Plasma Assisted Combustion and Chemical Processing

Plasma Assisted Combustion and Chemical Processing provides an introduction to the opportunities of plasma-assisted combustion and chemical processing for green energy conversion using renewable electricity.

Covering the fundamentals of combustion and plasma physics and chemistry, it details the mechanisms and technologies of plasma-enhanced combustion, chemical process, materials manufacturing and recycling, and pollutant control. Addressing future challenges and opportunities of plasma-assisted combustion and electrified green manufacturing, this book covers the state-of-the-art methods of modeling and diagnostic tools to optimize the process design.

This book offers graduate students and researchers a comprehensive review of the fundamentals and research frontier in this emergent field.

Plasma Assisted Combustion and Chemical Processing

Yiguang Ju and Andrey Starikovskiy

CRC Press is an imprint of the
Taylor & Francis Group, an **informa** business

Designed cover image: Yiguang Ju and Andrey Starikovskiy

First edition published 2025
by CRC Press
2385 NW Executive Center Drive, Suite 320, Boca Raton FL 33431

and by CRC Press
4 Park Square, Milton Park, Abingdon, Oxon, OX14 4RN

CRC Press is an imprint of Taylor & Francis Group, LLC

© 2025 Yiguang Ju and Andrey Starikovskiy

Reasonable efforts have been made to publish reliable data and information, but the author and publisher cannot assume responsibility for the validity of all materials or the consequences of their use. The authors and publishers have attempted to trace the copyright holders of all material reproduced in this publication and apologize to copyright holders if permission to publish in this form has not been obtained. If any copyright material has not been acknowledged please write and let us know so we may rectify in any future reprint.

With the exception of Chapter 5, no part of this book may be reprinted or reproduced or utilised in any form or by any electronic, mechanical, or other means, now known or hereafter invented, including photocopying and recording, or in any information storage or retrieval system, without permission in writing from the publishers.

Chapter 5 of this book is freely available as a downloadable Open Access PDF at http://www.taylorfrancis.com under a Creative Commons Attribution-Non Commercial-No Derivatives CC-BY-NC-ND 4.0 license.

Any third party material in this book is not included in the OA Creative Commons license, unless indicated otherwise in a credit line to the material. Please direct any permissions enquiries to the original rightsholder.

For permission to photocopy or use material electronically from this work, access www.copyright.com or contact the Copyright Clearance Center, Inc. (CCC), 222 Rosewood Drive, Danvers, MA 01923, 978-750-8400. For works that are not available on CCC please contact mpkbookspermissions@tandf.co.uk

Trademark notice: Product or corporate names may be trademarks or registered trademarks and are used only for identification and explanation without intent to infringe.

ISBN: 978-1-032-06610-3 (hbk)
ISBN: 978-1-032-06611-0 (pbk)
ISBN: 978-1-003-20306-3 (ebk)

DOI: 10.1201/9781003203063

Typeset in Times
by codeMantra

Contents

About the Authors xii
Nomenclature xiii

Chapter 1 Introduction 1

    1.1   Energy Consumption and Its Impact on Climate Change 1
    1.2   Challenges in Transportation Fuels and Internal Combustion Engines 5
    1.3   Industrial Manufacturing Processes and Carbon Emissions 10
    1.4   Hydrogen Production and Carbon Emissions 11
    1.5   Methane Emissions and the Impact on Climate Change 12
    1.6   Renewable Electricity and Challenges in Intermittency and Electricity Storage 13
    1.7   Transition from Fossil Energy to Electron Energy and the Opportunities of Nonequilibrium Plasma-Assisted Combustion and Chemical Manufacturing 14
    1.8   Plasmas Properties and Reactivity 15
    1.9   Opportunities and Challenges of Plasma-Assisted Combustion 18
    1.10 Challenges and Opportunities of Nonequilibrium Plasma-Assisted Chemicals and Materials Manufacturing 20
          1.10.1 Hydrogen Production 22
          1.10.2 CO_2 Reduction 24
          1.10.3 Ammonia Synthesis and Nitrogen Fixation 28
          1.10.4 Materials Manufacturing and Waste Materials Recycling 30
    1.11 Aim of this Book 30
    References 31

Chapter 2 Low-Temperature Plasma Generation and Recombination 37

    2.1   Introduction 37
          2.1.1 Chemistry under Non-Boltzmann Conditions 37
          2.1.2 Temperature of Electrons in the Plasma 38
          2.1.3 Discharge Energy Distribution 40
          2.1.4 Major Pathways in PAC 40
          2.1.5 Mechanisms of PAC 43
    2.2   Discharges Used for Combustion and Ignition Control 43
          2.2.1 Streamer Discharge 45
          2.2.2 Dielectric Barrier Discharge 51
          2.2.3 Atmospheric Pressure Glow Discharge 55
          2.2.4 Pulsed Nanosecond Discharges 59
          2.2.5 Pulsed Spark Discharges with Fast Gas Heating 62
          2.2.6 Optical Discharges 66
          2.2.7 RF and MW Discharges 70
    2.3   Streamer Discharge Development: Branching and Streamer Head Structure 72
          2.3.1 Experimental Study of Streamer Discharge 72
          2.3.2 Single Channel Development 73
          2.3.3 Streamer Branching 77

		2.3.4 Numerical Model of Streamer Discharge	78
2.4	Fast Ionization Wave		81
	2.4.1	FIW Velocity and Electron Energy Distribution	81
	2.4.2	Spatial Uniformity of FIW	82
	2.4.3	Electric Field and Electron Density in FIW	85
2.5	Models for Nonequilibrium Plasma Recombination		86
	2.5.1	Plasma Decay in Molecular Nitrogen	86
	2.5.2	Plasma Decay in Molecular Oxygen	88
	2.5.3	Plasma Decay in CO_2	91
	2.5.4	Plasma Decay in Water Vapor	92
	2.5.5	Plasma Decay at Elevated Gas Temperatures	94
2.6	Excitation of Internal Degrees of Freedom in Various Types of Gas Discharges		96
	2.6.1	Glow Discharge	96
	2.6.2	RF Discharge	97
	2.6.3	Barrier Discharges	97
2.7	Ionization Waves in Different Types of Discharges		99
	2.7.1	History of Pulsed Discharge Research	99
	2.7.2	Ionization Waves in Townsend Mechanism of Breakdown	101
	2.7.3	Ionization Waves in Streamer Mechanism of Breakdown	101
2.8	FIW – Pulsed Discharge at High Overvoltage		103
	2.8.1	FIW Modeling	104
	2.8.2	EEDF Relaxation and Local Field Approximation	106
	2.8.3	Theoretical Description of FIWs	107
2.9	Numerical Model of FIW Development		109
	2.9.1	Cross Sections of Electron Interaction with N_2	114
	2.9.2	EEDF Relaxation Dynamics	117
	2.9.3	Gas Excitation in Rapidly Changing Electric Field	121
	2.9.4	EEDF Formation and Gas Excitation in FIW	122
2.10	Pulse-Periodic Gas Discharge at Atmospheric Pressure Conditions		128
	2.10.1	Discharge Dynamics and Active Particle Production in the "Needle-Plane" Geometry	129
	2.10.2	Active Particles Generation in Discharge Gap	132
	2.10.3	Streamer Propagation Velocity	133
	2.10.4	Peak Concentrations of Active Particles in Streamer Head	134
	2.10.5	Integral Active Particles Production	135
	2.10.6	Electric Field in the Region of Effective Excitation of Electronic Levels	137
	2.10.7	Velocity of Ionization Wave Propagation	141
	2.10.8	Efficiency of Active Particles Generation	143
	2.10.9	Transition to Spark Breakdown	146
2.11	Theoretical Study of Pulsed Discharges at High Pressure		148
	2.11.1	Hydrodynamic Model of Pulsed Gas Discharge	148
	2.11.2	The Role of Photoprocesses in the Development of Cathode-Directed Streamer	154
	2.11.3	Analytical Model of Cathode-Directed Streamer	159
2.12	Uncertainty in the Rates of Elementary Processes		167
References			168

Contents   vii

Chapter 3   Energy Exchange and Chemical Reactions in Nonequilibrium Regimes in Chemically Active Plasmas 181

    3.1   Vibrational Relaxation in Chemically Active Gas 181
        3.1.1   Mode Approximation 181
        3.1.2   "Ladder" Approximation 182
        3.1.3   Diffusion Approximation 183
        3.1.4   Processes of Vibrational Energy Exchange in the C–N–O–H System 186
    3.2   Chemical Reaction Rates and Vibrational Excitation of Products in Nonequilibrium Conditions 191
        3.2.1   Change of Reaction Threshold and Transition Probability under Excitation of Reactants 192
        3.2.2   Evaluation of Transition Probability for Selected Levels 193
    3.3   Dependence of Reaction Rates on Vibrational Energy of Reagents and Products 195
        3.3.1   Monomolecular Decomposition at $T_{tr} > T_{vib}$ 198
        3.3.2   Reactions in N_2-O_2 Mixtures 199
        3.3.3   Reactions in H_2-O_2 System 200
    References 204

Chapter 4   Mechanisms of Plasma-Chemical Reactions 206

    4.1   Kinetics of Hydrogen Oxidation in a Stoichiometric Hydrogen-Air Mixture in a Pulsed Nanosecond Discharge 206
        4.1.1   Electron Energy Distribution Function and Gas Excitation by Electron Impact 207
        4.1.2   Energy Exchange and Chemical Reactions in the H_2-Air System under Pulsed Discharge Conditions 210
    4.2   Nitrous Oxide Non-Thermal Decomposition in a Pulsed High-Current Discharge 250
        4.2.1   Experimental Investigations of N_2O Decay in a Pulsed Discharge 250
        4.2.2   Numerical Model of Non-Thermal Decay of N_2O in Plasma 254
        4.2.3   Active Particle Fluxes and Main Stages of the Process Non-Thermal Decomposition of N_2O in a Pulsed Discharge 258
    4.3   Oxydation of Hydrocarbons in Nanosecond Discharge 261
        4.3.1   Nanosecond Discharge Formation in Chemically Active Gas 262
        4.3.2   Analysis of Gas Excitation in Repetitive Nanosecond Gas Discharge 263
        4.3.3   Slow Oxidation of Hydrocarbons in Nonequilibrium Plasma 267
        4.3.4   Kinetics of Alkanes Oxidation in Nanosecond Discharge 271
    4.4   Kinetic Model of CH_4, C_2H_6, C_2H_5OH Oxidation by Nanosecond Discharge 284
        4.4.1   Electron Impact Processes in the Discharge 284
        4.4.2   Positive Ions Kinetics 284
        4.4.3   Production of Radicals by Electron Impact 288
        4.4.4   Kinetic Model for Low-Temperature Nonequilibrium Oxidation 289

		4.4.5 Numerical Analysis of Dynamics of Hydrocarbons Oxidation 290

- 4.5 Acetone, Acetylene, and Ethyl Alcohol Oxidation under Pulsed Discharge Excitation 293
  - 4.5.1 Comparison of Oxidation of Different Fuels 293
  - 4.5.2 Oxidation Kinetics of Acetylene 295
  - 4.5.3 Kinetics of C_2H_6 Oxidation 299
- 4.6 Carbon Monoxide Oxidation in Mixtures with Oxygen and Water 300
  - 4.6.1 Major Pathways of CO Oxidation 300
- 4.7 Kinetics of Neutrals in Hydrocarbon-Oxygen Mixtures 302
- 4.8 Ignition of H_2-O_2-Ar Mixtures by Nanosecond Gas Discharge 308
  - 4.8.1 Theoretical Analysis of Ignition Efficiency 308
  - 4.8.2 Plasma Shock Tube 312
  - 4.8.3 Ignition Delay Time Change in H_2-O_2 Mixture by Pulsed Discharge 315
- 4.9 Ignition Delay Time Change in H_2-Air Mixture by Pulsed Discharge 317
  - 4.9.1 Experiments in H_2: O_2: N_2: Ar Mixture 317
  - 4.9.2 Experiments in H_2: O_2: N_2 Mixture 318
- 4.10 Comparison with the Computational Model 320
  - 4.10.1 Gasdynamic Model of the Shock Tube 320
  - 4.10.2 Kinetic Model 321
- 4.11 Ignition Delay Time Measurements in CH_4: O_2: N_2: Ar Mixture 322
- 4.12 Gas Discharge and Ignition Homogeneity Measurements 323
- 4.13 Plasma-Assisted Ignition of C1–C4 Hydrocarbons 324
- 4.14 Kinetics of Plasma-Assisted Kinetics of C1–C5 Hydrocarbons in Gas Mixtures 330
  - 4.14.1 Experimental Analysis of the Plasma-Assisted Ignition. Mixtures with Hydrocarbons 332
  - 4.14.2 Simulation of Production of Atoms and Radicals 333
  - 4.14.3 Simulation of Autoignition and Plasma-Assisted Ignition. Comparison with the Experiments 336
- References 339

Chapter 5 Plasma-Assisted Combustion: Chemistry 346

- 5.1 Elementary Reactions in Combustion 346
  - 5.1.1 Chain-Initiation, Propagation, and Termination Reactions in Combustion 346
  - 5.1.2 Elementary Kinetics of Hydrogen 347
  - 5.1.3 Elementary Kinetics of Methane 351
  - 5.1.4 Reaction Kinetics of Ammonia 354
  - 5.1.5 Reaction Kinetics of Large Alkanes and Oxygenates at Different Temperatures 358
- 5.2 Chemical Kinetics in Plasma-Assisted Combustion 361
  - 5.2.1 Nonequilibrium Plasma Energy Transfer and Its Impact on Combustion 361
  - 5.2.2 Elementary Reactions of Plasma 363
  - 5.2.3 Elementary Reactions of Plasma-Assisted Hydrogen Combustion 365
  - 5.2.4 Kinetics of Plasma-Assisted CH_4 Combustion 368
  - 5.2.5 Kinetics of Plasma-Assisted NH_3 Combustion and NO_x Emissions 373

Contents ix

		5.2.6	Kinetics of Plasma-Assisted Large Hydrocarbon and Oxygenate Fuel Combustion 377
	5.3	Impact of $O(^1D)$, NO_x, and Ozone Production in Plasma on Combustion Kinetics 383	
		5.3.1	$O(^1D)$ Reactions with Saturated and Unsaturated Hydrocarbons and Oxygenates 383
		5.3.2	Plasma-Produced NO_x and the Impact on Combustion Kinetics 387
		5.3.3	Plasma-Produced, Ozone-Assisted Combustion Kinetics 389
	References 394		

Chapter 6 Plasma-Assisted Combustion: Dynamics 402

    6.1 Plasma-Assisted Ignition 402
        6.1.1 Ignition and Ignition Delay Time 402
        6.1.2 Plasma-Assisted Ignition 405
        6.1.3 Plasma-Assisted Ignition and Extinction 409
        6.1.4 Control of Non-equilibrium Plasma for Ignition Enhancement 410
    6.2 Plasma-Assisted Flame Propagation 412
        6.2.1 Flame Propagation Speed 412
        6.2.2 Effect of Electric Field and Plasma on Flame Propagation Speed 415
        6.2.3 Ignition-Assisted Flame Propagation 420
        6.2.4 Extinction and Flammability Limits of Stretched Flames 422
        6.2.5 Plasma-Assisted Flame Propagation and Stabilization by Ozone and Singlet Oxygen 429
    6.3 Minimum Ignition Energy 438
        6.3.1 The Minimum Ignition Energy and the Critical Radius 438
        6.3.2 Impact of Plasma Properties on the Minimum Ignition Energy and Flame Initiation 443
    6.4 Plasma-Assisted Hot and Cool Diffusion Flames 449
        6.4.1 Extinction Limits of Hot Diffusion Flames 449
        6.4.2 Extinction Limits of Cool Diffusion Flames 458
        6.4.3 Extinction Limits of Warm Flames 459
    6.5 Plasma-Assisted Control of Detonation and Deflagration to Detonation Transition 462
        6.5.1 Detonation 462
        6.5.2 Deflagration to Detonation Transition 464
        6.5.3 Control of Detonation Using Plasma-Enhanced Combustion Chemistry 468
    References 472

Chapter 7 Plasma-Assisted Combustion: Applications 479

    7.1 Supersonic Combustion, Scramjet Engines, and Detonation Engines 479
        7.1.1 Plasma Torch and Arc Jet 480
        7.1.2 Gliding Arc 482
        7.1.3 Nanosecond Discharge 485
        7.1.4 Microwave and RF Discharge 486
        7.1.5 Laser Ignition and Electron Beam 487

	7.2	Detonation Engines	488
	7.3	Extension of Blowoff and Lean-Burn Limits	492
	7.4	Gas Turbine Engines and Thermal Acoustic Instability Control	495
	7.5	Internal Combustion Engine Applications	498
	7.6	Emission Control	503
		7.6.1 NO_x Emission Control	503
		7.6.2 Soot Emission Control	505
	References		507

Chapter 8 Electrified Non-Equilibrium Chemical Manufacturing 512

    8.1 Electrified Transient Non-Equilibrium Chemical Manufacturing 512
        8.1.1 Chemical Equilibrium 512
        8.1.2 Concept of Non-Equilibrium Chemical Manufacturing 514
        8.1.3 Non-Equilibrium Chemical Manufacturing by Pulsed Electrical Joule Heating 515
    8.2 Kinetics of Catalysis Reactions 521
        8.2.1 Binding Energy and Its Impact on the Activation Energy of Catalytic Reactions 522
        8.2.2 Adsorption Dissociation Reactions 524
        8.2.3 Langmuir–Hinshelwood Mechanism 527
        8.2.4 Eley–Rideal Mechanism 527
        8.2.5 Mars-Van Krevelen Mechanism 528
    8.3 Plasma Catalysis for Non-Equilibrium Chemical Synthesis 529
        8.3.1 Ammonia Synthesis 529
        8.3.2 Nitric Oxide for Nitrogen Fixation 535
        8.3.3 Hydrogen Production from Methane Pyrolysis 537
        8.3.4 CO_2 Reduction 539
    8.4 Plasma-Assisted Chemical Looping 553
    8.5 Plasma for Materials Synthesis and Waste Materials Recycling ... 557
        8.5.1 Plasma-Assisted Waste Resources Recycling and Upcycling of Battery Electrode Materials 557
        8.5.2 Plasma-Assisted Materials Synthesis 559
    References 562

Chapter 9 Plasma Diagnostics 566

    9.1 Optical Emission Spectroscopy 566
        9.1.1 Optical Emission Spectroscopy Method 567
        9.1.2 Applications of OES in Plasma 568
    9.2 Laser Absorption Spectroscopy 572
        9.2.1 Introduction 572
        9.2.2 The Beer-Lambert Law 574
        9.2.3 Experimental Method of LAS and Applications 577
        9.2.4 Femtosecond LAS 582
    9.3 Faraday Rotational Spectroscopy and Elementary Reaction Rate Measurements 585
        9.3.1 Faraday Rotational Spectroscopy and Experimental Method 585
    9.4 Raman Scattering and Thomson Scattering 588
        9.4.1 Raman Scattering 589
        9.4.2 Thomson Scattering for Electron Property Measurements 594

Contents xi

| | | 9.4.3 | Application of Thomson Scattering for Electron Property Measurements in Plasma | 595 |

- 9.5 Electric Field Induced Second Harmonic Generation (E-FISH) 598
- 9.6 Hybrid Femtosecond/Picosecond Coherent Anti-Stokes Raman Scattering (fs/ps CARS) 601
  - 9.6.1 fs/ps CARS 601
  - 9.6.2 Applications of fs/ps CARS 603
- 9.7 Femotsecond Two-Photon Lasers-Induced Fluorescence 608
  - 9.7.1 TALIF and Calibration 608
  - 9.7.2 Femtosecond TALIF Applications in Plasma 610
- References 612

Index 617

About the Authors

Yiguang Ju is the Robert Porter Patterson Professor in the Department of Mechanical and Aerospace Engineering at Princeton University, the Director of the US Department of Energy (DOE) Energy Earthshot Research Center (EERC) for plasma-enhanced hydrogen production, and the head of Electromanufacturing Science and Associated Faculty at Princeton Plasma Physics Laboratory (PPPL). He received his B.S. in Engineering Thermophysics from Tsinghua University in 1986, and his Ph.D. in Mechanical and Aerospace Engineering from Tohoku University in 1994. Dr. Ju's research interests include combustion, green fuels, plasma, and electrified manufacturing. He has published more than 300 refereed journal articles. He is a fellow of the American Society of Mechanical Engineers (ASME) and American Institute of Aeronautics and Astronautics (AIAA). He is also an inaugural fellow of the Combustion Institute. He served as the chair of the US Section of the Combustion Institute and a board member of the Combustion Institute. He also served as an associate editor for Proceedings of Combustion Institute and AIAA Journal and a steering committee member of National Academies of Sciences Decadal Survey on Biological and Physical Sciences Research in Space of NASA. He received many honors including the NASA Director's Certificate of Appreciation award, the Friedrich Wilhelm Bessel Research Award by the Alexander von Humboldt Foundation, the International Prize of the Japanese Combustion Society, and the Alfred C. Egerton Gold Medal from the Combustion Institute. He is a co-founder of HiTNano Inc, Princeton NuEnergy Inc, Polymer-X Inc, and USPlasma Inc. He is also the founding President of Asian American Academy of Science and Engineering (AAASE).

Andrey Starikovskiy is a senior research specialist in the Department of Mechanical and Aerospace Engineering at Princeton University. He received his Ph.D. in Mechanics of Gases and Plasma from the Moscow Institute of Physics and Technology in 1991 and his Doctor of Science degree in Physics and Chemistry of Plasma, Thermophysics and Molecular Physics from the Institute for High-Temperature Studies, Russian Academy of Science in 2000. Dr. Starikovskiy's research interests include combustion, fuels, plasma, molecular energy transfer processes, electron and ion kinetics, plasma aerodynamics, and supersonic and hypersonic flows. He has published more than 250 refereed papers and 8 book chapters. He is the author of six patents. He is an Editor-in-Chief for *Plasma* journal. He is also a member of the editorial board for the journals *Energies* and *Plasma Research Express*. Dr. Starikovskiy is an associate fellow of the American Institute of Aeronautics and Astronautics. He is also a member of AIAA Plasmadynamics and Lasers Technical Committee.

Nomenclature

a: stretch rate or acoustic speed
A: the pre-exponential factor or a reactant or the Einstein coefficient for spontaneous emission
B: the Einstein coefficient for stimulated absorption or rotational constant or reaction rate constant
c: light speed
C: mole concentration or constant
C_p: specific heat at constant pressure
C_v: specific heat at constant volume
d: diameter or height
D: diffusivity or gap distance or correction term in rotation energy
D_a: Damköhler number
E: electric field or energy
E_a: activation energy
f: a function or frequency or focal length
F: fuel
g^0: Gibbs energy at the standard condition
G: Gibbs energy of the mixture or the experimental gain
h: enthalpy or Planck coefficient
H: enthalpy of a system
I: intensity or moment of inertia
k: reaction rate or thermal conductivity or absorption coefficient
k_B: Boltzmann constant, 1.38×10^{-23} J/K
k_v: the spectral absorption coefficient
L: length of a channel
Le: Lewis number
J: rotational quantum number
m: mass burning rate (production of density and flame speed) or mass of a molecule
M: molecule or molecular weight or the third-body in reaction or Mach number or magnetic quantum number
m_e: mass of an electron
n: number density or mole number or a constant
N: molecular number or total species number or number of discharge per burst
p: pressure
P: product or probability or scattered power
q: charge
Q: heat release or partition function or quenching rate
q_e: charge of electron
r: distance
R: radicals or the universal gas constant
S: enthalpy of a reaction system or line strength or scattering signal
S_L: laminar flame speed
t: time
T: temperature
T_f: flame temperature
T_{ad}: adiabatic flame temperature
T_e: electron temperature
T_v: vibrational temperature
U: velocity

v: velocity or vibrational state
V: voltage
v_e: electron velocity
x: coordinate or anharmonicity constants
X: mole fraction
y: coordinate or anharmonicity constants
Y: mass fraction
Z: Zeldovich activation energy
α: energy loss coefficient or absorptivity or polarizability
δ_f: flame thickness
ε: small parameter of reaction zone or energy of a state or permittivity
ε_0: permittivity of free space
ϕ: equivalence ratio or line shape factor or line broadening or angle
γ: the ratio of the specific heat or the collisional halfwidth
χ: susceptibility
λ: mean free path or wave length
μ: mobility or chemical potential
$\vec{\mu}$: dipole momentum
ω: angular frequency or reaction rate
$\dot{\omega}$: reaction rate
ρ: density
σ: cross-sectional area or plasma conductivity or Stefan-Boltzmann constant
θ: angle or nondimensional variable
τ: time or transmissivity
τ_c: collision time
': plasma reaction
v: collision frequency, frequency or wave number
$\tilde{v}$: wave number
¯: dimensional parameter
0D, 1D, 2D, 3D: zero-, one-, two- and three-dimensional

AC:	alternatng current
ACI:	advanced compression ignition
AFRL:	Air Force Research Laboratory
APEC:	Asia-Pacific Economic Cooperation
ATDC:	after top dead center
Bcm:	billion cubic meter
BE:	Birkland-Eyde
CAEP:	Committee on Aviation Environmental Protection
CARS:	Coherent anti-Stokes Raman spectroscopy
CBS:	complete base set
CEAS:	cavity-enhanced absorption spectroscopy
CF:	cool flame
CFE:	cool flame extinction
CLC:	Chemical-looping combustion
CJ:	Chapman-Jouguet
CNT:	carbon nanotubes
CRDS:	cavity ring down spectroscopy
CRMs:	Collisional–Radiative Models
Da:	Damköhler number, the ratio of flow residence time to reaction time

DBD:	dielectric barrier discharge
DC:	direct current
DDT:	deflagration to detonation transition
DFB:	distributed-feedback
DFT:	density functional theory
DME:	dimethyl ether
DOE:	Department of Energy
DRM:	dry reforming of methane
EEDF:	electron energy distribution function
E-FISH:	electric field induced second harmonics
EI:	electron ionization or electron impact
EIA:	Energy Information Association
ELTC:	extreme low temperature chemistry
E/N:	reduced electric field
EPA:	Environmental Protection Agency
E-R:	Eley–Rideal
EV:	electric vehicles
FB:	fluidized bed
FIW:	fast ionization wave
FRS:	Faraday rotation spectroscopy
fs:	femtosecond
FT:	Fischer-Tropsch
FWHM:	full width at half maximum
GC:	gas chromatograph
Gt:	Giga-tons
GWP:	global warming potential
HB:	Haber–Bosch
HCCI:	homogenous charge compression ignition
HF:	hot flame
HFE:	hot flame extinction
HP-Mech:	high pressure mechanism
HTC:	high temperature chemistry
HTI:	high temperature ignition
ICAO:	International Civil Aviation Organization
ICCD:	intensified charge-coupled device
ICOS:	off-axis integrated cavity output spectroscopy
ICP:	inductively coupled plasma
INS:	inelastic neutron scattering
IPCC:	Intergovernmental Panel on Climate Change
IR:	infrared
ITI:	intermediate temperature ignition
JSR:	jet stirred reactor
KHP:	keto-hydroperoxide
Kp:	Planck mean absorption coefficient
LAS:	laser absorption spectroscopy
L-H:	Langmuir–Hinshelwood
LHCP:	left-handed circularly polarized
LiB:	lithium-ion battery
LIF:	laser-induced fluorescence
LSPI:	low speed pre-ignition

LTC:	low temperature chemistry
LTE:	local thermodynamic equilibrium
LTI:	low temperature ignition
MARCS:	multi-scale adaptive reduced chemistry solver
MCB:	magnetically induced circular birefringence
MCGA:	multi-channel gliding arc
MBMS:	molecular-beam mass spectrometer
MHD:	magneto-hydrodynamic discharge
MHX:	methyl hexanoate
MIE:	minimum ignition energy
MIP:	minimum ignition power
MRCI:	multireference configuration interaction
MSD:	multiple spark discharge
MTOE:	million tons of oil equivalent
MW:	microwave
MVK:	Mars-Van Krevelen
NAS:	National Academies of Sciences
NOAA:	National Oceanic and Atmospheric Administration
NPHFD:	nanosecond pulsed high frequency discharges
NSD:	nanosecond discharge
NRP:	nanosecond repetitive plasma discharge
NTC:	the negative temperature coefficient
OES:	optical emission spectroscopy
OPA:	optical parametric amplifier
PAC:	plasma-assisted combustion, plasma-assisted catalysis
PAI:	plasma-assisted ignition
PB:	packed bed
PET:	poly ethylene terephthalate
PDE:	pulse detonation engine
PIC-MCC:	particle-in-cell–Monte Carlo collision
PLIF:	planar laser-induced fluorescence
PM:	particulate matter
PHQ:	programmable heating and quenching
POZ:	primary ozonide
PP:	polypropylene
PPCI:	partially premixed compression ignition
PRF:	pulse repetition frequency
ps:	picosecond
QCL:	quantum cascading laser
QOOH:	hydrocarbon hydrogen peroxide
RCCI:	reactivity controlled compression ignition
RCM:	rapid compression machine
RDE:	rotating detonation engine
ReaxFF:	reactive molecular dynamics simulations
res:	flow residence
RET:	rotational energy transfer
RF:	radiofrequency
RGA:	rotating gliding arc
RH:	fuel molecule
RHCP:	right handed circularly polarized

RON:	research octane number
RTD:	flow residence time distribution
SACI:	spark-assisted compression ignition
SCRAM:	supersonic ramjet
SDBD:	surface dielectric barrier discharge
SEI:	specific energy input
SI:	spark ignition
SMR:	steam-methane reforming
SP-JSR:	supercritical pressure jet stirred reactor
STH:	spatiotemporal heating
Ta:	reduced temperature of the activation energy, E_a/R
TALIF:	two-photon laser-induced fluorescence
TDLAS:	tunable diode laser absorption spectroscopy
TOF:	the turnover frequency
UHC:	unburned hydrocarbons
USP:	Uniform stable plasma
UV:	ultraviolet
VOC:	volatile organic compounds
V-T:	vibration-rotational or translational energy transfer
V-V:	vibration-vibration energy transfer
WGS:	water gas shift
XRD:	X-ray diffraction analysis
ZND:	Zeldovich, von Neumann, and Doring

1 Introduction

1.1 ENERGY CONSUMPTION AND ITS IMPACT ON CLIMATE CHANGE

The recent surge of extreme hurricanes, draughts, and wildfires has raised increasing public concerns on climate change and CO_2 emissions in energy consumption. As shown in Figure 1.1, the daily CO_2 concentration measured by the National Oceanic and Atmospheric Administration (NOAA) at the Mauna Loa Observatory in Hawaii at 3400 m above sea level showed the monthly change of CO_2 at Mauna Loa and the global trend. This figure clearly shows that CO_2 has risen at a yearly averaged rate of 2.38 ppm/year during the last 10 years, reaching 413 ppm on September 1, 2021 from 389 ppm on September 1, 2011 [1], which is 53% increase than the CO_2 level of the preindustrial time. The last time the Earth's atmosphere had the same level of 400 ppm of CO_2 was about 3 million years ago when the Earth was 2°C–3°C degrees above preindustrial time and the sea levels were 12–25 m higher than today. More importantly, the CO_2 rise today is caused by mankind's carbon-based energy consumption, which is vastly different from that of 3 million years ago. For example, the world energy consumption by human activities in 2018 was 14,420 million tons of oil equivalent (MTOE), which produced 33.1 Gt of CO_2 [2] and is a large quantity compared to 3,230 Gt CO_2 present in the Earth's atmosphere at 421 ppm in February 2024.

The new 2021 Intergovernmental Panel on Climate Change (IPCC) report stated that climate change is widespread, rapid, and intensifying. It was estimated that the chances of crossing the global warming level of 1.5°C in the next two decades are high, and that unless there are immediate, rapid, and large-scale reductions in man-made CO_2 emissions, limiting warming to close to 1.5°C or even 2°C will be beyond reach [3]. A comparison of the Arctic ice caps between 1984 and 2016 published by NASA (Figure 1.2) shows clearly that the Arctic ice has not only shrunk in surface area dramatically but also become thinner as well. The accelerated global warming in the next

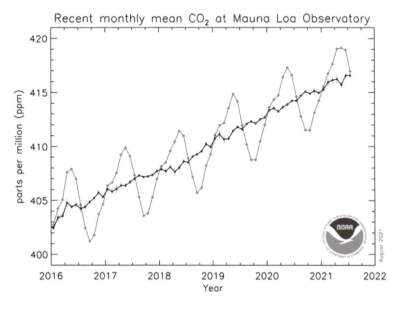

FIGURE 1.1 Monthly time history of CO_2 concentration measured at Mauna Loa observatory. (https://gml.noaa.gov/webdata/ccgg/trends/co2_trend_mlo.png, Sep. 1, 2021 [1].)

DOI: 10.1201/9781003203063-1

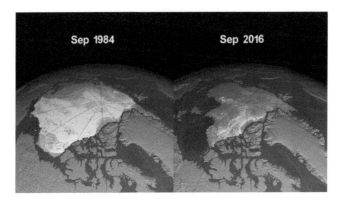

FIGURE 1.2 A comparison of the Arctic sea ice caps between 1984 and 2016 published by NASA (https://climate.nasa.gov/vital-signs/arctic-sea-ice/) [4].

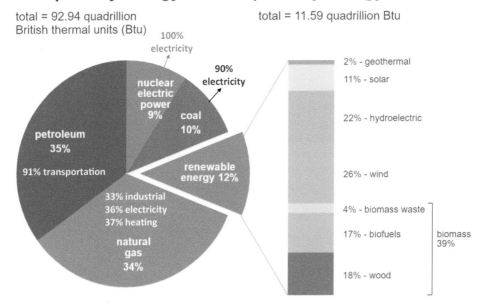

FIGURE 1.3 US primary energy consumption by sources and sectors published by EIA [5].

decades will further deteriorate the shrinking of the ice caps in both Arctic and Antarctic areas and cause serious concerns of sea level rise and flooding.

However, the current world energy consumption remains dominated by carbon-intensive fossil resources. Figure 1.3 shows the 2021 US primary energy consumption by energy sources and sectors published by the Energy Information Association (EIA) [5]. It is estimated that about 79% of the US energy consumption is still from fossil fuels (petroleum 35%, natural gas 34%, and coal 10%). More importantly, about 91% of petroleum was used for transportation, which accounted for 29% of CO_2 emissions by sectors. In addition, 90% of coal was used for electricity production, which contributed about 19% of CO_2 emissions and caused enormous emissions of particulates and NO_x and SO_x. Although 12% of the US energy consumption was from renewable resources, energy production from solar and wind were only 1.3% and 3.1%, respectively. Therefore, if energy conversion efficiency from fossil fuels can be improved by 10%, it will be equivalent to 7.9% of total energy consumption, which is even greater than the sum of renewable energy production from solar and wind (4.4%) in 2021. In addition, if water, biomass, waste plastics, and CO_2 emitted from fossil energy consumption can be used to produce E-fuels and chemicals using renewable electricity, a significant amount of carbon reduction can be achieved, especially in the industry sector.

If one looks at the CO_2 emissions per capita in different countries, one can see how much CO_2 emissions will continue to increase if everyone in the world consumes the same amount of energy as a person in the US. Despite the fact that China (10.3 Gt CO_2) and the US (4.7 Gt CO_2) are the No. 1 and No. 2 CO_2 emitters in the world, Figure 1.4 shows that the CO_2 emissions per capita in the US are twice higher than that in China, ten times higher than that in India, and more than hundred times higher than that in Uganda [6]. Given the huge populations and the future growth potential in Asia and Africa, it is expected that the world's CO_2 emissions may continue to increase as the economic development in Asia and Africa continues to improve. EIA recently published its 2019 outlook of future primary energy consumption. As shown in Figure 1.5 [7], by 2050 the world renewable energy consumption will likely reach 28% from today's 15%. The largest growth mostly comes from the electricity production from solar and wind. Nevertheless, the global consumption of natural gas will also increase by more than 40% from 2018 to 2050 due to the increasing demands in the industrial sector, electricity production, and chemicals and materials manufacturing. Moreover,

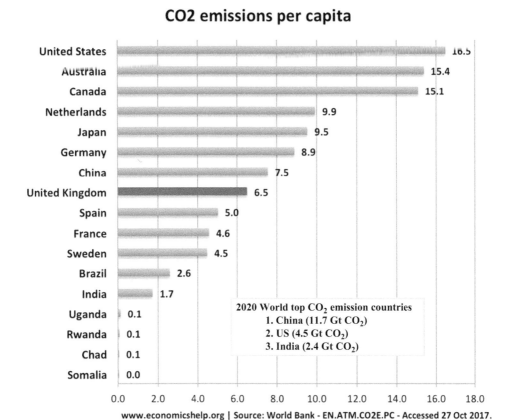

FIGURE 1.4 List of countries energy use per capita (https://www.economicshelp.org/blog/5988/economics/list-of-countries-energy-use-per-capita/) and the world top CO_2 emission countries (https://worldpopulationreview.com/country-rankings/co2-emissions-by-country).

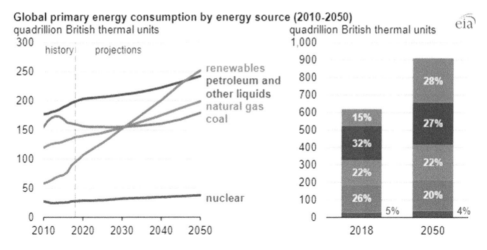

FIGURE 1.5 The global energy consumption outlook by EIA (2019) [7].

FIGURE 1.6 Images of Tiananmen in Beijing respectively taken on 2013.1.13 (Smog) and 2014.11.10 (APEC blue). (Photos by Xinhua News.)

it is also projected that both oil and coal consumption will either further increase or stay at a similar level. Therefore, the sustained needs for fossil energy in the next several decades will create enormous challenges in carbon and pollutant emissions. At the same time, the rapid increase of renewable electricity production and the rising impact of its intermittence on the electrical grids have also created a new challenge in GW scale electricity storage at different timescales.

In addition to carbon emissions, the consumption of fossil fuels also has created a significant impact on air pollution and smog [8]. In the spring of 2013, the air pollution caused by energy consumption of fossil fuels created heavy smog in Beijing [9]. Figure 1.6 shows a comparison of the air pollution in front of Tiananmen Square between 2013 and 2014. The level of particulate matter (PM2.5) in 2013 reached 150 μg/m^3. The concerns of combustion particulate emissions to human health were so serious that public schools, transportation, and people's daily activities were severely interrupted. To remedy the air pollution and hold the 2014 Asia-Pacific Economic Cooperation (APEC) economic forum in Beijing in November 2014, the Chinese government took a drastic measure to regulate the number of daily passenger vehicles in Beijing, limited the access of diesel trucks from neighboring provinces, shut down or relocate the polluting coal-burning plants, restaurants, and steel companies, and retrofitted coal power plants with natural gas power plants. This was an unprecedented experiment at a national government level at this scale to control air pollution from fossil fuel burning. The outcome is clearly shown in Figure 1.6. Before the 2014 APEC meeting, the sky of Beijing had become bluer and bluer day by day. This phenomenon was called the APEC Blue. The APEC Blue confirmed two things very clearly: the smog in Beijing was caused by man-made pollutions from burning fossil fuels and a fast and firm action by the government can reverse this consequence.

Introduction

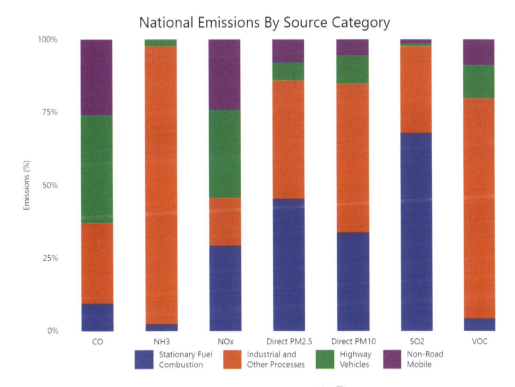

FIGURE 1.7 Pollutants emitted by a variety of sources published by EPA [10].

The 2021 US Environmental Protection Agency (EPA) report [10] analyzed the sources of air pollutants from stationary fuel combustion sources (such as electric utilities and industrial boilers), industrial and other processes (such as metal smelters, petroleum refineries, cement kilns, and dry cleaners), vehicles, and non-road mobile sources (such as recreational and construction equipment, marine vessels, aircraft, and locomotives). As shown in Figure 1.7, in the US although pollutants are emitted by a variety of sources, volatile organic compounds (VOCs) and NH_3 were mainly emitted from industrial and chemical processes. SO_2, particulate matter (PM) 2.5 and 10 were mainly from stationary power plants burning coal and oil. CO and NO_x were mainly contributed by transportation vehicles, stationary power plants including natural gas, and industrial processes.

Thanks to the progress of clean combustion technologies and the stringent regulations of air pollutants, the air pollutant emissions from 1990 have been decreasing steadily (Figure 1.8). PM2.5 and VOC emissions have dropped 38% and 48%, respectively. Moreover, NO_x and SO_2 emissions have decreased by 68% and 92%, respectively. This data clearly shows again that governmental regulations and technological advancement can play a critical role in reducing environmental pollutants. However, CO_2 emissions are different from air pollutants. It is more complicated and strongly dependent on energy consumption per capita, populations, regional energy resources, economic status, and energy conversion technologies.

1.2 CHALLENGES IN TRANSPORTATION FUELS AND INTERNAL COMBUSTION ENGINES

Transportation accounts for 29% of global CO_2 emissions by sector. Figure 1.9 shows the predicted world energy consumption by fuel type on energy consumption by Navigant Research [11]. It shows that the total road transportation energy consumption will grow from 81.1 quadrillion Btu in 2014 to 101.7 quadrillion Btu in 2035. Moreover, approximately 84% of transportation fuel will still be conventional fuels. Although the annual fuel consumption in the US, which is the largest fuel

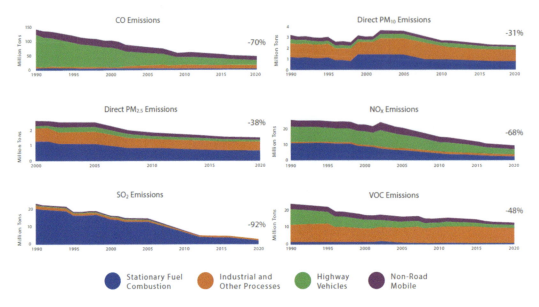

FIGURE 1.8 Changes of air pollutant emissions from 1990 to 2020 (EPA) [10].

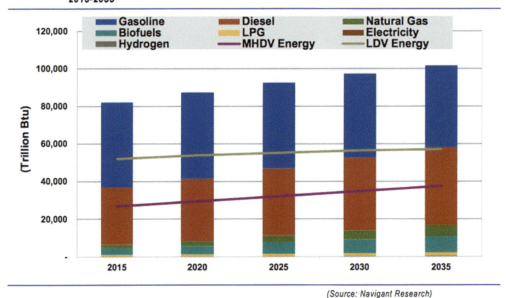

FIGURE 1.9 Annual road transportation sector energy consumption by fuel type, world markets: 2015–2035. (Navigant Research [11].)

consumption in transportation, will decrease due to the development of alternative fuels, fuel efficiency improvements, and electric vehicles, economic development in Eastern Europe, Middle East, Africa, and Asia-Pacific regions will still increase fossil fuel consumption. Brazil and India, and China will represent the largest increases due to the increase in number of vehicles. Although the recent rapid acceleration of electric vehicles (EVs) and renewable energy development in China could slow down the growth of conventional fuel consumption, the increase of coal power plant construction to low the electricity price for EVs is still propel the concerns of an increase of CO_2 emissions.

Table 1.1 shows the comparisons of energy return ratio, carbon index, and energy density of different fossil and alternative fuels [12]. It is seen that although conventional fuels have a high energy return ratio and energy density, they also have a high carbon index. In particular, the carbon indexes for coal and petroleum are, respectively, 105 and 96 gCO_2/MJ. Biofuels and hydrogen have a much less carbon index; however, their production costs remain high, and the energy return ratios are lower.

In 2005, President George W. Bush signed the America's Renewable Fuel Standard into law as part of the Energy Policy Act of 2005. The Renewable Fuel Standard called for an annual increase in production and use of biofuels as shown in Table 1.2. Over the past two decades, the Renewable Fuel Standard has successfully brought large government and industrial investment in the development of biofuels and displaced a significant portion of the use of gasoline fuels for transportation with ethanol. As shown in Table 1.2, with the government subsidy, the food-based first-generation bioethanol production in the US was more than tripled to 15 billion gallons per year from 2005 to 2015, which is about 10% of the gasoline consumption in the US. However, after 2015, the increase rate of food crop-based bioethanol production was reduced due to the impact of the decrease of government subsidy and the impact of COVID-19. Food crop-based first-generation ethanol competes directly with humans and animals and thus raises the price of food expenses. However, at the same time, as seen in Table 1.2, the non-food crop-based second-generation biofuel production is far below the projected production in 2005, despite significant research and commercial investment. It is seen that the projected biodiesel production in 2015 would be 1 billion gallons/year, but the

TABLE 1.1
Energy Return Ratio, Carbon Index, and Energy Density of Fuels [12]

Fuel	Energy Return Ratio	Carbon Index, gCO_2/MJ	Energy Density, MJ/kg
Coal	60	105	26
Petroleum	30–40	96	40
Shale gas	68	53	55
Hydroelectric	30–100		
Wind electric	20–40		
Solar electric	10–35		
Biodiesel	2.5	17–40	37
Corn ethanol	0.8–1.7	34–80	27
Algal oil	10		37
Cellulosic ethanol	6–36	20	27
Dry biomass	10–50	< 0	19
Hydrogen (solar)	1–5	10–20	120

TABLE 1.2
Forecasted and Actual Biofuel Production in the US from 2005 to 2012

Billion gallons/year	2005	2010	2015	2020	Status
Biodiesel, projected and (actual)	0	1	1 (0.12)	1 (0.16)	Low
First-generation bioethanol, projected and (actual)	4	12 (13)	15 (14.8)	15 (14)	On track
Second-generation biofuel, projected and (Actual)	0	0.5	4.5 < 0.1	14 < 0.1	Very low

actual production was only 0.12 and 0.16 billion gallons/year in 2015 and 2020, respectively, much lower than the forecasted yield. More disappointing was the production of the low yield non-food crop biofuel production. The projected second-generation biofuel production in 2015 was also 1 billion gallons/year, but the actual production was less than 0.1 billion gallons/year in 2015. The technological challenges in the second generation of biofuel production are much bigger than we anticipated. It will become more challenging with the current inflation. Therefore, it is necessary to improve efficiency and develop alternative methods such as E-fuels from CO_2 and H_2 for sustainable transportation.

To improve efficiency and reduce CO_2, NO_x, and particulate emissions, over the past decades, significant efforts have been made in combustion engines by optimizing combustion process and utilizing low carbon fuels (e.g., biofuels, E-fuels, syngas, hydrogen, and ammonia). In addition, high-pressure, low-temperature, and lean-burn combustion strategies such as homogenous charge compression ignition (HCCI) [13–15], reactivity controlled compression ignition (RCCI) [15], partially premixed compression ignition (PPCI) [16–18], advanced compression ignition (ACI), and spark-assisted compression ignition (SACI) technologies have been developed [19,20]. For gasoline engines, for example, Figure 1.10 shows the progress of the lean-burn gasoline engine efficiencies achieved by the science innovation project of Japan [21]. It is seen that by using the super lean-burn technology to control engine knock and heat losses, the net engine thermal efficiency had been increased from 38.5% to 51.5% in 5 years from 2014 to 2019, suggesting a significant improvement of engine efficiency and carbon reduction. In addition, to take advantages of high efficiency of diesel engines which operate at a high compression ratio in gasoline engines using premixed mixtures, a new technology called SKYACTIV-X was developed by Mazda, in which a diesel-like premixed fuel-lean compression ignition was achieved and controlled by using a spark-assisted ignition. This is a breakthrough technology of gasoline engines which can potentially improve energy efficiency by 30%. However, due to the difficulty of ignition control at different loads and for different fuels, robust control of ignition using nonequilibrium plasma is needed. Recently, research progress in plasma-assisted cool flames and warm flames and plasma-assisted combustion enhancement have attracted significant attention [22]. The kinetic enhancement of low-temperature combustion and ignition in engines using plasma provides great opportunities in control emissions, engine knock, lean burn, and heat losses and further improvement of engine efficiency. In addition to engine

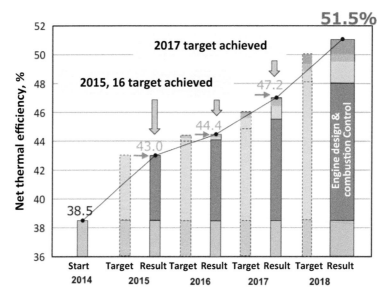

FIGURE 1.10 Achieved engine net thermal efficiency of lean-burn gasoline engines by the science innovation program in Japan [21].

development, many research efforts for the development of low carbon fuels such as E-fuels and biofuels by oil industry. The DOE CO-Optima project and the National Jet Fuels Combustion Program jointly led by the Federal Aviation Administration have been made for engine-fuel co-optimization [23–25]. Despite the successful demonstration of low carbon fuels in engine tests, the high cost of low carbon fuels remains a challenging issue.

To reduce the carbon emissions in ground power generation and air transportation, in 2019 the Department of Energy requested that the National Academies of Sciences (NAS) to convene an ad hoc committee to identify high-priority opportunities for improving and creating advanced gas turbine technologies that could be achieved by 2030. The committee reviewed the state-of-the-art gas turbine technologies and came out with 10 recommendations [26]. The top recommendation in this report was to enhance the foundational knowledge needed for low-emission combustion systems that can work in high-pressure and high-temperature environments and operate acceptably over a range of power settings and fuel compositions. As shown in Figure 1.11, since the 1990s, the gas turbine efficiency has been improved considerably from 34% to 44% and turbine inlet temperature has also increased from 1,300°C to 1,600°C. With the possibility of further increase of gas turbine inlet temperature with new high temperature materials, the challenges are how to reduce NO_x emissions at high temperature and how to increase the fuel flexibility such as hydrogen, biogas, ammonia, and other high hydrogen component fuels which have a broad distribution of heating values (Figure 1.12), flame speeds, ignition delay times, and emissions. New methods need to be developed to stabilize flames and reduce emissions with very low flame speeds and heating values but high NO_x and N_2O emissions such as ammonia and biogas.

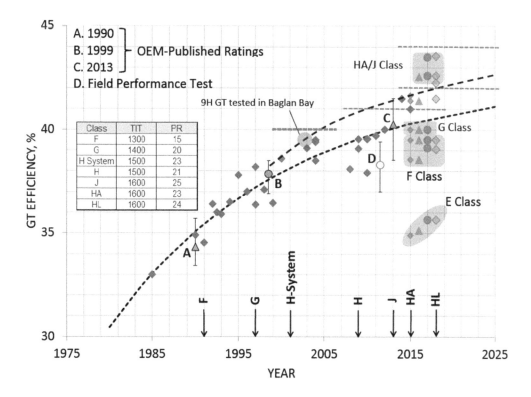

FIGURE 1.11 Gas turbine efficiency over years from National Academies of Sciences, Engineering, and Medicine. 2020. *Advanced Technologies for Gas Turbines*. Washington, DC: The National Academies Press [26].

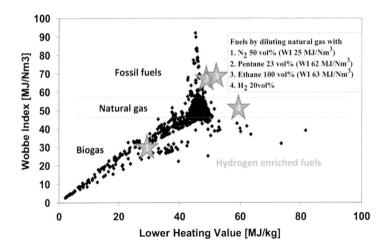

FIGURE 1.12 Heating values and the Wobbe Index of a variety of fuels for gas turbine engines. (Courtesy of Dr. Jenny Larfeldt at Siemens.)

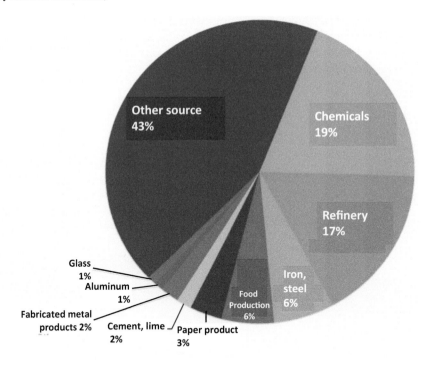

FIGURE 1.13 Energy-related CO_2 from industry (2020), EIA AEO 2021 [33].

1.3 INDUSTRIAL MANUFACTURING PROCESSES AND CARBON EMISSIONS

CO_2 emissions from various industrial sectors are shown in Figure 1.13. It is seen that total CO_2 emissions from chemical and material industrial processing are greater than that from other sources. Among them the chemical sector is the largest CO_2 contributor (19%), followed by refinery (17%), iron and steel manufacturing (6), food (6%), cement (2%) and glass (1%). The chemical industry accounts for 15% of total primary demand for oil on a volumetric basis and 9% (305 bcm) of natural gas. Chemical syntheses such as ammonia, methanol, and high-value chemicals such as ethylene and chlorine are the major sources of CO_2 emissions. In the 2020 IEA report [27], the estimated direct CO_2 emissions from primary chemical production were 880 $MtCO_2$ in 2018, a nearly 4% increase from the previous year, driven by growth in production.

Recently, ammonia has drawn significant attention as a hydrogen carrier for power generation and transportation because of its high energy density (compared to compressed and liquid hydrogen), easy storage, and safe transportation [28]. However, ammonia is difficult to ignite and has a very narrow burning limit, and ammonia combustion has high NO_x/N_2O emissions. Therefore, it is important to develop advanced technologies such as plasma-assisted ignition and combustion to enhance ammonia combustion to reduce ammonia slip and NO_x/N_2O emissions. Moreover, because the current ammonia production using the Haber–Bosch (HB) process requires high temperature (~500°C) for N_2 activation and high pressure (~250 atm) to maximize the ammonia yield, thus it is very energy and CO_2-intensive [29–32]. The estimated energy intensity of ammonia is 39.3 GJ/ton. Moreover, this process is limited to large-scale plants to be economically viable. Therefore, there has been an increasing need for a novel method using renewable resources for a distributed, efficient, and environmentally friendly NH_3 manufacturing. In addition, if H_2 is produced from methane reforming, 1.9 tons of carbon dioxide (CO_2) will be generated for producing one ton of NH_3. As such, it is also necessary to develop new technologies for efficient ammonia synthesis with a low carbon footprint using renewable electricity.

In high-value chemicals, ethylene is the major product in terms of production volume of the petrochemical. About 200 million metric tons were produced worldwide in 2020. Other simple organic chemicals are propylene, benzene, and chloride. These chemicals form the building blocks for many chemical products such as plastics (e.g., polyethylene, polypropylene, polystyrene, and PVC), resins, fibers, detergents, etc. Production of these chemicals is also very energy-intensive. Table 1.3 shows the estimated energy intensity of petrochemical production. The energy intensity of chlorine is the highest one because it uses electricity produced from fossil fuels.

Methanol is another important chemical for other chemicals and fuel products. Methanol's main end use is for formaldehyde to produce several specialized plastics and coatings. Methanol is mainly produced from natural gas reforming. The estimated energy intensity of methanol is 38 GJ/ton. Recently, methanol production grew at the fastest rate of all primary chemicals, with 6% growth in 2018. More recently, to reduce carbon footprint, methanol is planned to be produced from renewable electricity, CO_2, and H_2/H_2O and is used for producing E-fuels. E-methanol can be a key driver of growth in the future. Therefore, there is a critical need to develop innovative methods for green chemicals and E-fuels using renewable electricity.

1.4 HYDROGEN PRODUCTION AND CARBON EMISSIONS

Hydrogen has been considered as the ultimate clean energy carrier because it only produces water without any emissions when oxidized in internal combustion engines or fuel cells. However, hydrogen is not an energy resource because it does not exist freely on earth and must be produced from other sources of energy. There are many methods to produce hydrogen, but the carbon emissions and energy intensities are different. For example, as shown in Figure 1.14 [35], hydrogen can be produced directly from fossil fuels, biomass, and electrolysis from water. Today, the world commercial production of hydrogen from natural gas, oil, coal, and electrolysis accounts for 48%, 30%, 18%, and 4%, respectively. In the US, today about 95% of hydrogen is made from natural gas and

TABLE 1.3
Energy Intensity of Petrochemicals [34]

Chemicals	Ethylene	Chlorine	Ammonia	Methanol
Energy intensity	26 GJ/ton	47.8 GJ/ton	39.3 GJ/ton	38 GJ/ton

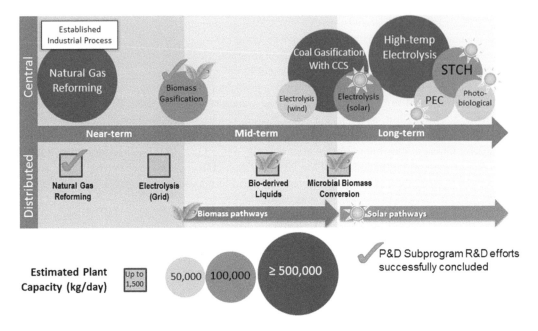

FIGURE 1.14 Hydrogen production processes in near, mid, and long term. (U.S. Department of Energy, Office of Energy Efficiency and Renewable Energy [35].)

coal reforming in large central plants. Therefore, in the near-term, hydrogen production from fossil fuels reforming or pyrolysis remains the dominant source of industrial hydrogen. However, hydrogen production from coal gasification technology [36] is not only energy inefficient but also very CO_2 intensive. Even without CO_2 capture, the energy efficiency in coal gasification process is only 63% [37]. Although the energy efficiency of hydrogen production from natural gas reforming is between 70% and 85%, the natural gas steam-methane reforming (SMR) and water gas shift (WGS) reactions require high temperatures (700°C–1,100°C) and high pressures (5–25 atm) to operate. The SMR process has a cost of \$2/kg H_2 and produces CO_2 at a level of 9–14 kg CO_2/kg H_2. In the mid-term, to reduce carbon emissions, as shown in Figure 1.14, biomass pathways and fossil fuel gasification with carbon capture will be used for hydrogen production. In the long term, solar and wind energy-based water electrolysis and photo-biological hydrogen synthesis will be employed to further reduce carbon emissions. However, without technology innovation, these pathways will significantly increase the cost of hydrogen production. Therefore, to enable hydrogen as an energy carrier, it is critical to develop an energy-efficient blue or green hydrogen production method from methane, shale gas, CO_2, H_2O, and biomass using renewable electricity.

1.5 METHANE EMISSIONS AND THE IMPACT ON CLIMATE CHANGE

Methane is another greenhouse gas, which accounts for accounted for about 10% of all US greenhouse gas emissions from human activities and is the second-most-prevalent greenhouse gas after carbon dioxide. In addition, methane is also emitted by natural sources such as natural wetlands and flared from oil and gas wells. More importantly, methane has strong thermal radiation absorption and is about 25 times the global warming potential (GWP) of CO_2 over 100 years [38]. Methane has a short lifetime in atmosphere (about 10 years), and it will be oxidized into CO_2 via photochemistry.

Globally, more than 50% of CH_4 emissions come from human activities [38]. Methane is mainly emitted from energy, industry, agriculture, land use, and waste management activities. Figure 1.15 shows the source distribution of methane emissions in 2019 in the US. Agriculture fermentation, landfill, wet land, and animal waste produced about 53% of methane emissions. At the same time,

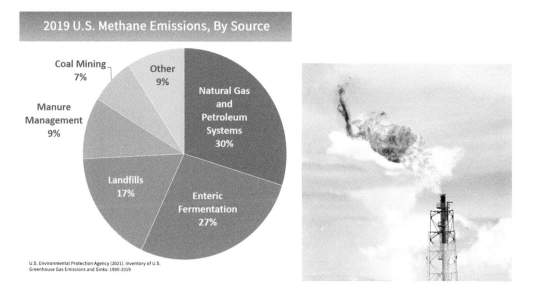

FIGURE 1.15 Total U.S. CO_2 Emissions in 2019 = 6,558 Million Metric Tons of CO_2 equivalent (excludes land sector) [38].

natural gas, petroleum, and coal mining contribute about 37% of total CH_4 emissions. Methane is emitted into the atmosphere during the production, processing, storage, distribution, and usage of natural gas, oil, and coal. Among the industrial methane emissions, natural gas flaring associated with oil extraction significantly contributes to the emissions of methane and CO_2. According to the 2020 EIA report, about 145 bcm of natural gas was flared just in 2018, which resulted in roughly 275 $MtCO_2$ emissions and a large amount of methane emissions from the leakage of flares. Russia, Iraq, Iran, Algeria, and the United States were responsible for more than half of the global flaring. To reduce greenhouse gas emissions from flaring, advanced technologies for (1) reduction of the methane leakage from the flaring by advanced combustion technologies, (2) on-site reforming of natural gas into liquid fuels, and (3) on-site electricity generation, need to be developed.

1.6 RENEWABLE ELECTRICITY AND CHALLENGES IN INTERMITTENCY AND ELECTRICITY STORAGE

To reduce carbon emissions in electricity generation, a significant increase of renewable electricity production has been made across the globe. For example, in 2020, the renewable electricity production in Norway, Brazil, and New Zealand reached respectively 98%, 84%, and 80%. The renewable electricity production from Germany and US are, respectively, 44% and 20%. Although the share of renewable energy production remains low, as shown in Figure 1.16, the renewable electricity production from wind and solar are increasing dramatically to 8.4% and 2.3% of the total US electricity production. The EIA forecasted that the US renewables share of electricity generation will be doubled by 2050 to 42%. The Biden administration announced in Sept of 2021 that the US will produce 50% of its electricity from solar power by 2050, a significant increase from the EIA prediction. The rapid increase of renewable electricity will also generate a transformative impact on the production of low carbon fuels, materials, and chemicals.

Germany is a world leader in renewable electricity production. The share of renewable electricity in Germany rose from just 3.4% in 1990 to 42.1% of consumption in 2019. However, renewable electricity production from wind and solar is very intermittent, which creates great challenges on electrical grids and energy storage. For example, the evolution of the electricity production in Germany for January 2017 is shown in Figure 1.17. The variation of sunshine and wind in days and

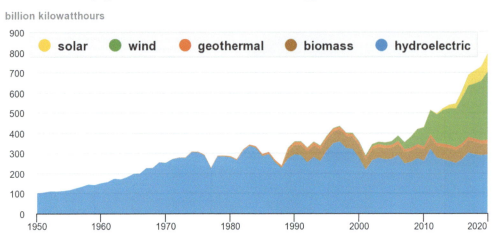

FIGURE 1.16 US electricity generation from renewable energy sources, EIA [39].

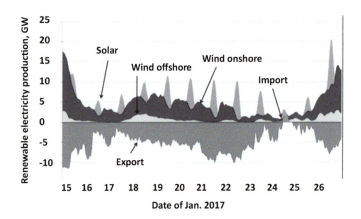

FIGURE 1.17 Time evolution of renewable electricity production during a dark and wind still period and compared to the electricity export for Germany (January 16–25, 2017) [40].

nights caused a large intermittency of renewable electricity production and the import and export of electricity to and from Germany. Especially, in the morning of the 24th of January 2017, the lack of wind and sunshine almost caused a collapse of the German renewable electricity supply. Therefore, during the time when the Sun does not shine and the wind does not blow, electricity production from E-fuels, hydrogen, and ammonia combustion is needed. On the other hand, if the renewable electricity production is at a time greater than the demand, it can be used to produce E-fuels, hydrogen, and ammonia.

1.7 TRANSITION FROM FOSSIL ENERGY TO ELECTRON ENERGY AND THE OPPORTUNITIES OF NONEQUILIBRIUM PLASMA-ASSISTED COMBUSTION AND CHEMICAL MANUFACTURING

As discussed in the above sections, today our energy is primarily fossil fuel (more than 80%) based energy production and consumption. Fossil fuels power our industry, transportation, heating and

Introduction

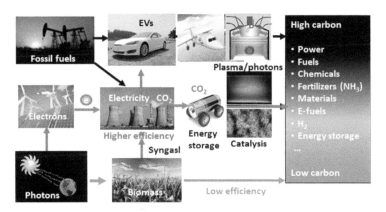

FIGURE 1.18 Transition from fossil energy to electron energy and opportunities of nonequilibrium plasma.

cooling, electricity production, and aerospace propulsion (Figure 1.18). At the same time, it produces enormous CO_2 (36.1 billion tons in 2022) and has raised the atmosphere CO_2 level from 389 ppm to 421 ppm (2024) in just 10 years. The rise of CO_2 level has triggered extreme weather, draughts, and wildfires, resulting in desperate concerns of climate change. To reduce carbon emissions from fossil energy use, over the last two decades, significant efforts have been made to develop biofuels and renewable energy electricity production. However, as discussed in Table 1.2, despite billions of dollars in investment in biofuel production from non-food crops for decades, the yield of the second-generation non-food crop-based biofuels remains low for the foreseeable future. However, at the same time, significant progress has been made in increasing energy conversion efficiency of combustion engines [21] and in wind and solar electricity production.

Note that the energy efficiency of renewable electricity production from solar and wind is much higher than that of biofuel production. Moreover, renewable electricity production from wind and solar does not consume water, and the electricity price has become comparable or even lower than that from fossil fuels. Therefore, in the foreseeable future, today's fossil fuel economy will be in transition to an electron energy economy. Electrons or electricity will be the primary energy carrier in world energy consumption. However, the biggest challenge of renewable electricity is its large intermittency and the lack of GW battery storage capacity.

Therefore, as shown in Figure 1.18, the recent development in renewable electricity provides an unprecedented opportunity to decarbonize the emissions in industrial sectors and enable distributed chemical manufacturing by using plasma-assisted combustion of low carbon or zero-carbon fuels and plasma-assisted chemical synthesis to produce hydrogen, ammonia, E-fuels, and other chemical feedstocks [41] from water, biomass, CO_2, and waste resources. For example, nonequilibrium plasma can extend the lean-burn limits and enable low-temperature combustion to improve energy efficiency. Plasma catalysis can create new reaction pathways and enable nonequilibrium reaction kinetics to accelerate chemical reactions at lower temperatures and increase the yield and selectivity by controlling the plasma reactivity via energetic electrons, excited states, and active radicals in plasmas [42–44].

1.8 PLASMAS PROPERTIES AND REACTIVITY

Plasma is a partially ionized mixture and the fourth state of matter (Figure 1.19a). The term 'plasma' was first introduced by Irving Langmuir in 1928 [45]. It contains energetic electrons, ions, electronically and vibrationally excited molecules, active radicals, and neutral molecules (Figure 1.19b). Plasmas are therefore more chemically active than neutral reactants and have the potential to enable new reaction pathways and reduce activation energies while increasing reactivity, energy efficiency, selectivity, and yield at low temperatures [44,46,47]. In plasma, the electrons are accelerated in an

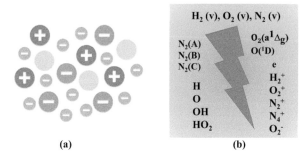

FIGURE 1.19 Schematic of plasma. (a) Plasma as a partially ionized mixture. (b) Electrons, ions, electronically and vibrationally excited species, and radicals in a hydrogen/air plasma.

electric field (E) and the kinetic energy is transferred to other molecules via electron impact collisions, causing ionization, dissociation, and electronic and vibrational excitations, and leading to active species formation and low temperature reaction pathways [42,48]. By using the Newton's law, the electron velocity, v_e, is governed by the Coulomb force and the collisional momentum loss,

$$m_e \frac{dv_e}{dt} = -q_e E - \frac{m_e v_e}{\tau_c} \quad (1.1)$$

where m_e is the electron mass, t the time, q_e the electrical charge of an electron, and τ_c the collision time between electrons and other molecules, respectively. Since the collisional time is proportional to the inverse of molecule number density (N), the increase of N will reduce the electron mean free bath between collisions and thus reduce the acceleration of electrons. Therefore, the electron velocity (v_e) or temperature (T_e) in a plasma is a function of the reduced electric field, $T_e = f(E/N)$. The electron energy will be transferred to other molecules via collisional excitations or by photons. The electron collisional impact on a neutral molecule will result in electronic excitation (M^*) or vibrational excitation, $M(v)$, of the molecules. These electron impact collisional excitation processes depend on the reduced electric field (E/N). Therefore, control of E/N in plasma by using different plasma sources and frequencies can change the plasma reactivity, electron temperature, and excited states.

Various plasmas such as sprites, lightening, glow discharge, gliding arc, laser, flame, corona, and dielectric discharge are generated in the universe or in the laboratory by using either static charges, direct current (DC), or alternating current (AC) (Figure 1.20). The AC frequency can be radio frequency (RF) or microwave (MW). The DC discharge can be pulsed and hybrid with AC discharge to change the electron number density and energy distribution in plasma. Moreover, plasma can also have different configurations such as dielectric barrier discharge (DBD), spark, corona, gliding arc,

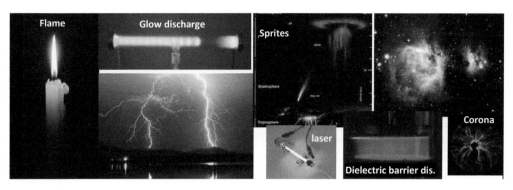

FIGURE 1.20 Various kinds of plasmas. (Courtesy from Mikhail Shneider.)

arc, microwave, and glow discharge [49]. Different plasmas have different electron energy, electron number densities, and reactivities (Table 1.4) [42], providing opportunities to tune chemistry pathways and reactivity in combustion and chemical synthesis.

Plasma can be either in thermodynamically nonequilibrium (nonthermal plasma) or equilibrium (thermal plasma). In a nonequilibrium plasma such as corona and nanosecond DBD, because there is not enough collisional energy transfer, the electron temperature, vibrational temperature, and the rotational temperature are not in equilibrium, $T_e > T_v > T$. By controlling the energy transfer timescales of electronic and vibrational excitations in comparison to vibrational-rotational energy transfer by increasing the discharge time or raising the discharge pressure, a nonequilibrium plasma can be transitioned to an equilibrium plasma, $T_e \approx T_v \approx T$. Although thermal plasma is in near equilibrium and produces enormous heating effect, it provides several advantages compared to traditional fossil fuel combustion heating. This includes its high temperature (3,000–20,000 K), high dissociation and ionization rates, and high power. As a result, thermal plasma is widely used for surface coating, fine powder synthesis, metallurgy, and the treatment of hazardous waste materials. Therefore, one can change the plasma properties and equilibrium such as the electron number density and electron temperature to manipulate the reactivity in plasma-assisted combustion and manufacturing.

A schematic diagram of the averaged electron temperature and electron number density is shown in Figure 1.21 for different types of plasmas. More detailed plasma characteristics are summarized in Table 1.4. Among different types of plasmas, spark and arc discharges are close to equilibrium

TABLE 1.4
Summary of Typical Characteristics of Different Types of Plasmas

	Arc Spark	NSD	RF	DBD	Corona	Streamer	MW
Pressure (atm)	Up to 20	Up to 5	10^{-3}–1	10^{-3}–1	0.1–10	0.1–1	0.1–10
Current (A)	1–10^5	50–200	10^{-4}–2	10^{-4}–10^{-3}	0.01–50	10^{-4}–10^{-3}	0.1–1
Voltage (kV)	0.01–0.2	1–100	0.5–2	1–10	0.1–50	10–100	0.1–100
E/N (Td)	0.5–2	100–1000	10–100	10–100	50–200	10–100	10–50
T_g (K)	3000–20000	300–1000	300–1000	300–500	500–1000	300–500	300–6000
T_e (ev)	0.5–2	5–30	1–5	1–5	1–5	1–3	1–5
n_e (m^{-3})	10^{21}–10^{22}	10^{17}–10^{19}	10^{17}–10^{19}	10^{17}–10^{19}	10^{12}–10^{15}	10^{17}–10^{18}	10^{15}–10^{23}

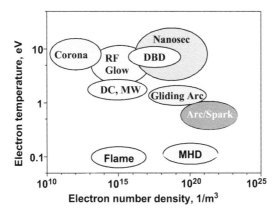

FIGURE 1.21 Schematic of electron temperature and number density for different discharges, Corona, direct current (DC) discharge, microwave (MW) discharge; dielectric barrier discharge (DBD), radio-frequency discharge (RF), glow discharge, gliding arc, nanosecond pulsed discharge (NSD), arc, magneto-hydrodynamic discharge (MHD), and flame.

plasmas. Spark and arc have a high neutral gas temperature and electron number density, but a low electron temperature. On the other hand, nonequilibrium plasmas such as corona, DC, glow, RF, DBD, nanosecond discharge (NSD), gliding arc, and MW all have higher electron temperature [42]. Among them, NSD can produce a very strong reduced electric field (E/N) and allow energy transfer mainly for electronic excitation and dissociation with less collisional energy loss. As such, NSD has a high electron temperature and high low temperature reactivity. Corona is a weakly ionized streamer discharge occurring near a sharp conducting electrode with locally high E/N. Ionization takes place only locally and produces glow. As shown in Table 1.4, the electron number density of corona is significantly low compared to streamer and spark (arc) discharges. On the other hand, gliding arc has a reasonably high electron temperature (0.1–3 eV), gas heating, and higher electron number density so that it has wide applications in plasma-assisted combustion and fuel reforming.

1.9 OPPORTUNITIES AND CHALLENGES OF PLASMA-ASSISTED COMBUSTION

About 80% of the US energy in 2021 was provided through combustion. Renewable energy from wind and solar is still only at 4.4%. Therefore, if we could improve combustion efficiency by 6%, the amount of carbon reduction would be equal to the entire contribution from wind and solar electricity sectors. As such, the development of low carbon fuels and increase of energy efficiency are both important.

In low carbon fuel development such as hydrogen, ammonia, biogas, and E-fuels, lean-burn ignition, low speed pre-ignition (LSPI), low-temperature and high-pressure combustion, near-limit flame stabilization, engine knock, fire safety and explosion, and NO_x and soot emissions have become new challenges. For example, in hydrogen combustion, due to the NO_x formation pathways from the NNH and N_2O mechanisms, the NO_x emissions of hydrogen combustion will increase. Moreover, hydrogen has a very high flame speed, it is easier to trigger deflagration to detonation transition and leads to increased fire safety concerns. For ammonia combustion, ammonia is difficult to ignite and has very low flame speeds. In addition, it may produce thousands ppm NO_x when burned at near stoichiometric conditions. For biogas, as shown in Figure 1.12, the volumetric heating value is very low. Therefore, it is difficult to ignite, and the flammability limits are very narrow. For E-fuels and bio-derived fuels, their ignition and combustion properties are different from conventional gasoline and diesel fuels. Thus, it is necessary to actively control their ignition, combustion, detonation, and emission properties.

The combustion engine efficiency is limited by the first and the second law of thermodynamics, i.e., the heat losses and the Carnot efficiency. For gas turbine engines, it is critical to raise the combustion temperature to increase efficiency (the Carnot efficiency) and to reduce NO_x emissions. However, internal combustion engines of gasoline and diesel cars, because of the engines high peak combustion temperature (~2,500 K), it is more important to reduce heat losses (lower peak combustion temperature) to improve engine efficiency (because the Carnot efficiency is not a limiting factor). In recent years, various new combustion engine technologies such as the Homogeneous Charge Compression Ignition (HCCI) engines [13–15], Partially Premixed Compression Ignition engines (PPCI) [16–18], and the Reactivity Controlled Compression Ignition (RCCI) engines [15] have been developed to achieve low-temperature and high-pressure combustion to reduce emissions and heat losses. However, there are several challenges to realize low-temperature combustion at high pressure and to reduce soot and NO_x. One example is that when the fuel's low-temperature reactivity is increased at high pressure, LSPI or engine knock is easier to occur. To suppress engine knock and LSPI, it is necessary to achieve lean burn. Unfortunately, ignition and combustion at fuel-lean conditions are difficult. An effective ignition and combustion method are needed to enable lean-burn technology. Another challenge is the large variation of fuel compositions in alternative fuels and engine loads. To enable fuel flexible efficient combustion with a wide range of engine loads, it is necessary to accurately control ignition timing and heat release rate in engines. Failure to accurately control ignition timing and heat release rate may lead to excessive unburned hydrocarbon emissions. Therefore, there is a great need to develop a new method for precise control of engine ignition.

Recently, there is also a renewed interest in supersonic and hypersonic propulsion. The development of airbreathing supersonic ramjet (SCRAM) engines requires fast ignition and combustion in a flow with speed five times or more than the sonic speed. Therefore, the ignition control and combustion enhancement in a scramjet engine is also critical. In addition, for ground power generation, fuel flexibility using low carbon or zero carbon fuels such as hydrogen, ammonia, biogas, high hydrogen content fuels and their mixtures is important. Therefore, it is necessary to develop novel reliable ignition and flame stabilization methods to address the fuel flexibility in gas turbine engines for power generation.

In addition to reducing carbon emissions and improving energy efficiency, reduction of NO_x, unburned hydrocarbons (UHCs), and particulate emissions in power generation and transportation is also important. Stringent emission standards for NO_x, UHCs, and particulates have been developed by the international Committee on Aviation Environmental Protection (e.g. CAEP-11), the Environmental Protection Agency (EPA) and the International Civil Aviation Organization (ICAO) (e.g., the Tier 6 standards) for aircraft engines and ground power gas turbine engines. Moreover, the emissions of UHCs from chemical plants and the methane leakage from flaring and abandoned gas well are also regulated. Therefore, it is necessary to develop novel and energy-efficient methods to reduce emissions of NO_x, particulates, and UHCs.

As shown in Figure 1.19b, plasma can generate many chemically active species such as electronically and vibrationally excited molecules and atoms, radicals, and intermediate species even at room temperature and nanosecond timescale. The rapid production of such reactive species can enhance ignition and combustion, especially at low-temperature and fuel-lean conditions. In addition, the radicals and intermediate species production can also be used to accelerate the oxidation of UHCs and particulates and reburning of NO_x, leading to reduction of emissions. Furthermore, plasma can raise temperature within a few nanoseconds to microseconds to enhance combustion and flame stabilization with precise time control. Moreover, plasma can induce turbulence via shock wave generation and ionic wind by the acceleration of the ion motion. The plasma induced shock wave, turbulence, and ionic wind can enhance fuel/air mixing and reduce soot, UHCs, and NO_x emissions. Therefore, as shown in Figure 1.22, plasma provides a promising technique to enhance ignition and combustion in scramjet engines, mild combustion, low-temperature combustion (cool flames) and enables the development of new engine technology with increased energy efficiency and fuel flexibility as well as low carbon and pollutant emissions.

Over the last 20 years, considerable progress has been made in plasma-assisted combustion using corona, DC, glow, RF, DBD, NSD, gliding arc, and microwave for ignition, flame stabilization, flame extinction, supersonic combustion, detonation, alternative fuels, low-temperature combustion, and emission control. Details of these studies can be found in several review papers [42,43,50–53]. The results showed that plasma can dramatically enhance ignition, flame stabilization, and lean-burn

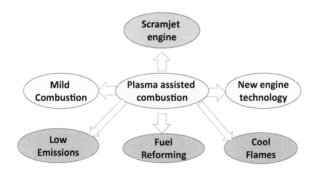

FIGURE 1.22 Schematic of plasma-assisted combustion and applications.

limit. In addition, it can promote low-temperature combustion and accelerate deflagration to detonation transition (DDT) [54]. Research also reported that plasma can reduce NO_x emissions in ammonia combustion and soot formation. In addition, many advanced laser diagnostic methods to probe plasma properties such as electron temperature and number density by using Thomson scattering [55,56], electric field using electric field induced second harmonic (E-FISH) generation [57–59], temperature and species using femto-second and picosecond (fs/ps) Coherent anti-Stokes Raman spectroscopy (CARS) [60–62], Raman scattering, Rayleigh scattering, femtosecond cavity-enhanced absorption spectroscopy (fs-CEAS) [63], UV and IR laser absorption spectroscopy (LAS), and two photon laser-induced fluorescence (TALIF) [64–66]. Such *in situ* diagnostic methods have provided valuable data for the validation of plasma-assisted combustion kinetic models for small hydrocarbons and n-dodecane [67]. In simulations of plasma-assisted combustion, multi-dimensional adaptive plasma-assisted combustion simulation methods such as the multiscale adaptive reduced chemistry solver for plasma-assisted combustion (MARCS-PAC) method [68,69] and a new unstructured, massively parallel code dedicated to low-temperature plasmas modeling [70] have been developed. These models can incorporate detailed combustion and plasma kinetic models with detailed transport properties for compressible reactive flow modeling. In addition, a new theory of plasma thermal-chemical instability to include the impact of plasma and combustion chemistry on plasma dynamics has been developed [71–73]. Such progress has significantly advanced the understanding of the physics and chemistry of plasma-assisted combustion and the development of predictive tools to optimize plasma-assisted combustion.

However, despite many efforts in plasma-assisted combustion research, quantitative or even qualitative understanding of the kinetic enhancement pathways has not been well-accomplished. Precise reactivity and discharge control at engine conditions remain difficult. Large knowledge gaps in both discharge development and fundamental understanding and modeling capabilities remain. In engineering applications, the development of efficient plasma discharge for large volume ignition at high pressure is the key to extend plasma-assisted combustion in engines. In fundamental research, understanding the energy transfer between different electronic, vibrational, and rotation excitations remains challenging. *In situ* diagnostics of different excited states is still challenging for many excited molecules. The existing plasma mechanisms have not been extensively validated in practical application conditions. There are many missing reaction pathways for plasma-assisted combustion modeling. Experimental data and cross-section areas for large hydrocarbon and oxygenated fuels are scarce. Accurate modeling of plasma-assisted combustion even in one-dimensional is still difficult due to the multi-timescale and nonequilibrium nature of plasma. To address these challenges, future efforts on experiments, kinetic mechanism, numerical modeling, and diagnostics are needed.

1.10 CHALLENGES AND OPPORTUNITIES OF NONEQUILIBRIUM PLASMA-ASSISTED CHEMICALS AND MATERIALS MANUFACTURING

With the rapid deployment of renewable electricity, there is an unprecedented opportunity to decarbonize the industrial sectors (Figure 1.13) for chemicals and materials manufacturing of steel, cements, hydrogen, E-fuels, and chemicals via plasma-assisted green manufacturing using CO_2, H_2O, N_2, waste plastics, and renewable resources (Figure 1.23). With precise control of the plasma reactivity and plasma-catalyst interaction using excited states of electrons, electron-hole pairs, phonons, ions, and molecules as well as surface charges and per-atom design of catalytic sites, not only we can reach the goal of net-zero carbon emissions but also increase the conversion energy efficiency and selectivity via manipulation of nonequilibrium interfacial chemical kinetics (Figure 1.24).

Compared to conventional thermal catalysis, nonequilibrium plasma provides several advantages to increase chemical conversion efficiency. Firstly, nonequilibrium plasma results in higher concentration of electronically and vibrationally species at low temperature due to the non-Boltzmann distribution of electron energy (Figure 1.25a). Such electronically and vibrationally excited molecules (Figure 1.24) can reduce the reaction activation energy by creating new reaction pathways and lower

Introduction 21

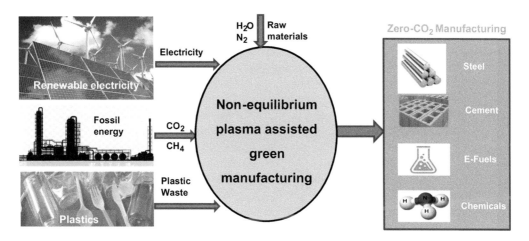

FIGURE 1.23 Schematic of nonequilibrium plasma-assisted green manufacturing and recycling for materials, chemicals, and E-fuels.

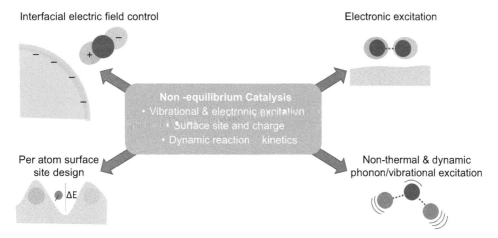

FIGURE 1.24 Schematic of manipulation of nonequilibrium interfacial chemical kinetic processes for energy-efficient plasma synthesis of chemicals and materials.

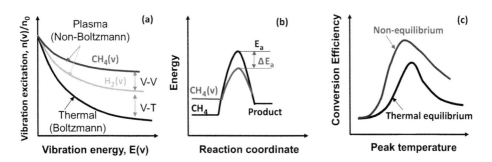

FIGURE 1.25 (a) Vibrational energy distributions in thermal equilibrium and nonequilibrium systems, (b) schematic of reduction in activation energy via excited molecules and new reaction pathways, (c) schematic of increase of conversion efficiency using nonequilibrium plasma process in comparison to thermal equilibrium process.

the reaction temperature. Secondly, plasma generates active radicals such as H, O, OH, and N. The production of active radicals will not only accelerate low-temperature reactions in gas phase reactions but also create new non-equilibrium Eley–Rideal and Mars-van Krevelen mechanism (MvK) reaction pathways on the catalyst surface in comparison to the dominant Langmuir–Hinshelwood catalytic pathways in a thermal equilibrium system. Thirdly, the surface charge created by plasma will change the polarization, transport, and adsorption dissociation energy of reactants on the catalyst surface. In addition, the surface charge also enhances the electric field near the surface and accelerates plasma and catalytic reactivities. Fourthly, the plasma-catalyst interaction will generate local hot spots on catalyst surface and enhance the electron excitation for molecules on the catalyst, thus changing the plasma catalysis process. Fifthly, the plasma-generated active species can also directly react with the catalysts and support materials to create surface vacancies and metastable catalysts (e.g. nitride, oxynitride, and hydride) to enable a plasma assisted Mars–van Krevelen mechanism for creating new catalysis pathways. Moreover, since plasma can be activated and quenched in a much shorter timescale (100 ns or less) in comparison to thermal catalysis (milliseconds to seconds), plasma can be manipulated to enable transient nonequilibrium manufacturing processes (shifting equilibrium) to further increase selectivity and conversion efficiency [74]. Therefore, as shown in Figure 1.25c, nonequilibrium plasma catalysis can increase chemical conversion at lower temperature, increase the peak conversion efficiency via new reaction pathways, and increase the yield and conversion efficiency even at high temperature by using the transient nonequilibrium processes.

Over the last decade, many advances have been made in the research of plasma-assisted chemicals and materials manufacturing in various areas such as hydrogen production, CO_2 reduction, ammonia synthesis, plastic and battery electrode materials recycling, and cement and steel manufacturing. Advanced diagnostic methods using LAS, E-FISH, CARS, and TALIF have been developed to probe time and species resolved nonequilibrium processes. Multiscale computational models using quantum chemistry, particle-in-cell–Monte Carlo collision (PIC–MCC) simulations, multi-fluid models, machine learning, and adaptive chemistry have been developed. Per-atom and high-entropy catalyst design, porous electrode, ferroelectric electrode, high-entropy catalysts, and high ion-conducting solid-state membranes have been made. Methods to control nonequilibrium plasma reactivity using hybrid nanosecond and RF/DC discharge and surface charge have been proposed and explored [75–77]. New plasma theories including thermal-chemical instability [71–73] were proposed and experimentally validated. These advances have led to extensive studies in plasma aided chemicals and materials manufacturing.

1.10.1 Hydrogen Production

Zero carbon energy carriers such as H_2 are crucial for the efficient integration of renewable electricity and distributed manufacturing [78]. However, the high cost and high carbon emissions in today's H_2 SMR and electrolysis production are among the great challenges of attaining net-zero carbon emissions. According to DOE's "Pathways to Commercial Liftoff: Clean Hydrogen" report [79], demand for clean H_2 production in the US could scale up by an order of magnitude to 10 MMT H_2/yr from the current <1 MMT H_2/yr by 2030. The report envisions the replacement of the majority, if not all, of the current carbon-intensive SMR H_2 processes. While H_2 generation from water electrolysis (green H_2) is ideal in the long term, it is currently not cost- or energy-competitive, especially compared with natural gas pyrolysis using renewable energy (turquoise H_2) [80]. The relatively low cost and abundance of natural gas and renewable electricity in the US makes methane, biomass, and plastic waste an appealing source for H_2 production with renewable electricity and enables carbon capture.

Table 1.5 tabulates the different clean H_2 production processes, including the current state-of-the-art SMR process and the competing green/blue/turquoise H_2 processes. SMR produces 4 moles of H_2, but also 1 mole of CO_2 [81]. However, the energy cost per kg H_2 production from methane pyrolysis is much less than that of water electrolysis per mole H_2 and produces no CO_2.

TABLE 1.5
Comparison of Standard Reaction Enthalpies per Mole of H_2 and CO_2 Emissions for Different H_2 Production Processes

H_2 Production Process	Pertinent Reaction(s)	Standard Reaction Ethalpy [kJ/mol H_2]	CO_2 emmited?	Reaction index
Steam methane reforming (SMR)	$CH_4(g) + H_2O(g) \rightarrow CO(g) + 3H_2(g)$ $CO(g) + H_2O(g) \rightarrow CO_2(g) + H_2(g)$ $CH_4(g) + 2H_2O(g) \rightarrow CO_2(g) + 4H_2(g)$	69 -41 41	no (CO) yes yes	R3
Water electrolysis	$2H_2O(l) \rightarrow O_2(g) + 2H_2(g)$	286	no	N/A
Methane pyrolysis	$CH_4(g) \rightarrow C(graphite) + 2H_2(g)$	37	no	R1
Methane coupling	$2CH_4(g) \rightarrow C_2H_2(g) + 3H_2(g)$ $2CH_4(g) \rightarrow C_2H_4(g) + 2H_2(g)$	125 101	no	R1
Plastic/polymer decomposition	$C_nH_m(s) \rightarrow xCH_4(g) + yC_pH_q(g) + zH_2(g)$ $C_pH_q(g) + pH_2O(g) \rightarrow pCO(g) + \frac{(2p+q)}{2}H_2(g)$	variable (> 0)	variable (CO)	R2
Dry methane reforming (DMR)	$CH_4(g) + CO_2(g) \rightarrow 2CO(g) + 2H_2(g)$	124	no (CO)	R3

The standard reaction enthalpies are for 298 K and 1 atm of gaseous molecules calculated from the enthalpies of formation.

In addition, methane pyrolysis using plasma is much more thermally efficient than SMR using combustion heating and can also convert carbon to valued carbon materials such as graphene and carbon nanotubes (CNTs), further increasing the product value and energy efficiency. An alternative method for H_2 and chemical production is biomass or waste plastic recycling for monomers, ethylene, and propylene. Polyethylene and polypropylene are the most manufactured plastics (60–70 million tons in 2019) [82,83]. Furthermore, ~70% of plastics end up in the environment (landfills and oceans) [84].

Different plasma discharges have been applied to hydrogen production from methane. Atmospheric pressure nonthermal plasmas such as DBD [85], corona discharge [86], and glow discharge [87] were used for methane conversion to hydrogen and syngas. However, the relatively low electron number density and low gas temperature in these weakly ionized plasmas make it difficult to achieve a high conversion efficiency at high flow rates [88,89]. To increase the electron number density and synthesis temperature, an intermediate temperature gliding arc has been used to produce hydrogen and carbon [88–90]. These experiments showed that gliding arc has significant advantages over other low-temperature plasmas in terms of CH_4 conversion. It was demonstrated that the maximum methane conversion reached 91.8% and the maximum hydrogen selectivity was

80.7%. The hydrogen production energy yield was 22.6 g/kWh. More recently, high-temperature microwave discharge was also used to produce hydrogen and carbon from methane in comparison to gliding arc in terms of methane conversion, product spectrum, and energy efficiency. The comparison revealed considerable differences between the performance of these two intermediate temperature plasma sources (~1,000–3,000 K). The gliding arc has higher methane conversion compared to the microwave discharge to produce hydrogen. Also, in the gliding arc, the selectivity toward acetylene was higher while the selectivity for hydrogen showed the opposite trend. The carbon produced with the gliding arc consisted of graphitic flakes, while the carbon produced with the microwave consisted of smaller spherical particles of amorphous carbon. These differences might be induced by the difference of temperature distribution and vibrational excitation. More recently, to further increase the temperature, studies to use thermal plasma such as arc and arc jet to produce hydrogen and valuable carbon materials such as carbon nanotubes were also conducted.

1.10.2 CO_2 Reduction

The conversion of CO_2 into value-added chemicals and E-fuels is one of the most interesting and active research areas of plasma-assisted chemical processing for decarbonization. CO_2 has strong carbon-oxygen bonds (783 kJ/mol) and thus is very stable. Breaking down a CO_2 molecule thermally is not effective and requires a large amount of energy (the standard formation enthalpy of CO_2 is −393.5 kJ/mol). As shown in Table 1.6, dissociation of CO_2 to CO and $1/2 O_2$ requires 283 kJ/mol. Moreover, the activation energy of this gas phase reaction is 532 kJ/mol, making this reaction more difficult to proceed without very high temperature. Therefore, an alternative method for CO_2 reduction is to react CO_2 with another reactant such as (e.g., H_2O, H_2, and CH_4) to produce syngas and other chemicals. The Gibbs energy changes of these reactions are shown in Table 1.6. Fortunately, by using catalysts, the activation energies of these reactions can be lowered so that the peak reaction temperature can be significantly reduced. For example, although the catalytic dry reforming of CO_2 with CH_4 still requires 247 kJ/mol, the lower activation energies for the catalytic CO_2 consumption over $Ce_{0.7}La_{0.2}Ni_{0.1}O_{2-\delta}$ catalyst is between 70.2 and 100.6 kJ/mol, which lowers the catalytic CO_2 dry reforming to 1,000 K [91]. Therefore, CO_2 reduction using nonequilibrium plasma-assisted catalytic reactions is very promising to further lower the activation energy and increase conversion efficiency by creating new reaction pathways and increasing the vibrational excitation population.

CO_2 dissociation ($CO_2 \rightarrow CO + 1/2O_2$): Recently, extensive studies have been conducted for CO_2 conversion using nonequilibrium plasmas. For direct CO_2 dissociation ($CO_2 \rightarrow CO + 1/2O_2$), both nonequilibrium and thermal equilibrium plasmas have been tried. For example, a coaxial DBD reactor with different catalysts [92] was used to examine the energy efficiency and CO_2 conversion. High-temperature MW and RF discharges, which have relatively higher electron density and temperature and lower reduced electric field to increase the thermal effect and the vibrational excitation of CO_2 were also tested [93]. To shift the equilibrium toward CO production, most microwave and RF studies were conducted at reduced pressures. The results showed that both DBD and microwave discharges have their own merits and demerits in CO_2 decomposition. DBDs can operate at

TABLE 1.6
Chemical Reactions for CO2 Reduction

$CO_2 \rightarrow CO + 1/2O_2$	$\Delta H = 283$ kJ/mol
$CO_2 + CH_4 \rightarrow 2CO + 2H_2$	$\Delta H = 247$ kJ/mol
$CO_2 + H_2O \rightarrow CO + H_2 + O_2$	$\Delta H = 525$ kJ/mol
$CO_2 + 2H_2O \rightarrow CH_3OH + 3/2O_2$	$\Delta H = 676$ kJ/mol
$CO_2 + 4H_2 \rightarrow CH_4 + 2H_2O$	$\Delta H = -165$ kJ/mol
$CO_2 + 3H_2 \rightarrow CH_3OH + H_2O$	$\Delta H = -41$ kJ/mol

atmospheric pressure but are not energy efficient. On the other hand, MW and RF showed high conversion efficiency, but these improvements were limited to lower pressures. To overcome these difficulties, gliding arcs were used to operate at atmospheric pressure and to achieve high efficiency at the same time. Most gliding arc experiments reported a maximum energy efficiency of 40%–50% or higher [94,95]. However, because of the thin channel of gliding arc, the CO_2 conversion is relatively low in comparison to microwave discharge. Figure 1.26 summarizes the energy conversion efficiency as a function of the CO_2 conversion for different plasma discharges in comparison with the limit of thermal equilibrium [96]. It is seen that low-temperature DBDs are not efficient for the conversion of CO_2 because of the low temperature and low electron number density. The gliding arcs at atmospheric pressure, however, showed capable of reaching energy efficiency of 60%, which is above the thermal equilibrium limit, demonstrating the advantage of nonthermal plasma in enhancing CO_2 conversion efficiency. However, the challenge is its low CO_2 conversion (less than 20%). It is also seen in this figure that although operating at reduced pressure, microwave discharge provides more promising capabilities than other plasma discharge. The CO_2 conversion can be higher than 40% and the conversion energy efficiency can reach 90%. However, optimization of plasma discharge is still needed to reach the targets of both high energy efficiency and high CO_2 conversion.

CO_2/CH_4 *dry reforming*: For CO_2 dry reforming with methane to syngas ($CO_2 + CH_4 \rightarrow 2CO + 2H_2$) and other oxygenate fuels and chemicals, various plasma catalysis studies have also been conducted [97–99]. By using DBD discharge in a fluidized-bed reactor and a packed-bed reactor, Chen et al. [99] showed that dry reforming was promoted dramatically in the fluidized-bed reactor due to an extended surface area of powdered catalysts and their interaction with plasma-generated reactive species and by vibrational excitation from DBD. In addition to syngas production, minor amounts of C_2-C_3 hydrocarbons such as C_2H_6, C_2H_4, C_2H_2, and C_3H_8 were also observed. A study of microwave discharge in dry reforming was also conducted [100]. The experiments reported CH_4 and CO_2 conversions of 71% and 69%, respectively, but the energy efficiency was quite low.

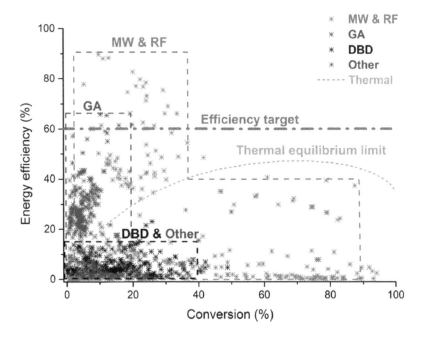

FIGURE 1.26 Comparison of all the data collected from the literature for CO_2 splitting in the different plasma types, showing the energy efficiency as a function of the conversion. The thermal equilibrium limit and the 60% efficiency target are also indicated [96].

CO_2 dry reforming was also conducted by using gliding arcs [101,102]. The results showed that gliding arc results in similar conversion to DBD, but the energy efficiency is significantly higher. To increase the electron energy, research using corona discharge, NSD, and glow discharge were also conducted. The results showed that CO_2 dry reforming is strongly affected by the CH_4/CO_2 ratio and has a broad distribution in products and conversion efficiency. A summary by Snoeckx and Bogaerts [96] on plasma-assisted dry reforming for energy cost as a function of conversion is shown in Figure 1.27. Like CO_2 decomposition, DBD and other weakly ionized plasma-assisted dry reforming do not allow the energy-efficient conversion of CO_2 and CH_4 into syngas, partly because of the low electron number density. However, it may have a higher yield of oxygenate formation due to the low-temperature process. Gliding arc demonstrates the best results of up to 40% conversion and higher energy efficiency than thermal equilibrium processes. By further raising the temperature using spark and arc discharges, conversions up to 85%–95% have already been achieved. Therefore, like CO_2 decomposition, high temperature and low electron energy-driven vibrational excitation play a critical role in plasma-assisted dry reforming.

Recently, CO_2 dry reforming of methane for oxygenates and hydrocarbon synthesis was also explored. Wang et al. [103] developed a DBD plasma process (Figure 1.28) for the direct activation of carbon dioxide and methane into oxygenated liquid fuels and chemicals (e.g., acetic acid, methanol, ethanol, acetone, and formaldehyde) at room temperature and atmospheric pressure with a water electrode. The results showed total selectivity of oxygenates of 50%–60% and a 40.2% selectivity to acetic acid. Scapinello et al. further investigated the catalytic effect of the electrode surface (Cu and Ni) on the conversion of CH_4 and CO_2 to $C_nH_{2n+1}COOH$ and CH_2O_2 at different discharge powers in a DBD reactor [104]. It was shown that CO_2 hydrogenation on the metal surface was enhanced when increasing the discharge power and that the selectivity of the end products was related to the CO_2/CH_4 molar ratio. Although these experiments showed that CO_2 conversion with methane is a promising process for the synthesis of oxygenates, more research is required to improve the selectivity and conversion efficiency of these value-added products.

CO_2 conversion with H_2O: Studies of CO_2 conversion to syngas and oxygenated fuels via reactions of $CO_2 + H_2O \rightarrow CO + H_2 + O_2$ and $CO_2 + 2H_2O \rightarrow CH_3OH + 3/2 O_2$ have also attracted research

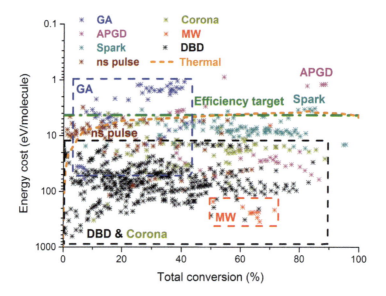

FIGURE 1.27 Comparison of all the data collected from the literature for dry reforming regarding the different plasma types, showing the energy cost as a function of the conversion. The thermal equilibrium limit and the target energy cost of 4.27 eV per molecule for the production of syngas (Corresponding to a 60% efficiency target) are also indicated [96].

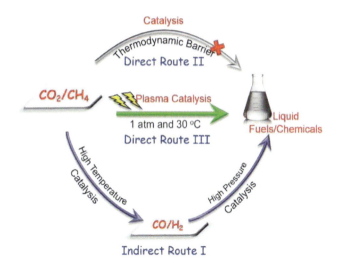

FIGURE 1.28 Dry reforming for the synthesis of liquid fuels and chemicals: single-step plasma catalysis versus two-step thermal catalysis [103].

interest because of the large amount of CO_2 and H_2O resources. However, these reactions are highly endothermic which require high temperatures. DBD plasmas [105] in a packed-bed reactor with $BaTiO_3$ pellets were used for CO_2+H_2O reactions. The results showed CO_2 conversion of 12.3% and yields of 12.4% H_2, 11.8% CO and 2.8% O_2. Mahammadunnisa et al. [106] later used a Ni/g-Al_2O_3 catalyst for this reaction. They obtained the conversion to syngas ratio of 24–36% for the partially reduced catalyst. In addition, the formation of CH_3OH and other compounds such as C_2H_2 was also detected. Microwave discharge was also tried for CO_2+H_2O reforming [107]. The production of oxygenated species such as methanol, oxalic acid, and H_2O_2 was observed. Recently, Chen et al. [108] investigated the dissociation of CO_2 and H_2O to syngas in a surface-wave microwave discharge. It was shown that syngas with a H_2/CO ratio close to unity could be produced at a CO_2/H_2O molar ratio of 1:1. In addition, gliding arc was also used for the same reaction [94]. However, the energy conversion efficiency was low. It was shown that even a small amount of water addition reduced the CO_2 conversion, possibly due to the vibrational energy losses to water molecules or because of the decrease of electron number density. Therefore, more kinetic understanding of energy transfer between CO_2 and H_2O excited states in nonequilibrium plasma is needed.

$CO_2 + H_2$ *hydrogenation*: With the rapid development of hydrogen, research in CO_2+H_2 hydrogenation reactions also has attracted significant interest. In this reaction, hydrogen can provide H radicals to accelerate CO_2 reduction to CO and CH_4. In addition, the formation of OH radicals can also provide reaction pathways for the formation of alcohols and other oxygenates.

The major products in plasma-assisted hydrogenation of CO_2 are CO and H_2O, although some amounts of methane, methanol, and other oxygenates can be formed by choosing different catalysts and catalytic reforming temperatures and pressures. Maya [109] employed a microwave discharge in a gaseous mixture of CO_2/H_2. The results showed that the major products were CO and H_2O although a small amount of formic acid was also observed. Reduction of CO_2 with hydrogen in a nonequilibrium microwave plasma reactor was also examined by Fuente et al. [110]. The main products were also CO and H_2O. The reported CO_2 conversion ratio was 65%. Recently, CO_2 hydrogenation for CO production was investigated in Pd_2Ga/SiO_2 alloy catalysts by Kim et al. [111] in comparison to thermal catalysts. It was shown that both thermal and plasma catalysis showed close to 100% CO selectivity. Plasma catalysis increased CO_2 conversion by more than 2-fold and broke the thermodynamic equilibrium limitation. The results demonstrated that vibrationally excited CO_2 reacts directly with hydrogen adsorbed on Pd sites and accelerates formate formation. This work revealed the potential of designing new CO_2 hydrogenation catalysts toward value-added chemicals synthesis using nonequilibrium plasma.

Methanation of CO_2 with hydrogen was another important research direction of CO_2 hydrogenation. CO_2 methanation in a DBD-packed Ni/zeolite catalyst pellets was investigated by Jwa et al. in the 1990s [112]. More than 95% conversion of CO_2 was observed. It was inferred that the rate limiting reaction was adsorbed CO dissociation to carbon and oxygen atom on the catalyst surface and that the reactive species generated in the plasma reactor can speed up the rate-determining-step of the catalytic hydrogenation. Nizo et al. [113] conducted hybrid plasma-catalytic methanation of CO_2 at low temperature and atmospheric conditions over ceria zirconia supported Ni catalysts (Ni-$Ce_xZr_{1-x}O_2$). The experiments showed a promising result that CO_2 conversion was as high as 80% with 100% selectivity toward methane at 90°C. More recently, Nozaki et al. investigated the nonthermal plasma effect on CO_2 methanation over Ru-based multi-metallic catalysts in a packed-bed DBD reactor at reduced pressure [114]. The experiments revealed that the low-temperature CO_2 methanation was improved by Ru-based multi-metallic catalyst and that Ru is essential for leading to 100% CH_4 selectivity. In addition, it concluded that methanation performance was improved by La because the adsorption of vibrationally excited CO_2 over La sites is enhanced. This was one of the first experiments demonstrating that vibrational excitation of CO_2 increased yield and reduced the activation energy.

Direct production of methanol and oxygenate synthesis from CO_2 hydrogenation is of interest for E-fuels. CH_3OH is an important chemical feedstock for fuels and chemicals. However, direct production of methanol from CO_2 hydrogenation using thermal catalysis often requires high temperatures and high pressure to activate CO_2 and shift chemical equilibrium, thus is energy-intensive. Nonequilibrium plasma offers a very promising technique for direct CO_2 hydrogenation to methanol under low temperatures and ambient pressure. Eliasson et al. [115] studied the hydrogenation of CO_2 to methanol in a DBD reactor with $CuO/ZnO/Al_2O_3$ catalyst. The methanol yield at high pressure (8 atm) increased to 40% and the selectivity was around 10%. Recently, Wang et al. reported a plasma-catalytic process for direct CO_2 hydrogenation to methanol at room temperature and ambient pressure [116]. A DBD reactor using a water ground electrode was used to shift the chemical equilibrium to low temperature. A methanol yield of 11.3% and selectivity of 53.7% was achieved with Cu/γ-Al_2O_3 catalyst in the plasma. They also found that the reactor structure greatly affected the production of methanol. As such, more improvement in methanol yield and selectivity is needed.

1.10.3 Ammonia Synthesis and Nitrogen Fixation

Ammonia (NH_3) is an important fertilizer and a potential energy carrier for hydrogen. Ammonia has several advantages over compressed and liquid hydrogen such as higher energy density, easier to transport and storage, an established distribution network, more difficult to ignition and higher flammability limit for fire safety. Currently, ammonia is made by the Haber–Bosch (HB) process [117] (Figure 1.29) using the conventional fossil fuel-based, high pressure, and high-temperature process from nitrogen and hydrogen: $N_2(g)+3H_2(g) \leftrightarrow 2NH_3(g)$, $\Delta H^{\circ}_{298} = -91.9$ kJ/mol [118]. This reaction is exothermic and has a decrease in the number of moles of molecules. Le Chatelier's principle suggests that low temperature and high pressure favor NH_3 production. However, a relatively high temperature is still necessary for N_2 activation because of its strong triple covalent bond. As a result, the industrial HB process is operated at large scale powered by fossil-fuel-based heating at moderately high temperature (700–800 K) and high pressure (> 200 bar) with catalysts. As a result, the HB process requires significant energy input (e.g., 8.3–9.7 kWh/kg NH_3) [117] and results in a significant amount of carbon dioxide emissions (Figure 1.29, responsible for ~1.5% of global CO_2 emissions, the highest of any commodity chemical [119]. It is therefore imperative to develop sustainable and distributed NH_3 synthesis approaches for H_2 storage and utilization using nonequilibrium plasma (Figure 1.29).

To enable electrified nonequilibrium manufacturing, pulsed electrical heating and plasma have been developed for NH_3 synthesis [30,31,46,76,120–124]. Studies by Mehta et al. [46] suggested

Introduction

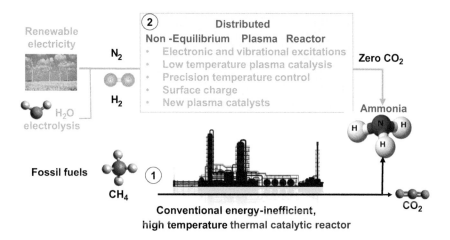

FIGURE 1.29 Schematic of ammonia synthesis from fossil fuel and nonequilibrium plasma green synthesis from renewable electricity.

that by populating vibrational excitation of nitrogen, the ammonia catalytic reaction rate (turnover frequency) can be increased by orders of magnitude. However, so far, the predicted drastic increase of ammonia reaction rate via vibrational excitations has not been experimentally validated. Instead, a recent experimental study showed that plasma-generated nitrogen atom plays a critical role in plasma-assisted ammonia synthesis [125]. Another DBD plasma study suggests that plasma-generated NH and NNH in gas phase kinetics could be another important pathway for plasma-assisted ammonia synthesis [47]. Both of these studies suggest that in addition to the Langmuir–Hinshelwood mechanism in thermal catalysis, Eley–Rideal mechanism and possibly the Mars–van Krevelen mechanism [126] via plasma-generated radicals may also play important roles. More recently, researchers also explored other mechanisms such as the surface charge effect and the surface porosity and temperature effects for plasma-assisted ammonia synthesis. Although it has been experimentally observed that plasma interaction with catalysts can enhance ammonia production at low temperature and atmospheric pressure, the underlying mechanisms are still far from understood. On the other hand, plasma-assisted ammonia synthesis has attracted much interest from researchers in various disciplines. Plasma physicists are designing hybrid plasmas and plasma instability theories to control plasma reactivity and plasma surface reaction. Material scientists are looking for new catalysts and catalyst structures to enhance plasma catalysis and stabilization of ammonia products in plasma. Various machine learning approaches and advanced laser diagnostic methods are also developed to understand the atomic scale reaction kinetics. At the same time, some chemical engineers are looking at the opportunity to use thermal plasma for nitrogen fixation by converting nitrogen and oxygen into NO [127].

Nevertheless, plasma catalysis is a very complicated multiphysics and multiscale problem. Several great challenges remain to be resolved before a large-scale commercial production can start. The mechanisms of non-plasma chemistry involving energy transfer between excited states, electrons, and radicals are not well understood. In addition, the interfacial kinetic mechanisms of plasma surface interactions involving active species reactions with catalysts and support materials remain difficult to explore. Moreover, *in situ* time-resolved diagnostic data and experimentally validated predictive plasma catalysis models for methane and plastics are not available. Moreover, since plasma reactivity is governed by electron energy distribution, electron number density, and temperature, it is necessary to control both electron temperature and electron number density as well as plasma instability at elevated pressure for commercial scale synthesis. There are numerous research opportunities in plasma-assisted chemical manufacturing.

1.10.4 Materials Manufacturing and Waste Materials Recycling

Electrified materials manufacturing: As shown in Figure 1.13, steelmaking and cement production globally accounted for 8% of CO_2 emissions [128] in 2021. The majority of CO_2 emissions in steelmaking arise from the coal-based reduction of ferrous oxide ores at 2,200°C in ironmaking [129]. Cement-making CO_2 emissions originate mostly from the byproducts of carbonate minerals in natural limestone and the use of fossil fuels for heating (~1,450°C) to form the "clinker phases" generated in furnaces from limestone, clay, and additives. Both processes rely on energy-intensive high-temperature reaction chambers that produce CO_2. Therefore, low-temperature plasma may not only remove CO_2 emissions but also have a great opportunity to lower the synthesis temperature by producing active species and higher heating flux. In addition, there is also an opportunity to apply carbon recycling to cement and steel manufacturing, where CO_2 generated from plasma-based synthesis of cement can offer key control over the flue gas composition.

Electrified materials recycling: Prolific plastic use has resulted in a surge of waste plastics and associated environmental concerns. To date, there exist ~8.3 billion metric tons of plastics globally [31], among which ~75% are wastes that can dramatically harm the environment if not properly treated. However, waste plastics offer alternative sources of high-value chemical products [130]. Existing steady-state plastic thermal pyrolysis-gasification via catalytic reforming at high temperature (~800°C) is not only carbon- and energy-intensive but also causes carbonaceous coke deposition on the catalyst due to chemical equilibrium. Thus, there is a need to develop a novel electrified nonequilibrium process for plastic pyrolysis using low-temperature ferroelectric plasma catalysis to increase the selectivity and efficiency and suppress coke formation.

Also with the rapid development of electric vehicles, By 2025 it is estimated that the global LiB market will reach 10 million tons of cathode materials [131]. Thus, these LiB batteries will need to be recycled. However, the current battery recycling technology heavily relies on the pyrolysis and mechanical-hydro-metallurgical processes and requires a series of thermal treatment and chemical leaching to break down the cathode materials into atomic form by using H_2SO_4. As such, these recycling processes are very energy and chemicals-intensive, and environmentally unfriendly. Therefore, there is an urgent need to develop a novel method using combined electric heating and plasma that directly upgrades the aged low energy density cathode materials without breaking down the materials structure and producing wastewater [132].

1.11 AIM OF THIS BOOK

Because of the need to reduce carbon emissions from fossil energy use, improvement of energy efficiency and electrified manufacturing have attracted significant attention. Nonequilibrium plasma powered by renewable electricity is a promising technology to enable nonequilibrium energy conversion to improve energy efficiency and selectivity. Recently, many exciting research works and innovative patents using nonequilibrium plasmas have been published. Unfortunately, plasma-assisted combustion and manufacturing is a young and multidisciplinary research area. It involves knowledge in plasma, combustion, fuels, chemistry, materials, physics, diagnostics, and computation. Moreover, there are many physics and chemistry in plasma-assisted combustion and manufacturing that are not well understood. Recently, there have several excellent review articles in plasma discharge [37,133], plasma kinetics [36], plasma-assisted combustion [42,43,51,52], and plasma catalysis [44,134,135]. However, there is not a book that presents plasma physics, chemistry, diagnostics, and modeling methods for applications in combustion and chemical manufacturing for graduate students and researchers. As such, this book aims to describe the fundamentals of plasma physics and chemistry as well as the state-of-the-art laser diagnostics and multiscale modeling methods for understanding the physics and chemistry in plasma-assisted combustion and manufacturing. In addition, it will summarize the research frontiers and recent progress in plasma-assisted combustion and manufacturing. We hope this book will serve as a gateway for graduate students, researchers, and engineers who are interested in research and technology development of plasma-assisted combustion and chemical manufacturing.

In this book, Ju contributed the writing of Chapters 1, 5–9. Starikovsky wrote Chapters 2–4. YJ would like to thank the encouragement and support to write this book from Prof. Chung K. Law at Princeton University, Prof. Yuji Suzuki at the University of Tokyo, his wife, Limei Zhu, and his two children, Alex Ju and Caren Ju. Ju is also indebted to the help from Drs. Timothy Chen, Zheng Chen, Ziqiao Chang, Joseph Lefkowitz, Aditya Lele, Ning Liu, Xingqian Mao, Timothy Ombrello, Aric Russo, Wenting Sun, Madeline Vorenkamp, Sanghee Won, Wenbin Xu, Hongtao Zhong, Arthur Dogariu, Richard Miles, Mikhail Shneider, and Gerard Wysocki who worked and collaborated with Ju in plasma at Princeton University for editing some chapters and sharing figures as well as those who have worked with Ju and contributed significantly to the research outcomes highlighted in this book. This book is not possible without their contributions. Ju also would like to thank his many distinguished collaborators in combustion and plasma research, especially Dr. Julian Tishkoff at Air Force Air Force Office of Scientific Research and Dr. Campbell Carter at Air Force Research Laboratory for funding and supporting his first research project in plasma assisted combustion. Ju also would like to thank the funding support from DOE-BES, DOE-EERC, DOE-FES, and NSF for the work published in this book.

REFERENCES

1. Pro Oxygen, Creator of CO2.Earth, 2024, *NOAA Monthly CO_2 Data*, https://www.co2.earth/monthly-co2. Email: aloha@co2.earth
2. IEA (2019), *Global Energy & CO_2 Status Report 2019*, https://www.iea.org/reports/global-energy-co₂-status-report-2019/emissions.
3. IPCC, 2021, *Climate Change Widespread, Rapid, and Intensifying – IPCC*, https://www.ipcc.ch/2021/08/09/ar6-wg1-20210809-pr/. 2021.
4. NASA, 2024, https://climate.nasa.gov/vital-signs/arctic-sea-ice/.
5. EIA, 2024, *U.S. Energy Facts Explained*, https://www.eia.gov/energyexplained/us-energy-facts/.
6. Tejvan Pettinger, 2020, *List of Countries Energy Use per Capita*, https://www.economicshelp.org/blog/5988/economics/list-of-countries-energy-use-per-capita/.
7. EIA, 2024, *Today in Energy*, https://www.eia.gov/todayinenergy/detail.php?id=41433.
8. Seinfeld, J.H. and S.N. Pandis, *Atmospheric Chemistry and Physics: From Air Pollution to Climate Change*. 2016: John Wiley & Sons.
9. Wikipedia, 2013, *2013 Eastern China Smog*, https://en.wikipedia.org/wiki/2013_Eastern_China_smog.
10. US EPA, 2020, *Our Nation's Air*, https://gispub.epa.gov/air/trendsreport/2021/. 2021.
11. Mike Millikin, Bioagemedia, *2024, Green Car Congress*, https://www.greencarcongress.com/2014/07/20140728-navigant.html. Email: mmillikin@bioagemedia.com
12. Sayre, R., *Next Generation Biofuel from Algae*. Los Alamos National Lab (LANL), Los Alamos, NM (United States) 2013. **LA-UR-13-24632**.
13. Lu, X., D. Han, and Z. Huang, Fuel design and management for the control of advanced compression-ignition combustion modes. *Progress in Energy and Combustion Science*, 2011. **37**(6): pp. 741–783.
14. Dec, J.E., Advanced compression-ignition engines—understanding the in-cylinder processes. *Proceedings of the Combustion Institute*, 2009. **32**(2): pp. 2727–2742.
15. Splitter, D., R. Reitz, and R. Hanson, High efficiency, low emissions RCCI combustion by use of a fuel additive. *SAE International Journal of Fuels and Lubricants*, 2010. **3**(2): pp. 742–756.
16. Kalghatgi, G., Developments in internal combustion engines and implications for combustion science and future transport fuels. *Proceedings of the Combustion Institute*, 2015. **35**(1): pp. 101–115.
17. Musculus, M.P., P.C. Miles, and L.M. Pickett, Conceptual models for partially premixed low-temperature diesel combustion. *Progress in Energy and Combustion Science*, 2013. **39**(2): pp. 246–283.
18. Hildingsson, L., et al., *Fuel Octane Effects in the Partially Premixed Combustion Regime in Compression Ignition Engines*. 2009, SAE Technical Paper 01-2648.
19. Ju, Y., et al., Multi-timescale modeling of ignition and flame regimes of n-heptane-air mixtures near spark assisted homogeneous charge compression ignition conditions. *Proceedings of the Combustion Institute*, 2011. **33**(1): pp. 1245–1251.

20. Manofsky, L., et al., *Bridging the Gap between HCCI and SI: Spark-Assisted Compression Ignition.* 2011, SAE Technical Paper 01-1179.
21. Iida, N., Research and development of super lean burn technology for efficient gasoline engines. *Public symposium on Innovative Combustion Technology*, 2019.1.28. https://www.jst.go.jp/sip/k01_kadai_siryo0129.html. 2019.
22. Ju, Y., et al., Dynamics of Cool Flames. *Progress in Energy and Combustion Science*, 2019. **75**(100787): p. 39.
23. Farrell, J., *Co-Optimization of Fuels & Engines (Co-Optima) Initiative: Recent Progress on Light-Duty Boosted Spark-Ignition Fuels/Engines.* 2017. National Renewable Energy Laboratory (NREL).
24. Heyne, J.S., et al., Year 3 of the national jet fuels combustion program: Practical and scientific impacts of alternative jet fuel research. in *2018 AIAA Aerospace Sciences Meeting*. Kissimmee, Florida, 2018.
25. Kohse-Höinghaus, K., et al., Biofuel combustion chemistry: From ethanol to biodiesel. *Angewandte Chemie International Edition*, 2010. **49**(21): pp. 3572–3597.
26. National Academies of Sciences, Engineering and Medicine, *Advanced Technologies for Gas Turbines.* 2020: National Academies Press.
27. Levi, P., et al., *Chemicals.* 2020, IEA. https://www.iea.org/reports/chemicals.
28. Bertagni, M.B., et al., Minimizing the impacts of the ammonia economy on the nitrogen cycle and climate. *Proceedings of the National Academy of Sciences*, 2023. **120**(46): p. e2311728120.
29. Kobayashi, H., et al., Science and technology of ammonia combustion. *Proceedings of the Combustion Institute*, 2019. **37**(1): pp. 109–133.
30. Gorky, F., et al., Plasma ammonia synthesis over mesoporous silica SBA-15. *Journal of Physics D: Applied Physics*, 2021. **54**(26): p. 264003.
31. Dong, Q., et al., Programmable heating and quenching for efficient thermochemical synthesis. *Nature*, 2022. **605**(7910): pp. 470–476.
32. Erisman, J.W., et al., How a century of ammonia synthesis changed the world. *Nature Geoscience*, 2008. **1**(10): p. 636.
33. U.S. EIA, *Annual Energy Outlook 2021*, https://www.eia.gov/outlooks/aeo/pdf/00%20AEO2021%20Chart%20Library.pdf. 2021.
34. Worrell, E., et al., Energy Use and Energy Intensity of the U.S. Chemical Industry. Lawrence Berkeley National Laboratory Report, 2000. **LBNL-44314**.
35. EIA, *Production of Hydrogen*, https://www.eia.gov/energyexplained/hydrogen/production-of-hydrogen.php. 2021.
36. Gnanapragasam, N.V. and M.A. Rosen, A review of hydrogen production using coal, biomass and other solid fuels. *Biofuels*, 2017. **8**(6): pp. 725–745.
37. Abbas, H.F. and W.M.A. Wan Daud, Hydrogen production by methane decomposition: A review. *International Journal of Hydrogen Energy*, 2010. **35**(3): pp. 1160–1190.
38. EPA, *Greenhouse Gas Emissions*, https://www.epa.gov/ghgemissions/overview-greenhouse-gases.
39. EIA, *US Electricity Generation from Renewable Energy Sources*, https://www.eia.gov/energyexplained/electricity/electricity-in-the-us.php. 2020.
40. Ongena, J., et al., *Hidden Consequences of Intermittent Electricity Production*, https://www.wind-watch.org/documents/hidden-consequences-of-intermittent-electricity-production/. 2017.
41. Kameshima, S., et al., Parametric analysis of plasma-assisted pulsed dry methane reforming over Ni/Al_2O_3 catalyst. *Plasma Processes and Polymers*, 2017. **14**(6): p. 1600096.
42. Ju, Y. and W. Sun, Plasma assisted combustion: Dynamics and chemistry. *Progress in Energy and Combustion Science*, 2015. **48**: pp. 21–83.
43. Starikovskaia, S.M., Plasma assisted ignition and combustion. *Journal of Physics D: Applied Physics*, 2006. **39**(16): p. R265.
44. Bogaerts, A., et al., The 2020 plasma catalysis roadmap. *Journal of Physics D: Applied Physics*, 2020. **53**(44): p. 443001.
45. Langmuir, I., Oscillations in ionized gases. *Proceedings of the National Academy of Sciences*, 1928. **14**(8): pp. 627–637.
46. Mehta, P., et al., Overcoming ammonia synthesis scaling relations with plasma-enabled catalysis. *Nature Catalysis*, 2018. **1**(4): p. 269.
47. Zhao, H., et al., *In situ* identification of NNH and N_2H_2 by using molecular-beam mass spectrometry in plasma-assisted catalysis for NH_3 synthesis. *ACS Energy Letters*, 2021. **7**: pp. 53–58.
48. Kimura, M. and Y. Itikawa, *Advances in Atomic, Molecular, and Optical Physics: Electron Collisions with Molecules in Gases: Applications to Plasma Diagnostics and Modeling.* 2000: Elsevier.

49. Puliyalil, H., et al., A review of plasma-assisted catalytic conversion of gaseous carbon dioxide and methane into value-added platform chemicals and fuels. *RSC Advances*, 2018. **8**(48): pp. 27481–27508.
50. Pilla, G., et al., Stabilization of a turbulent premixed flame using a nanosecond repetitively pulsed plasma. *IEEE Transactions on Plasma Science*, 2006. **34**(6): pp. 2471–2477.
51. Starikovskiy, A. and N. Aleksandrov, Plasma-assisted ignition and combustion. *Progress in Energy and Combustion Science*, 2013. **39**(1): pp. 61–110.
52. Lacoste, D.A., Flames with plasmas. *Proceedings of the Combustion Institute*, 2023. **39**(4): pp. 5405–5428.
53. Sun, W., et al., The effect of ozone addition on combustion: Kinetics and dynamics. *Progress in Energy and Combustion Science*, 2019. **73**: pp. 1–25.
54. Vorenkamp, M., et al., Plasma-assisted deflagration to detonation transition in a microchannel with fast-frame imaging and hybrid fs/ps coherent anti-Stokes Raman scattering measurements. *Proceedings of the Combustion Institute*, 2023. **39**(4): pp. 5561–5569.
55. Chen, T.Y., et al., Time-resolved characterization of plasma properties in a CH4/He nanosecond-pulsed dielectric barrier discharge. *Journal of Physics D: Applied Physics*, 2019. **52**(18): p. 18LT02.
56. Chen, T.Y., et al., Time-resolved *in situ* measurements and predictions of plasma-assisted methane reforming in a nanosecond-pulsed discharge. *Proceedings of the Combustion Institute*, 2021. **38**: pp. 1–9.
57. Chen, T.Y., et al., *Single Shot Burst Imaging of Electric Fields Measured by Split Excitation Electric Field Induced Second Harmonic Generation (SEE-FISH)*. 2020, Sandia National Lab (SNL-NM).
58. Raskar, S., et al., Spatially enhanced electric field induced second harmonic (SEEFISH) generation for measurements of electric field distributions in high-pressure plasmas. *Plasma Sources Science and Technology*, 2022. **31**(8): p. 085002.
59. Dogariu, A., et al., Species-independent femtosecond localized electric field measurement. *Physical Review Applied*, 2017. **7**(2): p. 024024.
60. Germann, G.J. and J.J. Valentini, CARS study of the OH (2Π) radical product of the photodissociation of H_2O_2 AT 266 nm. *Chemical Physics Letters*, 1989. **157**(1–2): pp. 51–54.
61. Roy, S., et al., Temperature measurements in reacting flows by time-resolved femtosecond coherent anti-Stokes Raman scattering (fs-CARS) spectroscopy. *Optics Communications*, 2008. **281**(2): pp. 319–325.
62. Chen, T.Y., et al., Simultaneous single-shot rotation–vibration non-equilibrium thermometry using pure rotational fs/ps CARS coherence beating. *Optics Letters*, 2022. **47**(6): pp. 1351–1354.
63. Liu, N., et al., Sensitive and single-shot OH and temperature measurements by femtosecond cavity-enhanced absorption spectroscopy. *Optics Letters*, 2022. **47**(13): pp. 3171–3174.
64. Stancu, G., et al., Atmospheric pressure plasma diagnostics by OES, CRDS and TALIF. *Journal of Physics D: Applied Physics*, 2010. **43**(12): p. 124002.
65. Dogariu, A., et al., Neutral Atomic-Hydrogen Measurements in a Mirror/FRC Plasma Device using fs-TALIF. in *APS Annual Gaseous Electronics Meeting Abstracts*. October 5–9, 2020, Virtual Conference, 2020.
66. Ding, P., et al., Temporal dynamics of femtosecond-TALIF of atomic hydrogen and oxygen in a NRP discharge-assisted methane-air flame. *Journal of Physics D: Applied Physics*, 2021. **54**(275201): pp. 1–9.
67. Zhong, H., et al., Kinetic studies of plasma assisted n-dodecane/O_2/N_2 pyrolysis and oxidation in a nanosecond-pulsed discharge. *Proceedings of Combustion Insittute*, 2021. **38**: pp. 6521–6531.
68. Mao, X., et al., Modeling of the effects of non-equilibrium excitation and electrode geometry on H_2/air ignition in a nanosecond plasma discharge. *Combustion and Flame*, 2022. **240**: p. 112046.
69. Mao, X., et al., Effects of inter-pulse coupling on nanosecond pulsed high frequency discharge ignition in a flowing mixture. *Proceedings of the Combustion Institute*, 2023. **39**: pp. 5457–5464.
70. Cheng, L., et al., *AVIP: A Low Temperature Plasma Code*. arXiv preprint arXiv:2201.01291, 2022.
71. Zhong, H., et al., Thermal-chemical instability of weakly ionized plasma in a reactive flow. *Journal of Physics D: Applied Physics*, 2019. **52**(48): p. 484001.
72. Zhong, H., et al., Dynamics and chemical mode analysis of plasma thermal-chemical instability. *Plasma Sources Science and Technology*, 2021. **30**: p. 035002.
73. Zhong, H., et al., Plasma thermal-chemical instability of low-temperature dimethyl ether oxidation in a nanosecond-pulsed dielectric barrier discharge. *Plasma Sources Science and Technology*, 2022. **31**(11): p. 114003.
74. Dong, Q., et al., Programmable heating and quenching for non-equilibrium thermochemical synthesis. *Nature*, 2022. **605**(7910): pp. 470–476.

75. Mao, X., et al., Numerical modeling of ignition enhancement of CH_4/O_2/He mixtures using a hybrid repetitive nanosecond and DC discharge. *Proceedings of the Combustion Institute*, 2019. **37**: pp. 5545–5552.
76. Sun, J., et al., A hybrid plasma electrocatalytic process for sustainable ammonia production. *Energy & Environmental Science*, 2021. **14**(2): pp. 865–872.
77. Orr, K., et al., Characterization and kinetic modeling of Ns pulse and hybrid Ns pulse/RF plasmas. in *AIAA SciTech 2021 Forum*. Jan. 11-21, Virtual meeting, 2021.
78. Dragoon, K., et al., Hydrogen as part of a 100% clean energy system: Exploring its decarbonization roles. *IEEE Power and Energy Magazine*, 2022. **20**(4): pp. 85–95.
79. DOE. *Pathways to Commercial Liftoff: Clean Hydrogen*, https://liftoff.energy.gov/wp-content/uploads/2023/03/20230320-Liftoff-Clean-H2-vPUB.pdf. 2023.
80. Diab, J., et al., Why turquoise hydrogen will Be a game changer for the energy transition. *International Journal of Hydrogen Energy*, 2022. **47**(61): pp. 25831–25848.
81. Chase, M.W. and N.I.S. Organization, *NIST-JANAF Thermochemical Tables*. Vol. 9. 1998: American Chemical Society.
82. Sánchez-Bastardo, N., R. Schlögl, and H. Ruland, Methane pyrolysis for zero-emission hydrogen production: A potential bridge technology from fossil fuels to a renewable and sustainable hydrogen economy. *Industrial & Engineering Chemistry Research*, 2021. **60**(32): pp. 11855–11881.
83. Dong, Q., et al., Depolymerization of plastics by means of electrified spatiotemporal heating. *Nature*, 2023. **616**(7957): pp. 488–494.
84. Li, H., et al., Expanding plastics recycling technologies: Chemical aspects, technology status and challenges. Green Chemistry, 2022. **24**: pp.8899–9002.
85. Zheng, X., et al., Silica-coated $LaNiO_3$ nanoparticles for non-thermal plasma assisted dry reforming of methane: Experimental and kinetic studies. *Chemical Engineering Journal*, 2015. **265**: pp. 147–156.
86. Ravari, F., et al., Kinetic model study of dry reforming of methane using cold plasma. *Physical Chemistry Research*, 2017. **5**(2): pp. 395–408.
87. Li, D., et al., CO_2 reforming of CH_4 by atmospheric pressure glow discharge plasma: A high conversion ability. *International Journal of Hydrogen Energy*, 2009. **34**(1): pp. 308–313.
88. Garduño, M., et al., Hydrogen production from methane conversion in a gliding arc. *Journal of Renewable and Sustainable Energy*, 2012. **4**: pp. 021202.
89. Zhang, H., et al., Plasma activation of methane for hydrogen production in a N_2 rotating gliding arc warm plasma: A chemical kinetics study. *Chemical Engineering Journal*, 2018. **345**: pp. 67–78.
90. Kreuznacht, S., et al., Comparison of the performance of a microwave plasma torch and a gliding arc plasma for hydrogen production via methane pyrolysis. *Plasma Processes and Polymers*, 2023. **20**(1): p. 2200132.
91. Pino, L., et al., Kinetic study of the methane dry (CO_2) reforming reaction over the $Ce_{0.70}La_{0.20}Ni_{0.10}O_{2-\delta}$ catalyst. *Catalysis Science & Technology*, 2020. **10**(8): pp. 2652–2662.
92. Ozkan, A., A. Bogaerts, and F. Reniers, Routes to increase the conversion and the energy efficiency in the splitting of CO_2 by a dielectric barrier discharge. *Journal of Physics D: Applied Physics*, 2017. **50**(8): p. 084004.
93. Spencer, L. and A. Gallimore, CO_2 dissociation in an atmospheric pressure plasma/catalyst system: A study of efficiency. *Plasma Sources Science and Technology*, 2012. **22**(1): p. 015019.
94. Nunnally, T., et al., Dissociation of CO_2 in a low current gliding arc plasmatron. *Journal of Physics D: Applied Physics*, 2011. **44**(27): p. 274009.
95. Li, L., et al., Magnetically enhanced gliding arc discharge for CO_2 activation. *Journal of CO_2 Utilization*, 2020. **35**: pp. 28–37.
96. Snoeckx, R. and A. Bogaerts, Plasma technology – a novel solution for CO_2 conversion? *Chemical Society Reviews*, 2017. **46**(19): pp. 5805–5863.
97. Uytdenhouwen, Y., et al., How process parameters and packing materials tune chemical equilibrium and kinetics in plasma-based CO_2 conversion. *Chemical Engineering Journal*, 2019. **372**: pp. 1253–1264.
98. King, B., et al., Comprehensive process and environmental impact analysis of integrated DBD plasma steam methane reforming. *Fuel*, 2021. **304**: p. 121328.
99. Chen, X., et al., CH_4 dry reforming in fluidized-bed plasma reactor enabling enhanced plasma-catalyst coupling. *Journal of CO_2 Utilization*, 2021. **54**: p. 101771.
100. Zhang, J., et al., Study on the conversion of CH_4 and CO_2 using a pulsed microwave plasma under atmospheric pressure. *Acta Chimica Sinica*, 2002. **60**(11): p. 1973.
101. Martin-del-Campo, J., S. Coulombe, and J. Kopyscinski, Influence of operating parameters on plasma-assisted dry reforming of methane in a rotating gliding arc reactor. *Plasma Chemistry and Plasma Processing*, 2020. **40**(4): pp. 857–881.

102. Chun, Y.N., et al., Hydrogen-rich gas production from biogas reforming using plasmatron. *Energy & Fuels*, 2008. **22**(1): pp. 123–127.
103. Wang, L., et al., One-step reforming of CO_2 and CH_4 into high-value liquid chemicals and fuels at room temperature by plasma-driven catalysis. *Angewandte Chemie*, 2017. **129**(44): pp. 13867–13871.
104. Scapinello, M., L.M. Martini, and P. Tosi, CO_2 hydrogenation by CH_4 in a dielectric barrier discharge: Catalytic effects of nickel and copper. *Plasma Processes and Polymers*, 2014. **11**(7): pp. 624–628.
105. Futamura, S. and H. Kabashima, Synthesis gas production from CO_2 and H_2O with nonthermal plasma, *Studies in Surface Science and Catalysis*. 2004, **153**: pp. 119–124.
106. Mahammadunnisa, S., et al., CO2 reduction to syngas and carbon nanofibres by plasma-assisted *in situ* decomposition of water. *International Journal of Greenhouse Gas Control*, 2013. **16**: pp. 361–363.
107. Ihara, T., M. Kiboku, and Y. Iriyama, Plasma reduction of CO_2 with H_2O for the formation of organic compounds. *Bulletin of the Chemical Society of Japan*, 1994. **67**(1): pp. 312–314.
108. Chen, G., et al., Simultaneous dissociation of CO_2 and H_2O to syngas in a surface-wave microwave discharge. *International Journal of Hydrogen Energy*, 2015. **40**(9): pp. 3789–3796.
109. Maya, L., Plasma-assisted reduction of carbon dioxide in the gas phase. *Journal of Vacuum Science & Technology A: Vacuum, Surfaces, and Films*, 2000. **18**(1): pp. 285–287.
110. de la Fuente, J.F., et al., Reduction of CO_2 with hydrogen in a non-equilibrium microwave plasma reactor. *International Journal of Hydrogen Energy*, 2016. **41**(46): pp. 21067–21077.
111. Kim, D.-Y., et al., Cooperative catalysis of vibrationally excited CO_2 and alloy catalyst breaks the thermodynamic equilibrium limitation. *Journal of the American Chemical Society*, 2022. **144**(31): pp. 14140–14149.
112. Jwa, E., et al., Plasma-assisted catalytic methanation of CO and CO_2 over Ni–zeolite catalysts. *Fuel Processing Technology*, 2013. **108**: pp. 89–93.
113. Nizio, M., et al., Hybrid plasma-catalytic methanation of CO_2 at low temperature over ceria zirconia supported Ni catalysts. *International Journal of Hydrogen Energy*, 2016. **41**(27): pp. 11584–11592.
114. Zhan, C., et al., Nonthermal plasma catalysis of CO_2 methanation over multi-metallic Ru based catalysts. *International Journal of Plasma Environmental Science & Technology*, 2022. **16**: p. c03006.
115. Eliasson, B., et al., Hydrogenation of carbon dioxide to methanol with a discharge-activated catalyst. *Industrial & Engineering Chemistry Research*, 1998. **37**(8): pp. 3350–3357.
116. Wang, L., et al., Atmospheric pressure and room temperature synthesis of methanol through plasma-catalytic hydrogenation of CO_2. *ACS Catalysis*, 2018. **8**(1): pp. 90–100.
117. Smith, C., A.K. Hill, and L. Torrente-Murciano, Current and future role of Haber–Bosch ammonia in a carbon-free energy landscape. *Energy & Environmental Science*, 2020. **13**(2): pp. 331–344.
118. Wikipedia, 2023, https://www.iipinetwork.org/wp-content/Ietd/content/ammonia.html.
119. The Institute for Industrial Productivity acknowledges Fraunhofer ISI, IREES, LBNL-China Energy Group, E3M Inc., ISR-UC, Holtec, Utrecht University, and FAI for their valuable contributions. *Ammonia*, https://www.iipinetwork.org/wp-content/Ietd/content/ammonia.html.
120. Nakajima, J. and H. Sekiguchi, Synthesis of ammonia using microwave discharge at atmospheric pressure. *Thin Solid Films*, 2008. **516**(13): pp. 4446–4451.
121. Peng, P., et al., Atmospheric pressure ammonia synthesis using non-thermal plasma assisted catalysis. *Plasma Chemistry and Plasma Processing*, 2016. **36**(5): pp. 1201–1210.
122. Kim, H.H., et al., Atmospheric-pressure nonthermal plasma synthesis of ammonia over ruthenium catalysts. *Plasma Processes and Polymers*, 2017. **14**(6): p. 1600157.
123. Hawtof, R., et al., Catalyst-free, highly selective synthesis of ammonia from nitrogen and water by a plasma electrolytic system. *Science Advances*, 2019. **5**(1): p. eaat5778.
124. Engelmann, Y., et al., Plasma catalysis for ammonia synthesis: A microkinetic modeling study on the contributions of Eley–Rideal reactions. *ACS Sustainable Chemistry & Engineering*, 2021. **9**(39): pp. 13151–13163.
125. Barboun, P.M., et al., Plasma-catalyst reactivity control of surface nitrogen species through plasma-temperature-programmed hydrogenation to ammonia. *ACS Sustainable Chemistry & Engineering*, 2022. **10**(48): pp. 15741–15748.
126. Tricker, A.W., et al., Mechanocatalytic ammonia synthesis over TiN in transient microenvironments. *ACS Energy Letters*, 2020. **5**(11): pp. 3362–3367.
127. Kim, H.-H. Sustainable Nitrogen Fixation Using Spark Discharge for Power-to-X. in WE-Heraeus-Semenar @Bad Honnef (26 April 2023). 2023.
128. Spreitzer, D. and J. Schenk, Reduction of iron oxides with hydrogen—a review. *Steel Research International*, 2019. **90**(10): p. 1900108.

129. Conejo, A.N., J.-P. Birat, and A. Dutta, A review of the current environmental challenges of the steel industry and its value chain. *Journal of Environmental Management*, 2020. **259**: p. 109782.
130. Wilk, V. and H. Hofbauer, Conversion of mixed plastic wastes in a dual fluidized bed steam gasifier. *Fuel*, 2013. **107**: pp. 787–799.
131. Rothermel, S., et al., Graphite recycling from spent lithium-ion batteries. *ChemSusChem*, 2016. **9**(24): pp. 3473–3484.
132. Koel, B., et al., *Recycling Lithium-Ion Batteries*. Andlinger Center for Energy and the Environment, Princeton University, March 3, 2022. https://www.princeton.edu/news/2022/03/01/better-way-recycle-lithium-batteries-coming-soon-princeton-startup
133. Raizer, Y.P., *Gas Discharge Physics*. 1991.
134. Bogaerts, A., et al., CO_2 conversion by plasma technology: Insights from modeling the plasma chemistry and plasma reactor design. *Plasma Sources Science and Technology*, 2017. **26**(6): p. 063001.
135. Chen, G., et al., Plasma pyrolysis for a sustainable hydrogen economy. *Nature Reviews Materials*, 2022. **7**(5): pp. 333–334.

2 Low-Temperature Plasma Generation and Recombination

2.1 INTRODUCTION

The use of plasma in thermal equilibrium to modulate combustion processes is a century-old concept that was first applied to internal combustion engines and spark ignition mechanisms [1–3]. These fundamental concepts continue to be relevant in modern efforts to improve combustion efficiency in a variety of fields. Recently, there has been a surge of interest in the use of nonequilibrium plasma for ignition and combustion optimization. This growing fascination is largely due to the introduction of novel approaches to enhance ignition and stabilize flames using plasma-assisted techniques. Over the past few decades, significant progress has been made in delineating the interactions between plasma and chemical processes, energy distribution, and combustion initiation under nonequilibrium conditions, with investigations involving a wide range of fuels and different types of plasma discharges [4].

Extensive reviews of experimental research on nonequilibrium plasma-assisted ignition (PAI) and combustion are documented in the scientific literature [4–10]. In recent years, remarkable progress has been made in unraveling the underlying mechanisms of plasma-assisted combustion (PAC) in a wide range of mixtures, particularly those containing hydrocarbons [5].

These mechanisms have been corroborated by experiments conducted under rigorously controlled conditions and subsequent comparison with numerical simulations of both discharge and combustion phenomena. This discourse explores the feasibility of generating chemically active discharge plasmas with optimal parameters and examines the impact of nonequilibrium plasmas on ignition and combustion processes through various mechanisms [11]. It also addresses the challenges encountered in these investigations and highlights unresolved issues. The proliferation of research groups working on PAC indicates a growing interest in this field. The purpose of this book is to summarize the key discoveries in the physics and chemistry underlying PAI and combustion, to identify barriers to further progress in the use of nonequilibrium plasma for combustion enhancement, and to present state-of-the-art diagnostic methods for the study and analysis of plasma processes.

2.1.1 CHEMISTRY UNDER NON-BOLTZMANN CONDITIONS

The fundamental difference between conventional combustion and PAC is the significant nonequilibrium excitation of gases in the discharge plasma [11]. This excitation results from an external electric field that accelerates electrons, leading to remarkably slow energy exchange between these electrons and the translational motion of the molecules due to their substantial mass discrepancy. Consequently, electron collisions predominantly transfer energy to the internal degrees of freedom of the molecules. In scenarios where the rate of internal energy relaxation remains modest, the distribution of excited states of the molecules diverges significantly from the equilibrium Maxwell-Boltzmann distribution [12,13]. The resulting excess of excited states, along with the thermally-nonequilibrium dissociation and ionization of molecules that can further induce UV generation and additional gas heating, enhances the reactivity of the system. This enhancement greatly facilitates ignition and flame propagation [14–18]. Therefore, in the context of plasma-stimulated chemistry, key questions revolve around the pathways by which the discharge energy is distributed among the various molecular degrees of freedom, the rate at which the system undergoes relaxation or thermalization, and how the chemically active entities of the system respond to such nonequilibrium excitations.

The population rate of excited states in an electrical discharge depends on the energy of the electrons. The excitation of molecular rotational degrees of freedom requires the least amount of electron energy, with a typical rotational quantum energy being only a few degrees Kelvin. Consequently, an electron energy of about 300 K (~ 0.03 eV) is sufficient to effectively excite molecular rotations. In contrast, the quantum energy required to excite vibrational degrees of freedom is in the range of 1–5 K, requiring a higher average electron energy for effective vibrational excitation, specifically in the range of 1–4 eV for air. Excitation of electronical degrees of freedom and molecular dissociation require even higher energies, typically between 3 and 10 eV. When the mean electron energy exceeds 10 eV, ionization of the gas becomes the dominant process in the plasma. Therefore, the ability to modulate electron energy provides a means to selectively direct energy deposition and excite different molecular degrees of freedom, thereby providing some control over the chemical and physical processes of the plasma [19].

2.1.2 Temperature of Electrons in the Plasma

The mean electron energy in a gas discharge is determined by the ratio of the electric field strength, E, to the gas density, n, the so-called reduced electric field E/n [20]. This relationship is illustrated in Figure 2.1, which shows the divergence of the characteristic electron energy, expressed as D/μ, from the molecular temperature, T, in various gases. In this context, D is the electrons' diffusion coefficient and μ is the electrons' mobility. A critical E/n value, at which a noticeable difference between D/μ and T occurs, leading to the establishment of a nonequilibrium electron energy distribution function (EEDF), is approximately $E/n \sim 0.1$ Td for atomic gases and $E/n \sim 1$ Td for molecular gases, where 1 Td is equal to 10^{-17} V×cm².

The nonequilibrium EEDF emerges as a solution to the Boltzmann equation. At its most fundamental level, this solution can be approximated as a steady-state function of energy that depends

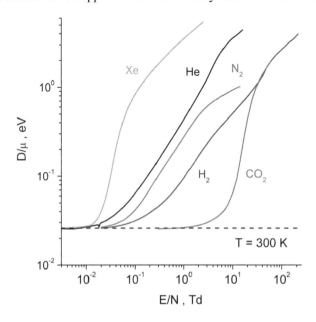

FIGURE 2.1 Characteristic electron energy [5]. He and Ar – [21]; H_2, N_2, and CO_2 – [22].

TABLE 2.1
Electron-Molecules Collisions Cross Sections

Atmospheric	Saturated	Unsaturated	Oxygenated	Isomers
N_2	H_2	C_2H_2	CO	Iso-propane
O_2	CH_4	C_2H_4	CH_3OH	Iso-butane
H_2O	C_2H_6	C_3H_6	C_2H_5OH	Neo-pentane
Ar	C_3H_8		CH_3OCH_3 (DME)	
CO_2	C_4H_{10}			
O_3	C_5H_{12}			
N_2O				

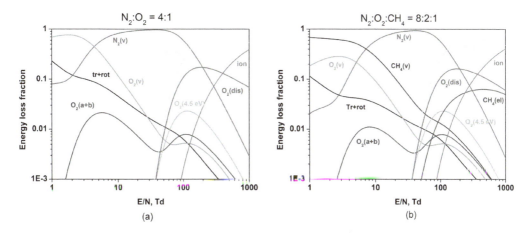

FIGURE 2.2 Fractional power dissipated by electrons into the different molecular degrees of freedom as a function of E/n. (a) Air; (b) methane-air stoichiometric mixtures [5].

only on the local electric field. This approximation can be further refined by using a two-term approximation, where the electron distribution function is written as $f(v) = f_0(v) + f_1(v)\cos(\theta)$ [20], where v represents the electron velocity and θ represents the angle between velocity vector v and the electric field vector E. This approach requires a comprehensive set of cross-sectional data for electron-molecule interactions. Table 2.1 summarizes the currently available self-consistent electron cross-section sets relevant to the modeling of PAC processes.

The derived EEDF allows quantification of the energy distribution among the various molecular degrees of freedom. This distribution is illustrated in Figure 2.2, which shows the distribution of energy among the internal degrees of freedom for certain gases at various E/n ratios in a discharge. At very low E/n values, about ~ 0.1 Td, rotational excitation predominates. The fast exchange of energy between rotational and translational degrees of freedom leads to thermally equilibrated heating of the gas within such discharge scenarios. An increase in the E/n value up to 0.4 Td changes the distribution of energy.

Beyond this threshold, the vibrational excitation of oxygen emerges as the primary pathway for electron energy dissipation. The superior efficiency in electron energy loss is attributed to the excitation of the vibrational states of nitrogen, as shown in Figure 2.2, in the reduced electric field ranging from 4 Td < E/n < 110 Td in air.

At low temperatures, vibrational-translational (VT) relaxation is slow, resulting in a scenario where the vibrational temperature within the discharge exceeds the translational temperature of the molecules. At the same time, within this specific range of E/n values, the lowest electronic level of oxygen, the $O(a^1\Delta)$ state could be efficiently excited in pure oxygen. On the other hand, the efficiency of this electronic excitation in the presence of nitrogen remains relatively low, about 2%. However, under certain conditions, the reduced quenching rate for singlet oxygen can cause an increase in its concentration.

2.1.3 Discharge Energy Distribution

The electric field threshold, $E/n \sim 120$ Td, plays a key role in air. Exceeding this threshold allows the electric field to ionize the air gases, allowing self-sustained propagation of the discharge. Below this critical value, an external ionization source is required to sustain the discharge. In the electric field strength range from $E/n \sim 140$ to 500 Td, the excitation of the electronic triplet states of nitrogen predominates as the primary mode of energy dissipation, facilitated by the elevated electron energy (ranging from 3 to 10 eV) within this E/n range. As a result, gas ionization within the discharge gap occurs rapidly. For E/n values ranging from 500 to 1,000 Td, activation of the nitrogen singlet states gains prominence, and for E/n values exceeding 1,000 Td, most of the electron energy is devoted to gas ionization.

The presence of fuel additives, as shown in Figure 2.2, does not significantly alter the overall dynamics of energy distribution. This minimal effect is primarily due to the relatively low concentration of fuel molecules present in the mixture under conventional combustion conditions. Specifically, Figure 2.2 illustrates the influence of methane on the distribution of electron energy across different degrees of freedom of the gas. In a stoichiometric CH_4-air mixture ($CH_4:O_2:N_2 = 1:2:8$), methane makes up approximately 9.1% of the mixture. The analytical results show that such additives only slightly adjust the energy distribution pattern, especially at moderate to high E/n values above 20 Td. This is where the excitation of molecular nitrogen dominates. Although the vibrational and electronic excitation and ionization of methane slightly tweak the energy distribution, the effect is more pronounced at lower E/n values, below 10 Td. Within this regime, the vibrational excitation of methane particularly affects the energy dissipation for E/n between 5 and 10 Td, enhancing the energy transfer to both rotational and translational molecular degrees of freedom, highlighting its significant role under conditions of low reduced electric field values.

2.1.4 Major Pathways in PAC

The electron multiplication process is most important for the kinetics of discharge development. Ionization, charge transfer, and recombination reactions of electrons should be included in the kinetic model (Figure 2.3).

Gas ionization coupled with electron-ion recombination culminates in rapid heating of the gas, the generation of translationally-hot atoms, the production of electronically excited radicals, and the formation of ion chains. While these phenomena are of minimal importance at high temperatures, they become critical at lower temperatures. At significantly lower temperatures, this pathway becomes even more important, as competing reactions involving neutral reactants are hindered and cannot proceed at temperatures around $T = 300$ K. For example, the reaction $O_2^- + H \rightarrow OH^- + O$ proceeds at a rate that is 11 orders of magnitude faster at $T = 300$ K than the reaction $O_2 + H \rightarrow OH + O$. The ion-involved processes are limited by the mechanisms of ion–ion and electron-ion recombination and the generation of negative ions by electron attachment, as illustrated by the reaction $O_2 + e^- + M \rightarrow O_2^- + O + M$.

Electron attachment to oxygen molecules changes the predominant recombination pathway from electron-ion to ion–ion recombination. This shift results in a significant reduction in the recombination rate due to the much lower mobility of heavy ions relative to electrons. Consequently, this adjustment in the recombination mechanism introduces a dependence on ambient pressure, as the efficacy of the process depends on the kinetics of the three-body attachment reaction.

Low Temperature Plasma Generation and Recombination

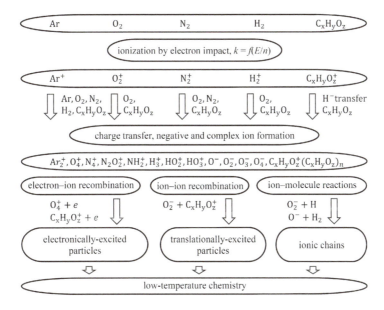

FIGURE 2.3 Ionization and charge transfer pathways.

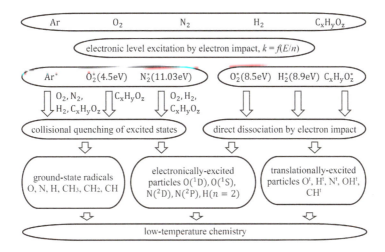

FIGURE 2.4 Electronic levels excitation and quenching mechanisms.

At low temperatures, the distinction between atomic oxygen in its ground state, $O(^3P)$, and in its electronically excited state, $O(^1D)$, becomes significant, as shown in Figure 2.4. At ambient conditions approaching $T = 300$ K, $O(^3P)$ tends to recombine, while $O(^1D)$ facilitates chain-branching reactions, with the reaction $O(^1D) + RH \rightarrow R^* + OH$ being particularly important. The main mechanisms producing atomic oxygen ($O_2 + e \rightarrow O + O + e$ and $N_2(C^3) + O_2 \rightarrow N_2 + O(^3P) + O(^1D)$) yield nearly equivalent amounts of O in both the ground (^3P) and excited (^1D) states. Therefore, at low temperatures, it is imperative to study all reactions involving (^1D). At elevated temperatures, $O(^3P)$ exhibits reactivity similar to $O(^1D)$, allowing their collective consideration as "atomic oxygen".

This phenomenon shows a dependence on pressure conditions. At low pressure, the radiative depopulation of N_2 (C^3): N_2 (C^3) $\rightarrow$ N_2 (B^3) $+ h\nu$ accompanied by photon emission changes the results of the quenching reaction, transforming N_2 (B^3) and O_2 into N_2 and two $O(^3P)$ atoms.

Conversely, at elevated pressures, the radiative lifetime of N_2 (C^3), approximately 37 ns, is excessively prolonged relative to collisional quenching, resulting in the formation of $O(^3P)$ and $O(^1D)$, as shown in Figure 2.3.

Molecular dissociation by electrons and photons is facilitated by repulsive energy surfaces of molecules, with excess energy being converted into translational motion of the products of dissociation. Consequently, initial collisions between these products and neighboring molecules occur with considerable energy, typically in the range of 1–5 eV. This energy level is sufficient to overcome reaction thresholds in all possible pathways, including direct dissociation in a single collision event. Therefore, the reactivity of such "hot" radicals is significantly enhanced; for example, the normally inert $O(^3P)$ (at $T = 300$ K), when endowed with additional kinetic energy, can engage in chain-branching reactions with N_2, H_2, and hydrocarbons. The effect of this phenomenon diminishes with increasing temperature, as atomic oxygen exhibits remarkable reactivity at elevated temperatures even without additional energy. Therefore, in low-temperature environments, it is crucial to consider translational nonequilibrium effects, which can increase the radical production rate by two to four times, depending on the specific conditions. At higher temperatures, however, their importance can be neglected.

The dependence of this phenomenon on concentration is remarkable. In dilute mixtures, the complete thermalization of hot atoms can be expected prior to their interaction with potential reactants. Conversely, in undiluted mixtures, this reaction pathway tends to dominate [5].

Vibrational excitation significantly accelerates reactions. At elevated temperatures, however, two factors contribute to a decrease in the importance of nonequilibrium vibrational excitation: (1) an increase in the reactivity of molecules in ground state and (2) an increase in the rates of VT relaxation. Consequently, at higher temperatures, the distribution of vibrational energy levels can be expected to align with the translational degrees of freedom, rendering deviations from the "nonequilibrium" state as insubstantial, as shown in Figure 2.5.

At low temperatures, reactions involving non-excited molecules proceed at a significantly slower rate. In such scenarios, two main vibrational mechanisms are observed. The first mechanism involves vibrational excitation of the reactants, which significantly increases the rate of chain-branching reactions. An example of this is the electron impact-induced vibrational excitation of hydrogen and its subsequent reaction with oxygen, where the reaction H_2 $(v=1)+O \rightarrow H+OH(v=1)$ occurs at a rate 3,000 times greater than its counterpart involving nonexcited hydrogen, H_2 $(v=0)$, at $T = 300$ K. The second mechanism involves vibrational excitation of nitrogen, also via electron impact, which significantly accelerates the decomposition of peroxides. This pathway is limited by the transfer of vibrational energy from N_2 (v) to HO_2, resulting in $N_2(v) + HO_2 \rightarrow N_2 + HO_2^*$, followed by the rapid decomposition of the vibrationally excited HO_2 and the re-generation of active radicals within

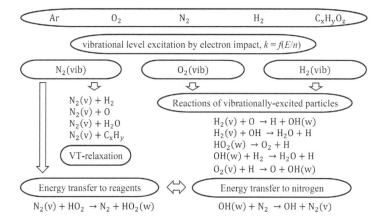

FIGURE 2.5 Vibrational levels excitation and vibrationally-activated reactions.

the mixture [23,24]. This mechanism has the potential to extend the chemical reaction chains by one to two orders of magnitude below the autoignition threshold, thus profoundly increasing the chemical energy release during the nonequilibrium phase of cold ignition.

2.1.5 Mechanisms of PAC

Several mechanisms influence gas behavior during combustion initiation or flame stabilization using discharge plasma. These mechanisms can be divided into two groups: (1) thermally-equilibrium heating and (2) non-thermal mechanisms.

The gas heating results from the thermalization of "hot" atoms along with the relaxation of vibrational and electronic energies, which subsequently accelerates chemical reactions [12,13]. Inhomogeneous heating induces flow perturbations that increase turbulence and mixing within the gas. Together, these thermal effects contribute to the modulation of combustion dynamics under the influence of the discharge plasma.

Among the mechanisms that influence the gas dynamics in the context of PAC, several non-thermal processes are worth mentioning. These include the ionic wind effect, which is characterized by momentum transfer from the electric field to the gas due to space charge, and ion and electron drift phenomena. The latter can lead to increased fluxes of active radicals within the nonuniform media in electric field.

In addition, excitation, dissociation, and ionization of the gas due to electron impact play a crucial role. These processes facilitate the production of nonequilibrium radicals and significantly alter the kinetic mechanisms underlying ignition and combustion processes [5].

These processes, whether operating in isolation or collectively, enable enhanced control of combustion, a feature essential for managing ultra-lean flames, high-speed flows, and the cold, low pressure conditions required for high-altitude gas turbine engines (GTEs) restart, initiating detonation in pulsed (PDE) or continuous (RDE) detonation engines, and facilitating distributed ignition control in homogeneous charge compression ignition (HCCI) engines, among other applications. Despite the existing gaps in the detailed understanding of these mechanisms, nonequilibrium plasma offers significant potential for managing ultra-lean, high-speed, low-temperature combustion scenarios. This positions it as a promising technology for a wide range of applications.

2.2 DISCHARGES USED FOR COMBUSTION AND IGNITION CONTROL

Ignition of fuel-oxygen and fuel-air mixtures is a central area of investigation in both fundamental and applied combustion research. The study of relatively simple fuels facilitates a comprehensive understanding of the precise mechanisms underlying combustion initiation. Conversely, complex fuel mixtures provide a closer approximation of potential real-world applications. Studies spanning several decades, as referenced in Refs. [25–27], have explored autoignition mechanisms within simple fuel-oxidizer mixtures, with these phenomena adequately characterized numerically by conventional kinetic models, such as the GRI-Mech mechanism [28]. The ignition delay, defined as the interval between the initiation of the process (either by gas heating or by mixing) and the observed increase in temperature, pressure, and emission intensity, is influenced by the rate of initial radicals production in dissociation reactions, which are slow at low-temperature conditions. The initial generation of atoms and radicals by external source of energy can significantly reduce the ignition delay. Currently, several strategies exist to accelerate ignition.

In the field of ignition studies under dynamically transient conditions, a key consideration is the interplay between various mechanisms and processes during the initial phase of ignition. The comparative importance of gas vibrational excitation, electron level excitation, ionization, and dissociation has been extensively analyzed [29,30]. Laser ignition techniques, while widely used, have certain limitations, most notably the inability to achieve uniform ignition initiation over large volumes of combustible gas. This limitation has led researchers to explore the concept of multipoint ignition [31]. To fully understand the dependence of the ignition process on system parameters, it is essential to investigate a spectrum of mechanisms, including laser-induced thermal processes, photoprocesses, and laser-induced spark evolution [32]. Experimental studies have been conducted over a range of wavelengths and system parameters, such as gas mixture compositions and pressure levels, to elucidate these dynamics [32,33].

A systematic evaluation was conducted to compare the ignition capabilities of three different lasers (ArF, KrF, and Nd:YAG) with traditional electric spark ignition in a standard four-stroke single-cylinder high-pressure combustion chamber [34]. The results showed that for a similar total energy input (approximately 50 mJ), both laser and spark ignition methods produce virtually identical pressure profiles. However, the ignition delay associated with electric sparks was observed to be 4–6 ms longer compared to laser-induced sparks. It is noteworthy that spark ignition inherently imposes significant constraints on the geometric configuration of the system and has been studied primarily in the context of gasoline engine ignition. Extensive two-dimensional (2D) modeling integrating hydrodynamic simulations with chemical kinetics [35], as well as comparative analyses with laser-induced fluorescence (LIF) measurements of OH radicals [36], have been documented in this area.

In addition, ignition by direct injection of an arc discharge plasma is of particular interest for hypersonic propulsion applications. The prototypical system configuration for such applications typically includes a turbofan equipped with a special fuel injection mechanism and a plasma torch system, where one electrode is constituted by a metallic nozzle. In particular, the ignition boundaries are significantly influenced by system geometric considerations, such as the positioning of the plasma torch relative to the fuel injectors and the penetration depth of the plasma torch, which has an inverse relationship with the flow velocity [37].

Amidst the diversity of ignition techniques, the challenge of achieving rapid and uniform ignition of combustible mixtures remains pressing. This chapter presents a method for homogeneous ignition that significantly exceeds the characteristic timescales of combustion processes and gas dynamics. This technique utilizes a nanosecond high-voltage discharge that manifests as a fast ionization wave (FIW) that generates atoms, radicals, and excited molecules. The genesis of the FIW occurs under conditions of extreme overvoltage—several hundred percent above nominal—on the electrode system, indicating that the voltage at the onset of discharge initiation is well above the threshold required to initiate a steady-state glow discharge. Experimentally, this means a rapid voltage increase rate of 1 kV/ns or greater, with pulse amplitudes ranging from tens to hundreds of kilovolts and durations of 10–100 ns. Characteristics of this type of discharge include high propagation velocity ($10^9 - 10^{10}$) cm/s, consistent reproducibility of discharge parameters, and spatial uniformity over large gas volumes, making it advantageous for a variety of applications, such as hypersonic combustion or ignition of lean fuel mixtures. A comprehensive review of this type of discharge is presented, for example, in Ref. [38], along with a detailed exposition of the nanosecond data acquisition methodology.

The elevated electric field before and after the FIW front facilitates efficient gas ionization, dissociation, and excitation without inducing significant changes in the translational temperature of the gas. The resulting nonequilibrium energy distribution enhances the effectiveness of the FIW as an ignition initiator, achieving high efficiency with minimal power input. Although the discharge yields relatively small amounts of active species such as O or H atoms, even a small degree of dissociation between 10^{-3} and 10^{-5} has a profound effect on the kinetic processes under high-temperature conditions. This chapter discusses the use of plasmas for PAI and combustion, focusing on both experimental and theoretical studies of key plasma parameters, including the reduced electric field, electron density, and energy distribution in various gas discharges. Discharge types such as streamers, pulsed nanosecond discharges, dielectric barrier discharges (DBDs), radio frequency (RF) discharges, and atmospheric pressure glow discharges are examined in detail.

In the context of combustion and ignition phenomena, the focus is on chemical interactions that occur at moderate to high gas densities and temperatures characteristic of the ignition and combustion phases in most of IC engines, typically in the range of 500–1,500 K. The gas densities of primary relevance to the onset of combustion are on the order of $10^{17} - 10^{18}$ cm^{-3}, consistent with conditions encountered during hypersonic flight. Such densities can escalate to as much as 10^{21} cm^{-3} in the high-pressure environments found in turbines or automobile engines. At standard ambient temperatures, these densities correspond to pressures ranging from a few Torr to tens of bar.

The spark discharge, widely used for ignition in various technical systems, is initiated by the application of a high voltage across a discharge gap typically several millimeters in size. Unlike the continuous evolution of an arc, the spark discharge has two discrete phases: an initial brief breakdown phase lasting 1–10 ns at atmospheric pressure, and a second phase lasting milliseconds due to the hydrodynamic expansion of the discharge volume, which subsequently reduces the current flow. The atmospheric pressure arc is a common and extensively studied form of dense, low-temperature equilibrium plasma maintained by an electric field [39]. Characterized by a decreasing volt-ampere relationship (i.e., voltage reduction with increasing current), this type of discharge exhibits relatively low-voltage drops across the discharge gap (below several hundred volts), along with substantial currents (ranging from fractions of an ampere to several kiloamperes, depending on external circuit impedances) and, most importantly, elevated gas temperatures within the arc (reaching thousands of Kelvins). Since the focus of this chapter is not on the thermal equilibrium plasma, reference is made only to one of the contemporary reviews [40] that delves into the intricacies of the arc plasma through an example of numerical modeling.

An arc plasma is developed from a spark discharge by applying a relatively long high-voltage pulse across two metal electrodes separated by a gaseous layer. Significant variations in plasma parameters can be achieved by modifying either the electrode configuration or the profile of the applied high-voltage pulse. Within the limits of this book, it is impractical to discuss each discharge variant in detail. Therefore, only those discharges that have historically been used to initiate or sustain combustion will be briefly discussed. For a full understanding of the associated plasmas, the reader is referred to the specialized literature on the subject (see, for example, [39]).

2.2.1 STREAMER DISCHARGE

The formation of a streamer discharge is observed when a high-voltage pulse, characterized by a steep front, breaks within a time interval comparable to the time required to cross the discharge gap (ranging from a few nanoseconds at atmospheric pressure through gaps of a few millimeters to a centimeter). In essence, a streamer discharge at atmospheric pressure represents the initial phase of spark breakdown. A streamer is characterized by the presence of two distinct spatially separated regions: the streamer head, where a high reduced electric field (on the order of hundreds of Townsends) is generated, and the streamer channel, which exhibits relatively low electric fields and enhanced conductivity. The initiation of a streamer discharge occurs when the electric field intensity at the streamer head equals or exceeds the external electric field in the gap, a criterion known as the Meek and Loeb criterion. The velocity of the streamer, the reduced electric field within the streamer head, and the electron density are predominantly determined by the processes occurring at the streamer head, with the primary function of the conductive channel being to maintain the electric potential of the head.

The primary generation of active species, crucial for the initiation of combustion and consisting mainly of electronically excited atoms and molecules, is localized within the streamer head. This phenomenon is supported by experimental monitoring of the electric field and emission from electronically excited states. The visual manifestation of a streamer discharge is characterized by the formation of thin filamentary channels extending between the electrodes, where the degree of branching is significantly influenced by the gas pressure within the discharge cell.

Occasionally, terminological ambiguities arise in the discourse on gas discharges, particularly with respect to the observed structural features. The optical observation of "filamentary" structures does not unambiguously indicate the formation of a streamer plasma. For example, in the propagation of microwave discharges at atmospheric pressure [41], a distinct filamentary spatial configuration is observed, supposedly representing slender channels of plasma approaching equilibrium conditions. Despite the occasional reference to these phenomena as "microwave streamers", there appears to be no substantial correlation with the conventional streamer mechanism.

A significant number of works devoted to streamer discharges elucidate the peculiarities of this phenomenon based on both experimental research and mathematical modeling. Several papers are now considered foundational texts in the field. Notable among these is the work of Galimberti and colleagues [42,43], who introduced a model for the numerical representation of cathode-directed streamers and explored the potential influence of accumulating vibrationally excited metastable N_2 molecules on streamer progression. Williams and coworkers [44] presented computational analyses of streamers in two-dimensional geometries using flux-corrected transport methods. In addition, Kunhardt and coworkers [45] studied the avalanche-to-streamer transition as an initial phase of streamer evolution, delineating the differences between cathode-directed and anode-directed streamers and the importance of photoionization and "run-away" electrons in the process.

In their study, the authors of Ref. [46] performed a comprehensive analysis of the processes occurring within the streamer channel. In particular, they reported that the electron density in the channel is primarily regulated by electron attachment and electron-ion recombination and observed that the gas heating within the streamer channel does not exceed a few tens of degrees Kelvin. They also developed a theoretical framework for streamer initiation and progression that includes a number of elementary processes, including ionization, attachment, recombination, electron diffusion, and photoionization [47].

The authors of Ref. [48] introduced a two-dimensional simulation model designed to characterize the propagation of a negative (anode-directed) streamer in a plane-to-plane geometry, achieving a spatial resolution of ≤ 5 μm, applicable to electrodes of arbitrary shape. At the same time, the dynamics of a positive streamer in a weak uniform electric field was computed in a two-dimensional framework [49].

These computational approaches have allowed the correlation of specific streamer channel properties, such as propagation velocity and channel radius, with the parameters of the external electric field and applied voltage. Currently, streamer simulations are being extended to a range of gases, different voltage supply configurations [50], and for extended discharge gaps exceeding 10 cm in length [51].

Streamer branching plays a crucial role in the characterization of streamer evolution, especially in multi-millimeter discharge gaps. This phenomenon has been the subject of extensive experimental research [52,53]. Figure 2.6 illustrates streamer branching within a discharge gap of a few millimeters, as observed in the air at atmospheric pressure. The observed emission is due to the second positive system of molecular nitrogen. The left part of the figure shows an image taken with an ICCD camera with a gate duration of 100 ns, while the right part of the figure shows a streak image with a duration of 50 ns.

ICCD images acquired with relatively long gate durations (5–50 μs, as shown in Figure 2.7) essentially provide an integral view of the discharge evolution, especially when considering the

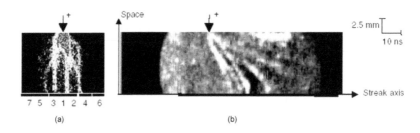

FIGURE 2.6 Example of still (a) and streak (b) images of a streamer emission [52].

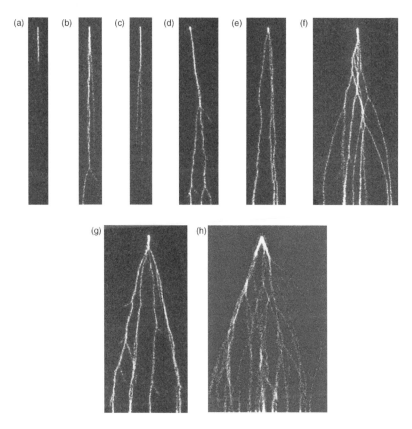

FIGURE 2.7 CCD photos of a point-plane discharge in air using the semiconductor switch. The applied voltage is (a) 9 kV, (b) and (c) 10 kV, (d)–(g) 12.5 kV and (h) 15 kV. The optical gate is (a)–(e) 5 μs, and (f)–(h) 50 μs [53].

use of long high-voltage pulses (on the order of hundreds of nanoseconds). This implies that a more nuanced physical interpretation of these images will require further studies with improved temporal resolution. Nevertheless, these images provide a vivid qualitative insight into the streamer evolution within gaps of tens of millimeters.

The image in Figure 2.8, taken at high temporal resolution (camera gate duration of 0.8 ns), clearly shows that the emission during a streamer flash is predominantly from the streamer heads. This observation further supports the notion that high-energy electronic degrees of freedom are primarily excited within the streamer head.

Despite various efforts to numerically simulate streamer branching in recent years [54,55], there remains skepticism as to whether these computational models accurately represent the real physical instabilities at the streamer front or whether the observed branching phenomena are artifacts of the specific numerical techniques employed.

The effect of pressure on discharge characteristics was investigated in Ref. [56]. The authors noted that discharges under several atmospheres of pressure are commonly used to initiate combustion in automotive engines, but their characteristics are not fully understood. This is attributed to the reduced size and duration of the discharges at elevated pressures, which presents challenges for experimental investigation. Their research included the study of discharge evolution at both atmospheric pressure and 6 bar.

To achieve similarity in discharge development, the authors maintained a constant $N \times t$ value, where N is the gas density and t is the voltage rise time. As a result, the voltage rose to 26 kV in 50 μs at 6 bar, compared to a rise of 300 μs at 1 atm. It is noteworthy that a significantly lower initiation

FIGURE 2.8 Photograph of streamer heads of a point-wire discharge in air taken with the ICCD with 0.8 ns gate. A pulse of 25 kV with a rise time of 20 ns is applied. The photo is taken 31 ns after the start of the pulse. Discharge gap is 25 mm, plane-to-plane geometry [53].

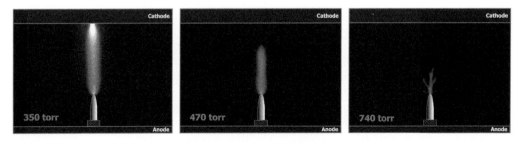

FIGURE 2.9 ICCD images of streamer in air at different pressures. Interelectrode gap is 30 mm, high-voltage amplitude is 20 kV, and pulse duration is 25 ns. According to Ref. [58].

voltage (by 6 kV) was observed under conditions of steeper voltage rise, specifically at 6 bar. After analyzing the data, the authors inferred the formation of a leader channel and the generation of a thermally equilibrated plasma. However, this conclusion seems questionable considering that the discharge current remained relatively modest, not exceeding the magnitudes typical for streamers (ranging from tens to hundreds of mA).

Another example of the modification of streamer behavior as a function of pressure is described in Refs. [57,58]. The authors showed that with decreasing pressure, streamers initially triggered by a voltage pulse of 22 kV amplitude and 20 ns duration at the high-voltage electrode cease to branch and instead propagate as a single channel. It was also observed that the diameter of the channel increases with decreasing pressure. Figure 2.9 shows ICCD images capturing streamer emission at different pressures, serving as a visual representation of this phenomenon.

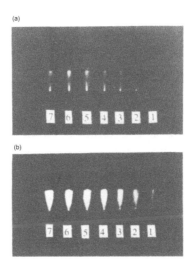

FIGURE 2.10 The effect of temperature on streamer propagation [59]. Numbers indicate the number of superimposed shots. (a) $T=290$ K and (b) $T=373$ K. The applied electric field was 500 kV/m with pulse voltage 4.25 kV and gap length 15 cm.

Studies of streamers generated by 50 ns voltage pulses of 10 kV at pressures ranging from 1 to 10 atm at ambient temperature show that the diameter of the streamer contracts with increasing pressure, while the number of branches increases. Experimental analysis of streamer discharges typically focuses on monitoring electrical parameters such as discharge current and voltage across the gap, in addition to assessing emission characteristics. High-resolution spatial and temporal control of the emission allows the determination of the optical diameter of the streamer and its propagation velocity.

Of particular note are the works that explore the use of standard measurement techniques to assess changes in the object, as they allow the parametric dependencies of various variables to be examined. For example, a study outlined in the paper [59] examined the evolution of streamers in gas preheated at constant pressure over a temperature range from ambient to 450 K. The results indicated an expansion of the diameter of the emitting region as the density decreased (see Figure 2.10). In addition, the study considered variations in the electric field required for streamer propagation in air and the velocity of streamer propagation.

In the context of plasma chemistry applications, including PAI and PAC, a select set of plasma parameters emerge as critical. These parameters include the magnitude of the electric field, the electron number density, and the densities of atoms, radicals, and excited species generated in the plasma, in addition to the gas temperature. Unfortunately, experimental data on these parameters are relatively scarce. This scarcity is due to the fast evolution of the discharges, characterized by a typical timescale in the nanosecond range, together with the spatial non-uniformity and the transient nature of the streamer flashes.

A technique utilizing the Pockels effect has been used to detect the electric field during the streamer-leader transition in air at atmospheric pressure [60]. The temporal responsiveness of this detector depends on the resolution of the optoelectronic converter, allowing its application even in the picosecond time frame. Due to its relatively small size (a few millimeters) and its dielectric composition, this detector induces fewer perturbations in the plasma compared to a conductive material. However, in cases of elevated electric fields, the polarization of the dielectric can cause localized field enhancement within the plasma. This, in turn, increases the ionization rate and alters the electric field in the vicinity of the detector, potentially distorting measurements.

Electric field control within streamer discharges can be achieved by non-contact methods, such as monitoring the emission intensity ratios between the first negative and second positive systems of molecular nitrogen. This approach is discussed in detail in Refs. [61,62]. Electric fields within the

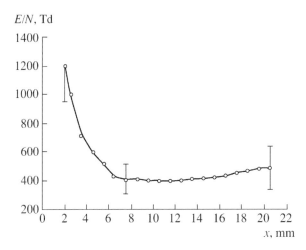

FIGURE 2.11 Reduced electric field in the streamer head *vs.* the distance from the high-voltage electrode for the interelectrode distance equal to 24 mm [63].

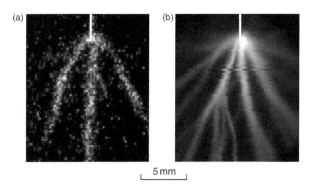

FIGURE 2.12 (a) Distribution of OH radicals at 5 μs observed by the two-dimensional LIF and (b) spontaneous emission of a positive discharge in $H_2O(2.4\%){:}N_2$ mixture [64]. Each image is obtained at a different discharge pulse. The white line represents the needle electrode. Discharge gap is 16 mm.

streamer head typically rise to the order of hundreds of Townsends. Figure 2.11 from Ref. [63] illustrates the behavior of the electric field during a streamer flash triggered by a high-voltage pulse of about 20 kV amplitude (at the high-voltage electrode) with a duration of 25 ns and a repetition frequency of about 1 kHz. The emission was captured through a narrow slit diaphragm oriented perpendicular to the slit axis, implying that the data represent a spatial average. Nevertheless, given the relatively short duration of the high-voltage pulse (75 ns at half-height), these measurements provide a credible approximation of the electric field within the streamer head during the propagation of a slightly branched streamer flash.

Assuming a streamer diameter of 1 mm, the density of excited nitrogen is calculated to be 10^{12} cm^{-3}, based on a given linear density of $\left[N_2\left(C^3\Pi_u\right)\right]$ approximately equal to $\approx 10^{10}$ cm^{-1}.

The study presented in Ref. [64] is one of the rare instances of direct radical density measurements. The researchers detailed the quantification of hydroxyl radicals produced by a pulsed corona discharge using LIF. This discharge, characterized by a 35 kV voltage and 100 ns pulse current, was initiated between needle and plate electrodes in $H_2O{:}O_2{:}N_2$ mixtures at atmospheric pressure. The OH density was estimated to be about $7 \cdot 10^{14}$ cm^{-3} in $H_2O(2.4\%){:}N_2$ mixture 10 μs after discharge. Unfortunately, these measurements relate only to OH concentration during the late afterglow of the discharge, which complicates the analysis of discharge kinetics. Conversely, spatially resolved absolute measurements of OH density (as shown in Figure 2.12) are recognized as a major strength of this investigation. In addition, the same team conducted a study of NO production facilitated by a similar pulsed discharge in a $N_2{:}O_2{:}Ar$ gas mixture [65].

Thus, the streamer discharge is recognized as a nonequilibrium, low-temperature plasma characterized by relatively high electric fields in the streamer head and lower fields in its channel. At moderate pressures and voltage amplitudes ranging from a few to tens of kilovolts across discharge gaps of several centimeters, streamer development occurs without branching, provided the voltage rise is fast enough and the pulse duration short enough to prevent the propagation of secondary streamers. As the pressure increases, the streamer diameter decreases and branching increases. Currently, two-dimensional (2D) and three-dimensional (3D) streamer simulation codes have advanced to the point where they can effectively model streamer propagation through gaps as large as 100 cm. However, the debate about the relative importance of the elementary processes that facilitate streamer initiation and propagation under different experimental conditions continues and is subject to ongoing refinement. The generation of electronically excited states capable of inducing dissociation occurs predominantly within the streamer head. Nevertheless, the potential contribution of superelastic collisions within the streamer channel, leading to additional dissociation through interactions with relatively low-energy excited states, is not universally dismissed by researchers.

2.2.2 Dielectric Barrier Discharge

A discharge occurring in an electrode system where at least one electrode is covered by a dielectric layer is called a dielectric barrier discharge [66]. This phenomenon is also known as a silent discharge, due to its noiseless operation [67], or alternatively as a dielectric corona discharge, reflecting the underlying physical principles [68]. When powered by a pulsed source characterized by a short nanosecond rise time and pulse durations of tens to hundreds of nanoseconds, the dielectric barrier acts to limit the current, thereby preventing the discharge from transitioning to an arc.

Turning our attention to the conventional configuration of the barrier discharge, it is appropriate to acknowledge some historical aspects. Barrier discharge has been intensively studied for about five decades, primarily because of its utility in ozone generation. To date, the mechanism of ozone production via DBD has been thoroughly studied and extensively implemented in various industrial applications. In particular, certain publications, such as [69,70], have provided comprehensive reviews of the synthesis of ozone using DBD technology.

Studies of the electrical properties of barrier discharges have elucidated the principal dynamics governing their behavior. Typically, such discharges are initiated in either planar or coaxial configurations, with one or both electrodes insulated by a dielectric material. The distance between the electrodes is kept relatively small, not exceeding a few millimeters. Gas flows through the system at a moderate rate, with provisions for additional electrode cooling. A sinusoidal high voltage of a few kilovolts peak-to-peak is applied, triggering the discharge twice per cycle.

Extensive analysis of the discharge structure [71,72] has shown that breakdown occurs at a specific voltage threshold and exhibits a filamentary microdischarge pattern. The typical microdischarge diameter is on the order of a few millimeters [73]. These observations suggest that the progression of DBD bears a physical resemblance to incomplete streamer breakdown. Figure 2.13 [74] shows a typical DBD emission. The reactor used had a knife-to-plate geometry, with the knife electrode 2.5 cm long and a gap distance of 1.7 mm. The grounded plate electrode was covered by a 1 mm thick quartz plate. The image captures the discharge under flowing dry nitrogen (flow rate of 3 L/min) at atmospheric pressure.

When a sinusoidal voltage is applied across a gap, the resulting current pulses typically do not exceed tens of nanoseconds in duration, a limit imposed by the charging of the dielectric layer. Figure 2.14a [75] shows characteristic current and voltage oscillograms for a barrier discharge in air at atmospheric pressure. Figure 2.14b and c show a time-resolved current pulse from a single microdischarge and from a series of microdischarges, respectively. The sinusoidal line shown in Figure 2.14a represents the voltage profile across the gap, with the peak-to-peak voltage reaching 13.2 kV. The electrode separation in these experiments ranged from 0.5 to 1 mm.

It is appropriate to acknowledge that spatially resolved kinetic measurements of active species within the DBD present significant challenges due to the pronounced spatial inhomogeneity and the

FIGURE 2.13 Short-time-exposure photographs of a filamentary DBD (anode, top; cathode, bottom) from a single voltage pulse (800 ns exposure time, N_2, 3 L/min, 1 atm, interelectrode distance 1.7 mm, repetition rate 2,000 Hz) [74].

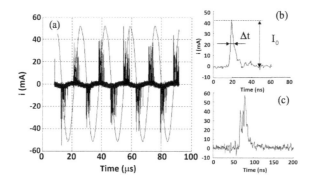

FIGURE 2.14 (a) Voltage and instantaneous current, (b) single current pulse and (c) series of current pulses, typical for DBD discharge [75].

short interelectrode distance. However, the consistent replication of discharge patterns makes such measurements more feasible than in "classical" streamer discharges. This accessibility probably accounts for the abundance of experimental studies focused on quantifying electric field characteristics, temperature, and particle densities in DBD's plasmas.

A comprehensive analysis of the electric field and electron density within individual microdischarges was performed in Ref. [76]. The researchers designed a discharge cell (shown in Figure 2.15a) to ensure the generation of a unique, consistently reproducible microdischarge. By measuring the emission from the second positive and first negative systems of molecular nitrogen, they determined the electric field values within the microdischarge. As shown in Figure 2.15b, as the microdischarge progresses, the peak electric field reaches several hundred Townsends, similar to the electric field intensity observed at the head of a streamer.

Within the discharge gap of a DBD, a certain amount of charged, excited, and dissociated species is generated as the DBD evolves. Papers [77,78] provide computational analyses of the gas composition following DBD evolution, based on empirical observations of electrical parameters within the system.

A subsequent publication [79] reports both experimental measurements and sophisticated calculations of the production of ozone and nitrogen oxides in DBD discharges by O_2/NO_x and $N_2/O_2/NO_x$ mixtures, with NO_x concentrations up to 2,000 ppm. The results indicate that the gas temperature within the discharge does not exceed 440 K, and typical ozone densities are in the vicinity of 10^4 ppm. This corresponds to an ozone concentration of up to 1%, which is in line with the performance characteristics of DBD-based ozonizers [70].

Direct measurements of dissociated particles within DBD microdischarges are of considerable value. In this context [80], the author presents the first measurements of spatial and temporal relative atomic density distributions of nitrogen at a single filament within a DBD reactor with submillimeter radial dimensions. Two-photon absorption laser-induced fluorescence (TALIF) spectroscopy using radiation at $\lambda = 206.7$ nm was used to map N atoms concentration. A specially designed multi-pin reactor facilitated 2D TALIF analysis by moving the discharge cell relative to the laser beam. A high-voltage pulse was applied, rising to 15 kV within 90 ns, with the pulse lasting approximately 200 ns at half-height. An N-atom image from Ref. [80] is shown in Figure 2.16. The authors

Low Temperature Plasma Generation and Recombination

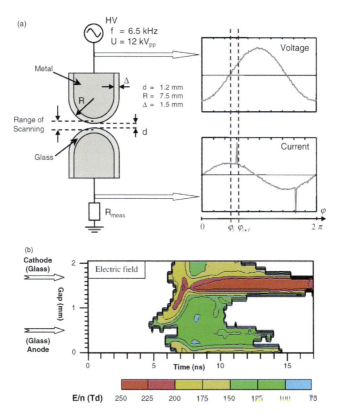

FIGURE 2.15 (a) Electrode arrangement with the indicated range of axial scanning and a typical example of the oscillograms of voltage and current; (b) determined distributions of electric field and electron density. The positions of the tips of the electrodes indicated by the arrows on the left [76].

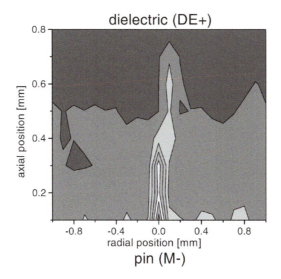

FIGURE 2.16 Density distribution of atomic nitrogen in a DBD at 950 mbar, containing 90% N_2 and 10% O_2; 8 kV voltage; positive dielectric (DE+); 2 μs after discharge initiation. Light colors correspond to low densities, dark colors to high densities [80].

used numerical modeling to compare the spatial distribution of nitrogen atoms with the experimental data. The calculated absolute density values of N atoms were in the range of $(4-6) \cdot 10^{14}$ cm^{-3}, reflecting the spatial distribution observed under experimental conditions.

The study presented in Ref. [81] elucidates the results of TALIF-based (zero-dimensional, yet time-resolved) assessments of the relative densities of atomic oxygen within the DBD, initiated in a point-to-plane configuration, across nitrogen-oxygen mixtures at varying levels of humidity (ranging from 0% to 2.4%). It was found that the decay of atomic oxygen density occurs at different rates over tens of microseconds, even when the production of atomic oxygen remains constant and the densities of oxygen atoms are identical in the initial phase of the afterglow (under 1 µs). The paper discusses possible chemical reactions responsible for the decay of oxygen atoms under different humidity conditions.

In addition, the gas temperature at the exit of the discharge cell is significantly influenced by the specifics of the system, especially when the distance between the electrodes is minimal (on the order of millimeters), making heat dissipation through the electrodes a critical consideration. Traditionally, spectroscopic methods are used to measure the gas temperature in the discharge zone.

In the study conducted by the authors of Ref. [82], the rotational temperature of molecular nitrogen ion emissions was measured to determine the gas temperature within the microdischarge zones and the surrounding gas environment of DBDs in helium/nitrogen mixtures. The DBD discharge was initiated using a sinusoidal voltage with a frequency of 19.7 kHz and a peak-to-peak amplitude of 3,500 V. The duration of the current pulses in this setup was significantly longer than those typically observed in filamentary discharges, which are about 10 ns, and comparable to pulses observed in atmospheric pressure glow discharges (APGD) in pure helium, about 1 µs. In contrast to the singular current pulse characteristic of uniform APGDs, the authors detected a series of pulses. They analyzed the rotational spectra of the transition $N_2^+ \left(B^2\Sigma_u^+ \to X^2\Sigma_g^+ \right)$ using a well-established reaction scheme for discharges in helium/nitrogen mixtures, which includes direct ionization of nitrogen by electron impact and Penning ionization. Consequently, they determined the temperature in the microdischarge region to be between 400 and 600 K and the ambient gas temperature within the discharge cell to be 300 ± 10 K.

The measurements of the gas temperature in a 30% N_2:70% Ar mixture subjected to short-pulsed ($\tau < 15$ ns) DBD discharges were presented in Ref. [83]. The temperature was determined by time-resolved diode laser absorption spectroscopy of metastable Ar* atoms and analysis of the rotationally resolved N_2^+ first negative and N_2 second positive systems emissions. The gas temperature inferred from the nitrogen emission spectra was derived from an unresolved rotational contour analysis of individual vibrational bands. Simultaneously, the temperature under identical conditions was inferred from the Doppler width of the metastable Ar absorption spectrum. Measurements showed that the gas temperature in low average power DBD, performed at pressures of 10, 30, and 100 Torr with pulse repetition rates ranging from 0.5 to 30 kHz, remained within 350–400 K and showed no variation with changes in pulse repetition rate.

Coherent anti-Stokes Raman scattering (CARS) technique was used to measure gas temperatures in DBD systems with different electrode geometries in N_2:O_2:NO mixtures, at pressures of 20 and 98 Pa, using bipolar high-voltage pulses of 10 kV at pulse repetition rates of 1 and 2 kHz [84]. The measurements yielded rotational (gas) temperatures close to room temperature and vibrational temperatures of about 800 K at atmospheric pressure, rising to 1,400 K at a pressure of 20 kPa.

In particular, in the 1970s, researchers emphasized the improved plasma-chemical efficiency that could be achieved in DBD systems by switching from conventional frequency sinusoidal power supplies to pulsed nanosecond voltage sources. Such modifications were shown to increase the efficiency of ozone synthesis in oxygen from 90 to 130 g/(kW.h) [85]. Similar results were presented in Ref. [74], where the authors used low repetition frequency, high-voltage unipolar square pulses with amplitudes up to 15 kV and rise and fall times below 20 ns to excite DBD. According to their experimental results, this excitation method for DBDs significantly improved the energy efficiency of ozone production to 8–9 eV per ozone molecule, an improvement of 30% over conventional sinusoidal excitation. The authors further investigated the transition of the discharge to a homogeneous form with decreasing pressure by conducting experiments in air at a pressure of 40 Torr and an interelectrode distance of 4 cm.

Over the past 10–20 years, theoretical and experimental research on DBDs has been driven primarily by their utility in applications such as surface treatment and plasma chemistry, including ozone generation and gas purification processes.

The paper [86] presents a comprehensive review of recent advances in the numerical modeling of non-thermal gas discharge plasmas in air at atmospheric pressure. The discussion includes the theoretical basis for the development of DBDs, methods and strategies for the numerical modeling of low-temperature plasmas at atmospheric conditions, and the analysis of streamers over various electrode configurations (point-to-plane, plane-to-plane) and interelectrode distances (ranging from millimeters to several centimeters). The significance of this approach lies in its emphasis on the universal physical principles governing the propagation of streamers and DBD microdischarges across the interelectrode gap.

Thus, DBD, as well as streamer discharge, represents a category of ionization waves characterized by current limitation due to the charging of the dielectric layer. In this type of discharge, the typical gas temperature is close to ambient conditions, the electric fields do not exceed several hundred Townsends—similar to those in streamer discharges—and the densities of dissociated and excited species are in the $10^{13} - 10^{15}$ cm^{-3} range, as reported in various studies. It is noteworthy that electron density measurements have not been performed experimentally in either streamers or DBD environments; however, numerical models have consistently yielded electron density values around 10^{14} cm^{-3} for streamers in atmospheric air.

2.2.3 Atmospheric Pressure Glow Discharge

Remarkably, two different types of discharges, studied under fundamentally different electrode configurations and power supplies, by different research groups, have been given the same name. Coincidentally, this nomenclature effectively encapsulates the physical essence of these discharges.

The first variant was realized using a sinusoidal power supply within an electrode system reminiscent of the DBD configuration. In 1988, the researchers [87] presented the observation of stable glow plasma at atmospheric pressure during plasma treatment experiments. Modifications were made to the electrode structure, the type of carrier gas used, and the power frequency. From their experimental observations, they outlined the following requirements for achieving a stable glow discharge in a DBD setup: (1) helium should be used as the carrier gas; (2) an insulating plate should be placed on the lower electrode; (3) a brush-type electrode is recommended for the upper, and simultaneously, high-voltage electrode (consisting of 25 fine wires of stainless steel or tungsten in their experiments); and (4) a continuous stable discharge should be facilitated by using a high-frequency power source. The authors used a RF power supply at 3,000 Hz, noting the inability to maintain a stable discharge at a frequency of 50 Hz. Using visual observations ("transform to arc" or "dark blue, stable") and examining the nuances of surface treatment for polyethylene terephthalate (PET) films with an $O_2:CF_4:He$ mixture, they articulated the differences between arc and glow discharge regimes.

Another study [88] examined APGD using a plane-to-plane electrode configuration commonly associated with DBDs. The researchers pointed out that APGDs were first observed in the 1930s using cooled metal electrodes without a dielectric barrier and in ambient air. They further observed that the evolution of a spatially uniform APGD is characterized by a single, prolonged current pulse within a voltage half-cycle, with a current amplitude of approximately 10 mA. The addition of He gas was found to prolong the transition phase from glow discharge to arc discharge.

Paper [89] presented the feasibility of maintaining a spatially homogeneous APGD in a plane-to-plane electrode configuration with a commercial (50 Hz) power frequency over different gases (air, argon, oxygen, and nitrogen). The researchers suggested that APGD stabilization could

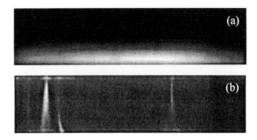

FIGURE 2.17 Typical 10 ns exposure time photographs of a 4 mm gas gap during (a) an APGD ($V = 11$ kV) and (b) a filamentary discharge ($V = 14$ kV) [91].

be facilitated by current limiting through a specific component of the electrode system, and proposed the use of a specially designed wire mesh electrode configuration. They referenced [90], which detailed the current transition during discharge evolution from dark Townsend to arc discharge, to support their findings. Paper [89] also described that a single current pulse within a voltage half-cycle is indicative of APGD, whereas a filamentary DBD is characterized by an irregular series of pulses (generally in the single-digit to ten-digit range).

Figure 2.17 clearly contrasts the appearance of APGD (top image) with that of filamentary DBD (bottom image) as captured by an ICCD camera with a 10 ns gate duration.

Later, considerable progress was made in the study of APGD in configurations similar to those found in DBDs. Complex kinetic models have been proposed to elucidate the mechanisms behind the suppression of filamentation and the transition to a uniform discharge attributed to the accumulation of metastable particles (see for example [91]). The electrical and spectroscopic properties of APGDs in planar geometries, with interelectrode distances up to 1 cm, have been extensively studied [92]. Typically, experimental studies are complemented by numerical simulations [93], which address detailed kinetics, including the role of electronically excited species [94]. Numerous studies have focused on the application of APGD DBD plasma for the treatment of various films, such as organosilicon polymer films [95] and polypropylene [96], exploiting the advantages of spatially homogeneous discharges at atmospheric pressure for efficiency and technical reliability.

The second type of discharge, with the same name as the first, was realized using a DC power supply with the implementation of electrode cooling. In this scenario, the gas temperature significantly exceeds that of the APGD DBD, reaching thousands of Kelvin, but remains significantly lower than the electron temperature, indicating a predominantly nonequilibrium plasma state. The introductory section of the paper [97] provides a concise yet comprehensive overview, evaluating research on APGD under various power supply configurations, including both early explorations in the 1930s [98] and more recent studies [99].

Studies [100,101] described DC APGD in nitrogen, stabilized by water cooling and gas flow across the discharge cell. The discharge was established between two platinum pins separated by 0.85 cm and mounted vertically on water-cooled stainless steel tubes. Nitrogen flowed through the discharge area at a velocity of 20 cm/s. The researchers reported an electron temperature of about $T_e \sim 10,000$ K and a gas temperature of $T_g \sim 2,000 - 4,000$ K for the generated plasma. Figure 2.18 shows a typical APGD discharge in air across a 3.5 cm discharge gap.

Spatial concentration profiles of N_2^+ have been quantified in an atmospheric pressure nitrogen glow discharge using cavity ring-down spectroscopy (CRDS) [100]. At discharge currents of about 100 mA, radial profiles with a radial half-maximum of about 1 mm were reported. Using a collisional-radiative model, ion concentrations were correlated with electron number densities, in agreement with spatially integrated electrical measurements. The electron density ranged from $4 \cdot 10^{11}$ to $2 \cdot 10^{12}$ cm^{-3} for discharge currents from 50 to 190 A, exceeding the densities predicted by the local thermal equilibrium (LTE) approach by up to five times at higher currents and by several orders of magnitude at lower currents [100].

FIGURE 2.18 DC glow discharge in air (1.4 kV/cm, 200 mA). Interelectrode distance is 3.5 cm [101].

Paper [101] presented several optical methods for assessing the temperature and concentrations of charged species in air and nitrogen plasmas, applicable to both equilibrium and nonequilibrium conditions over a wide range of parameters with submillimeter spatial and submicrosecond temporal resolution. Two equilibrium plasma scenarios (a 50 kW inductively coupled plasma torch with a flow velocity of 10 m/s and a recombining air or nitrogen plasma with a flow velocity of about 1 km/s through a water-cooled test section) were evaluated and contrasted with glow discharge experiments induced by a DC electric field in air at atmospheric pressure. Gas temperature was measured by medium-resolution emission spectroscopy of the rotational line structures of selected OH, NO, N_2, and N_2^+ vibrational bands. Electron densities greater than $5 \cdot 10^{13}$ cm^{-3} were deduced from Stark-broadened H_β line shapes. Ion concentrations were obtained from CRDS, with N_2^+ ion concentrations linked to electron concentrations assuming charge neutrality and known ion chemistry. Rotational temperatures were given as $4{,}850 \pm 100$ K for the recombining plasma and $2{,}200 \pm 50$ K for the APGD, with electron densities of 10^{15} cm^{-3} for the plasma jet and 10^{12} cm^{-3} for the APGD.

Paper [97] provides a comprehensive analysis of DC APGD plasma, drawing parallels to Townsend glow discharge at low pressures. A broad overview of discharge evolution in air, hydrogen, helium, and argon was provided, with rotational and vibrational temperatures determined by emission spectroscopy of the second positive system of molecular nitrogen. APG discharge zones, similar to those in glow discharge, were identified both visually (by photography) and by measuring voltage-current characteristics. The electric field was derived from electrical and spectroscopic experimental data, and the electron density was derived from the current. Some of the parameters elucidated by the authors are summarized in Table 2.2. The comparison between APGD and conventional glow discharge requires a recalibration of the parameters not in pressure units but in density units to account for density variations due to heating. Measurements were therefore made at an "effective pressure" $P_{eff} = PT_n/T$, where P is the pressure in Torr, T_n is the standard gas temperature (293 K), and T is the actual gas temperature. For example, the reduced electric field for low-pressure glow discharges is typically 10–30 V/(cm·Torr). At an "effective pressure" P_{eff}, electric fields at 760 Torr and 1,500 K should be in the range of 1.5–4.5 kV/cm, which agrees well with the findings for APGD in Ref. [97].

Upon examination of the data delineated in Table 2.1, it becomes evident that there can be a notable variance in the values of electron density n_e and that the electric field magnitude $(3-5) \cdot 10^{-16}$ V·Cm^2 = 30–50 Td is significantly lower than that observed in a streamer head, predominantly correlating with the excitation of vibrational degrees of freedom.

TABLE 2.2
Measured and Estimated Discharge Parameters in APG Discharge in Air at 0.4 and 10 mA Discharge Currents

Parameter	Current Is 0.4 mA	Current Is 10 mA
Electrode spacing, mm	0.05	0.5
Discharge voltage, V	340	380
Discharge power, W	0.136	3.8
Translational temperature, K	700	1,550
Vibrational temperature, K	5,000	4,500
E/n in positive column, V cm^2	$4.8 \cdot 10^{-6}$	$3 \cdot 10^{-16}$
n_e in negative glow, cm^{-3}	$3 \cdot 10^{13}$	$7.2 \cdot 10^{12}$
n_e in positive column, cm^{-3}	—	$1.3 \cdot 10^{14}$

Data are taken from Ref. [97].

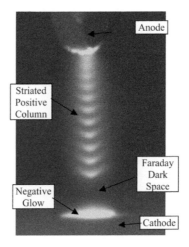

FIGURE 2.19 Image of the glow discharge in atmospheric pressure hydrogen. Positive column and negative glow are visible; standing striations are visible in the positive column [97].

The findings of Ref. [97] indicate that the developmental characteristics of APGD in helium, hydrogen, and argon bear resemblances to those observed in glow discharges under low pressures. Specifically, (1) each gas displays unique discharge colors and spectral lines indicative of various species; (2) in helium, the overall dimensions of the discharge are more expansive for equivalent current discharges in air; (3) hydrogen discharges exhibit standing striations within the primary column under certain conditions. Figure 2.19 showcases an instance of APGD in atmospheric pressure hydrogen, where regions characteristic of "common" glow discharges (such as negative glow, Faraday dark space, and positive column) are distinctly visible. Consequently, the authors proposed that DC APGD: (1) constitutes a normal glow discharge; (2) is thermally moderated by its dimensions, implying that electron density is governed by particle diffusion away from the discharge due to its constrained radius; (3) results in relatively low electric fields, facilitating the discharge's capacity to sustain a significant level of VT nonequilibrium.

Hence, APGD serves as an overarching term for discharges that facilitate nonequilibrium plasma conditions at atmospheric pressure. The physical characteristics of the plasma can exhibit considerable variability across a broad spectrum of parameters. This range extends from plasma bearing similarities to streamer-type phenomena in APGD DBDs, to plasma akin to that observed in glow discharges for direct current APGD configurations.

2.2.4 Pulsed Nanosecond Discharges

As previously described, the application of a high voltage, short duration pulse to a gas at atmospheric pressure and ambient temperature can result in two different types of discharge: a nonequilibrium streamer discharge for short pulse durations (ranging from units to tens of nanoseconds) or an equilibrium pulsed arc for longer pulse durations (ranging from tens to hundreds of nanoseconds). These two categories have fundamentally nonuniform spatial characteristics.

Discharge initiation with a steep voltage rise (about 1 kV/ns) at moderately low gas densities is of considerable scientific interest. Modern high-voltage generators, capable of delivering pulse amplitudes in the range of tens to hundreds of kilovolts, voltage rise times of units to tens of nanoseconds, pulse durations of tens to hundreds of nanoseconds, and repetition frequencies up to tens of kilohertz, allow the generation of spatially uniform discharges up to gas densities of $(6-8) \cdot 10^{18}$ cm^{-3} [102]. These conditions correspond to a temperature of about 1,000 K at 1 atm pressure, or 600 K at 0.6 atm pressure and are particularly relevant to the SCRAMjet engine conditions. Discharges can be generated within large volumes of gas, reaching liter-scale volumes.

It has been found that when the voltage across a discharge gap increases rapidly, a gas can sustain voltages exceeding the steady-state breakdown threshold [103]. Under such circumstances, both the overvoltage ratio $K = U/U_{br}$ (where U_{br} is the breakdown voltage) and the parameter $p \times d$ (where p is the gas pressure and d is the interelectrode distance) are critical in determining the breakdown mechanism. When the overvoltage is in the tens of percent range, the breakdown changes from a uniform glow Townsend to a streamer breakdown. Conversely, at overvoltages of hundreds of percent or more, the breakdown regains spatial uniformity, albeit by a completely different physical mechanism, as a subset of electrons attains high energy at the ionization wave front and enters the so-called "run-away" regime, providing uniform preionization in front of the ionization wave. Here, the discharge progresses from the high-voltage electrode to the low-voltage electrode at typical velocities of $10^9 - 10^{10}$ cm/s. The work [103] presents a demarcation between regions where either Townsend or streamer discharge occurs [104]. This demarcation, along with the threshold for the "run-away" electron regime indicated by a horizontal line, is shown in Figure 2.20 and serves as a criterion for initiating a uniform discharge.

The phenomenon of pulsed uniform nanosecond gas discharge has been observed for over a century, with the first observation attributed to J.J. Thomson in 1893. Thomson discovered [105] the propagation of a luminosity wave along a glass tube (15 m long and 5 mm in diameter) at a speed

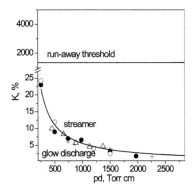

FIGURE 2.20 Regions of breakdown development per different mechanisms depend upon overvoltage in air [38].

that was at least half the speed of light in a vacuum. Since this discovery, there has been a periodic resurgence of interest in the topic every 20–30 years, probably in conjunction with advances in nanosecond technology, both in detection methods and in the generation of short pulses. The work presented in Ref. [106] provides a comprehensive review of research from the 1970s to the 1990s, emphasizing the importance of "run-away" electrons in the breakdown of short, overstressed gaps. A criterion for the transition to the run-away regime was established, confirming the critical role of run-away electrons in such breakdown mechanisms. Further exploration of this topic is available in the publication [107].

Research on nanosecond gas breakdown in long tubes conducted during the 1980s and 1990s was comprehensively reviewed in Ref. [108]. This review details the integral electrical parameters associated with the phenomena, including the amplitude and profile of the high-voltage pulse, the velocity of the ionization front, the signal attenuation during discharge propagation along the gap, the current (including the current of fast electrons), the delay in the onset of discharge, and the energy input. The discharge under study is considered promising for plasma chemistry applications due to the efficient excitation and ionization of gases facilitated by the high electric fields. Indeed, this review proposed the term "fast ionization wave" (FIW), which has since become the widely accepted nomenclature for this type of discharge.

The review [38] summarizes research on the detailed spatial and temporal dynamics of high-voltage pulsed nanosecond discharges, manifested as FIWs, during the last decade of the last century. It delves into the behavior of the electric field, electron concentrations, and excited states, using experimental data for analysis. The investigation, based on absolute time-resolved emission measurements of molecular bands, concluded that high-energy electrons significantly overpopulate the EEDF near the breakdown front. Conversely, it was found that the nonstationary and nonlocal characteristics of the EEDF could be neglected in the region behind the fast ionization wavefront. The energy distribution within the discharge was investigated, highlighting the potential of using FIWs as a source of uniform pulsed plasma with controllable energy deposition direction.

A brief review of the current understanding of homogeneous nanosecond discharges in air at moderate densities (up to 0.3–0.4 times the density at standard conditions) is given. The discharge propagates from the high-voltage electrode to the low-voltage electrode upon application of high voltage across the discharge gap. As the discharge progresses along the gap, the displacement currents close the circuit and the transition to the conductive currents occurs when the discharge contacts the low-voltage electrode. The spatial uniformity of the nanosecond discharge was demonstrated by extensive imaging using high-speed CCD cameras. An illustration of the development of a FIW in air, captured by an ICCD camera, is shown in Figure 2.21 [109]. Initiated by high-voltage positive polarity pulses with an amplitude of 11 kV in the cable, a duration of 25 ns at half-height, a rise time of 5 ns, and a repetition frequency of 40 Hz, the discharge occurred in a tube with a diameter of 5 cm and an electrode separation distance of 20 cm. Given the spectral sensitivity range of the optical system of 300–800 nm, the emission was predominantly from the second positive system of molecular nitrogen

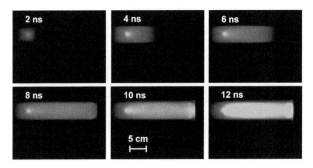

FIGURE 2.21 Subsequent ICCD images of nanosecond discharge in air. ICCD gate is equal to 1 ns, time moments from the discharge start are indicated. High-voltage electrode is on the left-hand side.

and accurately reflected the spatial evolution of the discharge over short time periods. Notable maxima were observed near the electrodes, although the discharge evolved uniformly overall. The propagation velocity of the front, estimated at 2.5 cm/ns, can be seen from a sequence of images.

The relationship between the velocity of nanosecond discharges and gas pressure, or more fundamentally, gas density, is often observed in experiments. Typically, this relationship shows a dome-shaped curve. It is noteworthy that nanosecond discharges can be initiated in a variety of gases, including both atomic and molecular types, as well as electronegative gases (such as SF_6). Under constant experimental conditions with increasing high-voltage amplitude, the velocity of the discharge with respect to pressure shifts toward higher gas densities. Maintaining a constant high-voltage pulse amplitude while increasing the gas density transforms the discharge into a non-homogeneous streamer configuration [110]. Conversely, there is evidence that this type of discharge can remain spatially uniform at atmospheric gas density [111], with energy input potentially reaching 80%–95% of the incident pulse [112]. Toward the end of the discharge, the current becomes significantly high, reaching into the hundreds of amperes.

Methods for estimating the electric field have been meticulously studied in Refs. [113–115], relying on measurements of the electric potential along the discharge gap or monitoring the emission from various molecular bands. The peak of high reduced fields (reaching several thousand Townsend) persists for only 2–3 ns before rapidly decaying to levels in the hundreds of Townsends. This brief surge in electric fields is critical for the spatially uniform progression of the discharge. Elevated electric fields generate high-energy electrons that pre-ionize the gas adjacent to the front, with most of the required electron concentration and excitation of upper electronic levels occurring behind the FIW front in residual fields. These residual fields are similar to those present in a streamer head (Figure 2.22).

The energy imparted to the plasma is predominantly used to excite electronic states, with the energy balance skewed toward high-energy level states, including excited ionic states. Figure 2.23 shows the energy distribution in air, oxygen and nitrogen subjected to a pulsed nanosecond discharge under parameters similar to those shown in Figure 2.21. The calculations were made from experimental data on the electric field distribution, using the two-term approximation of the Boltzmann equation to model the EEDF. This approach is applicable in the region behind the discharge front, where the discharge gap is fully spanned and the absence of steep electric field gradients is noted [116]. The data show that for all pressures and gases studied, a substantial fraction of the energy is allocated to the excitation of electronically excited states, consistent with expectations for electric field magnitudes on the order of hundreds of Townsends.

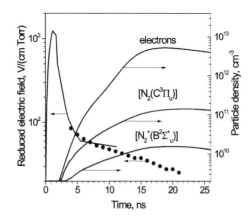

FIGURE 2.22 Temporal evolution of electric field, $N_2(C^3\Pi_u, v'=0)$ and $N_2^+(B^2\Sigma_u^+, v'=0)$ level densities and electron concentration in nitrogen at a distance 20 cm from the high-voltage electrode. Electric field represented with a solid line is determined from the electrical measurements; represented by symbols – from the spectroscopy. Nitrogen, $p = 4$ Torr, $U = -15.5$ kV [38].

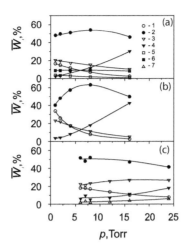

FIGURE 2.23 Fraction of energy contribution per pulse into inner degrees of freedom for various gases. (a) air (curves 1–4 correspond to N_2; 5 and 6 to O_2); (b) nitrogen; (c) hydrogen. Numbers designate. 1 and 5, ionization; 2 and 6, excitation of electronic terms; 3, dissociation; 4, excitation of vibration levels; 7, excitation of rotation levels [116].

Thus, from one perspective, a nanosecond discharge manifesting as a FIW represents a notable interest for the field of PAI/combustion due to several key attributes: (1) its evolution occurs within timescales significantly shorter than those typically associated with ignition processes, thereby facilitating temporal separation between discharge control and combustion kinetics; (2) it generates a spatially uniform plasma, effectively simplifying analysis to a zero-dimensional (0D) geometric consideration; (3) it excites predominantly high-energy molecular levels, leading to radicals production with minimal gas heating.

Conversely, the utility of this type of discharge at a given high-voltage pulse amplitude is limited by gas density (or equivalently, pressure). Increasing the pressure requires either increasing the amplitude of the high-voltage pulse—a solution that is not always technically straightforward—or employing special multielectrode systems designed to initiate the discharge in a quasi-uniform manner.

2.2.5 Pulsed Spark Discharges with Fast Gas Heating

To prevent glow discharges and other types of gas discharges from transitioning to arcs, a common strategy is to limit the discharge current by integrating large resistors or dielectric layers into the electrical circuit. The achievement of plasma maintenance at atmospheric pressure without transition to arc formation has also been realized by high-voltage, nanosecond repetitively pulsed discharges [117,118]. These discharges, which are characterized by their development under strong electric fields, generate a highly nonequilibrium plasma, with the short duration of the voltage pulse precluding arc formation. Active ionization occurs with each pulse, generating a significant amount of metastable particles that help sustain the discharge. In addition, rapid gas heating occurs, which is critical for flow control applications.

In Ref. [117], the application of high-voltage pulses, each lasting 10 ns, with an amplitude of 5 kV and a pulse repetition rate of 30 kHz – to a 1.5 mm gap demonstrated the feasibility of igniting a repetitively pulsed discharge in atmospheric air. Initially, the discharge characteristics resembled those of a pulsed streamer corona, but then transitioned to a regime characterized by a sharp increase in discharge current to 100–150 A, energy deposition per pulse escalating to 2 mJ, and a rise in gas temperature to $\tilde{3},000$ K. These conditions are indicative of discharge development at

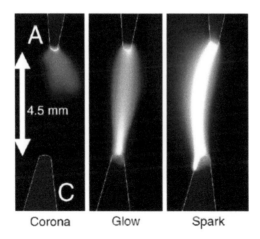

FIGURE 2.24 Photographs of a nanosecond pulsed-periodic discharge (corona, glow discharge, spark) in atmospheric pressure air at $T = 1,000$ K and frequency $f = 10$ kHz [121]. The length of the discharge gap length $d = 4.5$ mm. The anode and cathode are designated by the letters A and C, respectively. The voltage across the gap is 5, 5.5 and 6 kV (from left to right).

elevated E/n ratios, where fast gas heating becomes particularly effective [119,120]. This mode of discharge development is similar to a nanosecond spark regime.

In experiments in which air was heated to 2,000 K [118] and 1,000 K [121], a novel regime for sustaining a repetitively pulsed nanosecond discharge, termed the glow regime, was identified. Figure 2.24 illustrates the various forms of nanosecond discharge in heated air, including corona, glow mode, and spark, upon application of a repetitively pulsed voltage U with a frequency of 10 kHz, a duration of 25 ns, and a voltage rise and fall time of approximately 5 ns across a 4.5 mm gap between point electrodes [121]. The discharge was initiated in atmospheric air at a gas temperature of 1,000 K. The progression of different modes with increasing voltage U mirrored the corona-glow discharge-arc sequence observed in DC discharges [122]. The transition from corona to glow discharge mode occurred as U increased from 5 to 6 kV, followed by a transition to spark. Radiation within the air plasma in all nanosecond discharge modes was dominated by the second positive nitrogen system, indicating elevated reduced electric field E/n values within the discharge plasma. Radiation in the corona was confined to the vicinity of the high-voltage electrode, whereas in the glow mode the emission became diffuse and filled the entire gap. The transition to sparking was characterized by a sharp increase in radiation intensity and discharge current, from 1 A in the corona and glow modes to 20–40 A during spark formation. The energy deposition per pulse also escalated, from <1 µJ in the corona to 1–10 µJ in the glow mode, and reaching 200 µJ to 1 mJ in the spark [121]. Given the substantial E/n values (150 – 600 Td) in all regimes, significant heating was observed; emission spectroscopy measurements indicated a temperature rise of 200 K in the initial regimes and $\Delta T \sim 2,000 - 4,000$ K in the spark. The detection of intense emission from the second positive band of N_2^+ and the first negative band of N_2^- confirmed the presence of high electric fields. The mechanisms underlying rapid heating in such strong fields are consistent with those previously discussed. At the observed E/n parameters, energy release during the quenching of N_2 (A,B,C) triplet states by O_2 molecules emerges as one of the primary pathways for air heating.

The electron density within the discharge plasma was estimated to be approximately 10^{13} cm^{-3} in the glow mode and escalated to 10^{15} cm^{-3} in the spark mode [123], indicating a much higher emission intensity in the spark mode compared to the glow discharge mode. The emission radius of the plasma channel was measured to be about 1 mm in both scenarios. Specifically, in the spark mode, the current density reached 1 kA/cm^2, consistent with values observed for DC arc discharges [122].

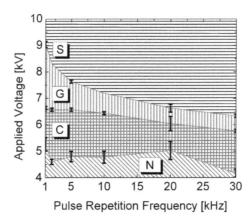

FIGURE 2.25 Regions of different modes of nanosecond pulse-periodic discharge in the plane "voltage-frequency of pulses" for air atmospherical pressure air at $T = 1,000$ K and $d = 5$ mm [125]. S - spark, G - glow discharge, C - corona, N - no discharge.

In glow mode, a cathode-directed streamer initially emanated from the anode. As the gap bridging, it triggered a reverse potential wave that homogenized the electric field across the discharge gap [121]. The voltage ceased before the formation of a cathode layer on the cathode, clearly distinguishing this discharge from DC glow discharges. A divergent scenario manifested itself in a nanosecond spark [123], where optical analysis revealed uniform discharge evolution driven by avalanche ionization throughout the volume. This breakdown mechanism differs from that of conventional sparks induced by a single voltage pulse in atmospheric air, where spark channel formation typically follows gap bridging by a streamer discharge propagating from one of the electrodes [124].

The uniform propagation of a nanosecond spark under repetitively pulsed voltage is attributed to the increased density of seed electrons ($n_e(0) \sim 10^{11}$ cm^{-3}) that persists from the end of one pulse to the beginning of the next. Optical measurements indicated rapid heating of several thousand degrees within the spark mode of a nanosecond repetitively pulsed discharge in the air within a mere ~ 30 ns duration [123], suggesting potential applications for flow/flame propagation and deflagration-to-detonation transitions via nanosecond discharge plasma.

In Ref. [125], a detailed investigation of the transition conditions between different modes of a nanosecond repetitively pulsed discharge in heated atmospheric air was performed. Experimentally defined ranges for different discharge modes, depending on voltage amplitude and frequency, are shown in Figure 2.25 for an air temperature of 1,000 K and an interelectrode gap of 5 mm. Reducing the pulse repetition rate from 30 to 1 kHz slightly widens and shifts the corona and glow domains toward higher voltages.

In the paper [125], variations in the frequency and amplitude of the voltage pulses were studied along with variations in the values of T (temperature) and d (gap size). Figure 2.26 shows the regions where different discharge modes materialize as a function of voltage amplitude and gas temperature, given a pulse frequency $f = 30$ kHz and an electrode gap $d = 5$ mm. It was observed that as T decreases, the voltage thresholds required to initiate a corona discharge escalate due to the increase in gas density at constant pressure. In particular, below 650 K, only corona and spark modes were sustainable under the specified experimental conditions, with the glow discharge mode manifesting exclusively at temperatures above 650 K. Variations in the electrode gap d at a fixed temperature $T = 1,000$ K and frequency $f = 30$ kHz revealed that the glow mode is attainable within a gap range of $3 < d < 9$ mm, beyond which only corona and spark modes were detectable. A pertinent question is the feasibility of achieving a glow discharge mode at ambient temperature conditions. Analytical calculations presented in Ref. [125] suggest a positive prognosis for this question, provided that the duration of the high-voltage pulse is extended and/or the radius of curvature of the electrode tips is reduced.

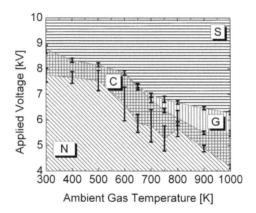

FIGURE 2.26 Areas of different modes of nanosecond pulse-periodic discharge in the voltage-temperature gas plane for atmospheric pressure air at $f = 30$ kHz and $d = 5$ mm [125]. S - spark, G - glow discharge, C - corona, N - no discharge.

Subsequent experimental studies of a nanosecond repetitively pulsed discharge in air elucidated the plasma characteristics in the temperature range of $T = 300 - 1,000$ K [126,127]. These studies, conducted in the nanosecond spark mode, revealed key findings including up to 50% dissociation of O_2, the density of electronically excited N_2 molecules, and the dynamics of gas heating. Insights into the fraction of energy rapidly converted to heat in air plasma at various E/n ratios were gained from analyses of heating dynamics and energy deposition within the discharge.

A groundbreaking numerical analysis of a nanosecond repetitively pulsed discharge in air was performed using a 1D model in Ref. [128], aligned with the experimental setup described in [117], where a nanosecond spark developed under standard conditions (1 atm, $T = 300$ K). The computational results showed qualitative agreement with the experimental observations and elucidated the primary mechanisms underlying the initiation and maintenance of the discharge, as well as the mechanisms of gas heating. The analysis underscored the importance of rapid gas heating in a strong electric field conditions ($E/n = 150 - 300$ Td), resulting mainly from the quenching of electronically excited N_2 states by O_2. Consequently, an escalation of the reduced field E/n occurred during the voltage pulse, due to a reduction of n between pulses, increasing the ionization rate, further heating the gas, and culminating in the transition to spark mode.

In Ref. [126], a 2D parametric numerical simulation was performed under conditions mirroring those of the experimental study [123], with a focus on the glow mode. The results indicate that the realization of the glow mode depends on the voltage pulse duration slightly exceeding the gap-bridging time of the streamer discharge. In addition, the post-gap-bridging phase is characterized by significant energy dissipation and gas heating within the channel, driving the discharge toward a transition to spark. This phenomenon was readily achieved in a 5 mm gap at $T = 1,000$ K, but proved challenging at ambient temperatures, confirming the observations of [123]. Further simulations in Ref. [127], examining the sequential pulse evolution in a glow mode for atmospheric pressure air at $T = 1,000$ K, showed that the discharge reached a quasi-stationary state, consistent with experimental findings on nanosecond repetitively pulsed discharges. Here, the density of residual charged particles at the onset of successive pulses was critical to the discharge evolution, with an increased density in heated gas due to slowed plasma decay at elevated T facilitating the discharge development.

An alternative form of pulsed spark discharge is the pulsed arc, which is the basis for the development of the localized arc filament plasma actuator (LAFPA). This discharge variant, in contrast to the nanosecond repetitively pulsed discharge, has received considerable attention in the field of plasma aerodynamics [129,130], but its evolutionary mechanisms have been less thoroughly explored. The proposal to use this type of discharge for gasdynamic flow control was first made in

Ref. [131]. The discharge mechanism involves generating a discharge in air between two electrodes separated by a few millimeters by applying a repetitively pulsed voltage with frequencies in the range of 10–300 kHz and pulse durations ranging from 1 μs to 1 ms. In order to prevent the plasma from being carried away by the high-speed flow, a groove was created on the surface of the body within the flow path to guide the discharge development. When voltage was applied, a breakdown occurred in the gap, resulting in spark formation. After the breakdown, the gap voltage dropped to several hundred volts, a level that was maintained until the pulse ended. During this phase, the electric field in the plasma channel was well below the breakdown threshold, heating the gas to temperatures between 1,000 and 2,500 K, indicating that the discharge was maintained as a pulsed arc. The key feature of the discharge, particularly advantageous for gasdynamic flow control, was fast gas heating, especially pronounced during the nanosecond spark phase [129], where elevated electric field values E/n were recorded. This fast heating mechanism was governed mainly by energy release upon quenching of excited N_2 (A,B,C) states by O_2 molecules. Gasdynamic perturbations caused by rapid local gas heating are used to modulate the gas flow. Thus, despite significant differences from the aforementioned nanosecond repetitive pulsed discharge, both share similarities in facilitating fast gas heating mechanisms.

Therefore, the application of pulsed spark discharges in atmospheric pressure air enables plasma generation in compact volumes at high reduced electric fields E/n and orchestrates fast gas heating. This process induces gasdynamic perturbations capable of influencing the gas flow, flames, and detonation initiation.

2.2.6 Optical Discharges

Compared to alternative energy deposition methods in gases, laser beams offer several distinct advantages. Characterized as a non-invasive energy deposition technique, lasers enable the achievement of high specific energy deposition at targeted locations. The process of focusing the laser beam ionizes the gas, resulting in spark formation, a phenomenon that has been validated over a wide range of wavelengths using various types of lasers, including CO_2 and excimer gas lasers, as well as solid state ruby and neodymium lasers [132,133].

The generation of a laser-induced spark, which facilitates the absorption of laser beam energy by the gas, is initiated by focusing the beam into a small focal volume. This process unfolds through a series of sequential phases as shown in Figure 2.27: initiation of electron production via multiphoton ionization, avalanche ionization within the focal zone, absorption of laser energy by the gaseous plasma, rapid plasma expansion, shock wave generation, and subsequent propagation of the shock wave through the adjacent gas [132,133].

In the context of gas ionization under laser radiation, two main electron multiplication mechanisms have been identified [134]. The first mechanism, multiphoton ionization, involves a neutral particle simultaneously absorbing a requisite number m of photons to facilitate ionization. This process can be expressed as

$$mh\nu + M = e + M^+$$

Here M represents a neutral particle, $h\nu$ represents a quantum of light, and $mh\nu$ represents the cumulative energy absorbed during this process. For ionization to occur, the energy absorbed must exceed the ionization potential, i.e. $mh\nu > I$. In this scenario, the electron density exhibits a linear increase with time. Since the ionization potential of molecules in air exceeds 12 eV and the quanta of the visible and near-infrared (IR) spectrum are typically equal to or <1 eV, ionization in a laser beam predominantly involves multiphotonic interactions.

The alternative ionization mechanism relies on the absorption of laser radiation by free electrons via inverse bremsstrahlung scattering processes, which serve as the inverse of the bremsstrahlung light quantum emission observed when high-energy electrons scatter off neutral particles.

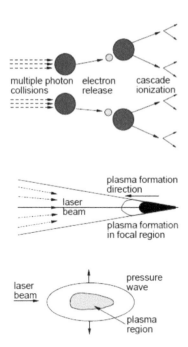

FIGURE 2.27 Processes controlling the gas breakdown by laser radiation [135].

When electrons acquire energy that exceeds the ionization potential of neutral particles, the resulting collisions can produce collisional ionization, described by the equation

$$e + M = e + e + M^+$$

This reaction initiates avalanche ionization, in which the electron density escalates exponentially over time as all newly generated electrons are also heated and subsequently involved in ionization collisions.

As avalanche ionization progresses, the resulting plasma begins to effectively absorb the laser beam energy. Initially, avalanche ionization is manifested within the small focal area characterized by the peak emission intensity (Figure 2.27). However, absorption of the laser radiation extends beyond the focal zone, the site of primary avalanche ionization initiation. Once the ionization level within the plasma reaches a significant value, avalanche ionization also begins in areas of lower emission intensities surrounding the plasma region. This newly ionized region, transitioning from transparency, begins to absorb the laser radiation. As a result, the energy-absorbing region continuously shifts toward the laser beam, creating a wave of gas ionization/heating. The thermal radiation emanating from the intensely heated region is absorbed by the cooler gas, which in turn initiates light emission [132]. This process, known as the radiative ionization wave propagation mechanism, allows the laser energy to be distributed over a larger area, thereby reducing the peak gas temperature. In addition to thermal plasma radiation, other mechanisms contribute to gas ionization adjacent to the established plasma, including molecular thermal conductivity and gas heating by a shock wave. These plasma propagation mechanisms can manifest themselves in various combinations, depending on specific conditions [132].

In the final phase of energy absorption, the laser pulse ceases, leading to plasma decay via electron-ion recombination. The focal and near-focal regions undergo significant heating, resulting in an increase in pressure and a decrease in gas density as shock and rarefaction waves propagate. After an optical breakdown, a toroidal vortex can form at the breakdown site, inducing a gas flow in the direction opposite to the propagation direction of the laser beam.

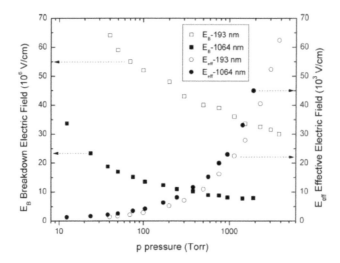

FIGURE 2.28 Dependence of the breakdown threshold (E_B) and effective electric field (E_{eff}) in air on its pressure for laser radiation with wavelengths of 193 nm [136] and 1,064 nm [137].

The threshold for laser-induced gas breakdown depends on a variety of factors [132,133], including the characteristics of the gaseous medium (such as gas composition, pressure, and presence of impurities, including aerosols and solid microparticles), the characteristics of the radiation (wavelength, pulse duration), and the dimensions of the focal zone. For illustration, Figure 2.28 shows the experimental results of the threshold electric field E_B for air breakdown over a wide range of pressures when exposed to UV radiation (193 nm from an ArF laser with a 20 ns pulse duration) [136] and IR radiation (1,064 nm from an Nd:YAG laser with a 6 ns pulse duration) [137]. The figure also shows the effective field, which describes the efficiency of electron heating in an alternating electric field.

$$E_{\mathrm{eff}} = \frac{E_B v_c}{\left(v_c^2 + \omega^2\right)^{1/2}}$$

Here, v_c represents the transport collision frequency of electrons with other particles (neutrals and ions), and ω represents the frequency of the electromagnetic field. Given the scenario where $\omega \gg v_c$, the effective field E_{eff} is approximately equal to $E_B \frac{v_c}{\omega}$. As shown in Figure 2.28, the threshold for gas breakdown increases with decreasing pressure and emission wavelength. The increased breakdown thresholds at reduced pressures are attributed to the decreased frequency of electron collisions with neutral particles under these circumstances [132]. For molecular ionization to occur, electrons must reach the ionization threshold energy by laser field heating, with electron heating resulting from collisional processes (inverse bremsstrahlung).

A tenfold decrease of the wavelength leads to an approximately tenfold increase of the breakdown threshold. At the same time, the field E_{eff} is almost independent of the radiation wavelength. Therefore, the increase of the threshold field E_B with decreasing wavelength correlates with the decreased efficiency of electron heating within the laser field at higher frequencies. The relationships between the gas breakdown threshold and various properties of the gas and the laser beam, together with their physical justification and the methods for theoretical numerical simulations, are described in detail in Refs. [132,133].

Figure 2.29 shows images capturing the sequence of events during laser-induced breakdown in air at atmospheric pressure, initiated by an ArF laser (193 nm) with an energy of 135 mJ and a pulse duration of 20 ns. These images were captured with a 10 ns gate of ICCD camera simultaneously

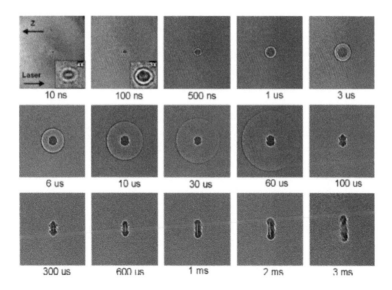

FIGURE 2.29 Dynamics of the plasma created in air by a laser beam with an energy of 135 mJ [iLs82]. The size of each image is 1.3×1.3 cm^2.

illuminated by a continuous wave (CW) laser. The laser beam traverses the image from left to right, with the time frame starting at the beginning of the laser pulse. Immediately after the breakdown ($t < 25$ ns), prominent bremsstrahlung from the plasma can be seen. In the initial phase ($t \leq 100$ ns), the heated region begins to expand from the focal point, causing the plasma to adopt a slightly elongated shape in the direction of the laser beam propagation. When 1 μs is reached, a shock wave detaches from the heated plasma core and propagates independently. Observation of the expansion of the shock wave continues up to 60 μs, at which point the wave diminishes in intensity and leaves the observable region. At the same time, noticeable deformation occurs within the radiating plasma zone, indicating that rapid gas dynamics significantly influence the stability of the expanding heated plasma core. As the wave moves sufficiently away from the plasma core, cooler air enters the heated region along the trajectory of the laser beam. The subsequent expansion of the plasma zone eventually leads to the emergence of a toroidal vortex-like configuration around the 2 ms mark. This evolutionary trajectory of the plasma zone and the associated gas jet as the gas flow progresses can significantly affect its properties.

The generation of a laser spark in a gaseous medium and the subsequent gasdynamic events represent a complex interplay of several physical mechanisms, culminating in the formation of a fully or partially ionized equilibrium plasma within nanoseconds, with temperatures ranging from 10^4 to 10^5 K [132,133]. Theoretical frameworks typically focus on one or two phases of laser spark evolution, while either neglecting or oversimplifying the rest. For example, simulations of the gasdynamic effects of laser spark energy deposition in an gas flow often assume instantaneous energy distribution within a predefined gas volume [138].

Laser radiation energy can be delivered to the gas not only by a single pulse but also by a more complex approach. Studies [139,140] introduced a novel methodology for laser energy delivery involving the synchronization of two laser pulses—one in the UV spectral range (266 nm) and the other in the near IR band (1,064 nm)—to simultaneously control both energy deposition and gas heating while avoiding optical breakdown. Initial electron generation was facilitated by multiphoton ionization by the UV pulse, followed by electron heating by an IR pulse after a predetermined delay (~ 10 ns) to ensure avalanche ionization of the gas. The resulting dual-pulse plasma exhibits nonequilibrium properties and requires significantly less energy to generate than the equilibrium plasma produced by a single laser pulse. In addition, this dual-pulse strategy allows fine-tuning of plasma density and gas temperature, offering the ability to generate plasma with different ionization levels ($10^{-4} - 10^{-2}$) and over a wide temperature spectrum (400–10,000 K) [141], thus proving beneficial for gas flow modulation tasks and fostering advances in PAI and combustion processes [140,142].

With the advancement of femtosecond lasers [143], recent proposals have emerged for modulating supersonic flows via thin plasma filaments generated during air breakdown by terawatt femtosecond laser radiation [144]. This method generates highly nonequilibrium plasmas of elevated density ($\sim 10^{17}$ cm^{-3}), whose decay triggers pulsed gas heating in the nanosecond regime [145], offering new possibilities for aerodynamic flow control.

2.2.7 RF AND MW DISCHARGES

The filamentary pulsed microwave discharge represents an innovative approach to gasdynamic flow control through rapid localized heating of air at atmospheric pressure [138,129]. This discharge is characterized by a conglomerate of plasma filaments in which gas temperatures rapidly increase to several thousand degrees within nanosecond time scale [146,147]. The rapid temperature rise within these filaments induces gasdynamic perturbations, such as weak shock waves, that can alter the gas flow properties. The principles underlying the rapid heating of gas in this plasma configuration are similar to those observed in other types of discharges discussed previously. A distinguishing feature of microwave (and, more broadly, optical) discharges, in contrast to other discharge methods for pulsed gas heating, is their ability to provide remote heating, free of electrode constraints, and effective over substantial distances from the electromagnetic wave source.

In their experimental work, the researchers in Ref. [148] used RF discharges to ignite fuel-oxidizer gas flows at stagnation pressures ranging from 30 Torr to 0.5 atm and flow velocities up to 70 m/s. Typically initiated within the frequency spectrum of 1 to 100 MHz, RF discharges are classified as either inductively coupled or capacitively coupled based on the method of excitation. In inductively coupled discharges, gas is confined within an inductive coil carrying RF current, creating a closed-loop vortex electric field. Conversely, in capacitive coupling, the gas is placed between electrodes to create a direct, potential electric field across them. The dynamics and principal processes governing capacitively coupled RF discharges, particularly their utility in CO_2 laser excitation (within a pressure range of 1–100 Torr) and surface treatments (pressure range of $10^{-3} - 1$ Torr), are thoroughly reviewed in Ref. [149].

This discussion focuses primarily on elevated pressure conditions where discharge uniformity is maintained either by electrode-based heat removal or by gas circulation through the discharge cell. In the absence of effective heat exchange stabilization, discharges tend to contract into brightly illuminated channels bridging the electrodes. The apparent channel diameter varies for different gases, with the vicinity of the electrodes exhibiting diffusive uniformity for molecular gases (N_2, CO_2, O_2), whereas atomic gases (Ar, Xe) exhibit filamentary patterns.

In both uniform and contracted configurations, the electron temperature reaches several eVs, although other parameters show significant variation. For contracted discharges, gas temperatures rise to 3,000–4,000 K with dissociation degrees of 90%–95%. Electron temperatures are significantly higher-by one to two orders of magnitude-compared to uniform RF discharge, while electric fields are lower. The characteristics of contracted RF discharges are very similar to those of arc discharges. It's worth noting that RF torches, generated by strong RF signals (ranging from hundreds of watts to kilowatts) at atmospheric pressure, have potential for various applications. These torches, when activated, produce elongated (more than 10 cm) hot plasma jets emanating from a single electrode that merge into the surrounding space by displacement currents.

At pressures in the tens to hundreds of Torr, coupled with effective heat removal, RF discharges produce a nonequilibrium plasma similar to glow discharges. RF voltage amplitudes are typically in the hundreds to thousands of volts, with gas temperatures not exceeding 1,000 K, electron densities around $n_e \sim 10^{11}$ cm^{-3}, and electric fields ranging from a few to hundreds of Townsend. The energy distribution is primarily aimed at low-energy electronic and vibrational states, mirroring the energy distribution seen in glow discharges. Some studies [150] use microwave discharges at moderate gas pressures (up to 300 Torr) to initiate combustion in both subsonic and supersonic gas flows. These discharges, when initiated in free space, manifest a complex spatial structure consisting of two distinct formations.

The first formation consists of thin, bright, highly conductive and hot filaments. The second consists of halo structures or "plasmoids" surrounding the hot filaments, characterized by nonequilibrium plasma, higher electric fields than the filaments, and relatively low conductivity. This intricate "skeleton" of hot, bright filaments surrounded by dimly glowing halos results from the interaction between microwaves and the gas flow. For an in-depth understanding of the physical nature of microwave discharges at moderate pressures, the references [41,151] are recommended.

To summarize this review of discharges relevant to PAI/PAC, a table is provided summarizing the discussion of typical plasma parameters associated with the discussed discharges (Table 2.3), where "vibr." represents vibrational degrees of freedom and "electr." represents electronic degrees of freedom. The wide range of values for microwave (MW) discharges reflects the variety of discharge forms under different experimental conditions. This summary serves as a general indicator that partially reflects the most commonly encountered discharges and plasma parameters in practice. Changing certain experimental conditions, such as frequency or flow velocity, can significantly alter the plasma characteristics while keeping the main system parameters unchanged (Table 2.3).

TABLE 2.3
A Summary Table of the Parameters of Plasmas Used for PAI/PAC

Parameter	Arc	Glow	Streamer	DBD	FIW	RF	MW
Pressure	Up to 10–20 atm	0.01–100 Torr	0.1–1 atm	1 Torr to 1 atm	0.01–200 Torr	10^{-3}–100 Torr	0.1 Torr to 1 atm
Current, A	$1-10^5$	$10^{-4}-10^{-1}$	$10^{-4}-10^{-3}$	$10^{-4}-10^{-3}$	50–200	$10^{-4}-10^{-1}$	0.1–1
Voltage	10–100 V	100–1,000 V	10–100 kV	1–10 kV	5–200 kV	500–5,000 V	0.1–10 kV
E/n or E	1–100 V/cm	10–50 V/(cm·Torr)	30–100 V/(cm·Torr) in a head	30–100 V/(cm·Torr)	30–100 V/(cm·Torr) behind the front	10–100 V/(cm·Torr)	1–1,000 V/cm
T_g, K	3,000–10,000	300–600	300–400	300–600	300–400	300–1,000	300–6,000
T_e	5,000–10,000 K	1–3 eV	1–3 eV	1–5 eV	1–10 eV	1–5 eV	1–5 eV
n_e, cm^{-3}	$10^{15}-10^{16}$	$10^{11}-10^{12}$	$10^{11}-10^{12}$	$10^{11}-10^{12}$	$10^{11}-10^{13}$	10^{11}	From 10^9 to 10^{17}
Uniformity in space	Non–uniform	Uniform (if not stratified)	Non–uniform	Non–uniform or uniform (APGD)	Uniform	Uniform (low pressure) or filamentary (high pressure)	Uniform (halo) and filamentary
Main energy input	Heating	Vibr.	Electr.	Electr.	Electr.	Vibr. or electr.	Heating, vibr., electr.

2.3 STREAMER DISCHARGE DEVELOPMENT: BRANCHING AND STREAMER HEAD STRUCTURE

The transition from self-sustained discharges to arc discharges, characterized by low charged particle energy and a high rate of energy transfer to gas heating, is generally unfavorable for effective plasma-chemical applications. Consequently, one of the most viable methods for generating highly nonequilibrium plasmas is to employ pulsed discharges with capacitive coupling, such as barrier discharges and their variants, where one or both electrodes are insulated with a dielectric layer. This dielectric barrier limits the total charge that can propagate through the gap.

Typically, at moderate to high pressures, the development of a barrier discharge unfolds in two phases: first, the discharge gap is bridged by a streamer flash, followed by the establishment of a non-zero electric current until the surface of the dielectric layer is charged. The significant energy contribution occurs in the latter phase of the discharge evolution, while the initial phase dictates the timing of gap bridging and the uniformity of distribution of the filamentary structure within the gap, a critical consideration in the design of practical devices [152].

A lot of studies, both experimental and theoretical, have examined the streamer corona, allowing discussion of the quantitative agreement between theoretical models and experimental observations, particularly with respect to individual filamentary streamers [51].

The primary mechanisms that influence the characteristics and parameters of these streamers as they advance are similar. These include preionization of the gas ahead of the discharge front by fast electrons and ionizing radiation, and ionization by electron impact in a relatively strong electric field at the front, as opposed to the relatively weak electric field behind it [153–155].

2.3.1 Experimental Study of Streamer Discharge

The typical experimental apparatus and diagnostic tools are depicted schematically in Figure 2.30 [57,156]. A thyristor-based generator, PAKM, equipped with magnetic pulse compression, served as the source of pulsed voltage. High-voltage pulses were delivered via a long 50 Ω coaxial cable to the discharge system's high-voltage input. To monitor the pulse parameters, a calibrated back-current shunt was installed at the midpoint of the coaxial cable's shield, 30 m away from the discharge setup. The voltage pulses had a positive polarity, with an amplitude ranging between 10 and 21 kV, a full width at half-maximum (FWHM) of 25 ns, a rise time of 5 ns, and a variable repetition rate from 0.5 to 100 Hz.

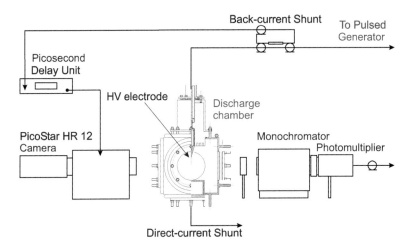

FIGURE 2.30 Experimental setup for streamer discharge investigations.

The experimental setup consisted of a 20 cm side cubic stainless steel vacuum chamber. Optical observations were facilitated by 100 mm diameter quartz windows. The discharge configuration used a point-plane geometry, with interelectrode distances ranging from 20 to 60 mm. The high-voltage electrode (cathode), made of brass, was disk-shaped and 80 mm in diameter. The grounded anode, also disk-shaped and made of aluminum, was 100 mm in diameter. A needle, 0.8 mm in diameter and 8 mm in height, mounted centrally on the grounded electrode, served to initiate the cathode-directed streamer.

When subjected to a high-voltage pulse of negative polarity, the electric field at the needle tip significantly exceeded the uniform field between the electrodes, triggering streamer development from the grounded electrode. This electrode configuration also allowed direct measurement of the streamer flash current by connecting the needle to ground through a 50-Ω resistor.

Emission spectroscopy was used to analyze the cathode-directed streamer discharge, with active particles identified by absolute emission spectroscopy. The optical setup included photomultipliers with sensitivity ranges 170–830 nm and 300–800 nm and UV and VIS monochromators with operating ranges 200–2,000 nm and 200–800 nm.

The spatiotemporal discharge characteristics were monitored using a PicoStar HR12 ICCD camera (LaVision) with a spectral sensitivity range of 200–800 nm and a minimal gate of 200 ps. The ICCD camera was aligned to capture the needle on the grounded electrode, ensuring that both electrodes were within the frame. The spatial resolution of the images was 0.1 mm. Simultaneously with the camera sync pulse, the discharge current and voltage across the gap were recorded using a Tektronix TDS-3054 oscilloscope. The ICCD camera, voltage and current measurements were synchronized to capture the same streamer flash.

Experiments were conducted in a N_2-O_2 (4:1) mixture over two discharge gap lengths: 30 and 50 mm. For the 30 mm gap, the chamber pressure varied from 320 to 1,300 Torr, while for the 50 mm gap, the pressure varied from 90 to 410 Torr. The minimum pressure for each gap length was determined by the threshold at which the streamer transition to the spark form was observed.

2.3.2 Single Channel Development

At pressures lower than 590 Torr authors [57] could register a branching structure or a single streamer channel from run to run under the same experimental conditions. At pressures lower than 470 Torr they observed only a single streamer channel.

2.3.2.1 Streamer Discharge Development at Different Voltage

The development of a streamer channel is determined by the density of seed electrons in front of the streamer's leading ionization wave. These electrons originate from photoionization by VUV photons emitted from the streamer head and background ionization from external radiation sources (such as cosmic rays). The streamer's advance through space is facilitated by its enhanced electrical potential relative to its surroundings. As the streamer moves from the high-voltage electrode to the low-voltage counterpart, the length of the streamer channel, and thus the voltage across it, increases, resulting in a decrease in the potential of the streamer head.

A decrease in pressure, given a fixed configuration of the electric field, results in an increase in the reduced electric field E/n. This, in turn, increases the rate of electron impact ionization while increasing the photoionization range due to the decreased gas density and decreased collisional quenching rate of electronically excited states. As a result, both the velocity of the streamer and its diameter increase.

The anode current of a streamer directed toward the cathode increases as the gas pressure decreases. Figure 2.31 shows the anode current profiles for different pulse voltages in air for a discharge gap length of 30 mm. The relationship between the velocity of the cathode-directed streamer and the applied gap voltage is shown in Figure 2.32. Experimentally measured velocity values for various pressures are shown as points, while numerical simulation results are shown as curves.

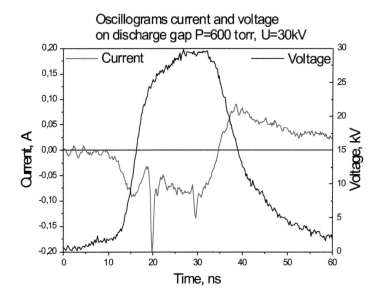

FIGURE 2.31 Voltage and anode current measurements. Discharge gap is 30 mm (air).

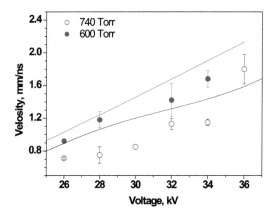

FIGURE 2.32 Velocity dependence on the voltage. $U = 24$ kV. Discharge gap is 30 mm (air). Cathode-directed streamer. Dots are experimental results, curves are numerical modeling.

The average streamer velocity across the gap for pressures of 600 and 740 Torr is highlighted. The observed velocities and current values show a good correlation between experimental data and computational predictions.

The streamer current was measured at the center of the current pulse plateau. The streamer head position is close to the center of the gap at this moment (Figure 2.33).

In theoretical analysis, the term "streamer radius" often refers to the electrodynamic radius that coincides with the field maximum. In the experiment, however, the streamer radius is derived from imaging techniques that capture the streamer's emission distribution.

In Ref. [57] the authors measured the streamer's emission distribution and defined its emission radius based on the half-maximum of the emission profile. The emission radius differs from the electrodynamic radius associated with the plasma-occupied region. For a comprehensive comparison with numerical simulations, it's advantageous to determine both the electrodynamic and the emission radius. The electrodynamic radius can be inferred from the weak extremities of the emission distribution at the streamer head, which requires the identification of the peak intensity within these

Low Temperature Plasma Generation and Recombination 75

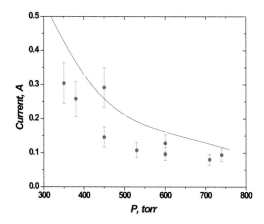

FIGURE 2.33 Current dependence on the Pressure. $U = 24\,\text{kV}$. Discharge gap is 30 mm (air). Cathode-directed streamer. Dots are experimental results, curve is numerical modeling.

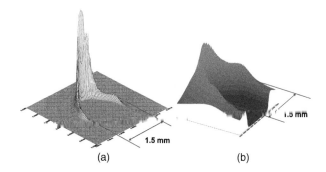

FIGURE 2.34 Calculated emission (a) and electric field (b) distributions. $U = 24\,\text{kV}$. Discharge gap is 30 mm (air). Cathode-directed streamer.

"tails". This determination cannot be made from an aggregate emission image; rather, an instantaneous image of the streamer head is required to analyze the spatial distribution of the emission.

Figure 2.34 shows both the calculated emission and electric field distributions, highlighting the discrepancies between their spatial distributions. However, the plasma domain (electrodynamic radius) remains consistent with emission measurements and can be determined from the emission distribution analysis.

In theoretical works when speaking about streamer radius usually means electrodynamical radius, which corresponds to field maximum. But during experiments, the streamer radius is determined using image, which is obtained using this or that technique.

It is observed that the calculated emission diameter is smaller than the electrodynamic diameter, as shown in Figure 2.35. A comparison between experimental results and numerical simulations shows that this discrepancy increases at lower pressures. Thus, the experimental derivation of the electrodynamic radius provides an essential parameter for comparison with the numerical model.

This methodology requires that the emission peak coincides with the field peak. In fact, the emission intensity of the 2^+ system of nitrogen is directly proportional to the concentration of excited nitrogen molecules. This concentration, in turn, depends on the electron impact excitation rate coefficient and the electron concentration. The excitation rate has an exponential dependence on the electron temperature and consequently on the electric field strength. Therefore, by analyzing the emission distribution across the streamer head section, one can determine the electrodynamic radius.

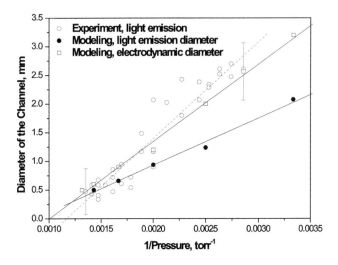

FIGURE 2.35 Comparison of the measured and calculated streamer channel diameter. $U = 24$ kV. Discharge gap is 30 mm (air). Cathode-directed streamer.

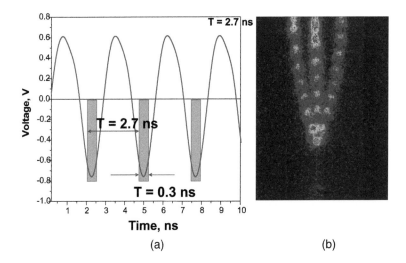

FIGURE 2.36 Voltage applied (a) and image of the streamer in the stroboscopic mode (b).

The investigation of the structure and velocity of the streamer head using stroboscopic imaging with an ICCD camera was performed in Ref. [57]. The experiment used a sinusoidal signal with a frequency of 300 MHz as the triggering mechanism for the camera amplifier. The signal shape is shown in Figure 2.36. In this particular mode of operation, the emission intensity received by the CCD matrix exhibits an exponential dependence on the applied voltage. It is postulated that the camera gate aligns with the lower part of the synchronization signal, which has a duration of 200–400 ps. By analyzing the interval between intensity peaks within the streamer image in stroboscopic mode, it is possible to calculate both the average velocity of the streamer and its velocity over different sections of the discharge gap.

To reconstruct the emission distribution within the streamer head, an image taken with a 300-picosecond gate duration was used. The analysis focused on the variation of the emission intensity over the radius perpendicular to the streamer propagation axis. Up to ten emission profiles were extracted from each streamer head image for detailed study (Figure 2.37).

Low Temperature Plasma Generation and Recombination

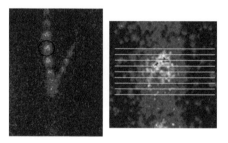

FIGURE 2.37 Emission distribution in the streamer head. Camera gate is 300 ps.

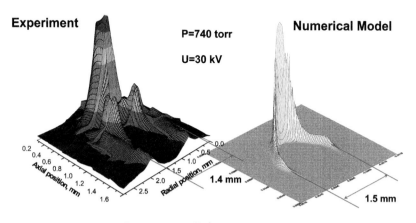

FIGURE 2.38 Streamer channel electrodynamic diameter. $U = 30$ kV, air.

The profile was generated by integrating the intensity of a two-dimensional, axisymmetric, cylindrical object. To reconstruct the emission pattern over the radius of the streamer head, the inverse Abel transformation was applied. By determining the distance between these profiles, the emission distribution within the streamer head could be reconstructed. The distance between the emission peaks indicates the diameter of the streamer. The diameter of the streamer head, derived from picosecond images and identified by emission peaks, was measured to be 1.4 mm (Figure 2.38). In addition, numerical simulation results were compared with the experimental emission pattern. The figure shows the emission distribution for the second positive system of N_2 calculated under experimental conditions. The electrodynamic diameter determined by the numerical model was 1.5 mm. It is observed that in the experimental distribution, the length of the emission zone along the trajectory of the streamer exceeds the calculated value, amounting to 0.3 mm. This discrepancy can be attributed to the movement of the streamer head during the camera's intensifier gate and to the spatial resolution of the intensifier.

For an interelectrode gap of 30 mm in a pressure range of 380–350 Torr, the streamer channel bridges the discharge gap and transforms into a spark as the pressure decreases below this limit.

Simultaneous synchronized measurements of the electrical parameters of the discharge show that as the pressure decreases, the streamer current, charge transferred and energy input increase monotonically until the transition to the spark.

2.3.3 Streamer Branching

Within the pressure range of 680–1,300 Torr, a branched streamer flash is consistently observed for all applied voltages. As the pressure is reduced, the frequency of branching decreases; in particular,

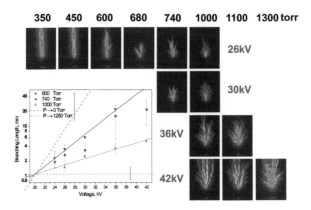

FIGURE 2.39 ICCD image of multichannel flash for different pressures and voltages. Discharge gap is 30 mm (air).

within the pressure range of 620–650 Torr, three or fewer branches are typically observed. Integral imaging of the cathode-directed streamer reveals that the number of branches escalates with increasing pressure and similarly increases with increasing voltage at constant pressure. The streamer branching phenomenon exhibits a pronounced dependence on both gas pressure and applied voltage. Experimental evidence suggests a direct proportionality between the number of branches and the gas pressure.

Figure 2.39 shows the typical dependence of streamer flash morphology on both pressure and voltage. In addition, the Figure shows the relationship between branch length and voltage applied at pressures of 600, 740 and 1,000 Torr for cathode-directed streamers. In particular, it is observed that the branching length increases exponentially with increasing voltage.

2.3.4 Numerical Model of Streamer Discharge

The goal of this section is to give a self-consistent description of the two-dimensional formulation of the problem studied in Refs. [51,157] and to model the dynamics of streamer discharges in long gaps.

2.3.4.1 Main Equations

The modeling of the streamer discharge was performed in the hydrodynamic approximation for the two-dimensional geometry. The numerical model includes the following balance equations for charged particles:

$$\frac{\partial n_e}{\partial t} + \text{div}\left(\vec{v}_e \cdot n_e\right) = S_{\text{ion}} + S_{\text{photo}} - S_{\text{att}} - S_{\text{rec}}^{\text{ei}} \tag{2.1}$$

$$\frac{\partial n_p}{\partial t} = S_{\text{ion}} + S_{\text{photo}} - S_{\text{rec}}^{\text{ei}} - S_{\text{rec}}^{\text{ii}} \tag{2.2}$$

$$\frac{\partial n_n}{\partial t} = S_{\text{att}} - S_{\text{rec}}^{\text{ii}} \tag{2.3}$$

where n_e is the electron density; n_p and n_n are the concentrations of positively and negatively charged ions; $\vec{v}_e$ is the drift velocity in a local electric field $\vec{E}$; and S_{ion}, S_{photo}, S_{rec}, and S_{att} are, respectively, the rates of ionization, photoionization, electron-ion (ei) and ion–ion (ii) recombination, and electron attachment.

The electric field $\vec{E}$ in the discharge gap is described by the potential φ, whose distribution over the gap was determined by solving Poisson's equation with prescribed boundary conditions and distribution of charged particles of different species:

$$\vec{E} = -\nabla \varphi \qquad (2.4)$$

$$\Delta \varphi = -\frac{e}{\varepsilon_o}\left(n_p - n_e - n_n\right) \qquad (2.5)$$

where e is the elementary charge and ε_o is the dielectric permittivity of the medium.

2.3.4.2 Photoionization Processes

The rate of gas photoionization, S_{photo}, is described on the basis of the model [157,158]:

$$S_{photo} = \frac{1}{4\pi} \cdot \frac{p_q}{p + p_q} \int_V d^3\vec{r_1} \frac{S_{ion}(\vec{r_1})}{|\vec{r}-\vec{r_1}|^2} \Psi\left(|r-r_1|\cdot p\right) \qquad (2.6)$$

where $1/4\pi$ is the normalization constant, p is the gas pressure, and $\Psi(|r-r_1|\cdot p)$ is the absorption coefficient of the ionizing radiation in the medium. The quenching pressure of the emitting states p_q was set to 60 Torr for pure nitrogen and 30 Torr for air [158] and was recalculated for any percentage of oxygen in nitrogen gas (2.6) (Figures 2.40 and 2.41).

To elucidate the effects of streamer branching, a simplified model capable of qualitatively assessing the influence of streamer branching on its propagation characteristics was developed in Ref. [157]. In an axially symmetric framework, streamer branching was conceptualized as a conical charge distribution ($\alpha = 0.5$ Rad) enveloping the primary streamer channel. The charge within this conical region was assumed to be uniformly distributed, and its total magnitude varied according to the desired number of branches to be evaluated (Figure 2.42). It was assumed that after branching, each streamer branch is identical in nature and carries an equal fraction of the total charge. Using

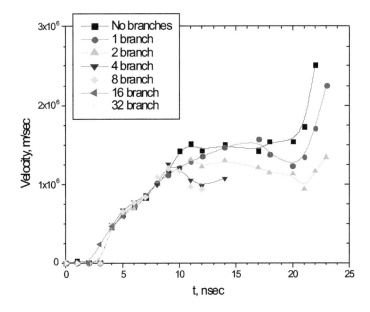

FIGURE 2.40 Dependence of the streamer velocity from number of branches discharge gap is 30 mm (air).

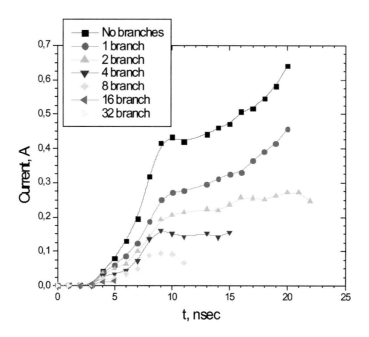

FIGURE 2.41 Dependence of the streamer current from number of branches discharge gap is 30 mm (air).

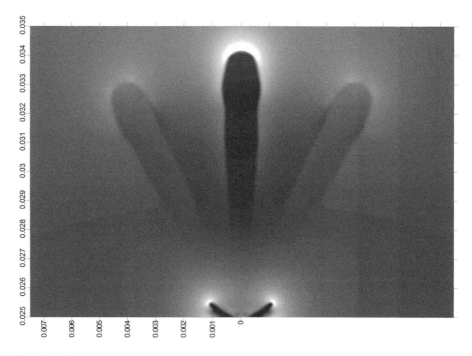

FIGURE 2.42 Electric field distribution in the gap ($\lg(E/n)$). Four additional "conical branches" are added.

this model, the authors of Ref. [157] attempted to approximate the effects of streamer branching on various streamer parameters.

To address the question of the influence of streamer branching, a simplified model was developed to qualitatively assess the effect of streamer branching on its properties. In the axially symmetric model, the branching was represented as a cone (with an angle of $\alpha = 0.5$ Rad) of charge surrounding the primary streamer. The charge was assumed to be uniformly distributed within the volume of the cone, with its quantity varying according to the desired number of branches to be simulated (Figure 2.42). It was assumed that after branching, each streamer branch is identical and carries the same amount of charge. Therefore, the above model was used to roughly estimate which parameters are affected by streamer branching.

For the baseline configuration, a streamer channel with a radius of $R_{streamer} = 0.282$ mm, a current of $I = 14$ mA, a total charge in the discharge gap of $Q = 7.4 \times 10^{-10}$ C, and a streamer velocity of $V = 5.06 \times 10^5$ m/s was observed. For the propagation of two streamer branches simultaneously, the results were $I = 10$ mA, $Q = 5.5 \times 10^{-10}$ C, and $V = 4.85 \times 10^5$ m/s, demonstrating a significant effect of streamer branching on the current. The numerical data indicate that, at the pressures and voltages specified, the velocity of a streamer is minimally affected by the presence or absence of branching. Conversely, branching predominantly affects the streamer current. It was found that the velocity of a single streamer in a plane gap and the velocity of streamers in a flash are closely related. This result indirectly supports the conclusion from the simplified model that streamer branching has little effect on the propagation velocity of the branches, and that the branches have minimal effect on each other's velocities.

The numerical model for a multichannel flash predicts a slight increase in flash current as the number of channels increases and a very slight decrease in velocity. The dependence of streamer velocity and current on the number of branches is shown in Figures 2.40 and 2.41.

2.4 FAST IONIZATION WAVE

Plasma chemistry research covers a wide range of topics, from the creation of chemical lasers and ozone generators to the treatment of liquid and solid wastes. These applications fall into two primary categories: those that use plasma to heat the gas [159,160] and those where the role of excited and charged particles is more important than the temperature increase [159,161,162]. A key factor influencing the effectiveness of plasma sources in the latter category is the degree of nonequilibrium of the plasma, essentially the proportion of energy directed to the internal degrees of freedom of the gas (excluding rotational motion of the molecules) relative to the total energy input. The reactivity of highly nonequilibrium gas mixtures exceeds that of their equilibrium counterparts by several orders of magnitude for equivalent total energy contributions. Especially in terms of chemical reactivity, discharges that direct most of the energy input to higher excited states are most beneficial. For many plasma-chemical applications, ensuring plasma homogeneity is critical, e.g. for uniform initiation of combustion processes.

Therefore, the creation of sources capable of producing highly nonequilibrium, homogeneous plasmas with efficient energy utilization is receiving considerable attention. Research in recent decades has highlighted the effectiveness of nanosecond pulsed power supplies, attributing this to the high reduced electric fields in the discharge, its spatial uniformity, and the resulting low gas temperature of the produced plasma. A promising method for generating highly nonequilibrium plasma is through discharges manifesting as FIWs [38,108,116]. This section reviews experimental analyses focusing on the kinetics of electron density decay and the spatial uniformity of nanosecond pulsed discharges.

2.4.1 FIW Velocity and Electron Energy Distribution

The velocity of the ionization wave propagating in the discharge tube was determined from the velocity of the electrical signal detected by capacitance dividers. It was found that there is an optimal pressure range conducive to the development of a uniform nanosecond breakdown, characterized by a smoothly varying ionization wave velocity with a notable peak within this range [163–168]. Consequently, tracking the wavefront velocity in each series of experiments facilitated the establishment of a relationship with the corresponding pressure conditions. For illustration, Figure 2.43 shows the velocities at which ionization waves of negative polarity propagate at each cross-section in air, nitrogen, and hydrogen [116].

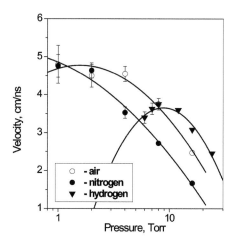

FIGURE 2.43 FIW front velocity upon pressure at a point $x = 20$ cm from the high-voltage electrode for different gases. $U = -13.5$ kV, pulse duration is 25 ns, pulse front duration is 3 ns, $f = 40$ Hz, discharge tube diameter is 1.8 cm, tube length is 60 cm.

The study of the propagation of the ionization wave of negative polarity with different configurations of high-voltage electrodes showed that, beyond a certain distance from the high-voltage electrode, the shape of the electrode does not significantly affect the breakdown evolution. However, the use of a pointed, e.g. conical, electrode was found to stabilize the initiation of the ionization wave at lower pressures.

The energy input into the gas was quantified using a conventional method that examines the current pulses directed to the discharge gap, those reflected from it, and those passing through. In our experimental setup, the current pulse reflected from the discharge gap upon reaching the generator is reflected back to the discharge tube. Within the optimal pressure range for ionization wave propagation (10–20 Torr in oxygen and its mixtures with nitrogen under experimental conditions of Ref. [116]), the energy of the electric pulse is typically absorbed by the plasma within the time frame of two pulse reflections at the tube. At pressures both lower and higher than this range, plasma absorption of the energy occurs over more reflections.

It has been found that there are regimes where at least 80% of the pulse energy is transferred to the discharge gap. Through extensive measurements in N_2-O_2 mixtures, it was confirmed that the pressure dependence of the energy input exhibits a bell shape similar to the pressure-velocity dependence, with the notable difference that the peak of the maximum is significantly shifted to lower pressures. For illustration, Figure 2.44 shows the energy input into the gas in the first pair of pulses W_I and the energy responsible for a single reflection W_{II} in pure oxygen in a 20 cm long and 5 cm diameter discharge tube.

2.4.2 Spatial Uniformity of FIW

A specific range of gas densities allows the FIW propagation at high speed. In order to investigate the spatial structure of the discharge and the homogeneity of the combustion initiated by this discharge, a series of experiments were performed in which the emission intensity (over the wavelength range of 300–800 nm) was recorded using a PicoStar HR12 La Vision ICCD camera. The gate duration of the camera was set to 1 ns, which allowed observation of the discharge progression from the high-voltage electrode to the low-voltage electrode with a 1 ns interval between frames, using the periodic regime of the discharge. These results are illustrated in Figure 2.45 for different high-voltage pulse polarities.

Low Temperature Plasma Generation and Recombination

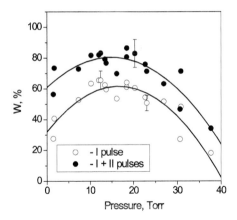

FIGURE 2.44 Energy contributed into gas in the first pair of pulses W_I and the energy, taking into consideration one reflection W_{II} in oxygen. $U = -15$ kV, pulse duration is 20 ns, pulse front duration is 8 ns, $f = 82$ Hz, discharge tube diameter is 4.7 cm, tube length is 20 cm.

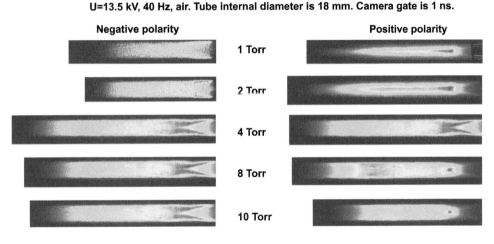

FIGURE 2.45 The imaging of nanosecond discharge propagation in long tube for different polarities of a high-voltage pulse. $U = \pm 13.5$ kV, pulse duration is 25 ns, pulse front duration is 3 ns, $f = 40$ Hz. Discharge tube diameter is 18 mm; interelectrode distance is 60 cm.

The ICCD system maintained a constant sensitivity across all images. Within the sensitivity spectral range (300–800 nm), the dominant emission was from the second positive system of molecular nitrogen (N_2, $C^3\Pi_u$ to $B^3\Pi_g$). With a lifetime of 40 ns, this system is a reliable indicator of plasma uniformity. For negative polarity, intense emission is observed around a conical electrode (cathode). At lower pressures (1–2 Torr), the discharge velocity is low and the attenuation along the tube is pronounced, but the emission remains relatively uniform. At the same pressures under positive polarity, higher propagation velocities are observed, but the discharge develops predominantly along the discharge tube axis, with the most intense emission near the electrode tip. At higher pressures (4–10 Torr), the discharge shows relative uniformity for both polarities, but with distinctly different evolution characteristics. In particular, for the positive polarity, the shape of the emission front changes from convex to concave as the pressure increases.

The region corresponding to the maximum velocity of the FIW, specifically 4–8 Torr under the experimental parameters of Ref. [116], denotes a region where the discharge evolution is spatially

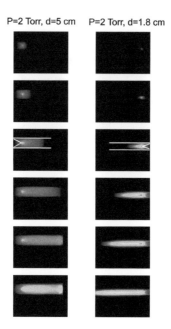

FIGURE 2.46 The imaging of nanosecond discharge propagation in tubes of different diameters. Positive polarity, $U = +13.5$ kV, pulse duration is 25 ns, pulse front duration is 3 ns, $f = 40$ Hz. Discharge tube diameter is 18 mm; interelectrode distance is 60 cm.

uniform. This pressure range at room temperature corresponds to gas densities of $(1.3 - 2.6) \cdot 10^{17}$ cm^{-3}. By increasing the amplitude of the high-voltage pulse, the uniform discharge region can be extended into denser gaseous media.

The analysis of the dynamics of nanosecond discharges in discharge tubes of different diameters provides interesting insights. Figure 2.46, which delineates the electrode and tube walls with white lines, provides a comparative view for discharges under positive pulse polarity. Clear differences can be seen between discharges in tubes of 2 and 5 cm diameter when subjected to identical electrical pulses. In the narrower diameter tube, there is a tendency to form a beam-like structure emanating from the high-voltage conical electrode (anode). Conversely, in the broader 5 cm diameter tube, the discharge has a more uniform emission profile with a much less pronounced bright zone at the electrode tip.

The evolution of a nanosecond discharge in large volumes was investigated in Ref. [110]. It was found that the discharge propagates quasi-spherically from the high-voltage electrode, at least as long as the field strength at the ionization front remains in the order of $E/n > 2,000 - 3,000$ Td. If the pulse amplitude and duration are sufficiently high, the discharge gap is bridged, resulting in a uniform plasma region near the low-voltage electrode. As the pressure is increased, the wavefront stops translationally closer to the high-voltage electrode. Under these conditions, a second wave of positive polarity is initiated from the surface of the low-voltage electrode toward the initial wave. This counter-wave is created by the effective preionization in front of the FIW and a significant potential difference between the low-voltage electrode and the FIW front. As the pressure continues to increase, the diameter of the region on the surface of the low-voltage electrode from which the opposing FIW originates begins to decrease. At a certain pressure level, the counter-wave loses its uniformity and transforms into a flash of streamers that close the discharge circuit. Subsequent increases in pressure lead to a reduction in the number of opposing streamers, eventually allowing the observation of a stalled spherical ionization wave. Finally, at elevated pressures, the FIW transitions to a pulsed corona emerging from the high-voltage electrode. A comparable sequence is observed when the amplitude of the high-voltage pulses is reduced at constant pressure.

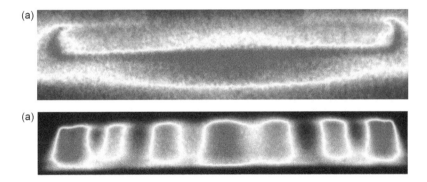

FIGURE 2.47 Emission image from a nanosecond discharge taken by LaVision ICCD camera. Camera gate is equal to 1 ns. Air, +5 kV pulse amplitude and 12 ns pulse duration. (a) −0.6 Torr; (b) −20 Torr.

The evolution of nanosecond discharges between two parallel electrodes in air, with the relatively short interelectrode gap was performed in Ref. [169]. The study spanned a pressure range from 0.6 to 20 Torr over various discharge repetition frequencies from 3 to 30 kHz. High-voltage pulses of positive polarity were used, each with an amplitude of 5 kV. The emission from the second positive system of molecular nitrogen was monitored using an ICCD camera with a 1 ns gate and 1.3 ns frame intervals. The plane electrodes were separated by 3 cm, with the top electrode having a diameter of 12 cm. Figure 2.47a shows the emission 14.1 ns after application of a high-voltage pulse at $p = 0.6$ Torr and repetition frequency $f = 30$ kHz. As the pressure increased, a series of discrete channels were observed starting at the electrode boundary and bridging the gap. These "streamers" were not stationary but moved along the boundary (see Figure 2.47). Thus, a remarkable phenomenon is highlighted for consideration in the use of pulsed nanosecond discharges in various applications: discharge uniformity in "free space" can be improved by the use of special electrode systems that avoid discharge destabilization due to boundary effects. The effect of frequency on discharge uniformity is modest; changing the frequency in the range of 3 Hz to 3 kHz resulted in a marginal shift of a few Torr in the threshold between uniform and nonuniform discharges. It's noteworthy that a much more pronounced dependence is observed with changes in pulse amplitude. By subjecting the gas to a high-voltage pulse with an amplitude of about 100 kV, it was possible to achieve uniform discharges up to gas densities of $\approx 0.5 \cdot 10^{18}$ cm^{-3}, corresponding to a pressure of about 150 Torr at ambient temperature.

2.4.3 Electric Field and Electron Density in FIW

For a comprehensive kinetic analysis of the FIW, it was essential to determine the temporal and spatial variations of the electric field and the electron concentration. These parameters can be derived from the dynamics of the spatial charge per unit length of the discharge tube, which can be measured by a capacitance sensors placed along the tube. Paper [116] has introduced a methodology for reconstructing the excess charge density that utilizes signal captures from capacitance sensors and solves the inverse problem while taking into account the spatial sensitivity function of the sensor.

The dynamics of the longitudinal electric field component and the electron concentration were inferred from the established charge distribution along the axis of the discharge tube. The analysis made the assumptions that the charge distribution is quasi-stationary and that the excess charge is located predominantly near the walls of the discharge tube. This latter assumption inherently suggests that the minimum possible value of the longitudinal electric field component at the wavefront was estimated. The electron concentration was evaluated using the drift approximation, relying on charge and field dynamics data. The methods for deriving field strengths and electron concentrations from electrical signals were discussed in detail in Refs. [113,114].

Illustrative reconstructions of the electric field for nitrogen and hydrogen are shown in Figure 2.48. For all scenarios, the longitudinal electric field profile exhibits a pronounced but short peak lasting

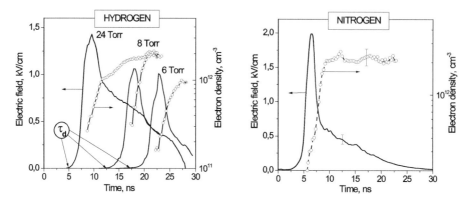

FIGURE 2.48 Electric field and electron density, typical temporal behavior. $U = -13.5$ kV, $x = 20$ cm from the cathode. Hydrogen pressure is pointed in the figure. Nitrogen pressure is equal 4 Torr.

only 2–4 ns. At this peak, the electric field reaches extremely high levels up to ~ 1 kV/(cm·Torr) or $E/n \sim 2-3$ kTd. The field then decays rapidly, stabilizing at tens to hundreds of V/(cm·Torr) for the duration of the pulse, 20–25 ns. In hydrogen, the reduction in total field peak duration with decreasing pressure is attributed to an increased delay in breakdown initiation, extending to 17 ns at a hydrogen pressure of 6 Torr.

It is important to emphasize that under identical experimental conditions in nitrogen, when a high-voltage pulse of positive polarity is applied, the temporal behavior of the electric field is similar, but its magnitude is several times greater; however, behind the front, the electric field decreases to a fraction of what is observed for the FIW of negative polarity [115]. Within the pressure range that produces a dome-shaped velocity profile of the ionization wave, the absolute value of the field increases slightly, by no more than 1.5–2 times, while the value of the reduced electric field decreases by about an order of magnitude. This significant variation suggests a profound redistribution of energy among the internal degrees of freedom of the gas with the pressure variations.

2.5 MODELS FOR NONEQUILIBRIUM PLASMA RECOMBINATION

Typical low-temperature noneuilibrium plasma has the ionization degree from 10^{-5} to 10^{-3}. The electron density observed in these experiments ranges from 10^{12} to 10^{14} cm^{-3}, while the density of neutral species ranges from 10^{16} to 10^{18} cm^{-3}. Based on observations from low-temperature experiments, the typical time frame for electron density decay is on the order of a few microseconds. In contrast, reaction times between neutral species are significantly longer. Consequently, reactions involving neutral species could be excluded from the kinetic scheme of plasma decay.

Variations in electron density are due to attachment/detachment processes and recombination phenomena. Recombination can be further divided into two main types: three-body recombination and two-body dissociative recombination. The latter is more prevalent at lower pressures, while three-body recombination plays a dominant role at elevated pressures (close to 1 atm). Ion-electron and ion–ion recombination mechanisms must also be considered. Under conditions of high electron density, three-body recombination involving electrons as a third body becomes a critical process.

2.5.1 Plasma Decay in Molecular Nitrogen

In the context of this analysis, the authors of Refs. [170–179] focus on positive ions, including N^+, N_2^+, N_3^+, N_4^+, in addition to neutral species N and N_2. The proposed kinetic scheme, based on Ref. [94], includes reactions with rate coefficients expressed in units of cm^3/s for bimolecular reactions and cm^6/s for trimolecular reactions, as follows:

$$e + N_2^+ \Rightarrow 2 \cdot N \quad k_{ea} = 2.8 \cdot 10^{-7} \cdot \left(\frac{300}{T_e}\right)^{0.5} \quad [94]$$

$$e + N_4^+ \Rightarrow 2 \cdot N_2 \quad k_{eb} = 3.7 \cdot 10^{-6} \cdot \left(\frac{300}{T_e}\right)^{0.5} \quad [94]$$

This reaction rate coefficient is multiplied by a factor of 1.7 comparing to [94].

$$2e + N_2^+ \Rightarrow e + N_2 \quad k_{eea} = 1 \cdot 10^{-19} \cdot \left(\frac{300}{T_e}\right)^{4.5} \quad [94]$$

$$e + N_3^+ \Rightarrow N_2 + N \quad k = 2 \cdot 10^{-7} \cdot \left(\frac{300}{T_e}\right)^{0.5} \quad [94]$$

$$2e + N^+ \Rightarrow e + N \quad k = 1 \cdot 10^{-19} \cdot \left(\frac{300}{T_e}\right)^{4.5} \quad [94]$$

$$e + N_2^+ + N_2 \Rightarrow 2 \cdot N_2 \quad k = 6 \cdot 10^{-27} \cdot \left(\frac{300}{T_e}\right)^{1.5} \quad [94]$$

$$e + N^+ + N_2 \Rightarrow N + N_2 \quad k = 6 \cdot 10^{-27} \cdot \left(\frac{300}{T_e}\right)^{1.5} \quad [94]$$

Conversion reactions:

$$N_2^+ + 2 \cdot N_2 \Rightarrow N_4^+ + N_2 \quad k_c = 5 \cdot 10^{-29} \quad [94]$$

$$N_2^+ + N + N_2 \Rightarrow N_3^+ + N_2 \quad k = 0.9 \cdot 10^{-29} e^{\frac{400}{T}} \quad [94]$$

$$N^+ + 2 \cdot N_2 \Rightarrow N_3^+ + N_2 \quad k = 0.9 \cdot 10^{-29} e^{\frac{400}{T}} \quad [94]$$

$$N^+ + N + N_2 \Rightarrow N_2^+ + N_2 \quad k = 1 \cdot 10^{-29} \quad [94]$$

$$N_2^+ + N \Rightarrow N^+ + N_2 \quad k = 2.4 \cdot 10^{-15} \cdot T \quad [94]$$

$$N_3^+ + N \Rightarrow N_2^+ + N_2 \quad k = 6.6 \cdot 10^{-11} \quad [94]$$

$$N_4^+ + N \Rightarrow N^+ + 2 \cdot N_2 \quad k = 1 \cdot 10^{-11} \quad [94]$$

The comparison of empirical observations with theoretical predictions were performed assuming a gas temperature of $T = 300$ K over a pressure spectrum of $p = 1 - 10$ Torr. At lower pressures (1 – 2 Torr), the observed electron densities fall below those predicted by the model. This discrepancy can be attributed to an uneven distribution of the discharge across the tube cross-section at these particular pressures. Conversely, at intermediate to high pressures (2.5 – 10 Torr), there is satisfactory agreement between experimental results and theoretical predictions. For example, at a pressure of $p = 5$ Torr, the comparison reveals that the initial segment (denoted as region N_1) is predominantly consistent with electron-ion recombination involving N_2^+ ions. The pronounced decrement is consistent with electron-ion recombination involving N_4^+ ions, while the noticeable bend in the curve is explained by the ion–ion transition from N_2^+ to N_4^+.

2.5.2 Plasma Decay in Molecular Oxygen

Here one should consider positive ions(O^+, O_2^+, O_3^+, O_4^+), negative ions (O^-, O_2^-, O_3^-, O_4^-) and neutrals (O, O_2, O_3):

Ion conversion:

$$O_2^+ + 2 \cdot O_2 \Rightarrow O_4^+ + O_2 \quad k_c = 2.4 \cdot 10^{-30} \cdot \left(\frac{300}{T_e}\right)^{3.2} \quad [94]$$

$$O_4^+ + O_2 \Rightarrow O_2^+ + 2 \cdot O_2 \quad k = 3.3 \cdot 10^{-6} \cdot \left(\frac{300}{T_e}\right)^4 \cdot e^{-\frac{5{,}030}{T}} \quad [94]$$

$$O^+ + O + M \Rightarrow O_2^+ + M \quad k = 1 \cdot 10^{-29} \quad [94]$$

$$O^+ + O_3 \Rightarrow O_2^+ + O_2 \quad k = 1 \cdot 10^{-10} \quad [94]$$

$$O_4^+ + O \Rightarrow O_2^+ + O_3 \quad k = 3 \cdot 10^{-10} \quad [94]$$

$$O_2^- + O \Rightarrow O_2 + O^- \quad k = 3.3 \cdot 10^{-10} \quad [94]$$

$$O_2^- + O_3 \Rightarrow O_2 + O_3^- \quad k = 4 \cdot 10^{-10} \quad [94]$$

$$O_2^- + O_2 + M \Rightarrow O_4^- + M \quad k = 3.5 \cdot 10^{-31} \cdot \left(\frac{300}{T_e}\right) \quad [94]$$

$$O^- + O_2 + M \Rightarrow O_3^- + M \quad k = 1.1 \cdot 10^{-30} \cdot \left(\frac{300}{T_e}\right) \quad [94]$$

$$O^- + O_3 \Rightarrow O + O_3^- \quad k = 5.3 \cdot 10^{-10} \quad [94]$$

$$O_3^- + O \Rightarrow O_2^- + O_2 \quad k = 3.2 \cdot 10^{-10} \quad [94]$$

$$O_4^- + M \Rightarrow O_2^- + O_2 + M \quad k = 1 \cdot 10^{-10} \cdot e^{-\frac{1{,}044}{T}} \quad [94]$$

$$O_4^- + O \Rightarrow O_3^- + O_2 \quad k = 4 \cdot 10^{-10} \quad [94]$$

$$O_4^- + O \Rightarrow O^- + 2 \cdot O_2 \quad k = 3 \cdot 10^{-10} \quad [94]$$

Detachment/attachment:

$$e + 2 \cdot O_2 \Rightarrow O_2^- + O_2^- \quad k_{at} = 1.4 \cdot 10^{-29} \cdot \frac{300}{T_e} \cdot e^{-\frac{600}{T}} \quad [94]$$

$$O_2^- + O_2 \Rightarrow 2 \cdot O_2 + e \quad k_{at} = 2.7 \cdot 10^{-10} \cdot \left(\frac{T_e}{300}\right)^{0.5} \cdot e^{-\frac{5,590}{T}} \quad [94]$$

The following reaction rate coefficients were proposed in the work [179]:

$$e + O_2^+ + O_2 \Rightarrow O_2^- + O_2^+ \quad k_r^1 = 2.5 \cdot 10^{-24}$$

$$e + O_4^+ + O_2 \Rightarrow O_2^- + O_4^+ \quad k_r^2 = 2.5 \cdot 10^{-24}$$

Electron-ion recombination:

$$e + O_4^+ \Rightarrow 2 \cdot O_2 \quad k_{eb} = 4.2 \cdot 10^{-6} \cdot \left(\frac{300}{T_e}\right)^{0.5}$$

This reaction rate coefficient is multiplied by a factor of 1.7 compared to Ref. [94].

$$e + O_2^+ \Rightarrow 2 \cdot O \quad k_{ea} = 2 \cdot 10^{-7} \cdot \frac{300}{T_e} \quad [94]$$

$$2 \cdot e + O_2^+ \Rightarrow e + O_2 \quad k_{eea} = 1 \cdot 10^{-19} \cdot \left(\frac{300}{T_e}\right)^{4.5} \quad [94]$$

$$e + O_2^+ + M \Rightarrow O_2 + M \quad k = 6 \cdot 10^{-27} \cdot \left(\frac{300}{T_e}\right)^{1.5} \quad [94]$$

$$2 \cdot e + O^+ \Rightarrow e + O \quad k = 1 \cdot 10^{-19} \cdot \left(\frac{300}{T_e}\right)^{4.5} \quad [94]$$

$$e + O^+ + M \Rightarrow O + M \quad k = 6 \cdot 10^{-27} \cdot \left(\frac{300}{T_e}\right)^{1.5} \quad [94]$$

Ion–ion recombination:

$$O_2^- + O_4^+ + M \Rightarrow 3 \cdot O_2 + M \quad k = 2 \cdot 10^{-25} \quad [94]$$

$$O_2^- + O_2^+ + M \Rightarrow 2 \cdot O_2 + M \quad k = 2 \cdot 10^{-25} \quad [94]$$

$$O_2^- + O_4^+ \Rightarrow 3 \cdot O_2 \quad k = 1 \cdot 10^{-7} \quad [94]$$

$$O_2^- + O_2^+ \Rightarrow 2 \cdot O_2 \quad k = 2 \cdot 10^{-7} \cdot \left(\frac{300}{T_e}\right)^{0.5} \quad [94]$$

$$O_2^- + O_2^+ \Rightarrow O_2 + 2 \cdot O \quad k = 1 \cdot 10^{-7} \quad [94]$$

$$O_2^- + O + M \Rightarrow O_3 + M \quad k = 2 \cdot 10^{-25} \cdot \left(\frac{300}{T_e}\right)^{2.5} \quad [94]$$

$$O_3^- + O_2^+ \Rightarrow O_3 + O_2 \quad k = 2 \cdot 10^{-7} \cdot \left(\frac{300}{T_e}\right)^{0.5} \quad [94]$$

$$O_2^- + O^+ \Rightarrow O_2 + O \quad k = 2 \cdot 10^{-7} \cdot \left(\frac{300}{T_e}\right)^{0.5} \quad [94]$$

$$O^- + O^+ \Rightarrow \cdot O \quad k = 2 \cdot 10^{-7} \cdot \left(\frac{300}{T_e}\right)^{0.5} \quad [94]$$

$$O_3^- + O^+ \Rightarrow O_3 + O \quad k = 2 \cdot 10^{-7} \cdot \left(\frac{300}{T_e}\right)^{0.5} \quad [94]$$

$$O^- + O_2^+ \Rightarrow O + O_2 \quad k = 2 \cdot 10^{-7} \cdot \left(\frac{300}{T_e}\right)^{0.5} \quad [94]$$

$$O^- + O_2^+ \Rightarrow O + 2 \cdot O \quad k = 1 \cdot 10^{-7} \quad [94]$$

$$O_3^- + O_2^+ \Rightarrow O_3 + 2 \cdot O \quad k = 1 \cdot 10^{-7} \quad [94]$$

$$O^- + O_4^+ \Rightarrow O + 2 \cdot O_2 \quad k = 1 \cdot 10^{-7} \quad [94]$$

$$O_3^- + O_4^+ \Rightarrow O_3 + 2 \cdot O_2 \quad k = 1 \cdot 10^{-7} \quad [94]$$

$$O_4^- + O^+ \Rightarrow 2 \cdot O_2 + O \quad k = 1 \cdot 10^{-7} \quad [94]$$

$$O_4^- + O_4^+ \Rightarrow 4 \cdot O_2 \quad k = 1 \cdot 10^{-7} \quad [94]$$

$$O_2^- + O^+ + M \Rightarrow O_2 + O + M \quad k = 2 \cdot 10^{-25} \cdot \left(\frac{300}{T_e}\right)^{2.5} \quad [94]$$

$$O^- + O_2^+ + M \Rightarrow O + O_2 + M \quad k = 2 \cdot 10^{-25} \cdot \left(\frac{300}{T_e}\right)^{2.5} \quad [94]$$

$$O^- + O^+ + M \Rightarrow 2 \cdot O + M \quad k = 2 \cdot 10^{-25} \cdot \left(\frac{300}{T_e}\right)^{2.5} \quad [94]$$

$$O^- + O_2^+ + M \Rightarrow O_3 + M \quad k = 2 \cdot 10^{-25} \cdot \left(\frac{300}{T_e}\right)^{2.5} \quad [94]$$

$$O^- + O^+ + M \Rightarrow O_2 + M \quad k = 2 \cdot 10^{-25} \cdot \left(\frac{300}{T_e}\right)^{2.5} \quad [94]$$

Chemical reactions with O–atom:

$$O + O_3 \Rightarrow 2 \cdot O_2 \quad k = 2 \cdot 10^{-11} \cdot e^{-\frac{2,300}{T}} \quad [94]$$

$$O + 2 \cdot O_2 \Rightarrow O_3 + O_2 \quad k = 6.9 \cdot 10^{-34} \cdot \left(\frac{300}{T_e}\right)^{1.25} \quad [94]$$

At pressures ($p = 1-3$ Torr), the recorded electron densities show lower values compared to those predicted by the model, suggesting the possibility of a non-homogeneous distribution of the discharge within the cross-section of the discharge tube at these specified pressures. In the range of moderate to high pressures ($p = 3.5-10$ Torr), a satisfactory agreement is found between the experimental results and the theoretical predictions. Upon analysis, it was found that the initial phase (identified as region N_1) corresponds predominantly to electron-ion recombination with O_2^+ ions. The further decrease is consistent with electron-ion recombination involving O_4^+ ions, while the inflection point of the curve is attributed to ion–ion conversion from O_2^+ to O_4^+.

2.5.3 Plasma Decay in CO_2

In CO_2 plasma the following positive ions (CO_2^+, $C_2O_4^+$) and neutrals (O, CO, CO_2) should be considered:

Conversion reactions:

$$CO_2^+ + 2 \cdot O_2 \Rightarrow C_2O_4^+ + CO_2 \quad k_c = 3 \cdot 10^{-28} \quad [180]$$

Recombination:

$$e + C_2O_4^+ \Rightarrow 2 \cdot CO_2 \quad k_{eb} = 4.2 \cdot 10^{-6} \quad [181]$$

$$e + CO_2^+ \Rightarrow CO + O \quad k_{ea} = 3.8 \cdot 10^{-8} \cdot \left(\frac{300}{T_e}\right)^{0.5} \quad [182]$$

$$2 \cdot e + CO_2^+ \Rightarrow e + CO_2 \quad k_{eea} = 1 \cdot 10^{-19} \cdot \left(\frac{300}{T_e}\right)^{4.5} \quad [94]$$

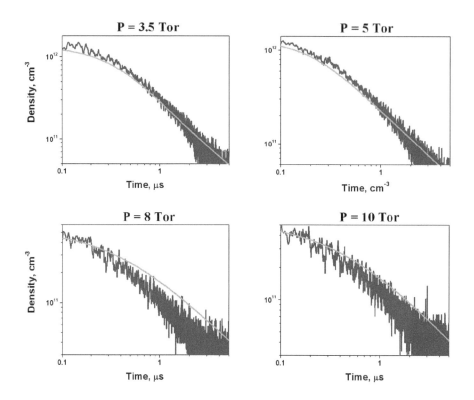

FIGURE 2.49 Electron density kinetic curve. Comparison of experiments (blue curves) and calculations (green curves). $T = 300$ K, CO_2.

$$e + CO_2^+ + CO_2 \Rightarrow 2 \cdot CO_2 \quad k_{eam} = 0.58 \cdot 10^{-24} \quad [180]$$

$$e + C_2O_4^+ + CO_2 \Rightarrow 3 \cdot CO_2 \quad k_{ebm} = 0.58 \cdot 10^{-24} \quad [180]$$

The typical result of the comparison between experimental observations and theoretical simulations is shown in Figure 2.49, where the numerical analysis was performed assuming that the gaseous medium is at a temperature of $T = 300$ K, spanning a pressure range of $p = 1 - 8$ Torr. In the graphical representation, the green lines indicate the results from the theoretical models, while the blue lines indicate the experimental results.

Analogous to the behavior observed in N_2 and O_2, the initial segment of the curve is predominantly characterized by electron-ion recombination processes involving the CO_2^+ ion. The pronounced dip in the curve is attributed to electron-ion recombination with the CO_4^+ ion. Furthermore, the inflection point of the curve is explained by the ion–ion conversion process from CO_2^+ to CO_4^+.

2.5.4 Plasma Decay in Water Vapor

The kinetic scheme, based on papers [94,180,182–184,] includes positive ions (H_2O^+, H_3O^+, $H_5O_2^+$, $H_7O_3^+$, $H_9O_4^+$) and neutrals (H, OH, H_2O, H_3O, H_5O_2, H_7O_3, H_9O_4) consists of the following steps:

Ion conversion:

$$H_2O^+ + H_2O \Rightarrow H_3O^+ + OH \quad k_{ca} = 1.8 \cdot 10^{-9} \quad [183]$$

$$H_3O^+ + 2 \cdot H_2O \Rightarrow H_5O_2^+ + H_2O \quad k_{cb} = 3.4 \cdot 10^{-27} \cdot \left(\frac{300}{T_e}\right)^4 \quad [180]$$

$$H_5O_2^+ + 2 \cdot H_2O \Rightarrow H_7O_3^+ + H_2O \quad k_{cc} = 2.3 \cdot 10^{-27} \cdot \left(\frac{300}{T_e}\right)^4 \quad [180]$$

$$H_7O_3^+ + 2 \cdot H_2O \Rightarrow H_9O_4^+ + H_2O \quad k_{cd} = 2.4 \cdot 10^{-27} \cdot \left(\frac{300}{T_e}\right)^4 \quad [180]$$

Electron-ion recombination:

$$e + H_2O^+ \Rightarrow H + OH \quad k_{ea} = 3.15 \cdot 10^{-7} \cdot \left(\frac{300}{T_e}\right)^{0.5} \quad [182]$$

$$e + H_3O^+ \Rightarrow H + H_2O \quad k_{eb} = 3.15 \cdot 10^{-7} \cdot \left(\frac{300}{T_e}\right)^{0.5} \quad [182]$$

$$e + H_3O^+ + H_2O \Rightarrow H + 2 \cdot H_2O \quad k_{ebm} = 2.7 \cdot 10^{-23} \quad [180]$$

$$e + H_5O_2^+ \Rightarrow H + 2 \cdot H_2O \quad k_{ec} = 2.5 \cdot 10^{-6} \quad [184]$$

$$e + H_5O_2^+ + H_2O \Rightarrow H + 3 \cdot H_2O \quad k_{ecm} = 2.7 \cdot 10^{-23} \quad [180]$$

$$e + H_7O_3^+ \Rightarrow H + 3 \cdot H_2O \quad k_{ed} = 4.5 \cdot 10^{-6} \quad [184]$$

$$e + H_7O_3^+ + H_2O \Rightarrow H + 4 \cdot H_2O \quad k_{edm} = 2.7 \cdot 10^{-23} \quad [180]$$

$$e + H_9O_4^+ \Rightarrow H + 4 \cdot H_2O \quad k_{ee} = 6.5 \cdot 10^{-6} \left(\frac{300}{T_e}\right)^{0.5} \quad [184]$$

$$e + H_9O_4^+ + H_2O \Rightarrow H + 5 \cdot H_2O \quad k_{eem} = 2.7 \cdot 10^{-23} \quad [180]$$

$$2 \cdot e + H_2O^+ \Rightarrow e + H_2O \quad k_{eea} = 1 \cdot 10^{-19} \cdot \left(\frac{300}{T_e}\right)^{4.5} \quad [94]$$

$$2 \cdot e + H_3O^+ \Rightarrow e + H_3O \quad k_{eeb} = 1 \cdot 10^{-19} \cdot \left(\frac{300}{T_e}\right)^{4.5} \quad [94]$$

$$2 \cdot e + H_5O_2^+ \Rightarrow e + H_5O_2 \quad k_{eec} = 1 \cdot 10^{-19} \cdot \left(\frac{300}{T_e}\right)^{4.5} \quad [94]$$

$$2 \cdot e + H_7O_3^+ \Rightarrow e + H_7O_3 \quad k_{eed} = 1 \cdot 10^{-19} \cdot \left(\frac{300}{T_e}\right)^{4.5} \quad [94]$$

$$2 \cdot e + H_9O_4^+ \Rightarrow e + H_9O_4 \quad k_{eee} = 1 \cdot 10^{-19} \cdot \left(\frac{300}{T_e}\right)^{4.5} \quad [94]$$

A comparison of the experimental data with the calculations was performed at gas temperature $T = 300$ K within pressure range $p = 1-8$ Torr. For all plots, green curves represent the results of the calculations, while blue curves represent the experimental data. The red line indicates the time of the reflected signal returning to the discharge cell. It can be concluded that there is reasonable agreement between the theoretical predictions and the second observed experimental peak [170–179]. However, the comparison of the first peak is challenging due to the suboptimal signal-to-noise ratio encountered under the prevailing experimental conditions.

2.5.5 Plasma Decay at Elevated Gas Temperatures

The core data set was acquired using a shock tube equipped with a discharge cell, where experiments were performed behind the incident and reflected shock waves [185]. This apparatus included a shock tube with a discharge cell, a discharge initiation system, and diagnostic tools. The shock tube, with a square cross-section of 25×25 mm^2, had a working channel of 1.6 m. A high pressure chamber of 60 cm length was used. Optical diagnostic capabilities were facilitated by two pairs of windows mounted along the stainless steel channel of the shock tube. The discharge segment of the shock tube, which maintains the same square cross-section, was fabricated from 40 mm thick Plexiglas and equipped with eight optical windows made of quartz and MgF$_2$. The metal end-plate of the tube acted as a high-voltage electrode, while the main steel section of the shock tube served as a ground electrode.

The nanosecond discharge was synchronized with the arrival of the incident or reflected shock wave at the predetermined observation location. The timing of the initiation was calibrated based on a preset delay after the activation signal from the initial Schlieren sensor. High-voltage pulses were generated by a Marx-type high-voltage generator, with the ferrite output line resulting in a voltage rise rate of approximately 8 kV/ns. This allowed the gas discharge to operate as a FIW within the dielectric portion of the tube, achieving front velocities in the range of 10^9 to 10^{10} cm/s, depending on specific experimental geometries. Parameters such as gas density (ρ_5), pressure (p_5), and temperature (T_5) after the reflected shock wave were derived based on ideal gas theory, taking into account the composition of the gas mixture, the initial pressure, and the velocity of the incident shock wave. The microwave diagnostics consisted of microwave (MW) transmitters and receivers, a waveguide assembly with a horn antenna, and an angular reflector and allowed the measurements of the electron density dynamics in nanosecond timescale [185–188].

The results of experiments conducted at elevated temperatures are shown in Figure 2.50, along with the corresponding results from numerical modeling. The discharge plasma, including the afterglow phase, was assumed to be spatially uniform. The kinetic scheme included processes considered significant over microsecond durations, omitting dynamics involving neutral species and interactions with the walls of the discharge cell.

For the high-temperature model of plasma recombination included the following reactions:
Dissociative recombination

$$O_2^+ + e \Rightarrow 2 \cdot O \quad k = 8.66 \cdot 10^{-7} \cdot \left(\frac{300}{T_e}\right)^{0.5}$$

$$N_2^+ + e \Rightarrow 2 \cdot N \quad k = 8.66 \cdot 10^{-8} \cdot \left(\frac{300}{T_e}\right)^{0.5}$$

$$e + NO^+ \Rightarrow N + O \quad k = 1.15 \cdot 10^{-6} \cdot \left(\frac{300}{T_e}\right)^{1.5}$$

Low Temperature Plasma Generation and Recombination

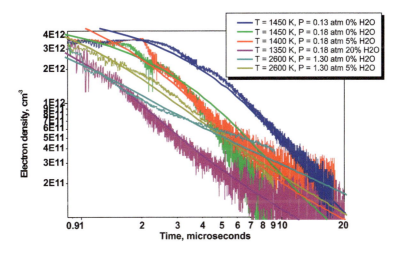

FIGURE 2.50 Results of high-temperature experiments (curves with the experimental noise) and their comparison with preliminary calculations (smooth curves). Gas mixture $CO_2:O_2:N_2 = 9:5:86$ with different admixtures of water.

$$H_2O^+ + e \Rightarrow H + OH \quad k = 5.77 \cdot 10^{-7} \cdot \left(\frac{300}{T_e}\right)^{0.5}$$

$$H_3O^+ + e \Rightarrow H + H_2O \quad k = 5.77 \cdot 10^{-7} \cdot \left(\frac{300}{T_e}\right)^{0.5}$$

Charge transfer

$$N_2^+ + O_2 \Rightarrow N_2 + O_2^+ \quad k = 8.5 \cdot 10^{-12}$$

$$N_2^+ + NO \Rightarrow NO^+ + N_2 \quad k = 1 \cdot 10^{-9}$$

$$O_2^+ + NO \Rightarrow NO^+ + O_2 \quad k = 1 \cdot 10^{-9}$$

$$N_2^+ + H_2O \Rightarrow N_2 + H_2O^+ \quad k = 4 \cdot 10^{-12}$$

$$H_2O^+ + H_2O \Rightarrow H_3O^+ + OH \quad k = 1.7 \cdot 10^{-9}$$

$$H_2O^+ + NO \Rightarrow NO^+ + H_2O \quad k = 1 \cdot 10^{-9}$$

Three-body recombination

$$N_2^+ + e + H_2O \Rightarrow N_2 + H_2O \quad k = 1.07 \cdot 10^{-21} \cdot \left(\frac{300}{T_e}\right)^{4.5}$$

$$N_2^+ + e + CO_2 \Rightarrow N_2 + CO_2 \quad k = 5.2 \cdot 10^{-28} \cdot \left(\frac{300}{T_e}\right)^{0.5}$$

$$O_2^+ + e + H_2O \Rightarrow O_2 + H_2O \quad k = 2.49 \cdot 10^{-20} \cdot \left(\frac{300}{T_e}\right)^{4.5}$$

$$O_2^+ + e + CO_2 \Rightarrow O_2 + CO_2 \quad k = 1.15 \cdot 10^{-26} \cdot \left(\frac{300}{T_e}\right)^{0.5}$$

$$H_2O^+ + e + H_2O \Rightarrow H_2O + H_2O \quad k = 2.49 \cdot 10^{-21} \cdot \left(\frac{300}{T_e}\right)^{4.5}$$

$$H_2O^+ + e + CO_2 \Rightarrow H_2O + CO_2 \quad k = 1.15 \cdot 10^{-27} \cdot \left(\frac{300}{T_e}\right)^{0.5}$$

$$NO^+ + e + H_2O \Rightarrow NO + H_2O \quad k = 1.78 \cdot 10^{-22} \cdot \left(\frac{300}{T_e}\right)^{4.5}$$

$$NO^+ + e + CO_2 \Rightarrow NO + CO_2 \quad k = 1.44 \cdot 10^{-27} \cdot \left(\frac{300}{T_e}\right)^{0.5}$$

This model describes the experimental data obtained, but the lack of data for rate coefficients for high-temperature regimes forces us to further verify the kinetic model. Nevertheless, the proposed model could serve as a good starting point.

2.6 EXCITATION OF INTERNAL DEGREES OF FREEDOM IN VARIOUS TYPES OF GAS DISCHARGES

The growing interest in the study of energy distribution on internal degrees of freedom in gas is dictated, first of all, by the expanding application of plasma in various fields of industry. Technological processes of surface treatment and generation of active particles can be significantly optimized by understanding the kinetics of elementary processes in low-temperature nonequilibrium plasma of gas discharges.

2.6.1 Glow Discharge

Studies of the behavior of the electric field and electron concentration in the glow discharge have the richest history, so it is not surprising that a large number of works are devoted to the study of the distribution of the energy by the internal degrees of freedom of the gas.

Glow discharge is widely used in spectroscopy to create radiation sources of various types, in analytical chemistry to analyze materials, in etching and sputtering, in membrane technologies, etc. The analysis and calculation of the EEDF in various gases (inert gases, metal vapors, nitrogen, oxygen, air, CO, CO_2) in an electric field, performed in [189], should be included among the works that have become classical. The calculated energy distributions over different degrees of freedom as a function of the reduced electric field are known for a wide set of gases. For nitrogen and air in the range $E/n = 10^{-16} - 3 \cdot 10^{-15}$ V·cm^2 they are given in Ref. [190]. Extensive experimental and theoretical material on the kinetics and mechanisms of excitation, ionization and dissociation of molecules, recombination of particles and chemical reactions in nonequilibrium low-temperature glow discharge plasma, accumulated up to 1980, is summarized in the books [191,192].

The excitation of atoms and molecules in a stationary glow discharge cannot be considered separately from the kinetics of heavy particles. At reduced electric field $E/n \approx (1-6) \cdot 10^{-16}$ V·cm^2, characteristic of the positive column of the glow discharge, the maximum amount of energy invested in the discharge in air (up to 90%) goes into nitrogen vibrations excitation. A collection of reviews [193] is devoted to the problems of vibrational kinetics in molecular gases in an electric field. In some cases, it is possible to significantly change the electron energy distribution in a glow discharge by changing the configuration of the system. Thus, a glow discharge with a hollow cathode is characterized by a large number of electrons with high energies, including the presence of run-away electrons [194,195]. Recent works devoted to gas excitation in a glow discharge are characterized by a multiparametric experimental study of the kinetics of the processes accompanied by detailed numerical modeling [196,197]. Illustrative in this sense is the work of Ref. [198], in which the glow discharge in argon was modeled using a hybrid model: the hydrodynamic approach for slow electrons in the real geometry of the discharge gap combined with the Monte Carlo method for fast electrons; the results of calculations were compared with experimentally obtained two-dimensional pictures of the distribution of various excited particles and probe measurements in the plasma.

2.6.2 RF Discharge

Radio frequency discharges (the frequency range used is 1–100 MHz) are divided into RFC (radio frequency capacitive) and radio frequency induction (RFI) discharges, depending on how the RF field is excited in the discharge volume. Induction discharge is excited inside a solenoid, powered by the RF generator. Capacitive discharge is initiated by applying RF voltage to the electrode system. A significant advantage of RF discharge from the point of view of clean plasma-chemical technologies is the possibility of organizing an electrodeless discharge when all metal electrodes are removed from the discharge volume. Medium-pressure (10–100 Torr) RFC discharge is used to create the active medium of CO_2 lasers; low-pressure (10^{-3} – 1 Torr) RFC discharge is used in technological processes of etching, sputtering, deposition of thin films, etc. For RFC discharges a large experimental material on the formation of active particles and temporal behavior of the electric field in various spatial regions of the plasma has been accumulated [199]; the monograph [200] summarizes approaches and methods to the study of HFE discharges, discusses the experimental technique, diagnostics and numerical simulation of the discharge.

RFI discharge has recently been actively studied at low (~mTorr) pressures due to the application of this type of discharge in microelectronics and surface treatment technologies. Its advantage is homogeneity in a large volume (characteristic dimensions are tens of centimeters) and the possibility of creating a plasma with a high concentration of electrons and ions due to the fact that plasma recombination at such pressures is slow. Typical for studies of this discharge are works with a detailed comparison of experimentally obtained spatial distributions of excited particles and the results of numerical modeling. Thus, in Ref. [201] the functions of electron energy distribution, spatial distribution of potential, electron density, and average ion energy calculated by a nonlocal model are compared with those experimentally measured for different pressures of argon; a model describing the energy distribution of electrons and the plasma behavior in the virtually collisionless regime was proposed.

2.6.3 Barrier Discharges

A barrier discharge is a discharge that occurs in a gas under the action of a voltage applied to the electrodes, if at least one of the electrodes is covered with a dielectric [66]. This discharge is also referred to in the literature as a "silent" discharge [67], because the discharge glows almost silently, or—according to the physics of the process—as a corona discharge with dielectric [68].

Barrier discharge has been intensively studied for more than 30 years in connection with its use for ozone generation. By now the process of ozone formation in barrier discharge has been studied in sufficient detail and has found wide application in industry. Practically all industrial ozone generators use this type of discharge. A detailed review of works on ozone synthesis in barrier discharge is given in Ref. [70].

The study of the electrical characteristics of the barrier discharge allowed us to determine the main peculiarities of its development. When a sinusoidal voltage is applied to the electrodes, the discharge in the gap is ignited twice per period.

A detailed study of the discharge structure [72,202] showed that the breakdown of the gas gap at reaching a certain voltage value at the electrodes occurs in the form of a series of microdischarges, the diameter of the microdischarge channel reaches several millimeters [73], that is, physically the picture of the barrier discharge development resembles an incomplete streamer breakdown. After the discharge passes through the discharge gap, a plasma with a certain amount of charged, excited, and dissociated particles is formed in the microdischarge channel.

In Refs. [77,78], the detailed gas composition after the end of the pulse is calculated on the basis of experimental data on the electrodynamics of the discharge development. It should be noted that due to strong spatial inhomogeneity and small interelectrode distance (as a rule, it does not exceed several millimeters for atmospheric pressure discharges), spatially resolved kinetic measurements of the behavior of active particles in a barrier discharge are difficult. For this reason, the results of calculations are usually compared with some values averaged over space (and often over time).

Streamer corona is used in plasma chemistry for the generation of active particles and surface treatment when homogeneity is critical only on a macroscopic scale: for example, in fabric treatment in the textile industry. Experimental methods for the study of streamer discharge are very limited. Usually, the volt-ampere characteristics are recorded in the experiment, giving, among other things, the value of the energy input into the system. To determine the spatial and temporal characteristics of the discharge development, as a rule, optical imaging of the discharge using EOPs and high-speed cameras is used. As a result of processing the data from such imaging, the velocity of the luminescence front propagation attributed to gas excitation processes in the streamer head can be obtained [203]. When studying the production of active particles, the integral yield of a certain substance is most often monitored after the flow of the treated gas passes through a flow reactor with a streamer corona in a coaxial [204] geometry.

The most significant parameters of the streamer discharge are considered to be the electron concentration and electric field strength in the streamer head and channel and the associated level of initial preionization upstream of the ionization wave.

Difficulties standing in the way of experimental methods for studying streamer discharges with the kinetic approach are obvious: high rates of processes (requiring subnanosecond time resolution at pressures close to atmospheric), small geometric dimensions, unpredictability of the propagation trajectory; as a rule, a high level of electromagnetic interference with recording equipment and signal cables. At the same time, the use of contact methods of diagnostics is practically impossible due to strong perturbations introduced by sensors into the field distribution in the gap.

In Ref. [205], measurements of the head diameter in a cathode-directed streamer in needle-plane geometry (height of the discharge gap $h = 21$ mm) in atmospheric air were carried out using high-speed optical emission spectroscopy. When constant voltage was used, the obtained value was $r = 170$ μm. During synchronous measurements of electrodynamic and optical parameters of the discharge, the start delay time from the moment of arrival of the high-voltage pulse at the electrode was studied. Experimentally, the effects of formation and propagation of a repeated streamer discharge, the transition of the streamer corona into arc and spark channels with increasing electric field in the gap or duration of the high-voltage pulse were detected.

In Ref. [206] the experimental work is analyzed and a model of the streamer discharge transition to the arc channel is constructed. The author states that in the case of bridging of a discharge gap in the air by a cathode-directed streamer, as a rule, no glow of the backward wave front is observed, but after its propagation, due to the development of attachment instability, there is the start of a secondary streamer propagating from the high-voltage electrode to the low-voltage electrode. The development of the secondary streamer precedes the formation of the arc channel. A detailed study of the streamer-arc transition stage in nitrogen at a pressure of 300 Torr in a uniform field (2 cm in plane-to-plane geometry) based on the analysis of the spatial and temporal variation of the emission spectrum in the range 300–860 nm is given in Ref. [207].

In Ref. [208], a systematic study of positive pulsed corona in a coaxial geometry (outer electrode diameter 56 mm, length 0.2 to 0.6 m) by multichannel emission spectroscopy in N_2, N_2 +NO, and N_2+O_2 mixtures at atmospheric pressure was carried out. The amplitude of the high-voltage pulse 100 kV, spectral range 200 to 500 nm. From the kinetic analysis performed by the authors, it follows that the population of $NO(A^2\Sigma^+)$ and $N_2(C^3\Pi_u)$ states during discharge (up to 200 ns) is determined by direct electron impact, and in the afterglow (up to 2 μs) by reactions involving excited molecular nitrogen in the metastable state $N_2(A^3\Sigma_u^+)$.

Thus, studies of the structure of the streamer flash have been limited to its general electrodynamic and, to a lesser extent, its radiative characteristics. The kinetics of the processes occurring in the gas discharge plasma of the streamer channel has been studied very poorly so far. An example of studies aimed at obtaining kinetic information about the streamer plasma is the work of Ref. [209].

In this work, the dependences of the emission of the second positive and first negative nitrogen systems on the length of the discharge gap were obtained experimentally in the mode of signal accumulation during the study of a positive streamer in the needle-plane geometry ($h = 20$ mm) in synthetic air N_2+O_2 (4:1) at atmospheric pressure. These results allowed, under the assumption of Maxwell's or Druyvestein's energy distribution of electrons, to obtain the dynamics of the average reduced field strength in the streamer head region. Unfortunately, the use of rough approximations for the EEDF and the use of relatively broadband light filters for radiation monochromatization somewhat reduce the value of the data obtained in the [209].

It should be noted that often in various applications, systems with a combined type of power supply combining several types of discharges are used. For example, it is known that superimposing a DC voltage of negative polarity when using a pulsed corona to clean flue gases significantly increases the cleaning efficiency [210]. Increasing the frequency of sinusoidal power supply in barrier discharge ozonators from 50 Hz up to kHz increases the ozone yield and the efficiency of utilization of the energy invested in the discharge, and the use of pulse power supply raises the efficiency of ozone synthesis in oxygen from 90 to 130 g/kWh [85]. By superimposing an additional electrical pulse of several hundred nanoseconds in length on the millisecond power pulse of the glow discharge unit, the plasma homogeneity [211] can be significantly improved.

Thus, in the study of low-temperature plasma of gas discharges, the approach analyzing the process of discharge development in time and space, based on a detailed consideration of the kinetics of electrons and related processes of population of active states is the most promising. To date, such an approach has been developed to a greater or lesser extent in most discharges actively used in plasma-chemical applications. At the same time, the consideration of a FIW as a type of gas breakdown has so far been limited to separately conducted electrodynamic studies and a few papers studying the effect of such a breakdown on the internal degrees of freedom of the gas. The main purpose of the next chapter is to fill this gap.

2.7 IONIZATION WAVES IN DIFFERENT TYPES OF DISCHARGES

2.7.1 History of Pulsed Discharge Research

The application of low-temperature plasma in technological processes began in the 1970s of the 20th century. In the last 20–30 years, the widespread use of plasma-chemical technologies, the development of new experimental methods of research and the fast growth of computational capabilities have led to a rethinking of the approach to the description of gas discharges.

The development of plasma-chemical technologies is largely determined by the degree of understanding of the processes occurring in plasma. Particularly important from the point of view of applications are the processes of active particle production in the discharge, which are closely related to the distribution of electric fields and the kinetics of electrons. It is the detailed study of the elementary processes occurring in the gas discharge that can both improve the understanding of the fundamental problems of low-temperature plasma physics and lead to the optimal solution of technological problems. Low-temperature plasma physics covers a wide range of gas discharge phenomena: from quasi-stationary (glow discharge) to extremely fast (ps- and ns-pulsed, HF, microwave discharges) processes, from homogeneous to sharply varying in space (streamer breakdown) structure.

Pulsed discharges occupy a special place in the study of low-temperature plasma properties. On the one hand, their study requires more advanced and, accordingly, more expensive experimental equipment. On the other hand, in pulsed discharges it is easier to separate the influence of different processes on the excitation and deactivation of active particles, to exclude the influence of processes on the walls of the discharge chamber, and to obtain information *in situ* on the changes in the composition and properties of the plasma when the electric field changes.

A special place among pulsed discharges is occupied by a high-voltage nanosecond discharge, which develops in the form of a FIW and creates a highly excited spatially homogeneous plasma in the discharge gap on characteristic time scales of nanoseconds.

A complete description of this type of discharge—both development from the high-voltage electrode and propagation—does not exist at present. Experimental data obtained by different authors give the following picture: when a voltage pulse is applied to the high-voltage electrode of a long—up to several meters—dielectric discharge tube surrounded by a metal screen, an ionization wave front starts from the high-voltage electrode after a certain delay time (typically few nanoseconds). The propagation velocity, having typical values of $\sim 10^9$ cm/s, is determined by the amplitude and shape of the high-voltage pulse, the geometry of the discharge cell, and the type of gas. As the pressure changes, the velocity passes through a maximum. For typical amplitudes of voltage pulse 15–30 kV, leading edge steepness 3–5 kV/ns, and discharge tube diameter a few centimeters, the range of pressures optimal for propagation of the ionization wave is from fractions to hundreds of Torr.

Motion of the potential front along the discharge tube is accompanied by the propagation of the glow front. Studies of discharge homogeneity using time-resolved emission spectroscopy of short-lived states have shown that a FIW can develop homogeneously over the volume in discharge cells with a characteristic diameter of at least tens of centimeters.

The values of the reduced electric field at the breakdown front can reach several kV/(cm·Torr), falling behind the front to hundreds of V/(cm·Torr). Active ionization, excitation, and dissociation of the gas occur. Changing the parameter E/n makes it possible to selectively excite certain degrees of freedom of the gas. The resulting plasma is characterized by a significant separation of the electron temperature from the gas temperature, homogeneity in volume over a wide range of pressures. The above properties of high-voltage nanosecond breakdown make it attractive from the point of view of studying elementary processes in nonequilibrium decaying plasma, and also allow us to hope for the possibility of using this type of discharge as an effective generator of chemically active atoms, radicals, and excited particles.

As the authors of the review [212,213] emphasize, the term "ionization waves" is often encountered in gas discharge physics. During the breakdown of long [214] tubes, as well as during the growth stage of electron avalanches [215], ionization waves are observed to propagate with a characteristic velocity of $10^5 - 10^7$ cm/s, which corresponds to the drift velocity of electrons in the applied field.

The propagation of streamer breakdown is also commonly described using the term "ionization wave". In this case, the ionization wave is understood as a [216] sprouting of the plasma channel into the region of a weak external field.

2.7.2 IONIZATION WAVES IN TOWNSEND MECHANISM OF BREAKDOWN

In the case of breakdown development by the Townsend mechanism, the volume charge of the single electron avalanche is so small that it does not distort the electric field in the gap. In addition, the condition of self-sustainability of the discharge is fulfilled, i.e., as a result of secondary processes on the cathode caused by the development of a single electron avalanche and the subsequent current of positively charged ions on the cathode or by the photo-effect, at least one electron must appear, giving rise to another electron avalanche. When speaking about the development of discharge by the Townsend mechanism, several temporal stages are usually distinguished (see, for example, [103]): the stage of avalanche generation, at which the current increase is caused by the development of successive electron avalanches (the concentration of charged particles at this stage does not exceed 10^{11} cm^{-3}); the propagation of ionization waves accompanied by a glow front and equalizing the concentration of charged particles along the length of the gap (at this stage, the concentration of charged particles increases by two orders of magnitude and is up to 10^{13} cm^{-3}) and the phase of the volumetric discharge development.

The stage of development of ionization waves in the glow discharge was studied both experimentally by measuring the discharge radiation [217] and by numerical simulation [218]. In the detailed analysis of these two works, given in Ref. [103], it is stated that the outcome of the studies was a qualitative description of the stage of ionization waves during the development of a glow discharge.

In the one-dimensional numerical simulation of the development of a glow discharge in hydrogen [218] at a pressure of 500 Torr, an interelectrode gap of 2 cm, a field strength of 19.84 kV/cm, and an overvoltage of 0.2%, the continuity equations for the electronic and ionic components were solved together with the Poisson equation and the external electric circuit equation. The initial number of initiating electrons was 10^2, the secondary emission coefficient was $\gamma = 8.34 \cdot 10^{-4}$, and the electron multiplication factor was $\mu = \gamma(\exp(\alpha d) - 1)) = 1.111$. It has been shown that during the first 30 μs after the electric field is turned on, the field is virtually undistorted by space charge. Then it begins to weaken near the anode and to strengthen in other regions. After 30 μs, a field enhancement near the anode is observed, leading to the propagation of a cathode-directed ionization wave. The velocity of the ionization wave was $2.8 \cdot 10^7$ cm/s, which is five times higher than the drift velocity of electrons in the external electric field. After the propagation of one or more ionization waves, a volumetric glow discharge is ignited. For example, four ionization waves were observed in Ref. [217].

2.7.3 IONIZATION WAVES IN STREAMER MECHANISM OF BREAKDOWN

In conditions when the breakdown develops by the streamer mechanism, there is no unambiguous opinion as to which stage of development to call the ionization wave. In the monograph [103], based on the processing of the results of [219,220] for breakdown in hydrogen at a pressure of 460 Torr, a gap length of 2 cm, and a reduced electric field of 23.5 V/(cm·Torr), a scheme of the luminescence front propagation of streamer breakdown in time is constructed, reproduced in Figure 2.51. The primary avalanche develops from the cathode with a drift velocity of $8.5 \cdot 10^6$ cm/s. After 180 ns, the number of charge carriers in it reaches a critical value and anode-directed and cathode-directed streamers propagate toward the electrodes of the discharge system. After the streamer channel bridges the gap, a series of ionization waves are excited, equalizing the conductivity along the length of the channel. The velocity of subsequent waves is greater than the velocity of previous waves and can reach values of 10^9 cm/s.

In the book [216], the ionization wave is understood as the streamer propagation in general, i.e., the germination of the plasma channel in the region of a weak external field. The authors argue this approach by the fact that the radius of the propagating streamer channel is much smaller than its length and the structure of the streamer head changes with time slowly enough for one to speak of its quasi-stationary change. The main processes determining the propagation of such a wave are ionization and electron drift in the field, providing ion exposure, space charge formation, and electric field redistribution.

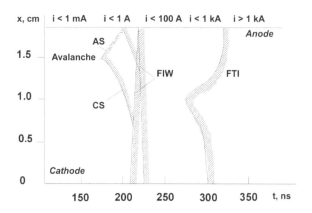

FIGURE 2.51 Scheme of propagation of discharge luminescence showing the different stages of streamer breakdown in hydrogen. AS, CS – anode- and cathode-directed streamers; FIW – fast ionization waves; FTI – front of thermal ionization wave.

According to Ref. [216], the streamer propagation as an ionization wave in a simple one-dimensional approximation can be described by a system of equations including the continuity equations for electrons and ions and the Poisson equation for the electric field:

$$\frac{\partial n_e}{\partial t} + \mathrm{div}\left(-n_e \mu_e \vec{E} - D_e \,\mathrm{grad}\, n_e\right) = v_i n_e - \beta n_e n_+ \tag{2.7}$$

$$\frac{\partial n_+}{\partial t} + \mathrm{div}\left(n_+ \mu_+ \vec{E} - D_+ \,\mathrm{grad}\, n_+\right) = v_i n_e - \beta n_e n_+ \tag{2.8}$$

$$\vec{E} = -\mathrm{grad}\,\varphi; \qquad \Delta\varphi = \frac{-\rho}{\varepsilon_0} \tag{2.9}$$

Here $\varepsilon_0 = 8.85 \cdot 10^{-12}$ F/m is the vacuum permittivity. Assuming that the front of the streamer head and the region in front of the wave front are spherical, the equation (2.9) can be written as follows

$$\frac{1}{r^2}\frac{d(r^2 E)}{dr} = \frac{e}{\varepsilon_0}(n_+ - n_e); \qquad \frac{dE}{dr} = \frac{e\Delta n}{\varepsilon_0} - \frac{2E}{r}, \tag{2.10}$$

where $\Delta n = n_+ - n_e$. The maximum value of the field E_m is reached at such a distance r_m from the center where $dE/dr = 0$. The value of the maximum field is related to the effective radius of the head r_m and the space charge density at this point $\rho = e\Delta n_m$ by the relation

$$E_m \approx \frac{e\Delta n_m r_m}{2\varepsilon_0} \tag{2.11}$$

Using the initial assumption of quasi-stationarity of the wave and assuming $E = E(x - v_c t)$; $n_e = n_e(x - v_c t)$ (here v_c is the wave velocity), let us write the Eq. 2.7 as follows:

$$-\frac{d}{dx}\left(n_e(v_c + v_e)\right) = v_i v_e; \qquad -v_c \frac{dn_+}{dx} = v_i n_e, \tag{2.12}$$

where $v_e = \mu_e E$ is the electron drift velocity modulus. The equation for positive ions (2.8) can be rewritten in a similar way. From (2.12) and the analogous equation for n_+, the law of conservation of total current follows:

$$\frac{d}{dx}(en_e - e\Delta n v_c) = 0; \qquad \Delta n = n_+ - n_e. \qquad (2.13)$$

Given that there are no electrons or bulk charge at large distances in front of the wave, the total current density is zero:

$$j_{\text{summ}} = en_e v_e - e\Delta n v_c = 0; \qquad \Delta n v_c = n_e v_e. \qquad (2.14)$$

Then from Eqs. 2.11 and 2.14, one can find the wave speed v_c:

$$v_c = \frac{e\mu_e n_m r_m}{2\varepsilon_0}, \qquad (2.15)$$

where $\mu_e = v_e/E$. Let us take the typical streamer electric field value in atmospheric pressure air $E_m = 300$ kV/cm, channel radius $r_m \approx 0.1$ cm, and initial electron concentration $n_0 = 10^6$ cm^{-3}. The electron mobility is $\mu_e \approx 270$ cm^2/(V·s); the value $n_m \approx 2 \cdot 10^{13}$ cm^{-3} can be estimated by integrating (2.12) from the point r_0, where $n = n_0$, to r_m. The final estimate for the ionization wave velocity (i.e., for the evolving streamer) gives $v_c \approx 5 \cdot 10^8$ cm/s.

As the authors [216], this estimate has a clear physical meaning: the bulk charge in the plasma dissolves over the time $\tau_M = \varepsilon_0/\sigma$. That part of the charge, which is located in front of the ionization wave, must disappear at this place, having moved to a distance of the order of r_m with velocity v_c. Then $\tau_M \approx \varepsilon_0/en_m\mu_e \sim \Delta t \sim r_m/v_c$.

2.8 FIW – PULSED DISCHARGE AT HIGH OVERVOLTAGE

What is the place of FIWs in pulsed discharges? Hereinafter we will understand by the word "breakdown" the processes of formation of charged particles in the gap and propagation of the ionized region from one electrode to another when a constant or pulsed voltage is applied to the electrodes of the discharge system. In this case, three physically different types of breakdown can be distinguished in the range of medium pressures (from fractions of Torr to hundreds of Torr).

Let us apply a voltage to the electrodes of the gas-filled gap. If the voltage is increased slowly, at a certain voltage value a glow discharge will ignite in the gap. This discharge develops by the Townsend breakdown mechanism [221], which is determined primarily by the efficiency of ionization and secondary emission of electrons from the cathode surface. Townsend breakdown typically diffusely fills the entire volume of the discharge gap.

In pulsed breakdown, the gap can withstand an overvoltage greater than the breakdown voltage. In fact, under these conditions, the overvoltage $K = U/U_{\text{br}}$ (U_{br} – breakdown voltage), along with the parameter pd (p – pressure, d – length of the interelectrode gap), determines the breakdown mechanism. In the case of overvoltages of tens of percent, the spatial charge of a single electron avalanche increases so much that the field inside the avalanche becomes comparable to the external field, and the field at the head and tail of the avalanche turns out to be amplified [103]. As a result, the breakdown develops according to the streamer mechanism: weakly conducting formations of small diameter propagate to one of the electrodes (or simultaneously to two electrodes) with a velocity of $10^7 - 10^8$ cm/s.

The release of energy into a narrow channel bridging the gap leads to the formation of a spark discharge. In the book [103], a curve separating the regions of the development of breakdown in the air by Townsend and streamer mechanisms is given [222]. This threshold is reproduced in Figure 2.20.

The above dependence can be obtained as follows [103]: when the avalanche diffuses, its radius is defined as $r = (4Dt)^{1/2}$, where $D = 2\varepsilon_{everage}\mu_e/3e$ – electron diffusion coefficient. The exponentially increasing number of charge carriers in the avalanche $n = n_0 \exp(\alpha d)$ is determined by the Townsend ionization coefficient $\alpha = A \times p \times \exp(-B/(E/p))$.

The field strength at the avalanche head is $E_1 = q_e \times n/(4\pi\varepsilon_0 r^2)$. The external field is $E = U/d$.

At a fixed size of the interelectrode gap $d = 1$ cm in the pressure range $p = 250 - 2500$ Torr overvoltage is found as follows: $K = (U/d - E_{crit} p/760) \cdot 100/(E_{crit} p/760)$. The dependence of $E_1(K)$ with the parameter pd defines the curve shown in Figure 2.20 as the overvoltage value at which a slight increase in overvoltage leads to a sharp increase in E_1.

In the case of higher overvoltages (hundreds of percent), the breakdown again shows a diffuse character of luminescence, but for other physical reasons: at sufficiently high strengths of the reduced electric field in the breakdown front, part of the electrons will go into a mode of continuous acceleration (the so-called "run-away" electrons), contributing to homogeneous volume predionization in the front. In this case, the breakdown will develop from the high-voltage electrode to the low-voltage electrode with a characteristic rate of cm/ns or more.

It should be noted that there is no clear boundary between streamer and spatially homogeneous nanosecond discharge: while Townsend discharge is distinguished by the presence of secondary emission from the cathode, the main elementary processes responsible for the development of both streamer and FIW breakdown are photoionization of the gas, in the case of sufficiently high fields – predionization by fast electrons, and ionization by electron impact in the ionization wave front. The horizontal line in Figure 2.20 marks the electron run-away threshold. It is determined [103,223], from the electron energy balance. For a nonrelativistic electron, its braking force in a gas is determined [224] by the density of gas molecules N_0, the number of electrons in the molecule Z, their kinetic energy $\varepsilon = mv^2/2$, and the average inelastic energy loss I:

$$F(\varepsilon) = \frac{2\pi e^4 N_0 Z}{\varepsilon} \ln \frac{2\varepsilon}{I}. \tag{2.16}$$

The energy balance in an electron having kinetic energy $\varepsilon > I/2$ will be written as

$$\frac{d\varepsilon}{dx} = eE - \frac{2\pi e^4 N_0 Z}{\varepsilon} \ln 2\varepsilon I \tag{2.17}$$

The function $F(\varepsilon)$ has a maximum F_m at energy $\varepsilon_m \approx 2.72 I/2$. As follows from (2.17), if the electric field exceeds the critical value $E_{crit} = F_m/e$, the electron starts to gain energy continuously as it moves along the x-axis. The critical field is defined as

$$E_{crit} = 1\pi e^3 N_0 Z/(2.72 I) = 3.88 \cdot 10^3 Z/I. \tag{2.18}$$

For nitrogen [103] $Z = 14$, $I = 80$ eV, $E_{crit}/p \approx 590$ V/(cm·Torr).

In Ref. [225], the existence of an upper bound on the parameter E/n for the streamer mechanism of breakdown in gases is shown on the basis of the analysis of the dependence of the energy loss per unit length on the electron energy. As an estimate of the critical value of the average field strength, $E_{crit} = (1/f)(d\varepsilon/dx)_{tot}^m$ is proposed, where $(d\varepsilon/dx)_{tot}^m$ is the maximum of the electron energy loss curve, f is the field enhancement factor near the space charge. In this case, in nitrogen of atmospheric density one can obtain $E_{crit} \approx 90$ kV/cm.

2.8.1 FIW Modeling

Numerical modeling based on the known peculiarities of the FIW, in comparison with experimental data obtained under the same conditions, allows us to significantly improve the understanding of the physics of the development of the pulse breakdown and its effect on the gas. The physical picture of the development of a FIW is close to that of the streamer; the differences in this case are quantitative rather than qualitative: in both cases, ionization before the front is due to photoionization and ionization by high-energy electrons. Higher values of the electric field and, correspondingly, a larger number of high-energy electrons lead to spatial homogeneity of the FIW, which is not characteristic of streamer breakdown.

We will not dwell here on the experimental study of streamers: a review of experimental work is detailed in the monograph [216]. Let us only note that accurate kinetic studies of the behavior of individual components in streamer plasma are difficult due to spatial inhomogeneity and poor reproducibility of the discharge development. The computational approaches modeling the streamer propagation and analyzing the processes responsible for its development have been actively developed during the last few decades. In contrast to the streamer, neat kinetic measurements are possible in a FIW. At the same time, numerical modeling of FIW is much less developed, primarily because of the need to take into account nonlocal effects or to select conditions when they cease to be significant.

Approaches to numerical modeling of the development and propagation of high-voltage pulse breakdown in gases can be divided into two classes. The purpose of the first of these is to describe the macroscopic characteristics of the breakdown, such as current, propagation velocity, total energy contribution, etc. Such models, as a rule, are limited to the consideration of electron kinetics in the average energy approximation, without asking about the energy distribution over the internal degrees of freedom of the gas and without trying to describe in detail the excitation and relaxation processes. The second, more detailed approach, involves analyzing the EEDF and the kinetics of elementary processes at various stages of breakdown; in this case, it is desirable to compare with the kinetic same experimental data on the rate of formation of individual excited states, atoms, and radicals.

During the last years, the calculations carried out in the works [226,227] have demonstrated the possibilities of computer modeling of streamer breakdown in homogeneous strong electric fields in a two-dimensional geometry. Almost all these works were based on the hydrodynamic drift-diffusion model for the description of electron motion, which was solved together with the Poisson equation to calculate the self-consistent electric field. There are only a few works [228–230] in which the Monte Carlo method was used to describe the motion and heating of electrons by the electric field. In spite of the fact that Monte Carlo methods have a much wider area of applicability and allow us to obtain the correct result even in the presence of very large values of electric fields and gradients in the solution, when hydrodynamic approaches lose their validity, their computational labor intensity is so great that so far such methods have been able to calculate only the initial stages of the avalanche-to-streamer transition [228].

In the work [231], the equations of conservation of momentum and energy for electrons were taken into account in the description of the streamer propagation. Electrons were considered as massless particles in local equilibrium with the electric field. The results obtained using this model gave results close to the results of calculations in the drift-diffusion approximation [232–236].

To solve the problem of calculating streamer propagation in long gaps, so-called "one and a half-dimensional" [237] models have been proposed. These models include one-dimensional conservation equations for charged particles and two-dimensional equations for the calculation of electric fields. Such models have been widely used to describe the propagation of streamers in weak fields and to simulate plasma-chemical processes in them [238,239]. Unfortunately, such models contain a fitting parameter—the streamer channel diameter, which almost completely determines the streamer dynamics and its properties. By fitting this parameter according to the experiment, one can achieve a reasonable degree of agreement between computational and experimental results. At the same time, in mixtures and under conditions where no experimental data are available, the results of calculations using such models will not be very reliable.

An interesting development of this approach is the [240,241] models taking into account the ionization expansion of the streamer channel. These models are somewhat less dependent on arbitrariness in the choice of the initial channel radius and their only disadvantages are the averaging of all plasma parameters over the streamer channel diameter and the assumption of uniform preionization of the gas.

Comparison of numerical simulations of streamer discharge with experiment has been carried out mainly for such parameters as propagation velocity, current, and energy contribution in the discharge gap. In recent works by Refs. [227,242] the calculated G-factors of active particle production (the number of active particles produced in the discharge per unit of energy of the discharge) are compared with those obtained from experiment. Unfortunately, these are integral quantities that give practically no information on the internal structure and spatial distribution of the plasma parameters.

2.8.2 EEDF Relaxation and Local Field Approximation

The EEDF is one of the most important characteristics of a gas discharge. The kinetic Boltzmann equation for the EEDF is a particle balance equation in phase space. In the unsteady case for electrons in an external field, the Boltzmann equation has the form [243]:

$$\frac{\partial f}{\partial t} + v\nabla f + \frac{eE}{m_e}\nabla_v f = S_{eM}^{el} + S_{eM}^{nel} + S_{ei} + S_{ee} + S_{ee} \qquad (2.19)$$

The EEDF in nonequilibrium plasma can be essentially non-Maxwellian even at relatively weak electric fields. Historically, the calculations of nonequilibrium EEDF were started with respect to electron beams [244,245]. In these works, an equation describing the degradation spectrum of electrons in the medium was derived and a calculation of the propagation of relativistic electrons through aluminum and lead was carried out. Reviews of works on degradation spectra are can be found in Refs. [246,247].

The most important task in calculations of the EEDF is the correct choice of cross sections of elementary processes of interaction of electrons with gas particles. Thus, the calculation of the EEDF in molecular gases, as compared to atomic gases, is considerably complicated by the necessity to take into account the excitation of vibrational levels of molecules by electron impact. The calculation of the stationary nonequilibrium EEDF in oxygen, nitrogen, and air under excitation by an electron beam, taking into account a sufficiently complete set of cross sections, is performed in Ref. [248].

It is necessary to present the difference between the nonequilibrium EEDF as a result of the influence of an electron beam and a gas discharge on the gas. When excited by an external source of electrons, the plasma nonequilibrium is determined by the relaxation processes of high-energy electrons; in the discharge, however, the energy is provided by the external electric field and the nonequilibrium is determined by the value of the reduced electric field strength E/n. This is manifested in the characteristic shape of the EEDF: in the discharge the distribution function, above the thresholds of inelastic excitation, shows the fast exponential decrease; in the e-beam plasma, however, the EEDF decays according to the power law at these energies.

A review of theoretical calculations of the EEDF in atomic and molecular gases in the case of a constant electric field is given in Ref. [189]. This review analyzes the two-term approximation used in solving the Boltzmann equation; the conditions under which electron-electron collisions are significant; and analytical methods for solving the Boltzmann equation for atomic and molecular gases.

Conditions when the EEDF depends not only on E/n, but also on density, degree of vibrational excitation, and degree of ionization are considered. A method for calculating the transport coefficients (drift velocity, diffusion coefficient, and Townsend ionization coefficient) in a gas in the presence of an electron concentration gradient is discussed.

The EEDF calculation for air was first performed in Ref. [249] for ionospheric physics problems. In Refs. [250,251], the Boltzmann equation for oxygen was numerically solved. By fitting the solution to the experimentally obtained kinetic coefficients of the gas, the cross sections of the inelastic interaction of electrons with O_2 were determined.

Experimental measurements of the EEDF, as already mentioned, are based mainly on the double differentiation of the voltammetric characteristic of the electric probe [252]. However, the application of this technique encounters fundamental difficulties at characteristic times shorter than 10^{-6} s. The EEDF in the case of characteristic for nanosecond discharge rates of change E/n requires a mandatory consideration of the $\partial f/\partial t$ term in the Boltzmann equation.

The numerical solution for the unsteady EEDF in nitrogen is presented in Ref. [253]. In this work, the problem of the relaxation of the EEDF under excitation of the gas by a pulsed source and the establishment of the distribution function when a constant source of primary electrons is switched on are solved. It is shown that the establishment of the rate of a certain inelastic process is faster the higher the characteristic energies for the cross-section of the considered process in the region of the maximum are. When the degradation spectrum of electrons is established, first of all the ionization rate is established, then the excitation rates of electronic levels and only then the excitation rates of vibrational levels of molecules. When a gas is excited by a pulsed source of high-energy electrons, the rates of inelastic processes at times shorter than the relaxation time of the EEDF can differ significantly from the stationary values. However, the total fraction of the source energy contributed by the degradation spectrum of electrons to the inelastic process for the full relaxation time appears to be close to its stationary value.

In Ref. [254] the non-standard Boltzmann equation for run-away electrons (with energies $\varepsilon \geq 100$ eV) is solved in the course of solving the problem of theoretical study of the breakdown of a neutral gas by ionizing waves of the potential gradient. Low-energy electrons were described in the hydrodynamic approximation. The system of charged particle balance equations included recombination, application, birth of low-energy electrons in the process of braking of run-away electrons, and development of an ionization cascade. The possibility of avalanche development and the probability of electrons transitioning to the run-away regime were taken into account. On the basis of the solution of the equation for unsteady EEDF, the main peculiarities of the development of ionizing waves were explained. Since the purpose of the [254] work was to demonstrate the influence of fast electrons on the propagation of the ionization wave, the refusal to consider the behavior of unsteady EEDF in the region of threshold energies of inelastic processes was quite justified. However, in complex molecular gases under short ($\sim 10^{-9}$ s) pulse action, the solution of the unsteady Boltzmann equation for the EEDF is necessary to determine the dynamics of energy relaxation in the plasma and the final energy balance.

2.8.3 Theoretical Description of FIWs

The theoretical description of FIWs has so far been limited mainly to models based on the hydrodynamic approximation. As a rule, integral parameters were compared with experiment: the wave velocity and its dependence on pressure, the current of high-energy electrons, etc.

In most theoretical works, it is assumed that the FIW propagates along the pre-ionized plasma, thus removing the question of preionization before the front. The problem is considered in a one-dimensional formulation. For this purpose, the plasma parameters are averaged over the cross-section of the discharge cell [255–257]. In Refs. [258,259] for transition to a one-dimensional problem an additional condition was imposed: the relation between longitudinal and transverse vectors of electric field, which follows from the theory of surface waves in plasma waveguides, was set. The solution of the problem under the assumption of the potential electric field, the dependence of the electron temperature on the local reduced electric field, and the automodality of the solution (it was assumed that the velocity of the FIW front does not change during wave propagation along the tube) allowed [258,259] to explain the shape of the current pulse at the high-voltage electrode and the different behavior of the discharge at different polarities of the pulse.

A more general approach in the theoretical consideration of the propagation of ionization waves is the solution of the system of the so-called telegraph equations for currents and potential:

$$\frac{\partial \varphi}{\partial x} = -\frac{\partial Li}{dt} - iR$$
$$\frac{\partial i}{\partial x} = -\frac{\partial C\varphi}{\partial t} - G\varphi, \qquad (2.20)$$

where φ is the potential, C, L is the capacitance and inductance of a unit length of the line, respectively (the long-wave approximation used here assumes that the transverse dimensions of the system are substantially smaller than the characteristic longitudinal dimension), i is the current, R is the resistance, and G is the conductivity of the line due to imperfect insulation (a value characterizing leakage through corona currents).

In the case of consideration of discharge in a tube, $G = 0$ is usually assumed. At wave velocities much less than the speed of light, the inductive term [256,257] is neglected in the equations (2.20). Then the system of telegraph equations reduces to the equation of nonlinear diffusion of potential along the plasma column:

$$\frac{\partial \varphi}{\partial t} - \frac{1}{R_0 C} \frac{\partial}{\partial x}\left(n_e \frac{\partial \varphi}{\partial x}\right) = 0, \qquad (2.21)$$

where $R = R_0/n_e$. No quantitative comparison with the experiment was performed in the [256]; in the [257] the calculated results were compared with the velocity of motion of the maximum of the potential gradient along the discharge tube at a fixed gas pressure.

At high wave velocities, as well as in the case when the characteristic scale of parameter changes in the front is comparable to the transverse dimensions of the system, the consideration of the system inductance and bias currents is necessary [255,260]. In this case the charged particle balance equations are solved together with the system of telegraph equations and the total current:

$$\frac{\partial \varphi}{\partial x} + \frac{\partial Li}{dt} + E(x,t) = 0$$
$$\frac{\partial i}{\partial x} + C\frac{\partial \varphi}{\partial t} = 0 \qquad (2.22)$$
$$i = \pi r^2 \left(\varepsilon_0 \frac{\partial E}{\partial t} + \sigma E\right),$$

where $E(x,t)$ is the longitudinal electric field, σ is the plasma conductivity, and r is the radius of the discharge tube. Such a consideration allowed us to Ref. [255] the variation of the velocity and amplitude of the electric field as it moves along the discharge cell. No quantitative comparison with experiment was made in Ref. [255]; in Ref. [260] the dependencies of the wave velocity on the initial concentration of electrons at different polarities of the pulse and of the wave velocity on the rate of voltage buildup $\partial U/\partial t$ were compared with experiment.

Generally, the assumption of a dependence between the electron temperature and the local reduced field strength E/n and the hydrodynamic consideration of the electron motion allows us to analyze only the most general electrodynamic characteristics of the FIW.

The first attempt to describe the electron distribution function in a FIW was made in the paper [254] mentioned above. The electrons were divided into two groups: "plasma" (low-energy electrons with isotropic velocity distribution) and "run-away" (high-energy electrons with preferential direction of motion). The energy at which the elastic scattering cross-section compares with the total inelastic loss cross-section (≈ 100 eV for air) was taken as the interface between the two groups.

Low-energy electrons were considered in the hydrodynamic approximation; for high-energy electrons, the Boltzmann equation was solved with the inelastic loss integral in the diffusion approximation. The EEDF at times of 2–4 ns for electron energies of $100-10^6$ eV were calculated. The dependences of the current, the velocity of the FIW, and the amplitude damping factor on gas pressure were compared with experiment.

A somewhat more complicated one-dimensional model of the FIW is developed in Ref. [261]. This work is also based on the idea of large energy groups of electrons developed in Ref. [262] to describe the electron avalanche at high overvoltages. Electrons are divided into three energy groups. The first includes a nearly isotropic distribution of electrons with energies below some ε_{max} value. For these electrons, the Boltzmann equation for the EEDF is solved in the Lorentz approximation. The value of ε_{max} is compared to the position of the maximum energy loss of the electron per unit path length. The second group of electrons has energies $\varepsilon_{max} < \varepsilon < \varepsilon_{th}$, where ε_{th} is the electron run-away threshold. A special algorithm based on a simplified statistical approach is developed to find the distribution of electrons in the second group. To describe the kinetics of run-away electrons, a deterministic approach is used, which is based on the equations of motion and the charge conservation equiation. The time dependence of the total conduction current in the discharge gap and the voltage on the gap are compared with experiment. In addition, the time profile of the fast electron current and the EEDF of run-away electrons at the anode were calculated.

The [263] work is based on the [257] model considering the nonlinear potential diffusion equation in a system with constant capacitance. The range of considered electron energies was divided into three regions; in each of them, the electron distribution was assumed to be Maxwellian with its mean energy. The balance of heavy particles in the system was determined by a system of kinetic equations including 18 reactions. A comparison with experimental results on the dependence of the wave front velocity and the energy contribution to the gas on pressure was carried out; in the proposed numerical model, the energy contribution was defined as the integral of the Joule losses over the pulse time and over the length of the discharge gap.

2.9 NUMERICAL MODEL OF FIW DEVELOPMENT

The availability of a set of consistent experimental data concerning the development of a FIW under identical experimental conditions, namely-the temporal behavior of the electric field, electron concentrations, and a number of excited particles-can serve as a reliable basis for the construction of a numerical model that would allow us to identify the peculiarities of the development of FIW compared to other types of discharges. For the calculation of pulsed nanosecond discharges characterized by the presence near the ionization wave front of superstrong electric fields reaching values of thousands of Townsends, the well-proven approaches developed for modeling glow discharges cannot be used. In particular, the two-term approximation of the Boltzmann equation, which well describes the dependence of the EEDF on the magnitude of the reduced electric field at small E/n, ceases to be valid in principle [264].

On the other hand, it is the presence of the region of ultrahigh fields and "run-away" electrons that distinguishes FIWs into a separate class of discharges. Therefore, the modeling of discharges of this type should be carried out using the most general models, which imposes very serious requirements both on the initial information (in particular, on the availability of data on differential and double-differential cross sections of various processes involving electrons in a wide range of energies) and on the numerical methods used to implement these models.

An overview of the various numerical methods used to solve the Boltzmann equation can be found, for example, in Ref. [265]. In the numerical description of the EEDF in discharges in strong electric fields, two approaches based on statistical modeling are usually used. The first approach-the molecular dynamics method or the method of test particles-assumes that the numerical scheme "tracks" the change in the characteristics of a single particle (in this case, an electron). As a rule, Newton's equations describing the motion of a charged particle in an electric field are used to calculate the trajectory of electrons, and the Monte Carlo method is used to model the collision of an electron with a molecule. The second approach (the method of direct statistical modeling, also called the Bird method [266]) consists in the numerical solution of the unsteady Boltzmann equation:

$$\frac{\partial f}{\partial t} + \vec{v} \cdot \nabla f - \frac{e\vec{E}}{m} \cdot \nabla_v f = \left(\frac{\delta f}{\delta t}\right)_c, \quad (2.23)$$

at which the motion of the whole ensemble of modeling particles is traced. According to the authors of Ref. [267], there is no reason to give a clear preference to any of the two mentioned methods, at the same time, their parallel use provides the possibility of mutual control of the correctness of the solution.

The numerical solution of Eq. 2.23 by direct modeling is possible using a certain number of constraints [266]. These constraints are to use a finite number of particles (while the distribution function in Eq. 2.23 can, in general, describe an infinite number of particles); to partition the physical space of the problem into a finite number of discrete cells; and to split the motion of molecules and collisions between them over a time interval Δt. To describe collisions between particles in this approach, the Monte Carlo method in one or another of its modifications is also used. Here are some examples of the use of the Monte Carlo method in problems of low-temperature plasma physics.

In Ref. [268], the Monte Carlo method was used to describe the collision of electrons with atoms in the study of the one-dimensional quasi-stationary near-cathode region of a glow discharge by the molecular dynamics method. The authors applied the method of "null" collisions proposed in Refs. [269,270] for solving problems of solid state physics to describe the collision of electrons with heavy particles. The essence of the method consists in the introduction of an additional formal mechanism of null-collisions, which does not change the direction of motion and energy of the electron, but allows to unify and carry out in one step two operations: the choice of "there is a collision – no collision" and the choice of the type of collision. For this purpose, the "null collision" is assigned a certain cross-section and this process is considered simultaneously with the processes of elastic and inelastic scattering of the electron by gas molecules. The "null-collision" cross-section is chosen so that the collision frequency or free path length is constant over the entire energy range.

In Ref. [271], the processes of selecting the presence – absence of collisions and selecting the collision type are implemented in a consistent manner, but the author notes, that when calculating the probability of an electron-molecule collision $P_{coll} = 1 - \exp(-\Delta s \Sigma(n\sigma_{coll}(\varepsilon)))$ at the spatial step Δs, the main requirement is that Δs be small compared to the characteristic scales of change in gas density n and electron energy ε, which essentially corresponds to the null-collision model. In recent works, the energy and spatial characteristics of electrons in strong electric fields have been calculated by the Monte Carlo method. The interaction of a stationary electron beam with a nitrogen jet was modeled in the [272,273]. The excitation rates of $B^2\Sigma_u^+, v' = 0$ and $C^3\Pi_u, v' = 0$ states of nitrogen under the conditions of interaction of a beam with a gas of non-homogeneous density have been calculated and compared with experiment in the approximation of isotropic electron scattering and a non-degenerate electron beam; the efficiency of various excitation mechanisms has been analyzed; the case of interaction of an electron beam with a supersonic gas jet in a three-dimensional geometry has been considered.

The EEDF in a high-voltage beam-type discharge is calculated in the [274]. The behavior of the electric field along the length of the discharge gap was set on the basis of an experiment in which the degree of linear polarization of radiation for a set of dipole allowed and forbidden transitions from the same level was measured with a time resolution of 25 ns [275]. The experiments were carried out in helium at a discharge gap length of 0.19 cm, pressure of 15 Torr, anode voltage of 3.1 kV, and current pulse width of 0.5 µs. Calculation of the stationary EEDF in one-dimensional geometry was carried out for the spatial field distribution corresponding to the time moment at which the amplitude value of the anode voltage is reached; for the linear decline of the field across the gap;

and for the case of a constant field. The chapter concludes that the electric field distribution has a significant influence on the type of EEDF, in particular, on the ratio of the "beam" component and low-energy electrons. It is shown that for both fast and slow electrons, the distribution function has a significant anisotropy: for both groups, the main contribution to the distribution is made by electrons scattering in the range $\Theta = 15° - 20°$. The numerical model presented below describes the development and propagation of a nanosecond high-voltage pulsed discharge in the form of an ionization wave in nitrogen. The calculated temporal and spatial behavior of the EEDF fully correlates with the data obtained from experimental studies. The calculations were performed for the conditions of the experiment to determine the electric field in nitrogen at a pressure of 4 Torr. The experiment was carried out in a discharge tube with a diameter of 1.75 cm and a length of 60 cm. A trapezoidal-shaped voltage pulse with an amplitude of 13.5 kV, a duration of 25 ns at half-height, and a rise time of 3 ns was used to initiate the breakdown.

The one-dimensional (1D3V) non-self-consistent problem of the development of a FIW when initiated by a negative polarity voltage pulse was solved by the direct simulation method. The behavior of the electric field $E(x,t)$ was set on the basis of the data obtained in the experiment [276]. In the process of solving the unsteady Boltzmann equation (2.23), the method of separation by physical processes was used: collisions, heating of the electron ensemble in the field, and transport were considered independently. To obtain the solution $f(\vec{x}, \vec{v}, t)$ at the next time instant $t' = t + \delta t$, the equations defining the function $f_\Sigma = f_{1\,\text{step}} + f_{2\,\text{step}} + f_{3\,\text{step}}$ were successively solved at each substep:

$$\frac{\partial f_{1\text{step}}}{\partial t} = \left(\frac{\delta f_\Sigma}{\delta t}\right)_c \tag{2.24}$$

$$\frac{\partial f_{2\text{step}}}{\partial t} = \frac{e\vec{E}}{m} \cdot \nabla_v f_\Sigma \tag{2.25}$$

$$\frac{\partial f_{3\text{step}}}{\partial t} = -\vec{v} \cdot \nabla f_\Sigma \tag{2.26}$$

In this case, in step 1, the Monte Carlo method is used to solve the problem of the birth and change of motion of electrons in velocity space when they collide with molecules. In step 2, the energy exchange between electrons and the field is considered, and in step 3, the electron transfer in the physical space of the problem (along the coordinate x) is considered. The transport in the momentum (step 2) and coordinate (step 3) spaces of the problem is calculated by an explicit finite-difference scheme of the "angle" type.

The block diagram of the numerical code is shown in Figure 2.52. The INI subroutine sets the initial conditions and the main parameters of the problem: the initial gas density, the mixture composition, the number of points in the coordinate and phase space, the time step, the time of outputting intermediate results to the file t_{print}, the maximum counting time t_{max}. The time step was set as $dt = \alpha/(N_0 v \Sigma \sigma_i)$, where N_0 is the density of the neutral gas, v is the maximum possible velocity in the considered velocity range, σ_i is the cross-section of the i-th process, α is the crossflow parameter (see below for a detailed explanation) introduced for more efficient mixing of electrons in phase cells. For the conditions used, the time step was $\sim 10^{-13}$ s. Note that for the characteristic cross-section $\sigma = 10^{-15}$ cm^2, electron velocity $v \sim 10^7$ cm/s, and gas density $n \sim 10^{17}$ cm^{-3} the time between collisions is $\tau_{\text{st}} = 1/(nv\sigma) \sim 10^{-9}$ s, which significantly exceeds the time step.

A self-consisted set of the cross sections of the considered collision processes of electrons with molecules was used for these calculations [264]. Elastic collisions, excitation of rotational, vibrational and electronic levels, dissociation and ionization were taken into account. Electron-electron, electron-tron-ion, superelastic collisions and step ionization have not been considered. The cross sections are presented in detail in 2.9.

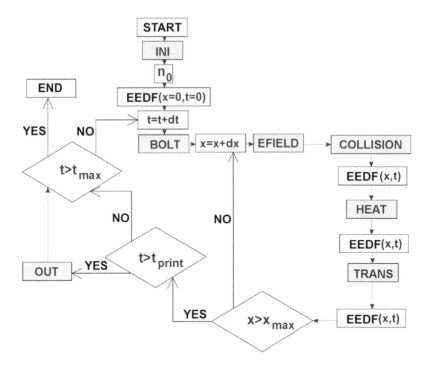

FIGURE 2.52 Block diagram of the direct simulation program for the Boltzmann equation.

It was assumed that at time $t = 0$ a non-zero electron concentration $n_e = 10^8$ cm^{-3} is present only in the first spatial cell. The electron energy distribution at time $t = 0$ was given by a zero-energy delta function. At each subsequent time step dt (subroutine BOLT), the change of the distribution function along the coordinate x was calculated. The BOLT subroutine included the COLLISION, HEAT, and TRANS subroutines.

First of all, based on the experimental data, the electric field at the considered point $E(x,t)$ was set (call of the EFIELD subroutine). Then, the change of coordinates of each phase space cell in the process of collisions of electrons with molecules was played out (subroutine COLLISION). The structure of this subroutine is presented in more detail in Figure 2.53.

In each cell of the phase space, provided there are a non-zero number of electrons, a set of events that change the energy of the electrons in the cell is played out. It is assumed that the coordinates and velocity components of the modeling electron are equal to the average values of these quantities for all electrons in the cell. At the current time step, the type of collision is determined based on the result of the comparison of a random number and an "event segment" of the form: "null collision" "elastic" "inelastic 1" "inelastic 2" "inelastic i" "ionization". The scattering probability of an electron at time dt in a j-type process colliding with a k-type molecule is calculated by the relation:

$$P_j^k = v_e \sigma_j^k(v_e) N_k dt, \qquad (2.27)$$

with the condition $\sum_{k,j} P_j^k = P_{\text{coll}} < 1$ being controlled at each step during the computation.

The probability of the electron moving without collision (null-collision) on the time interval δt is then $P_{\text{free}} = 1 - P_{\text{coll}}$. The interval from 0 to 1 was divided into subintervals according to the probability of a particular type of collision. The sum of the probabilities given null-collisions equals one. Depending on which of the intervals a randomly chosen number falls into, the type of process was selected.

Low Temperature Plasma Generation and Recombination

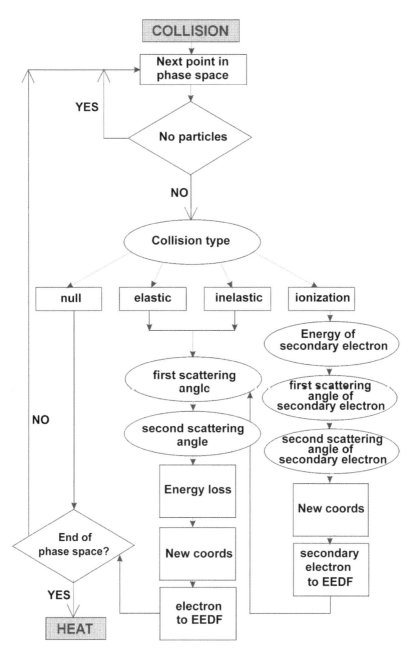

FIGURE 2.53 Block diagram of the program describing the collision process. The comparison with a randomly selected number is marked with a shaded oval.

Next, the parameters of the electrons after the collision were calculated. If it was a null collision, neither the energy nor the coordinates of the electrons were changed; the calculation continued for the next point in phase space. For ionization, the energy of the secondary electron was first of all played out. This is the only process for which the dependence of the double differential cross-section (DDCS) of ionization on energy was taken into account, which is necessary for careful consideration of processes involving high-energy electrons in the wave front; the directional distribution of the secondary electron velocity vector was assumed to be isotropic. The probability of birth of a secondary electron with energy u_2 in an ionizing collision involving a primary electron with energy u_1 was determined by the expression [277]:

$$P_{u_1,u_2} = \frac{1}{u \arctan((u_1 - \Delta u)/2u)} \cdot \frac{u^2}{u^2 + u_2^2} \qquad (2.28)$$

where $\Delta u = u_2 + 15.6$ – the energy threshold of the ionization act with the birth of an electron with energy u_2; $u = 13$ eV, $I = 15.6$ eV. The secondary electron energy u_2 was determined by matching a random number in the interval [0;1] with the electron birth probability $P(u_2)$. Next, the deflection angle and azimuthal angle were played out; the coordinates of the secondary electron were calculated and the resulting electron was added to the distribution function, after which the parameters of the scattered primary electron were calculated.

In the case of an inelastic or elastic collision, the scattering angle and azimuthal angle were played out; then the energy loss in the corresponding collision and the new coordinates of the electron were calculated. Once the new position of the velocity vector in phase space is determined, some fraction $0 < \alpha < 1$ of electrons from the original cell are transferred to the corresponding phase space cell. The introduction of the parameter α is due to the fact that this algorithm considers not the collision of a single electron with a molecule, which would lead to a significant increase in the calculation time, but the change in the parameters of the phase cell—energy and direction of motion—as a whole (in the limit $\alpha \sim n_e^{-1}$, a single electron is transferred in each collision). This parameter allows us to increase the size of the phase space of the problem during the calculation. The collision probability, as noted above, is increased by a factor of $1/\alpha$ to correctly reproduce the time dynamics. As a result of considering in the collision procedure not a single electron, but a phase cell, taking into account the weighting factor, the calculated time changes; the physical time of the problem remains the same. Such introduction of the weighting factor, as mentioned in Ref. [266], is used to equalize the sample volume when modeling the flow of gas mixtures with large differences in component concentrations or when it is necessary to introduce nonuniform computational mesh with large variations in cell volumes.

After determining the energy and coordinates of scattered electrons, the EEDF is recalculated. If the entire phase space is calculated, the algorithm proceeds to the HEAT subroutine. At a given point in space, the heating of electrons in an electric field varying along the coordinate x (Eq. 2.25) is calculated for each cell in the phase space. The resulting cell-to-cell flux is calculated, taking into account the energy conservation and total particle number, and the EEDF is adjusted accordingly.

The TRANS subroutine is then called to calculate the electron transfer from cell to cell at a given spatial and temporal step (Eq. 2.26), which results in another adjustment of the EEDF. The COLLISION-HEAT-TRANS sequence is repeated for the next point along the discharge axis. The BOLT subroutine calculation is repeated until the maximum calculation time is reached.

2.9.1 Cross Sections of Electron Interaction with N_2

In the work [264], a set of Refs. [278,279] cross sections describing the following processes was used:

1. Elastic scattering (Figure 2.54)
2. Excitation of rotational levels (Figure 2.54) (energy loss $\Delta E = 0.02$ eV)
3. Excitation of vibrational levels (Figure 2.55)
    - $v' = 1, \Delta E = 0.29$ eV
    - $v' = 2, \Delta E = 0.59$ eV
    - $v' = 3, \Delta E = 0.88$ eV
    - $v' = 4, \Delta E = 1.17$ eV
    - $v' = 5, \Delta E = 1.47$ eV

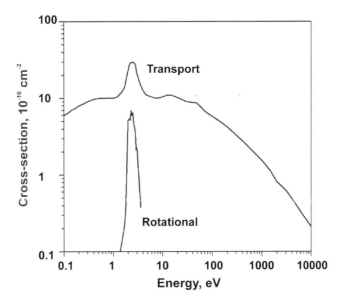

FIGURE 2.54 Transport cross-section and rotational level excitation cross-section.

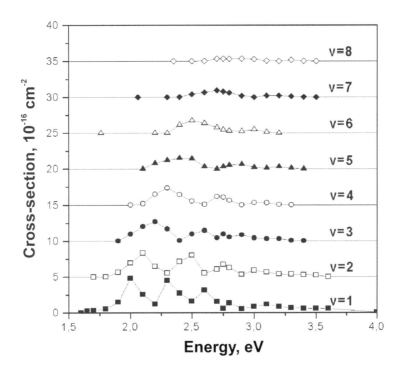

FIGURE 2.55 Excitation cross sections of vibrational levels of the ground electronic state N_2.

- $v' = 6, \Delta E = 1.76$ eV
- $v' = 7, \Delta E = 2.06$ eV
- $v' = 8, \Delta E = 2.35$ eV
4. Excitation of electronic levels (Figures 2.56 and 2.57)
    - $A^3\Sigma_u^+, v = 0 - 4, \Delta E = 6.17$ eV

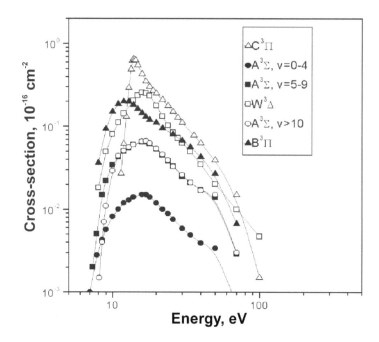

FIGURE 2.56 Excitation cross sections of N_2 electronic levels.

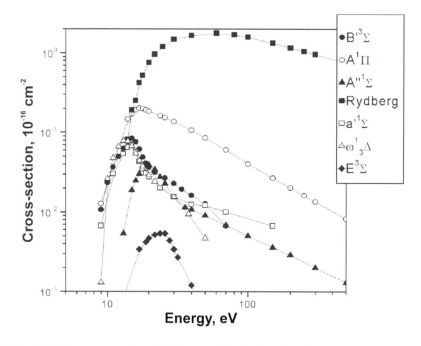

FIGURE 2.57 Excitation cross sections of N_2 electronic levels (continued).

- $A^3\Sigma_u^+, v = 5 - 9, \Delta E = 7.00$ eV
- $A^3\Sigma_u^+, v > 10, \Delta E = 7.80$ eV
- $B^3\Pi_g \; \Delta E = 7.35$ eV
- $W^3\Delta_u \; \Delta E = 7.36$ eV
- $B'^3\Sigma_u^- \; \Delta E = 8.16$ eV; $a'^1\Sigma_u^-, \Delta E = 8.40$ eV

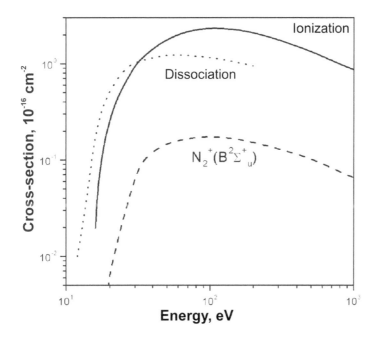

FIGURE 2.58 Ionization and dissociation cross sections of N_2 by electron impact.

- $a^1\Pi_g$, $\Delta E = 8.55$ eV
- $w^1\Delta_u$, $\Delta E = 8.89$ eV
- $C^3\Pi_u$, $\Delta E = 11.03$ eV
- $E^3\Sigma_g^+$, $\Delta E = 11.88$ eV
- $a'^1\Sigma$, $\Delta E = 12.25$ eV, Rydberg, $\Delta E = 12.487$ eV

5. Dissociation and ionization by electron impact (Figure 2.58)
- $N(^4S^0) + N(^4S^0)$, $\Delta E = 9.76$ eV
- N_2^+, $\Delta E = 15.60$ eV
- $N_2^+(B^2\Sigma_u^+)$, $\Delta E = 18.80$ eV

2.9.2 EEDF Relaxation Dynamics

The numerical modeling was done in the regimes given in Table 2.4 for different reduced electric field strengths while varying the internal parameters of the problem. The electric field in all cases was switched on instantaneously, and the time of the EEDF reaching a stationary value, the shape and absolute value of the EEDF were analyzed. The results of the calculations were compared with the known calculations of the stationary EEDF in the framework of the two-term approximation of the Boltzmann equation [280,281]. Note that the set of cross sections of electron-molecule collision processes [278] used in this problem, in general, differs from both the sets of cross sections [280,281]. However, the good agreement between the EEDFs calculated by the independent codes [280,281] gives reason to hope that the comparison is correct in our case as well.

The coincidence between the results of calculation 1 and calculation 2 shows that the formation of secondary electrons during electron impact ionization does not significantly affect the EEDF shape in relatively low electric fields.

Figure 2.59 shows the difference between the steady-state stationary EEDF calculated by the Monte Carlo method and the EEDF calculated in the framework of the two-term approximation of the Boltzmann equation. It can be seen that in the whole low-energy part of the EEDF the difference

TABLE 2.4
Test Regimes

Run #	E/n, Td	Varying Problem Parameter
Case 1	150	Baseline
Case 2	150	Secondary electrons[a]
Case 3	600	Secondary electrons
Case 4	150	Secondary electrons, fine mesh
Case 5	150	Secondary electrons, $\alpha = 0.25$

[a] The formation of secondary electrons in the ionization process is considered an energy loss; secondary electrons are not added to the EEDF.

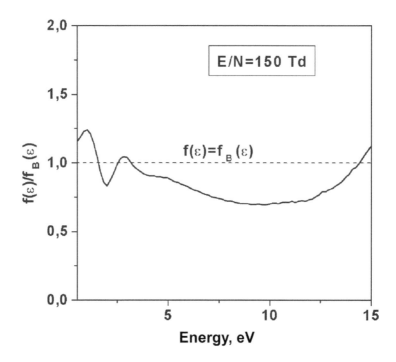

FIGURE 2.59 Difference of stationary EEDFs when calculated by the Monte Carlo method ($f(e)$) and when solved within the two-term approximation of the Boltzmann equation [280].

does not exceed 30%. Establishment of the EEDF parameters occurs for a characteristic time of 1–2 ns (Figure 2.60). At the same time, as a consequence of the difference in the EEDF shapes, the average energy of the obtained EEDF is slightly lower (by 15% for the case 1) than the average energy calculated in the framework of the two-term approximation.

A proportional reduction of the grid scale by an order of magnitude (calculation 4) had practically no effect on the physical relaxation rate and the form of the EEDF. The change of the crossflow parameter α should be considered separately. Reducing α from 0.5 to 0.25 led to a significant—several times—increase in the calculation time, but practically did not change the form of the EEDF. The distribution functions obtained for different values of α are shown in Figure 2.61.

Since one of the main characteristic features of the behavior of the unsteady EEDF is a significant overpopulation of the high-energy region (see below), it is necessary to compare the magnitude

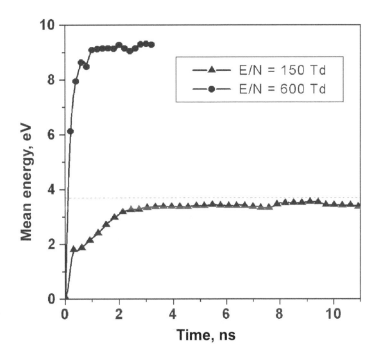

FIGURE 2.60 Relaxation of the average energy at instantaneous field activation. The dotted line indicates the mean energy value obtained in the [280] for $E/n = 150$ Td.

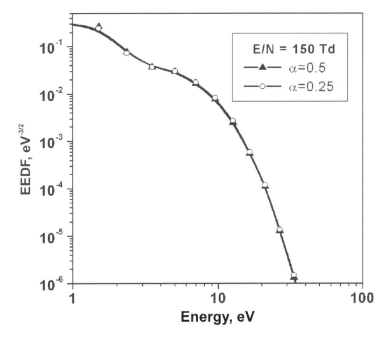

FIGURE 2.61 Electron energy distribution functions for different α-flow parameters.

of the deviation of the unsteady and stationary distribution functions obtained in this problem from the solution of the Boltzmann equation [280,281].

Figure 2.62 shows the results of EEDF calculations for four variants at a value of the reduced electric field of $E/n = 600$ Td. These are two independent solutions for the stationary distribution

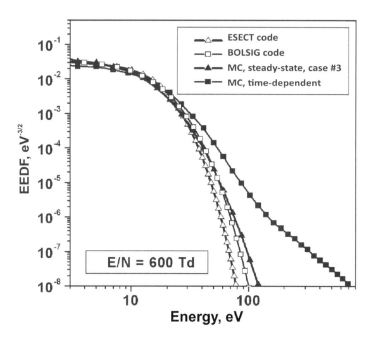

FIGURE 2.62 Comparing the unsteady EEDF corresponding to the real electric field variation in the problem and the stationary distribution functions. ESECT is the [281] calculation, BOLSIG is the [280].

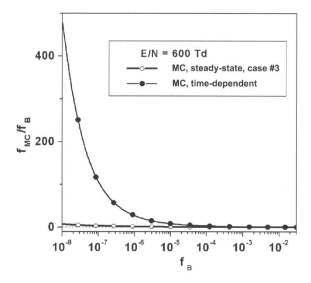

FIGURE 2.63 Scale of deviation of the steady-state and non-steady-state EEDF obtained in this problem (f_{MC}) from the distribution function f_B [280]. The ratio f_{MC}/f_B for different values of f_B is presented.

function [280,281], a Monte Carlo calculation for the stationary EEDF, and a calculation corresponding to the solution of the zero-dimensional problem, which takes into account the real behavior of the field, at a time of 5.3 ns at the current value of $E/n = 600$ Td (see section (2.9.3)). At energies higher than 20 eV, the "tail" of the unsteady EEDF is noticeably overpopulated. The magnitude of the deviation significantly exceeds the difference of the stationary EEDF for the different approaches, as clearly demonstrated by Figure 2.63.

2.9.3 GAS EXCITATION IN RAPIDLY CHANGING ELECTRIC FIELD

As a first step, zero-dimensional calculations in three-dimensional velocity space were performed. These calculations do not allow us to reveal the role of the effects of the spatial inhomogeneity of the discharge on the EEDF dynamics and the associated rates of excitation of molecules by electron impact. On the other hand, this approach allows us to estimate the scale of nonstationary effects associated with the finite relaxation time of the EEDF and the degree of deviation of the rates of various electron-molecular processes from "equilibrium" values.

The simulations were performed for conditions corresponding to the experiments presented in the [276]. In Figure 2.64 the data on the dynamics of electric field strength, electron density, and particle concentration in the states $N_2^+(B^2\Sigma_u^+, v = 0)$ and $N_2(C^3\Pi_u, v = 0)$ are reproduced in the front of the FIW (from the data of Ref. [282]). The data were obtained by optical emission spectroscopy (OES) spectral measurements in a cross-section, which is 20 cm away from the high-voltage electrode.

The electric field strength in the volume in the 0-D calculations was set from experimental measurements, and the initial electron concentration was assumed to be $n(e_0) = 10^8$ cm^{-3}. A nonuniform grid over the velocity space was used in the calculations, with the position of the cell center in the velocity space depending quadratically on its number. This distribution of nodes in the computational grid allowed to compensate for the low density of energy states (increasing in the velocity space as v^3) at low-energy values and to provide good energy resolution in the region of thresholds of inelastic processes. The total number of nodes of the computational grid was $N = 80 \times 80 \times 80 \times 80$, the minimum velocity step was $\Delta v = 1.0 \times 10^4$ m/s (corresponding to an energy resolution of $\Delta e = 1.4 \times 10^{-4}$ eV).

The results obtained are shown in Figures 2.65–2.69. Figure 2.65 shows the changes in the reduced electric field strength, electron concentration, electron average velocity, and energy in

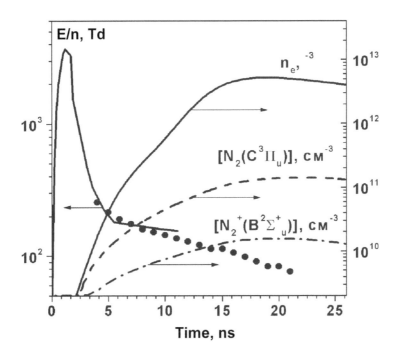

FIGURE 2.64 Temporal evolution of experimentally obtained electric field, electron concentration, and electronically excited particle concentration at levels $N_2(C^3\Pi_u, v = 0)$ and $N_2^+(B^2\Sigma_u^+, v = 0)$ at $p = 4$ Torr in nitrogen.

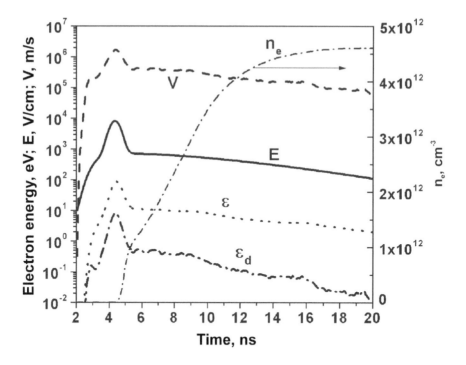

FIGURE 2.65 Electric field E, electron concentration n_e, directional velocity v, mean electron energy ε, and energy ε_d corresponding to velocity v. Nitrogen, pressure 4 Torr, 0D model.

nitrogen at a pressure of 4 Torr. The field strength profile, as already emphasized, is taken from the results of experiments (Figure 2.64), the other curves are the result of calculation in the 0D-3V geometry. The dynamics of the electron concentration is in good agreement with the dependences observed in experiments: the main production of ions and electrons occurs behind the ionization wave front in relatively weak electric fields.

2.9.4 EEDF Formation and Gas Excitation in FIW

Note that the low rate of change of the magnitude of the reduced electric field behind the wave front leads to the fact that the values of the mean energy and mean velocity practically coincide at this location with the values calculated using the two-term approximation of the Boltzmann equation, and slowly decrease together with the magnitude of the reduced field (Figure 2.65). This is explained by the relatively high relaxation rate of the high-energy electrons, noticeably exceeding in this region the rate of change of the reduced field strength. On the contrary, in the region of the field peak at the front of the ionization wave, the average electron energy changes with a characteristic time delay of $\delta t \sim 0.5$ ns relative to the electric field profile.

The dynamics of EEDF is shown in Figure 2.66. All stages of the EEDF formation are clearly visible. First, the ensemble of electrons is heated by the field and the population wave propagates in the energy space at times $\tau < 3$ ns. Then the energy of electrons becomes high enough and active ionization of the gas begins. High values of the reduced field strength in the wave front cause strong heating of the electrons at times $\tau \simeq 5$ ns. The cooling of the EEDF as E/n decreases is accompanied by continued gas ionization, and the secondary electrons' energy distribution affects the overall appearance of the EEDF for almost the entire duration of the voltage pulse.

The influence of the finiteness of the EEDF relaxation rate in the high-energy region on the shape of the distribution function is clearly demonstrated in Figure 2.67. The figure shows the EEDF dynamics calculated using the Monte Carlo approach and the EEDF calculation performed using

Low Temperature Plasma Generation and Recombination 123

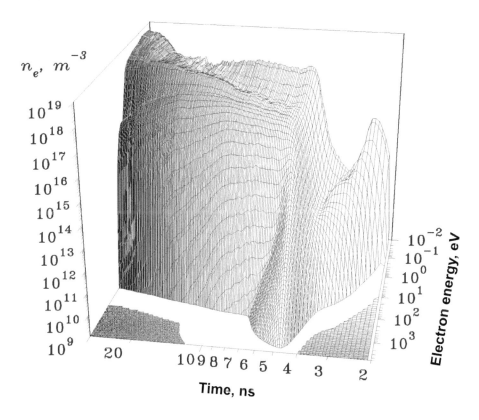

FIGURE 2.66 The electron energy distribution function. Nitrogen. Pressure 4 Torr.

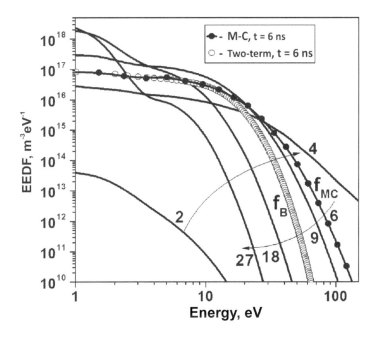

FIGURE 2.67 EEDF calculated from the 0D – model for different time moments (time moments are indicated by numbers near the corresponding curves). The curves f_B and f_{MC} represent the EEDFs calculated within the two-term approximation of the Boltzmann equation and within the presented 0D – model at time 6 ns.

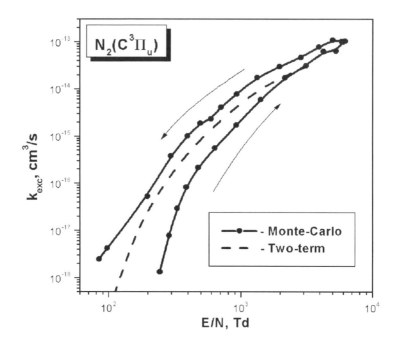

FIGURE 2.68 Comparison of the full excitation velocities of N_2 ($C^3\Pi_u$). Symbols – Monte Carlo calculations using the 0D – model; curve without symbols – calculation within the two-term approximation of the Boltzmann equation. At the starting point $E/n = 250$ Td; the direction of traversal (increasing time) is indicated by arrows.

the two-term approximation of the Boltzmann equation at the same instant of time. It can be seen that, in good agreement with the result obtained, the EEDFs match each other well in the energy range below 10 eV, and differ significantly at higher energies. This overpopulation of the high-energy "tail" of the distribution function is due to the relatively low relaxation rate of the EEDF. When the electric field decreases rapidly after the wave front has passed, the high-energy part of the EEDF loses its equilibrium with the field.

From the point of view of the elementary processes occurring at the front of a FIW and in its immediate vicinity, unsteady effects lead to a marked deviation of the rates of gas excitation, ionization, and dissociation from the values in equilibrium with a local electric field. Figures 2.68 and 2.69 show a comparison of the excitation rates of $C^3\Pi_u$ state of the nitrogen molecule and $B^2\Sigma_u^+$ state of the N_2^+ ion calculated by the Monte Carlo method and using the two-term approximation of the Boltzmann equation. It can be seen that at the leading edge of the wave, due to the finite relaxation time of the EEDF, the excitation rate is noticeably smaller than the value obtained in the two-term approximation. On the other hand, in the region of field decay, these velocities exceed the corresponding "local" values for the entire time of discharge development.

In calculations in one-dimensional geometry, it is possible to analyze the development of the ionization wave structure and the role of nonlocal effects, including those caused by the appearance of run-away electrons at the wave front. The conditions of calculations in velocity space corresponded to those in the 0D model, and the distribution in physical space was calculated on a grid containing 20 cells of size $\Delta x = 2$ cm. Note that the characteristic scale of field variation at a wave velocity of 3–4 cm/ns is about 6–8 cm.

Figures 2.70, 2.71, 2.73–2.75 show the initial stage of the FIW development—from the moment of the beginning of the electric field increase to the start of the wave and the initial stage of the movement of the point corresponding to the maximum of the mean electron energy along the discharge gap.

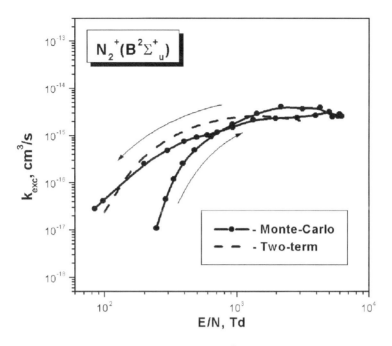

FIGURE 2.69 Comparison of full excitation velocities $N_2^+(B^2\Sigma_u^+)$; notations are the same as in Figure 2.68.

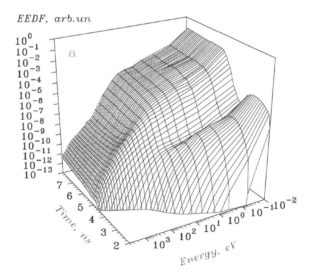

FIGURE 2.70 Behavior of electron energy distribution function. 1D 3 V–calculation. Nitrogen, pressure 4 Torr, distance from the beginning of the discharge gap is 1 cm.

The behavior of the electron distribution function in time at a small distance from the beginning of the discharge gap, where the spatial effects are still weak, is similar to the behavior of the EEDF calculated in the 0D geometry (Figure 2.66). At the initial moment of time, the EEDF is given by a delta function at zero energy. Then the field begins to give energy to the electrons. When the energies corresponding to the thresholds of inelastic processes are reached, the EEDF is somewhat

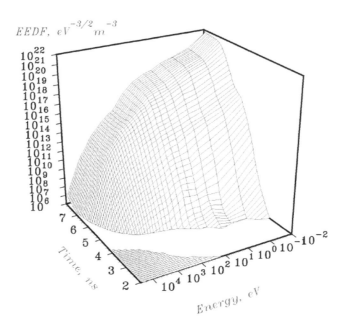

FIGURE 2.71 Behavior of electron energy distribution function. 1D 3 V–calculation. Nitrogen, pressure 4 Torr, distance from the beginning of the discharge gap is 4 cm.

"blurred"; the continued intense growth of the field leads to strong heating of the electrons. After passing the high-field peak, the formation of the EEDF is determined both by the continued energy gain of electrons in the field and by the degradation of high-energy electrons. Figure 2.71 shows the EEDF dynamics at a distance of 4 cm from the beginning of the discharge gap. Relatively low-energy electrons are the first to arrive at this point in time. This is caused by low values of the reduced field strength at the initial stage of the discharge development, correspondingly low values of the average electron energy, and a small value of the anisotropy of the EEDF. The increasing field strength leads to heating of the electron ensemble and a rapid increase in the electron energy near the wave front. Propagating in front of the wave, high-energy electrons cause ionization of the gas. The born secondary electrons have relatively low energy and zero average velocity. This leads to some decrease in the value of the average energy and velocity of electrons in front of the ionization wave compared to the maximum possible values ($\varepsilon_{max} \simeq 2U_0$). At the same time, the energy of electrons and the velocity of their directed motion in the wave noticeably exceed the analogous parameters at the front.

The main difference between the results of 1D modeling and the zero-dimensional model is the possibility of controlling the changes in the plasma parameters as the ionization wave front advances along the discharge gap. The nonequilibrium nonlocal character of the EEDF in front of the wave front is clearly demonstrated by Figures 2.73–2.75, which show the average velocity of electrons, their average energy, and the EEDF as a function of the distance from the high-voltage electrode at time $\tau = 6$ ns.

It is interesting to compare them with Figure 2.72, which shows the spatial distribution of the characteristic electric field in the FIW, which is given as initial data in the computational problem. The field is taken on the basis of experimental data obtained in a section 20 cm from the high-voltage electrode. It is assumed that there is no wave attenuation, and the given field profile $E(t)$ moves at a velocity corresponding to the wave propagation velocity along the discharge gap.

Figures 2.72–2.74 show three-dimensional representations of the electric field, average electron energy, average velocity as a function of time and coordinate along the discharge gap, and, in addition, a two-dimensional representation of each quantity as isolines on the $x-t$ plane. Heating of the distribution function leads to the appearance of a flux of high-energy electrons with average

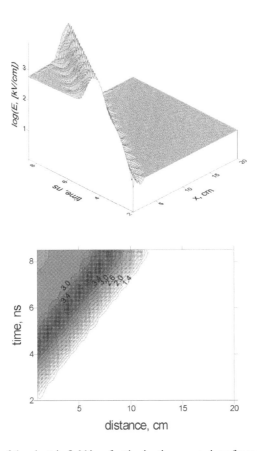

FIGURE 2.72 Dynamics of the electric field in a fast ionization wave given from experimental data. Pressure $p = 4$ Torr.

energies up to 1 keV (Figure 2.73) in front of the wave. The energy of electrons in the FIW front is significantly lower and is 110–150 eV, slowly increasing with increasing distance from the high-voltage electrode. The increase in energy is apparently due, first of all, to the process of formation of the EEDF, which at the initial moment of time is set by the delta peak at zero energy. The propagation velocity of the front corresponding to the second maximum of the average energy (labeled in the figure as the "main wave") is ~ 1.5 cm/ns and is set from the outside by the external electric field profile. The propagation velocity of the high-energy electron front is approximately 3 times faster.

A comparative analysis of Figures 2.73 and 2.74 illustrates the relationship between the energy and momentum relaxation lengths of electrons. Indeed, the characteristic momentum relaxation time of electrons is $\tau_{imp} \approx v_{el}^{-1} = 1/(n\sigma v)$, where v_{el} is the frequency of elastic collisions of electrons with gas molecules, n is the density of heavy particles, σ is the elastic transport cross-section of electron scattering, and v is the average thermal velocity of the electron. Let the gas pressure be 4 Torr, corresponding to a particle density $\sim 10^{17}$ cm^{-3}; for an electron energy of 10 eV $\sigma \approx 10^{-15}$ cm^2, the electron velocity $v \approx 2 \cdot 10^8$ cm/s. Then the momentum relaxation time is $\tau_{imp} \approx 5 \cdot 10^{-11}$ s.

As can be seen from Figure 2.73, the velocity of directed motion drops sharply with distance from the FIW front, i.e., the electron velocity quickly becomes isotropic when leaving the zone of high electric fields. Energy relaxation is determined by energy transfer in inelastic collisions; its characteristic time is 1–2 orders of magnitude longer than τ_{imp}. The behavior of the average electron energy in the front corresponding to the first electron energy maximum (denoted as "fast electrons" in Figure 2.73) is largely determined by the behavior of the electric field in the main wave and the predionization of the gas by fast electrons arising in the high-field front.

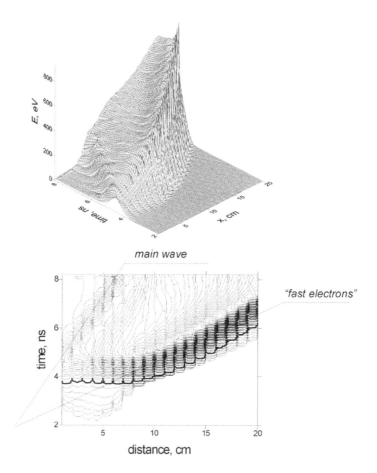

FIGURE 2.73 Spatial and temporal distribution of the average electron energy in a fast ionization wave. The distance between the isolines in the lower figure is 12.5 eV in the region bounded below by the bold line; in the region of small t and small x, the field values on successive isolines decrease from the bold line as 12.5/N, where N is the line number. Nitrogen, pressure 4 Torr, 1D 3 V - calculation.

The EEDF at large distances in front of the ionization wave becomes significantly overpopulated in the high-energy part, which is illustrated in Figure 2.75. The general view of the EEDF becomes close to the so-called "degradation" spectrum characteristic of the case of cooling in the atmosphere of an electron beam; the average energy of the electrons increases with distance from the front of the main wave, which at a time of 6 ns is at a distance of 3–4 cm from the beginning of the discharge gap.

2.10 PULSE-PERIODIC GAS DISCHARGE AT ATMOSPHERIC PRESSURE CONDITIONS

The main objective of this chapter is to describe the development of the spatiotemporal structure of the streamer corona of positive polarity in the range of values of the interelectrode gap corresponding both to the mode of partial bridging of the gap by the streamer, which does not transform into spark breakdown to spark breakdown, as well as to the mode of free streamer propagation. It is obvious that for various plasma-chemical applications an effective production of the corresponding kind of chemically active particles, which was estimated by the production of electronically excited states $NO\left(A^2\Sigma^+, v = 0\right)$, $N_2\left(C^3\Pi_u, v = 0\right)$, and $N_2^+\left(B^2\Sigma_u^+, v = 0\right)$. The excitation thresholds of these states from ground states are 5.42, 11.2, and 18.6 eV, respectively.

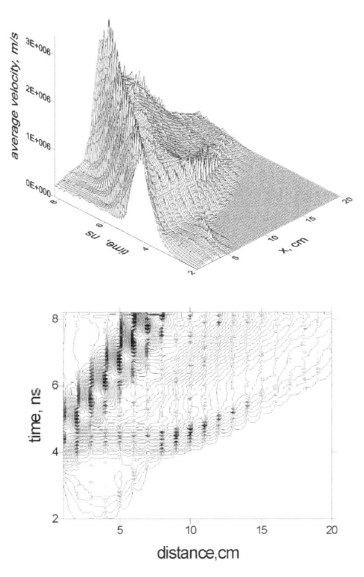

FIGURE 2.74 Spatial and temporal distribution of the average velocity of directional motion of electrons in a fast ionization wave. The distance between the isolines in the lower figure is 5×10^4 m/s. Nitrogen, pressure 4 Torr, 1D 3 V - calculation.

2.10.1 Discharge Dynamics and Active Particle Production in the "Needle-Plane" Geometry

Figure 2.76 shows a general view of the experimental setup and diagnostic system. High-voltage pulses were transmitted by means of the 60 m long 50-Ω coaxial cable to the high-voltage electrode of the discharge cell. A calibrated back-current shunt is used to control the parameters of the electrical pulse. The amplitude of the incident pulse was $U_{max} = 9$ kV (positive polarity), half-height duration $t_{1/2} = 75$ ns, rise time $t_{inc} = 25$ ns at a repetition rate $f = 1.2$ kHz.

The experimental setup consists of a needle-plane discharge gap with an adjustable size from 0 to 30 cm. The high-voltage electrode is designed as a 300 mm long cone with a tip radius of 0.5 mm, while the low-voltage electrode is an aluminum disk 8 mm thick and 550 mm in diameter. Active

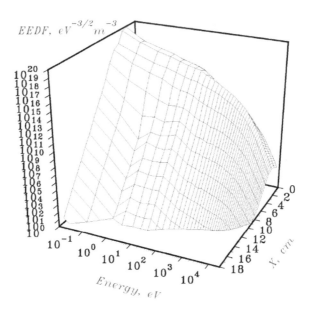

FIGURE 2.75 Spatial behavior of EEDF at time 6 ns. Nitrogen, pressure 4 Torr, 1D 3 V - model.

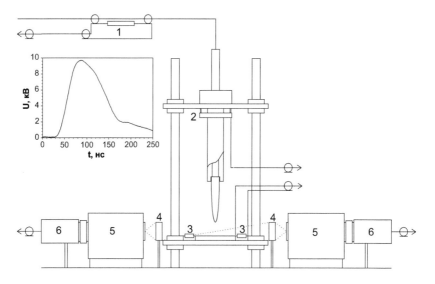

FIGURE 2.76 Scheme of the discharge diagnostics. **1**—Back-current shunt. **2**—Rogowski coil. **3**—Direct current shunt. **4**—Quartz lens. **5**—UV and VIS monochromators. **6**—PEMs.

particles in the discharge are detected by OES. Calibration of the optical system for absolute emission intensity measurements was performed using a deuterium arc lamp (200–500 nm) and a tungsten filament lamp (350–700 nm) as standard illumination sources. A signal accumulation mode with an adjustable time constant between 0.1 and 1.0 s was used to capture the discharge emission characteristics averaged over the pulse duration. With a pulse repetition rate of 1.2 kHz, the light signals are averaged over 100–1,000 pulses, improving measurement accuracy and facilitating identification of weak spectral bands and lines. Spatial and temporal analysis was performed using a 5 mm high dielectric slit aperture, which pozition can be adjusted along the axis of the discharge cell. Under the experimental conditions ($U_{gap} \sim 18$ kV, and $\tau_{1/2} \sim 75$ ns), reducing the discharge gap to

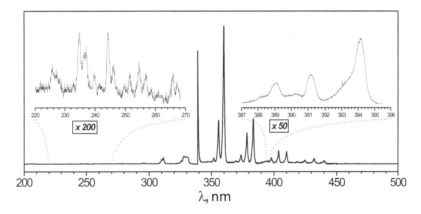

FIGURE 2.77 Streamer flash emission spectrum, $L = 24$ mm.

TABLE 2.5
Parameters of the Investigated Transitions of 2^+, 1^- Systems N_2 and γ-Band NO

	2^+, $0 \to 0$	1^-, $0 \to 0$	γ, $0 \to 1$
λ, nm	337.1	391.4	236.3
τ_0, ns	41	62	196
$A_{v'v''}$	0.523	0.718	0.298

$L \leq 16$ mm resulted in spark channel formation in the gap, preceded by a secondary streamer from the anode to the low-voltage electrode, as noted in Ref. [206]. For discharge gap lengths $L \geq 20$ mm, streamer flashes did not consistently lead to gap breakdown, although streamers spanned gaps up to $L \sim 24$ mm. Discharge spectra in streamer flash mode for gap lengths $L \geq 20$ mm exhibited consistent characteristics as shown in Figure 2.77, with identifiable molecular bands of nitrogen (2^+ system: $N_2(C^3\Pi_u, v') \to N_2(B^3\Pi_g, v'')$ transition), nitrogen ion (1^- system: $N_2^+(B^2\Sigma_u^+, v') \to N_2^+(X^2\Sigma_g^+, v'')$ transition), and nitrogen monoxide (γ bands: $NO(A^2\Sigma^+, v') \to NO(X^2\Pi, v'')$ transition) with distinct vibrational structures.

Spectral analysis in experiments was focused on specific molecular bands:

- 2^+ of the nitrogen system: 0–0 transition $\lambda = 337.1$ nm
- 1^- nitrogen system: 0–0 transition $\lambda = 391.4$ nm
- γ -nitric oxide bands: 0–1 transition $\lambda = 236.3, 237.0$ nm

The selection of specific bands for spectral measurements is guided by their distinct intensity and the clarity of their spectral transition, ensuring no overlap with other strong bands. The relationship between the concentration of excited particles and their radiative intensity, assuming no self-absorption in the band, is defined by the transition frequency $v_{v'v''}$, the concentration of particles in the excited state $[N^*]$, the radiative lifetime τ_0, and the probability of the vibrational transition $A_{v'v''}$. Relevant parameters are summarized in the Table 2.5. The Frank-Condon factors $q_{v'v''}$ are taken from Ref. [283], while the radiation lifetimes for the levels $N_2(C^3\Pi_u, v = 0)$, $N_2^+(B^2\Sigma_u^+, v = 0)$, and $NO(A^2\Sigma^+, v = 0)$ are taken from Refs. [284–286]. Consequently, this approach allows a direct correlation between the instantaneous concentrations of the excited states and the intensities of the respective transitions.

2.10.2 ACTIVE PARTICLES GENERATION IN DISCHARGE GAP

In the time-resolved mode, the dynamics of the total concentration of excited particles in the discharge gap was investigated. For this purpose, a system of diaphragms (with slit heights of 5 mm) was used, which cut out radiation from a certain region of the discharge gap. The positioning of the aperture block was performed with the help of a micrometer screw and provided accuracy not worse than 0.1 mm. When moving the diaphragm to a new position along the height of the discharge gap, the emission intensities of the studied transitions were measured of the studied transitions were measured. At the same time, the rise and fall times of the radiation correspond to the time of the streamers passing through distance equal to the length of the most actively emitting region, and the width of the radiation pulse at half-height—the time for the streamer to pass the height of the slit. Estimates made on the basis of the data obtained (see below) show that the radiation region is a narrow zone about 3 mm long. Hence an important conclusion—the main active particle production in the studied states occurs in the streamer head and the area immediately adjacent to it. All subsequent analyses and conclusions concern this region of the discharge, where the overwhelming number of excited particles is produced. The presence of a time reference point, which was used as the moment of arrival of a high-voltage pulse on the electrode, allowed us to synchronize the individual oscillograms in time and to construct diagrams of streamer motion through the discharge gap. Figure 2.78 shows the field of the radiation field (synchronous radiation profiles in different cross sections of the interelectrode gap) of the second positive nitrogen system at the value of the interelectrode gap $L = 24$ mm.

In the time-resolved mode, the study focused on the dynamics of the overall concentration of excited particles within the discharge gap. This was facilitated by the use of a diaphragm system (with slit heights of 5 mm) that isolated radiation from specific segments of the discharge gap. By moving the diaphragm to a new vertical position within the discharge gap, it was possible to measure the emission intensities along the axis of the discharge gap. Consequently, the rise and fall

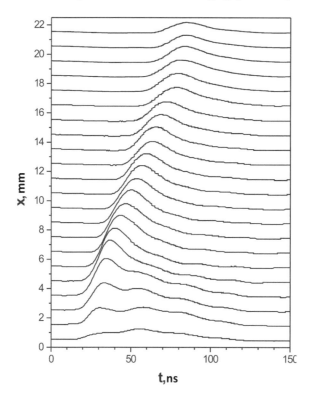

FIGURE 2.78 The radiation dynamics of the second positive nitrogen system, $L = 24$ mm.

times of the radiation were indicative of the time taken by the streamer to travel a distance corresponding to the most actively emitting section. In addition, the width of the radiation pulse at half-maximum provided a measure of the time required for the streamer to cross the slit height. Based on the data obtained, it was concluded that the region of radiation is a narrow zone in the streamer head approximately 3 mm in length. This leads to a significant insight that the main generation of active particles occurs within the streamer head and its immediate vicinity. The following analysis and conclusions were restricted to this discharge segment, where the majority of the excited particles are generated. The incorporation of a temporal reference point, defined by the instant at which a high-voltage pulse reaches the electrode, facilitated the synchronization of individual oscillographs in time, allowing the construction of diagrams depicting the streamer progression through the discharge gap. Figure 2.78 shows the radiation field (synchronous radiation profiles at different interelectrode gap cross sections) for the second positive system of nitrogen at an interelectrode gap of $L = 24$ mm.

2.10.3 Streamer Propagation Velocity

The dependence of the streamer velocity on the distance to the high-voltage electrode is represented by the so-called $x - t$ -diagram, i.e., the time for the emission to reach the level 0.1 of the maximum value as a function of the coordinate. The analysis allowed us to distinguish three characteristic regions of streamer propagation:

1. The anode region of streamer formation is characterized by a very high speed of the front movement ($l \leq 7$ mm) in a strong external field.
2. Central zone: the streamer velocity remains practically constant for gap lengths not exceeding $L = 42$ mm and slightly decreases for longer gaps.
3. Cathode region: the streamer velocity increases when approaching the low-voltage electrode. (Observed only when the gap is completely bridged by the streamer.)

Thus, for the linear portion of the motion, one can plot the dependence of the streamer velocity on the length of the discharge gap (Figure 2.79). As can be seen, the streamer velocity V increases as

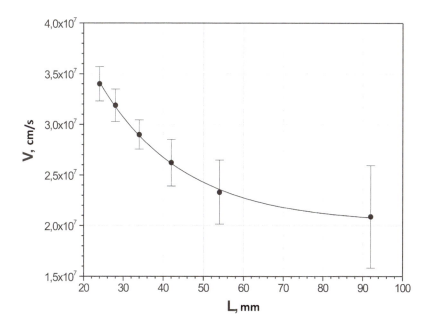

FIGURE 2.79 The dependence of streamer propagation velocity on the size of the interelectrode gap.

the gap L between the electrodes decreases, and varies (for our pulse parameters and geometry of the setup) in the range $(2 – 3.5) \times 10^7$ cm/s.

The oscillograms of streamer flash radiation in a given cross-section are characterized by the rise time and the half-width of the emission pulse. These characteristics are closely related to the length of the radiating zone near the streamer head and the velocity of its motion. It was found that the characteristic half-width of the pulse reflects the time of the streamer head passing the aperture of the measuring section, and the signal rise time is determined by the length of the emitting zone. The signal rise time was in the range of 10 – 20 ns, and the half-height width was in the range of 20 – 30 ns. The measured values of the streamer propagation velocity allow us to conclude that the length of the emitting zone is of the order of $l \simeq \langle v \times \Delta t \rangle \sim 3 – 4$ mm, corresponding to the size of the streamer head and the radial expansion zone immediately behind it.

2.10.4 Peak Concentrations of Active Particles in Streamer Head

Figure 2.80 shows the dependence of the peak concentration of states $N_2(C^3\Pi_u, v = 0)$, $N_2^+(B^2\Sigma_u^+, v = 0)$, $NO(A^2\Sigma^+, v = 0)$ (corresponding to the moment when the streamer head passes each of the gap sections) as a function of the distance from the high-voltage electrode.

From the data obtained, it can be concluded that the formation of a streamer corona occurs at a distance of up to $l_1 = 6$ mm from the high-voltage electrode (anode). At this stage, the motion of the luminescence front is controlled by local electric fields near the high-voltage electrode

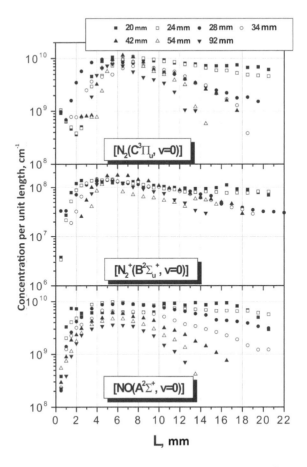

FIGURE 2.80 Peak concentration of active particles in the streamer head region as a function of distance from the high-voltage electrode for different size of the gap.

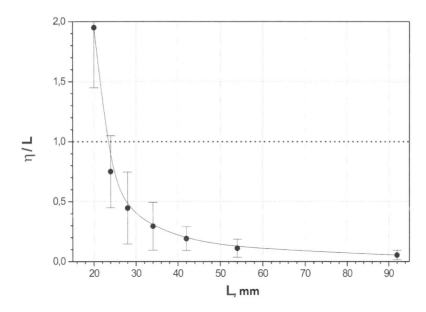

FIGURE 2.81 Dependence of the decrement of emission attenuation on the size of the interelectrode gap.

and almost does not depend on the distance to the low-voltage electrode, i.e., on the average electric field in the gap. At this stage, the maximum concentration of excited particles along the discharge gap is reached. Typical values are $\left[N_2\left(C^3\Pi_u, v=0\right)\right]_{max} \simeq (0.5-1.0) \times 10^{10}$ cm^{-1}, $\left[N_2^+\left(B^2\Sigma_u^+, v=0\right)\right]_{max} \simeq (1.0-2.0) \times 10^8$ cm^{-1}, $\left[NO\left(A^2\Sigma^+, v=0\right)\right]_{max} \simeq (0.3-1.0) \times 10^{10}$ cm^{-1}.

At small interelectrode distances, when the streamer channel overlaps the interelectrode gap, the concentration of active particles generated in the discharge varies weakly along the length of the discharge gap. For large gaps, the concentration of active particles decreases sharply with distance from the high-voltage electrode. As can be seen from Figure 2.80, the larger the interelectrode gap, the smaller the effective length of the active particles generation at the same voltage at the electrodes. The decrease in active particle concentration along the length of the discharge gap is approximated with good accuracy by a decaying exponential function:

$$[N^*](x) = [N^*]_{max} \exp\left(-\frac{(x-x_0)}{\eta}\right) \quad (2.29)$$

where $[N^*]_{max}$ is the peak concentration of excited particles reached at x_0 and η is the radiation attenuation decrement, which is a measure of the homogeneity of the active particle production along the length of the streamer corona. As can be seen from Figure 2.81, when the streamer overlaps the discharge gap, the reduced attenuation decrement (to the value of the interelectrode gap) $\eta/L > 1$ and the generation of active particles occurs almost uniformly along the entire gap. This is observed for gaps up to 24 mm long, the average electric field in the gap is $E = 7.5$ kV/cm. For larger discharge gaps, the zone of homogeneity becomes smaller than the length, and the homogeneity of active particle production along the discharge region is disturbed ($\eta/L < 0.1$ at $L > 65$ mm).

2.10.5 Integral Active Particles Production

The volumetric generation of active particles in the discharge also depends strongly on the value of the interelectrode gap and the resulting average electric field in the gap. In this case, as shown by experiments in the space-time resolution mode, the main production of active particles in the electronically excited states occurs in the streamer head.

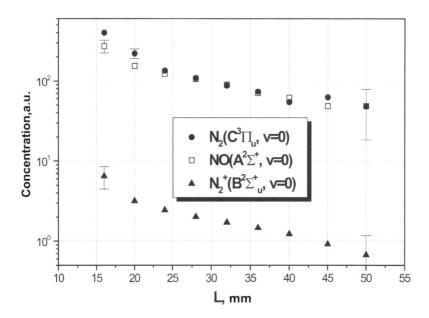

FIGURE 2.82 Integral yield of active particles in the discharge as a function of the length of the interelectrode gap.

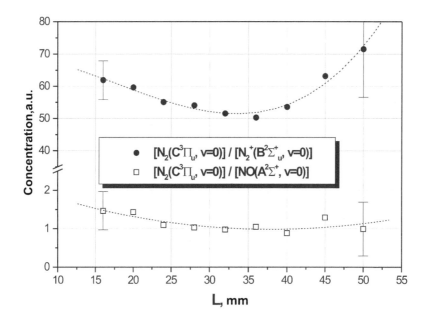

FIGURE 2.83 Relative concentrations of different excited states as a function of the length of the interelectrode gap.

The measurements performed in the range of gap lengths from 16 to 50 mm (Figures 2.82 and 2.83) have shown that the ratio of the production of different electronically excited particles in the discharge decreases sharply at the transition from the streamer gap-bridging mode ($L \simeq 20$ mm) to the mode corresponding to a free streamer corona ($L \simeq 50$ mm). Thus, the total production $N_2(C^3\Pi_u, v = 0)$ (Figure 2.81) decreases by an order of magnitude. Virtually the same behavior is shown by the electronically excited nitrogen ion $N_2^+(B^2\Sigma_u^+, v=0)$ and nitrogen monoxide $NO(A^2\Sigma^+, v=0)$. However, all these states differ significantly in excitation energy $(NO(A^2\Sigma^+, v=0)$—5.42 eV, $N_2(C^3\Pi_u, v=0)$—11.2 eV, $N_2^+(B^2\Sigma_u^+, v=0)$—18.6 eV). The effect of the different excitation thresholds is particularly clear in Figure 2.83, which shows the ratio of the concentrations of different electronically excited states as a function of the length of the discharge gap.

The value of $[N_2(C^3\Pi_u, v=0)]/[NO(A^2\Sigma^+, v=0)]$ is almost constant over the entire range of interval lengths and is 1–1.5. At the same time, the value of $[N_2(C^3\Pi_u, v=0)]/[N_2^+(B^2\Sigma_u^+, v=0)]$ varies in the range of 50–60 at decreasing and increasing of the interelectrode spacing lengths from 34 mm to 20 and 45 mm, respectively (Figure 2.83). Relatively weak dependences of the excitation ratio on the size of the gap can be explained by the fact that all states are mainly populated in the region of the strong fields at the streamer head, which depend weakly on the mean field in the gap.

The concentration ratios of the investigated states obtained in the integral mode (Figures 2.82 and 2.83) correspond to the values obtained during the space-resolved measurements (Figure 2.80), which is an additional confirmation of the fact of excitation of the investigated states (Figures 2.82 and 2.83) in the streamer head and negligibly small role of the production of active particles in the streamer channel for all sizes of the interelectrode gap up to the transition to spark breakdown.

No appreciable production of electronically excited components in the streamer channel could be detected at any size of the interelectrode gap, including the mode of partial bridging of the discharge gap by the streamer. Thus, the mechanism of a significant increase in the production of active particles in the streamer plasma channel has not been detected.

At the moment when the streamer touches the cathode surface, a cathode layer is formed, characterized by a high-voltage drop over a relatively short length. It is known that the near-cathode potential drop has a significant influence on the discharge development in the breakdown mode. The authors of Ref. [287] point out the strong dependence of the parameters of the cathode layer on the uncertainty of electron emission rates from the cathode (both photoemission and secondary electron emission). In spite of the fact that this result was obtained for the XeCl mixture, there are all reasons to expect similar high sensitivity of the plasma parameters to the uncertainty of electron emission rates in air.

2.10.6 Electric Field in the Region of Effective Excitation of Electronic Levels

The magnitude of the reduced electric field can be estimated by OES analyzing the population rates of electronically excited states of molecular nitrogen. The population of excited levels $N_2(C^3\Pi_u, v=0)$ and $N_2^+(B^2\Sigma_u^+, v=0)$ in strong electric fields is determined by the processes:

- excitation by direct electron impact from the ground state of nitrogen,

$$N_2(X^1\Sigma_g^+) + e \rightarrow N_2(C^3\Pi_u, v=0) \qquad k_{ex}^c = f(E/n)$$
$$N_2(X^1\Sigma_g^+) + e \rightarrow N_2^+(B^2\Sigma_u^+, v=0) \qquad k_{ex}^b = f(E/n) \qquad (2.30)$$

- radiative depopulation

$$N_2(C^3\Pi_u, v=0) \rightarrow N_2(B^3\Pi_g, v'') + h\nu_{2^+} \qquad \tau_0^c = 41 \text{ ns}$$
$$N_2^+(B^2\Sigma_u^+, v=0) \rightarrow N_2^+(X^2\Sigma_g^+, v'')) + h\nu_{1^-} \qquad \tau_0^b = 62 \text{ ns} \qquad (2.31)$$

- collisional quenching by heavy particles

$$N_2(C^3\Pi_u, v=0) + N_2 \rightarrow \text{products} \quad k_c^{N_2} = 0.09 \cdot 10^{-10} \text{ cm}^3/\text{s}$$
$$N_2(C^3\Pi_u, v=0) + O_2 \rightarrow \text{products} \quad k_c^{O_2} = 2.7 \cdot 10^{-10} \text{ cm}^3/\text{s}$$
$$N_2^+(B^2\Sigma_u^+, v=0) + N_2 \rightarrow \text{products} \quad k_b^{N_2} = 2.1 \cdot 10^{-10} \text{ cm}^3/\text{s} \quad (2.32)$$
$$N_2^+(B^2\Sigma_u^+, v=0) + O_2 \rightarrow \text{products} \quad k_b^{O_2} = 5.1 \cdot 10^{-10} \text{ cm}^3/\text{s}$$

- associative conversion of nitrogen ion

$$N_2^+(B^2\Sigma_u^+, v=0) + N_2 + M \rightarrow N_4^+ + M \quad k_b^M = 5 \cdot 10^{-29} \text{ cm}^6/\text{s} \quad (2.33)$$

Lifetimes and rate coefficients of collisional quenching (2.31 and 2.32) are taken from Ref. [284]. At atmospheric pressure, an essential role in the depopulation of the $N_2^+(B^2\Sigma_u^+, v=0)$ state is played by the process of ionic conversion (2.33), whose rate coefficient [288] at room temperature is rather large. In spite of the fact that the available data on this process refer to the nitrogen ion N_2^+ in the ground electronic state, the conversion rate of the $N_2^+(B^2\Sigma_u^+, v=0)$ state should not be significantly different from it because of the weak dependence of the of the polarization interaction responsible for the formation of the N_4^+ ion on the electronic excitation of the reactants.

The equation of the balance of excited particles $N^* = \{N_2(C^3\Pi_u, v=0), N_2^+(B^2\Sigma_u^+, v=0)\}$ is as follows:

$$\frac{d[N^*]}{dt} = k_{\text{ex}} \cdot n_e \cdot [N_2(X^1\Sigma_g^+)] - \frac{[N^*]}{\tau}, \quad (2.34)$$

where k_{ex} is the rate coefficient of level excitation by direct electron impact, n_e is the concentration of electrons, τ is the total lifetime taking into account radiative depopulation and collisional deactivation:

$$\tau_c = \left(1/\tau_0^c + k_c^{N_2}[N_2] + k_c^{O_2}[O_2]\right)^{-1}$$
$$\tau_b = \left(1/\tau_0^b + k_b^{N_2}[N_2] + k_b^{O_2}[O_2] + k_b^M[N_2][M]\right)^{-1} \quad (2.35)$$

To determine the magnitude of the electric field, let us consider the ratio of the rate coefficients of the population rates of levels $N_2(C^3\Pi_u, v=0)$ and $N_2^+(B^2\Sigma_u^+, v=0)$. Considering that the time derivative in the left part of the equation (2.34) is significantly smaller than the rate of population (right part), we have the relation:

$$\frac{k_{\text{ex}}^c}{k_{\text{ex}}^b} = \frac{[N_2(C^3\Pi_u, v=0)]}{[N_2^+(B^2\Sigma_u^+, v=0)]} \cdot \frac{\tau_b}{\tau_c} \quad (2.36)$$

On the one hand, the ratio of peak concentrations and inverse lifetimes of $N_2(C^3\Pi_u, v=0)$ and $N_2^+(B^2\Sigma_u^+, v=0)$ levels can be determined directly from experimental data:

$$f_{\text{exp}} = \frac{[N_2(C^3\Pi_u, v=0)]}{[N_2^+(B^2\Sigma_u^+, v=0)]} \cdot \frac{\tau_b}{\tau_c} \quad (2.37)$$

On the other hand, the ratio of the rate coefficients of the population rates of these levels can be obtained by solving the kinetic Boltzmann equation. Under such consideration, the excitation constants $k_{\text{ex}}^{c,b}$ are a function only of the reduced electric field E/n:

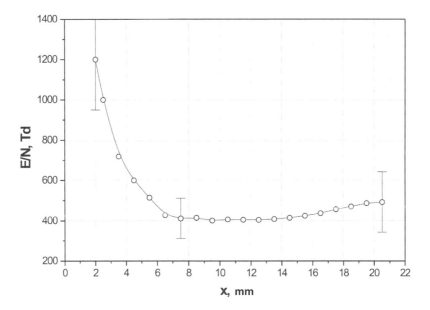

FIGURE 2.84 Electric field in the streamer head region as a function of distance from the high-voltage electrode, $L = 24$ mm.

$$f_{\text{calc}}(F/n) = \frac{k_{\text{ex}}^c}{k_{\text{ex}}^b} \qquad (2.38)$$

The electric field was estimated by comparing (2.37) and (2.38). We emphasize that the obtained values characterize the zone of the maximum excitation rate of states $N_2(C^3\Pi_u, v = 0)$ and $N_2^+(B^2\Sigma_u^+, v = 0)$, which represents the external boundary of the streamer head. With some degree of uncertainty, this value can be interpreted as the peak value of the electric field at the streamer's head.

A characteristic view of the distribution of the reduced electric field in the region of the streamer head along the length of the gap, corresponding to the region of the maximum rate of electronically excited molecules N_2, is presented in Figure 2.84 for a discharge gap $L = 24$ mm.

Near the high-voltage electrode, the electric field reaches its maximum value equal to $E/n \sim 10^3$ Td. At a distance of 7 mm from the high-voltage electrode, the formation of the streamer channel is completed, and the electric field decreases to 400 Td and further increases to 500 Td as the streamer head approaches the opposite electrode. For larger discharge intervals, the electric field dependence on the distance practically coincided with the one shown in Figure 2.84 up to the distance $x = 10$ mm from the high-voltage electrode. The range of variation of the reduced electric field outside the streamer formation region was, as in the previous case, $E/n \sim 400-500$ Td. For distances >12 mm from the high-voltage electrode in the case of large gaps, the intensity of the emission of the first negative nitrogen system became very low, preventing accurate absolute measurements of the population rate of the $N_2^+(B^2\Sigma_u^+, v = 0)$ state.

Using a similar methodology applied to the results of measurements in the integral regime (Figures 2.82 and 2.83), it is possible to estimate the average per pulse effective reduced electric field in the region of the streamer head as it moves along the gap during the whole impulse in the regions of effective excitation of electronic states of the molecules.

The obtained dependence of E/n on the size of the interelectrode gap L is shown in Figure 2.85.

As can be seen from the graph, the average field at a gap of $L = 24$ mm was $E/n \simeq 490$ Td. The obtained value agrees well with the results of spatial and time-resolved measurements, which give for the value of the reduced field strength in the middle of the discharge gap as $E/n \simeq 400-500$ Td. Thus, this approach can be used to estimate the electric field strength with an accuracy of up to 20%.

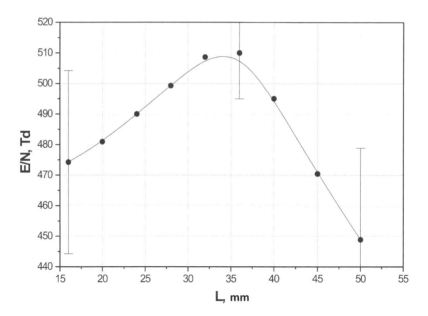

FIGURE 2.85 Effective volume-averaged reduced electric field in the streamer head region as a function of the length of the discharge gap.

This conclusion opens up possibilities for optimization of various discharges in conditions of industrial installations, when there are no variants to use diagnostics with nanosecond time resolution.

It should be noted that in estimating the electric field it is sufficient to know only the relative intensity of the emission of transitions, so the accuracy of absolute calibration ($\sim 15\%$) mainly affects only the obtained particle concentration. From Figure 2.85 it follows that the maximum of the average effective electric field is reached at the value of the interelectrode gap $L_{max} = 34 \pm 3$ mm.

In Figure 2.83, it is indicated that at intervals larger or smaller than L_{max}, the relative population of the $N_2^+\left(B^2\Sigma_u^+, v = 0\right)$ state (excitation threshold 18.6 eV) decreases more strongly than the relative population of $NO\left(A^2\Sigma^+, v = 0\right)$ molecules (5.42 eV), although the absolute concentrations of active particles decrease monotonically with increasing gap length. This effect is related to the shape of the high-voltage pulse applied to the gap (Figure 2.76). At small values of the interelectrode gaps, the streamer manages to cover the gap in the time less (or close) to the time of voltage rise on the high-voltage electrode. At larger gaps, the time of gap bridging increases, and the streamer moves in the average effective field is higher than in the case of a short gap. This fact reflects the average observed electric field in the streamer head (Figure 2.85). As the gap is further increased, the average field in the gap decreases and the streamer moves behind the formation zone with deceleration. The mean field near the head decreases slightly.

A detailed study of the distribution of active particle production along the streamer channel zone was carried out by the methods of time-resolved spectroscopy. The spatial resolution of the diagnostic system was set using a system of dielectric diaphragms at the edges of the discharge cell and quartz lenses and averaged over the discharge gap the value of the discharge gap was $\Delta L \simeq 10$ mm.

Figure 2.86 shows the radiation fields of the second positive, first negative nitrogen systems, and the γ-band of nitrogen monoxide obtained during the study of the spatial structure of the pulsed streamer corona in wire-plane geometry at a voltage of 80 kV at the high-voltage electrode.

Low Temperature Plasma Generation and Recombination

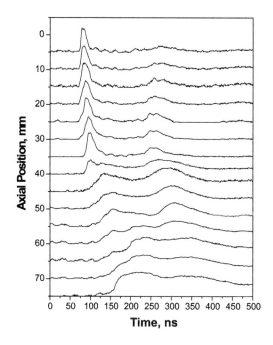

FIGURE 2.86 The radiation field of the second positive nitrogen system, $L = 110$ mm.

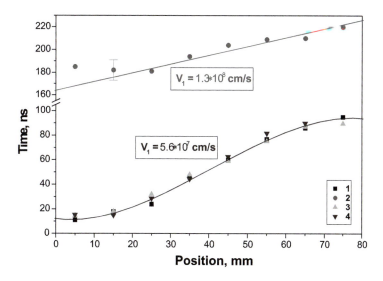

FIGURE 2.87 x-t radiation pattern of the 2^+ system N_2 at 0.5 of the maximum value as a function of the distance from the high-voltage electrode for different interelectrode spacings, $U = 80$ kV. **1** – 110 mm (first), **2** – 110 mm (second), **3** – 140 mm, **4** – 190 mm.

2.10.7 Velocity of Ionization Wave Propagation

In Figure 2.87, the $x - t$—diagram representing the dependence of the momentum of the arrival of the radiation front 2^+ of the nitrogen system into the measuring cross-section (at the level 1/2 of the maximum in the cross-section) for interelectrode gaps $L = 110, 140, 190$ mm. The propagation velocity of the radiation front is equal to the inverse derivative of the presented dependence.

As follows from the graph, the propagation velocity of the primary (and for gaps $L > 110$ mm the only one) front of ionization wave practically does not depend on the position of the low-voltage electrode. Following the region of formation $L \leq 20$ mm is followed by the region of propagation with almost constant velocity $V_1 = 5.6 \cdot 10^7$ cm/s up to $L \sim 60$ mm; then acceleration of the luminescence front is observed.

In the case of partial bridging of the gap ($L = 110$ mm), after a time of $\tau \simeq 150$ ns in the direction from the high-voltage electrode, a reverse luminescence front starts with a velocity $V_2 = 1.3 \cdot 10^8$ cm/s. In contrast to the primary front, the secondary front moves at a nearly constant velocity across the entire discharge gap.

2.10.7.1 Peak Concentrations of Active Particles

The distribution along the length of the gap of peak concentrations of the electronically excited states for different interelectrode spacings is shown in Figure 2.88. For the partial bridging mode ($L = 110$ mm), the peak concentrations are given separately, corresponding to the primary and secondary luminescence fronts.

As was shown in a series of experiments at low voltages in a single streamer mode, the discharge emission in the region of the formation is practically independent of the position of the low-voltage electrode; the influence of the interelectrode gap size is manifested on the profile of active particle production efficiency along the length of the gap.

In order to analyze the efficiency of gas excitation in the discharge gap, the total number of molecules of nitrogen molecules in the $N_2\left(C^3\Pi_u, v = 0\right)$ state along the discharge axis during the pulse is shown in Figure 2.89. These measurements indicate higher production efficiency and more homogeneous filling of the gas volume with active particles in the mode of partial gap bridging.

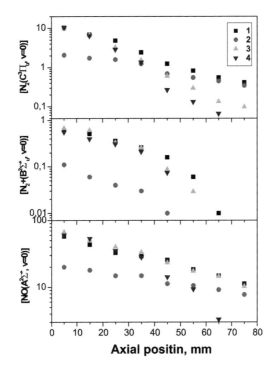

FIGURE 2.88 Peak concentration of active particles as a function of distance from the high-voltage electrode for different interelectrode spacing, $U = 80$ kV. **1** – 110 mm (first), **2** – 110 mm (second), **3** – 140 mm, **4** – 190 mm.

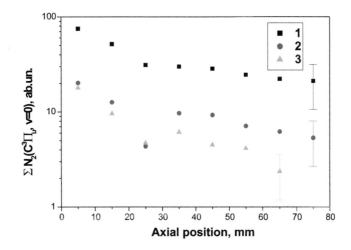

FIGURE 2.89 Distribution of the total operating time $N_2(C^3\Pi_u, v=0)$ in the streamer head region as a function of the distance from the high-voltage electrode for different interelectrode spacing, $U = 80$ kV. **1** – 110 mm, **2** – 140 mm, **3** – 190 mm.

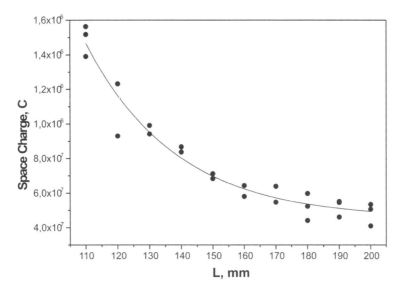

FIGURE 2.90 Charge carried into the discharge gap as a function of the size of the interelectrode gap, $U = 80$ kV.

2.10.8 Efficiency of Active Particles Generation

Figure 2.90 shows the amount of charge carried into the discharge gap as a function of the size of the interelectrode gap. The product of this function on the average effective voltage on the gap gives the value of total energy contribution to the discharge, which allows us to evaluate the relative efficiency of discharge at different positions of the high-voltage electrode.

The dependence of the charge Q pushed into a gap of length L can be approximated by a decreasing exponent as a function of distance from the high-voltage electrode:

$$Q(L) = Q_0 + Q_{\max} \exp\left(-\frac{(L - L_0)}{\eta_q}\right) \qquad (2.39)$$

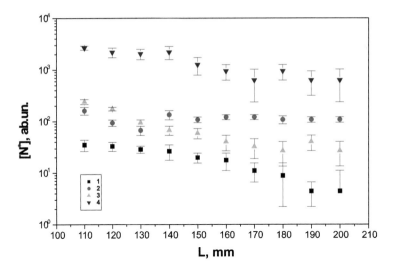

FIGURE 2.91 Peak lifetime of active particles over the whole volume for different spacings, $U = 80$ kV.

Here $Q_0 + Q_{max}$ is the charge carried into the gap at $L_0 = 110$ mm; η is the attenuation decrement (change of the interelectrode distance by the value η leads to decrease of the charge carried into the gap by a factor of e times).

The numerical values of the approximation parameters (2.39) for different values of the interelectrode distances are as follows $Q_0 = 4.5 \cdot 10^{-7}$ C, $Q_{max} = 1.0 \cdot 10^{-6}$ C, and $\eta_q = 28 \pm 4$ mm. It also follows that regardless of the position of the low-voltage electrode, under the action of strong electric fields near the high-voltage electrode, some minimum uncompensated charge Q_0 is carried into the discharge gap charge Q_0. This value can characterize the minimum energy contribution in the discharge from a high-voltage pulse generator.

In order to compare the efficiency of active particle production in the partial gap-bridging mode of the streamer, time-resolved spectral measurements and time-resolved spectral measurements and measurements of electrodynamic parameters. The optical systems were tuned in such a way as to collect radiation from the entire discharge gap along the height in the direction perpendicular to the axis of the extended electrode. Figure 2.91 shows the relative peak active particle yields $N_2(C^3\Pi_u, v = 0)$, $N_2^+(B^2\Sigma_u^+, v = 0)$, $NO(A^2\Sigma^+, v = 0)$ over the entire volume as a function of the interelectrode size depending on the size of the interelectrode gap. However, for plasma-chemical applications, the more important characteristic is the total number of generated active particles over the entire volume (2.40). This dependence of the total production of excited particles $N_2(C^3\Pi_u, v = 0)$, $N_2^+(B^2\Sigma_u^+, v = 0)$, and $NO(A^2\Sigma^+, v = 0)$) over the entire volume as a function of the size of the interelectrode gap is shown in Figure 2.92.

$$\Phi(L) = \iint [N^*](t, \vec{r})\, dt\, d^3\vec{r} \qquad (2.40)$$

The given dependence for convenience of comparison with the value of the charge carried out in the discharge gap can be approximated by an exponential decrease in the value of the interelectrode gap L:

$$\Phi(L) = \Phi_{max} \exp\left(-\frac{(L - L_0)}{\eta}\right) \qquad (2.41)$$

Here Φ_{max} is the maximum value of active particle yield achieved at $L_0 = 110$ mm; η – decrement of attenuation of active particle efficiency; this value characterizes the value of the interelectrode gap, at which the production efficiency decreases by a factor of e and can characterize the variation of active particle operating efficiency in different modes.

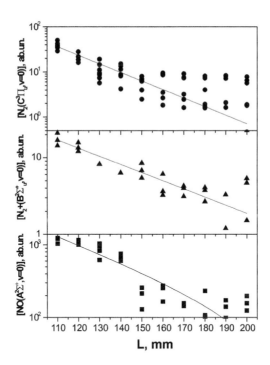

FIGURE 2.92 Integral lifetime of active particles in the discharge as a function of the size of the interelectrode gap, $U = 80$ kV.

TABLE 2.6
Decrement of Decrease in Efficiency of Active Particle Generation over the Discharge Gap from the Value of the Interelectrode Gap of the Interelectrode Gap $U = 80$ kV

	Φ_1	η, mm
$N_2(C^3\Pi_u, v = 0)$	$3.6 \cdot 10^{-5}$	23 ± 3
$N_2^+(B^2\Sigma_u^+, v = 0)$	$1.7 \cdot 10^{-5}$	41 ± 4
$NO(A^2\Sigma^+, v = 0)$	$1.4 \cdot 10^{-3}$	38 ± 9

The attenuation decrement obtained using this approximation for $N_2(C^3\Pi_u, v = 0)$, $N_2^+(B^2\Sigma_u^+, v = 0)$, and $NO(A^2\Sigma^+, v = 0)$ is given in Table 2.6. Since the decrement is 20 – 40 mm with a characteristic gap value of 100 – 200 mm, it is seen that the a strong sensitivity of the active particle yield at a fixed voltage pulse to the size of the interelectrode gap near the threshold. When the interelectrode gap decreases, both the active particle lifetime and the charge pushed to the gap (and with it the energy deposition to plasma) increase. The value of the charge carried into the discharge gap is less sensitive to the value of the mean field in the gap than the value of the integral production of particles in the $N_2(C^3\Pi_u, v = 0)$ state, and is more sensitive to it with respect to the production of particles in states $N_2^+(B^2\Sigma_u^+, v = 0)$ and $NO(A^2\Sigma^+, v = 0)$:

$$k_{N_2(C)} < k_q < k_{NO} < k_{N_2^+(B)}$$

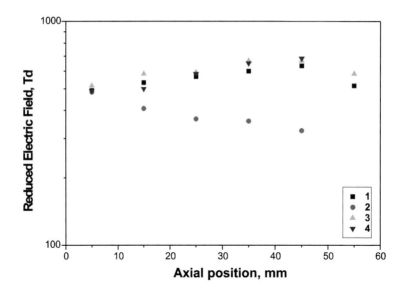

FIGURE 2.93 Electric field in the streamer head region as a function of distance from the high-voltage electrode for different electrode spacings $U = 80$ kV. **1** – 110 mm (first), **2** – 110 mm (second), **3** – 140 mm, **4** – 190 mm.

This means that when the interelectrode gap decreases, the energy cost of the formation of active particles in the $N_2(C^3\Pi_u, v = 0)$ state increases, and insignificantly increases the energy cost of formation of particles in $N_2^+(B^2\Sigma_u^+, v = 0)$ and $NO(A^2\Sigma^+, v = 0)$ states.

In general, these measurements allow us to conclude that the efficiency of the discharge in terms of production of the electronically excited states using the mode of partial bridging of the gap does not increase, although the total number of active particles produced in the discharge increases sharply in this mode.

Figure 2.93 shows the average effective electric field using the data from Figure 2.88. The obtained value varies from $E/n = 500$ Td to $E/n = 600$ Td at lengths from 5 to 75 mm in the primary streamer for all investigated interelectrode spacings and from $E/n = 500$ Td to $E/n = 350$ Td for the repeated discharge. This electric field corresponds to the region of effective gas excitation because this is the region where most electronically excited molecules are produced by electron impact.

2.10.9 Transition to Spark Breakdown

The transition to spark breakdown was accompanied by a jump-like increase of the conductive current (up to 150 – 300 A), flowing through the discharge gap. The transition from streamer discharge to pulse breakdown is also accompanied by significant changes in the emission spectrum. In the case of transition, the spectrum characteristic of an arc channel can be clearly seen, containing numerous atomic lines, including atoms and ions of the material of the electrodes, superimposed on a powerful continuum associated with the bremsstrahlung radiation and recombination emission of the quasi-equilibrium plasma of the arc channel Figure 2.94.

In order to find out the peculiarities of active particle generation in the pulsed corona and to select the region of optimal parameters, it is necessary to determine the size of the critical interelectrode gap, which corresponds to the transition to spark under given conditions.

Figure 2.95 presents the electrodynamic characteristics of the developed spark discharge. Voltage and conduction current profiles are given for an interelectrode gap of 5 mm and a high-voltage pulse with an amplitude of 13 kV. Figure 2.96 summarizes the dynamics of the energy and power input into the discharge gap.

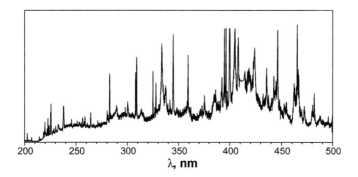

FIGURE 2.94 Emission spectrum of the discharge in the spark breakdown mode, $L = 10$ mm.

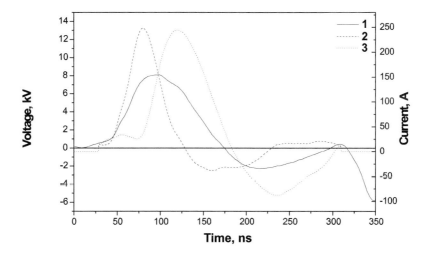

FIGURE 2.95 Electrodynamic characteristics of spark breakdown, $L = 5$ mm. **1** – voltage pulse shape in the supply line, **2** – voltage across the gap, conduction current through the high-voltage electrode.

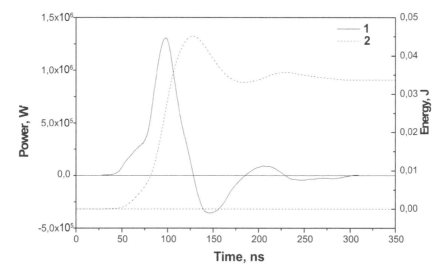

FIGURE 2.96 Electrodynamic characteristics of spark breakdown, $L = 5$ mm (continued). **1** – energy stored in the gap, **2** – the power of the embedded energy.

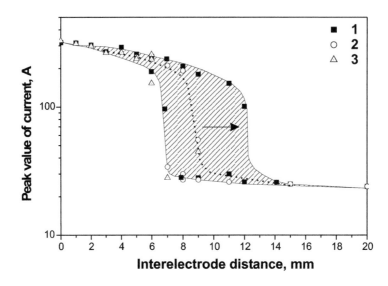

FIGURE 2.97 Peak value of spark current as a function of electrode gap value. **1** – positive polarity $f = 1.2$ kHz, **2** – positive polarity $f = 50$ Hz, **3** – negative polarity $f = 1.2$ kHz.

In the interelectrode gap range of 6–12 mm, the simultaneous appearance of both streamer and spark discharges is observed, i.e., in this range, the time of formation of the spark channel is comparable to the duration of the high-voltage pulse. Figure 2.97 shows the peak current values in this range of parameters. As can be seen from the figure, the minimum value of the gap at which the streamer (low-current) form of discharge can exist is 7 mm for both polarities of the high-voltage pulse, which corresponds to the average field $E = U_{max}/L \simeq 19$ kV/cm.

The maximum length of the gap at which spark breakdown can be realized is strongly dependent on the pulse repetition rate. Thus, at a pulse repetition rate of 50 Hz, this value was 9 mm, which corresponds to an average breakdown field of 15 kV/cm, but at a frequency of 1.2 kHz, the critical gap size increased to 12 mm, i.e., the breakdown field was 11 kV/cm. The change of the minimum gap size is related to the accumulation of active particles and partially with local heating of the gap at an increasing pulse repetition rate.

2.11 THEORETICAL STUDY OF PULSED DISCHARGES AT HIGH PRESSURE

2.11.1 Hydrodynamic Model of Pulsed Gas Discharge

2.11.1.1 Main Equations

Streamer discharges were modeled in the hydrodynamic approximation in 2D geometry for a long time [157,289–291]. Typical numerical model includes the charged particle balance equations (2.42–2.44) and the equation for electric field in the gap (2.45):

$$\frac{\partial n_e}{\partial t} + \text{div } j = S_{ion} + S_{photo} - S_{rec} - S_{att} \tag{2.42}$$

$$\frac{\partial n_p}{\partial t} = S_{ion} + S_{photo} - S_{rec} \tag{2.43}$$

$$\frac{\partial n_n}{\partial t} = S_{att} \tag{2.44}$$

$$\text{div } E = \frac{e}{\varepsilon_0}(n_i - n_e) \tag{2.45}$$

$$j = v_e \cdot n_e - D \cdot \nabla n_e$$

Here n_e, n_p, and n_n are the concentrations of electrons, positively and negatively charged ions, v_e, D is the drift velocity of electrons under the action of the local electric field E and the diffusion coefficient, S_{ion}, S_{photo}, S_{rec}, and S_{att} – the rates of ionization, photoionization, electron-ion recombination, and electron attachment to neutral gas particles, respectively.

2.11.1.2 Ionization and Transport of Charged Particles

The rate of ionization by electron impact was calculated using a local relation

$$S_{ion} = \alpha \mid v_e \mid n_e$$

More accurate [292,293] approaches take into account the effects of the nonlocal and nonstationary nature of ionization in the region of high spatial and temporal gradients. When modeling the streamer under similar conditions, the authors show a 10% difference from the local approach in the streamer propagation velocity and a negligible difference in the streamer current and active particle production efficiency.

The first Townsend coefficient α and electron drift velocity v_e were calculated by solving the kinetic Boltzmann equation in the two-term [294] approximation. A commercial implementation of the [281] and the freely available package **BOLSIG** (Kinema Software & CPAT) were used.

2.11.1.3 Photoionization

The most accurate is the photoionization model proposed in Ref. [158] for N_2-O_2 mixtures. The model is based on the assumption that the main contribution to the photoionization of the gas comes from the nitrogen emission in the spectral range 980–1,025 Å, where the absorption by nitrogen can be neglected. Here 1,025 Å is the photoionization threshold of O_2. Below 980 Å the radiation is essentially absorbed by nitrogen and makes no appreciable contribution to the photoionization processes of the air. Using the unresolved structure of the oxygen absorption spectrum in this interval, an integral expression for calculating the photoelectron birth rate in pulsed nitrogen-oxygen discharges was derived in Ref. [158].

In this case, the absorption length of photoionizing radiation, reduced to the partial pressure of oxygen, is a function of the product of distance and partial pressure of oxygen in the mixture: $\Psi/p_{O_2} = f(r \cdot p_{O_2})$.

In the introduction, the difficulties arising in accounting for photoionization processes for other gases have already been noted. However, to demonstrate qualitative effects and on the assumption that researchers often use these results for N_2-O_2 mixtures, the photoionization rate was taken from Ref. [295]. The collisional quenching of the emitting states at higher pressures [158] and the geometric factor should be taken into account:

$$S_{photo} = \frac{1}{4\pi} \cdot \frac{p_q}{p + p_q} \int_V d^3 r_1 \frac{S_{ion}(r_1)}{\mid r - r_1 \mid^2} \Psi(\mid r - r_1 \mid \cdot p) \tag{2.46}$$

Here $1/4\pi$ is the normalization constant, p is the gas pressure, p_q is the quenching pressure of photoionizing states, and $\Psi(\mid r - r_1 \mid \cdot p)$ is the absorption coefficient of ionizing radiation in the medium. Figure 2.98 shows the dependencies of the reduced radiation absorption length for pure nitrogen and N_2:O_2 mixtures based on the results of the [158,295].

The quenching pressure of photoionizing states was taken to be $p_q = 60$ Torr for pure nitrogen [296] and $p_q = 30$ Torr – for air [158].

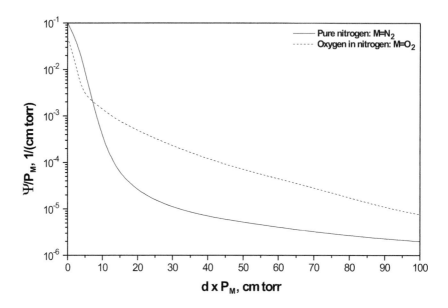

FIGURE 2.98 The photoionization rates of N_2 and O_2 molecules under excited nitrogen radiation for pure nitrogen and N_2-O_2 mixtures, respectively. (Based on the results of Ref. [295].)

2.11.1.4 Predionization

At the initial time moment, the gas in the gap was assumed to be completely quasi-neutral. The homogeneous spatial preionization was assumed to be $n_e \mid_{t=0} = n_i \mid_{t=0} = 0, 10^3, 10^7$ cm^{-3} for different situations. The [45] has demonstrated the strong sensitivity of the streamer formation time to a change in the ratio between the processes when photoionization and homogeneous spatial ionization are combined. Taking into account that: (1) the initial distribution of charged particles in the experimental conditions is poorly known; (2) a correct description of the near-electrode region is difficult to achieve in the hydrodynamic description, this calculation cannot claim to be a detailed description of streamer development. In this work, in order to reduce the calculation time, an initial preionization of the form:

$$n_e(z,r) \mid_{t=0} = n_i(z,r) \mid_{t=0} = n_0 \exp\left[-\left(\frac{r}{\sigma_r}\right)^2 - \left(\frac{z}{\sigma_z}\right)^2\right], \quad (2.47)$$

where $n_0 = 10^{14}$ cm^{-3}, $\sigma_r = \sigma_z = 5 \times 10^{-4}$ cm. Varying these parameters leads to a noticeable change in the streamer start delay without affecting the basic characteristics of the streamer in the gap. Thus, at single electron initiation, the discharge parameters differed markedly only at distances up to $\simeq 0.05$ cm from the anode, after which the streamer "forgets" the initial conditions.

2.11.1.5 Geometry of Discharge Gap

A gap $L = 1$ cm between a hyperbolic anode and a planar cathode was considered. The initial point of the cylindrical coordinate system (z,r) was located at the tip of the high-voltage electrode and the z-axis had a direction toward the low-voltage electrode. The geometry of the high-voltage electrode was given by the equation:

$$\left(\frac{z}{b}\right)^2 - \left(\frac{r}{a}\right)^2 = 1,$$

where $a = 0.18$ cm, $b = 1.0$ cm. The radius of curvature of the electrode tip is $a^2/b = 3 \times 10^{-2}$ cm.

2.11.1.6 Numerical Methods and Adaptive Meshes

The system of equations (2.42–2.45) was solved using the finite volume method on an adaptive space-time mesh. Equations (2.42–2.44) were solved by separation by physical processes. The transport of charged particles was calculated by an explicit scheme with first-order accuracy in time and space, and the right-hand sides by the Euler method with first-order accuracy in time. The Poisson equation was solved by the Gauss-Seidel fitting method with successive overrelaxation. The spatial distribution of the potential in the streamer head region was additionally shifted once per time step with the average speed of streamer motion—this technique allowed us to reduce the number of iterations in finding the electric fields by a factor of 1.5–2.5.

2.11.1.7 Algorithm of Adaptive Mesh Construction

The computational mesh changed as the maximum electric field in the gap was advanced (Figure 2.99). The radial position of the cell boundaries was chosen once and did not change during the calculation: up to $r_1 \leq 0.1$ cm the step in space was $\delta r_{min} = 5 \times 10^{-4}$ cm, and then exponentially increased to $\delta r_{max} \simeq 1$ cm at $R_{max} = 10$ cm. In the axial direction, there was a uniform mesh region of length $\Delta z = 0.2$ cm with a spacing of $\delta z_{min} = 5 \times 10^{-4}$ cm, beyond which the spatial spacing increased exponentially as one approached the boundary of the computational domain. The typical number of cells was values of $N_z \times N_r \simeq (700 - 900) \times (250 - 300)$.

A shift of the detailed mesh occurred each time the maximum of the electric fields was shifted by a distance of 0.05 cm. Thus, the regions of intense ionization and preionization in front of the streamer head were always on a uniform mesh of 400×200 cells with a space step of $\delta r_{min} = \delta z_{min} = 5 \times 10^{-4}$ cm. The time step was typically within $\delta t \simeq (1-5) \times 10^{-13}$ s and was significantly smaller than necessary for the stability of the scheme used. High spatial gradients of particle concentration and electric field impose limitations on the size of the mesh cells used. It was found that using mesh sizes larger than $\delta r_{min}, \delta z_{min} \simeq 10^{-3}$ cm in the streamer head region leads to a significant (10% or more) deviation of the main characteristics of the streamer discharge in pure nitrogen (peak electric field strength and electron concentration in the channel). When using photoionization constants corresponding to the nitrogen-oxygen mixture, this deviation was somewhat smaller, which is directly related to the larger radius of the streamer head. This effect was already mentioned in the chapter [297], the authors of which recommended the use of a grid with mesh sizes not larger than $\delta r_{min}, \delta z_{min} \simeq (5-10) \times 10^{-4}$ cm in pure nitrogen.

In order to exclude the influence of the peculiarities of the numerical realization on the studied effects, the calculations for all the cases presented in the chapter were performed on identical meshes and with identical values of time steps. These parameters were chosen according to the most "demanding" calculations.

To validate the described model, a simulation of the streamer discharge was performed under conditions close to the experimental conditions described in Section 2.10.1. The amplitude of the high-voltage pulse was $U_{max} = 18$ kV (positive polarity), half-height duration $t_{1/2} = 75$ ns, rise time $t_{inc} = 25$ ns, and interelectrode gap $L = 24$ mm.

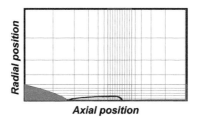

FIGURE 2.99 Principal scheme of adaptive mesh for modeling streamer discharge propagation.

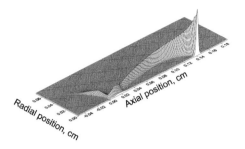

FIGURE 2.100 Spatial distribution of uncompensated charge. Air $t = 50$ ns. $L = 24$ mm.

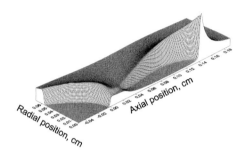

FIGURE 2.101 Spatial distribution of electric field strength. Air $t = 50$ ns. $L = 24$ mm.

2.11.1.8 Spatial and Temporal Characteristics of the Discharge

When a high-voltage pulse is applied to the anode after a certain time (start delay), a narrow channel with a well-defined region of high electric fields at the leading edge is formed from the high-voltage electrode to the low-voltage electrode (Figures 2.100 and 2.101). Figures 2.100 and 2.101 shows characteristic distributions of space charge and electric field strength at time $t = 50$ ns. For detailed analysis, 2D cross sections of the streamer are shown in Figure 2.102 with the main characteristics of the discharge.

2.11.1.9 Electric Fields and Active Particle Formation in the Streamer Head

The peak value of the electric field strength in the streamer head, according to the numerical simulation results, is in the range of 500–600 Td, except for the anode region, where fields as high as 1 kTd are realized at the streamer start. This value significantly exceeds the threshold of applicability of the hydrodynamic approximation using the two-term approximation of the Boltzmann equation for calculating the rates of elementary processes, but the deviation of the solution at such E/n in the air is $\sim 15\% - 30\%$ for the most critical parameter—the electron impact ionization rate [298].

Figure 2.103 shows the values of the peak electric field as the streamer head moves along the gap, obtained by calculation and on the basis of experimental measurements. As can be seen, this characteristic agrees with good accuracy in both cases.

The production rate of active particles in the streamer can be calculated from the known spatial and temporal distributions of the free electron concentration n_e and the electric field strength E/n (or rather from the rate coefficients of the processes responsible for the excitation of the particles). Thus, in order to compare the obtained values with the experimentally measured ones, the integral (in time) lifetime of the active particles per unit length of the discharge interval is given in Figure 2.104. The population rates of $N_2(C^3\Pi_u, v = 0)$ and $N_2^+(B^2\Sigma_u^+, v = 0)$ are obtained assuming direct electronic excitation from the nitrogen ground state $N_2(X^1\Sigma_g^+, v = 0)$. The excitation rate coefficients were calculated using the solution of the Boltzmann equation in the two-term approximation and

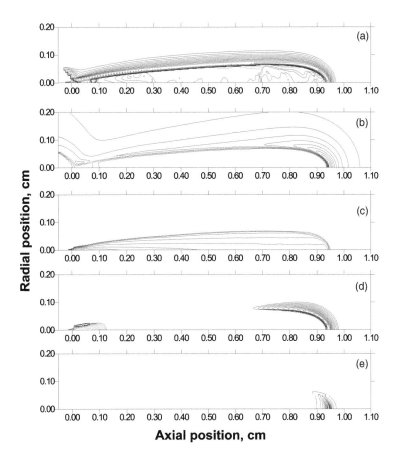

FIGURE 2.102 Streamer discharge dynamics in air at $t = 50$ ns and $L = 24$ mm. (a) Spatial charge (log scale). $3 \times 10^{-6} - 10^{-10}$ C/cm³, (b) electric field strengths up to 600 Td (steps of 50 Td); (c) electron concentration (log scale). $10^{12} - 5 \times 10^{14}$ cm⁻³; (d) ionization rate (log scale). $2 \times 10^{12} - 10^7$ cm⁻³/s; (e) photoionization rates up to 2×10^8 cm⁻³/s in uniform increments of 10^7 cm⁻³/s.

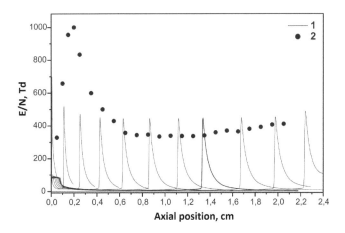

FIGURE 2.103 Peak electric field strength in the streamer head. Comparison of calculated (**1**) and experimental results (**2**).

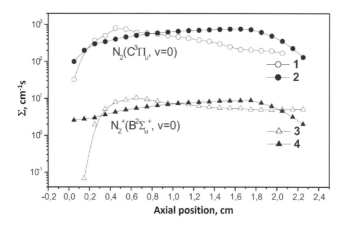

FIGURE 2.104 The production of active particles in streamer discharge. Comparison of calculation (**2,4**) and experiment (**1,3**) results.

the excitation cross sections of individual electron-vibrational transitions [299,300]. As can be seen, in the center of the discharge gap the concentration of the studied states is well described (with an accuracy of up to two times) within the framework of the hydrodynamic model used.

2.11.2 The Role of Photoprocesses in the Development of Cathode-Directed Streamer

This chapter paper [301] shows the example of simulating the streamer evolution for two cases:

- pure nitrogen (N_1 mode),
- a mixture of 99% N_2 + 1% O_2 (N_2 mode).

The addition of 1% oxygen makes no appreciable change in the drift velocity of electrons and ionization rates (up to the maximum electric fields in the problem). A completely different situation is observed for the photoelectron birth rate under the action of photoionizing radiation: a small admixture of oxygen does not lead to an appreciable increase in the total number of photoelectrons in space, but the path length of the photoionizing radiation increases strongly. Thus, the role of photoionization processes in streamer propagation should be explicitly demonstrated in a mixture of nitrogen and a small amount of oxygen.

It was found that for a constant voltage at the anode, the streamer velocity across the gap is almost constant—a consequence of the high conductivity of the streamer channel, which ensures that the streamer head potential is high compared to the surrounding space. However, in the near-electrode regions, the streamer head motion is essentially non-steady: this is due to both the increased electric field strength and the complex distribution of charged particles in the near-electrode regions. The spatial size of the regions of unsteady motion depends on the size of the streamer head and the path length of the photoionizing radiation and can be estimated as one or two radii of the streamer head.

2.11.2.1 Model Considering Gas Photoionization

Figure 2.105 shows $x - t$ – plots of the propagation of the electric field maximum along the length of the discharge gap for simulations in pure nitrogen and in a mixture of N_2:O_2 (99:1). The plot shows the time between the instant of voltage application to the high-voltage electrode and the arrival of the electric field maximum for each cross-section of the discharge gap. The slope of this dependence determines the local velocity of propagation of the streamer head.

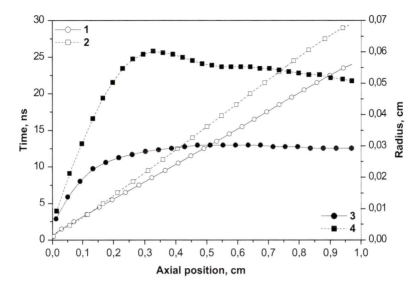

FIGURE 2.105 $x-t$ diagrams of the propagation of the maximum electric fields in the streamer head (**1,2**) and the streamer channel radius at half length (**3,4**). (**1,3**) – pure nitrogen, (**2,4**) – $N_2:O_2$ mixture. The modes are N_1 and N_2.

For the case of pure nitrogen (#1), the size of the near-electrode regions was $l_1 \simeq 0.05$ cm, and the velocity on this section was found to be $V_1 = 4.1 \times 10^7$ cm/s. For the $N_2:O_2$ (99:1) mixture (#2), the near-electrode region increased to $l_2 \simeq 0.1$ cm, and the streamer velocity decreased to $V_2 = 3.2 \times 10^7$ cm/s, i.e., a small oxygen admixture (1%) in pure nitrogen leads to a 20% decrease in streamer velocity. It should be noted that this deviation is controlled solely by the change in the path length of the photoionizing radiation and thus by the distribution of electrons in front of the streamer head.

To describe the geometry of the streamer channel in the transverse direction, the authors [301] consider the radius r_{st}. Due to the fact that the streamer channel is essentially a two-dimensional structure, the concept of "radius" cannot be introduced—such a property is quite distinct in each of the cross sections of the channel, but varies significantly along the length of the streamer channel. Because of this arbitrariness, r_{st} was defined as the point in the radial direction with the maximum electric field at the center of the streamer channel length. The introduced r_{st} is also shown in Figure 2.105. As can be seen, not only the streamer propagation velocity but also the transverse size of the channel undergoes a significant change when 1% oxygen is added to pure nitrogen, even though the drift and collision ionization rates used in both cases are the same.

The center of the gap (0.5 cm) is reached at times $t_1 = 12.7$ ns in pure nitrogen and $t_2 = 15.4$ ns in the $N_2:O_2$ mixture. Figure 2.106 shows the isolines (in steps of 10% of the maximum value) of the absolute value of the electric field strength E/n and the electron concentration n_e for both cases.

In Ref. [302], a modeling of streamer development in air was performed in a similar geometry and voltage at a high-voltage electrode. The author demonstrated the formation of a streamer "skirt" near the anode, similar to the one obtained in the $N_2:O_2$ mixture (Figure 2.106, $l \simeq 0.1$ cm), but much more pronounced. This effect increases with the increase of the path length of ionizing radiation and is not apparent when considering photoionization for pure nitrogen only. However, to all appearances, this effect has no physical basis, but is a consequence of simplification of calculations and, possibly, incorrect geometry of boundary conditions on the high-voltage electrode.

Due to the fact that the calculation of photoionization is a computationally intensive procedure, a number of simplifications are usually used to calculate the 2D integral (2.46) at each point of the problem. As a rule, one of these simplifications is the introduction of a restriction on the radiation path length, since the latter decreases extremely sharply with distance. It was found that the

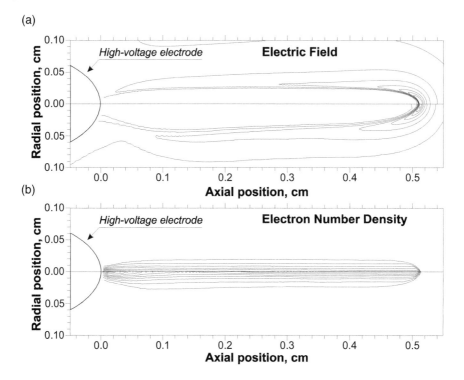

FIGURE 2.106 Comparison of streamer evolution in pure nitrogen (upper halves of the plots) and in the $N_2:O_2$ mixture (lower half). Isolines of electric field strength (a) and electron density (b). The spacing of the isolines in each figure is 10% of the maximum value. The length of the gap is 1 cm, and the anode voltage is 13 kV.

inflection in the streamer channel profile is directly related to such a restriction, and the curvature of such formation is determined by the considered path length of photoionizing radiation. Thus, for our conditions, the formation of a streamer "skirt" does not occur when taking into account the propagation of radiation for lengths up to 0.1–0.5 cm and longer. Note that this effect appears only at the streamer start in the near-electrode region and is relatively weak when the streamer propagates in the middle of the gap. It is important to mention this type of errors in the numerical modeling of pulsed discharges because they are quite typical and lead to significant errors.

It has been demonstrated above that an admixture of 1% oxygen to nitrogen markedly changes the propagation velocity and shape of the streamer discharge. In order to understand the nature of this effect, it is necessary to consider the 2D structure of the streamer head. The primary parameters of this model are the distributions of the reduced electric field strength E/n and electron concentration n_e. It is the distributions of these quantities before the front of high electric fields that determine the dynamics of further advancement of the streamer head [154,303–305].

Figure 2.107a–d shows the most important characteristics of the streamer head at the moment of time corresponding to the high electric field front reaching the middle of the gap shown in Figure 2.106. In Figure 2.107a the distribution of electric field E/n and electron concentration n_e along the axis of the discharge gap is given. It can be seen that the addition of 1% oxygen leads to a decrease in peak electric fields from 10^3 Td to 640 Td (40%) and electron concentration on the axis from 2.5×10^{14} cm^{-3} to 9×10^{13} cm^{-3} (60%). Figure 2.107b shows the radial distribution of electric field strength and electron concentration at the same moments in time in a cross-section located 0.1 cm away from the position of maximum electric fields. This cross-section is located behind the streamer head and corresponds to the distribution of the given quantities in the streamer channel. Figure 2.107c and d shows the corresponding spatial distributions of the electron birth rate in the head region under the action of collision ionization (S_{ion}) and photoionization (S_{photo}).

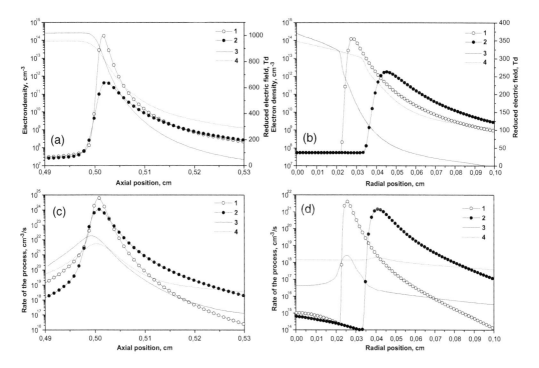

FIGURE 2.107 Spatial 2D structure of the streamer head in pure nitrogen (**1,3**) and in $N_2:O_2$ mixture (**2,4**) at the moment the streamer head reaches the middle of the gap. The axial (a,c) and radial distributions at a distance of 0.1 cm from the position of the maximum electric field (b,d) are given. (a,b) Electric field strength (**1,2**) and electron concentration (**3,4**); (c,d) collision ionization (**1,2**) and photoionization (**3,4**) rates. The N_1 and N_2 modes.

It can be seen that the distribution of electrons in front of the streamer head is significantly different in both cases. In turn, it is this distribution that sets all the characteristics of the discharge. In particular, the low level of seeding electrons in front of the electric field front in pure nitrogen leads to higher electric fields in the streamer head and, as a consequence, to a higher propagation velocity and smaller channel radius.

The above difference is noticeable throughout the entire process. The peak electric field strength at each cross-section of the discharge gap is given in Figure 2.108. This value is reached at each cross-section at different time points as the streamer head advances. The figure also shows the spatial distribution of the electron concentration on the axis in the streamer channel.

Thus, the addition of 1% oxygen in pure nitrogen leads to a marked change in all the characteristics of the cathode-directed streamer discharge. However, against the background of changes in propagation velocity (20%), electric field strength (40%), and electron concentration in the streamer channel (60%), the total channel conduction current ($J \simeq 250$ mA) in both cases differs by no more than 3%.

These results were obtained in the absence of homogeneous initial preionization over space. As indicated in Figure 2.107, the photoelectron flux in front of the high electric field region is quite intense: at a distance of the order of the streamer head radius, at the point where the collisional ionization and photoionization rates are equal, the electron concentration is of the order of $10^8 - 10^{10}$ cm^{-3}. On this basis, the effect of additional homogeneous preionization is to be expected at photoionization electron concentrations comparable to the figures given.

The modeling performed demonstrated that up to an initial predionization of 10^7 cm^{-3}, the presence of seed electrons has no appreciable effect on such characteristics as streamer velocity, electric field strength in the head, and electron concentration in the high electric field region and streamer channel. Thus, when the initial predionization was increased to 10^7 cm^{-3}, minor deviations (up to 3%) were observed only at the streamer formation length, i.e., $\simeq 0.05$ cm.

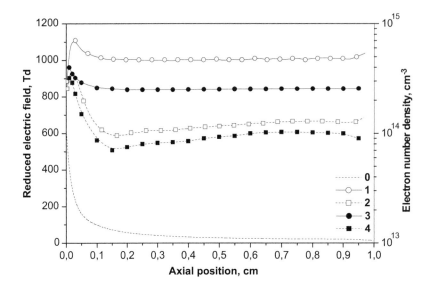

FIGURE 2.108 Dynamics of peak electric field strength in the head (**1,2**) and on-axis electron concentration in the streamer channel (**3,4**) at each section of the discharge gap. (**1,3**) – pure nitrogen, (**2,4**) – $N_2:O_2$ mixture. N_1 and N_2 modes. (**0**) – electric field distribution of the high-voltage electrode in the absence of streamer.

Further increase in "background" preionization leads to a decrease in the front propagation velocity and the magnitude of the peak electric field strength and an increase in the streamer head radius. Obviously, at a sufficiently high level of initial ionization in the gap, the streamer characteristics may differ significantly from those in the unperturbed gap. Based on the fact that the normal background concentration in pure nitrogen is $n_e \sim 10^3$ cm^{-3}, one can conclude that the "background" preionization under normal conditions can be neglected when considering the photoionization of the gas.

As stated in the introduction, a number of researchers, referring to the strong uncertainty in photoionization processes and the insignificant role of seed electrons in front of the high-field front, exclude from consideration the birth of photoelectrons in front of the streamer head. To demonstrate the consequences of this simplification, the streamer characteristics obtained with and without gas photoionization were compared.

To determine the effect of spatially uniform preionization, the streamer discharge in pure nitrogen was simulated for different levels of initial photoionization:

- $n_e |_{t=0} = n_i |_{t=0} = n_i |_{t=0} = 10^3$ cm^{-3} (N_3 mode),
- $n_e |_{t=0} = n_i |_{t=0} = n_i |_{t=0} = 10^7$ cm^{-3} (N_4 mode).

The first of these figures corresponds to the normal level of "background" electron concentration, while the second significantly exceeds it and can be used as an estimate from above. In this series of calculations, the photoelectron birth process was neglected and the electrons in front of the streamer head were set as initial homogeneous predionization. As before, to simplify the simulation of the starting moment, an initial region of inhomogeneous gas ionization (2.47) was set in front of the anode tip. The formation and propagation of streamer-like formation was observed for almost any level of "background" electron concentration. The variation of initial conditions demonstrated the possibility of changing all the streamer characteristics by varying the level of initial predionization, from which it should be concluded that by selecting the initial level of "background" electrons, the calculation results can always be reconciled with the experimentally measured characteristics.

Thus, the propagation velocity for the first case was $V_3 = 2.9 \times 10^7$ cm/s, and for the second – $V_4 = 3.7 \times 10^7$ cm/s (Figure 2.109). It follows from the figure that the streamer channel radius

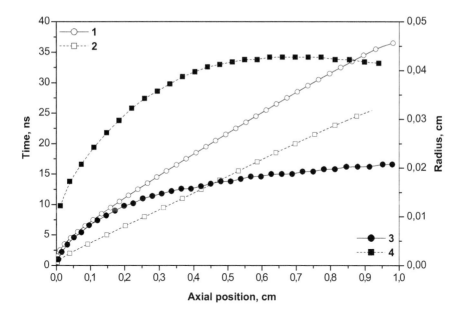

FIGURE 2.109 $x-t$ plots of the propagation of the maximum electric fields in the streamer head (**1,2**) and the streamer channel radius at half length (**3,4**) for different initial preionization levels n_0. (**1,3**) – $n_0 = 10^3$ cm^{-3}, (**2,4**) – $n_0 = 10^7$ cm^{-3}. Pure nitrogen. Regimes N$_3$ and N$_4$.

for background preionization $n_0 = 10^3$ cm^{-3} continues to increase as the maximum electric field advances up to the low-voltage electrode, while at $n_0 = 10^7$ cm^{-3} the channel radius reaches its maximum value at a length of $\simeq 0.6$ cm and then changes relatively little. In Figure 2.110 for both cases, the distribution of electric field strength and electron concentration at the moment the streamer reaches the middle of the gap is given. As can be seen, the streamer shape is also markedly different.

The dynamics of the peak electric fields are indicated in Figure 2.111. Almost throughout the entire interval, the peak values of the reduced electric field strength remain constant and are $(E/n)_3 \simeq 3 \times 10^3$ Td and $(E/n)_4 \simeq 1.1 \times 10^3$ Td for both modes, respectively. The electron concentration in the streamer channel increases proportionally to the peak fields (Figure 2.111). The streamer conduction currents were $J_3 \simeq 190$ and $J_4 \simeq 300$ mA for initial preionization of $n_0 = 10^3$ and $n_0 = 10^7$ cm^{-3}, respectively. Thus, by varying the initial predionization n_0, a reasonable match with almost any predetermined value can be achieved.

Comparing the obtained values with the results of the calculation that takes into account the birth of electrons under the action of VUV radiation from the region of intense ionization (Figure 2.108), it can be noted that the same properties of the streamer plasma without taking into account photoionization can only be obtained with an inexplicably high level of "background" preionization at $n_0 \sim 10^7$ cm^{-3}. However, there is a difference in both the propagation velocity (10%) of the streamer head and the transverse size of the streamer channel ($\simeq 30\%$).

2.11.3 Analytical Model of Cathode-Directed Streamer

As can be seen, the distribution of electrons ahead of the electric field front has a noticeable effect on the characteristics of the cathode-directed streamer. The theoretical consideration of this effect is complicated by the need to analyze the integro-differential equations (2.42–2.46). However, the role of the electron distribution ahead of the electric field front can be demonstrated at a qualitative level.

Consider the streamer as a perfectly conducting hemispherical head and a cylindrical channel. Assume the immobility of the heavy particles, i.e. the dominant influence of the electrons on the

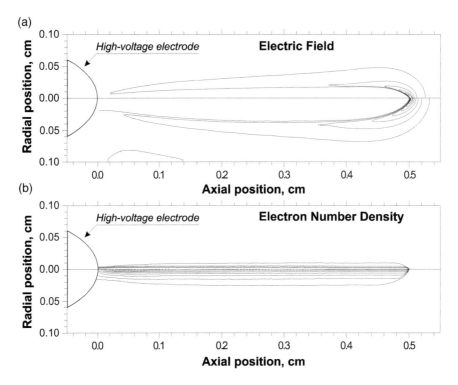

FIGURE 2.110 Comparison of streamer development in pure nitrogen excluding photoionization for initial homogeneous preionization $n_0 = 10^3$ cm^{-3} (upper halves of graphs) and for $n_0 = 10^7$ cm^{-3} (lower). Isolines of electric field strength (a) and electron density (b). The pitch of the isolines in each of the figures is 10% of the maximum value. The length of the gap is 1 cm, and the voltage at the anode is 13 kV. Pure nitrogen. Modes N_3 and N_4.

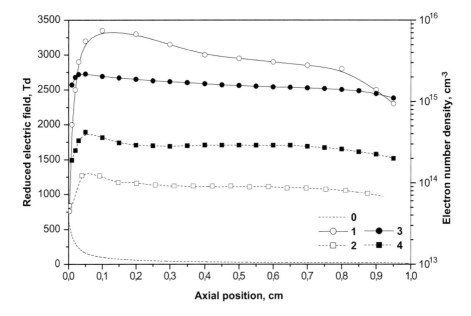

FIGURE 2.111 Dynamics of the peak electric field strength in the head (**1,2**) and the on-axis electron concentration in the streamer channel (**3,4**) at each section of the discharge gap. (**1,3**) – $n_0 = 10^3$ cm^{-3}, (**2,4**) – $n_0 = 10^7$ cm^{-3}. Pure nitrogen. Modes N_3 and N_4. (**0**) – electric field distribution of the high-voltage electrode in the absence of streamer.

streamer dynamics. Such a model is the most popular in the construction of analytical models due to the simplicity of the description of the electric field distribution and electron motion in the streamer head [306,307].

As follows from experimental studies and numerical simulations, at constant voltage at the cathode and not very large gas gaps (units of centimeters at atmospheric pressure) the streamer propagation is practically steady-state, except for the near-electrode regions. Thus, the departure of electrons into the streamer channel must be compensated by the flow of intensively multiplying electrons from the preionization region in front of the streamer head (the gas ionization in the streamer channel is neglected in this consideration).

Let us consider a single avalanche arising at some distance from the electric field front. Such an avalanche will reach the surface of the streamer head, producing on its way not only acts of ionization but also excitation of electronic levels generating the ionizing radiation. On the other hand, it is such photoionizing radiation from all avalanches (originating from different states) that is the source of photoelectrons in front of the streamer head. At the quasi-stationary development of the streamer, a balance between the drift motion of electrons into the head and their multiplication in front of it should be observed. If the described balance of electron fluxes is observed, the radius of the streamer head will be an eigenvalue of this balance.

A full two-dimensional analysis of the streamer structure is, for obvious reasons, extremely complicated and is possible only in numerical simulations. Therefore, to obtain qualitative results, the motion of charged particles only on the streamer axis will be considered. The origin of the coordinate system is associated with the center of the hemispherical streamer head. Let us consider the dynamics of the avalanche development from a photoelectron originating at the point x_0 on the symmetry axis. The rate of charge buildup in such an avalanche as it moves toward the streamer head is described by the equation:

$$\frac{dN_e}{dx} = -\alpha(x) \cdot \frac{v_e(x)}{v_{st} + v_e(x)} \cdot N_e$$

Here the following notations are adopted: $\alpha(x)$ – the first Townsend coefficient, $v_e(x)$ – the electron drift velocity, and v_{st} – the streamer propagation velocity. The fraction in the right-hand side is a correction to account for the mutual motion of the electrons and the streamer itself. The solution of this equation is a function of the form:

$$K(x_0, R) = \frac{N_R}{N_0} = \exp\left(\int_R^{x_0} \frac{\alpha(x) v_e(x)}{v_{st} + v_e(x)} dx\right) \qquad (2.48)$$

The resulting function reflects the propagation coefficient of an avalanche originating at point x_0 when it reaches the streamer head at distance R. In the stationary case, at the linear birth rate of such photoelectrons

$$\frac{\partial}{\partial t} \frac{\partial N0}{\partial x}\bigg|_{x_0},$$

the total secondary electron birth rate over the entire space in front of the streamer head will be:

$$\frac{\partial N_\Sigma}{\partial t} = \int_R^\infty \frac{\partial}{\partial t} \frac{\partial N_0}{\partial x}\bigg|_{x_0} \cdot K(x_0, R) dx_0 \qquad (2.49)$$

On the other hand, it is these secondary electrons that will give rise to photoionizing radiation and, consequently, new photoelectrons. Taking into account that the electric field increases rapidly when approaching the streamer head, and the rates of ionization and excitation of emitting levels depend on the electric field in an exponential manner, one can assume that photoionizing radiation is mainly generated in the streamer head, i.e., on a surface of radius R. Then the linear photoelectron birth rate can be written by representing the equation (2.46) in the form:

$$\frac{\partial}{\partial t} \cdot \frac{\partial N0}{\partial x}\bigg|_{x_0} = \frac{\partial N_\Sigma}{\partial t} \cdot G(x_0, R, p) \cdot 2\pi x_0^2 \qquad (2.50)$$

Here $G(x_0, R, p)$ is the absorption function of photoionizing radiation with distance in a given gas at a fixed pressure p. The computation of this function is done taking into account the hemispherical geometry of the emitting and absorbing layers and (2.46) can be used for the evaluation:

$$G(x_0, R, p) = \frac{1}{4\pi} \cdot \frac{p_q}{p + p_q} \cdot \frac{\Psi((x_0 - R) \cdot p)}{x_0^2} \qquad (2.51)$$

Combining the equations (2.49 and 2.50), we get the equality [304,305]:

$$2\pi \int_R^\infty G(x_0, R, p) \cdot K(x_0, R) \cdot x_0^2 \, dx_0 = 1 \qquad (2.52)$$

The resulting equality is an integral equation whose parameter is the desired radius R. As can be seen, the photoelectron distribution described by the function $G(x, R, p)$ plays no less role in the formation of the streamer head than the subsequent enhancement of avalanches under the action of collisional ionization in the electric field of the streamer head. Thus, the electron balance condition in the region in front of the positive streamer head can be written as:

$$\frac{1}{2} \cdot \frac{p_q}{p + p_q} \int_R^\infty \Psi((x_0 - R) \cdot p) \cdot \exp\left(\int_R^{x_0} \frac{\alpha(x) v_e(x)}{v_{st} + v_e(x)} dx\right) dx_0 = 1. \qquad (2.53)$$

To obtain the results, in addition to the velocity dependences of the elementary processes v_e, α, Ψ, it is necessary to determine the electric field distribution in the streamer head $E(x)$ and the velocity of the streamer itself v_{st}. Due to the proposed quasi-one-dimensional approach, this model cannot be closed self-consistently in principle, and some of the parameters must be introduced from outside. As noted in the introduction, the role of the conductive channel of the streamer is reduced to maintaining the head potential as close as possible to the anode potential. Then it is the value of the streamer head electric potential U that is one of the most logical parameters of the problem. Neglecting the small space charge in front of the streamer head, the Poisson equation (2.45) in this formulation leads to a space-quadratic decay of the electric field in front of the streamer head:

$$E(x) = \begin{cases} 0, & x < R \\ E_{max} \cdot (R/x)^2, & x \geq R \end{cases} \qquad (2.54)$$

However, the peak field $E_{max} = E(R)$ is different from that of the solitary sphere. As was shown in Ref. [306], accounting for the influence of charges in the cylindrical streamer channel leads to a reduction of the electric field by a factor of about two, i.e.,

$$E_{max} \simeq \frac{U}{2R}$$

Low Temperature Plasma Generation and Recombination

The advancement of the streamer head along the interelectrode gap is a consequence of the accumulation of space charge ahead of the electric field front, and, thus, the streamer velocity is determined by the dynamics of the generation of charged particles (both electrons and positively charged ions) in front of the streamer head. According to the results of numerical simulations of the streamer dynamics at small overvoltages, the streamer velocity v_{st} is smaller than the electron drift velocity at the electric field maximum $v_e|_{max} = v_e(R)$ by a factor of several. At high voltages, it appears that cases where $v_{st} > v_e(R)$ are possible. Therefore, keeping in mind that this approach is purely qualitative, one can assume that the streamer propagation velocity is equal to the electron drift velocity at the maximum of the electric field $v_{st} = v_e(R)$.

In view of the above, it seems possible to solve the equation (2.53) for different values of the head potential for the selected gas. The left part of the equation (2.53) (let us call it, for definiteness, $F = F(U,R,p)$) is, in fact, the ratio between the number of primary photoelectrons and secondary photoelectrons, i.e., generated by the primary ones when reaching the streamer head through photoionization of the gas. Figures 2.112 and 2.113 show the values of the left-hand side of this equation at different values of radius R for streamer head potentials $U = 10, 25, 50$ kV in pure nitrogen and air.

It can be seen that this dependence has a maximum at certain streamer head sizes away from this region, the efficiency of photoelectron generation drops due to the reduced gain of individual avalanches. To maintain the balance, this value must be equal to 1. The graph shows that at voltages below the critical $U_{min} \simeq 11.5$ kV for nitrogen and $U_{min} \simeq 4$ kV for air, the relation (2.53) is not realized at any values of the streamer head radius. This result is well known; at voltages below the critical voltage, streamer formation will not occur, but a slow ionization wave will propagate, controlled by the slow diffusion of electrons in the discharge gap. Thus, based on the generalization of extensive experimental material, the minimum value of the streamer head potential of 5–8 kV for discharge in air at atmospheric pressure is given in Ref. [308].

At voltages above critical voltages, there are two solutions of the equation (Figures 2.112 and 2.113), but one of them-smaller in value-is unstable and characterizes the minimum radius of streamer head curvature on the axis. The instability of such a solution follows from the fact that the angle of inclination of this dependence with respect to the axis is positive: at an arbitrary increase in the streamer radius, the number of photoelectrons produced increases, which, in turn, causes a further increase in the radius, and vice versa. The second solution is stable in this sense and represents the desired radius of the streamer head R.

In Figures 2.114 and 2.115, the values of radius and peak electric field strength for different streamer head potentials (up to 50 kV) in pure nitrogen and air are given. As the head potential increases, the radius increases almost linearly, while the electric field strength at its maximum decreases relatively rapidly to $E/n \simeq 600$ Td at $U \simeq 25$ kV, and then changes relatively little. For comparison, the figures show the results of numerical simulation in 2D geometry: radius R and peak applied electric field strength E/n for a number of voltages on the gap and values of the interelectrode gap $L = 10, 20,$ and 30 mm, respectively.

As can be seen from the above graph, the numerical simulation results are reproduced by the proposed model with good accuracy. Some differences, especially at high potentials, may be due to the following reasons: first, the shape of the streamer head can be represented by a hemisphere only with a certain accuracy; in the numerical simulation, especially at high overvoltages, the streamer head is elongated and resembles an ellipsoid of rotation (Figures 2.106 and 2.110), simplifying the calculation of gas photoionization in the form [51,106] may also make some difference, especially at short distances. The assumption that the streamer velocity is equal to the electron drift velocity at the maximum of the electric field does not make a significant difference when examined in more detail. As follows from the analysis of equation (2.53), the streamer velocity has a logarithmic weak effect on the streamer head radius, and if $v_{st} = 0$, this assumption only changes the streamer head radius by a factor of 2–3 depending on the head potential.

Thus, the proposed model correctly describes the streamer head radius and the value of the peak electric field strength depending on the collisional ionization and photoionization rates of a given gas

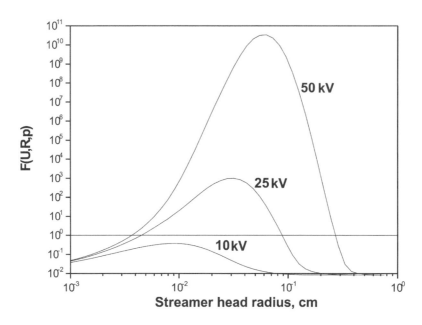

FIGURE 2.112 Dependence of the function $F(U,R,p)$ (53) on the streamer head radius in pure nitrogen at atmospheric pressure for voltages $U = 10, 25, 50$ kV.

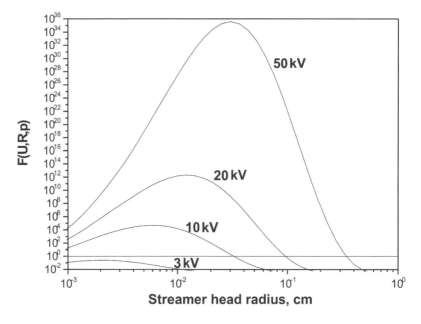

FIGURE 2.113 Dependence of the function $F(U,R,p)$ (53) on the radius of the streamer head in air at atmospheric pressure for voltages $U = 3, 10, 20, 50$ kV.

and is determined by a single parameter-the streamer head potential-which at not very large lengths of the interelectrode gap can be interpreted as the voltage applied to this gap. Figure 2.116 and 2.117 show the dependences of peak electric field and streamer head radius for values of streamer head potential 10, 25, 50 kV in the pressure range of $50 - 2 \times 10^3$ Torr in air. As can be seen from the graphs, these dependencies are linear dependencies in logarithmic coordinates over the entire

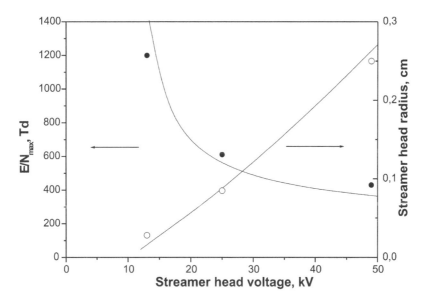

FIGURE 2.114 Dependence of radius (**1,2**) and peak electric field (**3,4**) on head potential. (**1,3**) are analytical model results, (**2,4**) are direct numerical simulations. Nitrogen.

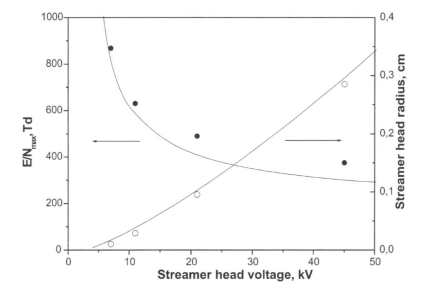

FIGURE 2.115 Dependence of radius (**1,2**) and peak electric field (**3,4**) on head potential. (**1,3**) are analytical model results, (**2,4**) are direct numerical simulation results. Air.

pressure range with good accuracy. At a fixed potential, the functional dependence of the streamer head radius on pressure can be represented as an inversely proportional dependence

$$R_{head} \propto 1/p, \qquad (2.55)$$

while the reduced electric field does not change faster when the pressure is varied than the

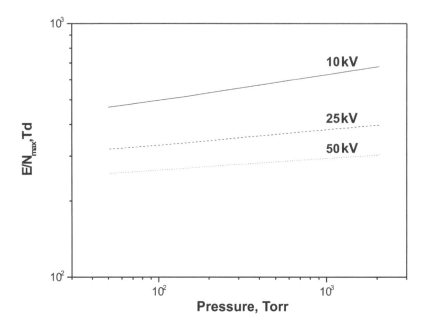

FIGURE 2.116 The peak electric field values as a function of air pressure for different values of streamer head potential.

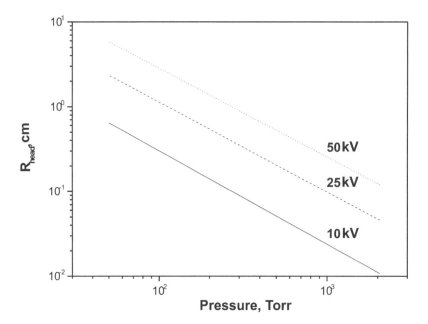

FIGURE 2.117 The values of streamer head radius as a function of air pressure for different values of streamer head potential.

$$E/n_{max} \propto p^{0.10 \pm 0.04}, \tag{2.56}$$

with the maximum sensitivity from pressure corresponding to small values of potential.

Recall that the obtained streamer parameters correspond to the quasi-stationary regime for the case when the electric field distribution can be represented as (2.54); the differences in the near-electrode regions can be significant. The streamer channel radius can be associated with the obtained streamer head radius at not too large intervals, but at large times, the deviation of these values from each other can be significant.

Thus, at large times, the channel radius is determined by ionization expansion and slower diffusion and gasdynamic expansion. Nevertheless, even at large values of the discharge gap, such streamer parameters as the electric field strength, the head radius, and the associated electron concentration in the streamer head are determined by processes only in the streamer head and, for a given gas, are unambiguously related to the head potential. As for the streamer propagation velocity and the channel conduction current, it was shown in Ref. [309] on the basis of direct numerical simulations in 2D geometry that they depend essentially on the gap geometry and are determined mainly by processes in the streamer channel. It should be added that the streamer parameters can undergo significant changes when additional photoelectrons are born at larger distances from the head.

2.12 UNCERTAINTY IN THE RATES OF ELEMENTARY PROCESSES

The influence of ionization by electron-molecular collisions and the process of photoionization of gas on the formation of the streamer head is of the same order (2.52), and therefore, it makes additional sense to consider the fairness of the used dependences of the elementary process rates.

The electron drift velocities v_e and the Townsend ionization coefficient α have been investigated many times. Experimentally measured dependences for pure gases (e.g. – [310,311]) have been complemented by theoretical works. The most widely developed approach for calculating the rates of elementary processes is the so-called two-term approximation of the Boltzmann equation [312,313].

As a rule, the dependences v_e and α are given for the local stationary case (when the EEDF depends on the local value of reduced electric field E/n), which obliges the study of pulsed discharges to take into account the nonlocal and nonstationary character of the formation of the electron energy distribution. In the literature, there are a number of works that take into account the effects formulated above by introducing "reverse" diffusion of electrons from the region of high electric fields [314,315]. The approach based on the solution of the Boltzmann equation in the multimomentum approximation is more commonly used [45,292,316,317]. However, the most closed approach, in our opinion, is the method of introducing corrections using the solution of the Boltzmann equation in the two-term approximation using perturbation theory [293,318,319].

The conclusions following from the latter two methods are in complete contradiction with each other. Thus, when modeling the streamer discharge, the works of Refs. [292,317] show that taking into account higher moments of the Boltzmann equation leads to significant changes in the characteristics of the streamer, while the results of Refs. [293,319] indicate a difference of up to 10% in the most critical parameter-the peak electric field strength. Nevertheless, this issue is widely discussed in the literature and is still under investigation.

In a more uncertain situation is the question concerning the rate of photoionization of gases. Due to the fact that the role of this process has been recently underestimated, there are only a few works devoted to this problem. Theoretical consideration of this process is extremely difficult [158,320], and experimental studies [295,321–323] are not very reliable, in particular, due to the use, as a rule, of gases of technical purity and require additional experimental studies and more detailed analysis. A complicating factor in modeling the streamer discharge is the need to know the dependence of the absorption function $\Psi(|r - r_1| \cdot p)$ at small values of the parameter (in particular, in this work, the values $5 - 20$ cm·Torr play the greatest role), where there is a strong dependence of the used approximations on the photoionization rate (Figure 2.98), i.e., in the region of maximum measurement error.

REFERENCES

1. Starikovskiy, A., High E/N discharges: Applications. In: *Encyclopedia of Plasma Technology*, edited by J. Leon Shohet. 2016: CRC Press, 33 pages. ISBN: 978-1-4665-0059-4.
2. Starikovskiy, A. and N. Aleksandrov, Plasma-assisted ignition and combustion. In: *Aeronautics and Astronautics*, edited by M. Mulder. 2011: Intechopen. ISBN 978-953-307-473-3.
3. Anikin, N.B., S.A. Bozhenkov, D.V. Zatsepin, E.I. Mintoussov, S.V. Pancheshnyi, S.M. Starikovskaia, and A. Yu.Starikovskiy, Pulsed nanosecond discharges and their applications. In: *Encyclopedia on Low-Temperature Plasma*, edited by V.E. Fortov. 2007: Chemistry of Low-Temperature Plasma, 87 pages.
4. Starikovskaia, S.M., Plasma-assisted ignition and combustion: Nanosecond discharges and development of kinetic mechanisms. *Journal of Physics D: Applied Physics*, 2014. **47**: p. 353001.
5. Starikovskiy, A.Y. and N.L. Aleksandrov, Plasma-assisted ignition and combustion. *Progress in Energy and Combustion Science*, 2013. **39**: pp. 61–110.
6. Starikovskiy, A.Y., Plasma supported combustion. *Proceedings of the Combustion Institute*, 2005. **30**: pp. 2405–2417.
7. Adamovich, I.V. and W.R. Lempert, Challenges in understanding and development of predictive models of plasma assisted combustion. *Plasma Physics and Controlled Fusion*, 2015. **57**: p. 014001.
8. Ju, Y. and W. Sun, Plasma assisted combustion: Dynamics and chemistry. *Progress in Energy and Combustion Science*, 2015. **48**: pp. 21–83.
9. Popov, N.A., Effect of a pulsed high-current discharge on hydrogen-air mixtures. *Plasma Physics Reports*, 2008. **34**: pp. 376–391.
10. Starikovskaia, S.M., and A.Y. Starikovskiy, Plasma assisted ignition and combustion. In: *Handbook of Combustion*, edited by M. Lackner, F. Winter, A.K. Agarwal. 2010: Wiley-VCH. Part 5: New Technologies. Paper 4.
11. Starikovskiy, A., Physics and chemistry of plasma-assisted combustion. *Philosophical Transactions of the Royal Society A*, 2015. **373**: 20150074. DOI: 10.1098/rsta.2015.0074.
12. Aleksandrov, N.L., A.A. Ponomarev, and A.Y. Starikovskiy. Monte Carlo simulation of the effect of "hot" atoms on active species kinetics in combustible mixtures excited by high-voltage pulsed discharges. *Combustion and Flame*, 2017. **176**: pp. 181–190.
13. Starikovskiy, A., On the role of "hot" atoms in plasma-assisted ignition. *Philosophical Transactions of the Royal Society A*, 2015. 373: p. 20140343.
14. Starikovskiy, A., N. Aleksandrov, and A. Rakitin. Plasma-assisted ignition and deflagration-to-detonation transition. *Philosophical Transactions of the Royal Society A*, 2012. **370**: pp. 740–773. DOI: 10.1098/rsta.2011.0344.
15. Starikovskaya, S.M., N.L. Aleksandrova, I.N. Kosarev, S.V. Kindysheva, and A.Y. Starikovskiy, Ignition with low-temperature plasma: Kinetic mechanism and experimental verification. *High Energy Chemistry*, 2009. **43**(3): pp. 213–218.
16. Aleksandrov, N.L., S.V. Kindysheva, I.N. Kosarev, S.M. Starikovskaia, and A.Y. Starikovskiy, Mechanism of ignition by non-equilibrium plasma. *Proceedings of the Combustion Institute*, 2009. **32**: pp. 205–212.
17. Kosarev, I.N., N.L. Aleksandrov, S.V. Kindysheva, S.M. Starikovskaia, and A.Y. Starikovskiy, Kinetic mechanism of plasma-assisted ignition of hydrocarbons. *Journal of Physics D: Applied Physics*, 2008. **41**(3):p. 032002.
18. Starikovskiy, A.Y., N.B. Anikin, I.N. Kosarev, E.I. Mintoussov, S.M. Starikovskaia, and V.P. Zhukov, Plasma-assisted combustion. *Pure and Applied Chemistry*, 2006. **78**(6): pp. 1265–1298.
19. Starikovskaia, S.M. and A.Y. Starikovskiy, Pulsed discharges in a wide density range: Plasma development and media excitation. In: *Runaway Electrons Preionized Diffuse Discharges*, edited by V.F. Tarasenko. 2014: NOVA Publishers. ISBN: 978-1-63321-883-3.
20. Raizer, Y.P., *Gas Discharge Physics*. 1991: Springer, pp. 373–420.
21. Dutton, J.J., A survey of electron swarm data. *Journal of Physical and Chemical Reference Data*, 1975. **4**: pp. 577–856.
22. Huxley, L.G.H. and R.W. Crompton, *The Diffusion and Drift of Electrons in Gases*. 1974: Wiley-Interscience.
23. Wu, L., J. Lane, N. Cernansky, D. Miller, A. Fridman, and A. Starikovskiy, Time resolved PLIF imaging of OH radicals in the afterglow of nanosecond pulsed discharge in combustible mixtures. *IEEE Transactions on Plasma Science*, 2011. **39**(11): pp. 2604–2605.

24. Wu, L., J. Lane, N.P. Cernansky, D.L. Miller, A.A. Fridman, and A.Y. Starikovskiy, Plasma assisted ignition below self-ignition threshold in methane, ethane, propane and butane-air mixtures. *Proceedings of the Combustion Institute*, 2010. **33**: pp. 3219–3224.
25. Cheng, R.K. and A.K. Oppenheim, Autoignition in methane-hydrogen mixtures. *Combustion and Flame*, 1984. **58**: p. 125.
26. Frenklach, M. and D.E. Bornside, Systematic optimization of a detailed kinetic model using a methane ignition example. *Combustion and Flame*, 1984. **56**: p. 1.
27. Burcat, A., R.W. Crossley, and K. Scheller, Shock tube investigation of ignition inethane-oxygen-argon mixtures. *Combustion and Flame*, 1972. **18**: p. 115.
28. Smith, G.P., D.M. Golden, M. Frenklach, et al., http://www.me.berkeley.edu/gri_mech/.
29. Starikovskaia, S.M., A.Y. Starikovskiy, and D.V. Zatsepin, *27th (International) Symposium on Combustion*, Boulder. 1998, p. 4.
30. Morris, R.A., A.A. Viggiano, S.T. Arnold, L.Q. Maurice, and E.A. Sutton, Effects of ionization on hydrocarbon-air combustion chemistry. *27th (International) Symposium on Combustion*, Boulder. 1998, p. 343.
31. Morsy, M.H., Y.S. Ko, S.H. Chuhg, and P. Cho, Laser-induced two-point ignition of premixture with a single-shot laser. *Combustion and Flame*, 2001. **125**: p. 724.
32. Phuoc, T.X., and F.P. White, Laser-induced spark ignition of CH_4/air mixtures. *Combustion and Flame*, 1999. **119**: p. 203.
33. Phuoc, T.X., Laser spark ignition: experimental determination of laser-induced breakdown thresholds of combustion gases. *Optics Communications*, 2000. **175**: p. 419.
34. Ma, J.X., D.R. Alexander, and D.E. Poulain, Laser spark ignition and combustion characteristics of methaneair mixtures. *Combustion and Flame*, 1998. **112**: p. 492.
35. Thiele, M., J. Warnatz, and U. Maas, Geometrical study of spark ignition in two dimensions. *Combustion Theory and Modelling*, 2000. **4**: p. 413.
36. Thiele, M., J. Warnatz, A. Dreizler, S. Lindenmaier, R. Schiessl, U. Maas, A. Grant, and P. Ewart, Spark ignited hydrogen/air mixtures, two dimensional detailed modelling and laser based diagnostics. *Combustion and Flame*, 2002. **128**: p. 74.
37. Takita, K., T. Uemoto, T. Sato, Y. Ju., G. Masuya, and K. Ohwaki, Ignition characteristics of plasma torch for hydrogen jet in an airstream. *Journal of Propulsion and Power*, 2000. **16**: p. 227.
38. Starikovskaia, S.M., N.B. Anikin, S.V. Pancheshnyi, D.V. Zatsepin, and A.Y. Starikovskiy, Pulsed breakdown at high overvoltage: development, propagation and energy branching. *Plasma Sources Science and Technology*, 2001. **10**: p. 344.
39. Raizer, Y.P., *Gas Discharge Physics*. 1991: Springer.
40. Gleizes, A., J.J. Gonzalez, and P. Freton, Thermal plasma modelling. *Journal of Physics D: Applied Physics*, 2005. **38**: pp. R153–R183.
41. Khmara, D. and Y. Kolesnichenko, Modeling of microwave filament origination. *44th AIAA Aerospace Sciences Meeting and Exhibit*, Reno, Nevada, 9–12 January 2006.
42. Gallimberti, I., A computer model for streamer propagation. *Journal of Physics D: Applied Physics*, 1972. **5**: 2179–2189.
43. Hartmannt, G. and I. Gallimberti, The influence of metastable molecules on the streamer progression. *Journal of Physics D: Applied Physics*, 1975. **8**: pp. 670–680.
44. Dhali, S.K. and P.F. Williams, Numerical simulations of streamer propagation in nitrogen at atmospheric pressure. *Physical Review A*, 1985. **31**: pp. 1219–1221.
45. Kunhardt, E.E. and Y. Tseng, Development of an electron avalanche and its transition into streamers. *Physical Review A*, 1988. **38**: pp. 1410–1421.
46. Gaivoronskii, A.S. and I.M. Razhanskii, Numerical model of the development of cathode–directed streamer in air gaps with non–uniform field. *Zhournal Tekchnicheskoi Fiziki*, 1986. **56**: pp. 1110–1116.
47. Morrow, R. and J.J. Lowke, Streamer propagation in air. *Journal of Physics D: Applied Physics*, 1997. **30**: pp. 614–627.
48. Vitello, P.A., B.M. Penetrante, and J.N. Bardsley, Simulation of negative–streamer dynamics in nitrogen. *Physical Review E*, 1994. **49**: pp. 5574–5598.
49. Babaeva, N.Y. and G.V. Naidis, Simulation of positive streamers in air in weak uniform electric fields. *Physics Letters A*, 1996. **215**: 187–190.
50. Babaeva, N.Y. and G.V. Naidis, Simulation of stepped propagation of positive streamers in SF_6. *Journal of Physics D: Applied Physics*, 2002. **35**: pp. 132–136.

51. Pancheshnyi, S.V. and A.Y. Starikovskiy, Two-dimensional numerical modelling of the cathode–directed streamer development in a long gap at high voltage. *Journal of Physics D: Applied Physics*, 2003. **36**: pp. 2683–2691.
52. Tardiveau, P., E. Marode, and A. Agneray, Tracking an individual streamer branch among others in a pulsed induced discharge. *Journal of Physics D: Applied Physics*, 2002. **35**: pp. 2823–2829.
53. van Veldhuizen, E.M. and W.R. Rutgers, Pulsed positive corona streamer propagation and branching. *Journal of Physics D: Applied Physics*, 2002. **35**: pp. 2169–2179.
54. Arrayas, M., U. Ebert, and W. Hundsdorfer, Spontaneous branching of anode–directed streamers between planar electrodes. *Physical Review Letters*, 2002. **88**(17): pp. 1–4.
55. Kulikovsky, A.A., The role of photoionization in positive streamer dynamics. *Journal of Physics D: Applied Physics*, 2000. **33**: pp. 1514–1524.
56. Tardiveau, P., E. Marode, A. Agneray, and M. Cheaib, Pressure effects on the development of an electric discharge in non–uniform fields. *Journal of Physics D: Applied Physics*, 2001. **34**: pp. 1690–1696.
57. Pancheshnyi, S., M. Nudnova, and A. Starikovskiy, Development of a cathode–directed streamer discharge in air at different pressures: Experiment and comparison with direct numerical simulation. *Physical Review E*, 2005, **71**: p. 016407 (1–12).
58. Nudnova, M.M., S.V. Pancheshnyi, and A.Y. Starikovskiy, Nonequilibrium plasma formation by high–voltage pulsed nanosecond gas discharge at different pressures. *42nd AIAA Aerospace Sciences Meeting and Exhibit*, Reno, Nevada, USA, 5–8 January 2004.
59. Allen, N.L. and A. Ghaffar, The variation with temperature of positive streamer properties in air. *Journal of Physics D: Applied Physics*, 1995. **28**: pp. 338–343.
60. Hidaka, K., T. Kouno, and I. Hayashi, Simultaneous measurement of two orthogonal components of electric field using a Pockels device. *Review of Scientific Instruments*, 1989. **60**(7): p. 1252.
61. Paris, P., M. Aints, M. Laan, and F. Valk, Measurement of intensity ratio of nitrogen bands as a function of field strength. *Journal of Physics D: Applied Physics*, 2004. **37**: pp. 1179–1184.
62. Paris, P., M. Aints, F. Valk, T. Plank, A. Haljaste, K.V. Kozlov, and H.E. Wagner, Intensity ratio of spectral bands of nitrogen as a measure of electric field strength in plasmas. *Journal of Physics D: Applied Physics*, 2005. **38**: pp. 3894–3899.
63. Pancheshnyi, S.V., S.V. Sobakin, S.M. Starikovskaya, and A.Y. Starikovskiy, Discharge dynamics and the production of active particles in a cathode–directed streamer. *Plasma Physics Reports*, 2000. **26**(12): pp. 1054–1065.
64. Ono, R. and T. Oda, Dynamics and density estimation of hydroxyl radicals in a pulsed corona discharge. *Journal of Physics D: Applied Physics*, 2002. **35**: pp. 2133–2138.
65. Ono, R. and T. Oda, NO formation in a pulsed spark discharge in $N_2/O_2/Ar$ mixture at atmospheric pressure. *Journal of Physics D: Applied Physics*, 2002. **35**: pp. 543–548.
66. Filippov, Y.V., Electrosynthesis of ozone. *Vestnik MGY, seria Khimia*, 1959. **4**: p. 153 (In Russian).
67. Bersis, D. and D. Katakis, Surface effects in the production of ozone in the silent discharge. *The Journal of Chemical Physics*, 1964. **40**(7): p. 1997.
68. Carlins, J.J. and R.G. Clark, Ozone generation by corona discharge. In: *Handbook of Ozone Technology and Applications*, edited by A. Netzer. 1982: Ann Arbor Science, p. 41.
69. Eliasson, B., M. Hirth, and U. Kogelschatz, Ozone synthesis from oxygen in dielectric barrier discharges. *Journal of Physics D: Applied Physics*, 1987. **20**: pp. 1421–1437.
70. Samoilovich, V.G., V.I. Gibalov, and K.V. Kozlov, *Phydical Chemistry of Barrier Discharge*. 1989: Moscow State University Publication (In Russian) 176 p.
71. Bagirov, M.A., N.E. Nuraliev, and M.A. Kurbanov, Investigation of discharge in air gap with dielectric and technique for determination of number of partial discharges. *Journal of Technical Physics*, 1972. **43**(3): p. 629 (In Russian).
72. Bagirov, M.A., K.S. Burziev, and M.A. Kurbanov, Investigation of energetic parameters of discharge in air between electrodes covered by dielectric at low pressures. *Journal of Technical Physics*, 1979. **49**(2): p. 339 (In Russian).
73. Tanaka, M., S. Yagi, and N. Tabata, Observations of silent discharge in air, oxygen and nitrogen by super–high sensitivity camera. *The Journal of the Institute of Electrical Engineers of Japan*, 1982. **102**A(10): p. 533.
74. Liu, S. and M. Neiger, Excitation of dielectric barrier discharges by unipolar submicrosecond square pulses. *Journal of Physics D: Applied Physics*, 2001. **34**: pp. 1632–1638.
75. Jidenko, N., M. Petit, and J.P. Borral, Electrical characterization of microdischarges produced by dielectric barrier discharge in dry air at atmospheric pressure. *Journal of Physics D: Applied Physics*, 2006. **39**: pp. 281–293.

76. Kozlov, K.V., H.E. Wagner, R. Brandenburg, and P. Michel, Spatio–temporally resolved spectroscopic diagnostics of the barrier discharge in air at atmospheric pressure. *Journal of Physics D: Applied Physics*, 2001. **34**: pp. 3164–3176.
77. Kline, L.E., Calculations of discharge initiation in overvolted parallel-plane gaps. *Journal of Applied Physics*, 1974. **45**(5): p. 2046.
78. Kline, L.E., Effect of negative ions on current growth and ionizing wave propagation in air. *Journal of Applied Physics*, 1975. **46**(5): p. 1994.
79. Stefanovic, I., N.K. Bibinov, A.A. Deryugin, I.P. Vinogradov, A.P. Napartovich, and K. Wiesemann, Kinetics of ozone and nitric oxides in dielectric barrier discharges in O_2/NOx and $N_2/O_2/NOx$ mixtures *Plasma Sources Science and Technology*, 2001. **10**: pp. 406–416.
80. Lukas, C., M. Spaan, V. Schulz–von der Gathen, M. Thomson, R. Wegst, H.F. Doebele, and M. Neiger, Dielectric barrier discharges with steep voltage rise: mapping of atomic nitrogen in single filaments measured by laser–induced fluorescence spectroscopy. *Plasma Sources Science and Technology*, 2001. **10**: pp. 445–450.
81. Ono, R., Y. Yamashita, K. Takezawa, and T. Oda, Behavior of atomic oxygen in a pulsed dielectric barrier discharge measured by laser–induced fluorescence. *Journal of Physics D: Applied Physics*, 2005. **38**: pp. 2812–2816.
82. Bibinov, N.K., A.A. Fateev, and K. Wiesemann, Variations of the gas temperature in He/N_2 barrier discharges. *Plasma Sources Science and Technology*, 2001. **10**: pp. 579–588.
83. Williamson, J.M., P. Bletzinger, and B.N. Ganguly, Gas temperature determination in a N_2/Ar dielectric barrier discharge by diode–laser absorption spectroscopy and resolved plasma emission. *Journal of Physics D: Applied Physics*, 2004. **37**: pp. 1658–1663.
84. Baeva, M., A. Dogan, J. Ehlbeck, A. Pott, and J. Uhlenbusch, CARS diagnostic and modelling of a dielectric barrier discharge. *Plasma Chemistry and Plasma Processing*, 1999. **19**: pp. 445–466.
85. Hosselet, L.M.L.F., Increased efficiency of ozone production by electric discharges. *Electrochimica Acta*, 1973. **18**: p. 1033.
86. Georghiou, G.E., A.P. Papadakis, R. Morrow, and A.C. Metaxas, Numerical modelling of atmospheric pressure gas discharges leading to plasma production. *Journal of Physics D: Applied Physics*, 2005. **38**: pp. R303–R328.
87. Kanazava, S., M. Kogoma, T. Moriwaki, and S. Okazaki, Stable glow plasma at atmospheric pressure. *Journal of Physics D: Applied Physics*, 1998, **21**: pp. 838–840.
88. Yokoyama, T., M. Kogoma, T. Moriwaki, and S. Okazaki, The mechanism of the stabilization of glow plasma at atmospheric pressure. *Journal of Physics D: Applied Physics*, 1990. **23**: pp. 1125–1128.
89. Okazaki, S., M. Kogoma, M. Uehara, and Y. Kimura, Appearance of stable glow discharge in air, argon, oxygen and nitrogen at atmospheric pressure using a 50 Hz source. *Journal of Physics D: Applied Physics*, 1993. **26**: pp. 889–892.
90. Kekez, M.M., M.R. Barrault, and J.D. Craggs, Spark channel formation. *Journal of Physics D: Applied Physics*, 1970. **3**: pp. 1886–1896.
91. Gherardi, N., G. Gouda, E. Gat, A. Ricard, and F. Massines, Transition from glow silent discharge to micro–discharges in nitrogen gas. *Plasma Sources Science and Technology*, 2000. **9**: pp. 340–346.
92. Navratil, Z., R. Brandenburg, D. Trunec, A. Brablec, P. St'ahel, H.E. Wagner, and Z. Kopecky, Comparative study of diffuse barrier discharges in Neon and Helium. *Plasma Sources Science and Technology*, 2006. **15**: pp. 8–17.
93. Brandenburg, R., V.A. Maiorov, Y.B. Golubovskii, H.E. Wagner, J. Behnke, and J.F. Behnke, Diffuse barrier discharges in nitrogen with small admixtures of oxygen: discharge mechanism and transition to the filamentary regime. *Journal of Physics D: Applied Physics*, 2005. **38**: pp. 2187–2197.
94. Kossyi, I.A., A.Y. Kostinsky, A.A. Matveyev, and V.P. Silakov, Kinetic scheme of the non-equilibrium discharge in nitrogen-oxygen mixtures. *Plasma Sources Science and Technology*, 1992. **1**: pp. 207–220.
95. Trunec, D., Z. Navratil, P. Stahel, L. Zajickova, V. Bursikova, and J. Cech, Deposition of thin organosilicon polymer films in atmospheric pressure glow discharge. *Journal of Physics D: Applied Physics*, 2004. **37**: pp. 2112–2120.
96. Guimond, S., I. Radu, G. Czeremuszkin, D.J. Carlsson, and M.R. Wertheimer, Biaxially Oriented Polypropylene (BOPP) surface modification by Nitrogen Atmospheric Pressure Glow Discharge (APGD) and by air corona. *Plasmas and Polymers*, 2002. **7**(1): pp. 71–88.
97. Staack, D., B. Farouk, A. Gutsol, and A. Fridman, Characterization of a dc atmospheric pressure normal glow discharge. *Plasma Sources Science and Technology*, 2005. **14**: pp. 700–711.
98. Fan, H.Y. The Transition from Glow Discharge to Arc. *Physical Review*, 1939. **55**: p. 769.

99. Yu, L., C.O. Laux, D.M. Packan, and C.H. Kruger, Direct–current glow discharges in atmospheric pressure air plasmas. *Journal of Applied Physics*, 2002. **91**: p. 2678.
100. Yalin, A.P., C.O. Laux, C.H. Kruger, and R.N. Zare, Spatial profiles of N_2^+ concentration in an atmospheric pressure nitrogen glow discharge. *Plasma Sources Science and Technology*, 2002. **11**: pp. 248–253.
101. Laux, C.O., T.G. Spence, C.H. Kruger, and R.N. Zare, Optical diagnostics of atmospheric pressure air plasmas. *Plasma Sources Science and Technology*, 2003. **12**: pp. 125–138.
102. Starikovskaia, S.M., E.N. Kukaev, A.Y. Kuksin, M.M. Nudnova, and A.Y. Starikovskiy, Analysis of the spatial uniformity of the combustion of a gaseous mixture initiated by a nanosecond discharge. *Combustion and Flame*, 2004. **139**: pp. 177–187.
103. Korolev, Y.D. and G.A. Mesyats, *Physics of Pulse Breakdown of Gases*. 1991: URO-Press, 224 pp.
104. Allen, K.R. and K. Phillips, Ozone generation in dielectric barrier discharge. *Electrical Review*, 1963. **173**: p. 779.
105. Thomson, J.J., *Recent Researches in Electricity and Magnetism*. 1893: Clarendon, p. 115.
106. Babich, L.P., T.V. Loiko, and V.A. Tsykerman, High voltage nanosecond discharge in dense gases at high overvoltages developing in the regime of electrons running away. *Physics – Uspekhi*, 1990. **49**: p. 49.
107. Babich, L.P., *High–Energy Phenomena in Electric Discharges in Dense Gases*, ISTC Science and Technology Series, Vol. 2. 2003: Futurepast.
108. Vasilyak, L.M., S.V. Kostyuchenko, N.N. Kudryavtsev, and I.V. Filyugin, High–speed ionization wave s at an electric breakdown. *Physics – Uspekhi*, 1994. **163**: p. 263.
109. Anikin, N.B. and N. Marchenko, Private Communication.
110. Starikovskaia, S.M., A.Y. Starikovskiy, and D.V. Zatsepin, Development of a spatially uniform fast ionization wave in a large discharges. *Journal of Physics D: Applied Physics*, 1998. **31**: pp. 1118–1124.
111. Rep'ev, A.G. and P.B. Repin, Dynamics of the optical emission from a high–voltage diffuse discharge in a rod–plane electrode system in atmospheric–pressure air. *Plasma Physics Reports*, 2006. **32**: 72–78.
112. Krasnochub, A.V. and L.M. Vasilyak, Dependence of the energy deposition of a fast ionization wave on the impedance of a discharge gap. *Journal of Physics D: Applied Physics*, 2006. **34**: pp. 1678–1682.
113. Anikin, N.B., S.V. Pancheshnyi, S.M. Starikovskaia, and A.Y. Starikovskiy, Breakdown development at high overvoltage: Electric field, electronic levels excitation and electron density. *Journal of Physics D: Applied Physics*, 1998. **31**: pp. 826–833.
114. Pancheshnyi, S.V., S.M. Starikovskaia, and A.Y. Starikovskiy, Population of nitrogen molecule electron states and structure of the fast ionization wave. *Journal of Physics D: Applied Physics*, 1999. **32**: pp. 2219–2227.
115. Anikin, N.B., S.M. Starikovskaia, and A.Y. Starikovskiy, Polarity effect of applied pulse voltage on the development of uniform nanosecond gas breakdown. *Journal of Physics D: Applied Physics*, 2002. **35**: pp. 2785–2794.
116. Anikin, N.B., S.M. Starikovskaia, and A.Y. Starikovskiy, Uniform nanosecond gas breakdown of negative polarity: initiation from electrode and propagation in molecular gases. *Journal of Physics D: Applied Physics*, 2001. **34**: p. 177.
117. Pancheshnyi, S.V., D.A. Lacoste, A. Bourdon, and C.O. Laux, Ignition of propane-air mixtures by a repetitively pulsed nanosecond discharge. *IEEE Transactions on Plasma Science*, 2006. **34**: p. 2478.
118. Kruger, C.K., C.O. Laux, L. Yu, D. Packan, and L. Pierrot, Nonequilibrium discharges in air and nitrogen plasmas at atmospheric pressure. *Pure and Applied Chemistry*, 2002. **74**: p. 337.
119. Nudnova, M.M., S.V. Kindysheva, N.L. Aleksandrov, and A.Y. Starikovskiy, Fast gas heating in N_2/O_2 mixtures under nanosecond surface dielectric barrier discharge: The effects of gas pressure and composition. *Philosophical Transactions of the Royal Society A*, 2015. **373**: p. 20140330. DOI: 10.1098/rsta.2014.0330.
120. Aleksandrov, N.L., S.V. Kindysheva, M.M. Nudnova, and A.Y. Starikovskiy, Mechanism of ultra-fast heating in a nonequilibrium weakly-ionized air discharge plasma in high electric fields. *Journal of Physics D: Applied Physics*, 2010. **43**: p. 255201 (19pp).
121. Pai, D.Z., G.D. Stancu, D.A. Lacoste, and C.O. Laux, Nanosecond repetitively pulsed discharges in air at atmospheric pressure—the glow regime. *Plasma Sources Science and Technology*, 2009. **18**: p. 045030.
122. Raizer, Y.P., *Gas Discharge Physics*. 1992: Springer-Verlag.
123. Pai, D.Z., D.A. Lacoste, and C.O. Laux, Nanosecond repetitively pulsed discharges in air at atmospheric pressure—the spark regime. *Plasma Sources Science and Technology*, 2010. **19**: p. 065015.
124. Bazelyan, E.M. and Y.P. Raizer, Spark Discharge. 1998: CRC.

125. Pai, D.Z., D.A. Lacoste, and C.O. Laux, Transitions between corona, glow, and spark regimes of nanosecond repetitively pulsed discharges in air at atmospheric pressure. *Journal of Applied Physics*, 2010. **107**: p. 093303.
126. Rusterholtz, D.L., D.A. Lacoste, G.D. Stancu, D.Z. Pai, and C.O. Laux, Ultrafast heating and oxygen dissociation in atmospheric pressure air by nanosecond repetitively pulsed discharges. *Journal of Physics D: Applied Physics*, 2013. **46**: p. 464010.
127. Xu, D.A., M.N. Shneider, D.A. Lacoste, and C.O. Laux, Thermal and hydrodynamic effects of nanosecond discharges in atmospheric pressure air. *Journal of Physics D: Applied Physics*, 2014. **47**: p. 235202.
128. Naidis, G.V., Simulation of spark discharges in high-pressure air sustained by repetitive high-voltage nanosecond pulses. *Journal of Physics D: Applied Physics*, 2008. **41**: p. 234017.
129. Russell, A., H. Zare-Behtash, and K.J. Kontis, Joule heating flow control methods for high-speed flows. *Electrostatics*, 2016. **80**: p. 34.
130. Samimy, M., N. Webb, and A. Esfahani, Reinventing the wheel: excitation of flow instabilities for active flow control using plasma actuators. *Journal of Physics D: Applied Physics*, 2019. **52**: p. 354002.
131. Samimy, M., I. Adamovich, B. Webb, J. Kastner, J. Hileman, S. Keshav, and P. Palm, Development and characterization of plasma actuators for high-speed jet control. *Experiments in Fluids*, 2004. **37**: p. 577.
132. Raizer, Y.P., *Laser Spark and Discharge Propagation*. 1974: Nauka.
133. Radziemski, L.J. and D. Cremers, *Lasers-Induced Plasmas and Applications*. 1989: Marcel Dekker.
134. Kandala, R. and G.V. Candler, Simulation of Laser-Induced Plasma Experiments for Supersonic Flow Control. *43rd AIAA Aerospace Science Meeting and Exhibit*, Reno, Nevada, January 10–13, 2005. . AIAA Paper 2005-205.
135. Starikovskiy, A.Y. and N.L. Aleksandrov, Gasdynamic flow control by ultrafast local heating in a strongly nonequilibrium pulsed plasma. *Plasma Physics Reports*, 2021, **47**(2), pp. 148–209.
136. Thiyagarajan, M. and J.E. Scharer, Experimental investigation of 193-nm laser breakdown in air. *IEEE Transactions on Plasma Science*, 2008. **36**: p. 2512.
137. Thiyagarajana, M., and S. Thompson, Optical breakdown threshold investigation of 1064 nm laser induced air plasmas. *Journal of Applied Physics*, 2012. **111**. p. 073302.
138. Knight, D.D., *Energy Deposition for High-Speed Flow Control*. 2019: Cambridge University Press.
139. Yalin, A.P., N. Wilvert, C. Dumitrache, S. Joshi, and M.N. Shneider, Laser plasma formation assisted by ultraviolet pre-ionization. *Physics of Plasmas*, 2014. **21**: p. 103511.
140. Dumitrache, C., R. Van Osdol, C.M. Limbach, and A.P. Yalin, Control of early flame kernel growth by multi-wavelength laser pulses for enhanced ignition. *Scientific Reports*, 2017. **7**: p. 10239.
141. Dumitrache, C., C.M. Limbach, and A.P. Yalin, Threshold characteristics of ultraviolet and near infrared nanosecond laser induced plasmas. *Physics of Plasmas*, 2016. **23**: p. 093515.
142. Mahamud, R., A.A. Tropina, M.N. Shneider, and R.B. Miles, Dual-pulse laser ignition model. *Physics of Fluids*, 2018. **30**: p. 106104.
143. Couairon, A., and A. Mysyrowicz, Femtosecond Filamentation in Air. In: *Progress in Ultrafast Intense Laser Science*, 2006. **84**: p. 235.
144. Elias, P.-Q., N. Severac, J.-M. Luyssen, J.-P. Tobeli, F. Lambert, R. Bur, A. Houard, A. Yves-Bernard, A. Sylvain, A. *Mysyrowicz, and I. Doudet, Experimental Investigation of Linear Energy Deposition Using Femtosecond Laser Filamentation in a M=3 Supersonic Flow*. AIAA Propulsion and Energy Forum, Cincinnati, Ohio, July 9–11, 2018. AIAA Paper 2018-4896.
145. Aleksandrov, N.L., S.B. Bodrov, M.V. Tsarev, A.A. Murzanev, Y.A. Sergeev, Y.A. Malkov, and A.N. Stepanov, Decay of femtosecond laser-induced plasma filaments in air, nitrogen, and argon for atmospheric and subatmospheric pressures. *Physical Review E*, 2016. **94**: p. 013204.
146. Kolesnichenko, Y., V. Brovkin, O. Azarova, V. Grudnitsky, V. Lashkov, and I. Mashek, Basics in Beamed MW Energy Deposition for Flow/Flight Control. *41st AIAA Aerospace Sciences Meeting and Exhibit*, Reno, NV, , 2004. AIAA Paper 2004-669.
147. Schaub, S.C., J.S. Hummelt, W.C. Guss, M.A. Shapiro, and R.J. Temkin, Electron density and gas density measurements in a millimeter-wave discharge. *Physics of Plasmas*, 2016. **23**: p. 083512.
148. Chintala, N., A. Bao, G. Lou, and I.V. Adamovich, Measurements of combustion efficiency in nonequilibrium RF plasma-ignited flows. *Combustion and Flame*, 2006. **144**(4): 744–756.
149. Raizer, Y.P., M.N. Shneider, and N.A. Yatsenko, *Radio–Frequency Capacitive Discharges*. 1995: CRC Press.
150. Leonov, A.B., D.A. Yarantsev, A.P. Napartovich, and I.V. Kochetov, Plasma-assisted ignition and flame-holding in high-speed flow. *44th AIAA Aerospace Sciences Meeting and Exhibit*, Reno, Nevada, January 9–12, 2006.

151. Kolesnichenko, Y.F., V.G. Brovkin, S.A. Afanas'ev, and D.V. Khmara, Interaction of high–power MW with DC, RF, SHF and laser created plasmas. *43rd AIAA Aerospace Sciences Meeting and Exhibit*, Reno, Nevada, January 10–13, 2005.
152. Van Veldhuizen, E.M. (ed), *Electrical Discharges for Environmental Purposes: Fundamentals and Applications*. 1999: Nova Science. ISBN 1-56072-743-8.
153. Starikovskiy, A., A. Nikipelov, and A. Rakitin, Streamer breakdown development in under-critical electric field. *IEEE Transactions on Plasma Science, Special Issue - Images in Plasma Science*, 2011. **39**(11): pp. 2606–2607.
154. Nudnova, M.M. and A.Y. Starikovskiy, Streamer head structure: role of ionization and photoionization. *Journal of Physics D: Applied Physics*, 2008. **41**(23): p. 234003.
155. Nudnova, M.M. and A.Y. Starikovskiy, Development of streamer flash initiated by HV pulse with nanosecond rise time. *IEEE Transactions on Plasma Science*, 2008. **36**(4), Part 1: pp. 896–897.
156. Krasnochub, A.V., M.M. Nudnova, and A.Y. Starikovskiy, Cathode-directed streamer development in air at different pressures. *43-rd AIAA Aerospace Sciences Meeting and Exhibit*, Reno, Nevada, 2005.
157. Pancheshnyi, S.V., S.M. Starikovskaia, and A.Y. Starikovskiy, Role of photoionization processes in propagation of cathode-directed streamer. *Journal of Physics D: Applied Physics*, 2001. **34**: p. 105.
158. Zheleznyak, M.B., A.K. Mnatsakanyan, and S.V. Sizykh, Photoionization of nitrogen–oxygen mixtures by emission from a gas discharge. *High Temperature* (Teplofizika Vysokikh Temperature), 1982. **20**: p. 423.
159. Polak, L.S., D.I. Ovsyannikov, D.I. Slovetskii, and F.B. Vurzel, *Theoretical and Applied Plasmachemistry*. 1975: Nauka.
160. Zanner, F.J. and L.A. Bertram, Behavior of sustained high-current arcs on molten alloy electrodes during vacuum consumable arc remelting. *IEEE Transactions on Plasma Science*, 1983. **11**: p. 223–232.
161. Boeuf, J.P., Plasma display panels: physics, recent developments and key issues. *Journal of Physics D: Applied Physics*, 2003. **36**: p. 53.
162. Hollenstein, C., The physics and chemistry of dusty plasmas. *Plasma Physics and Controlled Fusion*, 2000. **42**: p. 93.
163. Kosarev, I., A. Starikovskiy, and N.L. Aleksandrov, Development of high-voltage nanosecond discharge in strongly non-uniform gas. *Plasma Sources Science and Technology*, 2019. **28**(1): p. 015005.
164. Anokhin, E.M., M.A. Popov, A.Y. Starikovskiy, and N.L. Aleksandrov, Processes controlling properties of high-voltage nanosecond discharge plasma in combustible mixtures. *Journal of Physics: Conference Series*, 2018. **1058**(1): 012053.
165. Starikovskiy, A., Fast ionization wave development in atmospheric pressure air. *IEEE Transactions on Plasma Science, Special Issue - Images in Plasma Science*. 2011. **39**(11): pp. 2602–2603.
166. Rakitin, A.E. and A.Y. Starikovskiy, Streak images of pulsed discharge development at high overvoltage. *IEEE Transactions on Plasma Science*, 2008. **36**(4, Part 1), pp. 900–901.
167. Kirpichnikov, A.A. and A.Y. Starikovskiy, Nanosecond pulsed discharge: Always uniform? *IEEE Transactions on Plasma Science*, 2008. **36**(4, Part 1): pp. 898–899.
168. Anikin, N.B., N.A. Zavialova, S.M. Starikovskaia, and A.Y. Starikovskiy, Nanosecond: discharge development in long tubes. *IEEE Transactions on Plasma Science*, 2008. **36**(4, Part 1): pp. 902–903.
169. Kirpichnikov, A. and A. Starikovskiy, Nanosecond pulsed discharge: Always uniform? *IEEE Transactions on Plasma Science*, 2008. **36**(4, Part 1): pp. 898–899.
170. Aleksandrov, N.L., E.M. Bazelyan, A.A. Ponomarev, and A.Y. Starikovsky. Kinetics of charged species in non-equilibrium plasma in water vapor- and hydrocarbon-containing gaseous mixtures. *Journal of Physics D: Applied Physics*, 2022. **55**: p. 383002 (26pp).
171. Popov, M., I. Kochetov, A. Starikovskiy, and N. Aleksandrov, Repetitively pulsed nanosecond discharge plasma decay in propane-oxygen gas mixture in the presence of a heating electric field. *Combustion and Flame*, 2021. **233**: p. 111611.
172. Popov, M., E. Anokhin, I. Kochetov, A. Starikovskiy, and N. Aleksandrov, The effect of gas heating on the decay of plasma with hydrated ions after a high-voltage nanosecond discharge. *Plasma Physics Reports*, 2021. **47**(7): pp. 742–751.
173. Popov, M., I. Kochetov, A. Starikovskiy, and N. Aleksandrov, The effect of electron heating on plasma decay in $H_2:O_2$ mixture excited by a repetitively pulsed nanosecond discharge. *Journal of Physics D: Applied Physics*, 2021. **54**(33): p. 335201.
174. Popov, M.A., E.M. Anokhin, A.Y. Starikovskiy, and N.L. Aleksandrov, Plasma decay in hydrocarbons and hydrocarbon- and H_2O-containing mixtures excited by high-voltage nanosecond discharge at elevated gas temperatures. *Combustion and Flame*, 2020. **219**: pp. 393–404.

175. Popov, M.A., E.M. Anokhin, I.V. Kochetov, A.Y. Starikovskiy, and N.L. Aleksandrov, The effect of electron heating on hydrocarbon plasma decay after high-voltage nanosecond discharge. *Journal of Physics D: Applied Physics*, 2019. **52**(50): p. 505201.
176. Popov, M.A., I.V. Kochetov, A.Y. Starikovskiy, and N.L. Aleksandrov, Recombination of electrons with water cluster ions in the afterglow of a high-voltage nanosecond discharge. *Journal of Physics D: Applied Physics*, 2018. **51**(26), p. 264003.
177. Anokhin, E.M., M.A. Popov, A.Y. Starikovskiy, and N.L. Aleksandrov, The effect of fuel oxidation on plasma decay in combustible mixtures excited by high-voltage nanosecond repetitively pulsed discharge. *Combustion and Flame*, 2017. **185**: pp. 301–308.
178. Anokhin, E.M., M.A. Popov, I.V. Kochetov, A.Y. Starikovskiy, and N.L. Aleksandrov, Kinetic mechanism of plasma recombination in methane, ethane and propane after high-voltage nanosecond discharge. *Plasma Sources Science and Technology*, 2016. **25**(4): p. 044006.
179. Anokhin, E.M., M.A. Popov, I.V. Kochetov, N.L. Aleksandrov, and A.Y. Starikovskiy, Plasma decay in high-voltage nanosecond discharges in oxygen-containing mixtures. *Plasma Physics Reports*, 2016. **42**(1): pp. 59–67.
180. Smirnov, B.M., *Complex Ions*. 1983: Nauka.
181. Aleksandrov, N.L., A.M. Konchakov, L.V. Shackin, and V.M. Shashkov, Dissosiative antd triple electron-ion recombination in a CO_2 gas diacharge plasma. *Plasma Physics*, 1986. **12**: pp. 1218–1224.
182. Mitchell, J.B.A., The dissociative recombination of molecular ions. *Physics Reports*, 1990. **186**: p. 215.
183. McEwan, M.J. and L.F. Phillips, Chemistry of the atmosphere. Arnold, London, 1975.
184. Johnsen, R., Electron-temperature dependence of the recombination of $H3O+(H2O)n$ ions with electrons. *Journal of Chemical Physics*, 1993. **98**: p. 5390.
185. Aleksandrov, N.L., S.V. Kindysheva, I.N. Kosarev, and A.Y. Starikovskiy, Plasma decay in air and $N_2:O_2:CO_2$ mixtures at elevated gas temperatures. *Journal of Physics D: Applied Physics*, 2008. **41**(21): p. 215207.
186. Aleksandrov, N.L., E.M. Anokhin, S.V. Kindysheva, A.A. Kirpichnikov, I.N. Kosarev, M.M. Nudnova, S.M. Starikovskaia, and A.Y. Starikovskiy, Plasma decay in air and O_2 after a high-voltage nanosecond discharge. *Journal of Physics D: Applied Physics*, 2012. **45**: p. 255202 (10pp).
187. Aleksandrov, N.L., E.M. Anokhin, S.V. Kindysheva, A.A. Kirpichnikov, I.N. Kosarev, M.M. Nudnova, S.M. Starikovskaia, and A.Y. Starikovskiy, High-voltage nanosecond air discharge plasma decay. *Plasma Physics Reports*, 2012. **38**(2): pp. 200–208.
188. Aleksandrov, N.L., S.V. Kindysheva, A.A. Kirpichnikov, I.N. Kosarev, S.M. Starikovskaia, and A.Y. Starikovskiy, Plasma decay in N_2, CO_2 and H_2O excited by high-voltage nanosecond discharge. *Journal of Physics D: Applied Physics*, 2007. **40**(15): pp. 4493–4502.
189. Alexandrov, N.L. and E.E. Son, Energy distribution and kinetic coefficients of electrons in gases in an electric field. In: *Collected Works: Chemistry of Plasma*, edited by B.M.M. Smirnov. 1980: Atomizdat, p. 35.
190. Alexandrov, N.L., F.I. Vysikailo, R.S. Islamov, I.V. Kochetov, A.P. Napartovich, and V.G. Pevgov, Fuction of electron distribution in the mixture $N_2:O_2 = 4:1$. *TVT*, 1981. **19**: p. 22.
191. Polak, L.S., A.A. Ovsiannikov, D.I. Slovetsky, and F.B.M. Vurzel, *Theoretical and Applied Plasmochemistry*. 1975: Nauka.
192. Slovetsky, D.I., Dissociation of molecules by electron impact. In *Reviews of Plasma Chemistry*, Volume 1, edited by B.M. Smirnov. 1991: Kluwer Academic/Plenum Publishers, 334 pages.
193. Capitelli, M., *Non-Equilibrium Vibrational Kinetics*. 1989: Springer-Verlag, 392pp.
194. Bazhenov, V.Y., A.V. Ryabtsev, I.A. Soloshenko, V.V. Tsiolko, and A.I. Shchedrin, Features of physical kinetics in a glow discharge with a hollow cathode. *Proceedings of IX Conference on Gas Discharge Physics*, Ryazan, 1998. Part 2, p. 29.
195. Kolobov, V.I. and V.A. Godiak, Nonlocal electron kinetics in collisional gas discharge plasmas. *IEEE Transactions on Plasma Science*, 1995. **23**(4): p. 503.
196. Guerra, V. and J. Loureiro, Electron and heavy particle kinetics in a low-pressure nitrogen glow discharge. *Plasma Sources Science and Technology*, 1997. **6**: p. 361.
197. Nahorny, J., C.M. Ferreira, B.F. Gordiets, D. Pagnon, M. Touzeau, and M. Vialle, Experimental and theoretical investigation of a N_2–O_2 DC flowing glow discharge. *Journal of Physics D: Applied Physics*, 1995. **28**: p. 738.
198. Bogaerts, A., Mathematical modelling of a direct current glow discharge in argon. PhD thesis, Antwerpen, 1996.

199. Czanetzki, U., D. Luggenhlscher, and H.F. Dobele, Temporally and spatially resolved electric field measurements in helium and hydrogen RF–discharges. *Proceedings of XIV ESCAMPIG European Sectional Conference on the Atomic and Molecular Physics of Ionized Gases*, Malahide, Ireland, 1998. **22H**: p. 30.
200. Raiser, Y.P., M.N. Shneider, and N.A. Yatsenko, *High-Frequency Capacitive Discharge*. 1995: CRC Press, 320 pp.
201. Kortshagen, U., Modelling and diagnostics of low pressure inductively coupled plasmas. In: *Electron Kinetics and Applications of Glow Discharges*, edited by U. Kortshagen and L.D. Tsendin, NATO ASI Series, Series B: Physics, vol. 367. 2006: Springer, p. 329.
202. Suzuki, M. and Y. Naito, *Proceedings of the Japan Academy*, 1952. **28**: p. 469.
203. Reter, G., *Electron Avalanches and Breakdown in Gases*. 1968: Butterworths.
204. Ponizovsky, A.Z., L.Z. Ponizovsky, S.P. Kryuchkov, A.P. Shvedchikov, and E.V. Belousova, Use of the combined effect of corona discharge and UV irradiation for removal of ecologically harmful impurities from the air. *Proceedings of IX Conference on Gas Discharge Physics*, Ryazan, 1998. Part 1, p. 53.
205. Gilbert, A. and F. Bastien, *Journal of Physics D: Applied Physics*, 1989. 22: p. 1078.
206. Sigmond, R.S., *Journal of Applied Physics*, 1984. **56**(5): p. 1355.
207. Stritzke, P., I. Sander, and H. Raether, *Journal of Applied Physics*, 1977. **10**: p. 2285.
208. Simek, M., V. Babicky, M. Clupek, S. DeBenedictis, G. Dilecce, and P. Sunka, *Journal of Applied Physics*, 1998. **31**: p. 2591.
209. Spyrou, N. and C. Manassis, *Journal of Applied Physics*, 1989. **22**: p. 120.
210. Ponizovsky, A.Z., L.Z. Ponizovsky, S.P. Kryuchkov, A.P. Shvedchikov, and E.V. Belousova, Use of the combined effect of corona discharge and UV irradiation for removal of ecologically harmful impurities from the air. *Proceedings of IX Conference on Gas Discharge Physics*, Ryazan, 1998. Part 1, p. 53.
211. Beer, T.A., J. Laimer, and H. Stori, Investigation of the temporal and spatial evolution of a pulsed DC discharge used for plasma CVD. *Proceedings of XXIV International Conference on Phenomena in Ionized Gases (ICPIG XXIV)*, Warsaw, Poland, July 11–16, 1999. **1**: p. 63.
212. Asinovsky, E.I., L.M. Vasilyak, and V.V. Markovets, Wave breakdown of gas gaps. I. Fast stages of breakdown. *TVT*, 1983. **21**(2): p. 371.
213. Asinovsky, E.I., L.M. Vasilyak, and V.V. Markovets, Wave breakdown of gas gaps. II. Wave breakdown in distributed systems. *TVT*, 1983. **22**(3): p. 577.
214. Nedospasov, A.V. and A.E. Novik, Ionization front propagation rate at breakdown of long tubes. *ZHTF*, 1960. **30**(11): p. 1329.
215. Anikin, N.B., S.M. Starikovskaya, and A.Y. Starikovsky, *Proceedings of FNTP-98*, Petrozavodsk, 1998. CH.1, p. 198.
216. Bazelyan, E.M. and Y.P. Raiser, *Physics of Spark Breakdown*. 1997: Izd-vo of MIPT, 320pp.
217. Doran, A.A., The development of a Townsend discharge in N_2 up to breakdown investigated by image converter, intensifier and photomultiplier technique. *Physikalische Zeitschrift*, 1960. **208**(2): p. 427.
218. Köhrmann, W., Die zeitliche Entwicklung der Townsend: Entladung bis zum Durchschlag. *Zeitschrift für Naturforschung*, 1964. **19A**(7): p. 926.
219. Tholl, H., I. Sander, and H. Martinen, Eine automatische Apparatur zur örlich und zeilich aufgelosten Spectroscopic an Funkenentladungen. *Zeitschrift für Naturforschung*, 1970. **25A**(3): p. 412.
220. Koppitz, J., Die radiale und axiale Entwicklung des Leuchtens im Funkenkanal untersucht mit Wischkamera. *Zeitschrift für Naturforschung*, 1967. **22A**(11): p. 1089.
221. Raiser, Y.P., *Physics of Gas Discharge*. 1992: Nauka, 536pp.
222. Allen, K.R. and K. Phillips, Mechanism of spark breakdown. *Electrical Review*, 1963. **173**(3): p. 779.
223. Babich L.P., T.V. Loiko, and V.A. Tsukerman, High-voltage nanosecond discharge in dense gases at large overvoltages developing in the electron escape mode. *UVN*, 1990. **160**(7): p. 49.
224. Mott, N. and G. Messi, *Theory of Atomic Collisions*. 1969: Clarendon Press, 756pp.
225. Babich, L.P. and J.L. Stankevich, Criterion for the transition from the streamer mechanism of a gas discharge to continuous electron acceleration. *ZHTF*, 1972. **42**(8): p. 1669.
226. Dali, S. and P.F. Williams, *Journal of Applied Physics*, 1987. **62**: p. 4696.
227. Babaeva, N.Y. and G.V. Naidis, To calculation of plasmachemical efficiency of pulsed corona discharges. *Proceedings of the Conference on Low Temperature Plasma Physics FNTP-98*, Petrozavodsk, 1998, p. 637.
228. Kunhardt, E.E. and Y. Tzeng, Kinetics investigations of avalanche and streamer development. In: *Gaseous Dielectrics IV*, edited by L.G. Christophorou and M.O. Pace. 1984: Pergamon Press, p. 146.
229. Kline, L.E., Monte Carlo study of ionization zone electron kinetics in negative pin-plane coronas in atmospheric air. *Journal of Applied Physics*, 1985. **58**: p. 3715.

230. Satoh, K., H Y Ohmori, Y Sakai and H Tagashira, Computer simulation study of correspondence between experimental and theoretical electron drift velocities in CH_4 gas. *Journal of Applied Physics D*, 1991. **24**: p. 1354.
231. Guo, J.-M. and C.-H. Wu, Comparisons of multidimensional fluid models for streamers. In *Non-Thermal Plasma Techniques for Pollution Control*, Vol. 34A of NATO ASI, Series G, edited by M. Penetrante and E. Shultheis. 1993: Springer, p. 287.
232. Dali, S. and P.F. Williams, Two-dimensional studies of streamers in gases. *Journal of Applied Geophysics*, 1987. **62**: p. 4696.
233. Wu, C. and E.E. Kunhardt, Formation and propagation of streamers in N2 and N2-SF6 mixtures. *Physical Review A*, 1988. **37**: p. 4396.
234. Vitello, P.A., B.M. Penetrante, and J.N. Bardsley, Simulation of negative-streamer dynamics in nitrogen. *Physical Review E*, 1994. 49: p. 5574.
235. Kulikovsky, A.A., Positive streamer between parallel plate electrodes in atmospheric pressure air. *Journal of Physics D*, 1997. **30**: p. 441.
236. Kulikovsky, A.A., Positive streamer in a weak field in air: A moving avalanche-to-streamer transition. *Physical Review E*, 1998. **57**: p. 7066.
237. Davies, A.J. and C.J. Evans, Field distortion in gaseous discharges between parallel-plate electrodes. *Proceedings of IEEE*, 1967. 114: p. 1547.
238. Bayle, P. and B. Cornebois, Propagation of ionizing electron shock waves in electrical breakdown. *Physical Review A*, 1985. 31: p. 1046.
239. Morrow, R. and J.J. Lowke, Streamer propagation in air. *Journal of Physics D*, 1997. **30**: p. 614.
240. Alexandrov, N.L. and E.M. Bazelyan, Simulation of long-streamer propagation in air at atmospheric pressure. *Journal of Physics D*, 1996. **29**: p. 740–152.
241. Guo, J.M. and J. Wu, Streamer radius model and its assessment using two-dimensional models. *IEEE Transactions on Plasma Science*, 1996. 24: p. 1348.
242. Amirov, R.H., M.B. Zheleznyak, and E.A. Filimonova, Modeling of NOx, SO2, VOC removal by pulsed corona and dielectric barrier discharge. *Preprint of IVTAN N* 1-403. 1997, 61pp.
243. Lifshits, E.M. and L.P. Pitaevsky, *Physical Kinetics*, Volume 10, 1st Edition. 1981. Butterworth-Heinemann. 452 pages.
244. Fano, U., Degradation and range straggling of high energy radiations. *Physical Review*, 1953. **92**(2): p. 328.
245. Spencer, L.V. and U. Fano, Energy spectrum resulting from electron slowing down. *Physical Review*, 1954. **93**(6): p. 1172.
246. Nikerov, V.A. and T.V. Shomin, *Kinetics of Degradation Processes*. 1985: Energoatomizdat.
247. Konovalov, V.P. and E.E. Son, Degradation spectra of electrons in gases. In: *Chemistry of Plasma*, issue 14, edited by B.M. Smirnov. 1987: Energoatomizdat, C.196.
248. Konovalov, V.P. and E.V. Son, Electron distribution function and composition of a molecular plasma excited by an electron beam. *ZHTF*, 1980. **50**(2): p. 300.
249. Megill, L.R. and J.H. Cahn, The calculation of electron energy distribution functions in the ionosphere. *Journal of Geophysical Research*, 1964. **69**(23): p. 5041.
250. Lukas, J., D.A. Price, and J.L. Moruzzi, The calculation of electron energy distributions and attachment coefficient for electron swarms in oxygen. *Journal of Physics D*, 1973. **6**(2): p. 1503.
251. Islamov, R.S., I.V. Kochetov, and V.G. Pevgov, Analysis of the processes of interaction of electrons with an oxygen molecule. *FIAN Preprint*, 1977. 169: 24 pages.
252. Demidov, V.I., N.B. Kolokolov, and A.A. Kudryavtsev, *Probe Methods for Investigation of Low-Temperature Plasma*. 1996: Energoatomizdat, 240pp.
253. Konovalov, V.P., M.A. Skorik, and E.V. Son, Unsteady degradation spectrum of electrons in molecular nitrogen. *Plasma Physics*, 1992. **18**(6): p. 778.
254. Slavin, B.B. and P.I. Sopin, Breakdown of a neutral gas by ionizing waves of a negative polarity potential gradient. *TVT*, 1992. **30**(1): p. 1.
255. Golubev, A.I., A.V. Ivanovskii, A.A. Soloviev, V.A. Terekhin, and I.T. Shorin, One-dimensional model for description of fast breakdown waves in long discharge tubes. Voprosy atomnoy nauki i tekhnika. *Theoretical and Applied Physics*, 1985. **2**: p. 17.
256. Sinkevich, O.A. and Y.V. Trofimov, About the mechanism of the breakdown wave propagation through weakly ionized plasma in nanosecond discharges. *DAN USSR*, 1979. **249**: p. 597.
257. Sinkevich, O.A. and Y.V. Trofimov, About fast ionization of a long column of plasma by a secondary breakdown wave. *TVT*, 1980. **18**(5): p. 1088.
258. Lagarkov, A.N. and I.M. Rutkevich, *Waves of Electric Breakdown in a Confined Plasma*. 1989: Nauka.

259. Lagarkov, A.N. and I.M. Rutkevich, Ionizing waves of spatial charge. *DAN SSSR,* 1979. **249**: p. 593.
260. Slavin, B.B. and P.I. Sopin, Wave breakdown in long gas-filled tubes with pre-ionization. *TVT.* 1990. **28**(2): p. 243.
261. Babich, L.P. and I.M. Kutsyk, Numerical modeling of a nanosecond discharge in helium at atmospheric pressure developing in the electron escape mode. *TVT*, 1995. **33**(2): p. 191.
262. Kunhardt, E.E. and W.W. Buszewski, Development of overvoltage breakdown at high gas pressure. *Physical Review A*, 1980. 21: p. 2069.
263. Butin, O.V. and L.M. Vasilyak, Propagation of a high-speed ionization wave in long discharge tubes with preionization. *Plasma Physics*, 1999. **25**(8): p. 725.
264. Starikovskaia, S.M. and A.Y. Starikovskiy, Numerical modeling of the electron energy distribution function in the electric field of a nanosecond pulsed discharge. *Journal Physics D: Applied Physics*, 2001. **34**: pp. 3391–3399.
265. Belotserkovsky, O.M. and V.E. Yanitsky, Problems of numerical modeling of rarefied gas flows. 1979, Uspekhi mekhaniki **1** N1 C.2.
266. Bird, G., *Molecular Gas Dynamics.* 1981: Oxford University Press, 320 pp.
267. Belotserkovsky, O.M. and M.N. Kogan, Monte Carlo method in the dynamics of rarefied gases. In: *Molecular Gas Dynamics*, edited by G. Bird. 1981: Oxford University Press, C.303.
268. Boeuf, J.P. and E. Marode, A monte Carlo analysis of an electron swarm in a non-uniform field: The cathode region of a glow discharge in helium. *Journal of Physics D: Applied Physics*, 1982. **15**: pp. 2169–2187.
269. Skullerud, H.R., The stochastic computer simulation of ion motion in a gas subjected to a constant electric field. *Journal of Physics D: Applied Physics*, 1968. **1**: p. 1567.
270. Lin, S.L. and J.N. Bardsley, Monte Carlo simulation of ion motion in drift tubes. *The Journal of Chemical Physics*, 1977. **66**: p. 435.
271. Bogaerts, A., Mathematical modelling of a direct current glow discharge in argon. Ph.D. Thesis, 1996, Antwerpen University.
272. Vasenkov, A.V. and V.S. Malinovsky, Electron-beam plasma in a gas flow of inhomogeneous density. *Physics of Plasma*, 1995. **21**(12): pp. 1075–1081.
273. Vasenkov, A.V., R.G. Sharafutdinov, A.E. Belikov, and V.S. Malinovsky, Spatial and energy distribution of electrons generated as a result of interaction of an electron beam with a gas flow of inhomogeneous density. *Physics of Plasma*, 1996. **22**(12): pp. 1124–1133.
274. Demkin, V.P., B.V. Korolev, and S.V. Melnichuk, Calculation of the distribution function in strong electric fields. *Physics of Plasma*, 1995. **21**(1): pp. 81–84.
275. Demkin, V.P., et al., Electron energy distribution function in strong electric fields. *Atmospheric and Ocean Optics*, 1993. **6**: p. 263.
276. Anikin, N.B., S.V. Pancheshnyi, S.M. Starikovskaia, and A.Y. Starikovskiy, Breakdown development at high overvoltage: electric field, electronic levels excitation and electron density. *Journal of Physics D: Applied Physics*, 1998. **31**: p. 826.
277. Malinovsky, V.S., A.E. Belikov, O.V. Kuznetsov, and R.G. Sharafutdinov, Spatial and energy distribution of secondary electrons in an electron-beam plasma. *Physics of Plasma*, 1995. **21**(1): pp. 85–90.
278. Phelps, A.V. and L.C. Pitchford. Anisotropic scattering of electrons by N2 and its effect on electron transport. *Physical Review A*, 1985. **31**: p. 2932.
279. Jelenkovic, B.M. and A.V. Phelps. Excitation of N2 in dc electrical discharges at very high *E/n*. *Physical Review*, 1987. **36**: p. 5310.
280. Morgan, W.L., Kinema software (Software Solutions for Applied Physics and Chemistry) and J.-P. Boeuf, L.C.Pitchford CPAT (Center de Physique des Plasmas et Applications de Toulouse). BOLSIG Boltzmann solver (freeware), https://www.kinema.com.
281. Gordeev, O.A. and D.V. Khmara, Software package for modeling of kinetic processes in a gas discharge plasma in the reduced field approximation. *IX Conference on Gas Discharge Physics*, Ryazan, 1998. Abstracts, Ch.2, p. 91.
282. Pancheshny, S.V., S.M. Starikovskaya, and A.Y. Starikovsky, Dynamics of population of electronic states of molecular nitrogen and structure of high-speed ionization wave. *Plasma Physics*, 1999. **25**(4): p. 326.
283. Kuzmenko, N.E., L.A. Kuznetsova, and Y. Kuzyakov, *Franck-Condon Factors of Two-Atomic Molecules.* 1984: Moscow University Publishing House.
284. Pancheshnyi, S.V., S.M. Starikovskaia, and A.Y. Starikovskiy, Measurements of rate constants of the $N_2\left(C^3\Pi_u\right)$ and $N_2^+\left(B^2\Sigma_u^+\right)$ deactivation by N_2, O_2, H_2, CO and H_2O molecules in afterglow of the nanosecond discharge. *Chemical Physics Letters*, 1998. **294**: p. 523.

285. Pancheshnyi, S.V., S.M. Starikovskaia, and A.Y. Starikovskiy, Collisional deactivation of $N_2(C^3\Pi_u, v=0,1,2,3)$ states by N_2, O_2, H_2 and H_2O molecules. *Chemical Physics*, 2000. **262**: p. 349.
286. Okabe, D., *Photochemistry of Small Molecules*. 1981: Wiley.
287. Belasri, A., Boeuf, J.P., and L.C. Pitchford, Cathode sheath formation in a discharge-sustained XeCl laser. *Journal of Applied Physics*, 1993. **74**: p. 1553.
288. Grigoriev, I.S. and E.Z. Meilikhov, *Physical Quantities: Handbook*. 1991: Energoatomizdat.
289. Starikovskiy, A., N. Aleksandrov, and M. Shneider, Streamer self-focusing in external longitudinal magnetic field. *Physical Review E*, 2021. **103**(6): p. 063201.
290. Starikovskiy, A. and N. Aleksandrov, Blocking streamer development by plane gaseous layers of various densities. *Plasma Sources Science and Technology*, 2020. 29: p. 034002.
291. Starikovskiy, A.Y. and N.L. Aleksandrov, Gasdynamic diode: Streamer interaction with sharp density gradients. *Plasma Sources Science and Technology*, 2019. **28**(9): p. 095022.
292. Guo, J. and C. Wu, Two-dimensional simulation of the nonequilibrium fluid models for streamer. *IEEE Transactions on Plasma Science*, 1993. **21**: p. 684.
293. Naidis, G.V., Influence of nonlocal effects on streamer dynamics in positive corona discharges. *Letters in ZhTF*, 1997. **23**: p. 89.
294. Golant, V.E., A.P. Zhilinsky, and S.A. Sakharov, *Fundamentals of Plasma Physics*. 1977: Atomizdat.
295. Penney, G.W. and G.T. Hummert, Photoionization measurements in air, oxygen, and nitrogen. *Journal of Applied Physics*, 1970. **41**: p. 572.
296. Wu, C. and E.E. Kunhardt, Formation and propagation of streamers in N_2 and N_2–SF_6 mixtures. *Physical Review A*, 1988. **37**: p. 4396.
297. Vitello, P.A., B.M. Penetrante, and J.N. Bardsley, Multi-dimensional modelling of the dynamic morphology of streamer coronas. In *Non-Termal Plasma Techniques for Pollution Control*, edited by B.M. Penetrante and S.E. Schultheis. 1993: Springer Science & Business Media, p. 249.
298. Pinhao, N., Z. Donko, D. Loffhagen, M. Pinheiro, and E.A. Richley, Comparison of Monte Carlo and electron Boltzmann equation methods at low and moderate E/n field values. *Proceedings of XVth Europhysics Conference on Atomic and Molecular Physics of Ionized Gases*, Hungary, 2000, p. 156.
299. Skubenich, V.V. and I.P. Zapesochny, Excitation of two-atomic molecules in collisions with monoenergetic electrons. *High Energy Chemistry*, 1975. **9**: p. 387.
300. Borst, W.L. and E.C. Zipf, Cross-section for electron-impact excitation of the (0,0) first negative band of N_2^+ from thresholds to 3 keV. *Physical Review A*, 1970. **1**: p. 834.
301. Pancheshnyi, S.V., S.M. Starikovskaia, and A.Y. Starikovskiy, Role of photoionization processes in propagation of cathode-directed streamer. *Journal Physics D: Applied Physics*, 2001. **34**: p. 105.
302. Kulikovsky, A.A., Positive streamer in a weak field in air: A moving avalanche-to-streamer transition. *Physical Review E*, 1998. **57**: p. 7066.
303. Starikovsky, A.Y., N.L. Aleksandrov, and E.M. Bazelyan, The influence of humidity on positive streamer propagation in long air gap. *Plasma Sources Science and Technology*, 2022. **31**: p. 114009.
304. Starikovskiy, A., N. Aleksandrov, and M. Shneide, Simulation of decelerating streamers in inhomogeneous atmosphere with implications for runaway electron generation. *Journal of Applied Physics*, 2021. **129**: p. 063301.
305. Starikovskiy, A. and N. Aleksandrov, How pulse polarity and photoionization control streamer discharge development in long air gaps. *Plasma Sources Science and Technology*, 2020. **29**: p. 075004.
306. Bazelian, E.M. and Y.P. Reiser, *Spark Discharge*. 1997: Izd-v MIPT.
307. Raizer, Y.P. and A.N. Simakov, Hemispherical model of a streamer head. *Plasma Physics*, 1996. **22**: p. 668.
308. Bazelian, E.M. and Y.P. Raizer, *Lightning Physics and Lightning Protection*. 2000. CRC Press, 336 pages.
309. Babaeva, N.Y. and G.V. Naidis, Two-dimensional modelling of positive streamer dynamics in non-uniform electric fields in air. *Journal of Physics D: Applied Physics*, 1996. **29**: p. 2423.
310. Dutton, J., A survey of electron swarm data. *Journal of Chemical Physics*, 1975. **4**: p. 577.
311. Gallaher, J.W., E.C. Beaty, J. Dutton, and L.C. Pitchford, An annotated compilation and appraisal of electron swarm data in electronegative gases. *Journal of Chemical Physics*, 1983. **12**: p. 109–152.
312. Aleksandrov, N.L. and E.E. Son, Energy distribution and kinetic coefficients of electrons in gases in an electric field. In: *Chemistry of Plasma*, edited by B.M. Smirnov. 1980: Atomizdat Press. **7**: p. 35.
313. Phelps, A.V. and L.C. Pitchford, Anisotropic scattering of electrons by N_2 and its effect on electron transport. *Physical Review A*, 1985. **31**: p. 2932.
314. Odrobina, I. and M. Cernak, Numerical simulation of streamer-cathode interaction. *Journal of Applied Physics*, 1995. **78**: p. 3635.

315. Boeuf, J.P., Numerical model of rf glow discharges. *Physical Review A*, 1987. **26**: p. 2782.
316. Bayle, P. and B. Cornebois, Propagation of ionizing electron shock waves in electrical breakdown. *Physical Review A*, 1985. **31**: p. 1046.
317. Kanzari, Z., M. Yousfi, and A. Hamani, Modeling and basic data for streamer dynamics in N_2 and O_2 discharges. *Journal of Applied Physics*, 1998. **84**: p. 4161.
318. Aleksandrov, N.L. and I.V. Kochetov, Electron rate coefficients in gases under non-uniform field and electron density conditions. *Journal of Physics D: Applied Physics*, 1996. **29**: p. 1476.
319. Aleksandrov, N.L., E.M. Bazelyan, I.V. Kochetov, and A.M. Ohrimovsky, Velocities of inelastic processes in an alternating electric field in air. *Plasma Physics*, 1998. **24**: p. 662.
320. Lozanskii, E.D., Development of electron avalanches and streamers. *UVN*, 1975. **117**: p. 493.
321. Przybylski, A., The gas-ionizing radiation of a discharge in N2-O2 mixtures. *Zeitschrift für Naturforschung*, 1961. 16a: p. 1232.
322. Teich, T.H., Emission of gas-ionizing radiation from electron avalanches. I. Measurement set-up and procedure. Measurements in oxygen. *Physikalische Zeitschrift*, 1967. **199**: p. 378.
323. Seguin, H.J., J. Tulip, and D.C. Mc Ken, Ultraviolet photoionization in TEA lasers. *IEEE Journal of Quantum Electronics*, 1974. **QE-10**: p. 311.

3 Energy Exchange and Chemical Reactions in Nonequilibrium Regimes in Chemically Active Plasmas

3.1 VIBRATIONAL RELAXATION IN CHEMICALLY ACTIVE GAS

3.1.1 Mode Approximation

Modeling at the macroscopic level assumes the existence of local equilibrium on the degrees of freedom, i.e., the possibility of introducing the concept of "temperatures". The relaxation of vibrational energy, for example, in the N_2O - CO - Ar system, assuming the presence of a Boltzmann distribution over vibrational energy levels, can be described in the framework of the so-called "mode" approximation [1]:

$$N_2O(v_1) + M \xleftrightarrow{w_{12}^M} N_2O(v_2) + M, \tag{3.1}$$

$$N_2O(v_2) + M \xleftrightarrow{w_{20}^M} N_2O + M, \tag{3.2}$$

$$N_2O(v_3) + M \xleftrightarrow{w_{32}^M} N_2O(v_1, v_2) + M, \tag{3.3}$$

$$N_2O(v_3) + CO \xleftrightarrow{w_{34}} N_2O + CO(v=1), \tag{3.4}$$

where W_{ij}^M is the vibrational relaxation rate coefficient, M is the collision partner. Assuming that the vibrational temperatures of $v_1(T_{\text{vib}}(v_1))$ and $v_2(T_{\text{vib}}(v_1))$ modes of N_2O are equal to each other due to the fast $V-V$-exchange, one can write the following expression [1] for the vibrational energy relaxation:

$$-dx_i/dt = \sum_{j=1}^{3} a_{ij} x_j, \tag{3.5}$$

where $x_i = (\varepsilon_i(T_2) - \varepsilon_i)/\varepsilon_i(T_2)$, $i = 1,2,3$ corresponds to v_2, v_3 modes of N_2O and CO oscillations, ε_i and $\varepsilon_i(T_2)$ are the current and equilibrium number of vibrational quanta of a given mode. The coefficients a_{ij} are defined as follows:

$$a_{11} = \frac{1}{\varphi}\left(9\frac{\varepsilon_3(T_2)}{\varepsilon_2(T_2)}\frac{1}{\theta_2^2}w_{32} + \left[1 + \beta\psi_1\frac{C_{v2}}{C_{tr}}\left\{1 + \frac{2\vartheta_2}{\vartheta_1}\frac{C_{v1}}{C_{v2}}\right\}\right]w_{20}\right), \tag{3.6}$$

DOI: 10.1201/9781003203063-3

$$a_{12} = \frac{1}{\varphi}\frac{\varepsilon_3(T_2)}{\varepsilon_2(T_2)}\left(-3\frac{\theta_3}{\theta_2^{\ 3}}w_{32} + \beta\psi_1\frac{C_{v2}}{C_{tr}}\frac{\vartheta_3}{\vartheta_2}w_{20}\right), \tag{3.7}$$

$$a_{13} = \frac{1}{\varphi}\frac{\varepsilon_4(T_2)}{\varepsilon_2(T_2)}\frac{\vartheta_4}{\vartheta_2}\psi_2\frac{C_{v2}}{C_{tr}}w_{20}, \tag{3.8}$$

$$a_{21} = -\left(3/\theta_2^{\ 2}\right)w_{32}, \quad a_{22} = \left(\theta_3/\theta_2^{\ 3}\right)w_{32} + \psi_2(\varepsilon_4(T_2)/\varepsilon_2(T_2))w_{34}, \tag{3.9}$$

$$a_{23} = -(\theta_4/\theta_3)\psi_2(\varepsilon_4(T_2)/\varepsilon_3(T_2))w_{34}, \tag{3.10}$$

$$a_{31} = 0, \quad a_{32} = -(\theta_3/\theta_4), \quad a_{33} = -\psi_1 w_{34}, \tag{3.11}$$

$$\varphi = 1 + (C_{v1}/C_{v2})\cdot(2\vartheta_2/\vartheta_1)^2, \quad \theta_k = 1 - \exp(-\vartheta_k/T_2), \tag{3.12}$$

$$w_{ij} = w_{ij}^{N_2O}\psi_1 + w_{ij}^{CO}\psi_2 \tag{3.13}$$

Here ϑ_k, C_{vk} ($k = 1 \div 3$) are the characteristic vibrational temperatures of v_1, v_2 and v_3 modes of N_2O and their vibrational heat capacities, ψ_1 and ψ_2 are the mole fractions of N_2O and CO in the mixture.

This approach proved to be very fruitful in analyzing the vibrational relaxation of low-lying vibrational levels at high temperatures behind shock waves, when the main contribution to the energy exchange is given by several lowest (sometimes almost exclusively the first) vibrational levels, the distribution of which, as a rule, is close to the Boltzmann distribution with the current "vibrational temperature", and the energy distribution at high-lying vibrational levels practically does not play a role.

It is much more difficult to justify the validity of such an approach when analyzing chemical kinetics, where the energy distribution exactly on the upper states may play a dominant role. In particular, we considered the mechanism proposed in Ref. [2], based precisely on the change in the population distribution of high-lying vibrational states during bimolecular exchange reactions. The "runoff" of active highly excited molecules into the reaction can lead to depletion of the distribution at high vibrational levels and markedly change the reaction rate, practically without changing the distribution of low-lying states responsible for the main stock of vibrational energy in the system.

For this reason, the mode approximation is of little use for describing thermally nonequilibrium chemical kinetics.

3.1.2 "Ladder" Approximation

Several different approaches are being developed as an alternative to it. Among them, we will single out a class of so-called "Ladder" models, which approximate the real energy distribution over the states in a molecule by some piecewise continuous function defined at several energy points. The change in the population distribution of different levels occurs in collisions with the buffer gas and is characterized by the average portion of energy ΔE transferred per collision. The obvious disadvantages of this type of models include excessive coarsening of energy exchange kinetics by vibrational states, and poorly defined empirical parameters of energy exchange—the energy "step" of the ladder and the magnitude of the transferred energy portion.

3.1.3 Diffusion Approximation

Much more reasonable are models for describing the energy distribution of molecules by vibrational states, based on the view of vibrational energy exchanges as diffusion of the system in the energy phase space [3]. In this case, the distribution function of molecules by energy of vibrational degrees of freedom in a nonequilibrium reacting system can be described by a system of Fokker-Planck-type equations [4]:

$$\frac{\partial}{\partial t} f_i(E) + \frac{\partial}{\partial E} \left\{ j_{VT}^{(i)} + j_{VVV'}^{(i)} + j_{VVV}^{(0i)} + j_{VVV}^{(1i)} + j_R^{(i)} \right\} = 0 \tag{3.14}$$

Here the indices i define the sort of molecules, $j_{VT}^{(i)}$, $j_{VVV'}^{(i)}$, $j_{VVV}^{(0i)}$, $j_{VVV}^{(1i)}$ - fluxes, $j_R^{(i)}$ - the flux defining the loss or birth of vibrationally excited molecules in the reaction. The corresponding expressions for $j_{VT}^{(i)}$, $j_{VVV}^{(0i)}$, $j_{VVV}^{(1i)}$ and $j_{VVV'}^{(i)}$ are given, for example, in Ref. [4].

If a vibrationally excited molecule enters the reaction, the distribution function is distorted by the loss of highly excited molecules in the reaction [5]. In contrast, the energy release in the reaction to the internal degrees of freedom of the products leads to significant overpopulation of the upper levels. The presence of fast VV'exchanges in the system can lead to the appearance of energy feedback, which restores the distribution function of initial substances near the reaction threshold and accelerates chemical transformations.

The flux associated with monomolecular decay has a significant effect on the population of the upper levels of molecules. Bimolecular reactions can also lead to a marked change in the population distribution of vibrational states if the vibrational energy of the molecule can be utilized to overcome the energy threshold of the reaction. The formation of products at excited levels, on the contrary, leads to some overpopulation of the vibrational distribution.

In both cases, the reaction rate coefficient $k_R^{(i,j,k,l)}\left(e_i,e_j,e_k,e_l\right)$, which characterizes the reaction rate between vibrationally excited reactants i and j with the formation of products k and l in excited states with energies e_k, e_l, respectively, plays a determining role for the flux value.

Thus, the system of equations (3.1–3.8) allows us to describe the processes in a chemically reacting gas with a Boltzmann distribution of molecules by vibrational energy. As a parameter of the problem, the equation includes probabilities VT, VV, and VV' of the processes as well as rate coefficients of chemical transformations at $T_{vib} = T_{tr}$.

At all attractiveness of the system of equations of Fokker-Planck type (3.1–3.8), the description with its help of a real chemically reacting system has a serious disadvantage associated with the fundamentally discrete character of energy transfer of vibrational excitation. This is especially critical for the description of relaxation of molecules whose energy scale contains a small number of levels up to the dissociation threshold (e.g., H_2).

At the same time, small molecules can be described reasonably well using governing equations describing the relaxation of each individual energy level. Three-atomic molecules present some difficulty for this approach, since the number of states to be taken into account grows rapidly with increasing excitation energy. Nevertheless, this approach is the most free from restrictions and assumptions, the feasibility of which is not a priori obvious in a large number of cases.

For this reason, we will focus on the kinetics of vibrational relaxation in chemically reacting systems in the approximation of level-to-level kinetics.

Some justification for such a "brute force" approach to problem solving is the rapid growth in the power of computers and mathematical software, reducing the solution of an emerging system of several tens of thousands of kinetic equations to a routine operation. The obvious advantage of this approach is detailed information on the dynamics of the distribution of molecules by vibrational states, the obvious disadvantage—the need to introduce a huge number of processes, often with poorly or completely unknown rate coefficients. For this reason, experimental data on the probabilities of individual processes must, of necessity, be supplemented by various scaling relations. For vibrational relaxation, such a relation is the Schwartz–Slavsky–Herzfeld (SSH) model, which well describes the dependence of the relaxation rate of different energy levels of an anharmonic oscillator at a known relaxation rate of the lower levels.

3.1.3.1 Vibrational Energy Exchange in Collisions of Neutral Particles

Vibrational energy exchange in the collision of neutral particles plays an important role in the formation of the distribution of molecules by internal energy, competing with the processes of population and relaxation of vibrational states by electron impact, as a result of chemical reactions and energy transfer to electronic degrees of freedom.

Under the conditions of the temperature range characteristic of chemistry ($T = 300 - 3{,}500$ K), the vibrational relaxation of a large number of molecules can be described with good accuracy using the scaling relations given by the SSH theory. Calculation of the rate coefficient of vibrational VT-, VV-, and VV'- energy exchange for single-quantum transitions between different vibrational levels from given values of the rate coefficient of transitions between the first and zero levels in collisions involving two-atom and multi-atom molecules:

$$AB(v=n) + M \rightarrow AB(v=n') + M \tag{3.15}$$

$$AB(v=n) + CD(w=m) \rightarrow AB(v=n') + CD(w=m') \tag{3.16}$$

is carried out using [6] expressions:

The rate coefficients of vibrational energy exchange between molecules:

- *VT exchange*:

$$k_{n+1,n}(T) = (n+1)k_{10}(T)\exp(-\delta_{VT} n)\exp(n\bar{h}c\omega_e x_e / kT) \tag{3.17}$$

$$k_{n,n+1}(T) = k_{n+1,n}(T)\exp(-\Theta/T)\exp(2n\bar{h}c\omega_e x_e / kT) \tag{3.18}$$

$$k_{0,1}(T) = k_{1,0}(T)\exp(-\Theta/T) \tag{3.19}$$

where

$$\delta_{VT} = \begin{cases} \dfrac{4\pi}{3}\dfrac{\omega_e x_e}{\alpha}\sqrt{\dfrac{\mu}{2kT}} & \text{with } \gamma_n < 20 \\[2ex] \left(\dfrac{4\pi(\omega_e - 2\omega_e x_e)}{\alpha}\sqrt{\dfrac{\mu}{2kT}}\right)^{2/3}\dfrac{\omega_e x_e}{\omega_e - 2\omega_e x_e} & \text{with } \gamma_n \geq 20 \end{cases} \tag{3.20}$$

$$\gamma_n = \dfrac{\pi(\omega_e - 2(n+1)\omega_e x_e)}{\alpha}\sqrt{\dfrac{\mu}{2kT}} \tag{3.21}$$

$$\mu = m_{AB}m_M / (m_{AB} + m_M) \tag{3.22}$$

- *VV-exchange* with transfer of energy defect to translational degrees of freedom when identical molecules collide:

$$k_{n+1,n}^{m,m+1}(T) = (m+1)(n+1)k_{10}^{01}(T)\exp(-\delta_{VV}|n-m|)$$
$$\times \exp\left((m-n)\bar{h}c\omega_e x_e / kT\right)\left[\frac{3}{2} - \frac{1}{2}\exp(\delta_{VV}|n-m|)\right] \quad (3.23)$$

$$k_{n,n+1}^{m+1,m}(T) = k_{n+1,n}^{m,m+1}(T)\exp\left(2(n-m)\bar{h}c\omega_e x_e / kT\right) \quad (3.24)$$

where

$$\delta_{VV} = \frac{4\pi}{3}\frac{\omega_e x_e}{\alpha}\sqrt{\frac{\mu}{2kT}} \quad (3.25)$$

$$\mu = m_{AB}m_{CD}/(m_{AB} + m_{CD}) \quad (3.26)$$

- *VV'-exchange* with transfer of energy defect to translational degrees of freedom when different molecules collide:

$$k_{n+1,n}^{\prime m,m+1}(T) = (m+1)(n+1)k_{10}^{\prime 01}(T)\exp(-|\delta_{VV}'m - \delta_{VV}n + \delta_{VV}p|)$$
$$\times \exp\left((m\bar{h}c\omega_e' x_e' - n\bar{h}c\omega_e x_e)/kT\right)\exp(\delta_{VV}p) \quad (3.27)$$

$$k_{n,n+1}^{\prime m+1,m}(T) = k_{n+1,n}^{\prime m,m+1}(T)\exp\left(\bar{h}c\left[(\omega_e - \omega_e x_e(n+2)) - (\omega_e' - \omega_e' x_e'(m+2))\right]/kT\right) \quad (3.28)$$

$$p = \left[\omega_e - \omega_e' - 2(\omega_e' x_e' - \omega_e x_e)\right]/2\omega_e x_e \quad (3.29)$$

$$\delta_{VV} = \frac{4\pi}{3}\frac{\omega_e x_e}{\alpha}\sqrt{\frac{\mu}{2kT}} \quad (3.30)$$

$$\delta_{VV}' = \frac{4\pi}{3}\frac{\omega_e' x_e'}{\alpha}\sqrt{\frac{\mu}{2kT}} \quad (3.31)$$

$$\mu = m_{AB}m_{CD}/(m_{AB} + m_{CD}) \quad (3.32)$$

where δ_{VT}, δ_{VV}, δ_{VV}' are factors to account for the effect of anharmonicity in the coupling formulas for VT-, VV-, and VV'-exchange; m_{AB}, m_{CD}, m_M – masses of molecules AB, CD, and particle M, respectively; μ – reduced mass of colliding particles; Θ – characteristic vibrational temperature; γ_n – adiabatic factor for the transition $n+1 \to n$; α – the inverse radius of the Born-Mayer interaction; ω_e, ω_e' – the vibrational constant of the given molecule (without hyphen) and of the collision partner molecule (with hyphen); x_e, x_e' – the anharmonicity constant of the vibrations of the given molecule (without hyphen) and of the collision partner molecule (with hyphen). Accounting for anharmonicity leads to a marked increase in the values of the energy exchange rate coefficient with an increasing level number compared to the values obtained for the harmonic oscillator model.

3.1.4 Processes of Vibrational Energy Exchange in the C–N–O–H System

Here we present the data on the processes and relaxation rate coefficients of the lower vibrational levels for a typical C–N–O–H system [7] (Tables 3.1 and 3.2).

TABLE 3.1
Processes of Vibrational Energy Exchanges in a System of Molecules Containing C, N, O, H Atoms

Process	Rate
$CO_2(00^01) + N_2(v=0) \leftrightarrow CO_2(00^00) + N_2(v=1)$	$W_{3,4}$
$CO_2(00^01) + M \leftrightarrow CO_2(11^10, 03^10) + M$	$W_{3,\Sigma}^M$
$CO_2(00^01) + CO(v=0) \leftrightarrow CO_2(00^00) + CO(v=1)$	$W_{3,9}$
$CO_2(00^01) + N_2O(00^00) \leftrightarrow CO_2(00^00) + N_2O(00^01)$	$W_{3,13}$
$CO_2(00^00) + N_2(v=1) \leftrightarrow CO_2(11^10, 03^10) + N_2(v=0)$	$W_{4,\Sigma}$
$CO_2(00^00) + CO(v=1) \leftrightarrow CO_2(11^10, 03^10) + CO(v=0)$	$W_{9,\Sigma}$
$CO_2(00^00) + O_2(v=1) \leftrightarrow CO_2(10^00, 03^10) + O_2(v=0)$	$W_{5,1}$
$CO_2(10^00) + N_2O(00^00) \leftrightarrow CO_2(00^00) + N_2O(10^00)$	$W_{1,11}$
$CO_2(00^00) + N_2O(00^01) \leftrightarrow CO_2(11^10) + N_2O(00^00)$	$W_{13,\Sigma}$
$CO_2(10^00) + M \leftrightarrow CO_2(02^00) + M$	$W_{1,2}^M$
$CO_2(01^10) + M \leftrightarrow CO_2(00^00) + M$	$W_{2,0}^M$
$N_2(v=1) + CO(v=0) \leftrightarrow N_2(v=0) + CO(v=1)$	$W_{4,9}$
$N_2(v=1) + O_2(v=0) \leftrightarrow N_2(v=0) + O_2(v=1)$	$W_{4,5}$
$N_2(v=1) + H_2O(000) \leftrightarrow N_2(v=0) + H_2O(010)$	$W_{4,7}$
$N_2(v=1) + N_2O(00^00) \leftrightarrow N_2(v=0) + N_2O(00^01)$	$W_{4,13}$
$N_2(v=1) + NO(v=0) \leftrightarrow N_2(v=0) + NO(v=1)$	$W_{4,10}$
$N_2(v=1) + M \leftrightarrow N_2(v=0) + M$	$W_{4,0}^M$
$NO(v=1) + O_2(v=0) \leftrightarrow NO(v=0) + O_2(v=1)$	$W_{10,5}^M$
$H_2O(010) + O_2(v=0) \leftrightarrow H_2O(000) + O_2(v=1)$	$W_{7,5}^M$
$O_2(v=1) + N_2O(00^00) \leftrightarrow O_2(v=0) + N_2O(10^00)$	$W_{5,11}^M$
$O_2(v=1) + M \leftrightarrow O_2(v=0) + M$	$W_{5,0}^M$
$H_2O(001) + M \leftrightarrow H_2O(100) + M$	$W_{8,6}^M$
$H_2O(001) + M \leftrightarrow H_2O(020) + M$	$W_{8,7}^M$
$H_2O(100) + M \leftrightarrow H_2O(020) + M$	$W_{6,7}^M$
$H_2O(010) + M \leftrightarrow H_2O(000) + M$	$W_{7,0}^M$
$CO(v=1) + H_2O(000) \leftrightarrow CO(v=0) + H_2O(010)$	$W_{9,7}$
$NO(v=1) + H_2O(000) \leftrightarrow NO(v=0) + H_2O(010)$	$W_{10,7}$
$H_2(v=1) + H_2O(000) \leftrightarrow H_2(v=0) + H_2O(001)$	$W_{14,8}$
$OH(v=1) + H_2O(000) \leftrightarrow OH(v=0) + H_2O(100)$	$W_{15,6}$
$CO(v=1) + NO(v=0) \leftrightarrow CO(v=0) + NO(v=1)$	$W_{9,10}$
$CO(v=1) + O_2(v=0) \leftrightarrow CO(v=0) + O_2(v=1)$	$W_{9,5}$
$N_2O(00^01) + CO(v=0) \leftrightarrow N_2O(00^00) + CO(v=1)$	$W_{13,9}$
$CO(v=1) + M \leftrightarrow CO(v=0) + M$	$W_{9,0}^M$
$N_2O(00^01) + NO(v=0) \leftrightarrow N_2O(00^00) + NO(v=1)$	$W_{13,10}$
$NO(v=1) + M \leftrightarrow NO(v=0) + M$	$W_{10,0}^M$
$N_2O(00^01) + M \leftrightarrow N_2O(11^10, 03^10) + M$	W_{13,Σ_1}^M
$N_2O(00^01) + M \leftrightarrow N_2O(12^00, 20^00, 04^00) + M$	W_{13,Σ_2}^M
$N_2O(10^00) + M \leftrightarrow N_2O(02^00) + M$	$W_{11,12}^M$
$N_2O(01^10) + M \leftrightarrow N_2O(00^00) + M$	$W_{12,0}^M$
$H_2(v=1) + OH(v=0) \leftrightarrow H_2(v=0) + OH(v=1)$	$W_{14,15}^M$
$H_2(v=1) + M \leftrightarrow H_2(v=0) + M$	$W_{14,0}^M$

TABLE 3.2
Rate Coefficients of Vibrational Energy Exchange Processes for Molecules Containing C, N, O, H Atoms

Process	$W_{p,q}$ (cm³/s)	References
$W_{3,\Sigma}^{CO_2}$	$1.36T \times 10^{-16} / \exp(9.456 - 218.23T^{-1/3} + 1687.7T^{-2/3} - 3909.27T^{-1})$	[8]
$W_{3,\Sigma}^{N_2}$	$1.36T \times 10^{-16} / \exp(15.456 - 424.03T^{-1/3} + 3852.67T^{-2/3} - 10672T^{-1})$	[8]
$W_{3,\Sigma}^{O_2}$	$0.141T \times 10^{(-6.303-122.27T^{-1/3}+384T^{-2/3})}\theta$	[9]
$W_{3,\Sigma}^{H_2O}$	$4 \times 10^{-15} T\theta$	[8]
$W_{3,\Sigma}^{CO}$	$0.141T \times 10^{(-9.143-84.23T^{-1/3}+271.7T^{-2/3})}\theta$	[9]
$W_{3,\Sigma}^{NO}$	$\exp(-16.234 - 171.359T^{-1/3} + 453.455T^{-2/3})$, $T > 400$ K	[8]
	$W_{3,\Sigma}^{NO}(T = 400 \text{ K})$, $T \leq 400$ K	
$W_{3,\Sigma}^{H_2}$	$0.141T \times 10^{(-14.363-3.73T^{-1/3}+32T^{-2/3})}\theta$	[9]
$W_{3,\Sigma}^{OH}$	$W_{3,\Sigma}^{H_2O}$	[8]
$W_{3,\Sigma}^{O}$	$0.141T \times 10^{(-7.956-122.2T^{-1/3}+384T^{-2/3})}\theta$	[9]
$W_{3,\Sigma}^{H}$	$0.8 \cdot W_{3,\Sigma}^{H_2}$	[8]
$W_{3,\Sigma}^{He}$	$103.36T \times 10^{(-7.69T^{-1/3}-16.1)}\theta$	[9]
$W_{3,\Sigma}^{N_2O}$	$W_{3,\Sigma}^{CO_2}$	[8]
$W_{2,0}^{CO_2}$	$0.141T \times 10^{(-10.673-57.31T^{-1/3}+156.7T^{-2/3})}$	[10]
$W_{2,0}^{N_2}$	$0.141T \times 10^{(-9.987-68.78T^{-1/3}+188.5T^{-2/3})}$	[10]
$W_{2,0}^{O_2}$	$0.141T \times 10^{(-24.458+35.957T^{-1/3}-32.4623T^{-2/3})}$	[10]
$W_{2,0}^{Ar}$	$1.36 \cdot 10^{-10} \times T / 10^{(10.011-49.4T^{-1/3}+77.3T^{-2/3})}$	[10]
$W_{2,0}^{H_2O}$	$0.141T \times 10^{(-11.864-2.1\times10^{-3}T+0.705\times10^{-6}T^2)}$	[10]
$W_{2,0}^{CO}$	$0.141T \times 10^{(-11.04-51.65T^{-1/3}+121.8T^{-2/3})}$	[10]
$W_{2,0}^{NO}$	$0.141T \times 10^{(-14.57+2.8T^{-1/3})}$	[10]
$W_{2,0}^{H_2}$	$0.141T \times 10^{(-12.097-28.46T^{-1/3}+144.5T^{-2/3})}$	[10]
$W_{2,0}^{OH}$	$W_{2,0}^{H_2O}$	[8]
$W_{2,0}^{O}$	$0.141T \times 10^{(-10.989-49.4T^{-1/3}+77.3T^{-2/3})}$	[9]
$W_{2,0}^{H}$	$0.9 \cdot W_{2,0}^{H_2}$	[8]
$W_{2,0}^{He}$	$0.141T \times 10^{(-12.289+14.75T^{-1/3})}$	[10]
$W_{3,4}$	$4.16 \times 10^{-14}\sqrt{T} \exp(8.842 \times 10^{-7}T^2 - 2.072 \times 10^{-3}T)$	[8]
$W_{3,9}$	$0.141T \times 10^{(-11.856-31.91T^{-1/3}+103.5T^{-2/3})}$	[9]
$W_{3,13}$	$103.36T \times 10^{-21} \exp(-10.773 + 10.814T^{-1/3} - 11.128T^{-2/3})$	[8]
$W_{13,\Sigma}$	$W_{4,\Sigma}$	[8]
$W_{4,\Sigma}$	$0.28 \cdot W_{3,\Sigma}^{N_2}$	[8]
$W_{9,\Sigma}$	$0.141T \times 10^{(-10.292-69.94T^{-1/3}+203.3T^{-2/3})}$	[9]
$W_{4,9}$	$0.141T \times 10^{(-15.2+0.0865(T^{-1/3}-0.1)^2)}$	[8]
$W_{1,11}$	$W_{3,13}$	[8]
$W_{4,5}$	$\exp(-15.397 - 281.451T^{-1/2} + 902.709T^{-2/3} - 638.17T^{-1})$	[8]
$W_{4,7}$	$\exp(-17.658 - 149.023T^{-1/3} + 347.214T^{-2/3} - 881.093T^{-1})$	[8]
$W_{4,10}$	$\exp(-25.894 - 41.416T^{-1/3} - 166.29T^{-2/3})$	[8]

(Continued)

TABLE 3.2 (*Continued*)
Rate Coefficients of Vibrational Energy Exchange Processes for Molecules Containing C, N, O, H Atoms

Process	$W_{p,q}$ (cm³/s)	References
$W_{4,13}$	$103.36T \times 10^{(-2.094T^{-1/3}+0.732T^{-2/3}-16.559)}$	[11]
$W_{4,0}^{CO_2}$	$\exp(-19.414 - 134.727T^{-1/3} + 253.19T^{-2/3} + 2551.7T^{-1})$	[8]
$W_{4,0}^{N_2}$	$0.141T \times 10^{(-9.079-142.84T^{-1/3}+431.4T^{-2/3})}$	[9]
$W_{4,0}^{CO}$	$0.141T \times 10^{-21} \exp(24.81 - 220T^{-1/3})$	[8]
$W_{4,0}^{O_2}$	$W_{4,0}^{N_2}$	[9]
$W_{4,0}^{H_2O}$	$0.141T \times 10^{(-14.4-9.2T^{-1/3})}$	[9]
$W_{4,0}^{H_2}$	$0.141T \times 10^{(-10.44-72.63T^{-1/3}+171.3T^{-2/3})}$	[9]
$W_{4,0}^{NO}$	$9.2 \times 10^{-12} T \exp(-221T^{-1/3})$	[8]
$W_{4,0}^{He}$	$0.141T \times 10^{(-9.913-85.16T^{-1/3}+171.1T^{-2/3})}$	[9]
$W_{4,0}^{N_2O}$	$W_{4,0}^{CO_2}$	[8]
$W_{4,0}^{OH}$	$W_{4,0}^{H_2}$	[8]
$W_{4,0}^{H}$	$3.1 \cdot W_{4,0}^{H_2}$	[8]
$W_{4,0}^{O}$	$0.141T \times 10^{(-13.929-14.38T^{-1/3})}$	[9]
$W_{10,5}$	$\exp(-15.397 - 281.451T^{-1/3} + 902.709T^{-2/3} - 638.17T^{-1})$	[8]
$W_{7,5}$	$1.39 \times 10^{-11} T^{-1/2}$	[8]
$W_{5,11}$	$0.141T \times 10^{(-15.003-11.71T^{-1/3})}$	[12]
$W_{5,0}^{CO_2}$	$1.34 \times 10^{-12} T \exp(-146.5T^{-1/3})$	[8]
$W_{5,0}^{N_2}$	$6.8 \times 10^{-13} \exp(-132T^{-1/3})$	[8]
$W_{5,0}^{O_2}$	$1.11 \times 10^{-8} \exp(-157T^{-1/3})$	[8]
$W_{5,0}^{Ar}$	$1.36 \times 10^{-16} \times T / 10^{(-4.04+69.7T^{-1/3})}$	[13]
$W_{5,0}^{H_2O}$	$100.0 \cdot W_{5,0}^{O_2}$	[8]
$W_{5,0}^{CO}$	$6.8 \times 10^{-13} T \exp(-131.6T^{-1/3})$	[8]
$W_{5,0}^{NO}$	$7.5 \times 10^{-13} T \exp(-134T^{-1/3})$	[8]
$W_{5,0}^{H_2}$	$10^{-17} \exp(17.1 - 91.5T^{-1/3})$	[8]
$W_{5,0}^{OH}$	$W_{5,0}^{H_2O}$	[8]
$W_{5,0}^{N_2O}$	$W_{5,0}^{CO_2}$	[8]
$W_{5,0}^{He}$	$0.454 \times 10^{-13} \exp(-60.85T^{-1/3})$	[8]
$W_{5,0}^{H}$	$1.7 \cdot W_{5,0}^{H_2}$	[8]
$W_{5,0}^{O}$	$2.9T \times 10^{-13} \exp(-111T^{-1/3})$	[8]
$W_{7,0}^{CO_2}$	$0.5 \cdot W_{7,0}^{O_2}$	[8]
$W_{7,0}^{N_2}$	$0.141T \times 10^{(-14.407-8.065T^{-1/3}+20.51T^{-2/3})}$	[10]
$W_{7,0}^{Ar}$	$1.36 \times 10^{-10} \times T / 10^{(0.8935-10.657T^{-1/3}+31.339T^{-2/3})}$	[10]
$W_{7,0}^{O_2}$	$0.141T \times 10^{(-17.38-42.036T^{-1/3}+116.515T^{-2/3})}$	[10]
$W_{7,0}^{H_2O}$	$0.141T \times 10^{(-6.627-99.092T^{-1/3}+404.127T^{-2/3})}$	[10]
$W_{7,0}^{CO}$	$W_{7,0}^{N_2}$	[8]
$W_{7,0}^{NO}$	$0.141T \times 10^{-21} \exp(10.85 + 36.8T^{-1/3} + 2,300T^{-1})$	[8]
$W_{7,0}^{N_2O}$	$W_{7,0}^{CO_2}$	[8]

(*Continued*)

TABLE 3.2 (*Continued*)
Rate Coefficients of Vibrational Energy Exchange Processes for Molecules Containing C, N, O, H Atoms

Process	$W_{p,q}$ (cm³/s)	References
$W_{7,0}^{H_2}$	$W_{7,0}^{H_2O} \exp(-19.8 + 280T^{-1/3} - 1{,}117T^{-2/3})$	[10]
$W_{7,0}^{OH}$	$W_{7,0}^{H_2O}$	[8]
$W_{7,0}^{He}$	$0.35 \cdot W_{7,0}^{H_2O}$	[10]
$W_{7,0}^{O}$	2.2×10^{-12}	[8]
$W_{7,0}^{H}$	$1.4 \cdot W_{7,0}^{H_2}$	[8]
$W_{9,7}$	$W_{4,7}$	[8]
$W_{10,7}$	$W_{7,5}$	[8]
$W_{14,8}$	$1/N_A \exp(35.6 - 67T^{-1/3} + 62.1T^{-2/3})$	[8]
$W_{9,10}$	$1.62 \times 10^{-11} \sqrt{T} \exp(1.7 \times 10^{-3}T - 9.7)$	[8]
$W_{9,5}$	$1.72 \times 10^{-11} \sqrt{T} \exp(-1.56 - 106T^{-1/3})$	[8]
$W_{13,9}$	$103.36T \times 10^{(-15.34 - 1.32T^{-1/3} + 0.489T^{-2/3})}$	[11]
$W_{9,0}^{CO_2}$	$0.141T \times 10^{(-8.18 - 155.91T^{-1/3} + 450.5T^{-2/3})}$	[9]
$W_{9,0}^{N_2}$	$W_{9,0}^{CO}$	[9]
$W_{9,0}^{CO}$	$0.141T \times 10^{(-7.335 - 155.91T^{-1/3} + 450.3T^{-2/3})}$	[9]
$W_{9,0}^{O_2}$	$W_{9,0}^{CO}$	[9]
$W_{9,0}^{H_2O}$	$0.141T \times 10^{-14.3}$	[9]
$W_{9,0}^{H_2}$	$0.141T \times 10^{(-10.711 - 64.35T^{-1/3} + 153.8T^{-2/3})}$	[9]
$W_{9,0}^{N_2O}$	$W_{9,0}^{CO_2}$	[8]
$W_{9,0}^{NO}$	$W_{9,0}^{CO}$	[8]
$W_{9,0}^{OH}$	$W_{9,0}^{H_2O}$	[8]
$W_{9,0}^{He}$	$0.141T \times 10^{(-11.032 - 65.35T^{-1/3} + 113.6T^{-2/3})}$	[9]
$W_{9,0}^{O}$	$0.141T \times 10^{(-11.83 - 23.45T^{-1/3})}$	[9]
$W_{9,0}^{H}$	$0.141T \times 10^{(-13.34 - 1.3T^{-1/3})}$	[9]
$W_{13,10}$	$0.141T \times 10^{(-12.75 - 18.3T^{-1/3})}$	[13]
$W_{10,0}^{CO_2}$	$0.141T \times 10^{(-16.08 + T^{-1/3})}$	[8]
$W_{10,0}^{N_2}$	$1.8T \times 10^{-12} \exp(-165T^{-1/3})$	[8]
$W_{10,0}^{O_2}$	$2.2 \times 10^{-12} T \exp(-171T^{-1/3})$	[8]
$W_{10,0}^{Ar}$	$1.36 \times 10^{-16} \times T^2 / 10^{(-5.69 + 47.3T^{-1/3})}$	[13]
$W_{10,0}^{H_2O}$	$0.141T \times 10^{(-14.7 + T^{-1/3})}$	[8]
$W_{10,0}^{CO}$	$0.141T \times 10^{(-16.7 + T^{-1/3})}$	[8]
$W_{10,0}^{N_2O}$	$W_{10,0}^{CO_2}$	[8]
$W_{10,0}^{H_2}$	$W_{9,0}^{H}$	[8]
$W_{10,0}^{OH}$	$W_{10,0}^{H_2O}$	[8]
$W_{10,0}^{NO}$	$0.141T \times 10^{(-14.7 + T^{-1/3})}$	[8]
$W_{10,0}^{He}$	$0.141T \times 10^{(-16.4 + T^{-1/3})}$	[8]
$W_{10,0}^{O}$	3.6×10^{-11}	[8]
$W_{10,0}^{H}$	$W_{9,0}^{H}$	[8]

(*Continued*)

TABLE 3.2 (*Continued*)
Rate Coefficients of Vibrational Energy Exchange Processes for Molecules Containing C, N, O, H Atoms

Process	$W_{p,q}$ (cm³/s)	References
$W_{12,0}^{CO_2}$	$W_{12,0}^{N_2O}$	[8]
$W_{12,0}^{N_2}$	$0.141T\Psi \times 10^{\left(-12.62-16.7T^{-1/3}\right)}$	[10]
$W_{12,0}^{Ar}$	$1.36 \times 10^{-10} \times T / 10^{\left(7.584-8.619T^{-1/3}-47.62T^{-2/3}\right)}$	[10]
$W_{12,0}^{O_2}$	$0.141T\Psi \times 10^{\left(-12.545-17.55T^{-1/3}\right)}$	[10]
$W_{12,0}^{H_2O}$	$0.141T\Psi \times 10^{\left(-12.15-2.04\times10^{-3}T+0.892T^2\times10^{-6}\right)}$	[10]
$W_{12,0}^{CO}$	$0.141T\Psi \times 10^{\left(-9.601-65.345T^{-1/3}+196.47T^{-2/3}\right)}$	[10]
$W_{12,0}^{N_2O}$	$0.141T\Psi \times 10^{\left(-12.036-32.88T^{-1/3}+88.954T^{-2/3}\right)}$	[10]
$W_{12,0}^{H_2}$	$10^{-14}T\Psi$	[10]
$W_{12,0}^{NO}$	$0.141T\Psi \times 10^{\left(-13.92-1.6T^{-1/3}\right)}$	[10]
$W_{12,0}^{OH}$	$W_{12,0}^{H_2O}$	[8]
$W_{12,0}^{He}$	$0.141T\Psi \times 10^{\left(-13.15-5.75T^{-1/3}\right)}$	[10]
$W_{12,0}^{O}$	$4.9 \times 10^{-12}\sqrt{T}\exp(-1820/T)$	[8]
$W_{12,0}^{H}$	$0.9 \cdot W_{6,0}^{H_2}$	[8]
$W_{13,\Sigma_1}^{CO_2}$	$W_{13,\Sigma_1}^{N_2O}$	[8]
$W_{13,\Sigma_1}^{N_2}$	$W_{13,\Sigma_1}^{O_2}$	[11]
$W_{13,\Sigma_1}^{O_2}$	$103.36T \times 10^{\left(-13.99-45.34T^{-1/3}+96.81T^{-2/3}\right)} \cdot \phi$	[11]
$W_{13,\Sigma_1}^{H_2O}$	$3.1 \times 10^{-15}T\exp(190T^{-1}) \cdot \phi$	[11]
W_{13,Σ_1}^{CO}	$103.36T \times 10^{\left(-15.549-18.14T^{-1/3}\right)} \cdot \phi$	[11]
W_{13,Σ_1}^{NO}	$103.36T \times 10^{\left(-13.291-643T^{-1/3}+224.28T^{-2/3}\right)} \cdot \phi$	[11]
W_{13,Σ_1}^{Ar}	$1.36 \times 10^{-16} \times T / 10^{\left(-2.68+26.2T^{-1/3}\right)}$	[13]
$W_{13,\Sigma_1}^{H_2}$	$103.36T \times 10^{\left(-16.53-4.8T^{-1/3}\right)} \cdot \phi$	[11]
W_{13,Σ_1}^{OH}	$W_{7,\Sigma_1}^{H_2O}$	[8]
W_{13,Σ_1}^{O}	$W_{13,\Sigma_1}^{CO_2}\left(W_{3,\Sigma}^{O}/W_{3,\Sigma}^{CO_2}\right)$	[8]
W_{13,Σ_1}^{H}	$W_{13,\Sigma_1}^{CO_2}\left(W_{3,\Sigma}^{H}/W_{3,\Sigma}^{CO_2}\right)$	[8]
W_{13,Σ_1}^{He}	$103.36T \times 10^{\left(-12.086-82.26T^{-1/3}+260.27T^{-2/3}\right)} \cdot \phi$	[11]
$W_{14,0}^{CO_2}$	$0.07 \cdot W_{14,0}^{H_2}$	[8]
$W_{14,0}^{N_2}$	$0.08 \cdot W_{14,0}^{H_2}$	[8]
$W_{14,0}^{O_2}$	$0.08 \cdot W_{14,0}^{H_2}$	[8]
$W_{14,0}^{H_2O}$	$0.23 \cdot W_{14,0}^{H_2}$	[8]
$W_{14,0}^{CO}$	$0.08 \cdot W_{14,0}^{H_2}$	[8]
$W_{14,0}^{NO}$	$0.08 \cdot W_{14,0}^{H_2}$	[8]
$W_{14,0}^{N_2O}$	$W_{14,0}^{CO_2}$	[8]
$W_{14,0}^{H_2}$	$2.18 \times 10^{-9}\exp(-144.9T^{-1/3})$	[8]
$W_{14,0}^{OH}$	$W_{14,0}^{H_2O}$	[8]
$W_{14,0}^{O}$	2.2×10^{-12}	[8]
$W_{14,0}^{H_2}$	$1.73 \times 10^{-14}\sqrt{T}$	[8]
$W_{14,0}^{He}$	$1.363 \times 10^{-22}\exp(8.958-45.09T^{-1/3})$	[8]

3.2 CHEMICAL REACTION RATES AND VIBRATIONAL EXCITATION OF PRODUCTS IN NONEQUILIBRIUM CONDITIONS

The mechanism of vibrational energy exchanges allows us to calculate the distribution function of molecules on vibrational states. To describe chemical kinetics under nonequilibrium conditions, it is also necessary to know the dependence of the reaction rate coefficient on the vibrational excitation of the reactant. Let us consider an exchange-type reaction with energy release

$$AB + C \to A + BC + \Delta H \tag{3.33}$$

in which the molecules AB and BC may have a vibrational energy distribution different from the equilibrium (Boltzmann) distribution. Let the rate of this reaction at equilibrium energy distribution of the reactants be described by the law of acting masses, and let the experimentally measured rate coefficient have an Arrhenius dependence on temperature:

$$W = k[AB][C] \tag{3.34}$$

$$k = k_0 \exp\{-E_a/T\} \tag{3.35}$$

where W is the rate of the process, [AB] and [C] are the concentrations of the AB and C components, E_a is the activation energy. The integral reaction (3.33) proceeds through elementary channels of the form:

$$AB(v) + C \to A + BC(w) \tag{3.36}$$

v, w – the numbers of vibrational levels in which the molecules AB and BC are located, respectively. The rate of such a process will be determined by the ratio:

$$W^{v,w} = k^{v,w}[AB(v)][C] \tag{3.37}$$

where

$$k^{v,w} = k_0^{v,w} \exp\{-E_a^{v,w}/T\} \tag{3.38}$$

and $E_a^{v,w}$ is the activation energy of the process (3.36). For $k^{v,w}$, the normalization condition should be satisfied:

$$\sum_{v,w} W^{v,w} = W \tag{3.39}$$

at the equilibrium (Boltzmann) distribution of AB molecules in terms of vibrational energy. To determine the numerical values of $k_0^{v,w}$ and $E_a^{v,w}$, the vibronic term model was used [7,14,15]. In this approximation, the profile of the potential energy surface corresponding to the $AB(v) + C$ interaction is obtained by a parallel upward shift by the value E_{via}^{AB} of the surface corresponding to the AB+C state. The shape of the potential energy surface remains unchanged (Figure 3.1).

When the interaction between the states $AB(v) + C$ and $BC(w) + A$ is taken into account, the shape of the potential energy surface near the intersection point, generally speaking, changes. In this region, the true potential energy surface lies below the original surfaces, and the magnitude of the barrier lowering is determined by the interaction matrix element [16]. It was assumed [7] that

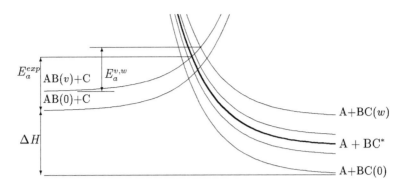

FIGURE 3.1 The structure of the potential energy curves of the reaction AB+C → A+BC. The bold line indicates the term corresponding to the velocity averaged over all vibrational levels of the products, for which $E_a = E_a^{exp}$ [15].

the interaction region of the states is small enough and lies near the barrier top, i.e. the potential energy surfaces everywhere, except, perhaps, for a narrow zone corresponding to either $AB(v) + C$ or $BC(w) + A$.

The application of this model is most justified for electronically non-adiabatic processes, when the interaction between the AB+C and A+BC states affects the potential curves of the corresponding states only in a small region near the point of transition from one state to another [14].

3.2.1 Change of Reaction Threshold and Transition Probability under Excitation of Reactants

One can approximate the initial terms by exponential dependences on the coordinate (r_{AB-C} and r_{A-BC}, respectively, with decreasing parameters r_1 and r_2), then the potential energy surfaces of the exothermic reaction (3.36) can be represented in the form of Ref. [15]:

$$U_1 = \Delta H + E_a^{00} \exp\{r / r_1\} \tag{3.40}$$

for reagents and

$$U_2 = \left(\Delta H + E_a^{00}\right) \exp\{-r / r_2\} \tag{3.41}$$

for the products, where E_a^{00} is the activation energy of the process.

$$AB(v = 0) + C \rightarrow A + BC(w = 0) \tag{3.42}$$

Within the framework of the made assumptions, the value of the barrier $E_a^{v,w}$ of the elementary process (3.36) is determined from the system of equations [15]:

$$E_{vib}^{AB}(v) + \Delta H + E_a^{00} \exp\{r^*/r_1\} = E_{vib}^{BC}(w) + \left(\Delta H + E_a^{00}\right)\exp\{-r^*/r_2\} \tag{3.43}$$

$$E_a^{v,w} = E_a^{00} \exp\{r^* / r_1\} \tag{3.44}$$

where r^* is the coordinate of the intersection point of the vibronic terms, $E_{vib}^{AB}(v)$ and $E_{vib}^{BC}(w)$ are the vibrational energies of molecules AB and BC in states with vibrational quantum numbers v and w,

Energy Exchange and Chemical Reactions in Nonequilibrium Regimes

respectively. In most cases, the interaction radii of reactants r_1 and products r_2 are close values. Assuming $r_1 = r_2$, the solution of this system has a simple form [15]:

$$E_a^{v,w} = \frac{1}{2}\left[\sqrt{\left(\Delta H + E_{\text{vib}}^{AB}(v) - E_{\text{vib}}^{BC}(w)\right)^2 + 4E_a^{00}\left(E_a^{00} + \Delta H\right)} - \left(\Delta H + E_{\text{vib}}^{AB}(v) - E_{\text{vib}}^{BC}(w)\right)\right] \quad (3.45)$$

3.2.2 Evaluation of Transition Probability for Selected Levels

The $k_0^{v,w}$ represents the probability of an electronically non-adiabatic transition corresponding to the rearrangement of the electronic structure of the A-B-C complex from the state correlated with the reactants to the state describing the reaction products occurring near the barrier top. The dependence of this quantity on v and w determines, together with $E_a^{v,w}$, the reaction rate coefficient (3.36) and the vibrational energy distribution of the products. The semi-classical [16] approximation was used to estimate the probability of such a transition.

In Ref. [16] it is shown that the [17] expression can be obtained for the transition probability (Landau-Ziner model):

$$P_{12} = |a_2(+\infty)|^2 \simeq \exp\left[\frac{2}{\hbar}\text{Im}\int^{t_c}\Delta U(t)dt\right], \quad (3.46)$$

Thus, in order to estimate the probability of transition P_{12} to a state correlated with the reaction products (the probability of the reaction proceeding provided that the energy of the system is sufficient to overcome the potential barrier), it is necessary to estimate the effective collision time t_c and the characteristic value of the energy gap during the transition to the new state ΔU.

As indicated in Ref. [18], the probability of a non-adiabatic transition is maximized in the region of maximum overlap of the translational wave functions near the turning point (see Figure 3.2). Assuming that most of the molecules entering the reaction have initial energies of translational motion that are only slightly higher than the reaction barrier $E_a^{v,w}$, in Ref. [7] it was assumed that the turning point of the system is located exactly at the top of the reaction barrier, i.e., at the point of intersection of the vibronic terms.

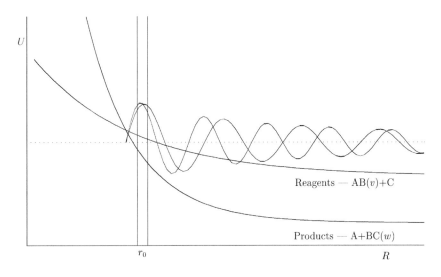

FIGURE 3.2 Region of maximum overlap of wave functions in AB - C collision [15].

The distance from the turning point to the region where the transition probability is maximized could be estimated as $r_0 \simeq \lambda/2$, where λ is the de Broglie wavelength of the system [15]. The value of λ near the turning point when moving along the surface U_1 in the coordinate system of the center of mass of the system can be estimated from the relation:

$$\lambda \simeq \frac{\hbar}{p} = \frac{\hbar}{\mu \bar{u}} \qquad (3.47)$$

where $\bar{u}$ is the average velocity in the interaction region, μ is the reduced mass of the system. Given the assumption made that the intersection point of the vibronic terms is a turning point and the relation $u = Ft/\mu$ [16], for $\bar{u}$ we have [15]:

$$\bar{u} = \frac{F}{2\mu} t_c \qquad (3.48)$$

which, in conjunction with (3.47):

$$\lambda \simeq \frac{2\hbar}{Ft_c} \qquad (3.49)$$

where t_c is the characteristic interaction time, λ is the de Broglie wavelength corresponding to the system near the turning point, $F = U'$ is the slope of the potential energy surface at that point.

Since the overlap of the wave functions is determined by both λ_1 and λ_2, the estimate for r_0 was obtained [15]:

$$r_0 \simeq \bar{\lambda}/2 = \frac{\lambda_1 + \lambda_2}{4} = \frac{\hbar}{2t_c} \left\{ \frac{F_1 + F_2}{F_1 F_2} \right\} \qquad (3.50)$$

The energy difference between the potential energy surfaces corresponding to the reactants and products at the point $r = r_0$ is equal to Ref. [15]:

$$\Delta U = |U_1(r) - U_2(r)| \simeq |F_1 - F_2| r_0 = \Delta F r_0 \qquad (3.51)$$

where

$$F_1(r) = U_1'(r) = E_a^{00}/r_1 \exp(-r/r_1), \qquad (3.52)$$

$$F_2(r) = U_2'(r) = (\Delta H + E_a^{00})/r_2 \exp(-r/r_2). \qquad (3.53)$$

Considering that the transitions between the terms U_1 and U_2 mainly occur in the region $r \simeq r_0$ [18], we neglect the transitions in the rest of the region. Hence for the transition probability P_{12} in Ref. [15] it was derived:

$$P_{12} \simeq \exp\left[\frac{2}{\hbar} \operatorname{Im} \int^{t_c} \Delta U(t) dt \right] \simeq \exp\left[-\left|\frac{F_1}{F_2} - \frac{F_2}{F_1}\right|\right] \qquad (3.54)$$

Thus, the value $\left|\dfrac{F_1}{F_2} - \dfrac{F_2}{F_1}\right|$ plays the role of the Messy parameter in this case and determines the transition probability from reactants to products. Taking into account the relation (3.54) for the value of $k^{v,w}$:

Energy Exchange and Chemical Reactions in Nonequilibrium Regimes

$$k^{v,w} = A\exp\left\{-\left|\frac{F_1}{F_2} - \frac{F_2}{F_1}\right|\right\}\exp\left(-E_a^{v,w}/kT\right) \qquad (3.55)$$

where

$$\frac{F_1}{F_2} = \frac{\left(E_a^{v,w}\right)^2}{\left(E_a^{00} + \Delta H\right)E_a^{00}}, \qquad (3.56)$$

A – the normalization multiplier.

Thus, the system of equations (3.45) and (3.55) determines the value of the rate coefficient of the process (3.36) at known values of A, E_a^{00} and ΔH, without requiring explicit consideration of other molecular constants [15].

Let us now turn to the problem of recovering the values of A and E_a^{00} from the experimentally measured in the quasi-equilibrium regime reaction activation energy (3.33) $E_a^{\exp}$ and the pre-exponential multiplier $k_0^{\exp}$, which are values averaged over all possible reaction channels (3.36). Obviously, the sum rate of the processes (3.36) under Boltzmann energy distribution of the reactants must be equal to the reaction rate (3.33) measured under thermally equilibrium conditions; their temperature derivatives must also coincide. This gives the two equations needed to determine A and E_a^{00} [15]:

$$\sum_{v,w}\left\{f_b^v(T)\times k_{v,w}\right\} = k_0^{\exp}\exp\left(-\frac{E_a^{\exp}}{kT}\right)$$

$$\sum_{v,w}\left\{f_b^v(T)\times k_{v,w}\right\}' = k_0^{\exp}\left\{\exp\left(-\frac{E_a^{\exp}}{kT}\right)\right\}' \qquad (3.57)$$

where "'" denotes the temperature derivative, $f_b^v(T)$ is the Boltzmann distribution function of the reactants over vibrational levels at a given temperature $T_{\text{vib}} = T_{\text{tr}} = T$.

3.3 DEPENDENCE OF REACTION RATES ON VIBRATIONAL ENERGY OF REAGENTS AND PRODUCTS

Thus, the equations (3.55) and (3.57) allow us to estimate the values of the rate coefficients of reactions between vibrationally excited reactants from the known reaction rate coefficient and its temperature dependence in quasi-equilibrium conditions [15].

However, it should be noted that the assumptions explicitly or implicitly introduced in the construction of this model (model of vibronic terms, size of the effective transition region, collision geometry) introduce uncertainty in the evaluation of the area of its possible application. With this in mind, it is necessary to treat the model (3.55) and (3.57) as a semiempirical model and to test it on a wide class of reactions at different ratios of the values of the translational and internal energy of the reactants.

Let us single out for analysis several cases important for practical application:

- Monomolecular decomposition at $T_{\text{tr}} > T_{\text{vib}}$:

$$N_2(v) + M \to N + N + M \qquad (3.58)$$

- Reactions in N_2-O_2 mixtures at $T_{\text{tr}} \neq T_{\text{vib}}$:

$$N_2(v) + O \rightarrow NO + N \tag{3.59}$$

$$N + O_2(v) \rightarrow NO + O \tag{3.60}$$

- Reactions in H_2-O_2 mixtures at $T_{tr} \ll T_{vib}$:

$$O + H_2(v) \rightarrow OH + H \tag{3.61}$$

$$OH + H_2(v) \rightarrow H_2O + H. \tag{3.62}$$

- The distribution of exchange reaction products by vibrational levels.

$$H + BrCl \rightarrow HBrr(w) + Cl. \tag{3.63}$$

$$Cl + F_2 \rightarrow ClF(w) + F \tag{3.64}$$

It should be emphasized that a significant part of the above reactions does not belong to the electronically non-adiabatic processes and the application of the model (3.55) and (3.57) to them should be considered as an analysis of the possibility of constructing an empirical methodology for calculating the rate of processes involving vibrationally excited reagents for the widest possible class of reactions.

Preliminary analysis has shown that the best match with the data available in the literature on the dependence of the rate coefficients of chemical reactions on the degree of nonequilibrium T_{vib}/T_{tr}, the distribution of products by vibrational degrees of freedom, and the ratio of reaction rates involving excited and unexcited reactants can be achieved by making a correction for the change in the shape of the potential energy surface at the growth of vibrational excitation of reactants and products.

In this case, the equation (3.43) changes to

$$E_{vib}^{AB}(v) + \Delta H + E_1 \exp\{r^*/r_1\} = E_{vib}^{BC}(w) + (\Delta H + E_2)\exp\{-r^*/r_2\} \tag{3.65}$$

$$E_a^{v,w} = E_1 \exp\{r^*/r_1\} \tag{3.66}$$

$$E_1 = \begin{cases} E_a^{00} - \gamma E_{vib}^{AB}(v), & \text{if } E_a^{00} - \gamma E_{vib}^{AB}(v) > \Delta E \\ \Delta E, & \text{if } E_a^{00} - \gamma E_{vib}^{AB}(v) \leq \Delta E \end{cases} \tag{3.67}$$

$$E_2 = \begin{cases} E_a^{00} - \gamma E_{vib}^{BC}(w), & \text{if } E_a^{00} - \gamma E_{vib}^{BC}(w) > \Delta E \\ \Delta E, & \text{if } E_a^{00} - \gamma E_{vib}^{BC}(w) \leq \Delta E \end{cases} \tag{3.68}$$

where the terms $\gamma E_{\text{vib}}^{\text{AB}}(v)$, $\gamma E_{\text{vib}}^{\text{BC}}(w)$ allow us to account for the deviation from the model of vibronic terms at large magnitude of vibrational excitation by specifying the relative decrease in the magnitude of the reaction energy barrier at vibrational excitation of reactants and products.

Then at $\gamma = 0$, equations (3.40) and (3.41) pass to the initial model of vibronic terms, and at $0 < \gamma < 1$ the relative height of the reaction energy barrier decreases with the growth of the vibrational level to some minimum level $\Delta E > 0$.

The equation (3.45) goes to

$$E_a^{v,w} = \frac{1}{2}\left[\sqrt{\left(\Delta H + E_{\text{vib}}^{\text{AB}}(v) - E_{\text{vib}}^{\text{BC}}(w)\right)^2 + 4E_1(E_2 + \Delta H)} - \left(\Delta H + E_{\text{vib}}^{\text{AB}}(v) - E_{\text{vib}}^{\text{BC}}(w)\right)\right] \quad (3.69)$$

of the equation (3.55) and (3.57) in the

$$F_1(r) = U_1'(r) = E_1/r_1 \exp(-r/r_1), \quad (3.70)$$

$$F_2(r) = U_2'(r) = (\Delta H + E_2)/r_2 \exp(-r/r_2). \quad (3.71)$$

and the equations (3.55) and (3.57) remain unchanged.

The condition $\Delta E > 0$ is required to ensure that the terms correlated with reactants and products overlap. In all the examples below

$$\Delta E = \frac{\left(\theta^{\text{AB}} + \theta^{\text{BC}}\right)}{250}$$

where θ^{AB}, θ^{BC} – the magnitude of the vibrational quantum of the reactant and reaction product, respectively. At such a choice of ΔE value the sensitivity of the calculation results to its variations by two to three times in any direction is practically absent.

The value of γ, which describes the relative lowering of the reaction energy barrier at high excitation levels, can be considered a parameter of the problem and can be selected in such a way as to ensure the best agreement of the model calculations (3.55) and (3.57) with the experimental data. The choice of a particular value of γ has little effect on reactions between unexcited components and becomes essential in the description of reactions with a large energy barrier between strongly excited reactants.

As a result of the analysis, it was obtained that the best agreement of the results of calculation by (3.55) and (3.57) with the data of other authors is achieved in a wide range of parameters at

$$\gamma = 0.3$$

This value was used in all the test calculations performed.

The results of approbation of the model (3.55) and (3.57) for some typical processes and their comparison with the data of other authors are presented below. The convolution of $k_{v,w}$ with the Boltzmann distribution function over vibrational levels at $T = T_{\text{vib}}$ was used for comparison with the results of other works, represented as a function of vibrational and translational temperatures:

$$k(T_{\text{tr}}, T_{\text{vib}}) = \sum_{v,w}\left\{f_b^v(T_{\text{vib}}) \times k_{v,w}(T_{\text{tr}})\right\} \quad (3.72)$$

where $f_b^v(T)$ is the Boltzmann distribution function of the reactants over vibrational levels at temperature $T = T_{\text{vib}}$.

3.3.1 Monomolecular Decomposition at $T_{tr} > T_{vib}$

Interest in monomolecular decay reactions is related to the problems of modeling aircraft re-entry and chemical reactions behind strong shock waves. A review of the numerous models developed to date to describe such processes is given, for example, in Ref. [19].

One of the most intensively studied reactions of monomolecular decomposition in the nonequilibrium regime is the process $N_2(v) + M \rightarrow N + N + N + M$ (3.60).

To model the dependence of the rate coefficient of monomolecular decay on the degree of vibrational nonequilibrium according to the model (3.55) and (3.57), it should be noted that the decay of molecule AB upon collision with partner C can be formally described by the scheme (3.36) with a quasi-continuous energy spectrum and zero lifetime for reaction product BC. Figure 3.3 shows the results of the reaction rate coefficient (3.58) at $T_{vib}/T_{tr} < 1$ obtained using different models.

It can be seen that up to the value of deviation from equilibrium $T_{vib}/T_{tr} = 0.5$, the results from the model (3.55) and (3.57) coincide well with the data obtained from the Macheret-Rich-Friedman model [23,24]. For larger values of deviation from equilibrium, the [23,24] model predicts higher dissociation rates than the (3.55) and (3.57) model. This is due to the fact that the [23,24] model takes into account multi-quantum VT-transitions, which form at large degrees of deviation from equilibrium an essentially non-Boltzmann distribution over the upper vibrational levels due to a decrease in the magnitude of the vibrational quantum and an increase in the rate of energy exchange with translational degrees of freedom.

The effective population temperature of the upper levels can reach the value of the translational one, which significantly accelerates the dissociation process (Figure 3.3). This effect fundamentally depends on the prehistory of the process (in particular, on the way of creating a nonequilibrium distribution) and is not universal.

Therefore, when modeling by the (3.55) and (3.57) model, the presence of a Boltzmann distribution over vibrational levels was assumed, which led to a decrease in the value of the rate coefficient at $T_{vib}/T_{tr} \ll 0.5$ compared to the [23,24] data. Taking into account the formation of the Boltzmann distribution over vibrational levels in the upper part of the energy spectrum, the model (3.55) and (3.57) gives a dependence of the dissociation rate coefficient on T_{vib}/T_{tr} similar to that of Refs. [23,24].

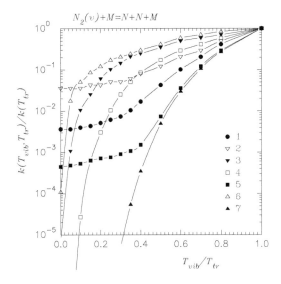

FIGURE 3.3 Dependence of the rate coefficient of monomolecular decay of N_2 on the degree of nonequilibrium T_{vib}/T_{tr} at $T_{tr} = 2 \times 10^4$ K. 1 – [20], $U = D/6k$; 2 – [20], $U = D/3k$; 3 – [21]; 4 – [22]; 5 – [23, 24]; 6 – [25]; 7 – models (3.55) and (3.57) [15,26].

The difference between the calculation results of Ref. [23,24] and (3.55) and (3.57) and the coarser Treanor [20] and Park [21,22] and Brun [25] models is quite large in the whole parameter region (Figure 3.3).

3.3.2 Reactions in N_2-O_2 Mixtures

The reaction $N_2 + O \rightarrow NO + N$ is one of the key reactions in describing the kinetics of chemical reactions in strong shock waves in air. The experimentally obtained estimate for the vibrational energy utilization factor in this process is $\alpha \simeq 0.51$ [27].

Figure 3.4 shows the results of the calculation of the reaction rate coefficient (3.59) as a function of the ratio between vibrational and translational temperatures T_{vib}/T_{tr} at a fixed translational temperature $T_{tr} = 15{,}000$ K for model (3.55) and (3.57), α-model [14] at $\alpha = 0.51$ and the Park model [28] at $s = 0.5$, which has become some benchmark of comparison for different models of description of high-temperature kinetics under thermally nonequilibrium conditions.

It can be seen that the (3.55) and (3.57) models provide good agreement between the dependence of the rate coefficient value on the vibrational excitation of the reactants and calculations based on the experimentally obtained value of $\alpha \simeq 0.51$ over a wide range of conditions.

At the same time, a sharper dependence of the calculation results for (3.55) and (3.57) on the vibrational temperature at $T_{vib} < T_{tr}$ compared to the data obtained from the Park model (Figure 3.4) is noticeable. Note that the α-model is not applicable in this range of parameters [19].

Another reaction of importance in describing nonequilibrium kinetics in air at high temperatures is the process $N + O_2(v) \rightarrow NO + O$, which has a much lower activation energy ($\Theta_{refr2} = 3{,}150$ K, $\Theta_{refr1} = 38{,}370$ K) and less sensitivity to the degree of vibrational excitation of the reactants (the experimentally measured value of the vibrational energy utilization factor $\alpha_{3.62} \simeq 0.24$ [27]). Figure 3.5 shows the results of the reaction rate coefficient (3.60) at translational temperature $T_{tr} = 300$ K with significant vibrational superheating of the reactants. The results obtained from the model (3.55) and (3.57) agree well with the calculations when using the experimentally measured value $\alpha = 0.24$ [14].

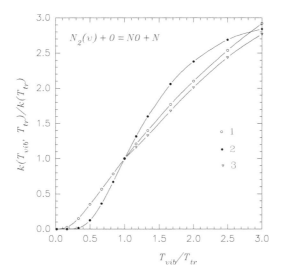

FIGURE 3.4 Dependence of the reaction rate coefficient $N_2(v) + O \rightarrow NO + N$ on the degree of nonequilibrium T_{vib}/T_{tr} at $T_{tr} = 1.5 \times 10^4$ K. 1 – model (3.55) and (3.57) [15]; 2 – [28], $s = 0.5$; 3 – α-model, [19], $\alpha = 0.51$ [26].

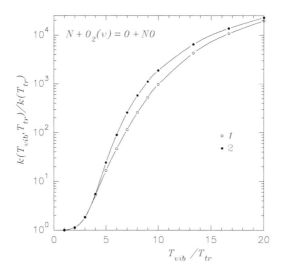

FIGURE 3.5 Dependence of the reaction rate coefficient $N+O_2(v) \to NO+O$ on the degree of nonequilibrium T_{vib}/T_{tr} at $T_{tr} = 300$ K. 1 – model (3.55) and (3.57) [15]; 2 – α-model, [19], $\alpha = 0.24$ [26].

3.3.3 Reactions in H_2-O_2 System

The reactions between excited hydrogen molecules $H_2(v)$ and radicals, the rate of which strongly depends on the magnitude of vibrational excitation of the gas, are of great interest for describing the processes of thermally nonequilibrium ignition.

One well-studied example of such reactions is the process $H_2(v) + O \to H + OH(w)$ [29]. It is shown in Ref. [30] that the ratio of specific reaction rate coefficients at $v = 1$ and at $v = 0$ is $k(v = 1)/k(v = 0) = 2,600$ at $T = 300$ K. A characteristic feature of this process is that the formation of OH radical occurs predominantly in the vibrationally excited [30] state:

$$O(^3P) + H_2(v = 1) \to OH(v = 1) + H, \quad k = \left(1.0^{+0.4}_{-0.6}\right) \times 10^{-14} \quad cm^3/s \tag{3.73}$$

$$O(^3P) + H_2(v = 1) \to OH(v = 0) + H, \quad k \leq 4.7 \times 10^{-15} \quad cm^3/s \tag{3.74}$$

The average vibrational energy utilization factor in this reaction from experimental measurements is $\alpha = 0.31$ [14]. Figure 3.6 shows the results of the calculation of the reaction rate coefficient (3.61) at translational temperature $T_{tr} = 300$ K for different vibrational temperatures at Boltzmann distribution of molecules by energy levels. The dependence calculated from (3.55) and (3.57) agrees well with the calculation from the α-model at $\alpha = 0.31$ at $T_{vib}/T_{tr} < 5$ (Figure 3.6). At higher values of T_{vib}/T_{tr}, the α-model calculation predicts a sharper increase in the reaction rate coefficient. It should be noted, however, that such degrees of nonequilibrium have not been investigated experimentally for this system, and the predictions of analytical models in this range of parameters are rather qualitative in nature.

At $T_{tr} = 300$ K, the model calculation (3.55) and (3.57) gives the ratio of specific reaction rate coefficients for $H_2(v = 1)$ and $H_2(v = 0)$ $k(v = 1)/k(v = 0) = 2,795$, which is in excellent agreement with experiment [30]. The ratio of the channels (3.73) and (3.74) at $T_{tr} = 300$ K is $k_{3.75}/k_{3.76} = 7.9$, which also agrees well with the [30] estimate.

In Ref. [31], an Arrhenius expression for the rate coefficient for reaction (3.73) was theoretically derived. The activation energy calculated in this work was $\Theta_{3.75} = 1,868$ K, which correlates well with the results of the model (3.55) and (3.57) predicting $\Theta_{3.75} = 2,160$ K.

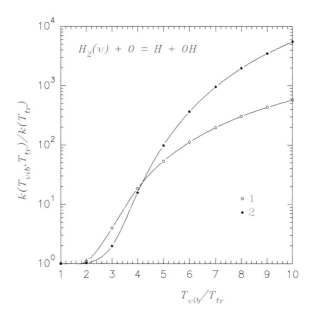

FIGURE 3.6 Dependence of the reaction rate coefficient $H_2(v)+O \rightarrow H+OH$ on the degree of nonequilibrium T_{vib}/T_{tr} at $T_{tr}=300$ K. 1 – model (3.55) and (3.57) [15]; 2 – α-model, [19], $\alpha=0.31$ [26].

A much more difficult case to consider are reactions in which multi-atomic molecules participate as reactants or products. Currently, existing analytical models do not allow us to estimate the relative efficiency of the contribution of different modes of vibrations in overcoming the activation barrier of the reaction and the distribution of energy released as a result of the reaction by different modes of vibrations.

In the paper [15], when modeling the reaction $OH+H_2(v) \rightarrow H_2O+H$, the assumption was made that as a result of the release of part of the energy into the internal degrees of freedom of the products, only the deformational (the lowest quantum energy) mode of vibrations of H_2O is excited. The dependence of the reaction rate coefficient (3.62) on the vibrational excitation obtained in this approximation is shown in Figure 3.7. This dependence coincides well with the calculation using the experimentally measured value $\alpha = 0.24$ [14]. In [32] an estimate for the ratio of the rate coefficients of the processes was obtained

$$OH + H_2(v=0) \rightarrow H_2O + H$$

and

$$OH + H_2(v=1) \rightarrow H_2O + H.$$

$k_{v=1}/k_{v=0} \leq 1,000$ at $T = 298$ K, which agrees well with the model-derived (3.55) and (3.57) estimate of $k_{v=0} = 2.9 \times 10^{-15}$ cm^{-3}s^{-1}, $k_{v=1} = 1.8 \times 10^{-12}$ cm^{-3}s^{-1}, giving for the ratio of these rate coefficients the value $k_{v=1}/k_{v=0} = 620$.

To investigate the possibility of describing the distribution of energy released in the process of chemical reactions by different degrees of freedom of the products, the following processes were considered:

$$H + BrCl \rightarrow HBr(w) + Cl$$

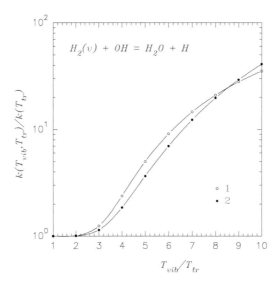

FIGURE 3.7 Dependence of the reaction rate coefficient $H_2(v)+OH \rightarrow H_2O+H$ on the degree of nonequilibrium T_{vib}/T_{tr} at $T_{tr}=300$ K. 1 – model (3.55) and (3.57) [15]; 2 – α-model, [19], $\alpha=0.24$ [26].

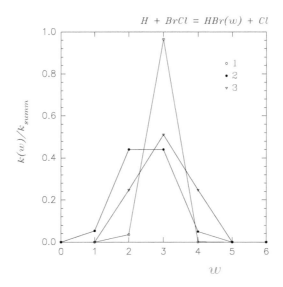

FIGURE 3.8 Distribution of HBr molecules by vibrational levels in the reaction $H+BrCl \rightarrow HBr(w)+Cl$ at $T_{tr}=300$ K. 1 – model calculation (3.55) and (3.57) [15]; 2 – classical trajectory method calculation [33]; 3 – experiment [34,35].

and

$$Cl + F_2 \rightarrow ClF(w) + F$$

in which the reaction products have a strong nonequilibrium Boltzmann distribution over vibrational levels [33].

The results of calculations of the product distribution function over vibrational levels f_w by the model (3.55) and (3.57) are presented in Figures 3.8 and 3.9.

For comparison, the same figures show the [33] data obtained by calculating the energy distribution of products in chemical two-channel reactions using Monte Carlo trajectory calculations and the results of Ref. [34] experiments.

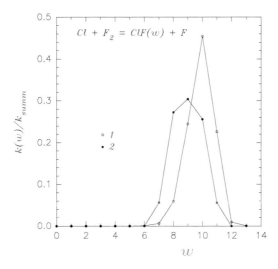

FIGURE 3.9 Distribution of ClF molecules over vibrational levels in the reaction $Cl+F_2 \to ClF(w)+F$ at $T_{tr} = 300$ K. 1 – model calculation (3.55) and (3.57) [15]; 2 – classical trajectory method calculation [33,35].

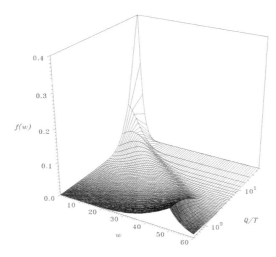

FIGURE 3.10 Change in the product distribution function of vibrational levels when the activation energy of reaction [26] is changed [15].

For all the processes studied, good agreement is obtained for the magnitude and the direction of vibrational energy allocated to vibrational degrees of freedom. The shape of the vibrational distribution function in the model calculations (3.55) and (3.57) is sharper than in the experiment [34] and trajectory calculations [33], but the position of the population maximum is reproduced with high accuracy (Figure 3.8). As the vibrational quantum of the products (reaction (3.64)) decreases, the population selectivity decreases and the half-width of the distribution function calculated from (3.55) and (3.57) becomes close to the results of trajectory modeling [33].

It is interesting to analyze the transformation of the product distribution function over vibrational levels when the activation energy of the reaction changes.

An increase in the activation energy leads to a more efficient population of the lower vibrational levels of the reaction products and at sufficiently large values (comparable to ΔH) the maximum population falls on the zero vibrational level of the product (Figure 3.10). On the contrary, a decrease in the activation energy leads to the fact that most of the energy released in the reaction goes to the vibrational degrees of freedom of the products.

The vibrational distribution function acquires a pronounced maximum, shifting with decreasing activation energy to the level for which $E_{\text{vib}} \simeq \Delta H$.

Thus, those processes for which the value of activation energy E_a^{00} is small will proceed with the greatest deviation of the product vibrational distribution function from the equilibrium value. The processes proceeding with a large activation energy, practically at any values of ΔH do not lead to an appreciable energy release into the vibrational degrees of freedom of the products and cannot cause a significant deviation of the energy of the internal degrees of freedom from the equilibrium values.

Thus, the model of thermally nonequilibrium reactions proposed here allows us to estimate the microconstants of the rates of the processes corresponding to certain vibrational states of the reactants $AB(v)$ and products $BC(w)$.

The values of the microconstants are completely determined by the magnitude and temperature dependence of the experimentally measured rate coefficient of the reaction $AB + C \rightarrow A + BC$ in thermally equilibrium conditions and the value of the energy release in the reaction ΔH, without requiring explicit consideration of other molecular constants.

Comparison of calculations using the proposed model with available experimental data and other models shows that it can be used for a wide range of processes and conditions.

REFERENCES

1. Zuev, A.P., S.S. Negodyaev, and B.K. Tkachenko, Measurement of V-T and V-V relaxation times of N2O in mixtures with CO by laser schlieren method. *Khimicheskaya Fizika*, 1985. **4**(10): p. 1303.
2. Zaslonko, I.S., Monomolecular reactions in shock waves and energy exchange of highly excited molecules. DS Thesis, Moscow, Institute of Chemical Physics, 1982.
3. Gordiets, B.F., A.I. Osipov, and L.A. Shelepin, *Kinetic Processes in Gases and Molecular Lasers*. 1980: Gordon and Breach Science Publishers.
4. Starikovsky, A.Y., Unsteady flows of reacting gases, reaction kinetics, ignition and detonation in systems containing nitrogen and carbon oxides. PhD Thesis. MIPT. Dolgoprudny, 1991.
5. Capitelli, M., *Non-Equilibrium Vibrational Kinetics*. 1989: Springer-Verlag.
6. Cherniy, G.G. and S.A. Losev, *Physico-Chemical Processes in Gas Dynamics*, Vol. 1. 1995: Moscow State University.
7. Starikovskiy, A.Y. and N.L. Aleksandrov, Plasma-assisted ignition and combustion. *Progress in Energy and Combustion Science*, 2013. **39**: pp. 61–110.
8. Salnikov, V.A. and A.M. Starik, Numerical analysis of energy characteristics of gas-dynamic lasers on hydrocarbon fuel combustion products. *TVT*, 1995. **33**(1): p. 121.
9. Achasov, O.V. and D.S. Ragozin, Constants of vibrational energy exchange in laser- active media with additives O_2, H_2, H_2O, CO. Preprint N 16. Minsk: ITMO, Academy of Sciences of BSSR, 1986, 53 p.
10. Zuev, A.P., S.A. Losev, A.I. Osipov, and A.M. Starik, Vibrational-translational energy exchange in collisions of triatomic molecules. *Chemical Physics*, 1992. **11**(1): p. 4.
11. Zuev, A.P., Laser schlieren method for VT relaxation measurements behind incident shock waves. *Chemical Physics*, 1983. **2**(7): p. 923
12. Zuev, A.P., VT relaxation of N2 at high temperatures. *Chemical Physics*, 1985. **4**(11): p. 1472.
13. Zuev, A.P. and B.K. Tkachenko, Vibrational energy exchange in N2-CO2 mixtures at elevated temperatures. *Chemical Physics*, 1988. **7**(11): p. 1451.
14. Rusanov, V.D. and A.A. Fridman, *Physics of Chemically Active Plasma*, Nauka, Moscow, 1985.
15. Starikovskiy, A. Y., State-to-state chemical reaction rate constants: Vibronic terms approximation. *Physical and Chemical Kinetics in Gas Dynamics*, 2003. **1**. https://www.chemphys.edu.ru/pdf/2003-12-25-001.pdf
16. Nikitin, E.E., *Theory of Elementary Atomic-Molecular Processes in Gases*. 1970: Nauka.
17. Landau, L.D. and E.M. Lifshits, *Quantum Mechanics*. 1983: Nauka.
18. Smith, Y., Vibrational energy transfer in collisions involving free radicals. In: *Nonequilibrium Vibrational Kinetics*, Ed. M. Capitelli. 1989: Springer-Verlag, pp. 113–157.
19. Cherniy, G.G. and S.A. Losev, *Physico-Chemical Processes in Gas Dynamics*. 1995: American Institute of Aeronautics and Astronautics.

20. Treanor, C.E. and P.V. Marrone, Chemical relaxation with preferential dissociation from excited vibrational levels. *Physics of Fluids*, 1963. **6**: p. 9.
21. Park, C., *Nonequilibrium Hypersonic Aerodynamics*. 1990: John Wiley & Sons.
22. Park, C., Assessment of a two-temperature kinetic model for dissociating and weakly ionizing nitrogen. *Journal of Thermophysics and Heat Transfer*, 1988. **2**: p. 1.
23. Macheret, S.O. and J.W. Rich, Theory of nonequilibrium dissociation rates behind strong shock waves. AIAA Paper 93-2860, 1993.
24. Macheret, S.O., A.A. Fridman, I.V. Adamovich, J.W. Rich, and C.E. Treanor, Mechanisms of nonequilibrium dissociation of diatomic molecules. AIAA Paper 94-1984, 1994.
25. Brun, R., and N. Belouaggadia, Dependence of N2 dissociation rate constant on vibrational temperature. *21st International Symposium on Shock Waves*, Paper 1620. Great Keppel Island, 1997.
26. Lashin, A.M. and A.Y. Starikovskiy, Ignition and detonation in N2O-CO-H2-He mixtures. *16th International Colloquium on the Dynamics of Explosions and Reactive Systems (ICDERS)*, Krakow, Poland, 1997, p. 603.
27. Levitsky, A.A., S.O. Macheret, L.S. Polak, et al., Role of atoms not in thermal equilibrium in the plasma chemical synthesis of nitrogen oxides. *High Energy Chemistry*, 1983. **17**: pp. 625–632.
28. Park, C., Assessment of two-temperature kinetic model for ionizing air. *Journal of Thermophysics and Heat Transfe*, 1989. **3**(3): p. 233.
29. Gardiner, W.C., *Combustion Chemistry*. 1984: Springer-Verlag.
30. Light, G.C., The effect of vibrational excitation on the reaction of O(3P) with H2 and the distribution of vibrational energy in the product OH. *The Journal of Chemical Physics*, 1978. **68**(6): pp. 2831–2843.
31. Johnson, B.R. and N.W. Winter, Classical trajectory study of the effect of vibrational energy on the reaction of molecular hydrogen with atomic oxygen. *The Journal of Chemical Physics*, 1977. 66: p. 4116.
32. Light, G.C. and J.H. Matsumoto, The effect of vibrational excitation in the reactions of OH with H_2. *Chemical Physics Letters*, 1978. **58**(4): pp. 578–581.
33. Konoplev, N.A., A.A. Stepanov, and V.A. Shcheglov, Numerical two-dimensional analysis of a ring DF-CO2 continuous chemical laser taking into account reagent mixing. In: *Dynamics of Elementary Atomic-Molecular Processes in Gas and Plasma*, Eds. V.A. Shcheglov, J.H. Misguich, G. Pelletier, and P. Schuck. 1991: Nova Science, **213**: pp. 34–60.
34. Polanyi, J.C. and W.J. Skrlac, Energy distribution among reaction products. X. H+ICl → HCl+I, H+ClBr → HCl+Br, HBr+Cl. *Chemical Physics*, 1977. **23**(2): pp. 167–194.
35. Lashin, A.M. and A.Y. Starikovskiy, On the chemical reactions rates and products vibrational distribution function calculation under nonequilibrium conditions. *26th Symposium (International) on Combustion*, Napoli, 1996, R.04-077.

4 Mechanisms of Plasma-Chemical Reactions

4.1 KINETICS OF HYDROGEN OXIDATION IN A STOICHIOMETRIC HYDROGEN-AIR MIXTURE IN A PULSED NANOSECOND DISCHARGE

In this part of this book, we describe the oxidation of hydrogen in a stoichiometric hydrogen-air mixture under the influence of a pulsed high-current nanosecond discharge.

The transition $H_2\left(a^3\Sigma_g^+ \to b^3\Sigma_u^+\right)$ leads to the appearance in the discharge emission spectrum of a continuum extending from 167 nm to the near-infrared region of the spectrum. The fact that the lower level of the transition is a repulsive term ensures the absence of self-absorption and the possibility to reconstruct by emission characteristics the population rate of the state $H_2\left(a^3\Sigma_g^+\right)$ by electron impact in the discharge [1]. The time course of the absolute concentration of molecular hydrogen in the mixture was reconstructed on the basis of the registration of the absolute emission intensity of this transition at a wavelength of $\lambda = 239$ nm with a spectral resolution of $\Delta\lambda = 0.24$ nm.

The total transition emission intensity and the concentration of H_2 molecules in the $a^3\Sigma_g^+$ excited state were determined using the distribution of the absolute transition intensity over wavelengths measured in the work of [2]. The wavelength dependence of the transition intensity was assumed to be the same as in [2], and normalization was performed according to the measured intensity value at 239 ± 0.12 nm.

Figure 4.1a shows the measured half-time decay to the steady-state concentration of molecular hydrogen in the $a^3\Sigma_g^+$ state in the discharge. Long times at pressures above 9 Torr corresponded to partial breakdown modes of the gap and will not be considered further. In the rest of the pressure range, after overlapping of the discharge gap by the ionization wave, a phase of quasi-stationary high-current homogeneous discharge with a duration of $\tau_1 \simeq 5-7$ ns was established for $\tau_2 \simeq 15-17$ ns, during which the main gas excitation occurred. The characteristic times of hydrogen conversion in this range of parameters are $\tau = 15-30$ s, which corresponds to $600-1,200$ discharge pulses. Note that despite the large total process time, the plasma-chemical transformations after passing the current pulse are practically completed in $10-20$ ms, which makes it possible to exclude the influence of diffusion on the experimental results.

Population of the upper excited level $H_2\left(a^3\Sigma_g^+\right)$ occurs by electron impact directly from the ground electronic state of the molecule:

$$H_2 + e \to H_2\left(a^3\Sigma_g^+\right) + e \tag{4.1}$$

which is due to the relatively small excitation of the gas during the current pulse and the complete relaxation of the electronic states between pulses. It turns out that the emission intensity of the transition $H_2\left(a^3\Sigma_g^+ \to b^3\Sigma_u^+\right)$ is directly proportional to the concentration of molecular hydrogen, the rate coefficient of the process (4.1), which depends on the reduced electric field strength, and the total electron concentration. Figure 4.1b shows data on the absolute concentration of excited hydrogen molecules $H_2\left(a^3\Sigma_g^+\right)$ at the initial moment of time as a function of the gas pressure in the gap.

The number of excited molecules varies within $3-7 \times 10^{11}$ cm^{-3}, which correlates well with the results of measurements of the concentration of electronically excited particles in the FIW [2,4]. The electron concentration in the discharge and the reduced electric field strength were determined from time-resolved measurements of the current through the discharge gap and the amplitude of the voltage pulse.

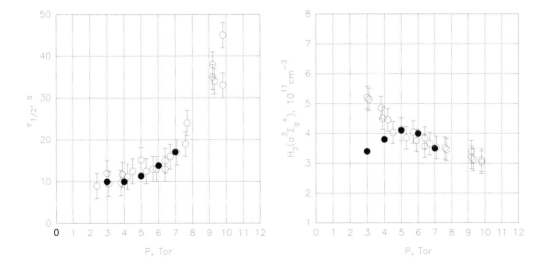

FIGURE 4.1 (a) Half-time to steady-state concentration of molecular hydrogen in the $a^3\Sigma_g^+$ state in the discharge as a function of pressure. Open symbols – experiment, solid symbols – calculation. (b) Maximum concentration of molecular hydrogen in the $H_2\left(a^3\Sigma_g^+\right)$ state in the discharge as a function of pressure. Open symbols – experiment, solid symbols – calculation [3].

The electric current in the discharge gap can be obtained as the sum of the incident and reflected pulses $I = I_{gen} + I^{ref}$. The pulse reflected from the discharge gap consists of two fundamentally different parts, according to the transmission line theory, which differ in sign: reflected from the ionization wave front propagating at a velocity of approximately $V_{fr} = 3 \times 10^9$ cm/s I_{fr}^{ref} (corresponding to reflection from the open end), and reflected from the discharge tube after the wave reaches the grounded electrode I_{sc}^{ref} (corresponding to reflection from the closed end). Because of the high velocity of the ionization wave front, the distortion of the pulse shape due to the Doppler effect must be taken into account for the part of the pulse reflected from the FIW front:

$$I_{fr}^{ref}(t) = \frac{1+\beta}{1-\beta} I_{fr}^{ref(measured)}\left(t\frac{1+\beta}{1-\beta}1+\beta\right), \qquad (4.2)$$

where $\beta = V_{fr}/c = 0.1$, c is the speed of light in vacuum. Figure 4.2 shows the incident I_{gen} (1), reflected I^{ref} (2), and passed I (3) current pulses. For convenience, the reflected current pulse (2) is shown in the figure with opposite sign, and its different parts are indicated by thick (transformed by the formula (4.2) I_{fr}^{ref}) and dashed (I_{sc}^{ref}) lines. The same figure shows a plot of the maximum values of the conduction current I as a function of pressure. It shows that in the pressure range 3–7 Torr, the amplitude of the current flowing through the discharge gap varies from $I = 300$ A at $p = 3$ Torr to $I = 155$ A at $p = 7$ Torr. The half-height current pulse duration remained constant over this range with good accuracy and was $\tau_{pul} \simeq 16$ ns.

4.1.1 Electron Energy Distribution Function and Gas Excitation by Electron Impact

The stage of quasi-stationary electric current flow through the discharge gap is characterized by two electrical parameters—the conduction current I and the voltage between the electrodes U. Under the conditions considered, the bridging of the discharge gap by the ionization wave occurs for a time significantly shorter than the total time of existence of the high-voltage pulse. This

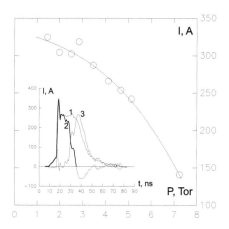

FIGURE 4.2 The pressure dependence of the discharge current amplitude and typical oscillograms of the incident *1*, reflected *2*, and passed *3* currents. The part of the pulse reflected from the ionization wave front is shown in bold line [3].

allows us to estimate the electric field strength in the discharge cell after bridging the gap as $E = U/l$, where l is the distance between the electrodes. Then, taking into account the homogeneity of the discharge over the cross-section S and the obvious relation

$$I = e n_e v_d S \quad (4.3)$$

it is possible to determine the electron concentration n_e in the discharge as a function of the conduction current through the discharge gap and the effective value of the reduced electric field E/n. The value of the drift velocity $v_d = f(E/n, \xi_i)$, where ξ_i are mole fractions of N_2, O_2, H_2, and H_2O, respectively, was determined from the solution of the quasi-stationary Boltzmann equation for the low-energy part of the electron energy distribution function (EEDF) in the two-term approximation [5].

It was assumed that the spatially homogeneous case of the EEDF, which is in equilibrium with the local electric field, is realized. This condition is violated in high fields near the FIW front [6], but after bridging of the discharge gap, the reduction of E/n to a few hundred Townsends makes the use of the local field approximation possible. Electron-electron and electron–ion collisions do not contribute significantly to the EEDF formation process, which is well fulfilled for relatively small gas ionization degrees ($\simeq 10^{-4}$) and total electron concentration $n_e \simeq 10^{12} - 10^{13}$ cm^{-3} under the conditions of this work. In this case, according to [7], the equation defining the stationary EEDF taking into account elastic and inelastic collisions of electrons in an electric field with intensity E, in a mixture of l gases, is of the form:

$$\frac{E^2 \varepsilon}{3 \sum_l N_l \sigma_{m,l}(\varepsilon)} + \sum_l \frac{2m}{M_l} \cdot N_l \varepsilon^2 \sigma_{m,l}(\varepsilon) \cdot \left\{ f(\varepsilon) + \frac{T}{e} \frac{df(\varepsilon)}{d\varepsilon} \right\} +$$

$$+ \sum_l B_{e,l} \cdot N_l \varepsilon \sigma_{rot,l}(\varepsilon) \cdot \left\{ f(\varepsilon) + \frac{T}{e} \frac{df(\varepsilon)}{d\varepsilon} \right\} = \quad (4.4)$$

$$= -\sum_l N_l \cdot \sum_{i,j} \int_\varepsilon^{\varepsilon + \varepsilon_{ij}} \varepsilon \sigma_{l,i,j}(\varepsilon') \cdot \varepsilon' f(\varepsilon') d\varepsilon'$$

where e, m, M are the charge, electron mass, and heavy particle mass; B_e is the rotational constant; $\sigma_m(\varepsilon)$, $\sigma_{rot}(\varepsilon)$ – transport cross-section of electron scattering and cross-section of excitation of rotational levels of molecules cross sections of inelastic collisions of electrons with gas molecules of l grade, T – gas temperature; summation is performed on l components of the gas mixture. The first summand in the left part describes the increase of electron energy in the field E, the second – energy losses in elastic collisions, and the third – losses for excitation of rotational levels (for molecules). The right part describes inelastic collisions of electrons in which their energy state changes (transition from state i to state j with a change in energy ε_{ij}).

In the calculations, it was assumed that the main components determining the nature of the EEDF and, consequently, the rates of all processes involving electrons are N_2, O_2, H_2, and H_2O. The list of processes taken into account in the calculation of the EEDF is given in Table 4.9. The transport cross sections were taken from [8–10]. The values of v_d calculated with the current chemical composition in the two-term approximation allow, using experimental data for the current density $j = I/S$, to obtain an estimate for the maximum electron concentration in the discharge (see expression 4.3).

Figure 4.3 shows plots of the dependence of the reduced electric field strength E/n and electron concentration n_e at different pressures corresponding to the initial mixture composition. In the investigated range of parameters, the value of the reduced electric field strength at the stage of the main current flow through the discharge cell at the initial moment is from $E/n = 560$ to 400 Td, which is close to the limit for the applicability of the two-term approximation of the Boltzmann equation for calculating the velocities of inelastic processes involving electrons. At the same time, E/n grows approximately by a factor of 1.4 with time due to the pressure growth in the system. It will be shown below that for the lowest investigated pressures this leads to appreciable deviations of the calculated and measured excitation rates of the electronic states of the gas. At the same time, in the range $p = 4 - 8$ Torr, as in the work of [2], the deviations of the calculated and experimentally measured excitation rates are insignificant.

The maximum electron concentration in the discharge during the current pulse varies in the range $n_e \simeq (1.3 \div 2.2) \times 10^{12}$ cm^{-3} (Figure 4.3), which agrees well with the results of measurements by [2,4] under similar conditions. The effect of changing the EEDF on the electron energy distribution between different processes in the H_2-O_2-N_2 mixture is shown in Figure 4.4. It can be seen that, under the conditions of the work [1], the maximum energy contribution of the discharge occurs in the electronic degrees of freedom of the gas, which, as will be shown later, determines the dominant

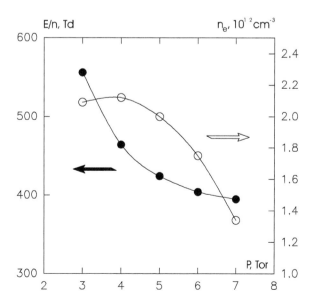

FIGURE 4.3 Dependence of maximum electron concentration and initial reduced electric field strength in the discharge cell at the stage of quasi-stationary high-current discharge on pressure [3].

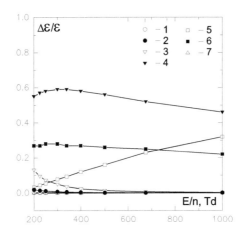

FIGURE 4.4 Effect of the magnitude of the reduced electric field strength E/n on the relative energy distribution of electrons between excitation of different degrees of freedom of a gas in a mixture H_2-O_2-N_2. The numbers indicate the fractions of energy going into translational *1*, rotational *2*, vibrational *3*, electronic degrees of freedom *4*, ionization *5*, gas dissociation *6*, and attachment *7* [3].

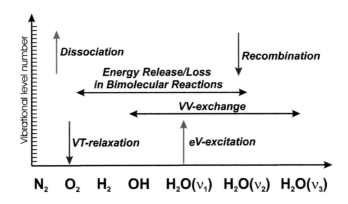

FIGURE 4.5 Schematic of energy exchanges with H_2, O_2, N_2, OH, H_2O system used in the analysis of vibrational kinetics [3].

role of processes involving electronically excited molecules in the hydrogen oxidation process. A somewhat smaller fraction of energy is accounted for the dissociation of the gas by electron impact and its ionization. Note that the role of ionization grows noticeably with increasing reduced field strength in the range of parameters of this work. The vibrational degrees of freedom in such strong fields are excited relatively weakly (Figure 4.4), and their role in chemical reactions will be discussed below.

4.1.2 Energy Exchange and Chemical Reactions in the H_2-Air System under Pulsed Discharge Conditions

4.1.2.1 VV and VT Relaxation Rate Coefficients in the Reacting H_2-O_2-N_2 System

In this part, the modeling of chemical reactions in the H_2-O_2-N_2 system took into account the level-by-level kinetics of vibrational energy exchanges between the components H_2, O_2, N_2, OH, and H_2O (Figure 4.5) will be discussed. The calculation of the rate coefficients $k_{n,n-1}^{n'-1,n'}(T)$ and $k_{n,n-1}(T)$ was carried out similarly to the above consideration for the N_2 molecule within the framework of SSH theory.

For the H_2O molecule, it was assumed that the relaxation of each mode (symmetric, bending and antisymmetric) occurs independently and can be described within the same model. The intermode energy exchange was described taking into account the anharmonicity of oscillations. In this case, the mixing of modes occurring at high levels of excitation [11] can be taken into account as an effective increase in the VV'-exchange rate coefficient.

In the framework of this consideration, the hypothetical effect of accelerating VT-relaxation at upper levels due to the appearance of low-frequency harmonics in the spectrum of the system at strong anharmonic interaction of different modes of oscillations is not taken into account in principle. The influence of this effect can lead in the conditions of the work [1] to the depletion of the vibrational distribution function of H_2O molecules at high levels. However, since H_2O is the main product of an almost irreversible reaction at $T = 300$ K, some increase in the decay rate associated with a possible underestimation of the VT-relaxation rate of the upper levels does not affect the calculation results. The vibrational relaxation rate coefficients (for the lower level transitions), which were used here in constructing the full set, were taken from reviews [12–15]. Their numerical values are summarized in Tables 4.1–4.4.

TABLE 4.1
VT Processes

Process			$\lg(k, [cm^3 s^{-1}])$
$H_2O(100)+O_2$	$\rightarrow$	H_2O+O_2	−13.24
$H_2O(100)+N_2$	$\rightarrow$	H_2O+N_2	−13.24
$H_2O(100)+H_2O$	$\rightarrow$	H_2O+H_2O	−8.84
$H_2O(100)+H_2$	$\rightarrow$	H_2O+H_2	−5.82
$H_2O(010)+O_2$	$\rightarrow$	H_2O+O_2	−8.76
$H_2O(010)+N_2$	$\rightarrow$	H_2O+N_2	−8.76
$H_2O(010)+H_2O$	$\rightarrow$	H_2O+H_2O	−5.84
$H_2O(010)+H_2$	$\rightarrow$	H_2O+H_2	−5.06
$H_2O(010)+OH$	$\rightarrow$	H_2O+OH	−5.84
$H_2O(010)+H$	$\rightarrow$	H_2O+H	−4.72
$H_2O(010)+O$	$\rightarrow$	H_2O+O	−6.84
$H_2O(001)+O_2$	$\rightarrow$	H_2O+O_2	−14.38
$H_2O(001)+N_2$	$\rightarrow$	H_2O+N_2	−14.38
$H_2O(001)+H_2O$	$\rightarrow$	H_2O+H_2O	−9.95
$H_2O(001)+H_2$	$\rightarrow$	H_2O+H_2	−6.66
$H_2(1)+H_2$	$\rightarrow$	H_2+H_2	−18.12
$H_2(1)+H_2O$	$\rightarrow$	H_2+H_2O	−18.76
$H_2(1)+OH$	$\rightarrow$	H_2+OH	−18.76
$H_2(1)+N_2$	$\rightarrow$	H_2+N_2	−19.22
$H_2(1)+O_2$	$\rightarrow$	H_2+O_2	−19.22
$H_2(1)+O$	$\rightarrow$	H_2+O	−11.65
$H_2(1)+H$	$\rightarrow$	H_2+H	−12.52
$N_2(1)+N_2$	$\rightarrow$	N_2+N_2	−26.49
$N_2(1)+N$	$\rightarrow$	N_2+N	−28.60
$N_2(1)+O_2$	$\rightarrow$	N_2+O_2	−26.49
$N_2(1)+H_2$	$\rightarrow$	N_2+H_2	−15.88

(Continued)

TABLE 4.1 (Continued)
VT Processes

Process			$\lg\left(k, \left[\mathrm{cm}^3\mathrm{s}^{-1}\right]\right)$
$N_2(1)+OH$	→	N_2+OH	−15.88
$N_2(1)+H$	→	N_2+H	−15.38
$N_2(1)+O$	→	N_2+O	−14.48
$N_2(1)+H_2O$	→	N_2+H_2O	−14.17
$O_2(1)+O_2$	→	O_2+O_2	−16.92
$O_2(1)+H_2$	→	O_2+H_2	−15.56
$O_2(1)+H$	→	O_2+H	−15.33
$O_2(1)+O$	→	O_2+O	−17.33
$O_2(1)+H_2O$	→	O_2+H_2O	−16.22
$O_2(1)+OH$	→	O_2+OH	−16.22
$O_2(1)+N_2$	→	O_2+N_2	−16.82
$OH(1)+H_2$	→	$OH+H_2$	−15.91
$OH(1)+H_2O$	→	$OH+H_2O$	−15.91

TABLE 4.2
VV-Exchange

Process			$\lg\left(k, \left[\mathrm{cm}^3\mathrm{s}^{-1}\right]\right)$
$N_2(1)+N_2$	→	$N_2+N_2(1)$	−13.06
$O_2(1)+O_2$	→	$O_2+O_2(1)$	−12.58
$H_2(1)+H_2$	→	$H_2+H_2(1)$	−12.56
$OH(1)+OH$	→	$OH+OH(1)$	−12.57
$N_2(1)+H_2O$	→	$N_2+H_2O(010)$	−15.30
$O_2+H_2O(010)$	→	$O_2(1)+H_2O$	−12.09
$OH+H_2O(001)$	→	$OH(1)+H_2O$	−11.59
$H_2(1)+H_2O$	→	$H_2+H_2O(001)$	−11.59
$H_2(1)+OH$	→	$H_2+OH(1)$	−11.59
$N_2+H_2(1)$	→	$N_2(1)+H_2$	−14.68
$N_2(1)+O_2$	→	$N_2+O_2(1)$	−17.70

TABLE 4.3
VV'-Exchange

Process			$\lg\left(k, \left[\mathrm{cm}^3\mathrm{s}^{-1}\right]\right)$
$H_2O(100)+H_2O$	→	$H_2O+H_2O(100)$	−7.50
$H_2O(010)+H_2O$	→	$H_2O+H_2O(010)$	−7.50
$H_2O(001)+H_2O$	→	$H_2O+H_2O(001)$	−7.50
$H_2O(100)+H_2O$	→	$H_2O+H_2O(020)$	−7.20
$H_2O+H_2O(001)$	→	$H_2O(100)+H_2O$	−6.41

TABLE 4.4
Electronic-Vibrational Energy Conversion

$H_2(v) + O_2(a^1\Delta_g)$	$\rightarrow$	$H_2 + O_2(b^1\Sigma_g^+)$
$N_2(v) + N_2(n)$	$\rightarrow$	$N_2 + N_2(a'^1\Sigma_u^-)$
$N_2(A^3\Sigma_u^+) + N_2(v)$	$\rightarrow$	$N_2(B^3\Pi_g) + N_2$
$N_2(B^3\Pi_g) + N_2(v)$	$\rightarrow$	$N_2(A^3\Sigma_u^+) + N_2(w)$
$N_2(v) + N_2(n)$	$\rightarrow$	$N_2(A^3\Sigma_u^+) + N_2$
$N_2(v) + N_2(n)$	$\rightarrow$	$N_2(B^3\Pi_g) + N_2$

4.1.2.2 Chemical Kinetics during the Discharge Stage

The kinetic scheme describing the processes at the stage of electric current flow consists of 9,500 processes involving 254 particles, including electronically excited and charged atoms and molecules, vibrationally excited molecules H_2, O_2, N_2, H_2O and OH; electrons, radicals, and unexcited components (Table 4.9). Associative and Penning ionization, recombination of positive ions and electrons, electron attachment and detachment, interaction between neutral unexcited components, interaction between neutral excited and neutral unexcited components, conversion of positive and negative ions, and recombination of positive and negative ions were taken into account (Table 4.10).

In the calculation, it was assumed that the cooling of the EEDF after the end of the voltage pulse is almost instantaneous, and until the next voltage pulse arrives in $\tau = 1/f = 25$ ms, the reactions of gas excitation by direct electron impact (see Table 4.9) can be excluded from the kinetic scheme; the calculation of chemical transformations in the afterglow of the discharge is performed without their participation. It was taken into account that in the process of hydrogen and oxygen conversion into water not only the chemical composition of the mixture changes but also the gas pressure, which together leads to the evolution of the electron energy distribution function and, consequently, to a change in the value of the electron drift velocity v_d, the rate of excitation of electronic and vibrational degrees of freedom, and the processes of ionization and dissociation.

4.1.2.3 Chemical Kinetics in the Discharge Afterglow

In modeling the oxidation process of H_2 in air under the action of a pulsed nanosecond discharge, processes involving vibrationally excited components and their inverse were taken into account (Figure 4.6).

In the H_2-O_2-N_2 system, it is necessary to take into account the dependence of the reaction rate on vibrational excitation for at least 15 exchange reactions and five monomolecular decomposition processes (Tables 4.5 and 4.6).

In addition to the processes involving uncharged particles, electron detachment and ionization processes involving vibrationally excited molecules were additionally taken into account to calculate the kinetics in the discharge afterglow (Tables 4.7 and 4.8).

Figures 4.7 and 4.8 show the vibrational level distributions for various molecules arising in the afterglow of the discharge. The strongly nonequilibrium character of the distributions is well observed. Thus, the distribution of H_2 molecules by vibrational levels up to $\tau \simeq 1\,\mu s$ has a pronounced local maximum at $n \simeq 10$, which is associated with the intensive population of these states during the radiative depopulation of singlet levels of molecular hydrogen, which are excited in the discharge (Figure 4.7).

The distribution over vibrational levels of the bending mode $H_2O(v_2)$ has a sharply pronounced inverse population near the dissociation threshold, which is due to the intense recombination flux to the upper states in the reaction

$$H_2 + O + M \rightarrow H_2O + M$$

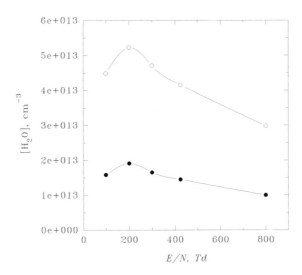

FIGURE 4.6 The amount of water formed by the arrival of the second pulse. Open symbols represent calculation including vibrationally excited components, solid symbols represent modeling without reactions between vibrationally excited particles [3].

TABLE 4.5
Bimolecular Exchange Reactions

$H_2(v)+N$	→	$NH+H$
$H_2(v)+O_2(n)$	→	HO_2+H
$H_2(v)+O_2(n)$	→	$OH(w)+OH(m)$
$H_2(v)+O$	→	$OH(w)+H$
$H_2(v)+OH(n)$	→	$H_2O(w_1,w_2,w_3)+H$
$N_2(v)+O$	→	$NO+N$
$O_2(v)+H$	→	$OH(w)+O$
$O_2(v)+N$	→	$NO+O$
$OH(v)+H$	→	$H_2(w)+O$
$OH(v)+N$	→	$NH+O$
$OH(v)+NO$	→	NO_2+H
$OH(v)+O$	→	$O_2(w)+H$
$OH(v)+OH(n)$	→	$H_2O+O(^1D)$
$OH(v)+OH(n)$	→	$H_2O(w_1,w_2,w_3)+O$
$OH(v)+OH(n)$	→	HO_2+H

TABLE 4.6
Monomolecular Decomposition Reactions

$H_2O(v_1,v_2,v_3)+M$	→	$H+OH+M$
$H_2(v)+M$	→	$H+H+M$
$OH(v)+M$	→	$H+O+M$
$O_2(v)+M$	→	$O+O+M$
$N_2(v)+M$	→	$N+N+M$

TABLE 4.7
Electron Detachment Processes Involving Vibrationally Excited Molecules

$O^- + N_2(v)$	$\rightarrow$	$O + N_2(w) + e$
$O_2^- + O_2(v)$	$\rightarrow$	$O_2 + O_2(w) + e$
$O_2^- + N_2(v)$	$\rightarrow$	$O_2 + N_2(w) + e$

TABLE 4.8
Ionization Processes

$N_2(a'^1\Sigma_g) + N_2(v)$	$\rightarrow$	$N_4^+ + e$
$N_2(A^3\Sigma_u^+) + N_2(v)$	$\rightarrow$	$N_4^+ + e$

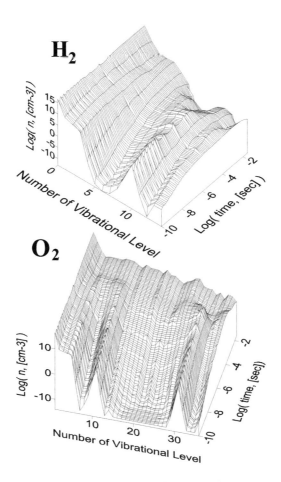

FIGURE 4.7 Distributions over vibrational energy levels for H_2 and O_2 molecules [3].

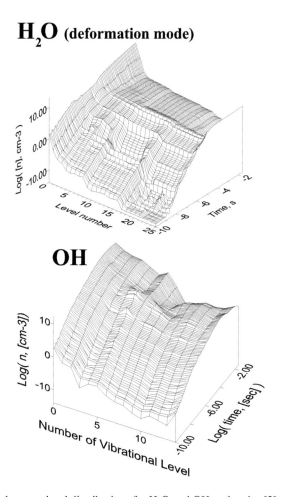

FIGURE 4.8 Vibrational energy level distributions for H_2O and OH molecules [3].

The distribution over vibrational levels of the OH radical, on the contrary, is significantly depleted in the region of levels $n > 5$ due to their rapid consumption in the reaction

$$OH(v) + OH \rightarrow HO_2 + H$$

The magnitude of the effect of reactions involving vibrationally excited components on the total rate of H_2 consumption and H_2O production is shown in Figure 4.6. It can be seen that accounting for the contribution of vibrational nonequilibrium to the kinetics leads to a strong (almost fivefold) increase in the reaction rate. Thus, the consideration of nonequilibrium vibrational kinetics in the conditions of this work is fundamental and leads to a sharp change in the rate of all processes in the chemically reacting plasma afterglow discharge.

In this connection, it becomes especially important to analyze the sensitivity of the kinetic scheme to the choice of rate coefficients of the main processes. As the results of the simulation show, the reaction

$$H_2(v) + O \rightarrow OH + H$$

which has an energy threshold $E_a = 4410$ K at $v = 0$ has the greatest influence on the acceleration of hydrogen oxidation in nonequilibrium plasma. The magnitude of this threshold decreases sharply with vibrational excitation of H_2 ($\Theta_{H_2} = 6{,}326$ K), with the experimentally measured increase in reaction rate at $T = 300$ K being $k_{v=1}/k_{v=0} = 2{,}600$ [16].

The model used in the work [1] to describe the kinetics of reactions between vibrationally excited components provides very good agreement on the rate coefficient of this process and its dependence on vibrational excitation (see the discussion above).

The total effect of vibrational excitation also depends on the rate of the processes VT, VVV, and VV'-relaxation of the H_2 molecule and the rate of populating the vibrational levels of H_2 in the discharge by electron impact. Because of the relatively low rate of vibrational relaxation in the range of pressures and temperatures used, the choice of specific values for the rate coefficients of these processes does not play a decisive role. The scatter of experimental data available in the literature (see, for example, the review in [13]) does not exceed a factor of 3 even for the slow VT-relaxation process, which does not affect the modeling results.

The sensitivity of the kinetics to the rate of population of vibrational levels of H_2 by electron impact is also small because it determines only a linear (proportional to the population of the corresponding states) dependence of the rate of H_2 oxidation. In addition, the rates of eV-processes depend weakly on the magnitude of the reduced electric field in the range of conditions of the work [1] $E/n = 400 - 800$ Td and are determined with high accuracy from the solution of the Boltzmann equation. In the calculation of $k_{eV}(H_2)$, the cross sections given in [17] were used.

Thus, the analysis allows us to assert that the proposed mechanism for describing chemical transformations under conditions of strong vibrational nonequilibrium shows that under conditions of pulsed high-current discharge the contribution of processes involving vibrationally excited components is fundamental and essentially determines the kinetics as a whole. The role of electronically excited, charged components and processes involving unexcited molecules, atoms, and radicals will be analyzed below.

4.1.2.4 The Role of Different Mechanisms in Hydrogen Oxidation in High E/n Discharge

The calculations were performed in "direct simulation" mode, i.e. all consecutive current pulses through the discharge gap were directly calculated, taking into account the changes in the mixture composition between and during the pulses.

Figure 4.9 shows the time-varying curves of the concentration of $H_2(a^3\Sigma_g^+)$ at different pressures obtained in experiment and calculation. Figure 4.9(1) shows the data corresponding to the pressure $p = 3$ Torr, and, consequently, the highest electric field strength E/n in the discharge gap during the high-voltage pulse. It is well seen that the calculated concentration of excited hydrogen molecules $H_2(a^3\Sigma_g^+)$ is almost one and a half times lower than that measured experimentally, despite the fact that the calculated hydrogen conversion time (Figure 4.1a) agrees well with the measured one. This result is explained by the decrease in the accuracy of the two-term approximation of the Boltzmann equation at very high values of the electric fields, which is particularly strong when calculating the rate coefficients for the population rate of high-bed states $\left(H_2(a^3\Sigma_g^+)\right)$. At the same time, the rate of population of the low-energy level $N_2(B^3\Pi_g)$ is determined more accurately.

$$N_2(B^3\Pi_g) + O_2 \rightarrow N_2 + O + O \tag{4.5}$$

As a result, the rate of quenching reaction of this level by molecular oxygen, which is one of the main suppliers of atomic oxygen in the system, is accurately calculated, resulting in a correct reproduction of the hydrogen conversion time in the discharge.

$$H_2 + O \rightarrow H + OH \tag{4.6}$$

Starting from pressure $p = 4$ Torr, a good agreement is observed both in the absolute value of the excited level concentration $H_2(a^3\Sigma_g^+)$ (Figure 4.1b) and in the hydrogen conversion time in the discharge (Figure 4.1a). The good coincidence of the calculated and measured $H_2(a^3\Sigma_g^+)$ concentration

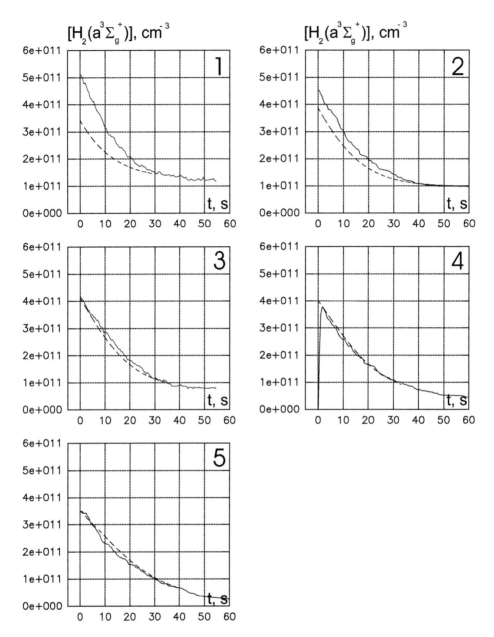

FIGURE 4.9 Profiles of the absolute concentrations of molecular hydrogen in the electronically excited state $a^3\Sigma_g^+$ as a function of pressure. Graphs 1–5 — initial mixture pressures $p = 3, 4, 5, 6$, and 7 Torr, respectively. Solid lines — experiment, dashed lines — calculation [3].

profiles (Figure 4.9(2)–4.9(5)) allows us to conclude that the description of the gas excitation process under these conditions using the two-term approximation of the Boltzmann equation is well accurate, and that the kinetics at the afterglow stage of the discharge is well reproduced.

In order to emphasize the most important processes occurring in the afterglow stage of the discharge, a rate analysis of the kinetic scheme was carried out under conditions corresponding to the propagation of the first current pulse through the discharge gap at a pressure of 5 Torr (initial mixture composition: H_2: O_2: $N_2 = 0.296$: 0.148: $0.556\$$).

Figure 4.10 shows the kinetic curves for the components involved in the most important processes in the afterglow of the first pulse. The time interval between the end of the first pulse and the beginning

Mechanisms of Plasma-Chemical Reactions

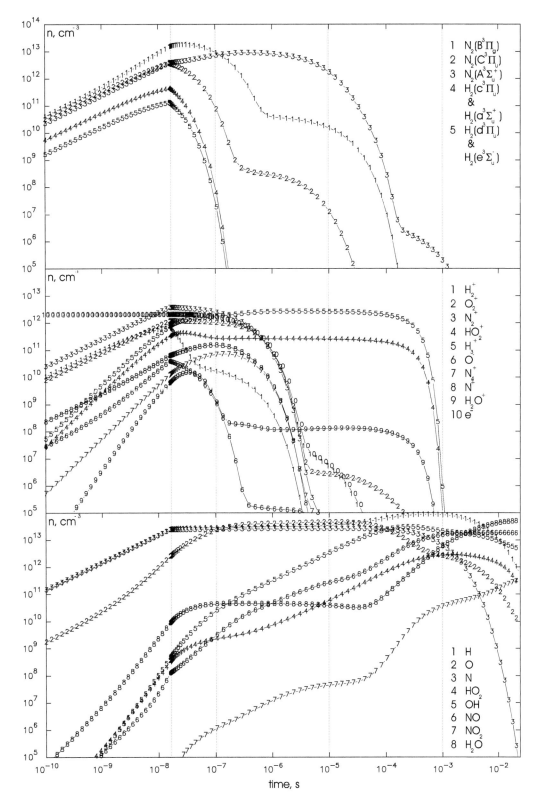

FIGURE 4.10 Calculated principal component profiles during the first current pulse and in the afterglow of the discharge until the arrival of the next pulse at $p=5$ Torr [3].

of the second pulse can be divided into four parts — the first — from 0 to 10^{-7} s, the second — from 10^{-7} to 10^{-5} s, the third — from 10^{-5} to 10^{-3} s, and the fourth — from 10^{-3} to 25×10^{-3} s.

The first of these intervals corresponds to the discharge stage (up to 16 ns) and the near-afterglow discharge zone, where excited particles, ions, and radicals are produced, and in the far-afterglow discharge zone, radiative scattering and collisional quenching of electronically excited states are active (Figure 4.10).

During the second stage, most ions recombine; only the concentration of $N_2(A^3\Sigma_u^+)$ remains significant from the electronically excited molecules. Reactions involving radicals practically do not take place.

The third stage is characterized by the conversion of the remaining metastables and ions into active radicals (Figure 4.10).

Finally, the fourth stage (time interval 10^{-3} to 25×10^{-3} s) is dominated by chemical processes involving free radicals. For each of these time intervals, diagrams of active particle fluxes for the fastest channels of chemical transformations were constructed (Figures 4.11 and 4.12). In the diagram, the thickness of the line and the number near it correspond to the rate of the process, and the

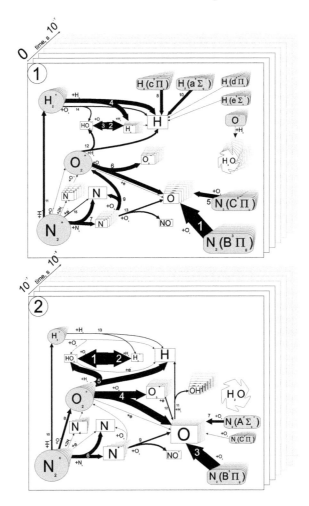

FIGURE 4.11 Active particle fluxes for the fastest channels of chemical transformations. The thickness of the line corresponds to the rate of the given process, and the number next to it corresponds to its place in the hierarchy in the rate analysis of the scheme. The dynamics of the concentration of a given chemical component on a given time interval is shown as a layer-by-layer increase or decrease of the corresponding field. (1) — time interval $\tau = 0 - 1 \times 10^{-7}$ s, (2) — $\tau = 1 \times 10^{-7} - 1 \times 10^{-5}$ s [3].

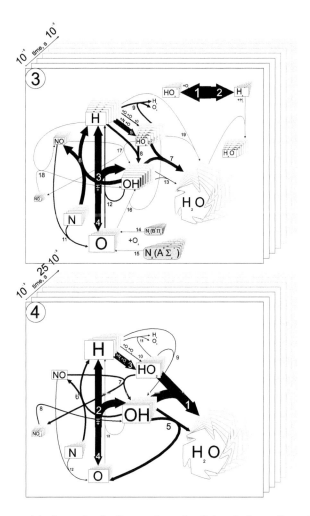

FIGURE 4.12 Active particle fluxes for the fastest channels of chemical transformations. The thickness of the line corresponds to the rate of the given process, and the number next to it corresponds to its place in the hierarchy in the rate analysis of the scheme. The dynamics of the concentration of a given chemical component at a given time interval is shown as a layer-by-layer increase or decrease of the corresponding field. (3) — time interval $\tau = 1\times 10^{-5} - 1\times 10^{-3}$ s, (4) — $\tau = 1\times 10^{-3} - 25\times 10^{-3}$ s [3].

"distribution" of a chemical component reflects the change in its amount over a given time interval. Let us consider the dominant processes in the discharge afterglow in each time interval.

1. 16×10^{-9} s — 10^{-7} s. The concentration of atomic oxygen continues to increase by more than an order of magnitude in the discharge afterglow stage, mainly due to the dissociative quenching of the electronically excited $NN_2(B^3\Pi_g)$ level by O_2 to form atomic oxygen. The dissociative quenching of the $N_2(C^3\Pi_u)$ state makes a much weaker contribution to the formation of atomic oxygen under these conditions.

    Next in the hierarchy is the reaction of conversion of H_2^+ to H_3^+ and H. It, as well as dissociation reactions of various electronic states of H_2, makes practically no contribution to the production of atomic hydrogen: the dominant amount of H atoms is produced in the discharge. The main channel of water production at this stage is the reaction of molecular hydrogen with the O^- ion formed in the discharge in the reaction $O_2 + e^- \rightarrow O + O^-$.

TABLE 4.9
Processes of Gas Excitation by Direct Electron Impact

Process			References
Excitation of H_2			
$e + H_2$	$\rightarrow$	$e + H_2(v=1)$	[17]
$e + H_2$	$\rightarrow$	$e + H_2(v=2)$	[17]
$e + H_2$	$\rightarrow$	$e + H_2(v=3)$	[17]
$e + H_2$	$\rightarrow$	$e + H_2(\text{rot})$	[17]
$e + H_2$	$\rightarrow$	$e + H_2(d^3\Pi_u)$	[18]
$e + H_2$	$\rightarrow$	$e + H_2(a^3\Sigma_g^+)$	[18]
$e + H_2$	$\rightarrow$	$e + H_2(b^2\Sigma_g)$	[18]
$e + H_2$	$\rightarrow$	$e + H_2(c^3\Pi_u)$	[18]
$e + H_2$	$\rightarrow$	$e + H_2(B'^1\Sigma_u^+)$	[18]
$e + H_2$	$\rightarrow$	$e + H_2(B^1\Sigma_u^+)$	[18]
$e + H_2$	$\rightarrow$	$e + H_2(E^1\Sigma_g^+)$	[18]
$e + H_2$	$\rightarrow$	$e + H_2(C^1\Pi_u)$	[18]
$e + H_2$	$\rightarrow$	$e + H_2(e^3\Sigma_u^+)$	[18]
$e + H_2$	$\rightarrow$	$e + H + H$	[17]
$e + H_2$	$\rightarrow$	$e + e + H_2^+$	[17]
$e + H_2$	$\rightarrow$	$H^- + H$	[19]
Excitation of N_2			
$e + N_2$	$\rightarrow$	$e + N_2(v=1)$	[20]
$e + N_2$	$\rightarrow$	$e + N_2(v=2)$	[20]
$e + N_2$	$\rightarrow$	$e + N_2(v=3)$	[20]
$e + N_2$	$\rightarrow$	$e + N_2(v=4)$	[20]
$e + N_2$	$\rightarrow$	$e + N_2(v=5)$	[20]
$e + N_2$	$\rightarrow$	$e + N_2(v=6)$	[20]
$e + N_2$	$\rightarrow$	$e + N_2(v=7)$	[20]
$e + N_2$	$\rightarrow$	$e + N_2(v=8)$	[20]
$e + N_2$	$\rightarrow$	$e + N_2(v=9)$	[21]
$e + N_2$	$\rightarrow$	$e + N_2(v=10)$	[21]
$e + N_2$	$\rightarrow$	$e + N_2(A^3\Sigma_u^+)$	[22]
$e + N_2$	$\rightarrow$	$e + N_2(B^3\Pi_g)$	[22]
$e + N_2$	$\rightarrow$	$e + N_2(C^3\Pi_u)$	[22]
$e + N_2$	$\rightarrow$	$e + N_2(W^1\Delta_u)$	[22]
$e + N_2$	$\rightarrow$	$e + N_2(W^3\Delta_u)$	[22]
$e + N_2$	$\rightarrow$	$e + N_2(a'^1\Sigma_u^-)$	[22]
$e + N_2$	$\rightarrow$	$e + N_2(a^1\Pi_g)$	[22]
$e + N_2$	$\rightarrow$	$e + N_2(a'''^1\Sigma_g)$	[22]

(Continued)

TABLE 4.9 (Continued)
Processes of Gas Excitation by Direct Electron Impact

Process			References
$e + N_2$	$\rightarrow$	$e + N_2(B'^3\Sigma_u^-)$	[22]
$e + N_2$	$\rightarrow$	$e + N_2(B^1\Pi_u)$	[23]
$e + N_2$	$\rightarrow$	$e + N_2(E^3\Sigma_g^+)$	[22]
$e + N_2$	$\rightarrow$	$e + N_2(\text{Rydberg})$	[22]
$e + N_2$	$\rightarrow$	$N_2^- \rightarrow e + N(^4S^0) + N(^4S^0)$	[24]
$e + N_2$	$\rightarrow$	$N + N$	[25,26]
$e + N_2$	$\rightarrow$	$e + e + N_2^+$	[27]
$e + N_2$	$\rightarrow$	$N(4s) + N^+(^3P) + e + e$	[28]
$e + N_2(j = 0)$	$\rightarrow$	$e + N_2(j = 2,4,6,8)$	[29]
	Excitation of O_2		
$e + O_2(j_1)$	$\rightarrow$	$e + O_2(j_2)$	[30]
$e + O_2$	$\rightarrow$	$e + O_2(v = 1)$	[18]
$e + O_2$	$\rightarrow$	$e + O_2(v = 2)$	[18]
$e + O_2$	$\rightarrow$	$e + O_2(v = 3)$	[18]
$e + O_2$	$\rightarrow$	$e + O_2(v = 4)$	[18]
$e + O_2$	$\rightarrow$	$e + O_2(a^1\Delta_g)$	[29]
$e + O_2$	$\rightarrow$	$e + O_2(b^1\Sigma_g^+)$	[30]
$e + O_2$	$\rightarrow$	$e + O_2(B^3\Sigma_u^-)$	[30]
$e + O_2$	$\rightarrow$	$e + O_2(A^3\Sigma_u^+)$	[30]
$e + O_2$	$\rightarrow$	$e + O_2(C^3\Delta_u)$	[31]
$e + O_2$	$\rightarrow$	$e + O_2(9.9 \text{ eV})$	[30]
$e + O_2$	$\rightarrow$	$e + O_2(\text{Rydberg})$	[30]
$e + O_2$	$\rightarrow$	$O_2^-(X^2\Pi_g) \rightarrow O^-(^2P^0) + O(^3P)$	[32]
$e + O_2$	$\rightarrow$	$e + O + O$	[30]
$e + O_2$	$\rightarrow$	$e + O^+ + O^-$	[32]
$e + O_2$	$\rightarrow$	$e + e + O_2^+$	[27]
$e + O_2$	$\rightarrow$	$e + e + O(^3P) + O^+(^4S)$	[28]
	Excitation of H_2O		
$e + H_2O$	$\rightarrow$	$H^- + OH$	[33]
$e + H_2O$	$\rightarrow$	$OH + H(^2P)$	[10,34]
$e + H_2O$	$\rightarrow$	$e + H_2O(7 \text{ eV})$	[10,34]
$e + H_2O$	$\rightarrow$	$e + H_2O(14 \text{ eV})$	[10,34]
$e + H_2O$	$\rightarrow$	$e + e + H_2O^+$	[27]
$e + H_2O$	$\rightarrow$	$e + e + OH^+ + H$	[35]
$e + H_2O$	$\rightarrow$	$e + H_2O\ (100+001)$	[10,34]
$e + H_2O$	$\rightarrow$	$e + H_2O\ (010)$	[10,34]

TABLE 4.10
The Reactions between Heavy Particles Included in the Kinetic Scheme. $A - 1/s$, cm^3/s, cm^6/s, $E_a - K$

Reaction			A^+	n^+	E_a^+, K	A^-	n^-	E_a^-, K	Ref$^+$	Ref$^-$
Ionization										
$N_2b(a'^1\Sigma_u^-) + N_2(a'^1\Sigma_u^-)$	$\rightarrow$	$N_4^+ + e$	5.0−11	0.0	0.0	0.0	0.0	0.0	[36]	—
$N_2(a'^1\Sigma_u^-) + N_2(a'^1\Sigma_u^-)$	$\rightarrow$	$N_4^+ + e$	2.0−10	0.0	0.0	0.0	0.0	0.0	[36]	—
$N(^2D) + N(^2P)$	$\rightarrow$	$N_2^+ + e$	1.0−12	0.0	0.0	0.0	0.0	0.0	[36]	—
$N(^2P) + O$	$\rightarrow$	$NO^+ + e$	1.0−12	0.0	0.0	0.0	0.0	0.0	[36]	—
$N(^2P) + N(^2P)$	$\rightarrow$	$N_2^+ + e$	5.0−12	0.0	0.0	0.0	0.0	0.0	[19]	—
$N_2(a'^1\Sigma_u^-) + N_2(a'^1\Sigma_u^-)$	$\rightarrow$	$N_2 + N_2^+ + e$	2.0−10	0.0	0.0	0.0	0.0	0.0	[37]	—
$N_2(a'^1\Sigma_u^-) + N_2(A^3\Sigma_u^+)$	$\rightarrow$	$N_2 + N_2^+ + e$	5.0−11	0.0	0.0	0.0	0.0	0.0	[37]	—
Recombination										
$N_2^+ + e$	$\rightarrow$	$N + N$	6.06−6	−0.50	0.0	1.41−14	1.0	6.77+4	[38]	[39]
$NO^+ + e$	$\rightarrow$	$N + O$	2.54−15	0.37	3.20+3	1.10−12	0.0	3.19+4	[40]	[39]
$O_2^+ + e$	$\rightarrow$	$O + O$	4.29−6	−0.60	0.0	0.0	0.0	0.0	[41]	—
$N_4^+ + e$	$\rightarrow$	$N_2 + N_2$	2.80−5	−0.50	0.0	0.0	0.0	0.0	[40]	—
$H_3^+ + e$	$\rightarrow$	$H + H_2$	4.0−6	−0.50	0.0	0.0	0.0	0.0	[40]	—
$H_2^+ + e$	$\rightarrow$	$H + H$	2.25−6	−0.40	0.0	0.0	0.0	0.0	[38]	—
$NH^+ + e$	$\rightarrow$	$H + N$	1.48−6	−0.50	0.0	0.0	0.0	0.0	[38]	—
$OH^+ + e$	$\rightarrow 0$	$H + O$	1.30−6	−0.50	0.0	0.0	0.0	0.0	[38]	—
$H^+ + e$	$\rightarrow$	H	2.28−10	−0.70	0.0	0.0	0.0	0.0	[41]	—
$N^+ + e$	$\rightarrow$	N	1.73−10	−0.70	0.0	0.0	0.0	0.0	[41]	—
$N_2O_2^+ + e$	$\rightarrow$	$NO + NO$	2.25−5	−0.50	0.0	0.0	0.0	0.0	[36]	—
$N_3O^+ + e$	$\rightarrow$	$N_2 + NO$	2.25−5	−0.50	0.0	0.0	0.0	0.0	[36]	—
$NO_3^+ + e$	$\rightarrow$	$NO + O_2$	2.25−5	−0.50	0.0	0.0	0.0	0.0	[36]	—
$N_2O_2^+ + e$	$\rightarrow$	$N_2 + O_2$	2.25−5	−0.50	0.0	0.0	0.0	0.0	[36]	—
$NO_2^+ + e$	$\rightarrow$	$NO + O$	3.46−6	−0.50	0.0	0.0	0.0	0.0	[36]	—

(Continued)

TABLE 4.10 (Continued)
The Reactions between Heavy Particles Included in the Kinetic Scheme. A – 1/s, cm^3/s, cm^6/s, E_a – K

Reaction			A^+	n^+	E_a^+, K	A^-	n^-	E_a^-, K	Ref$^+$	Ref$^-$
$N_2O^+ + e$	$\rightarrow$	$N_2 + O$	3.46−6	−0.50	0.0	0.0	0.0	0.0	[36]	—
$N_3^+ + e$	$\rightarrow$	$N + N_2$	3.46−6	−0.50	0.0	0.0	0.0	0.0	[36]	—
$N_2^+ + e + e$	$\rightarrow$	$N_2 + e$	1.40−8	−4.50	0.0	0.0	0.0	0.0	[36]	—
$O_2^+ + e + e$	$\rightarrow$	$O_2 + e$	1.40−8	−4.50	0.0	0.0	0.0	0.0	[36]	—
$NO^+ + e + e$	$\rightarrow$	$NO + e$	1.40−8	−4.50	0.0	0.0	0.0	0.0	[36]	—
$N^+ + e + e$	$\rightarrow$	$N + e$	1.40−8	−4.50	0.0	0.0	0.0	0.0	[36]	—
$O^+ + e + e$	$\rightarrow$	$O + e$	1.40−8	−4.50	0.0	0.0	0.0	0.0	[36]	—
$N_2^+ + N_2^+ + e$	$\rightarrow$	$N_2 + N_2$	9.35−21	−2.50	0.0	0.0	0.0	0.0	[36]	—
$N_2^+ + O_2^+ + e$	$\rightarrow$	$N_2 + O_2$	9.35−21	−2.50	0.0	0.0	0.0	0.0	[36]	—
$N_2^+ + NO^+ + e$	$\rightarrow$	$N_2 + NO$	9.35−21	−2.50	0.0	0.0	0.0	0.0	[36]	—
$N^+ + N_2^+ + e$	$\rightarrow$	$N + N_2$	9.35−21	−2.50	0.0	0.0	0.0	0.0	[36]	—
$N_2^+ + O^+ + e$	$\rightarrow$	$N_2 + O$	9.35−21	−2.50	0.0	0.0	0.0	0.0	[36]	—
$N_2^+ + O_2^+ + e$	$\rightarrow$	$N_2 + O_2$	9.35−21	−2.50	0.0	0.0	0.0	0.0	[36]	—
$O_2^+ + O_2^+ + e$	$\rightarrow$	$O_2 + O_2$	9.35−21	−2.50	0.0	0.0	0.0	0.0	[36]	—
$NO^+ + O_2^+ + e$	$\rightarrow$	$NO + O_2$	9.35−21	−2.50	0.0	0.0	0.0	0.0	[36]	—
$N^+ + O_2^+ + e$	$\rightarrow$	$N + O_2$	9.35−21	−2.50	0.0	0.0	0.0	0.0	[36]	—
$O^+ + O_2^+ + e$	$\rightarrow$	$O + O_2$	9.35−21	−2.50	0.0	0.0	0.0	0.0	[36]	—
$H_3O^+ + e$	$\rightarrow$	$H_2 + OH$	2.08−5	−0.50	0.0	0.0	0.0	0.0	[40]	—
$H_3^+ + e$	$\rightarrow$	$H + H + H$	4.0−6	−0.50	0.0	0.0	0.0	0.0	[38]	—
$O_2^+ + e$	$\rightarrow$	$O + O(^1S)$	1.93−7	−0.60	0.0	0.0	0.0	0.0	[41]	—
$O_2^+ + e$	$\rightarrow$	$O + O(^1D)$	2.11−6	−0.60	0.0	0.0	0.0	0.0	[41]	—
$N_2^+ + e$	$\rightarrow$	$N + N(^2D)$	3.46−6	−0.50	0.0	0.0	0.0	0.0	[36]	—
$NO^+ + e$	$\rightarrow$	$N(^2D) + O$	9.0−5	−1.0	0.0	0.0	0.0	0.0	[36]	—
$N_3^+ + e$	$\rightarrow$	$N + N_2(B^3\Pi_g)$	4.30−7	−0.50	0.0	0.0	0.0	0.0	[42]	—
$N_3^+ + e$	$\rightarrow$	$N + N_2(A^3\Sigma_u^+)$	4.30−7	−0.50	0.0	0.0	0.0	0.0	[42]	—

(*Continued*)

TABLE 4.10 (Continued)
The Reactions between Heavy Particles Included in the Kinetic Scheme. $A - 1/s$, cm^3/s, cm^6/s, $E_a - K$

Reaction		A^+	n^+	E_a^+, K	A^-	n^-	E_a^-, K	Ref+	Ref-
				Attachment					
N_2+O_2+e	$\rightarrow$	$N_2+O_2^-$	8.33−32	0.0	0.0	4.20−16	1.50	4.99+3	[39]	[39]
$O+O_2+e$	$\rightarrow$	$O+O_2^-$	2.76−32	0.0	0.0	1.0−16	1.50	5.33+3	[39]	[39]
O_3+e	$\rightarrow$	$O+O_2^-$	1.0−9	0.0	0.0	1.66−10	0.0	0.0	[36]	[39]
O_3+e	$\rightarrow$	O^-+O_2	1.0−11	0.0	0.0	5.0−15	0.0	0.0	[36]	[36]
NO_2+e	$\rightarrow$	$NO+O^-$	1.0−11	0.0	0.0	1.66−10	0.0	0.0	[36]	[36]
N_2O+e	$\rightarrow$	N_2+O^-	2.0−10	0.0	0.0	1.0−12	0.0	0.0	[36]	[43]
O_2+O_2+e	$\rightarrow$	$O_2+O_2^-$	4.0−30	0.0	0.0	9.0−15	1.50	4.99+3	[39]	[39]
$NO+NO+e$	$\rightarrow$	$NO+NO^-$	7.80−26	−1.50	9.40+2	3.30−10	0.0	1.22+3	[44]	[44]
$N_2O+NO+e$	$\rightarrow$	N_2O+NO^-	1.0−25	−1.50	9.70+2	4.30−10	0.0	1.24+3	[44]	[44]
N_2+O+e	$\rightarrow$	N_2+O^-	5.51−30	−0.50	0.0	4.0−13	1.0	1.69+4	[39]	[39]
H_2+NO+e	$\rightarrow$	H_2+NO^-	1.56−27	−1.50	6.80+2	6.30−12	0.0	9.60+2	[44]	[44]
NH_4+NO+e	$\rightarrow$	NH_4+NO^-	5.20−26	−1.50	4.10+2	2.30−10	0.0	6.80+2	[44]	[44]
$NO+O_2+e$	$\rightarrow$	$NO+O_2^-$	2.76−32	0.0	0.0	1.0−16	1.50	5.33+3	[39]	[39]
N_2+N_2O+e	$\rightarrow$	$N_2+N_2O^-$	3.50−39	2.0	0.0	0.0	0.0	0.0	[36]	—
NO_2+O_2+e	$\rightarrow$	$NO_2^-+O_2$	3.0−28	0.0	0.0	0.0	0.0	0.0	[36]	—
N_2+NO_2+e	$\rightarrow$	$N_2+NO_2^-$	8.0−28	0.0	0.0	0.0	0.0	0.0	[36]	—
$N_2(A^3\Sigma_u^+)+O^-$	$\rightarrow$	$N+NO+e$	1.0−10	0.0	0.0	0.0	0.0	0.0	[45]	—
O^-+O_3	$\rightarrow$	O_2+O_2+e	5.0−10	0.0	0.0	0.0	0.0	0.0	[45]	—
$O+O_2^-+e$	$\rightarrow$	O^-+O_2	4.0−30	0.0	0.0	0.0	0.0	0.0	[39]	—
O_2+O_3+e	$\rightarrow$	$O_2+O_3^-$	3.97−30	0.0	0.0	0.0	0.0	0.0	[39]	—
NO_2+O+e	$\rightarrow$	NO_2+O^-	8.33−31	0.0	0.0	0.0	0.0	0.0	[39]	—
$O+O+e$	$\rightarrow$	$O+O^-$	8.33−31	0.0	0.0	0.0	0.0	0.0	[39]	—
$NO+O+e$	$\rightarrow$	$NO+O^-$	8.33−31	0.0	0.0	0.0	0.0	0.0	[39]	—

(Continued)

TABLE 4.10 (Continued)
The Reactions between Heavy Particles Included in the Kinetic Scheme. A – 1/s, cm³/s, cm⁶/s, E_a – K

Reaction			A^+	n^+	E_a^+, K	A^-	n^-	E_a^-, K	Ref⁺	Ref⁻
					Detachment					
$N+O^-$	$\rightarrow$	$NO+e$	2.60–10	0.0	0.0	0.0	0.0	0.0	[36]	—
NO_2^-+O	$\rightarrow$	NO_3+e	1.0–12	0.0	0.0	0.0	0.0	0.0	[36]	—
$O+O^-$	$\rightarrow$	O_2+e	1.40–10	0.0	0.0	0.0	0.0	0.0	[39]	—
$N+O_2^-$	$\rightarrow$	NO_2+e	5.0–10	0.0	0.0	0.0	0.0	0.0	[43]	—
H_2+O^-	$\rightarrow$	H_2O+e	8.0–10	0.0	0.0	0.0	0.0	0.0	[43]	—
H_2O+O^-	$\rightarrow$	H_2O_2+e	6.0–13	0.0	0.0	0.0	0.0	0.0	[43]	—
$H+O_2^-$	$\rightarrow$	HO_2+e	1.20–9	0.0	0.0	0.0	0.0	0.0	[43]	—
$H+O_2^-$	$\rightarrow$	HO_2+e	1.20–9	0.0	0.0	0.0	0.0	0.0	[43]	—
$H+H^-$	$\rightarrow$	H_2+e	1.80–9	0.0	0.0	0.0	0.0	0.0	[43]	—
$OH+O^-$	$\rightarrow$	HO_2+e	1.80–9	0.0	0.0	0.0	0.0	0.0	[43]	—
$H+OH^-$	$\rightarrow$	H_2O+e	1.0–9	0.0	0.0	0.0	0.0	0.0	[43]	—
$O+O_2(a^1\Delta_g)$	$\rightarrow$	O_3+e	4.0–9	0.0	0.0	0.0	0.0	0.0	[19]	—
$O_2^-(a^1\Delta_g)+O_2^-$	$\rightarrow$	O_2+O_2+e	2.0–10	0.0	0.0	0.0	0.0	0.0	[44]	—
$O_2^-(b^1\Sigma_g^+)+O_2^-$	$\rightarrow$	O_2+O_2+e	3.60–10	0.0	0.0	0.0	0.0	0.0	[44]	—
$N_2(A^3\Sigma_u^+)+O_2^-$	$\rightarrow$	N_2+O_2+e	2.10–9	0.0	0.0	0.0	0.0	0.0	[44]	—
$N_2(a^1\Pi_g)+O_2^-$	$\rightarrow$	N_2+O_2+e	2.50–9	0.0	0.0	0.0	0.0	0.0	[19]	—
$N_2(a'^1\Sigma_u^-)+O_2^-$	$\rightarrow$	N_2+O_2+e	5.0–9	0.0	0.0	0.0	0.0	0.0	[19]	—
$N_2(B^3\Pi_g)+O_2^-$	$\rightarrow$	N_2+O_2+e	2.50–9	0.0	0.0	0.0	0.0	0.0	[44]	—
$O+O_2(b^1\Sigma_g^+)$	$\rightarrow$	$O+O_2+e$	6.90–10	0.0	0.0	0.0	0.0	0.0	[36]	—
$N_2(A^3\Sigma_u^+)+O^-$	$\rightarrow$	N_2+O+e	2.20–9	0.0	0.0	0.0	0.0	0.0	[36]	—
$N_2(B^3\Pi_g)+O^-$	$\rightarrow$	N_2+O+e	1.90–9	0.0	0.0	0.0	0.0	0.0	[36]	—
$O+O_3^-$	$\rightarrow$	O_2+O_2+e	1.40–10	0.0	0.0	0.0	0.0	0.0	[39]	—
$N+O_2^-$	$\rightarrow$	$NO+O+e$	4.0–10	0.0	0.0	0.0	0.0	0.0	[43]	—

(Continued)

TABLE 4.10 (Continued)
The Reactions between Heavy Particles Included in the Kinetic Scheme. $A - 1/s$, cm^3/s, cm^6/s, $E_a - K$

Reaction			A^+	n^+	E_a^+, K	A^-	n^-	E_a^-, K	Ref+	Ref−
			\multicolumn{6}{c}{Excitation and Quenching}		
$O+O+O_2$	→	$O_2+O_2(a^1\Delta_g)$	2.45−31	−0.63	0.0	4.07	−2.50	4.80+4	[36]	[46]
$O+O+O_2$	→	$O_2+O_2(b^1\Sigma_g^+)$	2.45−31	−0.63	0.0	4.07	−2.50	4.03+4	[36]	[46]
N_2+O+O	→	$N_2+O_2(a^1\Delta_g)$	2.76−34	0.0	−7.20+2	4.07	−2.50	4.80+4	[36]	[46]
N_2+O+O	→	$N_2+O_2(b^1\Sigma_g^+)$	2.76−34	0.0	−7.20+2	4.07	−2.50	4.03+4	[36]	[46]
$N+N+O_2$	→	$N_2(A^3\Sigma_u^+)+O_2$	8.27−34	0.0	−5.0+2	3.80+5	−3.50	4.16+4	[36]	[46]
$N+N+N_2$	→	$N_2+N_2(A^3\Sigma_u^+)$	8.27−34	0.0	−5.0+2	3.80+5	−3.50	4.16+4	[36]	[46]
N_2+O+O	→	$N_2(A^3\Sigma_u^+)+O_2$	2.76−34	0.0	−7.20+2	2.54−12	0.0	0.0	[46]	[36]
$O_2(b^1\Sigma_g^+)+O_3$	→	$O_2(a^1\Delta_g)+O_3$	6.60−12	0.0	0.0	6.60−12	0.0	7.66+3	[47]	[42]
$N+N_2(A^3\Sigma_u^+)$	→	$N(^2P)+N_2$	5.0−11	0.0	0.0	4.0−11	0.0	0.0	[48]	[42]
$N_2+N_2(B^3\Pi_g)$	→	$N_2+N_2(A^3\Sigma_u^+)$	2.0−12	0.0	0.0	3.0−16	0.0	0.0	[49]	[42]
$N_2(B^3\Pi_g)+NO$	→	$N_2(A^3\Sigma_u^+)+NO$	2.40−10	0.0	0.0	2.40−10	0.0	1.37+4	[36]	[42]
$N_2+N_2(a'^1\Sigma_u^-)$	→	$N_2+N_2(B^3\Pi_g)$	2.0−13	0.0	0.0	2.0−13	0.0	1.22+4	[36]	[42]
N_2+O+O	→	$N_2(a'^1\Sigma_u^-)+O_2$	2.75−34	0.0	−7.20+2	2.80−11	0.0	0.0	[46]	[36]
$N+N_2+O$	→	$N_2(a'^1\Sigma_u^-)+NO$	1.76−31	−0.50	0.0	3.60−10	0.0	0.0	[46]	[36]
$N_2+N_2(C^3\Pi_u)$	→	$N_2+N_2(a'^1\Sigma_u^-)$	1.0−11	0.0	0.0	1.0−11	0.0	3.05+4	[36]	[42]
$O+O_2+O_2$	→	$O_2(b^1\Sigma_g^+)+O_3$	8.60−31	−1.25	0.0	1.80−11	0.0	0.0	[46]	[36]
$N_2+O_2(b^1\Sigma_g^+)$	→	$N_2+O_2(a^1\Delta_g)$	2.30−15	0.0	0.0	2.30−15	0.0	7.66+3	[36]	[42]
$O_2+O_2(b^1\Sigma_g^+)$	→	$O_2+O_2(a^1\Delta_g)$	1.50−16	0.0	0.0	1.50−16	0.0	7.66+3	[36]	[42]
$O+O_2(b^1\Sigma_g^+)$	→	$O+O_2(a^1\Delta_g)$	8.0−14	0.0	0.0	8.0−14	0.0	7.66+3	[36]	[42]
$NO+O_2(b^1\Sigma_g^+)$	→	$NO+O_2(a^1\Delta_g)$	5.0−15	0.0	0.0	5.0−15	0.0	7.66+3	[50]	[42]
$H_2+O_2(b^1\Sigma_g^+)$	→	$H_2+O_2(a^1\Delta_g)$	8.30−13	0.0	0.0	8.30−13	0.0	7.66+3	[50]	[42]
$N_2(A^3\Sigma_u^+)+N_2(A^3\Sigma_u^+)$	→	$N_2+N_2(B^3\Pi_g)$	1.10−9	0.0	0.0	1.10−9	0.0	5.79+4	[48]	[42]

(Continued)

TABLE 4.10 (Continued)
The Reactions between Heavy Particles Included in the Kinetic Scheme. A – 1/s, cm^3/s, cm^6/s, E_a – K

Reaction		A^+	n^+	E_a^+, K	A^-	n^-	E_a^-, K	Ref$^+$	Ref$^-$
$N_2(A^3\Sigma_u^+) + N_2(B^3\Pi_g)$	$\rightarrow$	$N_2 + N_2(C^3\Pi_u)$	4.60−10	0.0	0.0	4.60−10	0.0	2.89+4	[42]	[42]
$N_2 + N_2(C^3\Pi_u)$	$\rightarrow$	$N_2 + N_2(B^3\Pi_g)$	3.30−11	0.0	0.0	3.30−11	0.0	4.27+4	[42]	[42]
$N(^2P) + N$	$\rightarrow$	$N + N$	1.60−12	0.0	0.0	1.13−12	0.0	4.15+4	[42]	a
$N_2(B^3\Pi_g) + O_2$	$\rightarrow$	$N_2(A^3\Sigma_u^+) + O_2(a^1\Delta_g)$	5.50−11	0.0	0.0	5.50−11	0.0	2.32+3	[51]	[42]
$N_2(B^3\Pi_g) + O_2$	$\rightarrow$	$N_2(A^3\Sigma_u^+) + O_2$	5.50−11	0.0	0.0	5.50−11	0.0	1.37+4	[51]	[42]
$N + O_2(a^1\Delta_g)$	$\rightarrow$	$NO + O$	2.0−14	0.0	6.0+2	0.0	0.0	0.0	[36]	—
$O + O_3$	$\rightarrow$	$O_2 + O_2(a^1\Delta_g)$	2.0−11	0.0	2.30+3	0.0	0.0	0.0	[36]	—
$N_2(A^3\Sigma_u^+) + NO$	$\rightarrow$	$N_2 + NO$	2.80−11	0.0	0.0	0.0	0.0	0.0	[48]	—
$N_2(B^3\Pi_g) + O_2$	$\rightarrow$	$N_2 + O + O$	3.0−10	0.0	0.0	0.0	0.0	0.0	[36]	—
$N_2(C^3\Pi_u) + O_2$	$\rightarrow$	$N_2 + O + O$	2.7−10	0.0	0.0	0.0	0.0	0.0	[52]	—
$N_2(A^3\Sigma_u^+) + O_2$	$\rightarrow$	$N_2 + O_2(b^1\Sigma_g^+)$	1.29−12	0.0	0.0	0.0	0.0	0.0	[36]	—
$N_2(C^3\Pi_u) + N_2$	$\rightarrow$	$N_2 + N_2$	9.0−12	0.0	0.0	0.0	0.0	0.0	[52]	—
$N_2(C^3\Pi_u) + H_2$	$\rightarrow$	$N_2 + H_2$	3.2−10	0.0	0.0	0.0	0.0	0.0	[52]	—
$N_2(C^3\Pi_u) + H_2O$	$\rightarrow$	$N_2 + H_2O$	3.9−10	0.0	0.0	0.0	0.0	0.0	[52]	—
$N_2(A^3\Sigma_u^+) + O$	$\rightarrow$	$N_2 + O$	2.10−11	0.0	0.0	0.0	0.0	0.0	[36]	—
$N + N_2(A^3\Sigma_u^+)$	$\rightarrow$	$N + N_2$	5.0−11	0.0	0.0	0.0	0.0	0.0	[36]	—
$N_2 + O_2(a^1\Delta_g)$	$\rightarrow$	$N_2 + O_2$	3.0−21	0.0	0.0	0.0	0.0	0.0	[50]	—
$O_2 + O_2(a^1\Delta_g)$	$\rightarrow$	$O_2 + O_2$	3.0−18	0.80	0.0	0.0	0.0	0.0	[50]	—
$O + O_2(a^1\Delta_g)$	$\rightarrow$	$O + O_2$	7.0−16	0.0	0.0	0.0	0.0	0.0	[36]	—
$NO + O_2(a^1\Delta_g)$	$\rightarrow$	$NO + O_2$	2.50−11	0.0	0.0	0.0	0.0	0.0	[36]	—
$N_2(A^3\Sigma_u^+) + O_2$	$\rightarrow$	$N_2O + O$	7.80−14	0.0	0.0	0.0	0.0	0.0	[36]	—
$N_2(A^3\Sigma_u^+) + O$	$\rightarrow$	$N(^2D) + NO$	7.0−12	0.0	0.0	0.0	0.0	0.0	[36]	—
$N_2(A^3\Sigma_u^+) + N_2O$	$\rightarrow$	$N + N_2 + NO$	1.0−11	0.0	0.0	0.0	0.0	0.0	[36]	—

(Continued)

TABLE 4.10 (Continued)
The Reactions between Heavy Particles Included in the Kinetic Scheme. A – 1/s, cm^3/s, cm^6/s, E_a – K

Reaction		A^+	n^+	E_a^+, K	A^-	n^-	E_a^-, K	Ref$^+$	Ref$^-$
$N_2(A^3\Sigma_u^+) + N_2(A^3\Sigma_u^+)$	$\rightarrow$	$N_2 + N_2(^3\Pi_u)$	5.54–4	−2.64	0.0	0.0	0.0	0.0	[53]	—
$N_2 + N_2(A^3\Sigma_u^+)$	$\rightarrow$	$N_2 + N_2$	3.0–18	0.0	0.0	0.0	0.0	0.0	[36]	—
$N_2(A^3\Sigma_u^+) + O_2$	$\rightarrow$	$N_2 + O_2(a^1\Delta_g)$	1.29–12	0.0	0.0	0.0	0.0	0.0	[36]	—
$O_2 + O_2(b^1\Sigma_g^+)$	$\rightarrow$	$O_2 + O_2$	4.0–17	0.0	0.0	0.0	0.0	0.0	[47]	—
$N_2(A^3\Sigma_u^+) + N_2^+$	$\rightarrow$	$N + N_3^+$	3.0–10	0.0	0.0	0.0	0.0	0.0	[36]	—
$O_2(a^1\Delta_g) + O_4^+$	$\rightarrow$	$O_2 + O_2 + O_2^+$	1.0–10	0.0	0.0	0.0	0.0	0.0	[36]	—
$O_2(b^1\Sigma_g^+) + O_4^+$	$\rightarrow$	$O_2 + O_2 + O_2^+$	1.0–10	0.0	0.0	0.0	0.0	0.0	[36]	—
$O + O_2(a^1\Delta_g)$	$\rightarrow$	$O + O_2^-$	1.0–10	0.0	0.0	0.0	0.0	0.0	[36]	—
$O_2(a^1\Delta_g) + O_4^-$	$\rightarrow$	$O_2 + O_2 + O_2^-$	1.0–10	0.0	0.0	0.0	0.0	0.0	[36]	—
$O_2(b^1\Sigma_g^+) + O_4^-$	$\rightarrow$	$O_2 + O_2 + O_2^-$	1.0–10	0.0	0.0	0.0	0.0	0.0	[36]	—
$H_2O + O_2(a^1\Delta_g)$	$\rightarrow$	$H_2O + O_2$	3.0–18	0.0	0.0	0.0	0.0	0.0	[50]	—
$H_2 + O_2(a^1\Delta_g)$	$\rightarrow$	$H_2 + O_2$	4.50–18	0.0	0.0	0.0	0.0	0.0	[50]	—
$O_2(a^1\Delta_g) + O_3$	$\rightarrow$	$O_2 + O_3$	4.0–15	0.0	0.0	0.0	0.0	0.0	[50]	—
$O_2(a^1\Delta_g) + O_2(a^1\Delta_g)$	$\rightarrow$	$O_2 + O_2(b^1\Sigma_g^+)$	2.0–17	0.0	0.0	0.0	0.0	0.0	[50]	—
$O(^1S) + O_2(a^1\Delta_g)$	$\rightarrow$	$O + O + O$	3.40–11	0.0	0.0	0.0	0.0	0.0	[50]	—
$N_2 + O_2(b^1\Sigma_g^+)$	$\rightarrow$	$N_2 + O_2$	2.30–15	0.0	0.0	0.0	0.0	0.0	[50]	—
$H_2O + O_2(b^1\Sigma_g^+)$	$\rightarrow$	$H_2O + O_2$	6.70–12	0.0	0.0	0.0	0.0	0.0	[50]	—
$H_2 + O_2(b^1\Sigma_g^+)$	$\rightarrow$	$H_2 + O_2$	7.90–13	0.0	0.0	0.0	0.0	0.0	[50]	—
$O_2(b^1\Sigma_g^+) + O_3$	$\rightarrow$	$O_2 + O_3$	1.80–11	0.0	0.0	0.0	0.0	0.0	[50]	—
$O + O_2(b^1\Sigma_g^+)$	$\rightarrow$	$O + O_2$	8.0–14	0.0	0.0	0.0	0.0	0.0	[50]	—
$O(^1S) + O_2(A^3\Sigma_u^+)$	$\rightarrow$	$O + O_2(a^1\Delta_g)$	1.30–10	0.0	0.0	0.0	0.0	0.0	[50]	—
$N_2 + N_2(a'^1\Sigma_u^-)$	$\rightarrow$	$N_2 + N_2$	6.0–14	0.0	0.0	0.0	0.0	0.0	[49]	—
$N + N + N_2$	$\rightarrow$	$N_2 + N_2(B^3\Pi_g)$	2.40–33	0.0	0.0	0.0	0.0	0.0	[49]	—

(*Continued*)

TABLE 4.10 (Continued)
The Reactions between Heavy Particles Included in the Kinetic Scheme A – 1/s, cm³/s, cm⁶/s, E_a – K

Reaction			A^+	n^+	E_a^+, K	A^-	n^-	E_a^-, K	Ref$^+$	Ref$^-$
$N(^2D)+N+N_2$	$\rightarrow$	$N_2+N_2(C^3\Pi_u)$	1.0−34	0.0	0.0	0.0	0.0	0.0	[42]	—
$N+N_2(B^3\Pi_g)$	$\rightarrow$	$N(^2P)+N_2$	1.0−10	0.0	0.0	0.0	0.0	0.0	[42]	—
$N+N_2(C^3\Pi_u)$	$\rightarrow$	$N(^2P)+N_2$	3.0−10	0.0	0.0	0.0	0.0	0.0	[42]	—
$N+N_2(a'^1\Sigma_u^-)$	$\rightarrow$	$N(^2P)+N_2$	3.0−10	0.0	0.0	0.0	0.0	0.0	[42]	—
$O_2(a^1\Delta_g)+O_3$	$\rightarrow$	$O+O_2+O_2$	5.20−11	0.0	2.84+3	0.0	0.0	0.0	[51]	—
$N_2(C^3\Pi_u)+O_2$	$\rightarrow$	$N_2(A^3\Sigma_u^+)+O_2$	1.35−11	0.0	0.0	0	0.0	0.0	[51]	—
$N_2(C^3\Pi_u)+O_2$	$\rightarrow$	$N_2+O+O(^1D)$	1.35−11	0.0	0.0	0.0	0.0	0.0	[51]	—
$O(^1S)+O_2(a^1\Delta_g)$	$\rightarrow$	$O+O_2(a^1\Delta_g)$	2.75−10	0.0	0.0	0.0	0.0	0.0	[54]	—
$O(^1S)+O_2(a^1\Delta_g)$	$\rightarrow$	$O(^2P)+O_2(a^1\Delta_g)$	2.75−10	0.0	0.0	0.0	0.0	0.0	[54]	—
$N_2+N_2(B^3\Pi_g)$	$\rightarrow$	N_2+N_2	1.0−12	0.0	0.0	0.0	0.0	0.0	[48]	—
$N_2(C^3\Pi_u)$	$\rightarrow$	$N_2(B^3\Pi_g)$	2.73+7	0.0	0.0	0.0	0.0	0.0	[55]	—
$N_2(B^3\Pi_g)$	$\rightarrow$	$N_2(A^3\Sigma_u^+)$	1.25+5	0.0	0.0	0.0	0.0	0.0	[55]	—
$N_2(A^3\Sigma_u^+)$	$\rightarrow$	N_2	5.26−1	0.0	0.0	0.0	0.0	0.0	[55]	—
$N_2(a'^1\Sigma_u^-)$	$\rightarrow$	N_2	2.0	0.0	0.0	0.0	0.0	0.0	[55]	—
$N+N+N_2$	$\rightarrow$	$N_2+N_2(A^3\Sigma_u^+)$	2.40−33	0.0	0.0	0.0	0.0	0.0	[56]	—
N_2+O+O	$\rightarrow$	$N_2+O_2(A^3\Sigma_u^+)$	2.10−37	0.0	0.0	0.0	0.0	0.0	[57]	—
$N_2(A^3\Sigma_u^+)+O_2$	$\rightarrow$	N_2+O_2	1.90−12	0.0	0.0	0.0	0.0	0.0	[48]	—
$N_2+N_2(C^3\Pi_u)$	$\rightarrow$	N_2+N_2	5.0−11	0.0	0.0	0.0	0.0	0.0	[19]	—
$N_2(C^3\Pi_u)+O_2$	$\rightarrow$	N_2+O_2	1.11−10	0.0	0.0	0.0	0.0	0.0	[19]	—
$N_2(C^3\Pi_u)+NO$	$\rightarrow$	N_2+NO	2.0−11	0.0	0.0	0.0	0.0	0.0	[19]	—
$O(^1D)+O_2$	$\rightarrow$	$O+O_2(a^1\Delta_g)$	1.0−12	0.0	0.0	0.0	0.0	0.0	[19]	—
$O+O^+$	$\rightarrow$	$O+O(^3P)$	5.0−8	0.0	0.0	0.0	0.0	0.0	[41]	—
$O+O^+$	$\rightarrow$	$O(^1D)+O(^5S)$	1.50−8	0.0	0.0	0.0	0.0	0.0	[41]	—

(Continued)

TABLE 4.10 (Continued)
The Reactions between Heavy Particles Included in the Kinetic Scheme. $A - 1/s$, cm^3/s, cm^6/s, $E_a - K$

Reaction			A^+	n^+	E_a^+, K	A^-	n^-	E_a^-, K	Ref$^+$	Ref$^-$
$N_2(A^3\Sigma_u^+)+NH_3$	$\rightarrow$	N_2+NH_3	8.20−11	0.0	0.0	0.0	0.0	0.0	[48]	—
$N_2+N_2(a'^1\Sigma_u^-)$	$\rightarrow$	$N_2+N_2(A^3\Sigma_u^+)$	2.0−13	0.0	0.0	0.0	0.0	0.0	[37]	—
$N+N+N_2$	$\rightarrow$	$N_2+N_2(C^3\Pi_u)$	1.0−34	0.0	0.0	0.0	0.0	0.0	[42]	—
$N+N_2(B^3\Pi_g)$	$\rightarrow$	$N+N_2(a''\Sigma_u^-)$	3.30−11	0.0	0.0	0.0	0.0	0.0	[58]	—
$N+N_2(B^3\Pi_g)$	$\rightarrow$	$N+N_2(A^3\Sigma_u^+)$	1.0−10	0.0	0.0	0.0	0.0	0.0	[58]	—
$N+N_2(b^3\Sigma_u)$	$\rightarrow$	$N+N_2(B^3\Pi_g)$	1.0−10	0.0	0.0	0.0	0.0	0.0	[58]	—
$N+N_2(C^3\Pi_u)$	$\rightarrow$	$N+N_2(B^3\Pi_g)$	3.30−11	0.0	0.0	0.0	0.0	0.0	[58]	—
$N+N_2(W^3\Delta_u)$	$\rightarrow$	$N+N_2(B^3\Pi_g)$	3.30−11	0.0	0.0	0.0	0.0	0.0	[58]	—
$N_2(B^3\Pi_g)$	$\rightarrow$	$N_2(W^3\Delta_u)$	7.46+4	0.0	0.0	0.0	0.0	0.0	[23]	—
$N_2(b^3\Sigma_u)$	$\rightarrow$	$N_2(B^3\Pi_g)$	2.20+4	0.0	0.0	0.0	0.0	0.0	[23]	—
$N_2(a^1\Pi_g)$	$\rightarrow$	N_2	8.70+3	0.0	0.0	0.0	0.0	0.0	[23]	—
$N_2(w^1\Delta_u)$	$\rightarrow$	N_2	4.0+3	0.0	0.0	0.0	0.0	0.0	[23]	—
$O(^1S)+O_2$	$\rightarrow$	$O+O_2$	2.45−12	0.0	8.60+2	4.61−13	0.0	4.95+4	[54]	a
$O+O(^1S)$	$\rightarrow$	$O+O$	2.50−11	0.0	3.10+2	8.49−12	0.0	4.88+4	[54]	a
$N(^2D)+O^+$	$\rightarrow$	N^++O	1.30−10	0.0	0.0	6.39−11	0.0	1.76+4	[36]	a
$N_2O+O(^1D)$	$\rightarrow$	$NO+NO$	7.20−11	0.0	0.0	1.06−12	0.0	4.09+4	[47]	a
$N_2O+O(^1D)$	$\rightarrow$	N_2O+O	1.0−12	0.0	0.0	1.13−12	0.0	2.28+4	[47]	a
$H_2O+O(^1D)$	$\rightarrow$	$OH+OH$	2.80−10	0.0	0.0	3.71−11	0.0	1.44+4	[59]	a
$H_2O+O(^1D)$	$\rightarrow$	H_2O+O	2.80−10	0.0	0.0	3.16−10	0.0	2.28+4	[59]	a
$H_2+O(^1D)$	$\rightarrow$	$H+OH$	1.10−10	0.0	0.0	5.30−11	0.0	2.18+4	[47]	a
$H_2+O(^1D)$	$\rightarrow$	H_2+O	1.10−10	0.0	0.0	1.24−10	0.0	2.28+4	[47]	a
$H_2O+O(^1D)$	$\rightarrow$	H_2+O_2	2.30−12	0.0	0.0	1.92−11	0.0	2.38+4	[47]	a
$O(^1S)+O_2(a^1\Delta_g)$	$\rightarrow$	$O(^1D)+O_2(b^1\Sigma_g^+)$	3.60−11	0.0	0.0	3.60−11	0.0	1.80+3	[50]	[43]
$N(^2D)+O_2$	$\rightarrow$	$NO+O$	4.30−13	0.50	0.0	2.85−13	0.50	4.37+4	[36]	a
$N(^2D)+NO$	$\rightarrow$	N_2+O	7.0−11	0.0	0.0	3.64−10	0.0	6.55+4	[48]	a

(Continued)

TABLE 4.10 (Continued)
The Reactions between Heavy Particles Included in the Kinetic Scheme. $A - 1/s$, cm^3/s, cm^6/s, $E_a - K$

Reaction			A^+	n^+	E_a^+, K	A^-	n^-	E_a^-, K	Ref+	Ref-
$N(^2D)+N_2O$	$\rightarrow$	N_2+NO	3.50−12	0.0	0.0	2.38−13	0.0	8.36+4	[48]	a
$N(^2D)+N_2$	$\rightarrow$	$N+N_2$	6.0−15	0.0	0.0	7.33−15	0.0	2.77+4	[36]	a
$N(^2P)+O_2$	$\rightarrow$	$NO+O$	2.60−12	0.0	0.0	9.96−13	0.0	5.75+4	[36]	a
$N(^2P)+NO$	$\rightarrow$	N_2+O	3.40−11	0.0	0.0	1.02−10	0.0	7.93+4	[36]	a
$N(^2P)+N_2$	$\rightarrow$	$N+N_2$	2.0−18	0.0	0.0	2.59−18	0.0	4.15+4	[36]	a
$N+N(^2P)$	$\rightarrow$	$N+N(^2D)$	1.80−12	0.0	0.0	1.04−12	0.0	1.38+4	[36]	a
$O(^1D)+O_2$	$\rightarrow$	$O+O_2(b^1\Sigma_g^+)$	1.70−5	−2.50	3.40+2	0.0	0.0	0.0	[19]	—
$O(^1D)+O_2$	$\rightarrow$	$O+O_2$	3.20−11	0.0	−5.70+1	7.89−12	0.0	2.29+4	[60]	a
$N_2+O(^1D)$	$\rightarrow$	N_2+O	3.20−11	0.0	1.07+2	3.61−11	0.0	2.29+4	[59]	a
$N_2O+O(^1D)$	$\rightarrow$	N_2+O_2	7.40−11	0.0	0.0	8.53−12	0.0	6.27+4	[59]	a
$N_2(B'^3\Sigma_u^-)+N_2$	$\rightarrow$	$N_2(B^3\Pi_g)+N_2$	1.0−11	0.0	0.0	0.0	0.0	0.0	[61,62]	—
$H_2(d^3\Pi_u)$	$\rightarrow$	$H_2(a^3\Sigma_g^+)$	1.5+7	0.0	0.0	0.0	0.0	0.0	[63]	—
$H_2(a^3\Sigma_g^+)$	$\rightarrow$	$H+H$	9.6+7	0.0	0.0	0.0	0.0	0.0	[64]	—
$H_2(c^3\Pi_u)$	$\rightarrow$	H_2	9.8+2	0.0	0.0	0.0	0.0	0.0	[65]	—
$H_2(B^1\Sigma_u^+)$	$\rightarrow$	H_2	1.0+8	0.0	0.0	0.0	0.0	0.0	[66]	—
$H_2(B^1\Sigma_u^+)$	$\rightarrow$	H_2	1.2+9	0.0	0.0	0.0	0.0	0.0	[67]	—
$H_2(C^1\Pi_u)$	$\rightarrow$	H_2	1.2+9	0.0	0.0	0.0	0.0	0.0	[67]	—
$O_2(B^3\Sigma_u^-)$	$\rightarrow$	O_2	1.3+7	0.0	0.0	0.0	0.0	0.0	[64]	—
$H(^2P)$	$\rightarrow$	H	2.0+8	0.0	0.0	0.0	0.0	0.0	[68]	—
$H(^4S)$	$\rightarrow$	H	1.0+8	0.0	0.0	0.0	0.0	0.0	[68]	—
$N+O_2$	$\rightarrow$	$NO+O$	1.59−12	0.50	3.73+3	5.31−15	1.	1.97+4	[39]	[39]
$N+O_3$	$\rightarrow$	$NO+O_2$	3.32−14	0.0	2.93+2	2.21−15	0.0	6.34+4	[39]	a
$N+NO$	$\rightarrow$	N_2+O	2.20−11	0.0	0.0	9.83−11	0.0	3.75+4	[39]	[39]
$N+NO_2$	$\rightarrow$	N_2O+O	7.97−12	0.0	0.0	1.10−8	−1.0	2.11+4	[39]	[39]
$N+NO_2$	$\rightarrow$	$NO+NO$	6.64−12	0.0	0.0	1.79−13	0.0	3.92+4	[39]	[39]

(Continued)

TABLE 4.10 (Continued)
The Reactions between Heavy Particles Included in the Kinetic Scheme. $A - 1/s$, cm^3/s, cm^6/s, $E_a - K$

Reaction			A^+	n^+	E_a^+, K	A^-	n^-	E_a^-, K	Ref$^+$	Ref$^-$
NO_2+O	$\rightarrow$	$NO+O_2$	9.97−12	0.0	0.0	3.29−10	−0.50	2.36+4	[69]	[39]
$NO+O_3$	$\rightarrow$	NO_2+O_2	8.0−13	0.0	1.20+3	1.10−12	0.0	2.51+4	[39]	[39]
$NO+NO_3$	$\rightarrow$	NO_2+NO_2	1.70−11	0.0	0.0	4.50−10	0.0	1.85+4	[36]	[70]
$NO_2+NO_3+O_2$	$\rightarrow$	$N_2O_5+O_2$	5.90−29	−1.27	0.0	1.74+8	−4.40	1.11+4	[36]	[47]
$N_2+NO_2+NO_3$	$\rightarrow$	$N_2+N_2O_5$	5.90−29	−1.27	0.0	2.80+12	−6.10	1.11+4	[36]	[71]
$NO+NO_2+O_2$	$\rightarrow$	NO_2+NO_3	0.0	0.0	−4.0+2	2.30−13	0.0	1.60+3	[71]	[36]
$N+N+N_2$	$\rightarrow$	N_2+N_2	7.44−32	−0.50	0.0	8.17−7	−0.50	1.13+5	[39]	[39]
$NO+O+O_2$	$\rightarrow$	NO_2+O_2	4.0−33	0.0	−9.70+2	2.61−7	0.0	3.63+4	[39]	[39]
$NO+NO_2+O$	$\rightarrow$	NO_2+NO_2	2.40−27	−1.80	0.0	7.30−8	0.0	3.30+4	[36]	[46]
$N+N+O_2$	$\rightarrow$	N_2+O_2	3.0−32	−0.50	0.0	3.32−7	−0.50	1.13+5	[39]	[39]
$N+N+NO$	$\rightarrow$	N_2+NO	3.03−32	−0.50	0.0	3.32−7	−0.50	1.13+5	[39]	[39]
$N+N+N$	$\rightarrow$	$N+N_2$	3.31−27	−1.50	0.0	3.62−2	−1.50	1.13+5	[39]	[39]
$N+N+O$	$\rightarrow$	N_2+O	3.03−32	−0.50	0.0	3.32−7	−0.50	1.13+5	[39]	[39]
N_2+O+O	$\rightarrow$	N_2+O_2	9.18−31	−1.0	0.0	6.71−4	−1.50	5.95+4	[39]	[39]
$O+O+O_2$	$\rightarrow$	O_2+O_2	4.08−30	−1.0	0.0	3.0−3	−1.50	5.95+4	[39]	[39]
$O+O+O$	$\rightarrow$	$O+O_2$	1.10−29	−1.0	0.0	8.0−3	−1.50	5.95+4	[39]	[39]
$N+O+O$	$\rightarrow$	$N+O_2$	8.27−33	−0.50	0.0	6.04−6	−1.0	5.95+4	[39]	[39]
$NO+O+O$	$\rightarrow$	$NO+O_2$	8.27−33	−0.50	0.0	6.04−6	−1.0	5.95+4	[39]	[39]
$N+N_2+O$	$\rightarrow$	N_2+NO	2.76−28	−1.50	0.0	6.74−4	−1.50	7.55+4	[39]	[39]
$N+O+O_2$	$\rightarrow$	$NO+O_2$	2.76−28	−1.50	0.0	6.74−4	−1.50	7.55+4	[39]	[39]
$N+N+O$	$\rightarrow$	$N+NO$	5.51−27	−1.50	0.0	1.35−2	−1.50	7.55+4	[39]	[39]
$N+O+O$	$\rightarrow$	$NO+O$	5.51−27	−1.50	0.0	1.35−2	−1.50	7.55+4	[39]	[39]
$N+NO+O$	$\rightarrow$	$NO+NO$	5.51−27	−1.50	0.0	1.35−2	−1.50	7.55+4	[39]	[39]
N_2+O+O_2	$\rightarrow$	N_2+O_3	3.50−35	0.0	−9.0+2	3.20−9	0.0	1.20+4	[39]	[39]
$O+O_2+O_2$	$\rightarrow$	O_3+O_3	1.50−34	0.0	−7.50+2	7.70−9	0.0	1.20+4	[39]	[39]
$O+O_2+O_2$	$\rightarrow$	O_2+O_3	4.10−35	0.0	−9.0+2	3.40−9	0.0	1.20+4	[39]	[39]
$O+O_3$	$\rightarrow$	O_2+O_2	1.40−12	0.0	1.50+3	1.10−11	0.0	4.96+4	[39]	[39]

(Continued)

TABLE 4.10 (Continued)
The Reactions between Heavy Particles Included in the Kinetic Scheme. $A - 1/s$, cm^3/s, cm^6/s, $E_a - $ K

Reaction		A^+	n^+	E_a^+, K	A^-	n^-	E_a^-, K	Ref$^+$	Ref$^-$
N_2+NO+O	$\rightarrow$	N_2+NO_2	4.0−33	0.0	−9.70+2	2.61−7	0.0	3.63+4	[39]	[39]
$NO+NO+NO$	$\rightarrow$	N_2O+NO_2	3.47−38	0.0	1.36+4	1.10−29	0.0	3.17+4	[71]	a
$HO_2+N_2+NO_2$	$\rightarrow$	$HO_2NO_2+N_2$	2.10−31	0.0	0	5.0−6	0.0	1.0+4	[59]	[47]
HO_2+NO	$\rightarrow$	$OH+NO_2$	3.70−12	0.0	−2.40+2	3.0−11	0.0	3.36+3	[47]	[72]
$NH+NO$	$\rightarrow$	$OH+N_2$	4.20−12	0.0	0.0	1.70−11	0.0	4.69+4	[73]	a
$OH+NH$	$\rightarrow$	H_2+NO	2.60−12	0.60	7.55+2	8.44−11	0.60	3.53+4	[73]	a
$NH+O$	$\rightarrow$	$H+NO$	8.30−12	0.0	0.0	1.15−10	0.0	3.35+4	[73]	a
$NH+NH$	$\rightarrow$	H_2+N_2	6.60−13	0.55	9.57+2	8.69−11	0.55	8.24+4	[73]	a
$N+NH$	$\rightarrow$	$H+N_2$	1.77−11	0.0	0.0	5.69−10	0.0	7.13+4	[73]	a
$H+NH$	$\rightarrow$	H_2+N	1.70−12	0.68	9.57+2	6.96−12	0.68	1.11+4	[73]	a
$NH+O$	$\rightarrow$	$OH+N$	1.40−11	0.70	5.0+1	2.45−11	0.70	9.15+3	[73]	a
$OH+NH$	$\rightarrow$	H_2O+N	2.60−12	0.56	7.55+2	3.87−11	0.56	1.83+4	[73]	a
$OH+NH_2$	$\rightarrow$	H_2O+NH	5.0−14	0.68	6.54+2	1.43−13	0.68	1.66+4	[73]	a
NH_2+O	$\rightarrow$	$OH+NH$	1.50−12	0.50	0.0	5.04−13	0.50	7.50+3	[73]	a
$H+NH_2$	$\rightarrow$	H_2+NH	2.30−13	0.67	2.16+3	1.81−13	0.67	1.07+4	[73]	a
NH_3+O	$\rightarrow$	$OH+NH_2$	1.30−12	0.50	0.0	6.03−14	0.50	−2.88+3	[73]	a
$OH+NH_3$	$\rightarrow$	H_2O+NH_2	6.60−14	0.68	5.54+2	2.60−14	0.68	6.07+3	[73]	a
$H+NH_3$	$\rightarrow$	H_2+NH_2	3.20−13	0.67	1.72+3	3.48−14	0.67	−1.60+2	[73]	a
$NO+NO+O$	$\rightarrow$	$NO+NO_2$	2.40−27	−1.80	0.0	9.87−21	−1.80	3.62+4	[36]	a
H_2+NO	$\rightarrow$	$H+HNO$	2.30−11	0.0	2.84+4	3.0−11	0.0	5.0+2	[72]	[72]
$H+NO_2$	$\rightarrow$	$OH+NO$	1.40−10	0.0	0.0	1.16−12	0.0	1.49+4	[72]	a
$HNO+O$	$\rightarrow$	$OH+NO$	3.0−11	0.0	0.0	9.69−12	0.0	2.67+4	[72]	a
$HNO+OH$	$\rightarrow$	H_2O+NO	8.0−11	0.0	5.0+2	2.20−10	0.0	3.56+4	[72]	a
$HNO+NO_2$	$\rightarrow$	$HONO+NO$	1.0−12	0.0	1.0+3	1.25−12	0.0	1.59+4	[72]	a
$HNO+HNO$	$\rightarrow$	H_2O+N_2O	1.40−15	0.0	1.55+3	9.52−14	0.0	4.53+4	[72]	a
$HONO+O$	$\rightarrow$	$OH+NO_2$	2.0−11	0.0	3.0+3	5.18−12	0.0	1.48+4	[72]	a
$H+HONO$	$\rightarrow$	H_2+NO_2	2.10−14	1.0	6.80+1	1.27−14	1.0	1.29+4	[72]	a

(Continued)

TABLE 4.10 (Continued)
The Reactions between Heavy Particles Included in the Kinetic Scheme. $A - 1/s$, cm^3/s, cm^6/s, $E_a - K$

Reaction			A^+	n^+	E_a^+, K	A^-	n^-	E_a^-, K	Ref$^+$	Ref$^-$
$H+N_2O$	$\rightarrow$	$OH+N_2$	6.30−10	0.0	9.80+3	6.10−13	0.0	3.91+4	[74]	a
$OH+N_2O$	$\rightarrow$	HO_2+N_2	1.40−11	0.0	5.0+3	1.31−12	0.0	1.93+4	[72]	a
$H+NO$	$\rightarrow$	$OH+N$	4.40−10	0.0	2.54+4	4.90−11	0.0	0.0	[75]	[47]
N_2+NO+O_2	$\rightarrow$	N_2+NO_3	1.41−26	−2.0	2.30+4	1.50−11	−2.0	2.32+4	[71]	[71]
N_2O+N_2	$\rightarrow$	N_2+O+N_2	8.47−6	−1.12	3.00+4	5.02−38	0.0	0.0	[76,77]	[78]
N_2O+O	$\rightarrow$	N_2+O_2	8.30−12	0.0	1.40+4	1.50−12	0.50	5.32+4	[79]	[39]
N_2O+O	$\rightarrow$	$NO+NO$	1.50−10	0.0	1.41+4	3.64−10	0.0	3.93+4	[79]	[39]
$NO+NO$	$\rightarrow$	N_2+O_2	5.64−3	−2.0	4.31+4	7.57	−2.50	6.46+4	[39]	[39]
N_2O+NO	$\rightarrow$	N_2+NO_2	4.17−10	0.0	2.52+4	1.91−10	0.0	4.19+4	[39]	a
$NO+NO+O_2$	$\rightarrow$	NO_2+NO_2	2.0−38	0.0	0.0	3.32−12	−1.80	1.36+4	[39]	[70]
$N_2O+NO+O$	$\rightarrow$	N_2O+NO_2	2.40−27	−1.80	0.0	9.87−21	0.0	3.62+4	[36]	a
$N_2+NO_2+NO_2$	$\rightarrow$	$N_2+N_2O_4$	9.0−28	−2.50	0.0	2.21−18	−2.50	6.45+3	[36]	a
$NO_2+NO_2+O_2$	$\rightarrow$	$N_2O_4+O_2$	9.0−28	−2.50	0.0	2.21−18	−2.50	6.45+3	[36]	a
$N_2O_4+NO_2+NO_2$	$\rightarrow$	$N_2O_4+N_2O_4$	9.0−28	−2.50	0.0	2.21−18	−2.50	6.45+3	[36]	a
$NO_2+NO_2+NO_2$	$\rightarrow$	$N_2O_4+NO_2$	9.0−28	−2.50	0.0	2.21−18	−2.50	6.45+3	[36]	a
$N_2O_5+NO_2+NO_3$	$\rightarrow$	$N_2O_5+N_2O_5$	5.90−29	−1.27	0.0	0.0	0.0	0.0	[36]	—
$NO+NO_2+NO_3$	$\rightarrow$	N_2O_5+NO	5.90−29	−1.27	0.0	0.0	0.0	0.0	[36]	—
$N+NO_2$	$\rightarrow$	N_2+O_2	7.0−13	0.0	0.0	0.0	0.0	0.0	[36]	—
$N+NO_2$	$\rightarrow$	N_2+O+O	9.10−13	0.0	0.0	0.0	0.0	0.0	[36]	—
NO_3+O	$\rightarrow$	NO_2+O_2	1.0−11	0.0	0.0	0.0	0.0	0.0	[36]	—
NO_2+O_3	$\rightarrow$	NO_3+NO_3	1.20−13	0.0	2.45+3	0.0	0.0	0.0	[36]	—
NO_3+NO_3	$\rightarrow$	$NO_2+NO_2+O_2$	5.0−12	0.0	3.0+3	0.0	0.0	0.0	[36]	—
NO_2+O+O_2	$\rightarrow$	NO_3+O_2	7.10−27	−2.0	0.0	0.0	0.0	0.0	[36]	—
N_2+NO_2+O	$\rightarrow$	N_2+NO_3	7.10−27	−2.0	0.0	0.0	0.0	0.0	[36]	—
N_3+O	$\rightarrow$	N_2+NO	1.0−11	0.0	0.0	0.0	0.0	0.0	[71]	—
$N+N_3$	$\rightarrow$	N_2+N_2	1.60−11	0.0	0.0	0.0	0.0	0.0	[71]	—
N_3+N_3	$\rightarrow$	$N_2+N_2+N_2$	1.40−12	0.0	0.0	0.0	0.0	0.0	[71]	—
$OH+N_2+NO$	$\rightarrow$	$HONO+N_2$	5.73−25	−2.40	0.0	0.0	0.0	0.0	[59]	—
$OH+N_2+NO_2$	$\rightarrow$	$HONO_2+N_2$	1.27−23	−2.70	0.0	0.0	0.0	0.0	[59]	—

(Continued)

TABLE 4.10 (Continued)
The Reactions between Heavy Particles Included in the Kinetic Scheme. $A - 1/s$, cm^3/s, cm^6/s, $E_a - K$

Reaction		A^+	n^+	E_a^+, K	A^-	n^-	E_a^-, K	Ref+	Ref-
$OH+HONO_2$	$\rightarrow$	H_2O+NO_3	9.50−15	0.0	−7.78+2	0.0	0.0	0.0	[47]	—
$OH+NO_2+O_2$	$\rightarrow$	$HONO_2+O_2$	3.36−23	−2.90	0.0	0.0	0.0	0.0	[47]	—
$HO_2+NO_2+O_2$	$\rightarrow$	$HO_2NO_2+O_2$	3.60−6	0.0	1.0+4	0.0	0.0	0.0	[47]	—
$OH+O$	$\rightarrow$	$H+O_2$	8.47−12	0.0	−3.75+2	3.10−10	0.0	8.46+3	[74,80]	[73]
$H+O_3$	$\rightarrow$	$OH+O_2$	1.40−10	0.0	4.80+2	1.18−12	0.0	3.92+4	[59]	a
H_2+OH	$\rightarrow$	$H+H_2O$	4.20−10	0.0	5.03+3	1.58−10	0.0	1.02+4	[73]	[73]
$OH+OH$	$\rightarrow$	H_2O+O	4.20−12	0.0	2.40+2	2.49−14	1.14	8.69+3	[60]	[75]
$OH+OH+N_2$	$\rightarrow$	$H_2O_2+N_2$	6.50−31	0.0	0.0	2.0−7	0.0	2.29+4	[59]	[75]
$OH+HO_2$	$\rightarrow$	H_2O+O_2	4.60−11	0.0	−2.30+2	3.25−10	0.0	3.40+4	[60]	a
$OH+O_3$	$\rightarrow$	HO_2+O_2	1.60−12	0.0	9.40+2	3.91−13	0.0	2.25+4	[60]	a
$OH+O_2+O_2$	$\rightarrow$	HO_2+O_3	3.76−21	0.0	1.39+4	1.10−14	0.0	5.0+2	[42]	[36,60]
H_2+O	$\rightarrow$	$H+OH$	2.50−17	2.0	3.8+3	2.27−11	0.0	2.30+2	[81]	a
$H+H+H_2$	$\rightarrow$	H_2+H_2	2.68−31	−0.60	0.0	2.14−26	−0.60	5.20+4	[75]	a
$H+H_2O+OH$	$\rightarrow$	H_2O+H_2O	3.86−25	−2.0	0.0	1.12−19	−2.0	5.94+4	[75]	a
$H+OH+N_2$	$\rightarrow$	H_2O+N_2	7.72−26	−2.0	0.0	3.30−8	0.0	5.29+4	[75]	[73]
$H+H_2+OH$	$\rightarrow$	H_2+H_2O	7.72−26	−2.0	0.0	3.30−8	0.0	5.29+4	[75]	[73]
$H+N_2+O$	$\rightarrow$	$OH+N_2$	1.66−31	0.0	−3.0+2	5.68−27	0.0	5.07+4	[75]	a
$H+H_2+O$	$\rightarrow$	H_2+OH	6.64−32	0.0	−3.0+2	2.27−27	0.0	5.07+4	[75]	a
$H+H_2O+O$	$\rightarrow$	H_2O+OH	4.32−31	0.0	−3.0+2	1.48−26	0.0	5.07+4	[75]	a
$H+HO_2$	$\rightarrow$	$OH+OH$	4.15−10	0.0	9.50+2	1.69−11	0.0	1.82+4	[75]	a
H_2O_2+OH	$\rightarrow$	H_2O+HO_2	3.30−12	0.0	2.0+2	9.69−12	0.0	1.79+4	[60]	a
$H+H_2O_2$	$\rightarrow$	H_2O+OH	1.66−11	0.0	1.80+3	5.65−13	0.0	3.62+4	[75]	a
$H+O_2+O_2$	$\rightarrow$	HO_2+O_2	1.77−29	−1.0	0.0	0.0	0.0	0.0	[47]	—
$H+O_2+O$	$\rightarrow$	H_2O+O	9.40−13	0.0	0.0	0.0	0.0	0.0	[59]	—
$H+N_2+O_2$	$\rightarrow$	HO_2+N_2	1.80−29	−1.0	0.0	0.0	0.0	0.0	[59]	—
HO_2+O	$\rightarrow$	$OH+O_2$	3.10−11	0.0	−2.0+2	0.0	0.0	0.0	[60]	—
H_2O_2+O	$\rightarrow$	$OH+HO_2$	2.70−12	0.0	2.10+3	0.0	0.0	0.0	[59]	—
H_2O_2+O	$\rightarrow$	H_2O+O_2	2.70−12	0.0	2.10+3	0.0	0.0	0.0	[59]	—
$H+H+N_2$	$\rightarrow$	H_2+N_2	1.34−31	−0.60	0.0	0.0	0.0	0.0	[75]	—

(Continued)

TABLE 4.10 (Continued)
The Reactions between Heavy Particles Included in the Kinetic Scheme. $A - 1/s$, cm^3/s, cm^6/s, $E_a - K$

Reaction		A^+	n^+	E_a^+, K	A^-	n^-	E_a^-, K	Ref$^+$	Ref$^-$
$H+H+H_2O$	$\rightarrow$	H_2+H_2O	1.34−30	−0.60	0.0	0.0	0.0	0.0	[75]	—
$H+HO_2$	$\rightarrow$	H_2+O_2	4.15−11	0.0	3.50+2	0.0	0.0	0.0	[75]	—
$H+H_2O_2$	$\rightarrow$	H_2+HO_2	2.82−12	0.0	1.90+3	0.0	0.0	0.0	[75]	—
HO_2+HO_2	$\rightarrow$	$H_2O_2+O_2$	3.32−12	0.0	0.0	0.0	0.0	0.0	[75]	—
$H_2+H_2O_2$	$\rightarrow$	$H_2+OH+OH$	5.0−7	0.0	2.29+4	0.0	0.0	0.0	[75]	—
$H_2O+H_2O_2$	$\rightarrow$	$H_2O+OH+OH$	3.25−6	0.0	2.29+4	0.0	0.0	0.0	[75]	—
Charge Transfer										
$N_2+N_2+N_2^+$	$\rightarrow$	$N_2+N_4^+$	5.0−29	0.0	0.0	2.10−16	0.0	0.0	[36]	b
$O_2+O_2+O_2^+$	$\rightarrow$	$O_2+O_4^+$	2.0−22	−3.20	0.0	2.67+4	−4.0	5.03+3	[36]	b
$N_2+N_2+O_2^+$	$\rightarrow$	$N_2+N_2O_2^+$	8.10−26	−2.0	0.0	1.48+7	−5.30	2.36+3	[36]	b
$N_2+N_2+NO^+$	$\rightarrow$	$N_2+N_3O^+$	1.60−20	−4.40	0.0	1.50+6	−5.40	2.45+3	[36]	b
N^++N_2+O	$\rightarrow$	N_2+NO^+	1.0−29	0.0	0.0	1.58−23	0.0	1.37+5	[36]	b
N^++O+O_2	$\rightarrow$	NO^++O_2	1.0−29	0.0	0.0	1.58−23	0.0	1.37+5	[36]	b
$N^++N^++N_2$	$\rightarrow$	$N_2+N_2^+$	1.0−29	0.0	0.0	1.40−11	0.0	0.0	[36]	b
$N+N^++O_2$	$\rightarrow$	$N_2^++O_2$	1.0−29	0.0	0.0	5.68−24	0.0	1.01+5	[36]	b
N_2+O+O^+	$\rightarrow$	$N_2+O_2^+$	1.0−29	0.0	0.0	3.92−24	0.0	7.77+4	[36]	b
$O+O+O_2$	$\rightarrow$	$O_2+O_2^+$	1.0−29	0.0	0.0	3.92−24	0.0	7.77+4	[36]	b
$N+N_2+O^+$	$\rightarrow$	N_2+NO^+	1.0−29	0.0	0.0	3.45−24	0.0	1.27+5	[36]	b
$N+O^++O_2$	$\rightarrow$	NO^++O_2	1.0−29	0.0	0.0	3.45−24	0.0	1.27+5	[36]	b
N^++O_2	$\rightarrow$	$NO+O_2^+$	2.30−10	0.0	0.0	4.49−10	0.0	2.84+4	[43]	b
N^++O_2	$\rightarrow$	NO^++O	5.0−10	0.0	0.0	4.29−10	0.0	7.73+4	[43]	b
N^++O_2	$\rightarrow$	$NO+O^+$	1.0−12	0.0	0.0	1.35−12	0.0	2.61+4	[43]	b
N^++O	$\rightarrow$	$N+O^+$	5.0−10	0.0	0.0	2.20−9	0.0	1.09+4	[39]	[39]
N^++O_3	$\rightarrow$	NO^++O_2	5.0−10	0.0	0.0	1.93−10	0.0	1.24+5	[36]	b
N^++NO	$\rightarrow$	$N+NO^+$	8.0−10	0.0	0.0	4.65−9	0.0	6.13+4	[43]	b
N^++NO	$\rightarrow$	N_2^++O	1.0−10	0.0	0.0	3.0−10	0.0	2.58+4	[39]	[39]
N^++NO	$\rightarrow$	N_2+O^+	1.0−12	0.0	0.0	1.06−11	0.0	4.79+4	[43]	b

(Continued)

TABLE 4.10 (Continued)
The Reactions between Heavy Particles Included in the Kinetic Scheme. $A - 1/s$, cm^3/s, cm^6/s, $E_a - K$

Reaction			A^+	n^+	E_a^+, K	A^-	n^-	E_a^-, K	Ref$^+$	Ref$^-$
N_2+O^+	$\rightarrow$	$N+NO^+$	3.0−12	0.0	0.0	1.10−12	0.0	1.22+4	[39]	[39]
O^++O_2	$\rightarrow$	$O+O_2^+$	2.0−11	0.0	0.0	6.0−12	0.0	1.65+4	[39]	[39]
O^++O_3	$\rightarrow$	$O_2+O_2^+$	1.0−10	0.0	0.0	9.63−12	0.0	6.54+4	[36]	b
$NO+O^+$	$\rightarrow$	NO^++O	2.0−11	0.0	0.0	3.0−11	0.0	5.01+4	[39]	[39]
$NO+O^+$	$\rightarrow$	$N+O_2^+$	3.0−12	0.0	0.0	4.34−12	0.0	2.30+3	[43]	b
N_2O+O^+	$\rightarrow$	$NO+NO^+$	2.30−10	0.0	0.0	3.81−12	0.0	6.93+4	[36]	b
N_2O+O^+	$\rightarrow$	$N_2+O_2^+$	2.0−11	0.0	0.0	8.74−13	0.0	5.82+4	[43]	b
$N_2^++O_2$	$\rightarrow$	$N_2+O_2^+$	1.24−7	−1.35	0.0	2.68−7	−1.35	4.10+4	[43]	b
N_2^++O	$\rightarrow$	$N+NO^+$	2.42−9	−0.50	0.0	6.71−9	−0.50	3.61+4	[39]	[82]
N_2^++O	$\rightarrow$	N_2+O^+	7.0−12	0.0	0.0	3.54−11	0.0	2.27+4	[43]	b
N_2^++NO	$\rightarrow$	N_2+NO^+	4.48−10	0.0	0.0	6.31−9	0.0	7.32+4	[39]	[39]
$N+N_2+NO^+$	$\rightarrow$	$N_2^++N_2O$	1.24−16	0.0	1.66+4	4.0−10	0.0	0.0	[42]	[36,43]
$N_2+O_2^+$	$\rightarrow$	$NO+NO^+$	9.96−16	0.0	0.0	5.31−16	0.0	1.19+4	[39]	[39]
$N+O_2^+$	$\rightarrow$	$NO+O$	1.80−10	0.0	0.0	1.05−10	0.0	4.89+4	[43]	b
$NO+O_2^+$	$\rightarrow$	$NO+O_2^+$	9.13−10	0.0	0.0	7.31−9	0.0	3.36+4	[39]	[39]
$NO_2+O_2^+$	$\rightarrow$	$NO+O_3$	1.0−11	0.0	0.0	2.94−11	0.0	9.05+3	[43]	b
$N_2+O_4^+$	$\rightarrow$	$N_2O_2^++O_2$	2.96−18	2.50	2.55+3	5.0−11	0.0	0.0	[36]	[36]
$O_2+O_2+O_2^-$	$\rightarrow$	$O_2+O_4^-$	3.0−31	0.0	0.0	1.0−10	0.0	1.04+3	[43]	[43]
$N_2+O_2+O_2^-$	$\rightarrow$	$N_2+O_4^-$	1.05−28	0.0	0.0	1.0−10	0.0	1.04+3	[36]	[36]
$O+O_2^-$	$\rightarrow$	O^-+O_2	8.0−11	0.0	0.0	3.45−12	0.50	1.62+4	[39]	[39]
$NO_2+NO_2^-$	$\rightarrow$	$NO+NO_3^-$	4.0−12	0.0	0.0	3.0−15	0.0	0.0	[36]	[36]
$H+O$	$\rightarrow$	$H+O^+$	3.80−10	0.0	0.0	2.30−11	0.50	0.0	[43]	[82]
$H_2+H_2^+$	$\rightarrow$	$H+H_3^+$	2.10−9	0.0	0.0	6.63−9	0.0	2.0+4	[43]	b
$H^++H_2+H_2$	$\rightarrow$	$H_2+H_3^+$	3.0−29	0.0	0.0	9.57−25	0.0	5.10+4	[43]	b
$H+NO$	$\rightarrow$	$H+NO^+$	1.90−9	0.0	0.0	2.62−9	0.0	5.09+4	[83]	b
N^++O^-	$\rightarrow$	$N+O$	4.50−6	−0.50	0.0	5.78−6	−0.50	1.51+5	[41]	b
$O_2^-+O_2^+$	$\rightarrow$	O_2+O_2	7.27−6	−0.50	0.0	1.25−5	−0.50	1.35+5	[41]	b

(Continued)

TABLE 4.10 (Continued)
The Reactions between Heavy Particles Included in the Kinetic Scheme. A – 1/s, cm^3/s, cm^6/s, E_a – K

Reaction		A^+	n^+	E_a^+, K	A^-	n^-	E_a^-, K	Ref$^+$	Ref$^-$
$N_2^+ + O_2^-$	$\rightarrow$	$N_2 + O_2$	2.77−6	−0.50	0.0	1.03−5	−0.50	1.76+5	[41]	b
$O^- + O^+$	$\rightarrow$	$O + O$	4.68−6	−0.50	0.0	1.31−6	−0.50	1.41+5	[41]	b
$N^+ + N_2 + O_2^-$	$\rightarrow$	$N_2 + NO_2$	3.10−19	−2.50	0.0	1.26−12	−2.50	2.15+5	[36]	b
$H_2O + O^+$	$\rightarrow$	$H_2O^+ + O$	2.30−9	0.0	0.0	4.79−10	0.0	1.06+4	[43]	b
$H_2 + O^+$	$\rightarrow$	$H + OH^+$	2.0−9	0.0	0.0	4.46−10	0.0	8.0+3	[43]	b
$H_2 + OH^+$	$\rightarrow$	$H + H_2O^+$	1.50−9	0.0	0.0	2.17−9	0.0	9.0+3	[43]	b
$H_2 + H_2O$	$\rightarrow$	$H_2 + H_2O^+$	3.60−9	0.0	0.0	6.43−9	0.0	3.13+4	[43]	b
$H_2 + N_2^+$	$\rightarrow$	$H_2^+ + N_2$	2.0−11	0.0	0.0	1.18−11	0.0	2.0+3	[43]	b
$H_2O + N_2^+$	$\rightarrow$	$H_2O^+ + N_2$	2.20−9	0.0	0.0	2.31−9	0.0	3.33+4	[43]	b
$N + N_2^+$	$\rightarrow$	$N^+ + N_2$	1.0−11	0.0	0.0	1.11−11	0	1.26+4	[43]	b
$N_2O + O^+$	$\rightarrow$	$N_2^+ + O_2$	2.0−11	0.0	0.0	4.05−13	0.0	1.72+4	[43]	b
$N^+ + N_2O$	$\rightarrow$	$N_2 + NO^+$	2.70−9	0.0	0.0	4.74−10	0.0	1.17+5	[43]	b
$NO + NO^+$	$\rightarrow$	$N_2^+ + O_2$	1.83−13	0.0	5.15+4	3.0−14	0.0	0.0	[39]	[39]
$N + N_2 + N_2^+$	$\rightarrow$	$N_2 + N_3^+$	9.0−30	0.0	4.0+2	0.0	0.0	0.0	[36]	—
$N + N_2 + N_2$	$\rightarrow$	$N_2 + N_3^+$	9.0−30	0.0	4.0+2	0.0	0.0	0.0	[36]	—
$N_2 + NO^+ + O_2$	$\rightarrow$	$N_2 + NO_3^+$	3.0−31	0.0	0.0	0.0	0.0	0.0	[36]	—
$NO^+ + O_2 + O_2$	$\rightarrow$	$NO_3^+ + O_2$	9.0−32	0.0	0.0	0.0	0.0	0.0	[36]	—
$NO_2 + O^+$	$\rightarrow$	$NO_2^+ + O$	1.60−9	0.0	0.0	0.0	0.0	0.0	[36]	—
$NO^+ + O_3$	$\rightarrow$	$NO_2^+ + O_2$	1.0−15	0.0	0.0	0.0	0.0	0.0	[36]	—
$NO + NO_2^+$	$\rightarrow$	$NO^+ + NO_2$	2.90−10	0.0	0.0	0.0	0.0	0.0	[36]	—
$O + O_4^+$	$\rightarrow$	$O_2^+ + O_3$	3.0−10	0.0	0.0	0.0	0.0	0.0	[36]	—
$N_2^+ + O_3$	$\rightarrow$	$N_2 + O + O_2^+$	1.0−10	0.0	0.0	0.0	0.0	0.0	[36]	—
$N_2O_5 + NO^+$	$\rightarrow$	$NO_2 + NO_2 + NO_2^+$	5.90−10	0.0	0.0	0.0	0.0	0.0	[36]	—
$N_3^+ + O_2$	$\rightarrow$	$N + N_2 + O_2^+$	2.30−11	0.0	0.0	0.0	0.0	0.0	[36]	—
$N_3^+ + NO$	$\rightarrow$	$N + N_2 + NO^+$	7.0−11	0.0	0.0	0.0	0.0	0.0	[36]	—
$N_4^+ + O$	$\rightarrow$	$N_2 + N_2 + O^+$	2.50−10	0.0	0.0	0.0	0.0	0.0	[36]	—

(Continued)

TABLE 4.10 (Continued)
The Reactions between Heavy Particles Included in the Kinetic Scheme. $A - 1/s$, cm^3/s, cm^6/s, $E_a - K$

Reaction		A^+	n^+	E_a^+, K	A^-	n^-	E_a^-, K	Ref$^+$	Ref$^-$
$N+N_4^+$	$\rightarrow$	$N^++N_2+N_2$	1.0−11	0.0	0.0	0.0	0.0	0.0	[36]	—
N_4^++NO	$\rightarrow$	$N_2+N_2+NO^+$	4.0−10	0.0	0.0	0.0	0.0	0.0	[36]	—
$N_2O_5+O_2^+$	$\rightarrow$	$NO_2^++NO_3+O_2$	8.80−10	0.0	0.0	0.0	0.0	0.0	[36]	—
$N_4^++O_2$	$\rightarrow$	$N_2+N_2+O_2^+$	4.0−10	0.0	0.0	0.0	0.0	0.0	[43]	—
N_2O^++NO	$\rightarrow$	N_2O+NO^+	2.90−10	0.0	0.0	0.0	0.0	0.0	[43]	—
$NO_2+O_2^+$	$\rightarrow$	$NO_2^++O_2$	6.60−10	0.0	0.0	0.0	0.0	0.0	[43]	—
$N_2^++N_2O$	$\rightarrow$	$N_2+N_2O^+$	5.0−10	0.0	0.0	0.0	0.0	0.0	[43]	—
N_3+O_2	$\rightarrow$	$N_2+NO_2^+$	6.0−11	0.0	0.0	0.0	0.0	0.0	[43]	—
$N+N_3^+$	$\rightarrow$	$N_2+N_2^+$	2.70−12	0.0	0.0	0.0	0.0	0.0	[43]	—
N_3^++NO	$\rightarrow$	$N_2+N_2O^+$	1.40−10	0.0	0.0	0.0	0.0	0.0	[43]	—
N_2O+O^+	$\rightarrow$	N_2O^++O	1.60−9	0.0	0.0	0.0	0.0	0.0	[43]	—
$NO+O_4^+$	$\rightarrow$	$NO^++O_2+O_2$	1.0−10	0.0	0.0	0.0	0.0	0.0	[36]	—
$O+O_2+O_2$	$\rightarrow$	$O_2+O_3^-$	8.0−31	0.0	0.0	0.0	0.0	0.0	[43]	—
$N_2O+O^-+O_2$	$\rightarrow$	$N_2+O_3^-$	3.30−28	0.0	0.0	0.0	0.0	0.0	[36]	—
$N_2+N_2+O_3^-$	$\rightarrow$	$N_2+N_2O_3^-$	1.50−31	0.0	0.0	0.0	0.0	0.0	[36]	—
$NO+O^-+O_2$	$\rightarrow$	$NO_2^-+O_2$	3.97−29	0.0	0.0	0.0	0.0	0.0	[39]	—
N_2+NO+O^-	$\rightarrow$	$N_2+NO_2^-$	2.04−32	0.0	0.0	0.0	0.0	0.0	[39]	—
$O_2^-+O_3$	$\rightarrow$	$O_2+O_3^-$	4.0−10	0.0	0.0	0.0	0.0	0.0	[36]	—
$NO_2+O_2^-$	$\rightarrow$	$NO_2^-+O_2$	7.16−10	0.0	0.0	0.0	0.0	0.0	[39]	—
$NO_3+O_2^-$	$\rightarrow$	$NO_3^-+O_2$	5.0−10	0.0	0.0	0.0	0.0	0.0	[36]	—
$N_2O+O_2^-$	$\rightarrow$	$N_2+O_3^-$	1.0−12	0.0	0.0	0.0	0.0	0.0	[36]	—
$O+O_3^-$	$\rightarrow$	$O+O_3^-$	5.30−10	0.0	0.0	0.0	0.0	0.0	[36]	—
NO_2+O^-	$\rightarrow$	NO_2^-+O	1.20−9	0.0	0.0	0.0	0.0	0.0	[36]	—
N_2O+O^-	$\rightarrow$	$NO+NO^-$	2.0−10	0.0	0.0	0.0	0.0	0.0	[36]	—
N_2O+O^-	$\rightarrow$	N_2O^-+O	2.0−12	0.0	0.0	0.0	0.0	0.0	[36]	—
$O+O_3^-$	$\rightarrow$	$O_2+O_2^-$	3.20−10	0.0	0.0	0.0	0.0	0.0	[36]	—

(Continued)

TABLE 4.10 (Continued)
The Reactions between Heavy Particles Included in the Kinetic Scheme. $A - 1/\text{s}$, cm^3/s, cm^6/s, $E_a - \text{K}$

Reaction		A^+	n^+	E_a^+, K	A^-	n^-	E_a^-, K	Ref$^+$	Ref$^-$
$NO + O_3^-$	$\rightarrow$	$NO_3^- + O$	1.0−11	0.0	0.0	0.0	0.0	0.0	[36]	—
$NO + O_3^-$	$\rightarrow$	$NO_2^- + O_2$	9.96−12	0.0	0.0	0.0	0.0	0.0	[39]	—
$NO_2 + O_3^-$	$\rightarrow$	$NO_2^- + O_3$	7.0−10	0.0	0.0	0.0	0.0	0.0	[36]	—
$NO_2 + O_3^-$	$\rightarrow$	$NO_3^- + O_2$	2.0−11	0.0	0.0	0.0	0.0	0.0	[36]	—
$NO_3 + O_3^-$	$\rightarrow$	$NO_3^- + O_3$	5.0−10	0.0	0.0	0.0	0.0	0.0	[36]	—
$NO + O_2^-$	$\rightarrow$	$NO + O_2^-$	5.0−10	0.0	0.0	0.0	0.0	0.0	[36]	—
$NO^- + NO_2$	$\rightarrow$	$NO + NO_2^-$	7.40−16	0.0	0.0	0.0	0.0	0.0	[36]	—
$N_2O + NO^-$	$\rightarrow$	$N_2 + NO_2^-$	2.80−14	0.0	0.0	0.0	0.0	0.0	[36]	—
$NO_2^- + O_3$	$\rightarrow$	$NO_3^- + O_2$	1.80−11	0.0	0.0	0.0	0.0	0.0	[36]	—
$NO_2^- + NO_3$	$\rightarrow$	$NO_2 + NO_3^-$	5.0−10	0.0	0.0	0.0	0.0	0.0	[36]	—
$N_2O_5 + NO_2^-$	$\rightarrow$	$NO + NO_3 + NO_3^-$	7.0−10	0.0	0.0	0.0	0.0	0.0	[36]	—
$O + O_4^-$	$\rightarrow$	$O_2 + O_3^-$	4.0−10	0.0	0.0	0.0	0.0	0.0	[36]	—
$NO + O_4^-$	$\rightarrow$	$NO_3^- + O_2$	2.50−10	0.0	0.0	0.0	0.0	0.0	[36]	—
$O + O_4^-$	$\rightarrow$	$O^- + O_2 + O_2$	3.0−10	0.0	0.0	0.0	0.0	0.0	[36]	—
$NO^- + NO_2^+$	$\rightarrow$	$NO + NO_2$	8.83−6	−0.50	0.0	0.0	0.0	0.0	[41]	—
$NO_2^- + O_2^+$	$\rightarrow$	$NO_2 + O_2$	7.10−6	−0.50	0.0	0.0	0.0	0.0	[41]	—
$NO^- + NO_3^+$	$\rightarrow$	$NO + NO_3$	5.89−7	−0.50	0.0	0.0	0.0	0.0	[41]	—
$NO^+ + NO_2^-$	$\rightarrow$	$NO + NO_2$	3.29−6	−0.50	0.0	0.0	0.0	0.0	[41]	—
$H^- + H^+$	$\rightarrow$	$H + H$	6.75−6	−0.50	0.0	0.0	0.0	0.0	[41]	—
$N^+ + O_2^-$	$\rightarrow$	$N + O_2$	3.46−6	−0.50	0.0	0.0	0.0	0.0	[36]	—
$O^+ + O_2^-$	$\rightarrow$	$O + O_2$	3.46−6	−0.50	0.0	0.0	0.0	0.0	[36]	—
$NO^+ + O_2^-$	$\rightarrow$	$NO + O_2$	3.46−6	−0.50	0.0	0.0	0.0	0.0	[36]	—
$NO_2^+ + O_2^-$	$\rightarrow$	$NO_2 + O_2$	3.46−6	−0.50	0.0	0.0	0.0	0.0	[36]	—
$N_2O^+ + O_2^-$	$\rightarrow$	$N_2O + O_2$	3.46−6	−0.50	0.0	0.0	0.0	0.0	[36]	—

(Continued)

TABLE 4.10 (Continued)
The Reactions between Heavy Particles Included in the Kinetic Scheme. $A - 1/s$, cm^3/s, cm^6/s, $E_a - K$

Reaction			A^+	n^+	E_a^+, K	A^-	n^-	E_a^-, K	Ref$^+$	Ref$^-$
$N_2^+ + O^-$	$\rightarrow$	$N_2 + O$	3.46−6	−0.50	0.0	0.0	0.0	0.0	[36]	—
$O^+ + O_2^-$	$\rightarrow$	$O + O_2$	3.46−6	−0.50	0.0	0.0	0.0	0.0	[36]	—
$NO^+ + O^-$	$\rightarrow$	$NO + O$	2.0−7	0.0	0.0	0.0	0.0	0.0	[39]	—
$NO_2^+ + O^-$	$\rightarrow$	$NO_2 + O$	3.46−6	−0.50	0.0	0.0	0.0	0.0	[36]	—
$N_2O^+ + O^-$	$\rightarrow$	$N_2O + O$	3.46−6	−0.50	0.0	0.0	0.0	0.0	[36]	—
$N_2^+ + O_3^-$	$\rightarrow$	$N_2 + O_3$	3.46−6	−0.50	0.0	0.0	0.0	0.0	[36]	—
$O_2^+ + O_3^-$	$\rightarrow$	$O_2 + O_3$	3.46−6	−0.50	0.0	0.0	0.0	0.0	[36]	—
$N^+ + O_3^-$	$\rightarrow$	$N + O_3$	3.46−6	−0.50	0.0	0.0	0.0	0.0	[36]	—
$O^+ + O_3^-$	$\rightarrow$	$O + O_3$	3.46−6	−0.50	0.0	0.0	0.0	0.0	[36]	—
$NO^+ + O_3^-$	$\rightarrow$	$NO + O_3$	3.46−6	−0.50	0.0	0.0	0.0	0.0	[36]	—
$NO_2^+ + O_3^-$	$\rightarrow$	$NO_2 + O_3$	3.46−6	−0.50	0.0	0.0	0.0	0.0	[36]	—
$N_2O^+ + O_3^-$	$\rightarrow$	$N_2O + O_3$	3.46−6	−0.50	0.0	0.0	0.0	0.0	[36]	—
$N_2^+ + NO^-$	$\rightarrow$	$N_2 + NO$	3.46−6	−0.50	0.0	0.0	0.0	0.0	[36]	—
$NO^+ + O_2^-$	$\rightarrow$	$NO + O_2$	3.46−6	−0.50	0.0	0.0	0.0	0.0	[36]	—
$N^+ + NO^-$	$\rightarrow$	$N + NO$	3.46−6	−0.50	0.0	0.0	0.0	0.0	[36]	—
$NO^+ + O^-$	$\rightarrow$	$NO + O$	3.46−6	−0.50	0.0	0.0	0.0	0.0	[36]	—
$NO^+ + NO^-$	$\rightarrow$	$NO + NO$	3.46−6	−0.50	0.0	0.0	0.0	0.0	[36]	—
$N_2O^+ + NO^-$	$\rightarrow$	$N_2O + NO$	3.46−6	−0.50	0.0	0.0	0.0	0.0	[36]	—
$N_2^+ + NO_2^-$	$\rightarrow$	$N_2 + NO_2$	3.46−6	−0.50	0.0	0.0	0.0	0.0	[36]	—
$N^+ + NO_2^-$	$\rightarrow$	$N + NO_2$	3.46−6	−0.50	0.0	0.0	0.0	0.0	[36]	—
$NO_2^+ + O^+$	$\rightarrow$	$NO_2 + O$	3.46−6	−0.50	0.0	0.0	0.0	0.0	[36]	—
$NO_2^- + NO_2^+$	$\rightarrow$	$NO_2 + NO_2$	3.46−6	−0.50	0.0	0.0	0.0	0.0	[36]	—
$N_2O^+ + NO_2^-$	$\rightarrow$	$N_2O + NO_2$	3.46−6	−0.50	0.0	0.0	0.0	0.0	[36]	—
$N_2^+ + NO_3^-$	$\rightarrow$	$N_2 + NO_3$	3.46−6	−0.50	0.0	0.0	0.0	0.0	[36]	—
$NO_3^- + O_2^+$	$\rightarrow$	$NO_3 + O_2$	3.46−6	−0.50	0.0	0.0	0.0	0.0	[36]	—
$N^+ + NO_3^-$	$\rightarrow$	$N + NO_3$	3.46−6	−0.50	0.0	0.0	0.0	0.0	[36]	—

(Continued)

TABLE 4.10 (Continued)
The Reactions between Heavy Particles Included in the Kinetic Scheme. A – 1/s, cm^3/s, cm^6/s, E_a – K

Reaction		A^+	n^+	E_a^+, K	A^-	n^-	E_a^-, K	Ref$^+$	Ref$^-$
$NO_3^- + O^+$	$\rightarrow$	$NO_3 + O$	3.46−6	−0.50	0.0	0.0	0.0	[36]	—
$NO_2^+ + NO_3^-$	$\rightarrow$	$NO_2 + NO_3$	3.46−6	−0.50	0.0	0.0	0.0	[36]	—
$N_2O^+ + NO_3^-$	$\rightarrow$	$N_2O + NO_3$	3.46−6	−0.50	0.0	0.0	0.0	[36]	—
$N_2^+ + N_2O^-$	$\rightarrow$	$N_2 + N_2O$	3.46−6	−0.50	0.0	0.0	0.0	[36]	—
$N_2O^+ + O_2^-$	$\rightarrow$	$N_2O + O_2$	3.46−6	−0.50	0.0	0.0	0.0	[36]	—
$N^+ + N_2O^-$	$\rightarrow$	$N + N_2O$	3.46−6	−0.50	0.0	0.0	0.0	[36]	—
$N_2O^+ + O^-$	$\rightarrow$	$N_2O + O$	3.46−6	−0.50	0.0	0.0	0.0	[36]	—
$N_2O^- + NO^+$	$\rightarrow$	$N_2O + NO$	3.46−6	−0.50	0.0	0.0	0.0	[36]	—
$N_2O^- + NO_2^+$	$\rightarrow$	$N_2O + NO_2$	3.46−6	−0.50	0.0	0.0	0.0	[36]	—
$N_2O^- + N_2O^+$	$\rightarrow$	$N_2O + N_2O$	3.46−6	−0.50	0.0	0.0	0.0	[36]	—
$N_2 + O^+ + O_2^-$	$\rightarrow$	$N_2 + O_3$	3.10−19	−2.50	0.0	0.0	0.0	[36]	—
$N_2 + NO^+ + O_2^-$	$\rightarrow$	$N_2 + NO_3$	3.10−19	−2.50	0.0	0.0	0.0	[36]	—
$N^+ + O_2 + O_2^-$	$\rightarrow$	$NO_2 + O_2$	3.10−19	−2.50	0.0	0.0	0.0	[36]	—
$O^+ + O_2 + O_2^-$	$\rightarrow$	$O_2 + O_3$	3.10−19	−2.50	0.0	0.0	0.0	[36]	—
$NO^+ + O_2 + O_2^-$	$\rightarrow$	$NO_3 + O_2$	3.10−19	−2.50	0.0	0.0	0.0	[36]	—
$N_2 + N_2^+ + O^-$	$\rightarrow$	$N_2 + N_2O$	3.10−19	−2.50	0.0	0.0	0.0	[36]	—
$N_2 + O + O_2^+$	$\rightarrow$	$N_2 + O_3$	3.10−19	−2.50	0.0	0.0	0.0	[36]	—
$N_2 + NO^+ + O^-$	$\rightarrow$	$N_2 + NO_2$	3.10−19	−2.50	0.0	0.0	0.0	[36]	—
$N_2 + O + O^+$	$\rightarrow$	$N_2 + O_2$	3.10−19	−2.50	0.0	0.0	0.0	[36]	—
$N^+ + N_2 + O^-$	$\rightarrow$	$N_2O + O_2$	3.10−19	−2.50	0.0	0.0	0.0	[36]	—
$N_2^+ + O^- + O_2$	$\rightarrow$	$O_2 + O_3$	3.10−19	−2.50	0.0	0.0	0.0	[36]	—
$O^- + O_2 + O_2^+$	$\rightarrow$	$NO_2 + O_2$	3.10−19	−2.50	0.0	0.0	0.0	[36]	—
$NO^+ + O^- + O_2$	$\rightarrow$	$O_2 + O_2$	3.10−19	−2.50	0.0	0.0	0.0	[36]	—
$O^- + O^+ + O_2$	$\rightarrow$	$NO + O_2$	3.10−19	−2.50	0.0	0.0	0.0	[36]	—
$N^+ + O^- + O_2$	$\rightarrow$	$N_2 + O_2 + O_2$	1.0−7	0.0	0.0	0.0	0.0	[36]	—
$N_2^+ + O_4^-$	$\rightarrow$	$O_2 + O_2 + O_2$	1.0−7	0.0	0.0	0.0	0.0	[36]	—
$O_2^+ + O_4^-$	$\rightarrow$								

(Continued)

TABLE 4.10 (Continued)
The Reactions between Heavy Particles Included in the Kinetic Scheme. $A - 1/s$, cm^3/s, cm^6/s, $E_a - K$

Reaction			A^+	n^+	E_a^+, K	A^-	n^-	E_a^-, K	Ref$^+$	Ref$^-$
$N^+ + O_4^-$	$\rightarrow$	$N + O_2 + O_2$	1.0−7	0.0	0.0	0.0	0.0	0.0	[36]	—
$O^+ + O_4^-$	$\rightarrow$	$O + O_2 + O_2$	1.0−7	0.0	0.0	0.0	0.0	0.0	[36]	—
$NO^+ + O_4^-$	$\rightarrow$	$NO + O_2 + O_2$	1.0−7	0.0	0.0	0.0	0.0	0.0	[36]	—
$NO_2^+ + O_4^-$	$\rightarrow$	$NO_2 + O_2 + O_2$	1.0−7	0.0	0.0	0.0	0.0	0.0	[36]	—
$N_2O^+ + O_4^-$	$\rightarrow$	$N_2O + O_2 + O_2$	1.0−7	0.0	0.0	0.0	0.0	0.0	[36]	—
$N_2^+ + N_2O_3^-$	$\rightarrow$	$N_2 + N_2 + O_3$	1.0−7	0.0	0.0	0.0	0.0	0.0	[36]	—
$N_2O_3^- + O_2^+$	$\rightarrow$	$N_2 + O_2 + O_3$	1.0−7	0.0	0.0	0.0	0.0	0.0	[36]	—
$N^+ + N_2O_3^-$	$\rightarrow$	$N + N_2 + O_3$	1.0−7	0.0	0.0	0.0	0.0	0.0	[36]	—
$N_2O_3^- + O^+$	$\rightarrow$	$N_2 + O + O_3$	1.0−7	0.0	0.0	0.0	0.0	0.0	[36]	—
$N_2O_3^- + NO^+$	$\rightarrow$	$N_2 + NO + O_3$	1.0−7	0.0	0.0	0.0	0.0	0.0	[36]	—
$N_2O_3^- + NO_2^+$	$\rightarrow$	$N_2 + NO_2 + O_3$	1.0−7	0.0	0.0	0.0	0.0	0.0	[36]	—
$N_2O^+ + N_2O_3^-$	$\rightarrow$	$N_2 + N_2O + O_3$	1.0−7	0.0	0.0	0.0	0.0	0.0	[36]	—
$N_2^+ + O_2^-$	$\rightarrow$	$N + N + O_2$	1.0−7	0.0	0.0	0.0	0.0	0.0	[36]	—
$O_2^- + O_2^+$	$\rightarrow$	$O + O + O_2$	1.0−7	0.0	0.0	0.0	0.0	0.0	[36]	—
$NO^+ + O_2^-$	$\rightarrow$	$N + O + O_2$	1.0−7	0.0	0.0	0.0	0.0	0.0	[36]	—
$NO_2^+ + O_2^-$	$\rightarrow$	$NO + O_2 + O_2$	1.0−7	0.0	0.0	0.0	0.0	0.0	[36]	—
$N_2O^+ + O_2^-$	$\rightarrow$	$N + NO + O_2$	1.0−7	0.0	0.0	0.0	0.0	0.0	[36]	—
$N_3^+ + O_2^-$	$\rightarrow$	$N + N_2 + O_2$	1.0−7	0.0	0.0	0.0	0.0	0.0	[36]	—
$N_4^+ + O_2^-$	$\rightarrow$	$N_2 + N_2 + O_2$	1.0−7	0.0	0.0	0.0	0.0	0.0	[36]	—
$O_2^- + O_4^+$	$\rightarrow$	$O_2 + O_2 + O_2$	1.0−7	0.0	0.0	0.0	0.0	0.0	[36]	—
$N_3O^+ + O_2^-$	$\rightarrow$	$N_2 + NO + O_2$	1.0−7	0.0	0.0	0.0	0.0	0.0	[36]	—
$NO_3^+ + O_2^-$	$\rightarrow$	$NO + O_2 + O_2$	1.0−7	0.0	0.0	0.0	0.0	0.0	[36]	—
$N_2O_2^+ + O_2^-$	$\rightarrow$	$NO + NO + O_2$	1.0−7	0.0	0.0	0.0	0.0	0.0	[36]	—
$N_2O_2^+ + O_2^-$	$\rightarrow$	$N_2 + O_2 + O_2$	1.0−7	0.0	0.0	0.0	0.0	0.0	[36]	—
$N_2^+ + O^-$	$\rightarrow$	$N + N + O$	1.0−7	0.0	0.0	0.0	0.0	0.0	[36]	—

(Continued)

TABLE 4.10 (Continued)
The Reactions between Heavy Particles Included in the Kinetic Scheme. $A - 1/\text{s, cm}^3/\text{s, cm}^6/\text{s}$, $E_a - K$

Reaction			A^+	n^+	E_a^+, K	A^-	n^-	E_a^-, K	Ref$^+$	Ref$^-$
$O + O_2^+$	$\rightarrow$	$O + O + O$	1.0−7	0.0	0.0	0.0	0.0	0.0	[36]	—
$NO^+ + O^-$	$\rightarrow$	$N + O + O$	1.0−7	0.0	0.0	0.0	0.0	0.0	[36]	—
$NO_2^+ + O^-$	$\rightarrow$	$NO + O + O$	1.0−7	0.0	0.0	0.0	0.0	0.0	[36]	—
$N_2O^+ + O^-$	$\rightarrow$	$N + NO + O$	1.0−7	0.0	0.0	0.0	0.0	0.0	[36]	—
$N_3^+ + O^-$	$\rightarrow$	$N + N_2 + O$	1.0−7	0.0	0.0	0.0	0.0	0.0	[36]	—
$N_4^+ + O^-$	$\rightarrow$	$N_2 + N_2 + O$	1.0−7	0.0	0.0	0.0	0.0	0.0	[36]	—
$O + O_4^+$	$\rightarrow$	$O + O_2 + O_2$	1.0−7	0.0	0.0	0.0	0.0	0.0	[36]	—
$N_3O^+ + O^-$	$\rightarrow$	$N_2 + NO + O$	1.0−7	0.0	0.0	0.0	0.0	0.0	[36]	—
$NO_3^+ + O^-$	$\rightarrow$	$NO + O + O_2$	1.0−7	0.0	0.0	0.0	0.0	0.0	[36]	—
$N_2O_2^+ + O^-$	$\rightarrow$	$NO + NO + O$	1.0−7	0.0	0.0	0.0	0.0	0.0	[36]	—
$N_2O_3^+ + O^-$	$\rightarrow$	$N_2 + O + O_2$	1.0−7	0.0	0.0	0.0	0.0	0.0	[36]	—
$N_2^+ + O_3^-$	$\rightarrow$	$N + N + O_3$	1.0−7	0.0	0.0	0.0	0.0	0.0	[36]	—
$O_2^+ + O_3^-$	$\rightarrow$	$O + O + O_3$	1.0−7	0.0	0.0	0.0	0.0	0.0	[36]	—
$NO^+ + O_3^-$	$\rightarrow$	$N + O + O_3$	1.0−7	0.0	0.0	0.0	0.0	0.0	[36]	—
$NO_2^+ + O_3^-$	$\rightarrow$	$NO + O + O_3$	1.0−7	0.0	0.0	0.0	0.0	0.0	[36]	—
$N_2O^+ + O_3^-$	$\rightarrow$	$N + NO + O_3$	1.0−7	0.0	0.0	0.0	0.0	0.0	[36]	—
$N_3^+ + O_3^-$	$\rightarrow$	$N + N_2 + O_3$	1.0−7	0.0	0.0	0.0	0.0	0.0	[36]	—
$N_4^+ + O_3^-$	$\rightarrow$	$N_2 + N_2 + O_3$	1.0−7	0.0	0.0	0.0	0.0	0.0	[36]	—
$O_3^- + O_4^+$	$\rightarrow$	$O_2 + O_2 + O_3$	1.0−7	0.0	0.0	0.0	0.0	0.0	[36]	—
$N_3O^+ + O_3^-$	$\rightarrow$	$N_2 + NO + O_3$	1.0−7	0.0	0.0	0.0	0.0	0.0	[36]	—
$NO_3^+ + O_3^-$	$\rightarrow$	$NO + O_2 + O_3$	1.0−7	0.0	0.0	0.0	0.0	0.0	[36]	—
$N_2O_2^+ + O_3^-$	$\rightarrow$	$NO + NO + O_3$	1.0−7	0.0	0.0	0.0	0.0	0.0	[36]	—
$N_2O_3^+ + O_3^-$	$\rightarrow$	$N_2 + O_2 + O_3$	1.0−7	0.0	0.0	0.0	0.0	0.0	[36]	—
$N_2^+ + NO^-$	$\rightarrow$	$N + N + NO$	1.0−7	0.0	0.0	0.0	0.0	0.0	[36]	—
$NO^- + O_2^+$	$\rightarrow$	$NO + O + O$	1.0−7	0.0	0.0	0.0	0.0	0.0	[36]	—
$NO^- + NO^+$	$\rightarrow$	$N + NO + O$	1.0−7	0.0	0.0	0.0	0.0	0.0	[36]	—

(Continued)

Mechanisms of Plasma-Chemical Reactions

TABLE 4.10 (Continued)
The Reactions between Heavy Particles Included in the Kinetic Scheme. $A - 1/s$, cm^3/s, cm^6/s, $E_a - K$

Reaction			A^+	n^+	E_a^+, K	A^-	n^-	E_a^-, K	Ref$^+$	Ref$^-$
$NO^- + NO_2^+$	$\rightarrow$	$NO + NO + O$	1.0−7	0.0	0.0	0.0	0.0	0.0	[36]	—
$N_2O^+ + NO^-$	$\rightarrow$	$N + NO + NO$	1.0−7	0.0	0.0	0.0	0.0	0.0	[36]	—
$N_3^+ + NO^-$	$\rightarrow$	$N + N_2 + NO$	1.0−7	0.0	0.0	0.0	0.0	0.0	[36]	—
$N_4^+ + NO^-$	$\rightarrow$	$N_2 + N_2 + NO$	1.0−7	0.0	0.0	0.0	0.0	0.0	[36]	—
$NO^- + O_4^+$	$\rightarrow$	$NO + O_2 + O_2$	1.0−7	0.0	0.0	0.0	0.0	0.0	[36]	—
$N_3O^+ + NO^-$	$\rightarrow$	$N_2 + NO + NO$	1.0−7	0.0	0.0	0.0	0.0	0.0	[36]	—
$NO^- + NO_3^+$	$\rightarrow$	$NO + NO + O_2$	1.0−7	0.0	0.0	0.0	0.0	0.0	[36]	—
$N_2O_2^+ + NO^-$	$\rightarrow$	$NO + NO + NO$	1.0−7	0.0	0.0	0.0	0.0	0.0	[36]	—
$N_2O_2^+ + NO^-$	$\rightarrow$	$N_2 + NO + O_2$	1.0−7	0.0	0.0	0.0	0.0	0.0	[36]	—
$N_2^+ + NO_2^-$	$\rightarrow$	$N + N + NO_2$	1.0−7	0.0	0.0	0.0	0.0	0.0	[36]	—
$NO_2^- + O_2^+$	$\rightarrow$	$NO_2 + O + O$	1.0−7	0.0	0.0	0.0	0.0	0.0	[36]	—
$NO^+ + NO_2^-$	$\rightarrow$	$NO + NO_2 + O$	1.0−7	0.0	0.0	0.0	0.0	0.0	[36]	—
$NO_2^- + NO_2^+$	$\rightarrow$	$NO + NO_2 + O$	1.0−7	0.0	0.0	0.0	0.0	0.0	[36]	—
$N_2O^+ + NO_2^-$	$\rightarrow$	$N + NO + NO_2$	1.0−7	0.0	0.0	0.0	0.0	0.0	[36]	—
$N_3^+ + NO_2^-$	$\rightarrow$	$N + N_2 + NO_2$	1.0−7	0.0	0.0	0.0	0.0	0.0	[36]	—
$N_4^+ + NO_2^-$	$\rightarrow$	$N_2 + N_2 + NO_2$	1.0−7	0.0	0.0	0.0	0.0	0.0	[36]	—
$NO_2^- + O_4^+$	$\rightarrow$	$NO_2 + O_2 + O_2$	1.0−7	0.0	0.0	0.0	0.0	0.0	[36]	—
$N_3O^+ + NO_2^-$	$\rightarrow$	$N_2 + NO + NO_2$	1.0−7	0.0	0.0	0.0	0.0	0.0	[36]	—
$NO_2^- + NO_3^+$	$\rightarrow$	$NO + NO_2 + O_2$	1.0−7	0.0	0.0	0.0	0.0	0.0	[36]	—
$N_2O_2^+ + NO_2^-$	$\rightarrow$	$NO + NO + NO_2$	1.0−7	0.0	0.0	0.0	0.0	0.0	[36]	—
$N_2O_2^+ + NO_2^-$	$\rightarrow$	$N_2 + NO_2 + O_2$	1.0−7	0.0	0.0	0.0	0.0	0.0	[36]	—
$N_2^+ + NO_3^-$	$\rightarrow$	$N + N + NO_3$	1.0−7	0.0	0.0	0.0	0.0	0.0	[36]	—
$NO_3^- + O_2^+$	$\rightarrow$	$NO_3 + O + O$	1.0−7	0.0	0.0	0.0	0.0	0.0	[36]	—
$NO^+ + NO_3^-$	$\rightarrow$	$N + NO_3 + O$	1.0−7	0.0	0.0	0.0	0.0	0.0	[36]	—
$NO_2^+ + NO_3^-$	$\rightarrow$	$NO + NO_3 + O$	1.0−7	0.0	0.0	0.0	0.0	0.0	[36]	—

(Continued)

TABLE 4.10 (Continued)
The Reactions between Heavy Particles Included in the Kinetic Scheme. $A - 1/s$, cm^3/s, cm^6/s, $E_a - K$

Reaction		A^+	n^+	E_a^+, K	A^-	n^-	E_a^-, K	Ref$^+$	Ref$^-$
$N_2O^+ + NO_3^-$	$\rightarrow$	$N + NO + NO_3$	1.0−7	0.0	0.0	0.0	0.0	0.0	[36]	—
$N_3^+ + NO_3^-$	$\rightarrow$	$N + N_2 + NO_3$	1.0−7	0.0	0.0	0.0	0.0	0.0	[36]	—
$N_4^+ + NO_3^-$	$\rightarrow$	$N_2 + N_2 + NO_3$	1.0−7	0.0	0.0	0.0	0.0	0.0	[36]	—
$NO_3^- + O_4^+$	$\rightarrow$	$NO_3 + O_2 + O_2$	1.0−7	0.0	0.0	0.0	0.0	0.0	[36]	—
$N_2O^+ + NO_3^-$	$\rightarrow$	$N_2 + NO + NO_3$	1.0−7	0.0	0.0	0.0	0.0	0.0	[36]	—
$NO_3^- + NO_3^+$	$\rightarrow$	$NO + NO_3 + O_2$	1.0−7	0.0	0.0	0.0	0.0	0.0	[36]	—
$N_2O_2^+ + NO_3^-$	$\rightarrow$	$NO + NO + NO_3$	1.0−7	0.0	0.0	0.0	0.0	0.0	[36]	—
$N_2O_2^+ + NO_3^-$	$\rightarrow$	$N_2 + NO_3 + O_2$	1.0−7	0.0	0.0	0.0	0.0	0.0	[36]	—
$N_2O_2^+ + NO_3^-$	$\rightarrow$	$N_2 + NO_3 + O_2$	1.0−7	0.0	0.0	0.0	0.0	0.0	[36]	—
$N_2^+ + N_2O^-$	$\rightarrow$	$N + N + N_2O$	1.0−7	0.0	0.0	0.0	0.0	0.0	[36]	—
$N_2O + O_2^+$	$\rightarrow$	$N_2O + O + O$	1.0−7	0.0	0.0	0.0	0.0	0.0	[36]	—
$N_2O + NO^+$	$\rightarrow$	$N + N_2O + O$	1.0−7	0.0	0.0	0.0	0.0	0.0	[36]	—
$N_2O + NO_2^+$	$\rightarrow$	$N_2O + NO + O$	1.0−7	0.0	0.0	0.0	0.0	0.0	[36]	—
$N_2O + N_2O^+$	$\rightarrow$	$N + N_2O + NO$	1.0−7	0.0	0.0	0.0	0.0	0.0	[36]	—
$N_2O + N_3^+$	$\rightarrow$	$N + N_2 + N_2O$	1.0−7	0.0	0.0	0.0	0.0	0.0	[36]	—
$N_2O + N_4^+$	$\rightarrow$	$N_2 + N_2 + N_2O$	1.0−7	0.0	0.0	0.0	0.0	0.0	[36]	—
$N_2O + O_4^+$	$\rightarrow$	$N_2O + O_2 + O_2$	1.0−7	0.0	0.0	0.0	0.0	0.0	[36]	—
$N_2O + N_3O^+$	$\rightarrow$	$N_2 + N_2O + NO$	1.0−7	0.0	0.0	0.0	0.0	0.0	[36]	—
$N_2O + NO_3^+$	$\rightarrow$	$N_2O + NO + O_3$	1.0−7	0.0	0.0	0.0	0.0	0.0	[36]	—
$N_2O + N_2O_2^+$	$\rightarrow$	$N_2O + NO + NO$	1.0−7	0.0	0.0	0.0	0.0	0.0	[36]	—
$N_2O + N_2O_2^+$	$\rightarrow$	$N_2 + N_2O + O_2$	1.0−7	0.0	0.0	0.0	0.0	0.0	[36]	—
$N_2O + O^+$	$\rightarrow$	$N + NO^+ + O$	2.30−10	0.0	0.0	0.0	0.0	0.0	[43]	—
$NH_3 + O_2^+$	$\rightarrow$	$NH_3^+ + O_2$	2.40−9	0.0	0.0	0.0	0.0	0.0	[43]	—
$H_2O + OH^+$	$\rightarrow$	$H_2O^+ + OH$	1.60−10	0.0	0.0	0.0	0.0	0.0	[43]	—
$O_2 + O_3^+$	$\rightarrow$	$O + O_3^+$	1.20−10	0.0	0.0	0.0	0.0	0.0	[43]	—
$H_2 + O_2^+$	$\rightarrow$	$H + HO_2^+$	4.0−11	0.0	0.0	0.0	0.0	0.0	[43]	—

(Continued)

TABLE 4.10 (Continued)
The Reactions between Heavy Particles Included in the Kinetic Scheme. $A - 1/s$, cm^3/s, cm^6/s, $E_a - K$

Reaction		A^+	n^+	E_a^+, K	A^-	n^-	E_a^-, K	Ref$^+$	Ref$^-$
$H_2O + O_4^+$	$\rightarrow$	$H_2O_3^+ + O_2$	1.20−10	0.0	−.0	0.0	0.0	0.0	[43]	—
$N_2O + O_4^+$	$\rightarrow$	$N_2O_3^+ + O_2$	2.50−10	0.0	−.0	0.0	0.0	0.0	[43]	—
$H_2^+ + O_2$	$\rightarrow$	$H + HO_2^+$	1.90−9	0.0	−.0	0.0	0.0	0.0	[43]	—
$H_3^+ + O_2$	$\rightarrow$	$H_2 + H_2^+$	1.30−10	0.0	−.0	0.0	0.0	0.0	[43]	—
$H_2O + O^- + O_2$	$\rightarrow$	$H_2O_2^- + O_2$	1.30−28	0.0	0.0	0.0	0.0	0.0	[43]	—
$H_2O + O_2^- + O_2^-$	$\rightarrow$	$H_2O_3^- + O_2$	2.20−28	0.0	0.0	0.0	0.0	0.0	[43]	—
$H + O_2^-$	$\rightarrow$	$OH^- + O$	1.20−9	0.0	0.0	0.0	0.0	0.0	[43]	—
$H + O_2^-$	$\rightarrow$	$H + O_2$	1.20−9	0.0	0.0	0.0	0.0	0.0	[43]	—
$H_2O + N_4^+$	$\rightarrow$	$H_2O^+ + N_2 + N_2$	1.90−9	0.0	0.0	0.0	0.0	0.0	[43]	—
$H_2O + H_3^+$	$\rightarrow$	$H_2 + H_3O^+$	5.90−9	0.0	0.0	0.0	0.0	0.0	[43]	—
$H_2O + OH^+$	$\rightarrow$	$H_3O^+ + O$	1.50−9	0.0	0.0	0.0	0.0	0.0	[43]	—
$H_2 + H_2O^+$	$\rightarrow$	$H + H_3O^+$	1.40−9	0.0	0.0	0.0	0.0	0.0	[43]	—
$N_3^+ + O_2$	$\rightarrow$	$N_2 + NO^+ + O$	1.0−10	0.0	0.0	0.0	0.0	0.0	[43]	—
$NO_2 + NO_2^+$	$\rightarrow$	$NO^+ + NO_3$	4.60−10	0.0	0.0	0.0	0.0	0.0	[43]	—
$NO + NO + NO^+$	$\rightarrow$	$N_2O_2^+ + NO$	5.0−30	0.0	0.0	0.0	0.0	0.0	[43]	—
$O + O_2 + O_2^-$	$\rightarrow$	$O_2 + O_3^-$	1.50−31	0.0	0.0	0.0	0.0	0.0	[39]	—
$NO + NO + O^-$	$\rightarrow$	$NO + NO_2^-$	2.04−31	0.0	0.0	0.0	0.0	0.0	[39]	—
$NO + O + O^-$	$\rightarrow$	$NO_2^- + O$	2.04−31	0.0	0.0	0.0	0.0	0.0	[39]	—

[a] Unless specifically stated, the relative efficiency of collisions with different partners is equal to $k_{Ar} = 1.5 \times k_{N_2} = 1.5 \times k_{O_2} = 3 \times k_{H_2} = 20 \times k_{H_2O} = 3 \times k_{NO} = 5 \times k_O = 5 \times k_H = 5 \times k_N$ (see [81]).
[b] Derived from the principle of detailed equilibrium.

2. 10^{-7} s — 10^{-5} s. The concentration of O atoms continues to increase due to the dissociative quenching of $N_2(B^3\Pi_g)$ and $N_2(A^3\Sigma_u^+)$. Complete recombination of H_2^+, O_2^+, and N_2^+ ions occurs, with the fastest reaction being the conversion of O_2^+ to form O and O_3^+; the latter recombines with an electron to form atomic oxygen O again. Another fast channel for the consumption of O_2^+ is its reaction with H_2, resulting in the formation of HO_2^+ and H. The rest of H_2^+ disappears in reactions with hydrogen to form H and H_3^+. The molecular nitrogen ion N_2^+ is a reservoir to feed H_2^+ and O_2^+ in charge-exchange reactions, but reacts even more rapidly with N_2 to form N and N^+, the latter in turn being efficiently converted to O. Up to now there has been an accumulation of atomic oxygen, and now it is gradually beginning to be converted into H and OH in reactions with hydrogen.
3. 10^{-5} s — 10^{-3} s. The process $(O+H_2 \rightarrow H+OH)$, along with the reverse process, continues to dominate until the beginning of the next pulse, being the main step in the chain of water formation. Nitrogen in the metastable state of $N_2(A^3\Sigma_u^+)$ in the dissociative quenching reaction by oxygen continues to energize the chemical system with O atoms, which in turn continue to increase the amount of H and OH. The trimolecular conversion reaction of atomic hydrogen with O_2 results in a very rapid buildup of HO_2, which, reacting with OH, forms water.
4. 10^{-3} s — 25×10^{-3} s. In the final stage, the sequence of water formation processes is as follows. Atomic oxygen reacts with H_2, producing H and OH. Atomic hydrogen H forms HO_2, which, reacting with OH, builds up H_2O. Another fast channel for water production is the OH+OH reaction.

Note that not only atoms and radicals involving O and H atoms play a significant role in reaction kinetics, but also processes involving N atoms and nitrogen oxides (Figures 4.11 and 4.12).

4.2 NITROUS OXIDE NON-THERMAL DECOMPOSITION IN A PULSED HIGH-CURRENT DISCHARGE

A key issue in the study of ignition under nonequilibrium conditions is the relation between the different mechanisms and processes at the initiation stage (see the 4.1 part of this book, where the relative role of excitation of vibrational, electronic degrees of freedom of the gas, ionization and dissociation of molecules in the H_2-air system was investigated).

For a better understanding of the peculiarities of the processes occurring in the phase of high-current discharge and its afterglow, it is necessary to study the excitation under such conditions of multiatomic molecules and reactions with their participation in the afterglow of the discharge. The decomposition of nitrous oxide has been investigated in sufficient detail both in the thermally equilibrium and in the "weakly nonequilibrium" regime behind shock waves. The excitation cross sections of nitrous oxides by electron impact are fairly well known and allow for detailed modeling of both the discharge phase of the process and the afterglow phase.

For these reasons, in this part we discuss the nonequilibrium mode of N_2O decomposition in the plasma of a nanosecond pulse discharge in order to isolate the influence of processes involving electronically excited, ionic, and vibrationally excited components on the kinetics of the process.

In the pressure range of 1–8 Torr at room temperature, the process of nitrous oxide decomposition in the conditions of exposure of the system to a high-speed ionization wave has been experimentally investigated, numerical simulation of such a process has been carried out, and the kinetics of plasma-chemical transformations in the system has been analyzed on the basis of comparison of the experimental results with the calculation [84].

4.2.1 Experimental Investigations of N_2O Decay in a Pulsed Discharge

The discharge cell was a quartz tube $l = 20$ cm long and 47 mm in diameter with a flat high-voltage electrode and a ring low-voltage electrode at the ends made of stainless steel. The low-voltage

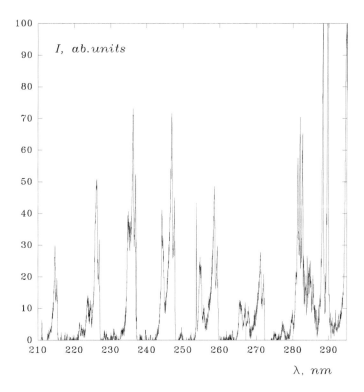

FIGURE 4.13 Spectrum of nanosecond discharge in N_2O at full pressure of 4 Torr 50 s after the start of the process.

electrode, short-circuited to the grounded shield of the supply cable by means of eight thick brass bars, has a CaF_2 window for radiation output. Negative polarity voltage pulses with amplitude $|U_{gen}| = 13$ kV, half-height duration 25 ns, and leading edge duration 2 ns were applied to the high-voltage electrode of the discharge cell from a pulse voltage generator with a repetition rate $f = 40$ Hz.

The amplitude and shape of the current pulse through the discharge section as the ionization wave propagated through it were recorded using a calibrated broadband reverse current shunt. The discharge emission was observed in the mode of signal accumulation from the end of the discharge cell using a UV-VIS monochromator. Measurements of the absolute pressure value during the decay of N_2O were carried out using a MD × 4C mechanotron. Figure 4.13 shows a part of the recorded near-ultraviolet spectrum. The bands corresponding to the γ-system of nitric oxide are well distinguished.

The complete set of spectroscopic data obtained in the experiment and the total pressure dynamics in the system are shown in Figure 4.14. The emission intensities of the second positive ($\lambda = 337.1$ nm, transition $C^3\Pi_u, v' = 0 \to B^3\Pi_g, v'' = 0$, Figure 4.24) and first negative ($\lambda = 391.4$ nm, transition $B^2\Sigma_u^+, v' = 0 \to X^2\Sigma_g^+, v'' = 0$) systems of molecular nitrogen—the main product of the N_2O decay reaction—were measured under the same initial conditions. This allowed us to determine the characteristic time of molecular nitrogen production in the N_2O decay process. The dynamics of the relative NO concentration during the decomposition process was measured from the emission intensity at wavelength $\lambda = 237.02 \pm 0.02$ nm (transition $NO(A^3\Sigma^+) \to NO(X^2\Pi)$). The upper level of this transition is populated by direct electron impact from the ground state of NO and in chemical processes involving electronically excited molecules, which significantly complicates the interpretation of these data. Figures in Figure 4.25 show the half-rise and half-decay times of the emission of the γ-system NO. Pressure measurements (Figure 4.24) provide important additional information on the rate of conversion of the triatomic reactant (N_2O) to bimolecular reaction products.

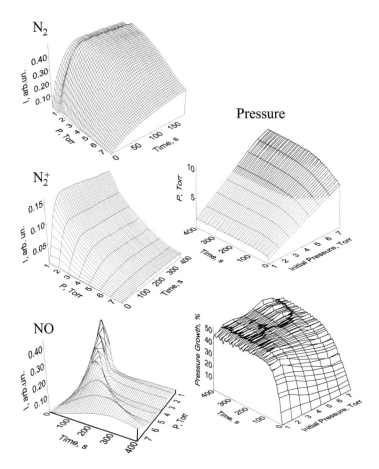

FIGURE 4.14 Experimental data. Temporal dynamics of pressure and emission changes of the second positive upper transition level $\left(N_2\left(C^3\Pi_u\right)\right)$, first negative $\left(N_2^+\left(B^2\Sigma_u^+\right)\right)$, and γ-systems $\left(NO\left(A^3\Sigma^+\right)\right)$ during the decay of N_2O depending on the initial conditions of the process development.

The electron concentration in the discharge and the magnitude of the reduced electric field were determined using time-resolved measurements of the dynamics of the current through the discharge gap and the voltage at the high-voltage electrode according to the technique described in detail in part 4.1.

The total current through the discharge gap can be obtained as the sum of the incident and reflected pulses. Figure 4.15 shows the measured incident, reflected, and calculated passed current pulses at initial gas pressure $p = 4.06$ Torr. The reflected pulse is shown with opposite sign for ease of comparison.

Figure 4.16 shows the dependence of the maximum value of current through the discharge gap I and its duration τ on the half-height of the current profile. In the investigated pressure range, the current amplitude in the discharge gap varied from $I = 210$ A at $p = 3$ Torr to $I = 110$ A at $p = 7.5$ Torr. Measurements of the discharge current and voltage drop across the discharge gap make it possible to estimate the electron concentration and the magnitude of the reduced electric field after overlapping of the discharge gap by a high-speed ionization wave (Figure 4.17).

The electron drift velocity v_d was calculated taking into account the current gas composition for a given pulse using the two-term approximation of the Boltzmann equation. Together with the data on the current density profile $j = I/S$ through the discharge gap, these data allow us to reconstruct the electron concentration profile at the high-current phase of the discharge $n_e = j/(ev_d)$, where e is the elementary charge.

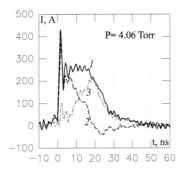

FIGURE 4.15 The incident (*1*), reflected (reverse sign) (*2*), and passed I (*3*) current pulses. The initial pressure $p = 4.06$ Torr.

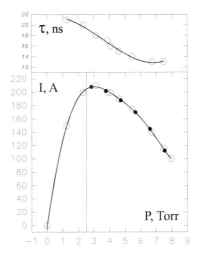

FIGURE 4.16 Average amplitude and duration of current pulse at different pressures.

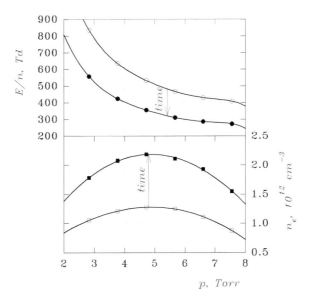

FIGURE 4.17 The magnitude of the reduced electric field and electron concentration as a function of initial pressure. Open symbols – initial values, solid symbols – values at the end of N_2O decay.

Figure 4.17 shows the dependences of the reduced electric field E/n and electron concentration n_e at different initial pressures of the mixture at the decay onset. In the range of parameters investigated in the work [84], the magnitude of the reduced electric field in the discharge gap varies from $E/n = 800$ to 300 Td, which is close to the limit of applicability of the two-term approximation of the Boltzmann equation in calculating the velocities of inelastic processes in electron-molecular collisions. However, as it was shown in [2], the error of such a calculation is insignificant when calculating the rates of population of low-lying electronic states under these conditions.

The maximum values of electron concentration lie in the range $n_e \simeq 0.9 - 2.2 \times 10^{12}$ cm^{-3} (Figure 4.17), which correlates well with the measurements made in [2,4].

4.2.2 Numerical Model of Non-Thermal Decay of N_2O in Plasma

The calculations were performed in the "direct modeling" mode, i.e., all consecutive current pulses through the discharge gap were calculated directly, taking into account the changes in the mixture composition between and during the pulses. The computational scheme was constructed from the results of the simulations carried out in part 4.1 of this chapter. In order to adequately describe the decay of N_2O in the presence of NO, N_2 and O_2, the processes of excitation of nitrous oxide and nitric oxide by electron impact were taken into account in the calculation of the electron energy distribution function. The cross sections of the corresponding processes were taken from the [85] review.

The scheme of kinetic equations included relaxation of vibrational and electronic states, ion-molecular and intermolecular processes in a similar way as was done in part 4.1 in describing the non-thermal oxidation of hydrogen in the H_2-air mixture. The results of the calculation of the energy distribution function for the vibrational states of various molecules realized in the decay process are shown in Figures 4.18–4.21. The distributions have a non-monotonic energy dependence, with strong overpopulation of the upper levels. Intermediate maxima on the distribution arise at selective scattering of the upper electronic states and due to recombination fluxes to the upper levels.

Figures 4.22 and 4.23 show the time-varying curves of the total pressure and relative concentration of $N_2(C^3\Pi_u)$ at initial pressure $p = 4.7$ Torr. Experimental data are indicated by solid line, calculation are indicated by dashed line. The horizontal dashed lines drawn on both graphs indicate the theoretical value of the total mixture pressure and relative concentration $N_2(C^3\Pi_u)$ at complete conversion of the initial gas into a mixture of N_2 and O_2 gases.

The good coincidence of the shape of the $N_2(C^3\Pi_u)$ yield curve in the calculation and experiment allows us to conclude that, under these conditions, the calculation using the two-term approximation

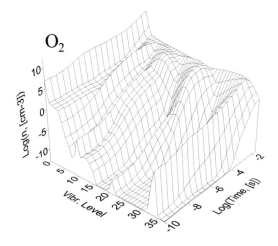

FIGURE 4.18 Calculation of vibrational state distribution function for O_2. $p = 2.8$ Torr. $U_{gen} = 13$ kV [86].

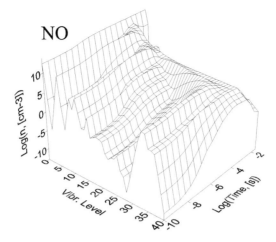

FIGURE 4.19 Calculation of the vibrational state distribution function for NO. $p = 2.8$ Torr. $U_{gen} = 13$ kV [86].

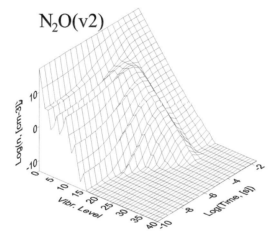

FIGURE 4.20 Calculation of the distribution function over vibrational states for $N_2O(v_2)$. $p = 2.8$ Torr. $U_{gen} = 13$ kV [86].

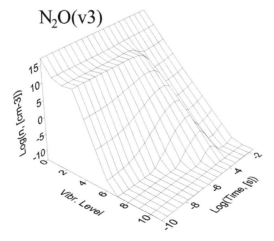

FIGURE 4.21 Calculation of the distribution function over vibrational states for $N_2O(v_3)$. $p = 2.8$ Torr. $U_{gen} = 13$ kV [86].

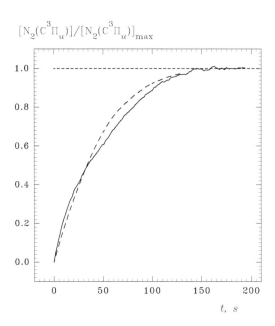

FIGURE 4.22 Dynamics of relative concentration $N_2(C^3\Pi_u)$ at initial pressure N_2O. $p=4.72$ Torr. $U_{gen}=13$ kV [86].

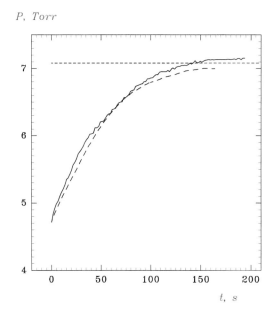

FIGURE 4.23 Dynamics of mixture pressure at initial pressure N_2O $p=4.72$ Torr. $U_{gen}=13$ kV [86].

of the Boltzmann equation correctly describes the process of gas excitation by electron impact, and the almost exact coincidence of the calculated and experimental pressure profiles indicates the correct reproduction of the N_2O decay kinetics at the afterglow stage of the discharge.

Figures 4.24 and 4.25 show the results of comparison of the calculated and experimental data. It is well seen that in the whole pressure range investigated, the proposed model describes both the conversion rate of triatomic nitrous oxide into two-atomic reaction products, which is expressed in the pressure growth as the main reagent is produced (Figure 4.24), and the time dependence of the population of the upper emitting states of molecular nitrogen and nitrogen ion.

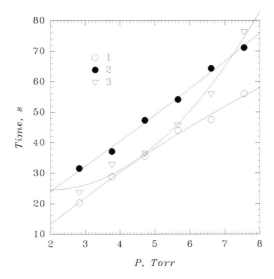

FIGURE 4.24 Major components production time. 1,2 – half-time of concentration $N_2(C^3\Pi_u)$ and $N_2^+(B^2\Sigma_u^+)$, 3 — half-time of pressure. Dots – experiment, curves – calculation.

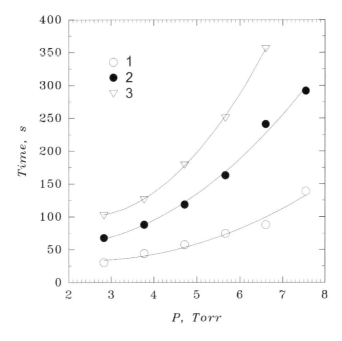

FIGURE 4.25 Major components production time. 1 – half-rise time, 2 – reaching the maximum, and 3 – half-decay of $NO(A^3\Sigma^+)$. Dots – experiment, curves – calculation.

The situation is much more complicated with the emission dynamics of the γ-band NO emission (Figure 4.14). It is well seen that the rise time of NO emission is much longer than that of $N_2(C^3\Pi_u)$ and $N_2^+(B^2\Sigma_u^+)$ (Figure 4.25). Thus, the maximum of $NO(A^3\Sigma^+)$ concentration is reached when the main decay of N_2O has already been completed and the NO concentration in the ground electronic state is small. This circumstance allows us to state that the population of $NO(A^3\Sigma^+)$ does not occur through the excitation of the NO molecule (e.g., in the processes $NO + e \rightarrow NO(A^3\Sigma^+) + e$ or $NO + N_2(A^3\Sigma_u) \rightarrow NO(A^3\Sigma^+) + N_2$, whose rates are proportional to the current NO concentration in the system), and through the direct formation of electronically excited nitric oxide NO. Unfortunately, information on such alternative channels for the formation of NO $\left(A^3\Sigma^+\right)$ is not available; in this work, several energetically-allowed channels for the formation of this state have been considered. The γ-bands closest to the experimentally obtained values in the emission profile were obtained under the assumption that the main population of NO $\left(A^3\Sigma^+\right)$ occurs in the process $N^+ + O_2^- \rightarrow NO\left(A^3\Sigma^+\right) + O$. The constantly increasing concentration of the negative oxygen ion O_2^- shifts the time to reach the maximum toward the end of the process, and the sharp drop in the N^+ concentration after the end of N_2O decay leads to a rapid decrease in the NO $\left(A^3\Sigma^+\right)$ population at large times, which correlates well with the experimentally observed dependences (Figure 4.25). Nevertheless, the question about the excitation channels of the γ-band under these conditions should, apparently, be considered open.

4.2.3 Active Particle Fluxes and Main Stages of the Process Non-Thermal Decomposition of N_2O in a Pulsed Discharge

To highlight the most important processes occurring in the afterglow of the discharge, a rate analysis of the kinetic scheme was carried out under conditions corresponding to the first current pulse through the discharge gap at a pressure of 4.72 Torr (initial composition: N_2O), and the 1,560th pulse (corresponding to 50% production of the initial component).

Figures 4.26 and 4.27 show the kinetic curves for the components involved in the most important processes in the afterglow of these pulses.

The characteristic differences between the kinetics at the initial phase of N_2O decay, when the concentration of molecular nitrogen and oxygen in the mixture is low, and at later stages, when the processes of excitation in the discharge of electronic levels of N_2 with their collisional dissociative quenching in collisions with N_2O become dominant, are well seen (Figures 4.26 and 4.27).

The time interval between the end of the first pulse and the beginning of the second pulse, as in the case of the reacting H_2-air system, can be divided into four parts: the first — from 0 to 10^{-7} s; the second — from 10^{-7} to 10^{-5} s; the third — from 10^{-5} to 10^{-3} s; and the fourth — from 10^{-3} to 25×10^{-3} s.

For each of these time intervals, diagrams of active particle fluxes for the fastest channels of chemical transformations are plotted. In the diagram, the thickness of the line and the number near

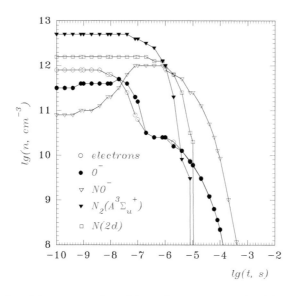

FIGURE 4.26 Dynamics of the concentration of the main components. The initial phase of decay. $([N_2O]/[N_2O]_0 = 1)$. Initial pressure N_2O $p = 4.72$ Torr. $U_{gen} = 13$ kV [87].

Mechanisms of Plasma-Chemical Reactions

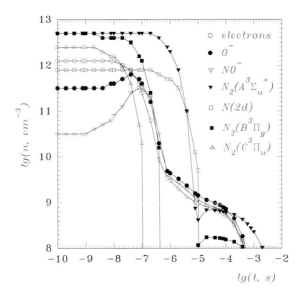

FIGURE 4.27 Dynamics of the concentration of major components. The final phase of decay. ($[N_2O]/[N_2O]_0=0.5$). Initial pressure N_2O $p=4.72$ Torr. $U_{gen}=13$ kV [86].

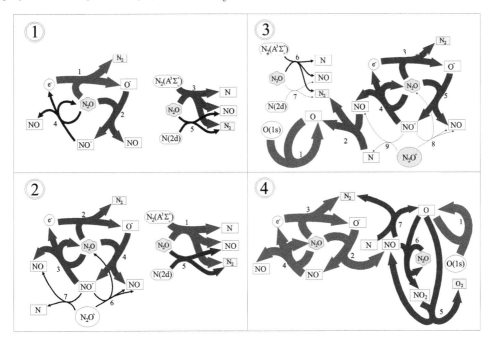

FIGURE 4.28 Active particle flux diagram. Initial phase of decay. ($[N_2O]/[N_2O]_0=1.0$). Initial pressure N_2O $p=4.72$ Torr. $U_{gen}=13$ kV [86].

it corresponds to the rate of the process. Let us consider in the afterglow of the first pulse the processes dominating in each time interval.

1. The fastest processes during the first 10^{-7} s after the current pulse is turned off (Figure 4.28(1)) are the formation of the charge transfer cycle

$$e^- + (N_2O) \to O^- \to NO^- \to e^-,$$

in which N_2O is actively decaying.

The second fast process is the dissociative quenching of the electronically excited level $N_2(A^3\Sigma_u^+)$ to N_2O. The decay process of N_2O on this time interval also yields the reaction $N(2d)+(N_2O) \rightarrow NO+N_2$.

2. During the second interval—from 10^{-7} to 10^{-5} s (Figure 4.28(2))—the cycle

$$e^- + (N_2O) \rightarrow O^- + (N_2O) \rightarrow NO^- + (N_2O) \rightarrow e^-,$$

significantly accelerates due to the accumulation of NO^-. However, the main role in the decay of N_2O is now played by the dissociative quenching process of $N_2(A^3\Sigma_u^+)$ to N_2O.

3. The third subinterval — from 10^{-5} to 10^{-3} s (Figure 4.28(3)) — is characterized by a decrease in the concentration of electronically excited nitrogen generated in the discharge and an increase in the relative role of the ionic mechanism of nitrous oxide decay. The relative contribution of secondary processes — reactions involving O and N atoms — increases.

4. At longer times—from 10^{-3} to 25×10^{-3} s (Figure 4.28(4))—the processes responsible for the production of N_2 and O_2 in the system are activated. The largest contribution here is made by processes involving NO:

$$NO + O + M \rightarrow NO_2 + M$$

$$NO_2 + O \rightarrow NO + O_2$$

and

$$NO + N \rightarrow N_2 + O$$

The general picture of the decomposition kinetics changes significantly when N_2O is partially decomposed and replaced in the mixture by molecular nitrogen, oxygen, and nitric oxide. The set of the main processes does not change, but there is a strong shift of emphasis to reactions involving electronically excited nitrogen molecules and oxygen ions (Figure 4.29).

The analysis allows us to speak about the principal role of processes involving ions and electronically excited particles in the decomposition of nitrous oxide in a high-current pulsed discharge at low temperatures. The relative role of processes involving O atoms is small because of the high-energy threshold in the process.

$$N_2O + O \rightarrow NO + NO$$

In contrast to the H_2-air system, where vibrational excitation of gas led to a sharp acceleration of the oxidation process, vibrational excitation of gas in the decomposition of N_2O does not cause a pronounced effect. The influence of vibrational excitation on the decomposition of nitrous oxide by the purely vibrational mechanism of accelerating monomolecular decomposition proposed in [87] cannot play any appreciable role under these conditions because of the high relaxation rates of energy distribution function at high values of vibrational numbers (in the deformation mode). A decrease in the reaction threshold for the process $N_2O + O \rightarrow NO + NO$, which is $\Theta \simeq 14,000$ K, also cannot lead to an appreciable (compared to other mechanisms) consumption of N_2O.

The analysis has shown that the contribution of vibrational nonequilibrium in this case manifests itself in a completely different way.

On the one hand, at vibrational excitation of the gas, the processes of collisional detachment of electrons from O^-, O_2^-, and NO^- ions are sharply accelerated. This leads to some acceleration of the recombination process of charged particles and slowing down of the ionic mechanism discussed above.

On the other hand, gas excitation leads to some increase in the average energy of electrons in the discharge due to the contribution of superelastic collisions and, consequently, to an increase in the

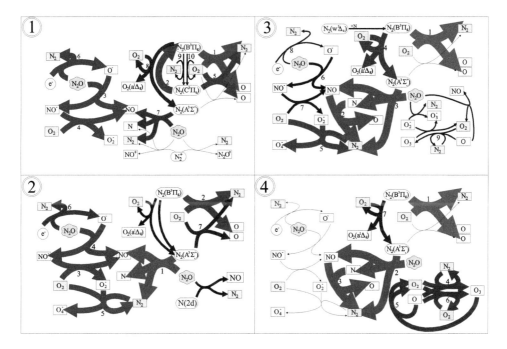

FIGURE 4.29 Active particle flux diagram. Final phase of decay. ([N₂O]/[N₂O]₀=0.5). Initial pressure N₂O $p=4.72$ Torr. $U_{gen}=13$ kV [86].

rate of $N_2(A^3\Sigma_u^+)$ population, leading to an increase in the rate of collisional dissociation of N₂O. Direct dissociation of N₂O by electron impact is also accelerated.

Both of these mechanisms only indirectly affect the integral rate of N₂O decomposition and practically do not change the rate of the process as a whole. Thus, on the basis of the data obtained, it can be stated that the thermally nonequilibrium decay of N₂O under conditions of high-current pulse discharge proceeds by the mechanism determined by reactions involving ions and electronically excited molecules.

4.3 OXYDATION OF HYDROCARBONS IN NANOSECOND DISCHARGE

A thorough experimental analysis was performed to study the slow oxidation processes of alkanes, ranging from methane to decane, in mixtures with oxygen and air under the influence of nanosecond uniform discharge [88]. The study meticulously examined the oxidation kinetics of alkanes in both stoichiometric and selected lean mixtures with oxygen and air at ambient temperature, facilitated by high-voltage nanosecond repetitive pulse discharge. The discharge mechanism employed involved the generation of high-voltage, negative polarity pulses with a magnitude of 10 kV and a duration of 25 ns at a repetition rate of 40 Hz. This process was performed in a shielded discharge tube measuring 5 cm in diameter and 20 cm in length. The initial pressures of the mixtures were varied over a range from 0.76 to 10.6 Torr, with increments of 0.76 Torr. Measurements of discharge current, electric field strength, and input energy were made with nanosecond precision. In addition, the study included the quantification of emission intensities from different molecular bands, including $NO(A^2\Sigma \to X^2\Pi, \delta v = 3)$, $N_2(C^3\Pi, v'=1 \to B^3\Pi, v''=7)$, $N_2(B^3\Pi, v'=6 \to A^3\Sigma, v''=3)$, $N_2^+(B^2\Sigma, v'=0 \to X^3\Sigma, v''=2)$, $CO_2^+(B^2\Sigma \to X^2\Pi, \delta v = 0)$, $CH(A^2\Delta, v'=0 \to X^2\Pi, v''=0)$, OH $(A^2\Sigma, v'=0 \to X^2\Pi, v''=0)$, and $CO(B^1\Sigma, v'=0 \to A^1\Pi, v''=2)$. The emission was measured in the integral regime (integration time of 2.2 s) and with nanosecond resolution. Methane concentration was measured by absorption of He-Ne laser emission at 3.3922 μm in the integral regime. Based on

TABLE 4.11
Investigated Mixture

Alkane	CH_4	C_2H_6	C_3H_8	C_4H_{10}	C_5H_{12}	C_6H_{14}
In mixture with O_2	33.3%	22.2%	16.6%	13.3%	11.1%	9.5%
In mixture with air	11.11%	–	–	–	3.03%	2.56%

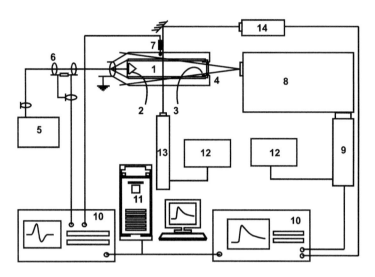

FIGURE 4.30 Experimental setup: 1 – discharge cell, 2,3 – electrodes, 4 – optical window (CaF2), 5 – high-voltage generator, 6 – back-current shunt, 7 – capacitive gauge, 8 – monochromator, 9 – photomultiplier, 10 – oscilloscopes S9-8 and TDS-380, 11 – computer, 12 – power supplies, 13 – He-Ne laser ($\lambda = 3.3922\,\mu m$), and 14 – photodetector Pb-Se.

optical measurements, the time of oxidation of alkanes was determined. The percentage of alkanes in all studied mixtures is presented in Table 4.11.

Additional research was conducted to study the slow oxidation processes of various hydrocarbon-containing molecules, specifically ethyl alcohol (C_2H_5OH), acetone (CH_3COCH_3), and acetylene (C_2H_2), in mixtures with oxygen and air, facilitated by nanosecond uniform discharge. The study focused on the experimental analysis of the kinetics of these substances in their respective mixtures with oxygen and in $CO:O_2$ mixtures containing small controlled additions of water vapor.

4.3.1 Nanosecond Discharge Formation in Chemically Active Gas

The experimental configuration is shown in Figure 4.30, with an arrangement in which electrical pulses are transmitted to the discharge cell via a 50-ohm RF coaxial cable 20 m in length. The core component, a discharge cell, consists of a thin-walled quartz tube with inner and outer diameters of 47 and 50 mm, respectively, surrounded by a metal screen and equipped with high and low-voltage electrodes. The shielding consists of eight brass rods, each 12 mm in diameter, uniformly spaced 7 cm from the central axis of the cell. The high-voltage electrode has a conical shape with an opening angle of 60°, while the low-voltage counterpart is a ring directly connected to the grounded shielding. The distance between the ends of the electrodes is set at 20 cm. The pressure in the mixture during the oxidation process is precisely monitored by a mechanotron MD×4C pressure gauge.

To capture the electrical characteristics of the discharge, measurements are made using a capacitive gauge that can be moved along the axis of the discharge cell, in conjunction with a back-current shunt integrated into the shield break of the coaxial supply cable. This back-current shunt is positioned at a distance sufficient to separate the incident and reflected current pulses.

For each mixture studied, the emission intensity of the discharge is quantitatively evaluated in both time-resolved and integral modes. Emission intensity measurements are made through a 3 cm diameter CaF_2 optical window embedded in the low-voltage ring electrode. A UV-VIS monochromator with a linear dispersion of 1.2 nm/mm and a UV-sensitive photomultiplier, the latter characterized by a signal rise time not exceeding 3 ns and a spectral sensitivity range of 200–800 nm, are used for these observations. The monochromator is positioned at a distance of 12 cm from the window to ensure that emission from the entire discharge volume is detected.

The measurements were performed at the beginning of the discharge and after the completion of the oxidation process, using an averaging regime over 128 pulses, resulting in an averaging duration of 3.2 s in this specific case. The evaluation of the temporal-spatial dynamics of the electric potential distribution was carried out only in the products.

For mixtures containing methane, in addition to the general set of parameters under investigation, the absorption of the He-Ne laser emission at a wavelength of 3.3922 µm was also quantified.

Time-resolved signals were acquired using a Tektronix TDS 380 oscilloscope with a bandwidth of 400 MHz. Conversely, integral measurements were performed with an oscilloscope model S9-8, using scales of 100, 200, and 400 s.

Mixtures of heavy hydrocarbons were prepared under conditions that ensured that the partial pressure of an alkane remained below its saturated vapor pressure at ambient temperature. A specified amount of alkane was injected into a 10-L chamber that had previously been subjected to vacuum treatment. The mass of the injected alkane was reduced by 20% from the mass corresponding to its saturated vapor pressure in that volume at room temperature. Monitoring the pressure within the chamber (which increased and stabilized as the alkane evaporated and warmed to ambient temperature) facilitated the determination of the alkane concentration. Oxygen was then added to achieve the desired pressure, and the gases were allowed to mix for 2 days.

4.3.2 Analysis of Gas Excitation in Repetitive Nanosecond Gas Discharge

4.3.2.1 Electric Current, Electric Field and Energy Input

The potential distribution along the discharge cell, denoted as $V_i(t)$, was determined using a capacitive gauge strategically placed at sections along the cell at distances x_i = 0.0, 3.6, 7.2, 10.8, 14.4, and 18.0 cm. The first section, x_0 = 0.0, was aligned with the edge of the conical electrode.

The methodology for measuring the current through the high-voltage electrode, $J_0(t)$, used standard techniques as described in [89]. A representative oscillogram of the signal from the shunt is shown in Figure 4.31a. The initial current pulse ("incident") is identified as the pulse propagating from the high-voltage generator to the discharge cell. Upon reflection from the discharge cell, this pulse returns within 100 ns to the section where the shunt is installed, forming the second pulse ("reflected") shown in Figure 4.31. The pulse then oscillates between the generator and the discharge cell, reflecting off each and returning to the shunt, continuing the cycle of pulse propagation. As a result, the current shunt captures pairs of pulses ("incident" and "reflected") cycling through it with a periodicity of 200 ns. The energy delivered to the gas during the initial pulse is quantified as the energy difference between the first pair of incident and reflected pulses. The current flowing through the discharge cell was derived from the charge continuity equation at the high-voltage electrode section (Figure 4.31), taking into account the directions of pulse propagation that affect the polarities of the signals detected by the back-current shunt.

$$J_0(t) = J_{sh}(t) + J_{sh}(t + 2\Delta t) \tag{4.7}$$

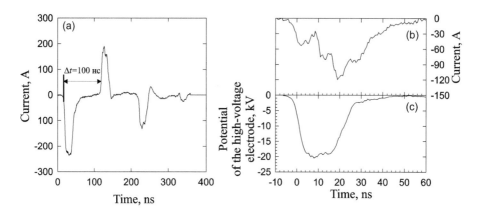

FIGURE 4.31 Characteristics measured by the current shunt: (a) the oscillogram of the current in the cable, (b) the current through the high-voltage electrode, and (c) the high-voltage electrode potential. The initial pressure of the mixture CH_4+air is 1.5 Torr.

where $J_{inc} = J_{sh}(t)$ represents the current within the incident pulse, while $J_{ref} = J_{sh}(t+2\Delta t)$ denotes the current within the reflected pulse, as shown in Figure 4.31. Here, Δt is the time required for the pulse to propagate from the current shunt to the high-voltage electrode. In addition, the current shunt measurements facilitate the determination of both the current flowing through the high-voltage electrode and the potential at the electrode, U_0 (Figure 4.31). The potential at the high-voltage electrode can be quantitatively correlated to the difference between the currents observed in the initial and subsequent pulses of the pair:

$$U_0(t) = R\left[J_{sh}(t) - J_{sh}(t+2\Delta t)\right], \tag{4.8}$$

where $R = 50$ Ohm is the impedance of the RF coaxial cable.

Telegraph equations were used to determine the electric field and current at various sections along the discharge cell, as described in previous chapters. The capacitance and inductance of the discharge setup were assumed to be uniform over the length of the discharge cell. Capacitance per unit length was calculated assuming that excess charge accumulates predominantly near the cell wall, resulting in a capacitance value of $C = 5.2$ pF/cm. Inductance per unit length was estimated based on the approximation of uniform current density throughout the cross-section of the discharge cell, yielding an inductance of $L = 2.8$ nH/cm. The resistance per unit length within the cell was assumed to vary both spatially and temporally.

It is postulated that the longitudinal component of the electric field, $E_l(x,t)$, can be expressed by the relationship $E_l(x,t) = J(x,t)R(x,t)$, where $E_l(x,t)$, hereafter referred to as E, denotes the longitudinal electric field.

$$\begin{cases} \dfrac{\partial V}{\partial x} + E = -L\dfrac{\partial J}{\partial t} \\ \dfrac{\partial J}{\partial x} = -C\dfrac{\partial V}{\partial t} \end{cases} \tag{4.9}$$

In this context, $J(x,t)$ is the current, $V(x,t)$ is the potential, and $R(x,t)$ is the resistance per unit length within the section located at x for a given time t.

Since the measurements were made at a discrete set of points, to solve the problem authors [88] use a linear approximation for the potential, $V_i(t)$, and the current, $J_i(t)$, between adjacent sections. Consequently, for the interval $i = (x_i, x_{i+1})$:

$$\begin{cases} V = V_i(t) + \dfrac{x - x_i}{x_{i+1} - x_i}(V_{i+1}(t) - V_i(t)) \\ J = J_i(t) + \dfrac{x - x_i}{x_{i+1} - x_i}(J_{i+1}(t) - J_i(t)) \end{cases} \quad (4.10)$$

Within the framework of the discrete approximation, the second equation contained in the system denoted by (4.10) is formulated as follows:

$$\frac{J_{i+1}(t) - J_i(t)}{x_{i+1} - x_i} = -C\frac{\partial V}{\partial t} \quad (4.11)$$

The right-hand side of the equation is dependent on the spatial variable x, whereas the left-hand side is independent of x. To facilitate the solution of this equation, the right-hand side is subjected to averaging over the interval $[x_i, x_{i+1}]$.

$$J_{i+1} = J_i - \frac{C(x_{i+1} - x_i)}{2}\frac{\partial}{\partial t}[V_i + V_{i+1}] \quad (4.12)$$

Consequently, the current within any given section, corresponding to the coordinate x_{i+1}, is described by the following relationship:

$$J_{i+1} = J_0 - \sum_{k=0}^{i} \frac{C(x_{k+1} - x_k)}{2}\frac{\partial}{\partial t}[V_k + V_{k+1}] \quad (4.13)$$

To determine the electric field E, Eq. 4.9 was averaged over the interval $i = (x_i, x_{i+1})$:

$$\frac{E_{i+1}(t) + E_i(t)}{2} = -\frac{V_{i+1}(t) - V_i(t)}{x_{i+1} - x_i} - \frac{L}{2}\frac{\partial}{\partial t}[J_i(t) + J_{i+1}] \quad (4.14)$$

Accordingly, by incorporating the inductance and capacitance characteristics of the discharge cell, the experimentally measured current through the high-voltage electrode and the potential distribution along the discharge cell were converted into distributions of current and electric field along the cell.

The derived values for $E(x,t)$ and $J(x,t)$ facilitated the calculation of the dynamics of the power contribution per unit volume, expressed as $P(x,t) = E(x,t)J(x,t)/S$, where S is the cross-sectional area of the discharge tube.

It is critical to recognize that the approximations used are insufficient to capture the dynamics of the electric field within the fast ionization wavefront, as indicated by [89]. This limitation arises both from the use of the delta function as the spatial sensitivity function of the capacitive gauge and from the omission of the radial components of the current and electric field. Nevertheless, these approximations are considered adequate for the phases beyond the wavefront and especially during the progression of the discharge toward the low-voltage electrode, as justified in [90]. The use of the delta function as the sensitivity function precluded accurate reconstruction of the electric field near the cathode. It is noteworthy that at pressures above 3 Torr, where the maximum energy contribution occurs under experimental conditions [90], most of the energy is transferred to the gas in later stages of the discharge, rendering the cathode potential fall insignificant. Therefore, the reconstruction methodology employed successfully delineates the distribution of the electric field during the time intervals critical for gas excitation.

4.3.2.2 Dynamics of Excitation of Electronical Degrees of Freedom and Discharge Emission

Under the specified experimental conditions, the complete oxidation of alkanes is facilitated by the application of several thousand pulses during tens to hundreds of seconds. Consequently, the integrated time signal derived from the photomultiplier tube contains comprehensive data on the hydrocarbon oxidation process. The integration time was $\tau_0 = CR_0 = 0.25$ s. The lifetimes of the investigated states, which are less than 10 μs, are significantly shorter than the interpulse interval, which is about $1/f \approx 25$ ms. Therefore, the influence of PEM signal from previous pulses can be ignored. With this in mind, the current generated within a pulse from the multiplier can be expressed as follows:

$$J(\delta t) = \frac{IAk_{exc}N_0'}{\tau_l} \exp\left[-\frac{\delta t}{\tau_l^{eff}}\right], \tag{4.15}$$

where I symbolizes the emission intensity at a particular wavelength λ, $A = A(\lambda)$ represents the spectral sensitivity of the combined optical system comprising the monochromator and the photomultiplier, N_0' denotes the concentration of non-excited molecules from which excited states are produced, k_{exc} is the excitation constant for the upper state, integrated over the excitation duration within a single high-voltage pulse, τ_l is the radiative lifetime of the upper state, τ_l^{eff} is the effective lifetime of the upper state, and δt is a variable indicating the time elapsed since the arrival of the last preceding pulse. The effective lifetime includes considerations for both radiative and collisional deactivation of the state.

$$\tau_l^{eff} = \frac{1}{1/\tau_l + N_0 \sum_i^M \alpha_i k_i^q}, \tag{4.16}$$

In this analysis, M is the number of stable components in a mixture, α_i is the fraction of component i, k_i^q is the quenching constant associated with the i-th component, and N_0 is the total concentration of stable components.

For mixtures characterized by slow reaction kinetics – where the characteristic time for composition changes significantly exceeds the integration time – the excitation rate of the state, k_{exc}, and the effective lifetime, τ_l^{eff}, show minimal variation from pulse to pulse.

Thus, the measured integral characteristics of the emission provide insight into the actual composition of the reactive mixture. The emission intensity was monitored over a number of spectral bands, noting that some bands overlapped. Table 4.12 lists the transitions monitored in all mixtures, while Table 4.13 details the transitions specifically investigated in mixtures with air, in

TABLE 4.12
Transitions Measured in all Mixtures

λ, nm	$\delta\lambda$, nm	Measured Band	Overlapped System of Bands
518.6 nm	2.8 nm	CO Angstrom, CO $(B^1\Sigma, v'=0 \to A^1\Pi, v''=2)$	–
430 nm	3 nm	$CH(A^2\Delta, v'=0 \to X^2\Pi, v''=0)$	$CO_2^+(A^2\Pi \to X^2\Pi)$, first negative and second positive systems of nitrogen
307.8 nm	3.4 nm	$OH(A^2\Sigma, v'=0 \to X^2\Pi, v''=0)$	$CO_2^+(A^2\Pi \to X^2\Pi)$ second positive systems of nitrogen
290 nm	2.8 nm	$CO_2^+(B^2\Sigma \to X^2\Pi, \delta v=0)$	$OH(A^2\Sigma \to X^2\Pi)$

TABLE 4.13
Transitions Measured in Air Mixtures

λ, nm	$\delta\lambda$, nm	Measured Band	Overlapped System of Bands
258.2 nm	3.7 nm	γ–system NO, NO$(A^2\Sigma \to X^2\Pi, \delta v = 3)$	–
490 nm	3.7 nm	$N_2(C^3\Pi, v' = 1 \to B^3\Pi, v'' = 7)$	$CO_2^+(A^2\Pi \to X^2\Pi)$
669.6 nm	3.7 nm	$N_2(B^3\Pi, v' = 6 \to A^3\Sigma, v'' = 3)$	$CO_2^+(A^2\Pi \to X^2\Pi)$
469.2 nm	3.3 nm	$N_2^+(B^2\Sigma, v' = 0 \to X^3\Sigma, v'' = 2)$	$CO_2^+(A^2\Pi \to X^2\Pi)$

addition to those outlined in Table 4.12. It is important to emphasize that the intensity levels of the transitions mentioned in Table 4.13 were sufficiently high to render the emission from overlapping bands negligible.

In scenarios where there was significant overlap between emitting spectral bands, identification of specific bands required time-resolved measurements of emission intensities. These measurements were made at two critical times: the onset of the discharge and the completion of the oxidation process, coinciding with the moment when the concentrations and the discharge parameters are stabilized at a steady-state value. In addition, the time-resolved evaluation of the emission intensities was useful in determining the quenching constant rates for both the initial and final composition of the mixture. By integrating these quenching rates, the excitation rates of higher energy states were accurately calculated.

4.3.2.3 Dynamics of Methane Concentration

The concentration of methane in the mixture was quantified by absorption of He-Ne laser emission at a wavelength of $\lambda = 3.3922\ \mu m$. This particular wavelength closely matches that of the vibrational transition associated with the asymmetric mode of the CH_3. The laser emission traversed the discharge cell in a direction perpendicular to its axis, eventually reaching a Pb-Se photodetector (whose spectral sensitivity band ranges from 2 to 4 μm). When the mixture was introduced into the discharge cell, the signals from both the mechanotron and the photodetector were recorded simultaneously. Using the known composition of the mixture and the established relationship between the photodetector signal and the mixture pressure $I(p)$, a calibration curve $I([CH_4])$ was derived for each experiment performed. The concentrations of other alkanes were not measured using this technique due to their minimal absorption coefficients at the 3.3922 μm wavelength and their relatively low concentrations in stoichiometric and lean mixtures.

4.3.3 SLOW OXIDATION OF HYDROCARBONS IN NONEQUILIBRIUM PLASMA

4.3.3.1 Electrodynamic Parameters of Discharge

The distributions of potential, current, reduced electric field, and power per unit volume along the discharge gap, derived from measurements using a capacitive gauge and shunt in a stoichiometric mixture of methane and air, are shown in Figure 4.32. To elucidate the behavior of the electrodynamic properties under high-pressure conditions, these properties were also measured at a pressure of 18 Torr, extending the standard parameter set.

The importance of considering the capacitance and inductance of the discharge cell is underscored by Eq. 4.9, from which it follows that

$$E = -\frac{\partial V}{\partial x} - L\frac{\partial J}{\partial t}$$

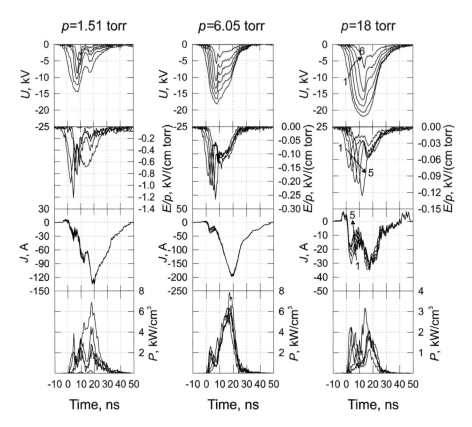

FIGURE 4.32 Electrodynamics characteristics in the first high-voltage pulse represented for three different pressures – 1.51, 6.05, and 18 Torr. Potential distributions $U(t)$ are represented in sections $x = 0.0, 3.6, 7.2, 10.8, 14.4, 18.0$ cm. Distributions of reduced electric fields $E(t)/p$, currents $J(t)$, and volumetric power densities $W(t)$ averaged over the intervals are represented for the intervals 0.0–3.6, 3.6–7.2, 7.2–10.8, 10.8–14.4 and 14.4–18.0 cm. Curves are enumerated from 1 to 6 for the potential and from 1 to 6 for other values consequently. Arrows in the figures show directions of curve shifts with the consequent change from the 1st to the 6th section for voltage and from the 1st to 5th interval for other values. The mixture is the product of the oxidation of CH_4 in the methane-air stoichiometric mixture.

Therefore, the inclusion of the inductance is critical in scenarios characterized by low electric fields and rapid changes in current. As shown in Figure 4.32, such conditions were encountered at the trailing edge of the high-voltage pulse. Specifically, at a pressure of 6.0 Torr, these conditions manifest themselves beyond the 25 ns time frame, as clearly shown in Figure 4.32. Consequently, at $t = 25$ ns, the rate of current rise reached 20 A/ns, and the inductive contribution to the electric field was calculated to be 60 V/cm, or 10 V/(cm × Torr), figures that are equivalent to the magnitude of the electric field itself.

The rapid development of the ionization wave is attributed to the charging of the capacitance of the discharge cell, which causes the current to decrease with increasing distance from the high-voltage electrode. In the later stages of the discharge, the effect of capacitance becomes insignificant.

The electrodynamic characteristics of the discharge within a relatively short discharge gap present an extensive profile that differs markedly from that observed in longer gaps, as detailed in [90]. The velocities of the fast ionization wave were typically observed to be between 2 and 3 cm/ns, with the propagation time of the wave through the discharge cell ranging from 7 to 15 ns. This duration is significantly shorter than that of the pulse. Because of this, a short-circuit regime appears in these experimental conditions. In this regime, the electric field and the current exhibit a uniformity across

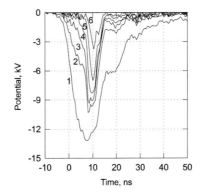

FIGURE 4.33 Distributions of the potential in the first high-voltage pulse for the pressure 0.76 Torr in sections $x = 0.0, 3.6, 7.2, 10.8, 14.4, 18.0$ cm, which are enumerated from 1 to 6 consequently. The mixture is the product of oxidation of CH_4 in the stoichiometric mixture with air.

the discharge cell, characterized by a comparatively gradual temporal evolution. For example, at a pressure of $p = 6.0$ Torr, this phase started at $t = 10$ ns.

At a pressure of 0.76 Torr, a precursor is observed, characterized as a fast ionization wave emerging against a background of a high cathode voltage drop and characterized by a minimal current through the cathode (as shown in Figure 4.33 for $t < 8$ ns). The dynamics of the precursor and the primary ionization wave in a long tube have been analyzed in detail in [90,91]. The initiation of the main wave facilitates the neutralization of positive charges in the vicinity of the cathode, leading to a significant reduction in the cathode drop.

Research has shown that the precursor is manifested in the early stages of the discharge, and its development is governed by the polarization current near the cathode, while the emission current remains comparatively low during this phase. This polarization effect near the cathode escalates the electric field intensity in this region, causing a rapid increase in the emission current and thus triggering the onset of the main wave. This main wave traverses the pre-ionized gas at a much higher speed than the precursor, eventually overtaking it. As the main wave merges with the precursor, a single ionization wave continues to propagate at a velocity greater than that of the precursor, but less than the velocity of the main wave prior to merging. The onset of the main wave results in the neutralization of excess positive charge near the cathode and a pronounced decrease in the cathode drop, to the extent that the cathode drop after the onset of the main wave falls below the threshold of experimental detection, as reported in [90].

At a pressure of 0.76 Torr, two distinct phenomena are observed: first, the precursor reaches the anode before the initiation of the main wave; second, a significant cathode voltage drop is maintained throughout the duration of the discharge. The occurrence of the first phenomenon can be attributed to the reduced length of the discharge tube used (20 cm), which is one-third of the length of the tube described in [90] (60 cm). Consequently, given roughly equivalent delay times for the initiation of the main wave and the precursor velocities, the precursor in these experiments has sufficient time to reach the anode before the start of the main wave. This dynamics results in the main wave passing through pre-ionized gas, allowing it to traverse the discharge cell in about 2 ns, with a velocity estimated to be 10^{10} cm/s. The second phenomenon is explained by the difference in the dimensions of the discharge cell, where both the length and the diameter are significantly different in the setup [90] – in particular, the tube diameter is 2.5 times larger than that used in [90]. For a comparable electron density, the resistance of the discharge cell (excluding the cathode layer) in this configuration was found to be 20 times lower than that reported in [90]. This difference leads to a proportionally higher voltage drop across the cathode layer in the experimental setup [90].

Under low-pressure conditions ($p < 4$ Torr), the evolution of the ionization wave and the appearance of the short-circuit stage are observed against a background of significant cathode voltage drop. It is important to note that the electric field value within the cathode layer at pressures below 4 Torr is significantly higher—by several orders of magnitude—than that shown in Figure 4.33. This discrepancy is due to the significant width (approximately the diameter of the tube's screen) of the half-width of the sensitivity function of the capacitive gauge, as discussed in [90]. The thickness of the cathode layer can be approximated by applying glow discharge theory. Under the conditions of these experiments, the cathode layer fits the characteristics of an anomalous glow discharge, characterized by a typical current density in the range of several to tens of A/cm², exceeding the normal current density by 3 to 4 orders of magnitude. Consequently, it is inferred that the parameter (pd) approximates the value $(pd)_{norm}/e$ [Torr· cm], where d represents the cathode layer width and $(pd)_{norm}$ denotes the normal reduced cathode layer width as per [92]. For an aluminum cathode in air, $(pd)_{norm}$ is about 0.25 [Torr· cm] according to [92], which leads to an estimated pd of about 0.09 [Torr· cm]. At a pressure of $p = 0.76$ Torr, this calculation yields a cathode layer width (d) of 1.2 mm, and at a pressure of $p = 9.83$ Torr, d is approximately 100 µm.

Examining the data at a pressure of $p = 1.51$ Torr with greater specificity, the thickness of the cathode layer was determined to be $d = 0.6$ mm. Analysis of Figure 4.32 indicates that a quasi-steady-state distribution of the electric field within the discharge cell was established from $t = 10$ ns. At this time, the voltage across the discharge cell was approximately 20 kV, as shown in Figure 4.31. At the same time, the voltage drop across the main part of the discharge cell, excluding the cathode, was about 10 kV, as evidenced by the observation that at $x = 3.6$ cm, the voltage reached 8 kV and the electric field intensity in the main part of the discharge cell was about 600 V/cm. Consequently, the electric field within the cathode layer is estimated to be about $E \sim 1.7 \cdot 10^5$ V/cm.

This value is sufficient to induce vacuum breakdown in a discharge gap with an aluminum cathode subjected to a pulsed discharge according to [93]. In the following 3 ns, the potential drop across the main part of the discharge column decreased to about 2 kV, while the potential at the high-voltage electrode remained virtually constant (decreasing to 19 kV). This scenario resulted in an increase in the electric field intensity within the cathode layer to approximately $E \sim 2.8 \cdot 10^5$ V/cm and a corresponding increase in the field emission current from the cathode. The escalation of the emission current culminated in a field emission breakdown of the cathode layer and the appearance of a wave of repeated breakdowns, causing a redistribution of potential between the cathode layer and the main body of the discharge column.

As shown in Figure 4.33, two ionization waves were observed at a pressure of 0.76 Torr ($t = 20$ ns and $t = 27$ ns); in the pressure range of $p = 1.51 - 3.02$ Torr and at higher pressures the cathode layer remained steady for the duration of the pulse (Figure 4.32).

The analysis performed reveals that at pressures below 3 Torr, a significant portion of the energy is deposited in the thin cathode layer. In addition, a significant amount of energy is deposited during the ionization wave propagation phase. Observations derived from the data indicate that from a pressure of 3.0 Torr, the majority of the energy is transferred to the gas during the short-circuit/high-current phase, while the energy input to the cathode layer during the entire discharge duration and to the main part of the discharge column during the wave propagation phase becomes minimal. For pressures above 10 Torr, the energy input during the short-circuit phase decreases with increasing pressure; however, even at 18 Torr, it remains three times greater than that deposited during the ionization wave propagation phase.

Previous observations have shown that a portion of the pulse is reflected back to the generator when it hits the discharge cell. Subsequently, this reflected pulse bounces off the generator and returns to the discharge cell. The discharge induced by these secondary pulses instantaneously forms the short-circuit phase because of high residual conductivity of the plasma. Notably, at pressures above 0.76 Torr, no cathode voltage drop was detected within these pulses, indicating a uniform distribution of energy contribution throughout these secondary discharge events.

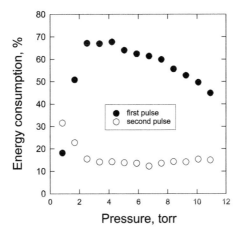

FIGURE 4.34 The energy input in the discharge for the first and the second pulses in the oxidized methane-air mixture in dependence on the initial pressure of the mixture.

The relationship between energy contribution and gas pressure in the discharge cell is shown in Figure 4.34. It is evident that at a pressure of 0.76 Torr, the energy contribution within the second pulse exceeds that of the first pulse. As the pressure escalates, the energy contribution experiences an increase, peaking at approximately 3.5 Torr. Conversely, the energy contribution attributed to the second pulse decreases and stabilizes at a constant value above $p = 3.5$ Torr. The energy imparted by the third and subsequent pulses is less than 2%. It is noteworthy that the increased energy contribution observed in the second pulse serves to mitigate the non-uniformity of the energy distribution at lower pressures.

Therefore, over the pressure range studied (0.76–10.6 Torr), the predominant energy input into the gas occurs during the short-circuit phase within either the first or second pulse. At the same time, the discharge non-uniformity induced by the cathode layer becomes insignificant. This feature facilitates the measurement of local electric fields and currents at the point of maximum energy contribution using only a back-current shunt, a methodology of particular relevance to the study of chemically active mixtures where mixture composition changes affect electrodynamic properties. Electrical measurements performed on oxidized stoichiometric mixtures ranging from ethane to hexane with either oxygen or air yielded results consistent with those described above.

Figure 4.35 shows the energy contributions in all the mixtures studied at the beginning of the discharge and after the oxidation process, with the total energy per pulse being approximately 60 mJ. Analysis of the figure shows that in the initial phase of oxidation, the energy contribution in any mixture containing oxygen at pressures above 3 Torr is consistently greater than that in mixtures with air. As the pressure escalates, this difference increases, peaking at a 12% difference at $p = 10.6$ Torr. Among the processed mixtures, the variations in energy contributions between different mixtures are minimal, indicating that even the addition of nitrogen has a negligible effect. Notably, the energy contribution in the methane-oxygen processed mixture is slightly higher compared to the other mixtures.

4.3.4 Kinetics of Alkanes Oxidation in Nanosecond Discharge

Emission spectra from discharges in both pure methane and stoichiometric methane-oxygen mixtures are shown in Figure 4.36. The spectrum from pure methane shows the repulsive continuum of hydrogen associated with the $H_2(a^3\Pi - b^3\Sigma)$ transition. The emission band of $CH(A^2\Delta \rightarrow X^2\Pi)$, corresponding to the $v' = 0 \rightarrow v'' = 0$ transition, is clearly visible. Molecular bands in the visible and

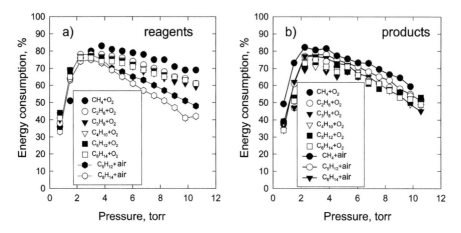

FIGURE 4.35 The sum of the energy input in the first and the second high-voltage pulses: (a) in the initial mixture, (b) in the mixture transformed by the discharge.

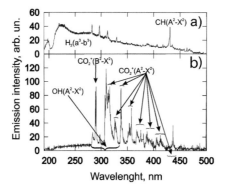

FIGURE 4.36 Spectrum of the discharge emission (a) in pure methane and (b) in the oxidized methane-oxygen mixture. The pressure is 4 Torr.

near-ultraviolet spectra appearing in the processed methane-oxygen mixture are mainly attributed to the $CO_2^+(A^2\Pi)$ molecule, with bands of the $OH(A^2\Sigma \rightarrow X^2\Pi)$ radical also prominently visible. In addition, some weak bands of the CO Angstrom system $(CO(B^1\Sigma \rightarrow A^1\Pi))$ have been detected and observed.

Figure 4.36 shows that the $CH(A^2\Delta \rightarrow X^2\Pi)$ band emission in methane overlaps significantly with the $CO_2^+(A^2\Pi \rightarrow X^2\Pi)$ emission in the mixture. Therefore, during the initial processing of the methane-oxygen mixture, it is crucial to take measures to differentiate these bands. The $OH(A^2\Sigma \rightarrow X^2\Pi)$ spectrum also overlaps with the CO_2^+ emission spectrum. In alkane-air mixtures, strong and pronounced bands of the first negative and the first and second positive nitrogen systems are observed within the spectrum, partially overlapping with the emission spectra of CH, CO_2^+, and OH bands. Within the Angstrom system, the band at 518.6 nm remains distinct and is not obscured by other bands. In addition, faint lines from the Balmer series of hydrogen were detected in all experimental setups.

The differentiation of overlapping spectral bands was accomplished by time-resolved experiments with nanosecond resolution, as mentioned above. Figure 4.37 illustrates time-resolved emission signals at 430 nm recorded at both the beginning and end of the oxidation process. Two emission peaks corresponding to two high-voltage pulses are clearly observed. The emission decay is accurately modeled by a composite of two exponential functions, each characterized by significantly

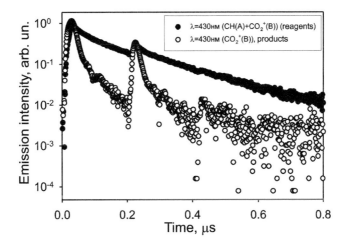

FIGURE 4.37 Time-resolved intensities of the emission on the wavelength $\lambda=430\,\text{nm}$ in the initial (CH_4+O_2) and the corresponded processed mixtures. The initial pressure is 3.78 Torr.

different time constants. The longer lifetime of the $CO_2^+\left(A^2\Pi\right)$ state ($\tau=124$ ns, as reported in [94]) is significantly larger than that of the $CH\left(A^2\Delta\right)$ state ($\tau=600$ ns, according to [95]). Thus, the longer characteristic time is associated with the $CH\left(A^2\Delta\right)$ state emission, while the shorter time constant is associated with the $CO_2^+\left(B^2\Sigma\right)$ state emission. It is clear from this figure that the emission intensity corresponding to the $CH\left(A^2\Delta \rightarrow X^2\Pi\right)$ transition decreases significantly during the oxidation process, while the emission intensity associated with the $CO_2^+\left(A^2\Pi \rightarrow X^2\Pi\right)$ transition increases significantly. The lifetimes and quenching constants determined for the "fast" emission closely match those for the CO_2^+ emission at the 290 nm wavelength. Consequently, the integrated time profile of the $CH\left(A^2\Delta \rightarrow X^2\Pi\right)$ emission intensity can be derived by subtracting the normalized emission curve at 290 nm from the measured emission curve at 430 nm. The normalization factor was adjusted to ensure congruence between the time-resolved emission signal at 290 nm and the "fast" emission signal at 430 nm.

Subsequent analysis allowed the delineation of an optimal wavelength interval for measuring $OH\left(A^2\Sigma\right)$ emissions, within which interference from both the nitrogen second negative system and carbon dioxide bands is minimal. In particular, the overlap of $CO_2^+(B^2\Sigma \rightarrow X^2\Pi)$ bands with $OH\left(A^2\Sigma \rightarrow X^2\Pi\right)$ bands corresponding to transitions from ($v'=1-3$) vibrational states is considered negligible at the 290 nm wavelength.

These molecular bands were consistently detected in all stoichiometric mixtures studied, regardless of the inclusion of oxygen or air, and were used as markers to monitor the hydrocarbon oxidation process in the discharge.

Incorporating the quenching constants and lifetimes of specific states determined by time-resolved experiments (a methodology detailed in [96]), the excitation rates of the emitting states were reconstructed, taking into account the cumulative duration of the discharge activity. It is noteworthy that the quenching constants for the ions $CO_2^+\left(B^2\Sigma\right)$, $CO_2^+\left(A^2\Pi\right)$ and $N_2^+\left(B^2\Sigma\right)$ by reactants and products show negligible variations. In contrast, the quenching constants for the $CO\left(B^1\Sigma\right)$, $CH\left(A^2\Delta\right)$ and $OH\left(A^2\Sigma\right)$ states are about 30%–50% higher in the products than in the reactants. This discrepancy is attributed to the formation of water in the products.

In stoichiometric alkane-air mixtures, the excitation rates for the $N_2\left(C^3\Pi, v'=1\right)$, $N_2\left(B^3\Pi, v'=6\right)$, and $N_2^+\left(B^2\Sigma, v'=0\right)$ states were determined from the time-resolved emission of the transitions $N_2\left(C^3\Pi, v'=1 \rightarrow B^3\Pi, v''=7\right)$, $N_2\left(B^3\Pi, v'=6 \rightarrow A^3\Sigma, v''=3\right)$, and $N_2^+\left(B^3\Sigma, v'=0 \rightarrow X^2\Sigma, v''=2\right)$. These rates were analyzed both at the beginning of the discharge and at the end of the oxidation, integrating these rates over the duration of the discharge.

It was observed that the integral excitation rates for the $N_2(C^3\Pi)$ and $N_2^+(B^2\Sigma)$ states, from the onset of the discharge to the completion of the oxidation, showed a negligible variation (within 10%). Conversely, the excitation rate for the $N_2^+(B^2\Sigma, v'=0)$ state decreased by about 25% during the oxidation process. According to [96], under similar conditions, the $N_2(C^3\Pi, v'=1)$ and $N_2^+(B^2\Sigma, v'=0)$ states are excited primarily by direct electron impact from the ground state of molecular nitrogen.

Since the nitrogen concentration remains essentially constant throughout the discharge, the integral rate coefficients for $N_2(C^3\Pi, v'=1)$ and $N_2^+(B^2\Sigma, v'=0)$ state excitations vary by 10% and 25%, respectively, during oxidation. It is noted that the $N_2(C^3\Pi, v'=1)$ and $N_2^+(B^2\Sigma, v'=0)$ excitations occur during high-voltage pulses, while the $N_2(B^3\Pi, v'=6)$ excitation occurs predominantly during the discharge afterglow, correlating with the emission from the $N_2(C^3\Pi, v'=1 \to B^3\Pi, v''=7)$ transition. This suggests that the excitation of the $N_2(B^3\Pi, v'=6)$ state results from the deactivation processes of the $N_2(C^3\Pi)$ state.

The excitation thresholds for the $N_2(C^3\Pi, v'=1)$ and $N_2^+(B^2\Sigma, v'=0)$ states by electron impact are about 11.2 and 18.6 eV, respectively. Since the excitation thresholds for all investigated states are below 19 eV, it is concluded that the integral rate coefficients for the excitation of these electronic states over the discharge emission duration do not vary by more than 25% throughout the oxidation process. Furthermore, lower excitation thresholds correlate with smaller variations in the rate coefficients during oxidation.

Consequently, changes in the composition of the mixture during the oxidation process, resulting from changes in quenching and excitation rates, affect the integral intensity of each monitored band by no more than 50%. Therefore, qualitative relationships of concentrations leading to excited state production can be derived from the integral emission intensity. In addition, these changes minimally affect the characteristic times derived from the integral emissions of the $CO(B^1\Sigma \to A^1\Pi)$, $CO_2^+(B^2\Sigma \to X^2\Pi)$, $OH(A^2\Sigma \to X^2\Pi)$ and $CH(A^2\Delta \to X^2\Pi)$ transitions.

The temporal evolution of the integral emission intensity in methane mixtures was accurately modeled using either the exponential decay function $\exp(-t/\tau(p))$ or its complementary function $1-\exp(-t/\tau(p))$, as shown in Figure 4.38. It is important to emphasize that complete oxidation of methane was observed in these experiments, as evidenced by the methane concentration curve $[CH_4](t)$ derived from laser absorption spectroscopy shown in Figure 4.38. The integral emission

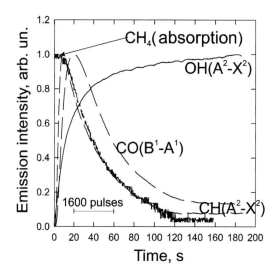

FIGURE 4.38 Integral over time emission intensities of transitions $CH(A^2\Delta \to X^2\Pi)$, $CO(B^1\Sigma \to A^1\Pi)$, $OH(A^2\Sigma \to X^2\Pi)$ and methane concentration, measured by the absorption of the He-Ne laser emission on the wavelength $\lambda=3.3922\mu$ m, in the discharge. The initial pressure of the mixture CH_4+O_2 is 7.6 Torr.

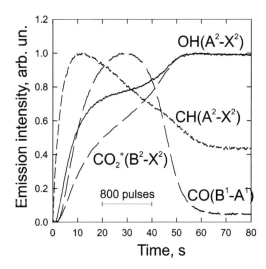

FIGURE 4.39 Integral over time emission intensities of transitions CH($A^2\Delta \to X^2\Pi$), CO($B^1\Sigma \to A^1\Pi$), CO$_2^+$ ($B^2\Sigma \to X^2\Pi$) and OH($A^2\Sigma \to X^2\Pi$) in the discharge. The initial pressure of the mixture $C_2H_6+O_2$ is 7.56 Torr.

of the CH$\left(A^2\Delta \to X^2\Pi\right)$ radical is also shown. It is noteworthy that from the onset of the discharge, the emission intensity curve of CH$\left(A^2\Delta \to X^2\Pi\right)$ closely follows the methane concentration profile. This observation suggests that the CH$\left(A^2\Delta \to X^2\Pi\right)$ emission serves as a reliable indicator of the hydrocarbons present in the discharge.

For mixtures of alkanes heavier than methane, the emission dependencies exhibit a more complex form due to the accumulation of intermediates, probably carbon monoxide (CO) and unsaturated hydrocarbons, during the alkane oxidation process and their subsequent further oxidation. An illustration of the integral oxidation of ethane at a pressure of 7.54 Torr is shown in Figure 4.39. It is clear from the figure that the curves cannot be characterized by simple mathematical functions and show significant divergence from each other.

Figure 4.40 shows the time required to achieve complete oxidation, derived from the emissions of different spectral bands, for alkane mixtures ranging from ethane to hexane when reacted with oxygen. The completion time of the oxidation process is determined at the point where the photomultiplier signal reaches a level of $0.95I_\infty$ for wavelengths (290 and 307.8 nm) associated with the OH$\left(A^2\Sigma \to X^2\Pi\right)$ and CO$\left(B^1\Sigma \to A^1\Pi\right)$ transitions, and a level of $1.05I_\infty$ for wavelengths (430 and 518.6 nm) corresponding to the CO$_2^+$($B^2\Sigma \to X^2\Pi$) and CH$\left(A^2\Delta \to X^2\Pi\right)$ transitions.

It should be emphasized that the oxidation times deduced from the emission of different bands under identical experimental conditions show remarkable disparities, due to the different effects of excitation and quenching processes on different states during the transformation of the mixture. However, the oxidation times derived from the emission of the same band in different mixtures are remarkably consistent and within the bounds of experimental accuracy. Given the minimal experimental scatter observed for the times determined from the CO$\left(B^1\Sigma \to A^1\Pi\right)$ band emission, these times are designated as the definitive duration for the complete oxidation of alkanes.

Figure 4.41 shows the times required for the complete oxidation of various alkanes. It is evident that the time required for complete oxidation of methane is about twice as long as that for other alkanes, both in mixtures with oxygen and in mixtures with air. The oxidation times for the other alkanes do not differ significantly within the limits of experimental dispersion. In mixtures with air, the times for complete oxidation are at least half of those observed for equivalent mixtures with oxygen. Considering that the total amount of alkanes in stoichiometric mixtures with air is less than that in mixtures with oxygen, the oxidation rates of alkanes in mixtures with air are observed to be twice as fast as those in mixtures with oxygen.

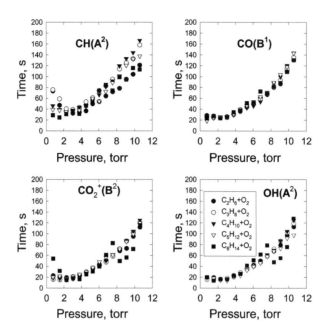

FIGURE 4.40 Dependencies of alkanes full oxidation times on initial pressures of stoichiometric mixtures, obtained from CH($A^2\Delta \rightarrow X^2\Pi$), CO($B^1\Sigma \rightarrow A^1\Pi$), CO_2^+ ($B^2\Sigma \rightarrow X^2\Pi$), OH($A^2\Sigma \rightarrow X^2\Pi$) bands emissions.

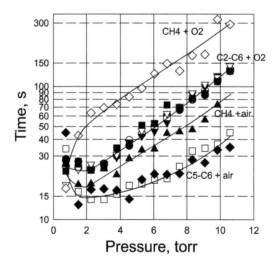

FIGURE 4.41 Dependencies of alkanes full oxidation times on initial pressures of stoichiometric mixtures, obtained from CO($B^1\Sigma \rightarrow A^1\Pi$) band emission. The data for the stoichiometric mixtures of oxygen with alkanes from ethane to hexane are denoted as "C2–C6+O_2", and data for the stoichiometric mixtures of oxygen with alkanes from ethane to hexane are denoted as "C5–C6+air" (see Tables 4.1 and 4.4).

Therefore, it has been found that stoichiometric mixtures of alkanes with air and oxygen subjected to a nanosecond homogeneous discharge at ambient temperature undergo complete oxidation. Oxidation occurs within the same time for all alkanes in stoichiometric mixtures starting with ethane. Methane, however, oxidizes at half the rate of the other alkanes studied.

4.3.4.1 C_4H_{10} Oxidation in Lean Mixtures with Oxygen

As mentioned above, the oxidation of alkanes starting from ethane, when subjected to nanosecond discharge, occurs over identical time under equivalent experimental conditions (initial pressure, energy input). This observation can be most rationally attributed to the relatively small effect of the excited and charged particles generated by electron impact on fuel molecules compared to their effect on oxygen molecules, intermediates, and water molecules. It is known that the electron impact dissociation thresholds for alkanes (from methane to propane) decrease by about 1 eV for each additional carbon atom. At the same time, the steric factor increases proportionally to the number of carbon atoms in alkane molecules, while the concentration of an alkane in a stoichiometric mixture decreases inversely to the number of carbon atoms. Therefore, it might be expected that changes in stoichiometric mixtures from lighter to heavier alkanes would result in a compensation between the change in alkane concentration and the change in steric factor. However, as the number of carbon atoms in an alkane molecule increases, the dissociation threshold due to electron impact decreases. This theoretically suggests a faster oxidation rate for heavier alkanes compared to lighter ones, a phenomenon not observed experimentally. This discrepancy invites consideration of possible explanations for this unexpected result.

1. Heavy alkanes undergo rapid decomposition under the influence of electron impact, resulting in the formation of an intermediate that is common to all alkanes starting with ethane, which then undergoes slow oxidation in plasma. In this scenario, the oxidation rate of the alkanes is determined by the oxidation rate of this intermediate. Ethane stands out among the hydrocarbons as a potential candidate for this intermediate role, since acetylene (C_2H_2) and ethylene (C_2H_4), along with alcohols and aldehydes, exhibit significantly lower thresholds for electron impact dissociation, making their decay within the discharge an unlikely rate-limiting step in oxidation. It is apparent that the dissociation of heavy alkanes does not result solely in the formation of ethane. This process also produces hydrocarbon radicals, with alkyl radicals being the most likely dissociation products. Although the recombination of these radicals could potentially produce ethane, this mechanism is not considered significant in the presence of oxygen. Instead, these radicals are rapidly converted to alkyl peroxy radicals, which precludes their recombination.
2. Consequently, there is no single hydrocarbon that can assume the role of a stable, unique intermediate. It becomes clear that this role could be attributed to CO, despite the observation that the emission of the $CH(A^2\Delta \rightarrow X^2\Pi)$ radical increases significantly at the onset of the discharge and then gradually decreases throughout the oxidation process. In particular, the intensity of the emission of this band shows minimal variation among different stoichiometric mixtures of alkanes (starting from ethane). Therefore, the hypothesis that heavy alkanes are rapidly oxidized to CO, which is then slowly oxidized to CO_2, is incorrect. Nevertheless, CO remains a major intermediate in the low-temperature oxidation process of alkanes. It is now postulated that the dissociation of alkanes occurs under conditions of a very high reduced electric field, where the average electron energy exceeds the electron impact dissociation thresholds for all alkanes starting from ethane. Under these conditions, the rate of dissociation is minimally affected by the thresholds and is more influenced by the steric factor of the molecule and the concentration of the alkane. Molecules of any heavy alkane are expected to dissociate into radicals, which are then oxidized to water and carbon dioxide.
3. The excitation, dissociation, and ionization of molecular oxygen, along with the potential excitation and dissociation of products (most important is water) and intermediates, dominate the oxidation of alkanes. It is important to note that within stoichiometric mixtures, the variation in the amount of oxygen from one alkane to another is minimal, resulting in the production of a consistent amount of water in all mixtures. The variation in the amount of carbon dioxide produced during the oxidation process in different mixtures is insignificant. For example, the oxygen concentration in a stoichiometric mixture with ethane is approximately 20% lower than that in a mixture with hexane, which corresponds to a proportional decrease in the amount of carbon dioxide produced during the oxidation process. It is also noteworthy that water is more susceptible to dissociation by electron impact

than carbon dioxide, since water has larger absolute cross sections for electron impact processes and significantly lower thresholds; moreover, radicals derived from water have greater reactivity.

The active oxidant particles generated during the discharge react with non-excited hydrocarbon molecules, while the excitation, dissociation, and ionization of the hydrocarbons themselves do not significantly affect the rate of oxidation. In addition, charge and energy transfers from oxygen atoms, molecules, ions, and products and intermediates to the hydrocarbons have minimal effect on the oxidation process. If this were not the case, the hypothesis of a universally slow oxidizing intermediate would have to be reconsidered. Thus, only chemical reactions between non-excited hydrocarbons and excited oxidants are likely to affect the oxidation rate, rendering the first explanation implausible. The second hypothesis, which requires an electron energy greater than 7 eV, also seems unlikely.

However, a number of additional experiments are needed to determine the most accurate explanation. Using the above methods, the oxidation of a lean mixture of butane with oxygen was studied. Figure 4.42 shows the energy input for stoichiometric and selected lean mixtures as a function of mixture pressure. It is apparent that variations in energy input relative to the equivalence ratio do not exceed 13% of the pulse energy, particularly in the range of maximum energy input for the initial mixture. This suggests that the discharge conditions governing the excitation processes do not vary significantly for mixtures with different equivalence ratios from 0 to 1. Figure 4.43 shows the dynamics of the $CO(B^1\Sigma \to A^1\Pi)$ integral emission in the studied mixtures, clearly indicating that a decrease in the equivalence ratio corresponds to a decrease in the time required for complete oxidation. If the duration of the oxidation were determined by the rates of excitation, dissociation and ionization of butane, then the time to complete oxidation would be invariant with respect to the

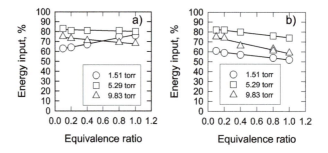

FIGURE 4.42 The sum of the energy input in the first and the second high-voltage pulses: (a) in the initial mixture, (b) in the mixture transformed by the discharge.

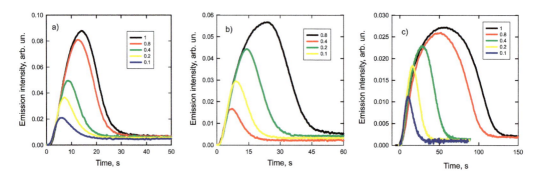

FIGURE 4.43 Integral over time emission intensities of transition $CO(B^1\Sigma \to A^1\Pi)$ in the discharge. The initial pressures of the mixtures are 1.51 Torr for (a), 5.29 Torr for (b) and 9.83 Torr for (c).

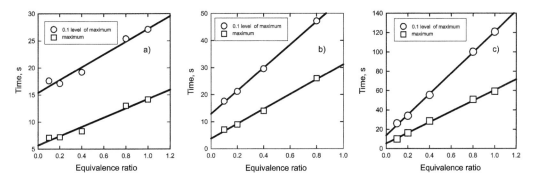

FIGURE 4.44 Time of full oxidation obtained from CO($B^1\Sigma \to A^1\Pi$) band emission and time of CO(B1$\Sigma \to$ A1Π) band emission maximum versus initial pressure of butane with oxygen lean mixtures [88]. The initial pressures of the mixtures are 1.51 Torr for (a), 5.29 Torr for (b) and 9.83 Torr for (c).

equivalence ratio, since a decrease in the equivalence ratio would proportionally reduce both the initial concentration of butane and the rate of oxidation. It is also worth noting that initial parts of the emission curves are remarkably similar.

The time required for complete fuel oxidation, as derived from the CO$\left(B^1\Sigma \to A^1\Pi\right)$ integral emission, and the timing of the CO$\left(B^1\Sigma \to A^1\Pi\right)$ emission peak are shown in Figure 4.44. It is observed that this duration decreases linearly with decreasing equivalence ratio, which is consistent with expectations in scenarios where the excitation, dissociation, and ionization of alkanes by the discharge have minimal effect. In fact, the oxygen content in a stoichiometric mixture varies by only 11% compared to a mixture with an equivalence ratio of 0.1, implying that fluctuations in the concentrations of excited molecules, ions, and oxygen atoms do not exceed 11%. Conversely, the concentration of butane changes tenfold, justifying the expectation that the dependence of the total oxidation time on the equivalence ratio should exhibit a linear correlation with considerable accuracy.

Thus, under the specified experimental conditions, the processes of excitation, ionization, and dissociation of an alkane by electron impact do not significantly affect the rate of alkane oxidation. Instead, the oxidation process is dominated by the generation of ions, excited molecules, and atoms from oxygen, water, carbon dioxide, and possibly from intermediates formed in the early stages of the hydrocarbon oxidation process.

4.3.4.2 $C_{10}H_{22}$ Oxidation in Lean Mixture with Oxygen

Modern fuels used in internal combustion engines are composed of hydrocarbons ranging from C8 to C16; therefore, it is imperative to verify the universality of the conclusions drawn so far by studying an alkane with a carbon chain length exceeding C6. For this purpose, n-decane ($C_{10}H_{22}$) was chosen. At 25°C, the saturated vapor pressure of n-decane is approximately 0.9 Torr, resulting in a maximum pressure of 18 Torr for the stoichiometric mixture with oxygen. The methodology used to prepare the mixture was analogous to that used for hexane and pentane mixtures. A specific amount of decane, less than the value corresponding to its saturated vapor pressure at 25°C in a 10-L volume, was introduced into the vacuum chamber. Oxygen was then added to the volume and the mixture was homogenized using a fan. Despite the application of forced mixing, the concentration of decane increased very gradually, as shown in Figure 4.45, a phenomenon attributed to the remarkably slow evaporation rate of decane under ambient temperature conditions.

The emission intensity of the OH$\left(A^2\Sigma \to X^2\Pi\right)$ radical in the oxidized stoichiometric mixture of butane with oxygen served as the reference for this study. The concentration of decane was monitored by the emission spectrum of the mixture that was compared with spectra of oxidized mixture of butane and oxygen at various equivalence ratios.

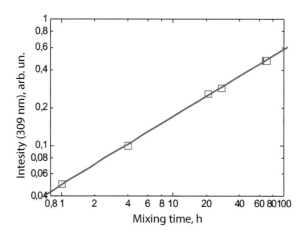

FIGURE 4.45 Intensity of OH($A^2\Sigma \to X^2\Pi$) band emission in the processed mixture of decane with oxygen versus time of mixing. The unit emission intensity level coincides to the emission intensity of OH($A^2\Sigma \to X^2\Pi$) in butane-oxygen stoichiometric mixture. The initial pressure of the mixture $C_{10}H_{22}+O_2$ is 4 Torr.

A detailed examination of the method used to determine the equivalence ratio is presented, with the oxidation of alkanes in a stoichiometric mixture with oxygen described by global reaction:

$$C_nH_{2(n+1)} + (3n+1)/2 O_2 = nCO_2 + (n+1)H_2O \tag{4.17}$$

Let us assume that N is the total number of molecules in the discharge volume, then the number of alkane molecules:

$$[C_nH_{2(n+1)}] = 2/3N/(n+1), \tag{4.18}$$

and quantity of oxygen molecules amounts to

$$[O_2] = N(n+1/3)/(n+1). \tag{4.19}$$

Further, the quantity of hydrogen atoms is

$$[H] = [C_nH_{2(n+1)}]/2/(n+1) = 4/3N, \tag{4.20}$$

and the quantity of carbon atoms is

$$[C] = [C_nH_{2(n+1)}]n = 2/3N/(n+1)n = 2/3N/(1+1/n). \tag{4.21}$$

Then quantity of carbon dioxide molecules in the processed mixture is

$$[CO_2] = 2/3N/(1+1/n), \tag{4.22}$$

and the quantity of water molecules is

$$[H_2O] = [H]/2 = 2/3N. \tag{4.23}$$

Consider the difference in the amount of carbon dioxide molecules in stoichiometric mixtures of butane and decane with oxygen.

$$[CO_2]_{C4} = 2/3N \times 0.8 \tag{4.24}$$

$$[CO_2]_{C10} = 2/3N \times 0.91 \tag{4.25}$$

Here and later the indices $C4$ and $C10$ indicate that a value corresponds to butane and decane mixtures with oxygen.

$$[CO_2]_{C10}/[CO_2]_{C4} = 1.14 \quad (4.26)$$

Then total pressure in oxidized mixtures is

$$p_{C4} = 2/3N \times 1.8 = 1.2 p_{in} \quad (4.27)$$

$$p_{C10} = 2/3N \times 1.91 = 1.27 \times p_{in} \quad (4.28)$$

$$P_{C10}/p_{C4} = 1.06 \quad (4.29)$$

where p_{in} is the pressure of the initial mixture.

The experiments revealed that the quenching rates of CO_2^+ are minimally affected by the type of quencher, leading to the observation that the quenching rate varies directly with pressure changes. Consequently, the emission intensities of the CO_2^+ bands in the mixture processed with decane are reduced by no more than 6% compared to those in the butane mixture. Given the small variations in energy input in different mixtures, it can be inferred that the excitation rates of the CO_2^+ bands are proportional to the CO_2 concentration.

Thus, in the mixture with decane, these rates show an increase of 14% relative to the butane mixture. Therefore, the emission intensity of a CO_2^+ band in the decane-containing mixture is expected to exceed that in the butane mixture by 8%–14%, depending on the initial gas pressure. It is well known that water acts as a more effective quencher than carbon dioxide, and that electron energy losses in electron collisions with water are significantly higher than with carbon dioxide.

In experiments with stoichiometric alkane-oxygen mixtures, it has been shown that the quenching rates of $OH(A^2\Sigma \rightarrow X^2\Pi)$ increase substantially (by 50%) throughout the mixture transformation under nanosecond discharge and show little dependence on the alkane type. The transition from butane-oxygen to decane-oxygen mixtures should minimally affect the change in emission intensity due to the equivalent amount of water in the oxidized mixtures, which plays a key role in both the quenching processes and the shaping of the electron energy distribution function. The emission intensity of OH during the conversion of stoichiometric mixtures of oxygen with ethane, propane, and butane is shown in Figure 4.46. It can be seen that the steady-state values of the emission intensity (after completion of the oxidation process) are practically identical for all mixtures.

The relationship between the emission intensity of the OH and CO_2^+ bands in processed butane-oxygen mixtures and the equivalence ratio, used to determine the concentration of decane,

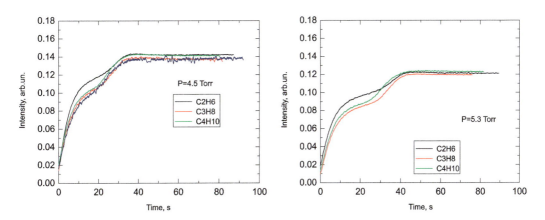

FIGURE 4.46 Intensities of $OH(A^2\Sigma \rightarrow X^2\Pi)$ band emission in mixtures of oxygen with different alkanes.

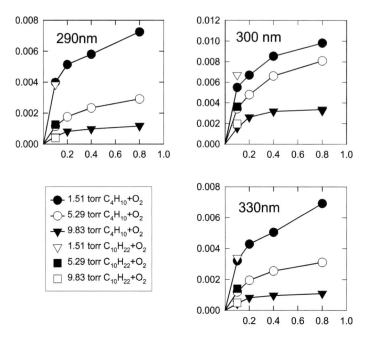

FIGURE 4.47 Integral over time emission intensities of transitions CO_2^+ ($B^2\Sigma \to X^2\Pi$) (290 nm), OH($A^2\Sigma \to X^2\Pi$) (306 nm), CO_2^+ ($A^2\Pi \to X^2\Pi$) (330 nm) in the discharge versus the equivalence ratio of the mixture. All points corresponded to the decane mixture are demonstrated at 0.1 equivalence ratio point [88].

TABLE 4.14
Equivalence Ratio

Pressure, Torr	1.51	5.29	9.83
$\lambda = 290$ nm	0.096	0.116	0.08
$\lambda = 306$ nm	0.205	0.114	0.135
$\lambda = 330$ nm	0.12	0.129	0.113

is shown in Figure 4.47. It is observed that the emission intensities corresponding to all the bands analyzed show a marked nonlinear dependence on the equivalence ratio. This phenomenon is attributed to the effect of water concentration on the electron energy distribution function, specifically the variation of the excitation constants of these states with changes in water concentration. In addition, this figure includes the emission intensities of processed decane-oxygen mixtures at identical wavelengths. The emission intensities for different bands were interpolated between measured points and linearly extrapolated to zero. These relationships facilitated the determination of the equivalence ratio for the decane mixture. The equivalence ratio values derived from measurements at different wavelengths and pressures are detailed in Table 4.14. The concentration of n-decane determined by this method reached 0.12 after 3 days of mixing, indicating that approximately 30 days are required to achieve a stoichiometric mixture of n-decane with oxygen.

Consequently, the preparation of a stoichiometric mixture of n-decane with oxygen at room temperature is considered impractical due to the length of time required for its preparation. Therefore, the oxidation process was studied in a mixture with a stoichiometric ratio of 0.12±0.01. Within this mixture, both the energy contribution and the $CO\left(B^1\Sigma \to A^1\Pi\right)$ emission intensity were monitored. The variation of the energy contribution with pressure is shown in Figure 4.48 for both the decane

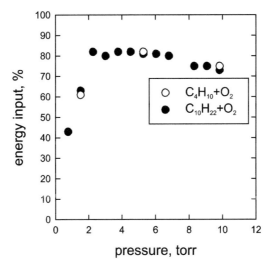

FIGURE 4.48 Energy input for butane and decane mixtures with oxygen. Equivalence ratio of the butane mixture is 0.1 and one of the decane mixture is 0.12±0.01.

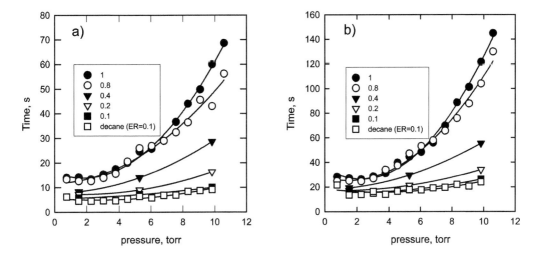

FIGURE 4.49 Time to reach the maximum $CO(B^1\Sigma \rightarrow A^1\Pi)$ emission — (a) and the time of full oxidation — (b) [88].

and butane mixtures. It is important to emphasize that the discrepancy between the energy contributions in the initial and processed mixtures is insignificant, due to the fact that oxygen is the primary component of these mixtures.

Observations indicate that the energy contribution across different mixtures remains the same within the limits of experimental data variability. Consequently, it can be inferred that gas excitation occurs with equivalent intensity in all mixtures.

Figure 4.49 shows the times required to achieve complete oxidation as determined by the emission intensities of the $CO(B^1\Sigma \rightarrow A^1\Pi)$ bands for both butane-oxygen mixtures and the decane-oxygen mixture. It is evident that the time required for the complete oxidation of n-decane is approximately the same as the time required for the complete oxidation of butane in a mixture with an equivalence ratio of 0.1. Therefore, the results obtained for C2 to C6 alkane mixtures can be extrapolated to heavier alkanes.

4.4 KINETIC MODEL OF CH_4, C_2H_6, C_2H_5OH OXIDATION BY NANOSECOND DISCHARGE

Relatively small molecules were chosen for the analysis of hydrocarbon oxidation kinetics. This choice facilitated the development of a concise kinetic scheme that accounts for the most important experimental observations [88]. The initial generation of radicals is assumed to be initiated by electron impact. Hereinafter, "primary radicals" refer to those radicals generated by electron impact from stable entities, with the most significant radicals being those derived from reactants and products, while those derived from stable intermediates are considered less critical.

It is recognized that radicals are generated by the excitation of repulsive or pre-dissociated electronic states. It is also recognized that almost all polyatomic molecules, with the exception of certain positive triatomic ions, do not emit radiation in the visible and ultraviolet spectrum, indicating that they possess only metastable, repulsive, or significantly pre-dissociated electronic states. It is important to note that the metastable states of almost all polyatomic molecules are predominantly pre-dissociated or repulsive, especially when the energies of these states exceed the first dissociation limit of the molecule [97].

It is pertinent to recognize that electron collisions result in the formation of entities other than uncharged radicals; positive and negative ions, as well as electronically and vibrationally excited particles, are also produced. Because the discharge operates in a pulsed-periodic mode, short intervals (~ 30 ns) of ion production within the discharge are interspersed with longer periods (~ 30 ms) of ion transformation and dissipation during the discharge afterglow [88]. The variety of positive ions produced in the discharge can be diverse; however, a significant fraction of the ions originate from stable molecules.

4.4.1 ELECTRON IMPACT PROCESSES IN THE DISCHARGE

In the previous section, the oxidation of alkanes was studied using experimental methods that included the measurement of electrical power and reduced electric fields in the discharge. As previously shown, a significant portion of the energy is uniformly distributed throughout the gas during the short-circuit phase of the discharge development. To evaluate the efficiency of radicals and ions production, the energy distribution during the discharge phase was calculated using the two-term approximation of Boltzmann equation [98]. The results of these calculations are shown in Figures 4.50 and 4.51, where the reduced electric fields measured in the experiments are shown. These field values correspond to the temporal peak of the energy input. It is clear from the experimental conditions (E/n values) that processes such as dissociative attachment, excitation of molecular oxygen metastables, and vibrational excitation of molecules by electron impact are negligible. The predominant excitation mechanisms are dissociation and ionization of molecules, with over 80% of the energy allocated to these processes.

It is pertinent to note that virtually all of the experiments were conducted at pressures below 11 Torr, so the contribution of vibrational excitation, which accounts for about 20% of the energy input at highest pressures in the reactive mixture, can be disregarded. Another reason to neglect the influence of vibrations is a fast VRT relaxation in mixtures with hydrocarbons due to low vibrational quantum energy and intense mixing of vibrational modes at higher levels.

In mixtures that have undergone partial oxidation, a substantial portion of the energy input could be devoted to the excitation of various degrees of freedom of stable intermediates. However, for the initial analysis, the excitation processes of a number of intermediates are not considered. In the following, we present a kinetic model [88] that allows the identification of key intermediates.

It is found that the main active species generated in the discharge are positive ions and radicals produced by electron impact from stable products (Figure 4.51).

4.4.2 POSITIVE IONS KINETICS

When studying the initial phase of the oxidation process, where the reactants are predominantly present, two main types of positive ions are produced within the discharge: O_2^+ and the initial

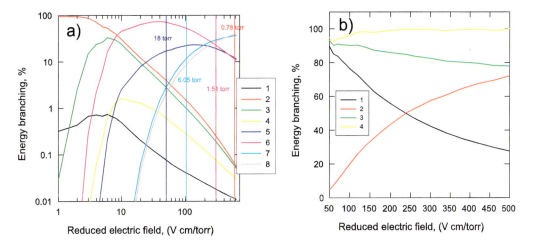

FIGURE 4.50 Energy branching in the discharge (stoichiometric mixture C_2H_6–O_2). (a) The curves indicate the percentage of energy directed to: (1) elastic collisions, (2) vibrational excitation (both C_2H_5OH and O_2), (3) singlet molecular oxygen excitation, (4) dissociative attachment (C_2H_6 and O_2), (5) C_2H_6 dissociation, (6) O_2 dissociation, (7) C_2H_6 ionization, (8) O_2 ionization. The vertical lines show the experimentally measured reduced electric field values corresponding to the maximum power in the discharge. (b) The curves indicate the percentage of energy directed to: (1) total (C_2H_6 and O_2) dissociation, (2) total (C_2H_6 and O_2) ionization, (3) total dissociation+0.7 total ionization, (4) total dissociation+total ionization.

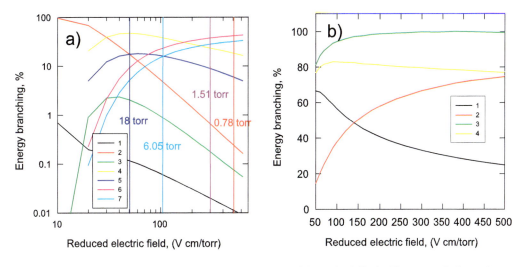

FIGURE 4.51 Energy branching in the discharge (mixture CO_2:H_2O=2:3). (a) The curves indicate the percentage of energy directed to: (1) elastic collisions, (2) vibrational collisions (both CO_2 and H_2O), (3) singlet molecular oxygen (both states), (4) dissociative attachment (CO_2 and H_2O), (5) CO_2 dissociation, (6) H_2O dissociation, (7) H_2O ionization, (8) CO_2 ionization. Vertical lines show measured experimentally reduced electric field values, which are corresponded to maximum power in the discharge. (b) The curves indicate the percentage of energy directed to: (1) total (CO_2 and H_2O) dissociation, (2) total (CO_2 and H_2O) ionization, (3) total dissociation+0.7 total ionization, and (4) total dissociation+total ionization [88].

hydrocarbon ion. The proton affinity of molecular oxygen is lower than the dissociation threshold of any of the above hydrogen-containing molecules, which means that under ambient temperature conditions the ions can be lost primarily by electron transfer reactions.

In the case of methane or hydrogen, this reaction is significantly endothermic, resulting in a negligibly small rate coefficient. Consequently, in hydrogen or methane-oxygen mixtures, O_2^+ can be converted to O_4^+. The rate of this conversion process is about $1.6 \cdot 10^{-13}$ cm^{3}s^{-1}. At an oxygen pressure of 3 Torr, this rate corresponds to a characteristic time of about 6 ms. Conversely, the rate of dissociative recombination is $2 \cdot 10^{-7}$ cm^{3}s^{-1}, and given a typical electron density for this type of discharge (about 10^{12} cm^{-3}) [90], the characteristic time for recombination is about 5 μs. It is evident that O_2^+ in mixtures with hydrogen or methane undergoes predominantly recombination without conversion, resulting in the formation of oxygen atoms $O(^3P)$ and $O(^1D)$.

The CH_4^+ ion has sufficient proton affinity to dissociate a methane molecule, thereby facilitating the proton transfer reaction at a much higher rate of $1.5 \cdot 10^{-9}$ cm^{3}s^{-1}. The electron transfer reaction from O_2 to CH_4^+ is also exothermic, with a significant reaction rate (about 10^{-9} cm^{3}s^{-1}). Therefore, CH_4^+ is rapidly converted to O_2^+ and CH_5^+ within tens of nanoseconds. The dominant dissociative recombination pathway for CH_5^+ yields CH_3 and 2H (70%). In hydrogen-oxygen mixtures, the main mechanism for the loss of H_2^+ involves charge transfer to molecular oxygen, since the proton affinity energy of hydrogen is substantially higher than its dissociation energy.

In the case of an ethane-oxygen mixture, the primary positive ions produced in the discharge are O_2^+ and $C_2H_6^+$. In these mixtures, a significant fraction of the O_2^+ ions are lost by electron transfer from C_2H_6, since the ionization potential of ethane is substantially lower than that of oxygen. The proton transfer process between $C_2H_6^+$ and C_2H_6 is exothermic and therefore fast. The product of this reaction, $C_2H_7^+$, has a dissociative recombination rate coefficient significantly higher (about 10^{-6} cm^{3}s^{-1}) than either of the initial ions. The ion recombination time corresponding to this process is about 1 μs. The recombination products can include C_2H_5 and 2H, indicating that a single ionization event in these mixtures can yield up to four radicals ($2C_2H_5$ and 2H).

For the ethyl alcohol–oxygen mixture, the predominant positive ions produced in the discharge are O_2^+ and $C_2H_5OH^+$. Similar to the ethane mixture, most of the O_2^+ ions are lost by electron transfer from C_2H_5OH. C_2H_5OH has the highest proton affinity, as shown by the data in Table 4.15. This implies that not only $C_2H_5OH^+$, but any molecular ion containing hydrogen, upon collision with C_2H_5OH, is susceptible to conversion to $H^+C_2H_5OH$ with a rate coefficient of about 10^{-9} cm^{3}s^{-1} (with a characteristic time on the order of tens of nanoseconds). Therefore, as long as the alcohol concentration remains sufficiently high, the primary positive ion is $H^+C_2H_5OH$. This ion can undergo molecular addition of either alcohol or water, but does not participate in proton or electron transfer in collisions with stable molecules. Beyond a certain threshold of molecules added, the growth of a cluster ion does not alter the products of its dissociative recombination, although the recombination constant may increase. The only exothermic channel of recombination of these ions is:

$$H^+(C_2H_5OH, H_2O)n + e \rightarrow H + n(C_2H_5OH, H_2O)$$

TABLE 4.15
Stable Molecules Ionization Potential, Affinity to Proton and R-H Bond Strength

Molecule	H_2	CO	CO_2	CH_4	H_2O	O_2
I, eV	15.43	14.01	13.79	12.98	12.61	12.08
P_a, eV	4.4	6.15	5.5	5.4	7.23	4.1
E_H, eV	4.478	–	–	4.51	5.12	–
Molecule	C_2H_6	C_2H_2	CH_3OH	C_2H_4	C_2H_5OH	H_2O_2
I, eV	11.50	11.41	10.85	10.51	10.47	10.47
P_a, eV	6.9	6.71	7.89	7.12	8.1	7.06
E_H, eV	4.3	4.9	4.47	4.6	4.4	3.8

TABLE 4.16
Main Loss Processes of $H^+(H_2O)_n$ Ions

Index	Process	Reaction		
c	Clusterization	$H^+(H_2O)_n + H_2O$	→	$H^+(H_2O)_{n+1}$
d	Decay	$H^+(H_2O)_n$	→	$H^+(H_2O)_{n-1} + H_2O$
r	Recombination	$H^+(H_2O)_n + e$	→	$H + nH_2O$

Thus, in the scenario described, about two radicals are produced per ionization event after about 1 µs – C_2H_5O resulting from the proton transfer from C_2H_5OH to $C_2H_5OH^+$ (or other radicals resulting from the proton transfer from another stable hydrogen-containing molecule), and H resulting from the dissociative recombination of ions. When a relatively small ion undergoes recombination, this process can yield more than two radicals per ionization event. However, dissociative recombination emerges as a critical mechanism that provides an initial concentration of radicals. In this reaction, the production of two or more radicals depends on the dynamics between the recombination and clusterization rates. While the ion kinetics in C_2H_5OH are not fully understood, it is plausible to suggest a similarity to the kinetics of water ions, as discussed below (Table 4.16).

The production of ions in oxidized mixtures was also considered in [88]. These mixtures consist almost only of carbon dioxide and water vapor. The main ions produced in the discharge are H_2O^+ and CO_2^+.

As described in Table 4.15, CO_2^+ can be converted to CO_2H^+ and H_3O^+ via electron and proton transfer reactions. This conversion is expected to occur rapidly (with an overall rate coefficient of about $2 \cdot 10^{-9}$ cm^{3}s^{-1}) due to the exothermic nature of both pathways. Given the significantly lower proton affinity energy of CO_2, CO_2H^+ is likely to transform into H_3O^+ within 1–2 collisions with H_2O. As a result, H_3O^+ emerges as the dominant positive ion tens of nanoseconds after the discharge stops. At ambient temperature and pressures of a few Torr or higher, this ion transitions to $H^+(H_2O)_2$ with a rate coefficient of $1 \cdot 10^{-10}$ cm^{3}s^{-1} (with a typical time scale of 100 ns). This ion further converts to $H^+(H_2O)_3$ with a nearly identical rate coefficient. The dissociative recombination rate coefficients for $H^+(H_2O)_2$ and $H^+(H_2O)_3$ are an order of magnitude larger than those for H_3O^+ ($3 \cdot 10^{-7}$ cm^{3}s^{-1}, [99]).

Table 4.17 lists the rate coefficients and characteristic times for various processes involving $H^+(H_2O)_n$ cluster ions, calculated for pressures at the lower and upper bounds of the investigated pressure range, at room temperature. These figures correlate with the processes identified in (Table 4.16). It is apparent that within the pressure range studied, H_3O^+ ions efficiently evolve into various cluster ions prior to recombination. The primary ions with sufficient lifetime for recombination under these conditions are $H^+(H_2O)_4$, $H^+(H_2O)_5$, and $H^+(H_2O)_6$ at pressures of 1 and 10 Torr, respectively.

These ions have only one endothermic channel of recombination with electrons:

$$H^+(H_2O)_n + e \rightarrow H + nH_2O$$

In oxidized mixtures, ionization results primarily in the formation of two products: H and OH. OH radicals are formed by proton transfer reactions, while H atoms are formed by dissociative recombination. For a typical electron density of $2 \cdot 10^{12}$ cm^{-3}, the characteristic recombination time is about 200 ns.

These conversion processes are remarkably fast, due to the significant permanent dipole moment and high polarization coefficient of water molecules. These properties ensure high rates of ion-molecular reactions. It is also worth noting that radicals also exhibit significant polarization and dipole moments. Considering ion-radical reactions under the given conditions, the radical concentration

TABLE 4.17
Kinetics of $H^+(H_2O)_n$ Ions

Ion	k^c	T_c	k^d	T_d	k^r	T_r
			Pressure is 1 Torr.			
$H^+(H_2O)_2$	$2.0 \cdot 10^{-10}$	$1.6 \cdot 10^{-7}$	1.7	0.6	$2.2 \cdot 10^{-6}$	$4.5 \cdot 10^{-7}$
$H^+(H_2O)_3$	$7.3 \cdot 10^{-11}$	$4.6 \cdot 10^{-7}$	440	$2.3 \cdot 10^{-3}$	$3.8 \cdot 10^{-6}$	$2.6 \cdot 10^{-7}$
$H^+(H_2O)_4$	$1.0 \cdot 10^{-11}$	$3.4 \cdot 10^{-6}$	$1.8 \cdot 10^{5}$	$5.5 \cdot 10^{-6}$	$4.9 \cdot 10^{-6}$	$2.0 \cdot 10^{-7}$
$H^+(H_2O)_5$	$1.2 \cdot 10^{-12}$	$2.8 \cdot 10^{-5}$	$5.8 \cdot 10^{4}$	$1.7 \cdot 10^{-5}$	$6.0 \cdot 10^{-6}$	$1.7 \cdot 10^{-7}$
$H^+(H_2O)_6$	$1.3 \cdot 10^{-13}$	$2.5 \cdot 10^{-4}$	$7.9 \cdot 10^{5}$	$1.3 \cdot 10^{-6}$	$> 6.0 \cdot 10^{-6}$	$< 1.7 \cdot 10^{-7}$
			Pressure is 10 Torr.			
$H^+(H_2O)_2$	$1.2 \cdot 10^{-9}$	$2.8 \cdot 10^{-9}$	10	0.1	$2.2 \cdot 10^{-6}$	$4.5 \cdot 10^{-7}$
$H^+(H_2O)_3$	$5.7 \cdot 10^{-10}$	$4.6 \cdot 10^{-9}$	3450	$2.3 \cdot 10^{-3}$	$3.8 \cdot 10^{-6}$	$2.6 \cdot 10^{-7}$
$H^+(H_2O)_4$	$9.5 \cdot 10^{-11}$	$3.4 \cdot 10^{-8}$	$1.3 \cdot 10^{6}$	$5.6 \cdot 10^{-7}$	$4.9 \cdot 10^{-6}$	$2.0 \cdot 10^{-7}$
$H^+(H_2O)_5$	$1.2 \cdot 10^{-11}$	$2.8 \cdot 10^{-7}$	$5.7 \cdot 10^{5}$	$1.7 \cdot 10^{-6}$	$6.0 \cdot 10^{-6}$	$1.7 \cdot 10^{-7}$
$H^+(H_2O)_6$	$1.3 \cdot 10^{-12}$	$2.5 \cdot 10^{-6}$	$7.5 \cdot 10^{6}$	$1.3 \cdot 10^{-7}$	$> 6.0 \cdot 10^{-6}$	$< 1.7 \cdot 10^{-7}$

Temperature is 300 K. Electron density is $2 \cdot 10^{12}$ cm^{-3}.

cannot exceed $1.5 \cdot 10^{14}$ cm^3. This estimate assumes that all of the energy from the electrical pulse is used to dissociate molecules, averaging ~ 8 eV per molecule. The rate coefficients for ion-radical reactions are less than $3 \cdot 10^{-9}$ cm^{3}s^{-1}, implying that the characteristic times for ion-radical reactions extend beyond 2 μs after discharge. Conversely, typical rate coefficients for ion-water reactions are of comparable magnitude, allowing ion-radical reactions to be omitted entirely. This is supported by the observation that the water concentration generated within the first ten discharge pulses is sufficient for this purpose.

Significant concentrations of CO, H_2O_2, or certain hydrocarbons may accumulate in partially-oxidized mixtures. Ions originating from these molecules are likely to be converted by electron or proton transfer reactions to form ions similar to those previously described. The kinetics in such scenarios are a composite of the two previously discussed cases. In addition, it is pertinent to recognize that the kinetics associated with water-ion interactions assume greater significance than those involving ions of the initial reactants due to the superior proton affinity energy of water compared to the reactants (with the exception of C_2H_5OH). In the specific case of C_2H_5OH, the most critical reaction is the water replacement reaction by C_2H_5OH in a $H^+(C_2H_5OH, H_2O)$ ion. This reaction is exothermic, with a rate coefficient of $3 \cdot 10^{-9}$ cm^{3}s^{-1}. The number of H atoms in C_2H_5OH is three times that in water, and the concentrations of water and alcohol in partially-oxidized mixtures equilibrate when only 1/4 of C_2H_5OH is consumed. Consequently, the kinetics of water-ion interactions become more relevant than those of the original reactant ions even at a relatively early stage of ethyl alcohol oxidation.

4.4.3 Production of Radicals by Electron Impact

As discussed earlier, one of the main mechanisms that occurs during discharge is dissociation of molecules induced by electron impact. This dissociation is facilitated by the excitation of repulsive and pre-dissociated electronic states or regions within these states. Table 4.18 outlines the dissociation channels for the most important stable molecules.

TABLE 4.18
Stable Molecules Dissociation Channels

Reaction			Importance of Channel	Reference	Degree of Dissociation
H_2+e	$\rightarrow$	$2H+e$	Main channel	Our estimation	α
CO_2+e	$\rightarrow$	$CO+O(^1D)+e$	Main channel	Our estimation	α
CH_4+e	$\rightarrow$	CH_3+H+e	70% main channel	Our estimation	α
CH_4+e	$\rightarrow$	CH_2+2H+e	30%	Our estimation	0
H_2O+e	$\rightarrow$	$H+OH+e$	Main channel	[100]	α
O_2+e	$\rightarrow$	$O+O(^1D)+e$	Main channel	[101]	α
O_2+e	$\rightarrow$	$O+O+e$	Minor channel	[101]	0
O_2+e	$\rightarrow$	$O+O(^1S)+e$	Minor channel	[101]	0
C_2H_6+e	$\rightarrow$	$2CH_3+e$	50%	Our estimation	$\alpha/2$
C_2H_6+e	$\rightarrow$	C_2H_5+H+e	50%	Our estimation	$\alpha/2$
C_2H_5OH	$\rightarrow$	C_2H_5O+H+e	Main channel	Our estimation	α

Hydrogen dissociation is primarily mediated by the excitation of a group of triplet states, which are connected to the repulsive $H_2(b^3\Sigma)$ state via radiative and collisional transitions. This repulsive state converges to a limit of 4.5 eV. Since the next dissociation limit is about 13 eV, dissociation leading to electronically excited H atoms can be neglected. In the process of VUV absorption, H_2O dissociates into H and OH with a 90% probability [100]. There is no compelling evidence that electron impact significantly alters the dissociation channels. As demonstrated earlier, dissociation of heavier alkanes (from C_2H_6 to C_6H_{14}) by electron impact does not significantly affect the oxidation process. Therefore, to obtain accurate results, it is sufficient to approximate the dissociation pathways of alkanes or to omit them from consideration.

4.4.4 Kinetic Model for Low-Temperature Nonequilibrium Oxidation

It is concluded that the discharge generates a spectrum of radicals that serve as initiators of the oxidation process. A simplified kinetic scheme has been developed for the analysis of hydrocarbon oxidation. To streamline the kinetics, it is postulated that the discharge generates radicals as described in Table 4.18 and identified under the "main channel". For C_2H_6 both identified channels are included. Reactions and species produced in these reactions with rate coefficient s below $1 \cdot 10^{-19}$ cm^{3}s^{-1} at ambient temperature were completely excluded from consideration ([100]).

To determine the initial concentration of primary radicals, the experimentally measured energy input, denoted as E_{exp}, was used. Since the energy input was measured at two different points, namely at the beginning and at the end of the oxidation process, the value of E_{exp} was determined by averaging these measurements. It was postulated that this average energy is assigned to a single dissociation event, with an energy threshold (E_{thr}) set at 8 eV. Furthermore, the main calculations assumed that 70% of the experimentally measured energy input is devoted to dissociation. Under these premises, the average degree of dissociation was defined as follows:

$$\alpha = 0.7 \frac{E_{exp}}{E_{thr} N \times V},$$

where N is the total concentration of stable molecules and V is the discharge volume. Radicals are generated at a frequency of 40 Hz from stable molecules in accordance with the dissociation channels outlined in Table 4.18. In the intervals between pulses, the "excited" mixture undergoes a progression without external perturbations, according to the kinetic scheme presented in Table 4.24.

It is important to note that ionization results in the dissociation of molecules, with a typical average energy consumption of about 12 eV per event. Consequently, it has been postulated that only 70% of the energy contributes to dissociation, as shown in Figures 4.50 and 4.51. Therefore, ionization has the potential to facilitate two dissociation events per ionization act, although, as previously discussed, this scenario occurs predominantly in the initial phase of oxidation. Subsequently, water complex ions undergo recombination, yielding only two radicals.

Thus, reactions 45, 172, 182, 185, 190, and 207 (refer to Table 4.24), which involve the formation of molecules containing more than two carbon atoms, are excluded from consideration. This exclusion is justified given the substantial concentration of molecular oxygen present in the mixtures. The reaction rates in this table are given in $cm^{-3}s^{-1}$ for bimolecular reactions and $cm^{-6}s^{-1}$ for trimolecular reactions, with all rate coefficient s provided for a temperature of 300 K.

4.4.5 Numerical Analysis of Dynamics of Hydrocarbons Oxidation

The kinetic profiles derived from this model are shown in Figure 4.52 for C_2H_6 and C_2H_5OH, and in Figure 4.53 for H_2 and CH_4. A notable difference between these profiles is the extended oxidation time for CH_4 and other substances, coupled with significant accumulations of CO and H_2O_2 during the oxidation of C_2H_6 and C_2H_5OH. It is observed that the oxidation times for C_2H_6 and C_2H_5OH are practically identical. Experimental results have shown that alkanes heavier than methane oxidize within the same time frame, and that the accumulation of stable intermediates significantly influences the oxidation processes. Accordingly, the computational model reflects these key features. The different kinetics of methane or hydrogen compared to ethane or ethyl alcohol are due to the different roles of OH radicals in their respective oxidation processes. OH radicals actively react with C_2H_6 and C_2H_5OH, whereas reactions with H_2 and CH_4 are not favored at room temperature (see Table 4.24, reactions 3, 17, 24, 59). The rate coefficient s for the reactions of OH with C_2H_6 or C_2H_5OH are 2–3 orders of magnitude larger than those with H_2 or CH_4. Consequently, in mixtures containing H_2 or CH_4, OH reacts predominantly with H_2O_2, CO, and radicals, whereas in mixtures of C_2H_6 or C_2H_5OH, a significant pathway for OH consumption involves reactions with the initial reactants.

To evaluate the accuracy of the proposed energy branching in the discharge, the full oxidation times for C_2H_6, C_2H_5OH, H_2, and CH_4 were calculated in stoichiometric mixtures with O_2. Figure 4.54 compare the experimental and calculated times of complete oxidation. The experimental

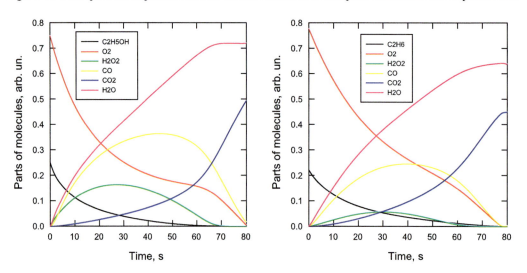

FIGURE 4.52 Calculated mole fraction of the molecules during the oxidation process of C_2H_5OH and C_2H_6 in stoichiometric mixtures with oxygen. The initial pressure of the mixture is 6.8 Torr. Dissociation degrees are corresponded to the 70% of the measured experimentally energy input in the discharge.

Mechanisms of Plasma-Chemical Reactions

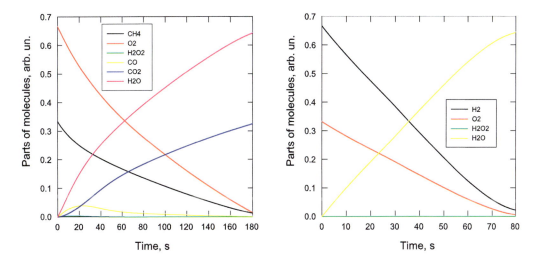

FIGURE 4.53 The calculated mole fractions of the molecules during oxidation process of CH_4 in stoichiometric mixture with oxygen. The initial pressure of the mixture is 6.8 Torr. Dissociation degrees correspond to 70% of the measured experimentally energy input in the discharge.

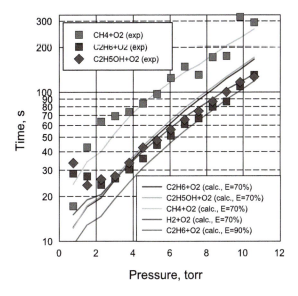

FIGURE 4.54 Calculated and measured oxidation times of C_2H_6, CH_4, H_2, C_2H_5OH molecules during the oxidation process in stoichiometric mixtures with oxygen. The degree of dissociation corresponds to 70% (with the exception of the cyan curve, 90%) of the experimentally measured energy input in the discharge.

durations were determined based on the point at which the $CO(B^1\Sigma \rightarrow A^1\Pi)$ emission intensity decreased to 10% of its peak value. Correspondingly, the calculated times for all hydrocarbons were defined as the interval required for the CO concentration to decrease to 10% of its peak. For H_2, the oxidation duration was determined by the time when the H_2 concentration decreased to 10% of its initial value, reflecting the excitation of the emitting state directly from the ground state of the H_2 molecule by electron impact.

The data show that the oxidation times for C_2H_6 and C_2H_5OH are nearly identical, with the oxidation time for C_2H_5OH slightly exceeding that of C_2H_6, in agreement with experimental observations.

Consequently, the oxidation time for CH_4 is observed to be twice as long as for other substances, again in agreement with experimental results.

At the upper limit of the pressure range studied, the calculated oxidation time exceeds the experimentally determined value, while at the lower limit, the calculated time falls short. Consequently, the oxidation time for C_2H_6 was recalculated under the assumption that 90% of the energy contributes to dissociation.

The calculated oxidation times, in this case, are consistently lower than those observed experimentally, a discrepancy that can be attributed to the differential allocation of energy input between dissociation and ionization under varying pressure conditions. At higher pressures, a significant fraction of the energy input is preferentially channeled directly into dissociation, whereas at lower pressures, a larger fraction of the energy is devoted to ionization. Subsequent dissociative recombination results in the production of different radicals than those produced by direct dissociation. Because of the different activities of these radicals, ionization may be less efficient for the oxidation process than direct dissociation. An alternative explanation may involve significant variability in the distribution among the dissociation channels, which may change substantially with pressure—calculations were performed assuming constant efficiency among the dissociation channels, independent of pressure variations. Thus, the effects of electron collisions with intermediates were not considered. Despite these limitations, the calculations successfully capture the main features (as shown in Figure 4.54) of the experimental data and facilitate the identification of the main intermediates (CO and H_2O_2).

Further analysis was performed to determine the influence of different dissociation channels. Figures 4.55 and 4.56 show the calculated kinetic curves for different degrees of dissociation through selected channels. It can be seen that the dissociation of CO_2 and hydrocarbons has a minimal effect on the overall process. Conversely, the dissociation of water and oxygen emerges as the most important channels (as shown in Figure 4.55).

The following figures confirm these observations, but illustrate that the accurate selection of the dissociation channel for O_2 is crucial mainly in the initial phase of the oxidation process. Beyond this early stage, the specific choice of dissociation channels becomes less critical. However, deactivation of either the O_2 or H_2O dissociation channels results in a significant slowing of the oxidation process.

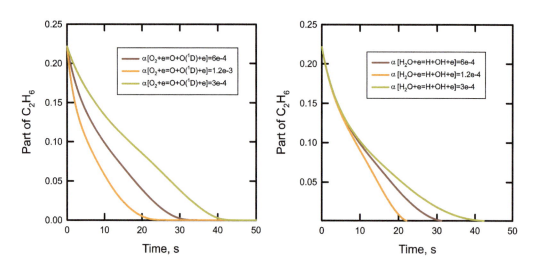

FIGURE 4.55 The calculated mole fractions of C_2H_6 molecules during the oxidation process in the stoichiometric mixture with oxygen. The initial pressure of the mixture is 3.78 Torr. The degrees of dissociation correspond to 70% of the experimentally measured energy input in the discharge.

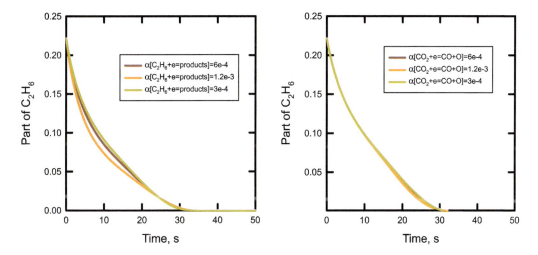

FIGURE 4.56 The calculated mole fractions of C_2H_6 molecules during the oxidation process in the stoichiometric mixture with oxygen. The initial pressure of the mixture is 3.78 Torr. Dissociation degrees correspond to 70% of the measured experimentally energy input in the discharge. Dissociation degrees of C_2H_6 and O_2 are varied.

Accordingly, the OH radical emerges as the most important species in the oxidation processes of the studied heavy hydrocarbons, with atomic oxygen O and $O(^1D)$ also playing significant roles.

It is worth noting that the rate coefficient for the reaction of C_2H_6 with OH (reaction 24) is about five times smaller than that for the reaction of C_2H_5OH with OH (reaction 59). Despite this substantial discrepancy, the difference in oxidation times is not proportionally large. This phenomenon can be attributed to the fact that the intermediates formed from the primary hydrocarbon during oxidation act as effective inhibitors, thereby establishing an efficient negative feedback mechanism within the system. Conversely, in systems involving H_2 or CH_4 with oxygen, such a feedback mechanism is absent.

In summary, a numerical model has been developed that accurately captures the key experimental features of hydrocarbons oxidation via high-E/n discharge. The critical roles of molecular oxygen and water dissociation induced by electron impact have been highlighted.

4.5 ACETONE, ACETYLENE, AND ETHYL ALCOHOL OXIDATION UNDER PULSED DISCHARGE EXCITATION

The experimental study of saturated hydrocarbons provides essential data for the development of kinetic models. At the same time, it is imperative to obtain additional experimental data on different types of hydrocarbons (see Table 4.19) to facilitate a comprehensive analysis of the kinetics. Accordingly, experiments focusing on the oxidation of acetone, acetylene, and ethyl alcohol were performed using the same technique (Figure 4.57).

4.5.1 COMPARISON OF OXIDATION OF DIFFERENT FUELS

Acetone and ethyl alcohol have been shown to oxidize with the same rate as heavy alkanes under identical experimental conditions. Consequently, all conclusions regarding the kinetics of alkane oxidation are applicable to these hydrocarbons. Furthermore, there is no reason to believe that the oxidation kinetics of other ketones and saturated alcohols would be significantly different. The results of these investigations are shown in Figure 4.58.

TABLE 4.19
Mixtures Investigated

Equivalence Ratio	C_2H_5OH	CH_3COCH_3	C_2H_2
1.0	25%	20%	16.7%
0.8	–	–	13.8%
0.4	–	–	7.4%
0.2	–	–	3.8%

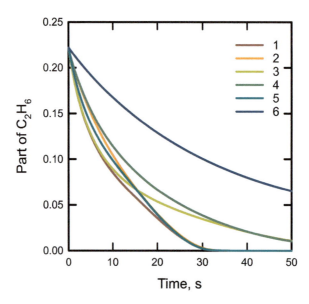

FIGURE 4.57 The calculated mole fractions of C_2H_6 molecules during the oxidation process in the stoichiometric mixture with oxygen. The initial pressure of the mixture is 3.78 Torr. The degrees of dissociation correspond to 70% of the experimentally measured energy input. Dissociation channels are changed or turned off in the following ways: (1) all dissociation channels are in accordance with Table 4.18, (2) O_2 dissociates to ground state O atoms only, (3) water dissociation is turned off, (4) O_2 dissociates to ground state O atoms only and water dissociation is turned off, (5) ethane dissociation is turned off, (6) oxygen dissociation is turned off.

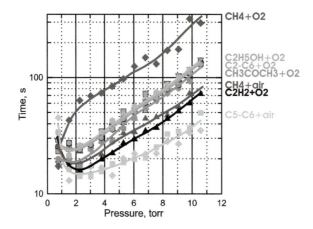

FIGURE 4.58 Oxidation time measured by $CO(B^1\Sigma \rightarrow A^1\Pi)$ transition intensity. C2–C6 and C5–C6 denote C_2H_6–C_6H_{14} and C_5H_{12}–C_6H_{14} correspondingly. These series of alkanes have practically the same oxidation times.

4.5.2 Oxidation Kinetics of Acetylene

The oxidation process for acetylene is completed in half the time required for the oxidation of other hydrocarbons studied under equivalent conditions (see Figure 4.58).

It is found that unsaturated hydrocarbons react directly with singlet molecular oxygen, unlike saturated hydrocarbons which act as quenchers. Singlet molecular oxygen is efficiently generated under a reduced electric field in the range of 10–15 V/(cm×Torr). Therefore, the contribution of singlet molecular oxygen within the nanosecond discharge is considered to be minimal because, as previously elucidated, most of the energy is imparted to the discharge at much higher electric field strengths.

To elucidate the effect of collisions with fuel and oxygen on the kinetics, experimental studies of acetylene oxidation in lean mixtures with oxygen were conducted in parallel with previous studies of butane kinetics. These studies demonstrated that the observed independence of oxidation time was due to the predominant role of fuel excitation—including dissociation, ionization, and other processes—induced by electron impact.

It was found that at pressures below 4 Torr, electron collisions with acetylene molecules dominate the oxidation process, as shown in Figure 4.59. The variation of acetylene oxidation time with changes in the equivalence ratio of the mixture at pressures above 4 Torr cannot be satisfactorily explained by the assumption that acetylene dissociation by electron impact plays a significant role. Consequently, a decrease in the proportion of acetylene in the mixture will result in a corresponding decrease in the rate of acetylene dissociation by electron impact, since the electron concentration generated during the discharge depends on the total molecular concentration in the given scenario. Therefore, under conditions of constant pressure, the oxidation time remains relatively unchanged with variations in acetylene content. This observation is particularly noticeable at low pressures, where the oxidation time does not depend on the proportion of acetylene in the mixtures. It can be concluded that at low pressures the dissociation of acetylene by electron impact is the critical determinant of the oxidation process. Beyond the threshold of 4 Torr, the impact of electron collisions on acetylene decreases with increasing pressure, and at about 10 Torr, alternative processes become dominant in influencing the oxidation dynamics.

The thresholds for electron impact excitation of the electronic states of alkanes are higher than the first dissociation limit of an alkane, significantly so for smaller alkanes. In addition, alkanes

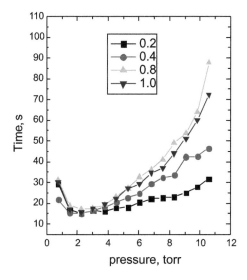

FIGURE 4.59 Oxidation times of acetylene measured from $CO(B^1\Sigma \rightarrow A^1\Pi)$ emission intensity in the different equivalence ratio mixtures. The equivalence ratio of the mixture varies from 1.0 to 0.2.

possess repulsive or highly pre-dissociated states, so that electron impact excitation of the electronic states of an alkane leads immediately to its dissociation. Dissociation occurs via several primary pathways because the excitation energy is more than sufficient for dissociation and the electronic states of alkanes are intricately connected through non-radiative transitions. The resulting fragments of dissociated molecules are produced with considerable external or translational energy.

Focusing on the electron impact excitation of acetylene, it is noteworthy that significant emission from acetylene has not been detected [97]. The first dissociation limit of acetylene is 4.9 eV [97], while the energy thresholds for the two lowest electronically excited states of acetylene are 1.9 and 5.1 eV [34]. The lowest state is identified as a metastable triplet state. The characterization of the second state is more nuanced, with very weak absorption bands of acetylene observed in the 200–240 nm range (6.25–5.2 eV) [97], with a well-defined vibrational and rotational structure indicative of stable higher transition states. Therefore, it can be postulated that the 5.1 eV state is also metastable due to its non-emissive nature. It is important to note that the efficiency of electron impact excitation of states with such low-energy thresholds is significantly higher at a pressure of 8 Torr compared to 4 Torr. Consequently, it can be inferred that the excitation of these lower energy states has a negligible effect on the rate of acetylene oxidation compared to the excitation of higher energy states. The subsequent electronic state of acetylene has an excitation threshold of 7.9 eV and exhibits a fourfold increase in the peak of the excitation cross-section [34]. This electronic term can be related to the dissociation through the channel:

$$C_2H_2 + e \rightarrow C_2H^* + H,$$

In this process C_2H^* is imparted with a substantial amount of chemical energy in its internal (vibrational and electronic) degrees of freedom. Dissociation through alternative channels requires additional energy. Since the absorption spectrum in the wavelength range below 200 nm (energy above 6.25 eV) has a diffuse character, it is concluded that states with threshold energies above 6.25 eV are pre-dissociated [97]. A system of distinct, diffuse bands is identified near 152 nm (8.2 eV) [97]. The intensity and diffuse nature of these bands, coupled with their absence in the emission spectra, strongly suggest that these bands are associated with a dipole-allowed transition to a significantly pre-dissociated upper state. Consequently, the state with a threshold of 7.9 eV [34] is conclusively linked to dissociation via this channel. Let us consider the C_2H-radical kinetics:

The right column of Table 4.20 details the lifetime of radicals in a stoichiometric acetylene-oxygen mixture at a pressure of 10 Torr. Under these experimental conditions and in the presence of oxygen, C_2H radicals are oxidized to CO and CO_2 in less than 1 ms, during which a variety of hydrogen-containing radicals are generated. It is pertinent to note that at lower pressures (below 4 Torr), the degree of dissociation of molecular oxygen and water per pulse is significantly less than that of acetylene. The cross-section peak of acetylene is several times larger than that of water and O_2. With increasing pressure, the dissociation degree of acetylene (with a threshold of 7.9 eV) decreases faster than that of water (threshold of 7 eV) because of E/n (and electron energy) decrease.

TABLE 4.20
Kinetics of C_2H Radical in Acetylene-Oxygen Mixtures

N				Constant Rate cm^{3}s^{-1}	Reaction characteristic time
1.	$C_2H + O_2$	=	$O + HCCO$	$1 \cdot 10^{-12}$	~10 μs
2.	$C_2H + O_2$	=	$CO + HCO$	$4 \cdot 10^{-12}$	~3 μs
3.	$HCO + O_2$	=	$CO + HO_2$	$5.5 \cdot 10^{-12}$	~2 μs
4.	$HCCO + O_2$	=	$CO_2 + CO + H$	$6 \cdot 10^{-13}$	~20 μs
5.	$HCCO + O_2$	=	$CO + CO + OH$	$6 \cdot 10^{-14}$	~200 μs

TABLE 4.21
Kinetics of OH Radical in Acetylene-Oxygen Mixtures

1.	$C_2H_2 + OH + M$	$\rightarrow$	$CH=CHOH + M$	$k_1 = 4 \cdot 10^{-29}$ cm^{6}s^{-1}
2.	$CH=CHOH + O_2$	$\rightarrow$	$OH + (CHO)_2$	$k_2 = 4.2 \cdot 10^{-12}$ cm^{3}s^{-1}
3.	$(CHO)_2 + OH$	$\rightarrow$	$H_2O + HC(O)CO$	$k_3 = 4.2 \cdot 10^{-11}$ cm^{3}s^{-1}
4.	$(CHO)_2 + HO_2$	$\rightarrow$	$H_2O_2 + HC(O)CO$	$k_4 = 5 \cdot 10^{-16}$ cm^{3}s^{-1}
5.	$HC(O)CO$	$\rightarrow$	$CO + HCO$	$k_5 = 3.7 \cdot 10^7$ s^{-1}
6.	$HC(O)CO + O_2$	$\rightarrow$	$2CO + HCO$	$k_6 = 1 \cdot 10^{-11}$ cm^{3}s^{-1}
7.	$HCO + O_2$	$=$	$CO + HO_2$	$k_7 = 5.5 \cdot 10^{-12}$ cm^{3}s^{-1}
8.	$CO + OH$	$\rightarrow$	products	$k_8 = 1.7 \cdot 10^{-13}$ cm^{3}s^{-1}

The dissociation degree of oxygen decreases most rapidly with pressure (the threshold of the primary dissociation channel is 8.4 eV). The observed experimental variation in oxidation time with equivalence ratio can be explained by the dominant role of electron impact dissociation of acetylene in the lower pressure range. At elevated pressures, the efficiency of C_2H_2 dissociation decreases, potentially making the dissociation process of water more important, similar to that observed for alkanes. It is noteworthy that in the case of acetylene, as with alkanes, the OH radical plays a predominant role in the oxidation process. However, unlike alkanes, the main reaction channel involves the addition of OH radicals to acetylene molecules in a third-order reaction (see Table 4.21).

For an initial pressure of 5 Torr, the rate coefficient, considering second-order kinetics, is $6 \cdot 10^{-12}$ cm^{3}s^{-1}, significantly higher than the rate coefficients associated with other OH radical loss processes. The newly formed OHCH=CH radical undergoes a rapid reaction with O_2 (2) (see Table 4.21). It is obvious that reactions (1) and (2) are a chain reaction mechanism. The primary termination mechanism for this chain is represented by the following reaction (3). The rate of OH radical loss in this chain termination reaction is equal to the rate of loss in the acetylene oxidation process when the concentration of $(CHO)_2$ (formyl) approximates that of acetylene. It is important to note that the formyl group, being a weaker bonded molecule compared to CO, is more readily oxidized by the HO_2 radical. This suggests that C_2H_2 is initially oxidized to CO by the specified reactions (see Table 4.21), followed by the oxidation of CO to CO_2. Since reaction (1) involves three participants, an increase in pressure increases the influence of OH. This effect is pronounced in experiments conducted with a stoichiometric mixture of acetylene and oxygen, where the discrepancy in oxidation times, as determined from $CO(B^1\Sigma \rightarrow A^1\Pi)$ and $CH(A^2\Delta \rightarrow X^2\Pi)$ emissions, increases with increasing pressure (see Figure 4.60), indicating that CO is oxidized significantly later than acetylene.

Comparative analysis of the emission intensities of different molecular bands in acetylene-oxygen (Figure 4.62) and ethane-oxygen (Figure 4.61) stoichiometric mixtures reveals distinctive characteristics. For the ethane-oxygen mixture, at the peak of the $CO(B^1\Sigma \rightarrow A^1\Pi)$ band emission, the intensity of the $CH(A^2\Delta \rightarrow X^2\Pi)$ band emission is approximately 0.6 of its maximum value. Conversely, for the acetylene-oxygen mixture, the peak of the $CO(B^1\Sigma \rightarrow A^1\Pi)$ band emission is only 0.35 of the $CH(A^2\Delta \rightarrow X^2\Pi)$ emission maximum. In addition, the emission intensity of the $OH(A^2\Sigma \rightarrow X^2\Pi)$ band in the acetylene-oxygen mixture increases more gradually than in the ethane-oxygen mixture, where the emission escalates rapidly to two-thirds of its maximum intensity and maintains this level until a sharp decline at the point of $CO(B^1\Sigma \rightarrow A^1\Pi)$ band emission decay. This variance highlights the different kinetics of the OH radical in these mixtures.

In the ethane-oxygen mixture, the OH radical predominantly drives ethane oxidation, with its concentration influenced by four main processes (see Table 4.22). Processes (1)–(4) produce OH, while process (5) involves the oxidation of C_2H_6, followed by the oxidation of C_2H_5 through a series of reactions culminating in the production of CO. The most effective oxidation of CO occurs in

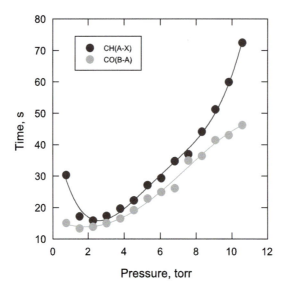

FIGURE 4.60 Oxidation times of acetylene determined from $CO(B^1\Sigma \rightarrow A^1\Pi)$ and $CH(A^2\Delta \rightarrow X^2\Pi)$ emission intensities.

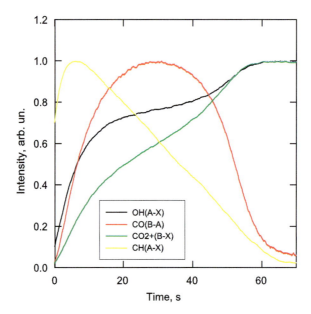

FIGURE 4.61 Emission intensities of $OH(A^2\Sigma \rightarrow X^2\Pi)$, $CO(B^1\Sigma \rightarrow A^1\Pi)$, $CO_2^+ (B^2\Sigma \rightarrow X^2\Pi)$, $CH(A^2\Delta \rightarrow X^2\Pi)$ bands. Initial pressure of ethane-oxygen stoichiometric mixture is 6.8 Torr.

process (6), which significantly reduces the concentration of OH radicals. Consequently, oxidation under discharge conditions exhibits negative feedback, as an increase in OH concentration leads to increased CO levels, which in turn consume the radical and reduce the oxidation rate. Thus, after an initial peak, the intensity of $OH(A^2\Sigma \rightarrow X^2\Pi)$ emission stabilizes while the intensity of $CO(B^1\Sigma \rightarrow A^1\Pi)$ band emission exhibits a prolonged maximum. Once the $CH(A^2\Delta \rightarrow X^2\Pi)$ emission stabilizes, indicating complete hydrocarbon oxidation, the $CO(B^1\Sigma \rightarrow A^1\Pi)$ band emission drops sharply to a low steady-state level, concurrent with a rapid rise of the $OH(A^2\Sigma \rightarrow X^2\Pi)$ band emission to a stable plateau. After ethane conversion, with the depletion of CO sources, residual CO rapidly oxidizes, resulting in the cessation of negative feedback within the system.

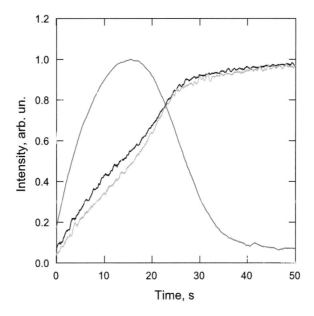

FIGURE 4.62 Emission intensities of OH($A^2\Sigma \rightarrow X^2\Pi$), CO($B^1\Sigma \rightarrow A^1\Pi$), CO_2^+ ($B^2\Sigma \rightarrow X^2\Pi$), CH($A^2\Delta \rightarrow X^2\Pi$) bands. Initial pressure of acetylene-oxygen stoichiometric mixture is 6.8 Torr.

TABLE 4.22
Kinetics of C_2H_6 Oxidation

1.	Electron impact to O_2	$e + O_2$	$\rightarrow$	$O(^1D) + O + e$
2.	Production in the process	$O(^1D) + H_2O$	$\rightarrow$	$2OH$
3.	Production in the process	$O(^1D) + C_2H_6$	$\rightarrow$	$C_2H_5 + OH$
4.	Electron impact to H_2O	$e + H_2O$	$\rightarrow$	$OH + H + e$
5.	Loss in the process of C_2H_6 oxidation	$C_2H_6 + OH$	$\rightarrow$	$C_2H_5 + H_2O$
6.	Loss in the process of CO oxidation	$CO + OH$	$\rightarrow$	products

Let us compare emission intensities of different molecular bands in acetylene-oxygen (Figure 4.62) and ethane-oxygen (Figure 4.61) stoichiometric mixtures. It is clearly seen that for ethane-oxygen mixture in the point of time when CO($B^1\Sigma \rightarrow A^1\Pi$) band emission is maximal the intensity of CH($A^2\Delta \rightarrow X^2\Pi$) band emission amounts to 0.6 of CH($A^2\Delta \rightarrow X^2\Pi$) emission maximum approximately. At the same time for the acetylene-oxygen mixture the point of CO($B^1\Sigma \rightarrow A^1\Pi$) band emission maximum corresponds to the point of 0.35 of CH($A^2\Delta \rightarrow X^2\Pi$) emission maximum. Also, it is necessary to note, that the emission intensity of OH($A^2\Sigma \rightarrow X^2\Pi$) band in the acetylene-oxygen mixture increase slowly relative to the emission in ethane-oxygen mixture, in which the emission rapidly increases to two-thirds of the maximum value and after that stays on the level up to sharp decreasing of CO($B^1\Sigma \rightarrow A^1\Pi$) band emission point of time. This difference clearly points to different kinetics of OH radicals in these mixtures.

4.5.3 Kinetics of C_2H_6 Oxidation

In the acetylene oxidation process, negative feedback mechanisms become prominent in later stages, coinciding with the significant accumulation of formyl radicals. Since formyl is oxidized faster than CO, its concentration peaks during these later stages. Consequently, the oxidation processes of acetylene and CO are temporally distinct, with the interval between these processes increasing with increasing pressure. This phenomenon can be attributed to the significant role played by the addition of OH to acetylene, where reaction (1) involves three participants.

It is important to note that the observed prolongation of the oxidation times with increasing proportions of acetylene in the mixture is consistent with the hypothesis that the OH radical exerts a decisive influence. A reduction in the initial concentration of acetylene, and thus a corresponding reduction in the concentration of water in the processed mixture, leads to a reduced generation of OH radicals by electron impact throughout the oxidation process. However, in lean mixtures, excess oxygen acts as a source of additional $O(^1D)$ states generated by electron collisions with molecular oxygen. This state, when reacting with water, produces two OH radicals instead of one as would be the case with hydrocarbon collisions. As a result, the rate of hydrocarbon oxidation shows little variation with changes in the equivalence ratio of the mixture.

Thus, in it was concluded that the main processes controlling acetylene oxidation are the dissociation of acetylene by electron impact and the addition of OH radical to an acetylene molecule. OH radicals are produced from water by electron impact and in collisions of water molecules with $O(^1D)$.

4.6 CARBON MONOXIDE OXIDATION IN MIXTURES WITH OXYGEN AND WATER

As previously shown, carbon monoxide contributes significantly to the oxidation processes of hydrocarbons. To study the kinetics of carbon monoxide in hydrocarbon-oxygen mixtures, paper [88] discussed the oxidation of CO in stoichiometric mixtures with O_2 and a small addition of water. The experimental apparatus and methods were the same as those used in previous experiments, although the experimental setup and the preparation of the mixtures differed considerably. A mixture containing 67% CO and 33% O_2 was premixed in a cylinder. Water vapor was also prepared in advance; water was injected into a vacuum vessel, and after evaporation, the water vapor was ready to be introduced into the discharge cell. After the water vapor was injected into the cell, the CO-O_2 mixture was added to the water vapor, mixing the water and CO-O_2 mixture over a period of 2 min. Since the total pressure did not exceed 11 Torr and the discharge cell had a length of 25 cm, this time was considered sufficient for thorough mixing of the water with the CO-O_2 premix, since the diffusion time is $t \sim L^2/D < 30$ s over the entire pressure range.

4.6.1 Major Pathways of CO Oxidation

The emission intensity of the $CO(B^1\Sigma \to A^1\Pi)$ band, the voltage across the discharge cell, and the electric current were monitored with nanosecond resolution and also in the integral regime. Experiments were performed with mixtures containing different water concentrations of 1, 2, 4, and 8%. The emission intensities of the $CO(B^1\Sigma \to A^1\Pi)$ band and their temporal derivatives as functions of time, at a pressure of 3.78 Torr, are shown in Figure 4.63 for the various water concentrations.

It is important to note that the emission intensities of the $CO(B^1\Sigma \to A^1\Pi)$ band can be directly correlated to the concentration of carbon monoxide, as the small addition of water is unlikely to have a significant effect on the overall emission. It is evident that a small addition of water significantly accelerates the oxidation process and significantly improves the final conversion of CO to CO_2. In Figure 4.64, the emission intensity of the $CO(B^1\Sigma \to A^1\Pi)$ band at $t = 160$ s (a marker, at a given initial pressure, for the level of CO conversion relative to the conversion level in the absence of water) and the peak value of the time derivatives of the emission $\frac{dI^{CO}(t)}{dt}$ (an indicator, at a given initial pressure, for the maximum rate of CO oxidation relative to the rate without water) versus initial pressure are plotted for various water vapor concentrations. It should be noted that the peaks of the derivatives can be generalized to all experimental data as follows:

Mechanisms of Plasma-Chemical Reactions

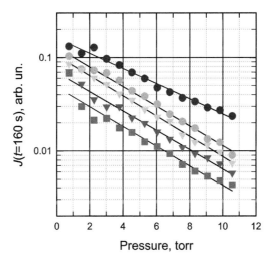

FIGURE 4.63 CO($B^1\Sigma \to A^1\Pi$) emission intensity after 6,400 high-voltage pulses. Stoichiometric mixture CO–O_2 with small addition of water vapor.

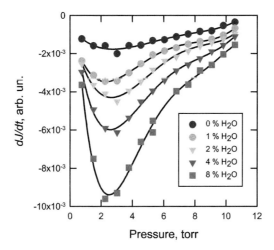

FIGURE 4.64 Peak values of CO($B^1\Sigma \to A^1\Pi$) emission derivatives. Stoichiometric mixture CO–O_2 with controlled addition of water vapor.

$$\frac{dI}{dT} = \frac{dI}{dT}\bigg|_{[H_2O]=0} + F(p)\ln[H_2O],$$

where $F(p)$ is a universal function independent of the H_2O concentration. Furthermore, it is observed that the emission intensities $I(t=160,p)$ have identical logarithmic derivatives with respect to pressure upon the addition of water:

$$(I(t=160,p)) = I_0\left(1 - \ln[H_2O]^{0.307}\right)\exp(-0.243p),$$

and substantially different without the addition of water:

$$I(t=160,p) \sim \exp(-0.187p).$$

Here [H_2O] is the percentage of water, p, [Torr] is the initial pressure.

TABLE 4.23
Kinetics of CO Oxidation in the Discharge

1.	Electron impact to CO	e+CO	→	C+O	$F(E/n), \delta\varepsilon \approx 12$ eV
2.	C atoms losses	$C+O_2$	→	$CO+O(^1D)$	$5 \cdot 10^{-11}$
		$C+H_2O$	→	CH+OH	$<1 \cdot 10^{-12}$
		$C+CO_2$	→	2CO	$<1 \cdot 10^{-15}$
3.	Electron impact to H_2O	$e+H_2O$	→	OH+H+e	$F(E/n), \delta\varepsilon \approx 7$ eV
5.	OH production in the process	$O(^1D)+H_2O$	→	2OH	$1.1 \cdot 10^{-10}$
6.	Loss in the processes of CO oxidation	CO+OH	→	CO_2+H	$1.7 \cdot 10^{-13}$

The pronounced correlation between the rate of oxidation and the degree of conversion of CO with water concentration is due to the predominant oxidation of CO by reactions with hydrogen-containing radicals (Table 4.23). In this context, any pathway that does not involve the OH radical is significantly less effective.

It is important to recognize the stability of a CO molecule, which is characterized by an ionization potential and a first dissociation limit of 14.01 and 11.09 eV, respectively. Consequently, dissociation or ionization of CO in plasma is not the primary pathway for energy deposition by discharge. However, a single event of CO dissociation initiates the production of O atoms and CO molecules, culminating in the production of two OH radicals through a chain of fast reactions (see Table 4.23). These OH radicals are then used to form two CO_2 molecules. Therefore, in the presence of water, the dissociation of CO by electron impact facilitates the fast and efficient conversion of CO molecules to CO_2.

4.7 KINETICS OF NEUTRALS IN HYDROCARBON-OXYGEN MIXTURES

Thus, slow oxidation of C_2H_2, CH_3COCH_3 and C_2H_5OH in stoichiometric and some lean mixtures with oxygen under the action of the nanosecond pulsed-periodic discharge has been investigated experimentally. CO oxidation under the action of the discharge in the stoichiometric mixture with O_2 in the presence of small additions of H_2 has been investigated experimentally. The kinetics of slow oxidation of H_2, CH_4, C_2H_6, C_2H_5OH, C_2H_2, CH_3COCH_3 under the action of the nanosecond pulsed-periodic discharge in mixtures with oxygen at room temperature under nanosecond uniform discharge have been considered. The main role of $O(^1D)$ and OH radical has been determined in the kinetics. Main paths of radicals and atoms creation in the discharge have been shown. The kinetic scheme for the H_2, CH_4, C_2H_6 and C_2H_5OH oxidation at room temperature under the nanosecond discharge action has been developed. The results of calculation using this scheme are in good agreement with experimental results.

The slow oxidation of C_2H_2, CH_3COCH_3 and C_2H_5OH in both stoichiometric and selected lean mixtures with oxygen subjected to nanosecond pulse-periodic discharge has been experimentally studied. In addition, the oxidation of CO under discharge in a stoichiometric mixture with O_2, supplemented with small additions of H_2, has been studied experimentally. The kinetics of slow plasma-assisted oxidation for H_2, CH_4, C_2H_6, C_2H_5OH, C_2H_2, CH_3COCH_3 in oxygen mixtures at ambient temperature have been analyzed. Computational results derived from this model show good agreement with the experimental data. The critical roles of $O(^1D)$ and the OH radical in these kinetic processes have been identified. The primary mechanisms for the generation of radicals and atoms in the discharge have been elucidated.

TABLE 4.24
Kinetics of Neutrals in Hydrocarbons (C_2H_6, CH_4, H_2, C_2H_5OH)–Oxygen Mixtures

N	Reaction			Rate Coefficients, cm^3/s
1.	$H_2 + O(^1D)$	=	$H + OH$	$1.1 \cdot 10^{-10}$
2.	$H_2 + O$	=	$H + OH$	$1 \cdot 10^{-17}$
3.	$H_2 + OH$	=	$H_2O + H$	$7 \cdot 10^{-15}$
4.	$H_2 + CHO$	=	$H + CH_2O$	$2.8 \cdot 10^{-26}$
5.	$H_2 + CH_3O$	=	$H + CH_3OH$	$3.7 \cdot 10^{-9}$
6.	$O_2 + O(^1D)$	=	$O_2 + O$	$4 \cdot 10^{-11}$
7.	$O_2 + O + M$	=	$O_3 + M$	$6 \cdot 10^{-34}$
8.	$O_2 + H + M$	=	$HO_2 + M$	$5.7 \cdot 10^{-32}$
9.	$O_2 + CH_3 + M$	=	$CH_3O_2 + M$	$4.5 \cdot 10^{-31}$
10.	$O_2 + C_2H_5 + M$	=	$C_2H_5O_2 + M$	$1.5 \cdot 10^{-28}$
11.	$O_2 + CHO$	=	$CO + HO_2$	$5.5 \cdot 10^{-12}$
12.	$O_2 + CHO$	=	$CO_2 + OH$	$1.1 \cdot 10^{-13}$
13.	$O_2 + CH_3O$	=	$CH_2O + HO_2$	$2 \cdot 10^{-15}$
14.	$O_2 + CH_3CO$	=	$CH_2O + CO + OH$	$3 \cdot 10^{-12}$
15.	$CH_4 + O(^1D)$	=	$CH_3 + OH$	$2.2 \cdot 10^{-10}$
16.	$CH_4 + O$	=	$CH_3 + OH$	$1 \cdot 10^{-17}$
17.	$CH_4 + OH$	=	$CH_3 + H_2O$	$6.3 \cdot 10^{-15}$
18.	$CH_4 + H$	=	$CH_3 + H_2$	$1 \cdot 10^{-17}$
19.	$C_2H_4 + O$	=	$H + CH_2CHO$	$6.24 \cdot 10^{-13}$
20.	$C_2H_4 + OH + M$	=	$HOCH_2CH_2 + M$	$1 \cdot 10^{-28}$
21.	$C_2H_4 + H + M$	=	$C_2H_5 + M$	$2 \cdot 10^{-30}$
22.	$C_2H_6 + O(^1D)$	=	$C_2H_5 + OH$	$6.3 \cdot 10^{-10}$
23.	$C_2H_6 + O$	=	$C_2H_5 + OH$	$5 \cdot 10^{-16}$
24.	$C_2H_6 + OH$	=	$C_2H_5 + H_2O$	$2.5 \cdot 10^{-13}$
25.	$CO + O(^1D)$	=	CO_2	$7.3 \cdot 10^{-11}$
26.	$CO + O(^1D)$	=	$CO + O$	$2.1 \cdot 10^{-10}$
27.	$CO + O + O_2$	=	$CO_2 + O_2$	$6 \cdot 10^{-35}$
28.	$CO + O + H_2O$	=	$CO_2 + H_2O$	$6 \cdot 10^{-35}$
29.	$CO + O + CO_2$	=	$2CO_2$	$3.5 \cdot 10^{-35}$
30.	$CO + OH$	=	$HOCO$	$1.5 \cdot 10^{-13}$
31.	$HOCO + O_2$	=	$CO_2 + HO_2$	$1.5 \cdot 10^{-12}$
32.	$CO + OH$	=	$CO_2 + H$	$1.7 \cdot 10^{-13}$
33.	$CO + H + M$	=	$CHO + M$	$1.5 \cdot 10^{-34}$
34.	$CO + CH_3 + M$	=	$CH_3CO + M$	$4.2 \cdot 10^{-36}$
35.	$CO + C_2H_5$	=	C_2H_5CO	$8.1 \cdot 10^{-17}$
36.	$CO_2 + O(^1D)$	=	$CO_2 + O$	$1.1 \cdot 10^{-10}$
37.	$CH_3CHO + O(^1D)$	=	$OH + CH_3CO$	$3 \cdot 10^{-10}$
38.	$CH_3CHO + O$	=	$CH_3CO + OH$	$4.3 \cdot 10^{-13}$
39.	$CH_3CHO + OH$	=	$CH_3CO + H_2O$	$1.45 \cdot 10^{-11}$
40.	$CH_3CHO + H$	=	$CH_3CO + H_2$	$1 \cdot 10^{-13}$
41.	$CH_3CHO + CH_3$	=	$CH_3CO + CH_4$	$5 \cdot 10^{-18}$
42.	$CH_3CHO + C_2H_5$	=	$CH_3CO + C_2H_6$	$1 \cdot 10^{-17}$

(Continued)

TABLE 4.24 (Continued)
Kinetics of Neutrals in Hydrocarbons (C_2H_6, CH_4, H_2, C_2H_5OH)–Oxygen Mixtures

N	Reaction		Rate Coefficients, cm³/s
43.	$CH_3CHO + CHO$	= $CHO + CO + CH_4$	$2 \cdot 10^{-17}$
44.	$CH_3CHO + CH_3O$	= $CH_3CO + CH_3OH$	$1 \cdot 10^{-14}$
45.	$CH_3CHO + CH_3CO$	= $CHO + C_2H_6CO$	$2.8 \cdot 10^{-13}$
46.	$CH_2O + O(^1D)$	= $OH + CHO$	$3 \cdot 10^{-10}$
47.	$CH_2O + O$	= $OH + CHO$	$1.7 \cdot 10^{-13}$
48.	$CH_2O + OH$	= $H_2O + CHO$	$9.4 \cdot 10^{-12}$
49.	$CH_2O + H$	= $H_2 + CHO$	$5.7 \cdot 10^{-14}$
50.	$CH_2O + CH_3O$	= $CH_3OH + CHO$	$1.1 \cdot 10^{-15}$
51.	$CH_3OH + O(^1D)$	= $CH_3O + OH$	$5.1 \cdot 10^{-10}$
52.	$CH_3OH + O$	= $CH_3O + OH$	$7 \cdot 10^{-15}$
53.	$CH_3OH + OH$	= $CH_2OH + H_2O$	$9.3 \cdot 10^{-13}$
54.	$CH_3OH + OH$	= $CH_3O + H_2O$	$1.4 \cdot 10^{-13}$
55.	$CH_3OH + H$	= $CH_2OH + H_2$	$1.3 \cdot 10^{-15}$
56.	$CH_3OH + CH_3O$	= $CH_2OH + CH_3OH$	$5.4 \cdot 10^{-16}$
57.	$C_2H_5OH + O(^1D)$	= $C_2H_5O + OH$	$5 \cdot 10^{-10}$
58.	$C_2H_5OH + O$	= $CH_3CHOH + OH$	$1 \cdot 10^{-13}$
59.	$C_2H_5OH + OH$	= $CH_3CHOH + H_2O$	$1.3 \cdot 10^{-12}$
60.	$C_2H_5OH + H$	= $CH_3CHOH + H_2$	$3 \cdot 10^{-15}$
61.	$CH_3OOH + O(^1D)$	= $CH_3O_2 + OH$	$5 \cdot 10^{-10}$
62.	$CH_3OOH + O$	= $CH_3O_2 + OH$	$1 \cdot 10^{-14}$
63.	$CH_3OOH + OH$	= $CH_3O_2 + H_2O$	$3.6 \cdot 10^{-12}$
64.	$CH_3OOH + OH$	= $CH_2OOH + H_2O$	$1.9 \cdot 10^{-12}$
65.	$CH_3OOH + H$	= $CH_2OOH + H_2$	$1 \cdot 10^{-15}$
66.	$CH_3OOH + H$	= $CH_3O + H_2O$	$4 \cdot 10^{-16}$
67.	$CH_3OOH + H$	= $CH_3O_2 + H_2$	$5 \cdot 10^{-16}$
68.	$C_2H_5OOH + O(^1D)$	= $C_2H_5O_2 + OH$	$5 \cdot 10^{-10}$
69.	$C_2H_5OOH + O$	= $C_2H_5O_2 + OH$	$1.2 \cdot 10^{-14}$
70.	$C_2H_5OOH + OH$	= $C_2H_5O_2 + H_2O$	$3.7 \cdot 10^{-12}$
71.	$C_2H_5OOH + OH$	= $CH_3CHO + H_2O + OH$	$2 \cdot 10^{-12}$
72.	$C_2H_5OOH + H$	= $CH_3CHO + H_2 + OH$	$1 \cdot 10^{-15}$
73.	$C_2H_5OOH + H$	= $C_2H_5O + H_2O$	$4 \cdot 10^{-16}$
74.	$C_2H_5OOH + H$	= $C_2H_5O_2 + H_2$	$5 \cdot 10^{-16}$
75.	$H_2O_2 + O(^1D)$	= $OH + HO_2$	$5.2 \cdot 10^{-10}$
76.	$H_2O_2 + O$	= $OH + HO_2$	$1.8 \cdot 10^{-15}$
77.	$H_2O_2 + OH$	= $H_2O + HO_2$	$1.7 \cdot 10^{-12}$
78.	$H_2O_2 + H$	= $H_2O + OH$	$4.2 \cdot 10^{-14}$
79.	$H_2O_2 + CH_3$	= $CH_4 + HO_2$	$5.5 \cdot 10^{-14}$
80.	$H_2O_2 + C_2H_5$	= $C_2H_6 + HO_2$	$2.8 \cdot 10^{-15}$
81.	$H_2O_2 + CH_3O$	= $CH_3OH + HO_2$	$1 \cdot 10^{-14}$
82.	$H_2O + O(^1D)$	= $OH + OH$	$2.2 \cdot 10^{-10}$
84.	$O(^1D) + O_3$	= $2O_2$	$1.2 \cdot 10^{-10}$
85.	$O(^1D) + O_3$	= $2O + O_2$	$1.2 \cdot 10^{-10}$
86.	$O + O + O_2$	= $O_2 + O_2$	$1 \cdot 10^{-32}$
87.	$O + O + H_2$	= $O_2 + H_2$	$2.6 \cdot 10^{-33}$

(Continued)

TABLE 4.24 (Continued)
Kinetics of Neutrals in Hydrocarbons (C_2H_6, CH_4, H_2, C_2H_5OH)–Oxygen Mixtures

N	Reaction			Rate Coefficients, cm³/s
88.	$O+O+H_2O$	=	H_2O+O_2	$1.7 \cdot 10^{-32}$
89.	$O+O+CH_4$	=	CH_4+O_2	$2.2 \cdot 10^{-33}$
90.	$O+O+CO$	=	$CO+O_2$	$1.9 \cdot 10^{-33}$
91.	$O+O+CO_2$	=	CO_2+O_2	$4 \cdot 10^{-33}$
92.	$O+O+C_2H_6$	=	$C_2H_6+O_2$	$3.3 \cdot 10^{-33}$
93.	$O+O_3$	=	$2O_2$	$8.3 \cdot 10^{-15}$
94.	$O+OH$	=	$H+O_2$	$3.5 \cdot 10^{-11}$
95.	$O+H+H_2$	=	$OH+H_2$	$9.2 \cdot 10^{-33}$
96.	$O+H+CH_4$	=	$OH+CH_4$	$9.2 \cdot 10^{-33}$
97.	$O+H+CO_2$	=	$OH+CO_2$	$9.2 \cdot 10^{-33}$
98.	$O+H+CO$	=	$OH+CO$	$6.9 \cdot 10^{-33}$
99.	$O+H+H_2O$	=	$OH+H_2O$	$2.8 \cdot 10^{-32}$
100.	$O+H+C_2H_6$	=	$OH+C_2H_6$	$1.4 \cdot 10^{-32}$
101.	$O+HO_2$	=	$OH+O_2$	$5.7 \cdot 10^{-11}$
102.	$O+CH_3$	=	$H+H_2+CO$	$3 \cdot 10^{-11}$
103.	$O+CH_3$	=	$H+CH_2O$	$1 \cdot 10^{-10}$
104.	$O+C_2H_5$	=	CH_3CHO+H	$1.3 \cdot 10^{-10}$
105.	$O+C_2H_3$	=	CH_2O+CH_3	$2.7 \cdot 10^{-11}$
106.	$O+CHO$	=	$H+CO_2$	$5 \cdot 10^{-11}$
107.	$O+CHO$	=	$OH+CO$	$5 \cdot 10^{-11}$
108.	$O+CH_3O$	=	CH_3+O_2	$2.5 \cdot 10^{-12}$
109.	$O+CH_3O$	=	CH_2O+OH	$2.4 \cdot 10^{-11}$
110.	$O+CH_3CO$	=	CO_2+CH_3	$3.3 \cdot 10^{-11}$
111.	$O+CH_3O_2$	=	CH_3O+O_2	$5 \cdot 10^{-11}$
112.	$O+C_2H_5O_2$	=	$C_2H_5O+O_2$	$7 \cdot 10^{-11}$
113.	O_3+OH	=	O_2+HO_2	$7.4 \cdot 10^{-14}$
114.	O_3+H	=	$OH+O_2$	$3 \cdot 10^{-11}$
115.	O_3+HO_2	=	$2O_2+OH$	$2.1 \cdot 10^{-15}$
116.	O_3+CH_3	=	CH_3O+O_2	$2.6 \cdot 10^{-12}$
117.	$O_3+C_2H_5$	=	$C_2H_5O+O_2$	$2.5 \cdot 10^{-11}$
118.	O_3+CHO	=	$H+O_2+CO_2$	$1 \cdot 10^{-12}$
119.	$O_3+CH_3O_2$	=	$2O_2+CH_3O$	$1 \cdot 10^{-17}$
120.	$O_3+C_2H_5O_2$	=	$2O_2+C_2H_5O$	$1 \cdot 10^{-17}$
121.	$OH+OH$	=	$O+H_2O$	$1.4 \cdot 10^{-12}$
122.	$OH+H+H_2O$	=	H_2O+H_2O	$4.3 \cdot 10^{-30}$
123.	$OH+H+H_2$	=	H_2O+H_2	$8.6 \cdot 10^{-31}$
124.	$OH+H+C_2H_6$	=	$C_2H_6+H_2O$	$3.5 \cdot 10^{-30}$
125.	$OH+H+O_2$	=	H_2O+O_2	$6.8 \cdot 10^{-31}$
126.	$OH+H+CO_2$	=	H_2O+CO_2	$9 \cdot 10^{-31}$
127.	$OH+H+CO$	=	H_2O+CO	$4.5 \cdot 10^{-31}$
128.	$OH+HO_2$	=	H_2O+O_2	$1.1 \cdot 10^{-10}$
129.	$OH+CH_3$	=	CH_2OH+H	$1.3 \cdot 10^{-11}$
130.	$OH+CH_3$	=	CH_3O+H	$1.6 \cdot 10^{-10}$
131.	$OH+CH_3$	=	CH_3OH	$1 \cdot 10^{-10}$

(Continued)

TABLE 4.24 (Continued)
Kinetics of Neutrals in Hydrocarbons (C_2H_6, CH_4, H_2, C_2H_5OH)–Oxygen Mixtures

N	Reaction			Rate Coefficients, cm³/s
132.	$OH+C_2H_5$	=	$C_2H_4+H_2O$	$4 \cdot 10^{-11}$
133.	$OH+C_2H_5$	=	C_2H_5OH	$1.2 \cdot 10^{-10}$
134.	$OH+CHO$	=	H_2O+CO	$1.7 \cdot 10^{-10}$
135.	$OH+CH_3O$	=	H_2O+CH_2O	$3 \cdot 10^{-11}$
136.	$OH+CH_3CO$	=	CH_2CO+H_2O	$2 \cdot 10^{-11}$
137.	$OH+CH_3O_2$	=	O_2+CH_3OH	$1 \cdot 10^{-10}$
138.	$OH+C_2H_5O_2$	=	$O_2+C_2H_5OH$	$1 \cdot 10^{-10}$
139.	$H+H+H_2$	=	H_2+H_2	$8.9 \cdot 10^{-33}$
140.	$H+H+CH_4$	=	H_2+CH_4	$1.8 \cdot 10^{-32}$
141.	$H+H+CO_2$	=	H_2+CO_2	$1.7 \cdot 10^{-32}$
142.	$H+H+C_2H_6$	=	$H_2+C_2H_6$	$2.8 \cdot 10^{-32}$
143.	$H+H+H_2O$	=	H_2+H_2O	$9.2 \cdot 10^{-32}$
144.	$H+HO_2$	=	H_2O+O	$2.4 \cdot 10^{-12}$
145.	$H+HO_2$	=	H_2+O_2	$5.6 \cdot 10^{-12}$
146.	$H+HO_2$	=	$2OH$	$7.2 \cdot 10^{-11}$
147.	$H+CH_3+C_2H_6$	=	$CH_4+C_2H_6$	$3 \cdot 10^{-28}$
148.	$H+CH_3+H_2O$	=	CH_4+H_2O	$6 \cdot 10^{-28}$
149.	$H+CH_3+H_2$	=	CH_4+H_2	$2 \cdot 10^{-28}$
150.	$H+CH_3+CH_4$	=	$2CH_4$	$2 \cdot 10^{-28}$
151.	$H+CH_3+CO$	=	CH_4+CO	$1.5 \cdot 10^{-28}$
152.	$H+CH_3+CO_2$	=	CH_4+CO_2	$2 \cdot 10^{-28}$
153.	$H+C_2H_5$	=	$2CH_3$	$1.25 \cdot 10^{-10}$
154.	$H+CHO$	=	H_2+CO	$1.13 \cdot 10^{-10}$
155.	$H+CH_3O$	=	H_2+CH_2O	$3 \cdot 10^{-11}$
156.	$H+CH_3CO$	=	$CHO+CH_3$	$3.6 \cdot 10^{-11}$
157.	$H+CH_3CO$	=	CH_2CO+H_2	$1.9 \cdot 10^{-11}$
158.	$H+CH_3O_2$	=	CH_3O+OH	$1.6 \cdot 10^{-10}$
159.	$H+C_2H_5O_2$	=	C_2H_5O+OH	$1.6 \cdot 10^{-10}$
160.	HO_2+HO_2+M	=	$O_2+H_2O_2+M$	$4.5 \cdot 10^{-32}$
161.	HO_2+HO_2	=	$O_2+H_2O_2$	$1.7 \cdot 10^{-12}$
162.	HO_2+CH_3	=	CH_3O+OH	$3 \cdot 10^{-11}$
163.	HO_2+CH_3	=	CH_4+O_2	$6 \cdot 10^{-12}$
164.	$HO_2+C_2H_5$	=	$O_2+C_2H_6$	$5 \cdot 10^{-13}$
165.	$HO_2+C_2H_5$	=	$H_2O_2+C_2H_4$	$3 \cdot 10^{-12}$
166.	$HO_2+C_2H_5$	=	$OH+C_2H_5O$	$5 \cdot 10^{-11}$
167.	HO_2+CHO	=	CH_2O+O_2	$5 \cdot 10^{-11}$
168.	HO_2+CH_3O	=	$CH_2O+H_2O_2$	$5 \cdot 10^{-13}$
169.	HO_2+CH_3CO	=	CH_3+CO_2+OH	$5 \cdot 10^{-11}$
170.	$HO_2+CH_3O_2$	=	O_2+CH_3OOH	$5.8 \cdot 10^{-12}$
171.	$HO_2+C_2H_5O_2$	=	$O_2+C_2H_5OOH$	$7.63 \cdot 10^{-12}$
172.	CH_3+CH_3	=	C_2H_6	$4.6 \cdot 10^{-11}$
173.	$CH_3+C_2H_5$	=	C_3H_8	$5.6 \cdot 10^{-11}$
174.	$CH_3+C_2H_5$	=	$C_2H_4+CH_4$	$1.9 \cdot 10^{-12}$
175.	CH_3+CHO	=	CH_4+CO	$4.4 \cdot 10^{-11}$

(Continued)

TABLE 4.24 (Continued)
Kinetics of Neutrals in Hydrocarbons (C_2H_6, CH_4, H_2, C_2H_5OH)–Oxygen Mixtures

N	Reaction			Rate Coefficients, cm³/s
176.	CH_3+CHO	=	CH_3CHO	$6.3 \cdot 10^{-12}$
177.	CH_3+CH_3O	=	CH_4+CH_2O	$4 \cdot 10^{-11}$
178.	CH_3+CH_3CO	=	$CO+C_2H_6$	$5.3 \cdot 10^{-11}$
179.	CH_3+CH_3CO	=	CH_2CO+CH_4	$1 \cdot 10^{-11}$
180.	$CH_3+CH_3O_2$	=	$2CH_3O$	$4 \cdot 10^{-11}$
181.	$CH_3+C_2H_5O_2$	=	$CH_3O+C_2H_5O$	$4 \cdot 10^{-11}$
182.	$C_2H_5+C_2H_5$	=	$C_2H_4+C_2H_6$	$2.4 \cdot 10^{-12}$
183.	$C_2H_5+C_2H_5$	=	C_4H_{10}	$1.9 \cdot 10^{-11}$
184.	C_2H_5+CHO	=	$CO+C_2H_6$	$2 \cdot 10^{-10}$
185.	$C_2H_5+CH_3O$	=	$CH_2O+C_2H_6$	$4 \cdot 10^{-11}$
186.	$C_2H_5+CH_3CO$	=	$C_2H_5COCH_3$	$3 \cdot 10^{-11}$
187.	$C_2H_5+CH_3O_2$	=	$C_2H_5O+CH_3O$	$4 \cdot 10^{-11}$
188.	$C_2H_5+C_2H_5O_2$	=	$2C_2H_5O$	$4 \cdot 10^{-11}$
189.	$CHO+CHO$	=	$CO+CH_2O$	$5 \cdot 10^{-11}$
190.	$CHO+CH_3O$	=	$CO+CH_3OH$	$1.5 \cdot 10^{-10}$
191.	$CHO+CH_3CO$	=	CH_3COCHO	$3 \cdot 10^{-11}$
192.	$CHO+CH_3CO$	=	$CH_3CHO+CO$	$1.5 \cdot 10^{-11}$
193.	$CHO+CH_3O_2$	=	$CH_3OOH+CO$	$5 \cdot 10^{-11}$
194.	$CHO+C_2H_5O_2$	=	$C_2H_5OOH+CO$	$5 \cdot 10^{-11}$
195.	$CH_3O+CH_3.O$	=	CH_3OH+CH_2O	$1 \cdot 10^{-10}$
196.	CH_3O+CH_3CO	=	CH_3OH+CH_2CO	$1 \cdot 10^{-11}$
197.	CH_3O+CH_3CO	=	$CH_3CHO+CH_2O$	$1 \cdot 10^{-11}$
198.	$CH_3O+CH_3O_2$	=	$CH_3OOH+CH_2O$	$5 \cdot 10^{-13}$
199.	$CH_3O+C_2H_5O_2$	=	$C_2H_5OOH+CH_2O$	$5 \cdot 10^{-13}$
200.	CH_3CO+CH_3CO	=	CH_2CO+CH_3CHO	$1.5 \cdot 10^{-11}$
201.	$CH_3CO+CH_3O_2$	=	$CH_3+CO_2+CH_3O$	$4 \cdot 10^{-11}$
202.	$CH_3CO+C_2H_5O_2$	=	$C_2H_5+CO_2+CH_3O$	$4 \cdot 10^{-11}$
203.	$CH_3O_2+CH_3O_2$	=	$2CH_3O+O_2$	$1.3 \cdot 10^{-13}$
204.	$CH_3O_2+C_2H_5O_2$	=	$CH_3O+C_2H_5O+O_2$	$2 \cdot 10^{-13}$
205.	$C_2H_5O_2+C_2H_5O_2$	=	$2C_2H_5O+O_2$	$4 \cdot 10^{-14}$
206.	CH_2OOH	=	CH_2O+OH	$5 \cdot 10^{-4}$
207.	$O_2+C_2H_5CO$	=	$CH_3CHO+CO+OH$	$3 \cdot 10^{-12}$
208.	$CH_3CHO+C_2H_5CO$	=	$C_3H8CO+CHO$	$2.8 \cdot 10^{-13}$
209.	O_2+CH_3CHOH	=	$CH_3CHO+HO_2$	$1.9 \cdot 10^{-11}$
210.	$CH_3CHOH+O$	=	$CH_3CHO+OH$	$3.2 \cdot 10^{-10}$
2.	$CH_3CHOH+H$	=	CH_3CHO+H_2	$8 \cdot 10^{-11}$
212.	CH_2CO+O	=	$2CHO$	$7 \cdot 10^{-13}$
213.	CH_2CO+OH	=	CH_2OH+CO	$1.2 \cdot 10^{-11}$
214.	$C_2H_5O+O_2$	=	$CH_3CHO+HO_2$	$1 \cdot 10^{-14}$
215.	$HOCH_2CH_2+H$	=	CH_3CHO+H_2	$8.3 \cdot 10^{-11}$
216.	$HOCH_2CH_2+O_2$	=	$CH_3CHO+HO_2$	$3 \cdot 10^{-12}$
217.	CH_2OH+O	=	$OH+CH_2O$	$1.5 \cdot 10^{-10}$
218.	CH_2OH+H	=	H_2+CH_2O	$5 \cdot 10^{-11}$
219.	CH_2OH+O_2	=	HO_2+CH_2O	$9.6 \cdot 10^{-12}$

(Continued)

TABLE 4.24 (Continued)
Kinetics of Neutrals in Hydrocarbons (C_2H_6, CH_4, H_2, C_2H_5OH)–Oxygen Mixtures

N	Reaction			Rate Coefficients, cm³/s
220.	$CH_2OH + OH$	=	$H_2O + CH_2O$	$9.6 \cdot 10^{-12}$
221.	$CH_2OH + HO_2$	=	$H_2O_2 + CH_2O$	$2 \cdot 10^{-11}$
222.	$CH_2OH + CHO$	=	$CO + CH_3OH$	$2 \cdot 10^{-11}$
223.	$CH_2OH + CHO$	=	$2CH_2O$	$3 \cdot 10^{-11}$
224.	$CH_2OH + CH_3O$	=	$CH_3OH + CH_2O$	$4 \cdot 10^{-11}$
225.	$CH_2OH + C_2H_5O$	=	$C_2H_5OH + CH_2O$	$4 \cdot 10^{-11}$
226.	$CH_2OH + CH_3$	=	$CH_4 + CH_2O$	$1.4 \cdot 10^{-10}$
227.	$CH_2OH + C_2H_5$	=	$C_2H_6 + CH_2O$	$4 \cdot 10^{-12}$
228.	$CH_2OH + C_2H_5$	=	$C_2H_4 + CH_3OH$	$4 \cdot 10^{-12}$
229.	$CH_2OH + H_2$	=	$CH_3OH + H$	$1 \cdot 10^{-13}$
230.	$CH_2CHO + O_2$	=	$CH_2O + OH + CO$	$3 \cdot 10^{-14}$
231.	$H + CO_2$	=	$CO + OH$	$1 \cdot 10^{-29}$

4.8 IGNITION OF H_2-O_2-AR MIXTURES BY NANOSECOND GAS DISCHARGE

4.8.1 Theoretical Analysis of Ignition Efficiency

The discharge generates active species as it develops in the combustible gas mixture [102]. In order to assess the impact of the fast ionization wave on the combustion dynamics, it is imperative to determine the density and sort of the active species generated by the discharge [103]. Since the duration of the discharge (10–100 ns in this context) is very short compared to the typical time frame of combustion processes (spanning microseconds), the discharge is treated as a preliminary stage. Subsequently, the dynamics within the discharge and the subsequent chemical reactions in the afterglow are analyzed as distinct segments. The concentrations of atoms and radicals generated during the discharge phase are used as initial conditions for subsequent high-temperature calculations in the discharge afterglow.

Extensive experimental studies of nanosecond discharges have shown [104] that an initial peak in the electric field within the ionization wavefront is observed during the first few nanoseconds, with its peak value approaching or exceeding the electron runaway threshold. Subsequently, over the next few tens of nanoseconds, the field decreases from levels conducive to active excitation of the internal gas degrees of freedom to zero. It has been found [96] that both the prolific generation of electrons and the population of electron states occur predominantly behind the ionization wavefront in "residual" fields (measured in hundreds of V/(cm·Torr)). For a relatively short discharge gap, the contribution of active particle production at the discharge front can be ignored and the focus can be placed on the generation of atoms and radicals after the bridging of the discharge gap [105]. The typical propagation velocity of a nanosecond discharge is in the range of 5–10 cm/ns. For a discharge gap of 20 cm, the transit time of the FIW from the high-voltage electrode to the low-voltage electrode is approximately 2–4 ns, which is significantly less than the typical duration of the high-voltage pulse.

To determine the densities of active species, it is imperative to compute the electron energy distribution function (EEDF) by solving the Boltzmann equation for electrons. In scenarios characterized by elevated electric fields, an accurate solution of this equation requires statistical modeling [106]. Excluding the FIW front, as discussed earlier, the two-term approximation of the Boltzmann equation can be used [107]. For discharge analysis, the standard software [19,98] was used. Both algorithms allow the calculation of a stationary electron energy distribution function and the distribution of electron energy over different processes. Inputs include a value of reduced electric field E/n (where E is the electric field and n is the gas density) and the gas composition, while outputs include the population rates of vibrational and electronically excited molecules, atoms, and radicals.

The set of processes used for EEDF calculations in a hydrogen-oxygen mixture is cataloged in Table 4.9. The excitation cross sections are taken from [8,17,108] for hydrogen and [9,28,30,31,109,110] for oxygen, with dissociation treated as in [101]. For the hydrogen molecule, dissociation takes place via $b^3\Sigma_u^+$ and $a^3\Sigma_g^+$ electronically excited states, while for oxygen via $A^3\Sigma_u^+$ and $B^3\Sigma_u^-$ states.

The concentrations of atoms and radicals produced in the discharge serve as the basis for subsequent calculations of high-temperature ignition. Analogous methods have been applied to other mixtures under investigation.

The kinetics of high-temperature plasma-assisted ignition was formulated using a kinetic model of Konnov [111] and the GRI-Mech 3.0 mechanism. For the mixtures studied at elevated temperatures, both kinetic schemes yielded congruent results. It is critical to recognize that the kinetics of electronically excited particles play a significant role in the afterglow of nonequilibrium gas discharges. The question of the influence of these excited particles on ignition processes at high temperatures will be addressed in subsequent discussions.

The electrical discharge was modeled as a square pulse characterized by a predefined amplitude and a duration of 40 ns. For preliminary analysis, the electric field strength was set to $E/n = 300$ Td ('Td' or Townsend denotes a unit of measurement in gas discharge physics, where 1 Td=0.33 V/(cm×Torr)=$10^{-17} \cdot V \cdot cm^2$ at 20°C), representative of the conditions that prevail in a high-current nanosecond discharge following the ionization wave front. First, the electron energy distribution function (EEDF), the energy distribution among different collision processes, and the densities of atoms and radicals produced during the discharge were calculated. Composition of the gas mixture, including the densities of atoms and radicals, and the gas temperature served as the initial point for ignition kinetics calculations. The ignition delay time was characterized as a point when a temperature rise reaches 200 K [103].

Numerical modeling has identified the parameter range within which the system exhibits an increased responsiveness to discharge influence. Figure 4.65 shows the results of numerical simulations evaluating the ignition delay time within a stoichiometric H_2-air mixture at a total pressure of $p=1$ atm. These simulations were performed for both autoignition scenarios and nanosecond discharge conditions. The zero excitation energy scenario is consistent with autoignition at a given temperature. The ignition delay time of the system shows significant sensitivity to the addition of

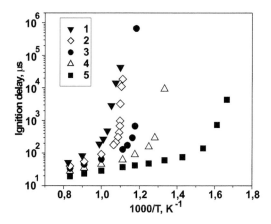

FIGURE 4.65 Calculated ignition delay time *vs* initial gas temperature. H_2: O_2: N_2=29.6: 14.8: 55.6 mixture. $p=1$ atm. Nonequilibrium excitation. $E/n=300$ Td. Different curves correspond to different energy consumption: 1 – 0 J/cm^3 (autoignition); 2 – $4 \cdot 10^{-4}$ J/cm^3; 3 – $4 \cdot 10^{-3}$ J/cm^3; 4 – $4 \cdot 10^{-2}$ J/cm^3; 5 – $4 \cdot 10^{-1}$ J/cm^3.

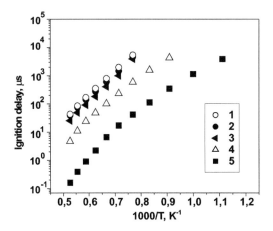

FIGURE 4.66 Calculated ignition delay time *vs* initial gas temperature. CH_4: O_2: N_2=9.4: 18.8: 71.8 mixture. $p=1$ atm. Nonequilibrium excitation. $E/n=300$ Td. Different curves correspond to different energy consumption: 1 – 0 J/cm³ (autoignition); 2 – 10^{-3} J/cm³; 3 – 10^{-2} J/cm³; 4 – 10^{-1} J/cm³; 5 – $4 \cdot 10^{-1}$ J/cm³.

nonequilibrium gas excitation. An excitation energy of $W = 4 \times 10^{-4}$ J/cm³ significantly reduces both the ignition delay time and the ignition temperature threshold. At an energy level of $W = 4 \times 10^{-1}$ J/cm³, the ignition threshold shift reaches a magnitude of 400 K. Consequently, variations in discharge energy offer the potential to adjust the ignition threshold and ignition delay time of a H_2-air mixture over a wide range.

In the CH_4-air mixture, the excitation induced by the pulsed discharge is significantly lower in intensity than in the H_2-air mixture. This disparity is attributed to the relatively modest cross-section of the methane molecule for electron impact dissociation. In addition, the methane concentration in a stoichiometric mixture with air is significantly lower than that of H_2, resulting in a minimal shift in the ignition threshold for CH_4-air mixtures at lower discharge energy levels. Conversely, at higher discharge energies ($W = 0.4$ J/cm³), the ignition threshold shift exceeds 400 K within the studied temperature range, as shown in Figure 4.66, where the energy release is denoted.

A comparison of these results with those obtained under conditions of equilibrium excitation, i.e. thermal heating, provides interesting insights. Figure 4.67 shows such an analysis for the H_2-air mixture. In particular, the ignition delay time changes even under equilibrium excitation, where all degrees of freedom are heated simultaneously. However, for the same magnitude of energy release, $W = 4 \times 10^{-3}$ J/cm³, gas excitation via a nonequilibrium discharge characterized by a reduced electric field $E/n = 300$ Td results in a more significant reduction of the ignition threshold. This difference in ignition efficiency between fast ionization wave (FIW) excitation and thermal heating depends on the magnitude of the energy input (as shown in Figure 4.68). Nonequilibrium excitation shows increased efficiency at lower energy inputs. High efficiency can be rationalized by recognizing that the generation of atoms and radicals, in the absence of additional systemic heating, increases the reaction rate and significantly alters the ignition delay. The effects of nonequilibrium energy input are equivalent to those of thermal heating at an energy input level of 0.1 J/cm³. In our experiments, the typical energy introduced by the nanosecond discharge was between 10^{-3} and $5 \cdot 10^{-2}$ J/cm³, placing it in the range where thermal ignition as a result of additional heating is less effective than the ignition facilitated by radicals' production in the gas discharge.

In addition, the effect of the electric field strength in the discharge on the ignition delay time was studied (as shown in Figure 4.69). It was deduced that the range of reduced electric fields considered most effective for gas excitation in terms of ignition initiation at elevated temperatures in typical air-fuel gas mixture is between 250 and 350 Td. This range is consistent with the values of reduced electric fields in the region of the maximum energy deposition in the fast ionization waves [90].

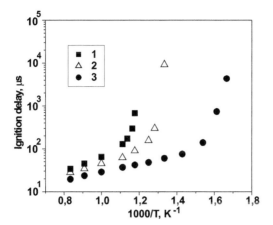

FIGURE 4.67 Calculated ignition delay time *vs* initial gas temperature for thermal and nonequilibrium plasma excitation. H_2: O_2: N_2 = 29.6: 14.8: 55.6 mixture. $p=1$ atm. Total energy release $W=4\times10^{-3}$ J/cm^3. Reduced electric field is $E/n=300$ Td. 1 – autoignition; 2 – equilibrium excitation; and 3 – nonequilibrium excitation.

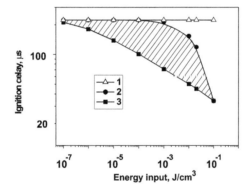

FIGURE 4.68 Calculated ignition delay time *vs* energy input for thermal and nonequilibrium plasma excitation. H_2: O_2: N_2 = 29.6: 14.8: 55.6 mixture. $p=1$ atm; $T_0=1{,}000$ K; $E/n=300$ Td. 1 – autoignition (corresponds to zero energy input); 2 – equilibrium excitation; 3 – nonequilibrium excitation ($E/n=300$ Td, $t=40$ ns).

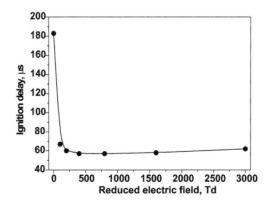

FIGURE 4.69 Calculated ignition delay time *vs* reduced electric field. H_2: O_2: N_2 = 29.6: 14.8: 55.6 mixture. $p=1$ atm; $T_0=1{,}000$ K. Energy release is $W=4\times10^{-3}$ J/cm^3; $t=40$ ns.

Thus, initial numerical modeling has predicted the range of temperatures and electric fields that are efficient for nonequilibrium plasma-assisted ignition. The temperature intervals of 800–1,000 K for hydrogen-containing mixtures and 1,200–1,800 K for methane-containing mixtures emerge as the focal point for experimental investigation. Such temperature ranges are readily achievable through the use of shock tube methods [112].

4.8.2 Plasma Shock Tube

The experimental setup used to study rapid homogeneous ignition in combustible mixtures, as shown in Figure 4.70, consists of a shock tube (ShT), a discharge cell (DC) connected to the shock tube, a discharge initiation mechanism, and a diagnostic system.

The stainless steel shock tube has a square cross-section of 25×25 mm^2 and a work channel length of 1.6 m. The high-pressure cell (HPC) is 60 cm long, with two pairs of diagnostic optical windows positioned along the stainless steel work channel. The final section of the shock tube is made of Plexiglas and has eight optical windows, six of which are quartz and two of which are IR-transparent CaF2$_2$ windows. The section has a 25×25 mm^2 internal square cross-section and measures 20 cm in length. A uniform nanosecond discharge, propagated as a fast ionization wave, was generated in the dielectric section; the end plate (EP) of the shock tube served as the high-voltage electrode for the discharge system, with the electrical circuit completed through the grounded stainless steel driven section of the shock tube.

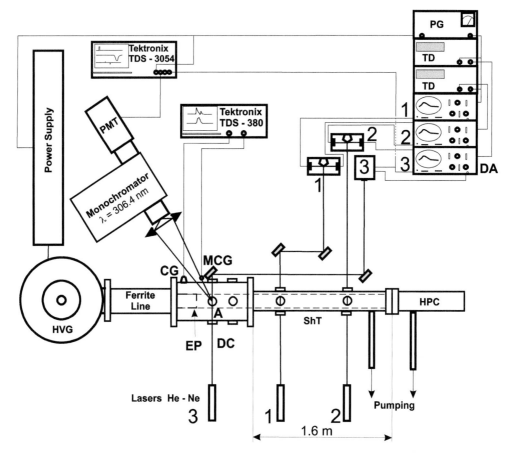

FIGURE 4.70 Scheme of the experimental setup.

4.8.2.1 Discharge Initiation

The initiation of the nanosecond discharge was synchronized with the arrival of the reflected shock wave at the observation point using a synchro-pulse generator (PG). A Marks-type generator was used to generate the high-voltage pulse. The high-voltage generator (HVG) consists of ten stages. To ensure spark gap operability, the HVG is pressurized with nitrogen at 3 atm, allowing a starting voltage range of 100- to 160 kV. A ferrite line of high nonlinear magnetic permittivity with an impedance of $Z = 40$ Ohm is used to refine the voltage pulse transmitted from the generator to the discharge gap. At the output of the ferrite line, the voltage rise rate reaches 16 kV/ns, allowing the generation of a gas discharge in the form of a fast ionization wave inside the dielectric segment of the shock tube. The propagation velocity of the ionization wave front is in the range of $10^9 - 10^{10}$ cm/s depending on the gas density.

4.8.2.2 Vacuum System and Mixtures Used

The gas mixtures used, consisting of either methane or hydrogen combined with oxygen and diluted with either Ar or He, were chosen to achieve specific pressure and temperature conditions after shock wave reflection (see Table 4.25 for details). The initial pressure within the driven section was varied between 5 and 40 Torr. The addition of Ar or He as diluents was used to increase the specific heat ratio, thus facilitating the achievement of higher temperatures after shock wave reflection. Helium, dry air or CO_2 were selected as the driving gas for these experiments. In some experiments, a special perforated flow plate was positioned between the high pressure and driven sections, in front of the diaphragm. This configuration was designed to moderate the velocity of the shock wave without changing initial gas pressure.

4.8.2.3 Diagnostic System

The diagnostic system is divided into three distinct components: one for monitoring shock wave parameters, another for observing the electrical parameters of the nanosecond discharge, and a third for registering ignition events.

The shock wave parameter monitoring component includes a system for measuring the velocities of the incident and reflected shock waves using the laser schlieren technique and a mechanism for controlling the gas pressure. The schlieren system consists of three He-Ne lasers positioned at different points along the shock tube and three pairs of photodiodes equipped with differential analyzers (DA). The time delays observed between points 2–1 and 1–3 along the length of the tube are recorded by time delay analyzers (TD). Using the known composition of the initial gas mixture, the initial pressure, and the velocity of the incident shock wave, the pressure, gas density, and temperature following the reflected shock wave are calculated by the application of conservation laws, assuming complete translational, rotational, vibrational and electronic relaxation and the absence of chemical reactions (see, for example, [113]).

The subsystem for monitoring the electrical parameters of the nanosecond discharge consists of a calibrated magnetic current gauge (MCG) for controlling the current pulse and a capacitance gauge (CG) placed above the high-voltage electrode to observe the shape and amplitude of the high-voltage pulse. An additional capacitance gauge was positioned near the observation point (12 cm from the first CG) to estimate the propagation velocity of FIW. The outputs from these electrical gauges

TABLE 4.25
Composition of the Gas Mixtures

H_2	O_2	He, %	Ar, %
12	6	—	82
12	6	82	—

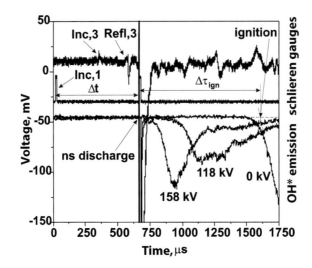

FIGURE 4.71 Typical behavior of synchronized signals from the schlieren system, delay generator, and photomultiplier at different amplitudes of high-voltage pulse. Mixture H_2: O_2: N_2: Ar=6: 3: 11: 80.

were recorded using a digital oscilloscope. To reduce high-frequency electrical noise, all cables were further shielded and the oscilloscopes were placed in a shielded chamber. Figure 4.71 shows the characteristic patterns of the electrode voltage, along with the voltage and current near the observation point. The propagation velocity of the fast ionization wave can be determined from the data obtained using capacitance gauges, with recorded velocities of approximately 10 cm/ns. It is evident that the maximum energy consumption occurs within the first 20–30 ns, characterized by a sufficiently high current. Under these conditions, the generation of active species in the discharge and the subsequent ignition/combustion processes can be analyzed sequentially, both in terms of numerical modeling and experiments.

The ignition process was studied by emission spectroscopy. Combustion-induced emissions were observed perpendicular to the shock tube axis, 55 mm from the end plate (referred to as point A), using a UV-VIS monochromator (dispersion 1.2 nm/mm), a UV photomultiplier, and a 1 GHz digital oscilloscope. The chosen observation point, relatively distant from the endplate of the shock tube, was intended to minimize the effects of boundary layers, although proximity to the endplate is typically preferred for kinetic measurements behind the reflected shock wave to reduce boundary layer effects. On the other hand, the homogeneity of the plasma increases with the distance from the high-voltage electrode (end plate of the shock tube). Thus, the location has been chosen to minimize disturbances both from the electric field side and from the gas dynamic side.

The OH emission ($\lambda = 306.4$ nm, $A^2\Sigma(v' = 0) \rightarrow X^2\Pi(v'' = 0)$ transition) was measured under a wide range of experimental conditions, including different mixture compositions and temperature ranges from 750 to 2,250 K, as well as pressure fluctuations between 0.3 and 2.3 atm. These experiments were also performed with high-voltage pulse amplitudes at the electrode ranging from 100 to 160 kV and pulse durations of approximately 30-40 ns at full width at half maximum (FWHM).

4.8.2.4 Synchronization

The characteristic synchronization of the signals from the third schlieren gauge, the synchro generator, and the photomultiplier over varying amplitudes of the high-voltage pulse is shown in Figure 4.71. For clarity, the signals from each detector are offset vertically on the y axis. The signal from the third schlieren gauge positioned at the observation point forms the top curve, with discernible peaks for both the incident (labeled "Inc,3") and reflected ("Refl,3") shockwaves. The second curve shows the sync signal, aligned with the arrival of the incident wave at the first schlieren gauge (labeled "Inc,1" in the figure). An adjustable delay Δt facilitates the precise timing of discharge activation, ensuring that the discharge begins after the reflected shock wave reaches reference point A. The interval between the gas heating by shock wave and discharge initiation is significantly shorter than the ignition delay observed in autoignition under equivalent temperature conditions. Typically, this time (from "Refl.,3" to "ns discharge" in the figure) is in the range of 20–40 µs.

The pronounced, narrow peak observed at 660 µs is due to electrical noise and OH emission resulting from the discharge. This peak is followed by the emission of OH radicals due to combustion processes. In all cases, the ignition delay time $\Delta \tau$ was quantified as the interval between the last physical intervention (namely the reflected shock wave signal from the third schlieren gauge in the case of autoignition, or the start of the discharge in the case of nanosecond discharge) and the onset of OH emission. The ignition delay is reduced under the influence of the fast ionization wave (FIW), with an inverse relationship between voltage magnitude and ignition delay. It is noteworthy that in this case the ignition delay time for autoignition exceeds the operating time of the shock tube ($t_{max} \approx 400-500$ µs), and this result is presented to show the wide range of effects of the discharge. All data points collected are shown in the figures, including those where the ignition delay time was $\tau_{ign} > t_{max}$, to provide baseline comparisons for ignition facilitated by nanosecond discharge.

4.8.3 Ignition Delay Time Change in H_2-O_2 Mixture by Pulsed Discharge

Representative data from the capacitive gauge and the magnetic current gauge are shown in Figure 4.72. The electric field was determined using the estimation $(U_2 - U_1)/l$, where l is the distance between the capacitive detectors. The energy input was calculated by integrating the product of $U \times I$ over time.

In regimes characterized by relatively low gas number densities, the ignition curve (the relationship between ignition delay time τ and temperature) is expected to shift toward lower temperatures. Experimental studies have been conducted using a variety of mixtures and pressures to achieve the particle number densities behind the reflected shock wave that would optimally facilitate nanosecond discharge development.

To illustrate the variations in gas number densities, Figure 4.73 shows the gas density behind the reflected shock wave for various mixtures. It is evident that system parameters can be adjusted to control the gas density, with the arrow indicating the direction of decreased gas number density. Modifications such as changing the high-pressure gas, adding a perforated flow plate, or changing the dilution gas can shift the curve $\rho_5(T)$. The minimum temperature achieved in these experiments was about 750 K, observed when the mixture H_2: O_2: He = 12: 6: 82 was ignited (as shown in Figure 4.74). Broadly speaking, there are two strategies for improving ignition efficiency in this

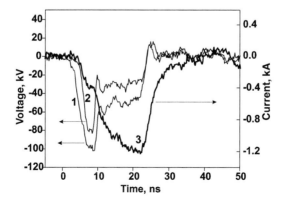

FIGURE 4.72 Typical voltage on the electrode (1), voltage (2), and current (3) at the observation point.

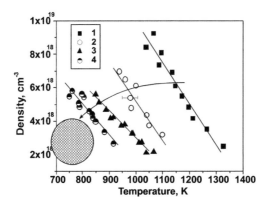

FIGURE 4.73 Densities behind the reflected shock wave, realized in different mixtures, vs temperature. 1 – mixture H_2: O_2: Ar = 12: 6: 82, high-pressure gas is air; 2,3 – mixture H_2: O_2: Ar = 12: 6: 82, high-pressure gas is CO_2, in the case of curve 3 special flow plate was used to decrease shock wave velocity; 4 – mixture H_2: O_2: He = 12: 6: 82, high-pressure gas is He, experiments with flow plate. Dashed circle corresponds to a region of densities where discharge development at $U = 160$ kV is optimal.

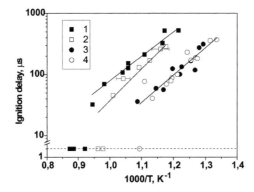

FIGURE 4.74 Measured ignition delay *vs* gas temperature for different voltage polarity. 1,2 – Mixture H_2: O_2: Ar = 12: 6: 82; 3,4 – Mixture H_2: O_2: He = 12: 6: 82. 1,3 – positive polarity of the electric pulse; 2,4 – negative polarity. $U = 160$ kV.

parameter range: reducing the density to the levels outlined by the dashed circle in Figure 4.73 or increasing the high-voltage amplitude.

The difference between the effect of positive and negative polarity pulses on ignition has been studied. It is recognized that the discharge propagates from the high-voltage electrode to the low-voltage one under both polarities [114], but the initial conditions, the velocity of ionization front propagation, the electric field distribution in the discharge, and other parameters are significantly affected by the polarity [115].

Figure 4.74 shows an example of ignition delays observed under varying electrical pulse polarities. For the H_2: O_2: Ar = 12: 6: 82 mixture, the difference in ignition threshold temperatures is between 30 and 40 K, while for the H_2: O_2: He = 12: 6: 82 mixture, no discernible difference was observed for discharges of positive and negative polarities. In order to determine which polarity is more preferable to ignition, a more detailed study involving a thorough analysis of the effect of the discharge on the kinetics of the system is required.

The ignition of the H_2: O_2: Ar = 0.12: 0.06: 0.82 mixture was analyzed numerically using the approach discussed earlier in this chapter. Figure 4.75 compares the calculated results with the experimental results. The voltage applied to the high-voltage electrode was 160 kV with a pulse

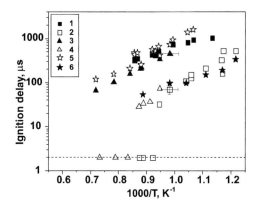

FIGURE 4.75 Comparison of the measured ignition delay time with calculated one. Mixture H_2: O_2: Ar = 12: 6: 82. 1 – autoignition, high-pressure gas is CO_2; 2 – with discharge, high-pressure gas is CO_2; 3 – autoignition, high-pressure gas is air; 4 – with discharge, high-pressure gas is air; 5 – numerical simulation for autoignition, GRI-Mech 3.0; 6 – numerical simulation for the shock wave and the discharge action, BOLSIG + GRI-Mech 3.0.

duration of 40 ns. The densities of O and H atoms were calculated using experimental estimates of the temporal electric field profile in the discharge $E(t)$ and a standard BOLSIG solver [98]. The GRI-Mech 3.0 mechanism was used to model the kinetics in the high-temperature ignition in the discharge afterglow [116]. The ignition delay time in these simulations was identified by a pronounced increase in the OH mole fraction.

The GRI-Mech 3.0 mechanism has been rigorously validated for this range of parameters, and the congruence observed between the experimentally measured autoignition delays and the computational results at temperatures $T < 1,200$ K and $\tau_{ign} < 600$ μs underscores the reliability of the experimental data. In regimes where $T > 1,200$ K, the initial pressure was maintained below 10 Torr, indicating that the discrepancies observed within this parameter range may be due to the pronounced influence of boundary layers on flow dynamics.

Notably, the agreement between experimental measurements and computational predictions is also sufficiently robust in cases involving discharge-initiated ignition. It is pertinent to mention that our analysis did not include the electronically excited species generated during the discharge; only the density of dissociated species was considered. The importance of these excited species for high-temperature ignition needs to be elucidated further. However, the favorable correlation between experimental and simulation results supports the utility of this simple kinetic model for evaluating the effectiveness of nanosecond discharges in igniting flammable mixtures at temperatures above the autoignition threshold.

4.9 IGNITION DELAY TIME CHANGE IN H_2-AIR MIXTURE BY PULSED DISCHARGE

4.9.1 Experiments in H_2: O_2: N_2: Ar Mixture

The relationship between ignition delay and temperature at a constant voltage of $U = 160$ kV was elucidated (see Table 4.26). The observed ignition delay times for the H_2-air-Ar mixture are shown in Figure 4.76. At $T = 1,000$ K, the energy contribution from the discharge reduces the ignition delay time by a factor of 4.8, from 860 to 140 μs. It is important to note that in this figure, as well as in subsequent figures detailing experimentally determined ignition delays, the dashed line at 2 μs represents a nominal value indicating the typical duration of electrical noise from the discharge and thus represents the minimum detectable limit of our measurements. If the ignition delay falls below this threshold, it exceeds the resolution capabilities of the experimental setup [103].

TABLE 4.26
Composition of the Gas Mixtures

H_2, %	O_2, %	N_2, %	Ar, %
6	3	11	80
16.7	16.7	66.6	—

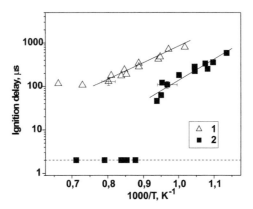

FIGURE 4.76 Measured ignition delay time *vs* gas temperature. Mixture H_2: O_2: N_2: Ar=6: 3: 11: 80. 1 – autoignition, 2 – with nanosecond discharge. $U=160$ kV.

4.9.2 Experiments in H_2: O_2: N_2 Mixture

The mixture H_2: O_2: N_2 exhibits distinct characteristics both in terms of discharge evolution and chemical reaction dynamics, due to the substantial presence of nitrogen. During the discharge phase, this leads to energy transfer to the vibrational degrees of freedom of N_2, as well as excitation of electronic levels of N_2 (e.g., A^3, B^3, C^3). In the afterglow phase, the electronically excited levels of N_2 are quenched by molecular oxygen, resulting in the formation of atomic oxygen. Consequently, in mixtures containing N_2, it is imperative to consider additional processes during both the discharge and afterglow phases.

The shock tube used in these studies was carefully designed to ensure a high repeatability of experimental conditions. Figure 4.77 shows the consistency of shock wave velocities across different experiments, coupled with the attenuation of the shock wave within the channel due to viscous dissipation.

To illustrate the variations in gas number densities, Figure 4.78 shows the gas density behind the reflected shock wave in various experimental regimes. It is evident that the parameters of the system can be modulated to effectively control the gas density. The voltage for this part of experiments was kept constant at 100 kV, which allowed to maintain high E/n values in the discharge. Variations in energy input, as shown in Figure 4.79, are attributed solely to changes in gas density.

The pressure range corresponding to the experimental conditions of this study is shown in Figure 4.80. The pressure after the reflected shock wave ranged from 0.1 to 0.6 atmospheres. Such conditions facilitated the development of uniform discharge in the discharge section.

The relationship between ignition delay time and temperature was established at a constant voltage of $U = 100$ kV. The observed time delays for ignition of the H_2-air mixture are shown in Figure 4.81. A notable difference between the undiluted H_2-air mixture and its analog—mixture diluted with Ar—is the specific heat ratio (γ). The H_2-air mixture has a γ of 1.4, while the H_2-air-Ar mixture has a γ of approximately 1.6. Lower specific heat ratio results in a pronounced interaction

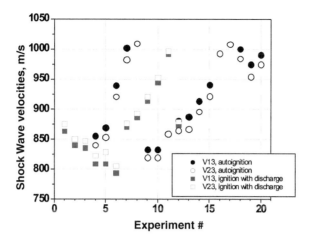

FIGURE 4.77 Velocity of the incident shock wave. H_2-O_2-N_2 = 16.7: 16.7: 66.6 mixture.

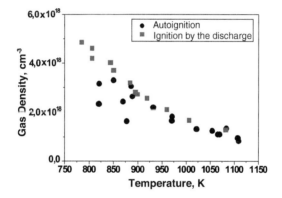

FIGURE 4.78 Gas density behind the reflected shock wave. H_2-O_2-N_2 = 16.7: 16.7: 66.6 mixture.

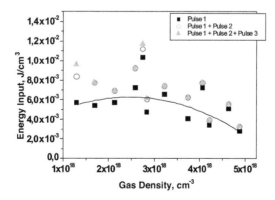

FIGURE 4.79 Energy of the discharge for different gas densities. H_2-O_2-N_2 = 16.7: 16.7: 66.6 mixture.

between the reflected shock wave and the boundary layer in the H_2-air mixture, leading to the formation of a so-called λ-shaped shock wave configuration. As a result, the gas parameters behind the reflected shock wave deviate significantly from one-dimensional theory predictions. This deviation contributes to a certain discrepancy between the measured and calculated ignition delay times for

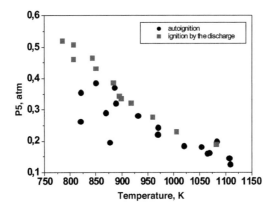

FIGURE 4.80 Pressure behind the reflected shock wave. H_2-O_2-N_2 = 16.7: 16.7: 66.6 mixture.

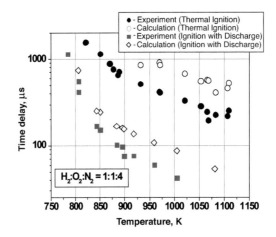

FIGURE 4.81 Comparison of the measured ignition delay time with calculated one. Mixture H_2: O_2: N_2 = 0.167:0.167:0.666. 1 – autoignition; 3 – with discharge; 2 – numerical simulation for autoignition, GRI-Mech 3.0, no gasdynamic correction; 4 – numerical simulation for the shock wave and the discharge action, BOLSIG+GRI-Mech 3.0, no gasdynamic correction. Log scale.

the H_2-air mixture, as shown in Figure 4.81. The nature and implications of this discrepancy will be discussed in the following sections.

4.10 COMPARISON WITH THE COMPUTATIONAL MODEL

4.10.1 Gasdynamic Model of the Shock Tube

The nonequilibrium flow behind the reflected shock wave has been numerically studied. Two-dimensional, axially symmetric simulations of the flow were performed. The MacCormack scheme with the Flux-Corrected Transport (FCT) correction was used to solve the fundamental equations governing chemistry, mass, momentum, and energy conservation:

$$\rho_t + (\rho u)_x + (\rho v)_y = 0$$

$$(\rho u)_t + (\rho u^2 + p)_x + (\rho u v)_y = 0$$

Mechanisms of Plasma-Chemical Reactions

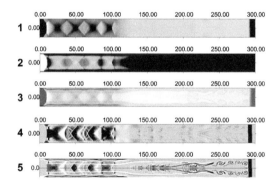

FIGURE 4.82 Gas dynamic fields. H_2-O_2-N_2 = 16.7: 16.7: 66.6 mixture. 1 – pressure field; 2 – v_x velocity field; 3 – temperature field; 4 – pressure gradients picture; 5 – temperature gradients picture. Incident shock wave velocity 900 m/s. $p_0 = 10$ Torr.

$$(\rho v)_t + (\rho v^2 + p)_y + (\rho u v)_x = 0$$

$$e_t + ((e+p)u)_x + ((e+p)v)_y = S_i h_i^0 w_i$$

$$(c_i \rho)_t + (c_i \rho u)_x + (c_i \rho u)_y = 0, \quad i = 1, n$$

Vibrational excitation and energy exchange in the N_2-O_2-H_2 plasma flow have been considered as well (Figure 4.82).

Figure 4.82 clearly illustrates the pronounced interaction between the reflected shock wave and the boundary layer, leading to the formation of a λ-shaped configuration and subsequent changes in gas parameters behind the reflected shock wave. The typical deviation in temperature from one-dimensional theory predictions in such scenarios is about $\Delta T \sim 100 - 150$ K during the ignition delay period. Furthermore, the calculations show that the magnitude of this deviation depends on the boundary layer regime (laminar or turbulent). Given the challenges in accurately reconstructing the real flow structure within the shock tube, the numerical results are considered to be of a semi-quantitative nature only. Actual deviations can be further elucidated by infrared emission measurements.

In this case, one will evaluate the flow parameters after the reflected shock wave and the discrepancies from the 1D model, using results from two-dimensional modeling along with experimental data on autoignition. The autoignition delay time of the H_2-O_2-N_2 mixture can be numerically predicted with high accuracy, serving as a "kinetic thermometer" for these conditions. The estimated temperature deviation due to gasdynamic effects, as determined by this "kinetic thermometer", is in close agreement with the predictions of the numerical model - ΔT is in the range of 120–160 K. This agreement allows the application of semi-quantitative adjustments to temperature measurements in both autoignition and plasma-assisted ignition experiments.

4.10.2 Kinetic Model

To quantify the densities of active species, it is essential to determine the electron energy distribution function. In scenarios characterized by elevated electric fields, an exact resolution of this equation requires statistical modeling [106]. Excluding the consideration of the fast ionization wave (FIW) front as mentioned above, the two-term approximation of the Boltzmann equation can be used [107]. For the purpose of discharge analysis, both the BOLSIG software [98] and the Gordeev

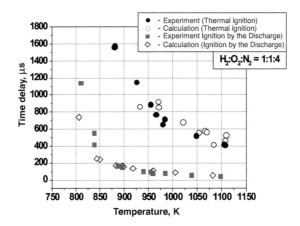

FIGURE 4.83 Comparison of the measured ignition delay time with calculated one. Mixture H_2: O_2: N_2=0.167: 0.167: 0.666. 1 – autoignition; 3 – with discharge; 2 – numerical simulation for autoignition, GRI-Mech 3.0, with gasdynamic correction; 4 – numerical simulation for the shock wave and the discharge action, BOLSIG+GRI-Mech 3.0, with gasdynamic correction.

solver [19] were used. These computational tools are designed to determine a steady-state electron energy distribution function along with the energy distribution over various collisional interactions of electrons with atoms and molecules.

The list of processes used for EEDF calculations in a hydrogen-air mixture is described above and in Table 4.9. Excitation cross sections have been taken from references [8,17,108] for hydrogen and [9,28,30,31,109,110] for oxygen, with dissociation channels adapted from [101]. For hydrogen molecules, dissociation proceeds through $b^3\Sigma_u^+$ and $a^3\Sigma_g^+$ electronically excited states, while for oxygen it proceeds through $A^3\Sigma_u^+$ and $B^3\Sigma_u^-$ states. The densities of atoms and radicals produced by the discharge served as the basis for subsequent high-temperature ignition analyses. Analogous methods were applied to all other mixtures studied.

The ignition of the H_2: O_2: N_2=0.167: 0.167: 0.666 mixture under specified experimental conditions was modeled using the previously described methodology. Figure 4.83 shows a comparison of the computational results with the experimental data. The applied voltage at the high-voltage electrode was set to 160 kV with a pulse duration of 40 ns.

The observed agreement between experimental measurements and computational predictions, particularly in cases involving plasma-assisted ignition (as illustrated in Figure 4.83), is sufficiently robust. This conclusion allows us to extrapolate the plasma-assisted ignition models from the mixtures diluted with Ar and He to fuel-air mixtures containing molecular nitrogen. The main difference between these two cases is the dynamics of excitation and depopulation of the internal degrees of freedom of nitrogen, which leads to a significant change in the channels of atomic oxygen production. For example, collisional quenching of nitrogen triplet states by molecular oxygen produces more oxygen atoms than direct oxygen dissociation by electron impact (see discussion in previous chapters).

4.11 IGNITION DELAY TIME MEASUREMENTS IN CH_4: O_2: N_2: AR MIXTURE

The measured ignition delay times for the CH_4:O_2:N_2:Ar (1:4:15:80) mixture are shown in Figure 4.84. At a temperature of 1,000 K, the energy introduced by the discharge reduces the ignition delay time by a factor of 4.8, from 860 to 140 μs. It is important to note that the minimum detectable ignition delay time was ~ 2 μs because of an electrical noise from the discharge [117].

Ignition delay as a function of temperature at constant voltage of $U = 160$ kV has been measured. The ignition delay time undergoes significant changes under the influence of the nanosecond

Mechanisms of Plasma-Chemical Reactions

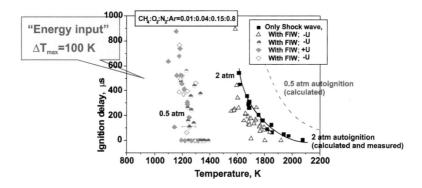

FIGURE 4.84 Measured ignition delay time *vs* gas temperature. Mixture CH_4: O_2: N_2: $Ar = 1: 4: 15: 80$. Pressure behind the reflected shock wave is indicated with numbers near appropriate curves. 1 – autoignition, 2 – with discharge, $W = 10^{-3}$ J/cm³; 3 – with discharge, $W = 3 \cdot 10^{-2}$ J/cm³; 4 – with discharge, $W = 5 \cdot 10^{-2}$ J/cm³.

discharge. In particular, ignition of a highly diluted methane-air mixture was achieved at a pressure of 0.3 atm and a temperature of 1,100 K with an energy release of 0.05 J/cm³ in the discharge. The figure also shows variations in the ignition delay time by nanosecond discharge at different pressures [117].

Curves corresponding to different pressure levels were derived by using different gases in the high-pressure chamber. Regime 1, achieved with helium, resulted in the maximum shock wave velocity and thus the highest temperature and density; regime 2 used dry air; and regime 3 was achieved with CO_2. Discharge development is efficient at gas densities $\sim 10^{18}$ cm⁻³. At this specific gas number density, energy of the discharge reaches its peak. Reducing the pressure p_5, and thus the particle number density from $\sim 10^{19}$ to 10^{18} cm⁻³, increases the energy consumption and significantly reduces the ignition delay. This phenomenon explains the tripartition observed in Figure 4.84.

In scenarios characterized by very low gas number densities, the ignition curve (the relationship between ignition delay time τ and temperature) is expected to shift toward the lowest possible temperatures. Experimental investigations have been conducted using a variety of mixtures, gases in the high-pressure chamber, and diaphragms to achieve optimal particle number densities after shock wave reflection for discharge efficiency.

The efficiency of discharge development, depending on the specified parameters of the generator, is maximized at gas densities of approximately 10^{18} cm⁻³. At such particle number densities, discharge energy consumption is at its peak. Reducing the pressure p_5, and thus the particle number density from about 10^{19} to 10^{18} cm⁻³, results in increased energy consumption and thus significantly reduces the ignition delay.

Returning to situations with relatively low gas number densities, it is predicted that adjustments can be made to the ignition curve to achieve as low a temperature as possible. Through experimentation with various mixtures, high-pressure chamber gases and diaphragms, efforts have been made to achieve particle number densities behind the reflected shock wave that are conducive to the most effective discharge development.

4.12 GAS DISCHARGE AND IGNITION HOMOGENEITY MEASUREMENTS

The PicoStar HR12 ICCD camera was used to capture images of discharge and ignition in a hot gas mixture with high temporal resolution. The wavelength sensitivity of the optical system ranged from 300 to 800 nm, allowing the acquisition of spectrally integrated images [102]. The synchronization scheme of the camera is shown in Figure 4.85, with the image intensifier gate set at 1 ns for nanosecond discharge and 30 μs for microsecond-range of combustion processes. The emissions were measured from a 2 cm diameter shock tube window.

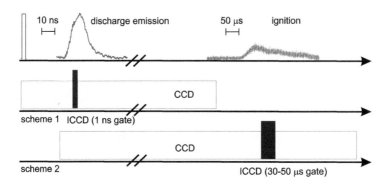

FIGURE 4.85 An example of a synchronization scheme for PicoStar Camera.

Figure 4.86 represents the different experiments for emission field measurement. These images allow to analyze the reproductivity of the gas discharge in the form of fast ionization wave, and homogeneity of mixture ignition. It is clearly seen that discharge remains spatially uniform up to pressures of about two atmospheres behind reflected shock waves. Ignition of the mixture is almost homogeneous. These experiments show the possibility of application of discharge in the fast ionization wave form to spatially uniform mixture excitation and ignition. It is clear from the experiments that the higher pressure the less uniform the discharge is. At relatively low pressure both discharge and combustion processes develop uniformly in space.

Figure 4.86 shows several experiments performed to evaluate the homogeneity of the gas discharge and ignition. These images show the high reproducibility of gas discharges in the form of fast ionization waves, as well as the homogeneity of mixture ignition. The discharge shows spatial uniformity up to about two atmospheres behind the reflected shock wave, and the ignition of the mixture appears nearly homogeneous as well. These observations underscore the feasibility of using fast ionization wave discharges for detailed kinetic analysis of plasma kinetics and plasma-assisted combustion.

4.13 PLASMA-ASSISTED IGNITION OF C1–C4 HYDROCARBONS

The uniform ignition of combustible gas mixtures is of great interest for both scientific research and technological applications. Fuel oxidation occurs by a fast chain mechanism, where the ignition delay time is limited by the rate of generation of active centers, typically by thermal dissociation. Consequently, the artificial initiation of a chain can increase the overall reaction rate by generating free radicals, usually achieved by breaking the weakest bond of the molecule [118].

When using a discharge to initiate combustion, it is important to consider the two primary mechanisms by which a discharge can affect a gas. In scenarios involving discharges that result in the formation of an equilibrium (or near-equilibrium) plasma, such as sparks and arcs, the predominant factor in reducing the ignition delay time is the localized heating of the gas. This heating accelerates the rate of thermal dissociation [101,119,120]. Conversely, in the context of nonequilibrium plasma, dissociation and molecular excitation by electron impact serve as the primary mechanisms for initiating chain reactions. The efficacy of using nonequilibrium plasmas for this purpose remains a subject of investigation. On the one hand, even modest concentrations of atoms and radicals (about 10^{-5} to 10^{-3} of the total number of gas particles) are capable of altering equilibria within the system and initiating chain reactions, suggesting potential efficiency. For example, Figure 4.87 illustrates the disparity in reaction rates, clearly indicating that electron impact dissociation in nonequilibrium plasma is significantly more efficient than thermal dissociation, which only becomes significant at high gas temperatures.

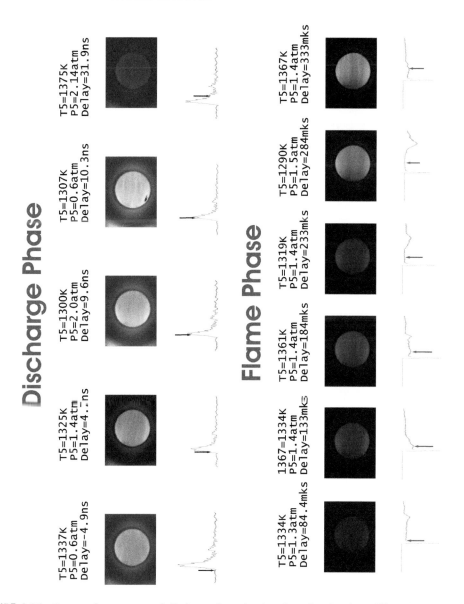

FIGURE 4.86 Image of a nanosecond discharge (gate 1 ns) and combustion (gate 30\mus) stages. Mixture $CH_4:O_2:N_2:Ar = 1:4:15:80$.

In addition, achieving a uniform distribution of active particles throughout the gas volume ensures that the resulting ignition will not lead to detonation. However, uniform ignition by a discharge over a substantial gas volume, especially when starting with a relatively high initial density of neutral particles, presents significant technical challenges.

The volumetric plasma-assisted ignition of hydrogen-air and methane-air mixtures diluted with either argon or helium has been studied in [103]. These mixtures were ignited by a spatially uniform nanosecond discharge. Experimental and computational analyses showed that for certain initial densities and temperatures, there was a significant reduction (up to 600 K) in the ignition temperature. A significant advantage of using high-voltage nanosecond discharges is their ability to almost instantaneously (within a few nanoseconds in experiments [103]) bridge a 20 cm discharge gap and subsequently generate the atoms and radicals necessary to initiate spatially uniform ignition. The

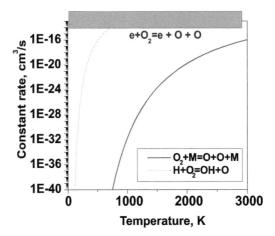

FIGURE 4.87 Comparison of typical reaction rates at different temperatures: solid line is a rate of reaction of thermal dissociation; dashed line is a rate of a typical reaction of chain propagation; square in the top of a plot – typical rate of reaction of dissociation by an electron impact at electric field values typical for the discharge.

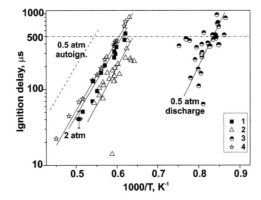

FIGURE 4.88 Ignition delay time vs temperature. Symbols: 1 – 2 atm, autoignition; 2 – 2 atm, plasma-assisted ignition; 3 – 0.5 atm, plasma-assisted ignition; dashed line – 0.5 atm, autoignition (calculated); 4 – 2 atm, autoignition (calculated).

duration of energy deposition in the gas could be extended to several hundred nanoseconds, limited by the transition of the discharge into an arc. Furthermore, it was shown [103] that the ignition delay time is significantly influenced by the discharge parameters and the gas density.

A comprehensive study of discharge parameters, as reported in the literature [106,121], facilitates the evaluation of the efficacy of plasma as a source of active particles. The primary objective of this part of the book is the analysis of plasma-assisted ignition kinetics of C1–C4 hydrocarbons, and discussion of the experiments aimed at determining the influence of dissociated and excited species on ignition processes (see also [117,122–127]). As baseline data, these papers analyzed results from experiments with a stoichiometric methane-synthetic air mixture diluted with argon. A graphical representation correlating plasma-assisted ignition with temperature in CH_4–N_2–O_2–Ar mixture is shown in Figure 4.88.

Complementing the experimental observations, the autoignition simulated with the GRI-Mech 3.0 mechanism [116] is presented. These simulations were performed under constant pressure conditions. Notably, there is a commendable agreement between the modeling (denoted by symbols 4)

and experimental (denoted by symbols 1) results for pressures approximately 2 atm behind the reflected shock wave. Autoignition parameters at a lower pressure of 0.5 atm were beyond the operational capability of the shock tube, with simulation results at $p_5 = 0.5$ atm shown by a dashed line. The introduction of a discharge reduces both the ignition time and the minimum temperature required for ignition. Specifically, at a pressure of 2 atm, the ignition temperature decreases by 100 K, while at a pressure of 0.5 atm, the ignition temperature decreases by 600 K.

For a detailed study of the influence of nanosecond discharge plasma on flammable mixtures, a thorough investigation of the electrical parameters of the discharge is essential. In general, the monitoring of the energy deposited in the gas serves as a diagnostic approach for the analysis of repetitive nanosecond discharges at ambient temperatures [90,128,129]. Typically, the relationship between the energy deposited in the gas and the gas density, denoted $w(n)$, exhibits a dome-shaped curve. The peak of $w(n)$ is closely aligned with the density range where the ionization wave propagation velocity has a maximum.

Figure 4.89a shows the energy input into the gas during a single-pulse discharge as a function of gas density at a baseline temperature of 300 K, with the pulse amplitude and duration set to 100 kV and about 50 ns, respectively. In Figure 4.89, curve (A) represents the energy input during the initial (primary) pulse (w_1), while curve (B) represents the cumulative deposited energy, including all subsequent reflections ($w_1 + w_2 + w_3$), with individual data points indicating variations in w_1, $w_1 + w_2$, and $w_1 + w_2 + w_3$. It is observed that almost all the energy is transferred during the main pulse and a subsequent reflection ($w_1 + w_2$). At a temperature of 300 K, the $w(n)$ function reflects the characteristic profile associated with discharges in the form of repetitive fast ionization waves [106]. The two lower trajectories in Figure 4.89b are consistent with the curves (A) and (B) measured at ambient temperature, while data points 1–3 represent the energy deposition in the gas over temperatures of 1,000–1,400 K and a high-voltage pulse of 100 kV. Despite a somewhat larger scatter in the experimental data, the energy deposition in the gas remains comparably consistent at both ambient and 1,000–1,400 K temperatures. Therefore, the effect of temperature on the total energy deposition in the gas is negligible even when heated to 1,000–1,800 K. This finding facilitates the extrapolation of experimental results on the discharge formation obtained at ambient temperature to higher temperatures.

Figure 4.89 further illustrates the total energy deposited into the gas, taking into account the main pulse and two subsequent reflections, plotted against gas density for a high-voltage pulse

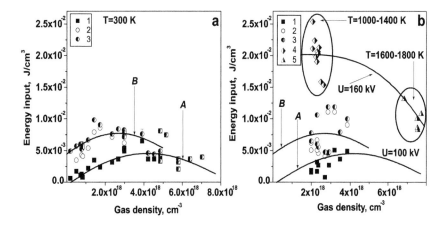

FIGURE 4.89 Energy deposited in the discharge: (a) – low temperature, (b) – high temperature. Curves A and B are discussed in the text. Symbols: $1 - w_1$, $U = 100$ kV; $2 - w_1 + w_2$, $U = 100$ kV; $3 - w_1 + w_2 + w_3$, $U = 100$ kV; $4, 5 - w_1 + w_2 + w_3$, $U = 160$ kV.

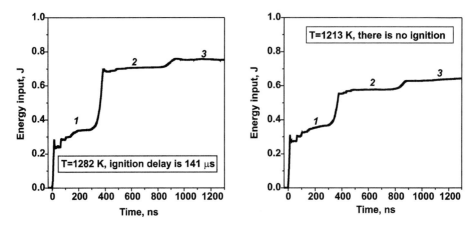

FIGURE 4.90 Energy input vs time. The plateau 1 corresponds to the energy W_1, the plateau 2 – to the W_1+W_2, the plateau 3 – to the $W_1+W_2+W_3$.

amplitude of 160 kV (represented by symbols 4 and 5). A one and a half times increase in voltage results in a four times increase in total energy deposition. Notably, two distinct clusters of data points encircled by ovals are observed. The cluster on the left is associated with temperatures between 1,000 and 1,400 K and pressures behind the reflected shock wave of about 0.5 atm, while the cluster on the right is associated with temperatures between 1,600 and 1,800 K and pressures of 2 atm. It is plausible to fit the $w(n)$ dependence with a dome-shaped curve, similar to the procedure used for measurements at $U = 100$ kV. Examination of two specific data points from the left cluster shows that about half of the energy (w_1) is deposited during the primary pulse, as shown in Figure 4.90. Despite the minimal variance (less than 10%) in deposited energy between the instances shown in Figure 4.90a and b, the ignition delay is markedly different: 141 μs at a temperature of 1,282 K compared to no ignition at a temperature of 1,213 K. Consequently, the sheer magnitude of the energy deposited (while indeed indicative of the characteristics of a gas discharge) does not necessarily reflect the kinetic behavior of the system, nor does it facilitate an analysis of the efficacy of the discharge as a plasmochemical activator.

Referring back to Figure 4.89b, it is observed that within the temperature range of 1,000–1,400 K, the specific energy deposition undergoes a significant variation—by a factor of 5 (ranging from 0.5×10^{-2} to 2.5×10^{-2} J/cm³)—as the amplitude of the high-voltage pulse is adjusted from 100 to 160 kV. At the same time, the temperature-dependent behavior of the ignition delay time, which includes all the referenced data points, shows a remarkably consistent trend (as shown in Figure 4.88). Analysis of the experimental discharge current and voltage data indicates that the high-voltage pulse amplitude has a minimal effect on the development of the discharge within the gas density range studied.

At elevated pressures (about 2 atm and above) and high temperatures (1,600–1,800 K), the discharge dynamics changes significantly. A detailed examination of the current, voltage, and emission intensity waveforms from the electronically excited molecular nitrogen states ($\lambda = 337.1$ nm, $C^3\Pi_u(v'=0) \to B^3\Pi_g(v''=0)$ transition) shown in Figure 4.91 reveals a distinct behavior of the discharge at lower gas densities (2.3×10^{18} cm⁻³, 0.5 atm, 1,000–1,400 K) compared to higher gas densities (7.5×10^{18} cm⁻³, 2 atm, 1,600–1,800 K). Specifically, at a pressure of 0.5 atm, the readings from both the current gauge and the photomultiplier show significant signal intensity. Conversely, at a pressure of 2 atm, the recorded current and emission intensity diminish, while a robust signal from the capacitance gauges indicates that the ionization front does not extend to the low-voltage electrode. Under these conditions, the discharge manifests itself primarily as a corona emanating from the high-voltage electrode, with atoms and radicals generated predominantly near the electrode surface. To increase the production rate of active particles, the output voltage of the high-voltage pulse generator must be increased to a level sufficient to bridge the discharge gap at elevated gas density.

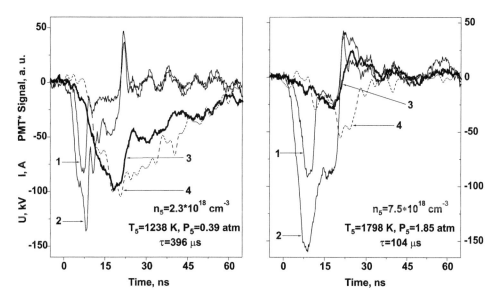

FIGURE 4.91 Typical waveforms of voltage (curves *1* and *2* are the signals from the different capacitive gauges), current (curve *3*) and emission (curve *4*).

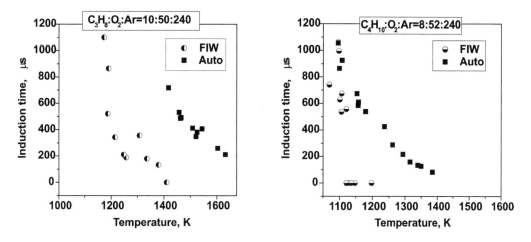

FIGURE 4.92 Ignition delay time *vs* temperature. (a) propane-containing mixture and (b) butane-containing mixture.

The same experiments were performed on a variety of hydrocarbons. Figure 4.92 shows the relationship between ignition delay times for both spontaneous autoignition and ignition initiated by nanosecond discharge in mixtures containing propane and butane. The results clearly show a significant difference in ignition delay in both cases. The divergence in ignition delay for C_4H_{10} is most pronounced in the temperature range of 1,100–1,200 K, while for C_3H_8 the shift remains relatively constant at about 200 K over the temperature range of 1,400–1,600 K.

Figure 4.93 shows the ignition delay for mixtures containing C3 and C4 hydrocarbons plotted against $1,000/T$ coordinates. Energy measurements were made for various mixtures, with an in-depth examination of the energy input for mixtures containing methane discussed earlier. Figure 4.94 shows the energy input observed in a series of experiments with mixtures containing propane. In particular, the magnitudes of the energy inputs are comparable to those recorded for mixtures containing methane, which means that the small admixture of hydrocarbons does not significantly affect the discharge evolution.

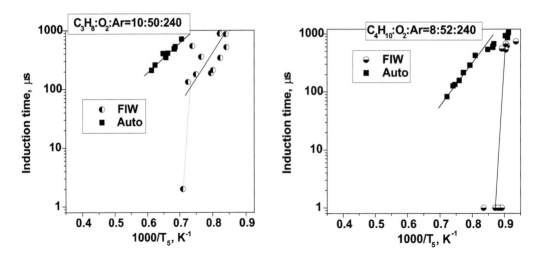

FIGURE 4.93 Ignition delay time *vs* inverse temperature. (a) propane-containing mixture and (b) butane-containing mixture.

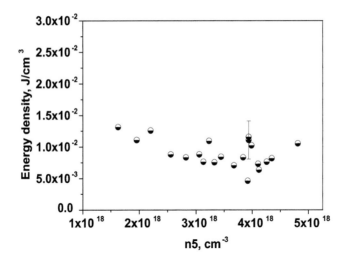

FIGURE 4.94 Energy input in a gas during nanosecond discharge action in dependence upon gas density behind the reflected shock wave.

4.14 KINETICS OF PLASMA-ASSISTED KINETICS OF C1–C5 HYDROCARBONS IN GAS MIXTURES

Elucidating the physical nuances underlying the ignition and combustion processes facilitated by nonequilibrium plasma from gas discharges is of paramount importance. Such nonequilibrium plasma-driven phenomena result from the intricate interplay of numerous physical mechanisms. To accurately characterize plasma-assisted ignition (PAI) and plasma-assisted combustion (PAC), it is imperative to integrate comprehensive insights from both gas discharge physics and chemical kinetics, encompassing both high-temperature dynamics and behavior influenced by excited and charged species [126,127].

Currently, several models have been established to describe the kinetics of fuel-air mixtures at elevated temperatures. These models account for hundreds of chemical species and thousands of reactions. Detailed sensitivity analysis is proving to be a critical approach to testing these models. In addition, even in "classic" combustion scenarios when Maxwell-Boltzmann equilibrium is reached, accurate replication of experimental results, especially for complex fuels, requires significant computational resources. The study of discharge kinetics at lower gas temperatures has been the subject of extensive research over the years, as documented in the literature [130,131].

The discourse surrounding the interplay between nonequilibrium plasma and combustion encompasses several perspectives, particularly with respect to discharges occurring at low reduced electric fields. In such scenarios, the predominant mechanisms involve low-energy excitations, such as vibrational excitations or the excitation of lower electronic states [132]. Consequently, a kinetic model has been postulated in which electronically excited oxygen molecules, specifically in the $a^1\Delta_g$ and $b^1\Sigma_g^+$ states, are identified as possible agents for increasing combustion efficiency [133]. This hypothesis was supported by findings suggesting the potential for initiating ignition in a $CH_4:O_2$ mixture at reduced temperatures through the excitation of these states by laser irradiation at wavelengths of 1.268 μm and 762 nm, respectively. Conversely, studies at high reduced electric fields underscore the paramount importance of high-energy electronic states that facilitate the dissociation of both fuel and oxygen molecules. This dissociation process can arise directly from electronic collisions or through interactions with electronically excited species, as exemplified by the reaction $N_2^* + O_2 = N_2 + O + O$ [130].

Additional perspectives were provided by other researchers [134], who examined the role of oxygen atoms dissociated by the discharge, focusing on hydrogen-oxygen kinetics at atmospheric pressure and varying initial temperatures of the mixture. This analysis included the artificial addition of oxygen atoms as an initial condition for a standard high-temperature ignition model. The minimum concentration of oxygen atoms required to initiate ignition was identified, highlighting that at lower temperatures, the predominant mechanism accelerating the ignition process is rapid heating due to radicals recombination. The contribution of oxygen atoms becomes increasingly important from several hundred degrees Kelvin, where they not only facilitate heating but also initiate chemical chain reactions, providing a more viable pathway for ignition acceleration.

The discussion was extended to emphasize that the influence of gas discharges goes beyond mere oxygen recombination, with nonequilibrium energy deposition in the gas leading to the generation of ions, as well as electronically and vibrationally excited species. This introduces "ion chains" and "energy chains", concepts familiar from laser physics, where ions or excited species, as opposed to radicals in ground states, act as intermediates. However, the presence of multiple chain mechanisms in the system does not lead to a cumulative effect due to the nonlinear nature of chain processes, and these mechanisms remain underexplored.

For example, studies and subsequent numerical analyses [135] revealed a plasma-chemical chain branching mechanism that amplifies the initial radicals produced in low-temperature hydrocarbon/air and CO/air plasmas. Conversely, experiments in ambient temperature hydrocarbon-air and hydrocarbon-oxygen mixtures [136] suggested that the role of the discharge in the low-temperature oxidation of alkanes ranging from CH_4 to C_6H_{14} is essentially as a radical source, with subsequent oxidation proceeding along established low-temperature oxidation pathways (e.g., mechanism [137]). Furthermore, some studies [138] have posited the importance of ions in ignition kinetics, demonstrating through numerical simulations that the addition of a small fraction (10^{-4}) of NO^+ ions/electrons to a mixture of i-$C_8H_{18}:O_2:NO:Ar$ at atmospheric pressure can significantly influence the ignition delay time.

Numerical simulations often evaluate the influence of the discharge on flammable mixtures by introducing various radicals at the beginning of the calculations. A study [139] numerically explored the pathways to radical-induced ignition in a methane-air mixture, demonstrating ignition initiation from temperatures as low as 300 K at atmospheric pressure by injection of CH_3, CH_2, CH, and C radicals, with radical concentrations reaching up to a few percent. The study identified the CH radical as critical in initiating combustion chains, primarily oxidizing to HCO. It was found that CH_2 could serve as a precursor for CH by reacting with H to form CH, while the role of C atoms was limited to inducing gas heating by conversion with O_2 to form CO. CH_3 was considered relatively unimportant for ignition. This led to the conclusion that radical-induced ignition is a promising technique, highlighting the variable radical distributions produced by real gas discharges under different experimental conditions.

A quantitative assessment by [140] investigated the effect of radical species (O, H, N, and NO_x) on the extinction limits of H_2 and CH_4 flames in a perfectly stirred reactor. The purpose of the study was to determine the differential effects of radical additions on combustion processes. Although the introduction of radicals into fuel mixtures increased the extinguishment thresholds for both H_2 and CH_4, this effect was less significant for extinguishment thresholds than for ignition thresholds. At low gas temperatures, where the initial radicals were quickly neutralized by recombination, the combustion enhancement was mainly attributed to the heat released by exothermic radical recombination. Conversely, at higher inlet temperatures, the effect of radical addition on the extinction limits was more pronounced.

Experimental evidence for the effectiveness of nonequilibrium plasma ignition/combustion enhancement is not always clear. Therefore, experiments that minimize uncertainties or compare the effects of different plasmas under identical conditions are particularly valuable.

The oxidation of ethylene by nanosecond discharge was reported in a study [141], where a high-voltage (16–18 kV) nanosecond pulse (20–30 ns duration) with a repetition frequency of up to 50 kHz was applied to electrodes within a pressure range of 70–100 Torr. The discharge power was maintained at a relatively low level of 70–115 W, with gas temperatures ranging from 100°C to 300°C. Notably, the research demonstrated the possibility of flameless oxidation via plasma or ignition by increasing the energy input to the discharge within the same experimental setup. The results led to several conclusions: first, that a significant fraction of the hydrocarbons can be oxidized by low-temperature plasma-chemical reactions before ignition is achieved; second, that the observed increase in gas temperature is due to the heat released during the exothermic plasma-chemical oxidation of the fuel; and third, that ignition occurs when the temperature increase due to the heat generated in the plasma-chemical oxidation process reaches an autoignition threshold. As articulated by the authors [141], the conversion of fuel by low-temperature plasma-chemical processes and the consequent heating of the flow effectively "open the door" to ignition.

4.14.1 Experimental Analysis of the Plasma-Assisted Ignition. Mixtures with Hydrocarbons

In the subsequent chapter, scenarios in which the effects of nanosecond pulsed discharge on ignition were temporally constrained, are explored. A methodology proposed in 1996 [142] suggested the use of shock wave techniques—typically used in autoignition studies—to evaluate the effectiveness of nanosecond discharge ignition. Using analytical models, it was shown that operating near the autoignition temperature threshold facilitates quantitative assessment of ignition threshold shifts. A nanosecond pulsed discharge characterized by a spatially uniform fast ionization wave was recommended for ignition initiation. This approach takes advantage of the ability of the shock tube to achieve predetermined pressure and temperature conditions behind the shock wave. By significantly diluting the flow with a monatomic gas, such as Ar or He, a quasi-one-dimensional flow is established, allowing the application of 1D shock tube theory to determine the gas temperature T_5, pressure p_5, and gas density n_5 behind the reflected shock wave. The nanosecond discharge maintains spatial uniformity down to gas densities of about 10^{19} cm^{-3} at high-voltage amplitudes, indicating that the plasma remains nonequilibrium over a substantial volume. These experimental results have been discussed in several studies [103,121,134].

The primary objective of this chapter is to present a computational model that quantitatively predicts the decrease in ignition delay time for hydrocarbon-oxygen mixtures, assuming that radical production during the nanosecond discharge and its immediate afterglow is the primary mechanism

Mechanisms of Plasma-Chemical Reactions

driving ignition. In addition, the chapter aims to contrast the results of the model with experimental observations in order of the model validation.

4.14.2 Simulation of Production of Atoms and Radicals

The numerical modeling of active particle generation in $Ar:O_2:C_nH_{2n+2}$ gas mixtures subjected to high-voltage nanosecond discharges involved two main processes: (1) the generation of active species (including electrons, ions, excited molecules, atoms, and radicals) induced by the high-energy electrons accelerated by the strong electric field present during the discharge, and (2) the transformation of these active particles as the plasma recombines following the intense phase of the discharge. The focus was on active species such as excited Ar atoms, O atoms, H atoms, hydrocarbon molecule radicals, electrons, and positive ions. Due to the elevated gas temperatures (> 1,000 K) in the discharge gap, the formation of negative ions and complex positive ions (such as Ar_2^+, O_4^+) was not considered. After plasma decay, the active particle ensemble was assumed to consist mainly of atoms and radicals, while the presence of long-lived excited states was neglected.

The primary pathways for the generation of active species via high-voltage nanosecond discharges exhibit the same characteristics across all gas mixtures studied. To illustrate, Table 4.27 details the main reactions considered in the modeling of active particle production in an $Ar:O_2:CH_4$ mixture.

Discharge dynamics were modeled using measured electric field dynamics in the discharge gap at successive points in time and space. Parameters such as electron drift velocity and rate coefficients for various electron impact processes-including dissociation (e.g., reactions (R1) and (R2) in Table 4.27), excitation (reaction (R3)), and ionization (reactions (R4)–(R6))-were determined by solving the electron Boltzmann equation using the classical two-term approximation. The input parameters for solving this equation included the electric field, neutral particle number density, gas temperature, and mixture composition.

The model considered only electron interactions with the atoms and molecules of the dominant species in the gas mixtures, excluding electron-electron interactions and those between electrons and newly created particles due to their low concentrations in the mixture. For electron collisions with Ar, O_2, CH_4, and C_2H_6, the existing literature provides self-consistent cross-sectional data sets that allow accurate prediction of transport and rate coefficients in these pure gases. In the absence of reliable data for C_3H_8, C_4H_{10}, and C_5H_{12}, it was simplistically assumed that their cross sections were similar to those for C_2H_6. This approximation could introduce the most significant inaccuracies in predicting the densities of atoms and radicals generated by electron impact dissociation of hydrocarbons during discharge. However, the following sections show that this influence is marginal under a wide range of conditions.

The simulation of the evolution of the active particle density in the high electric field discharge uses rate and transport coefficients derived from the Boltzmann equation. Due to the very short duration of the discharge phase, the model assumes negligible active particle loss, allowing for analytical determination of the particle density evolution based on balance equations.

At the onset of the voltage pulse (the ionization wave front), there is a very short peak of electric field, typically unresolved experimentally (see discussion in previous chapters). This peak in the electric field within the discharge gap is theorized to generate a significant density of charged particles, while the generation of neutral active particles remains minimal. To account for electron and ion production during this initial, less understood phase, the initial electron density at the onset of the main voltage pulses was treated as a tunable parameter. This parameter was determined by matching the simulated discharge current evolution to experimental data. For illustration, Figure 4.95 compares the calculated specific deposition energy with experimental values for the $Ar:O_2:CH_4$ mixture at 1.1 atm and 1,530 K, and for the $Ar:O_2:C_5H_{12}$ mixture at 0.52 atm and 1,390 K. Theoretical curves are derived from simulated discharge currents and observed electric fields in the gap, while experimental curves are derived directly from field and current measurements. It

TABLE 4.27
Processes Dominating Production of Atoms and Radicals by Gas Discharge in Ar:O_2:CH_4 Mixtures

Number	Reaction	Rate Coefficient, cm^3/s or cm^6/s	References
	Electron Impact Dissociation		
R1	$e+O_2 \rightarrow e+O+O$	$f(E/n)$	[143]
R2	$e+CH_4 \rightarrow e+CH_3+H$	$f(E/n)$	[144]
	Electron Impact Excitation		
R3	$e+Ar \rightarrow e+Ar^*$	$f(E/n)$	[143]
	Electron Impact Ionization		
R4	$e+Ar \rightarrow 2e+Ar^+$	$f(E/n)$	[143]
R5	$e+O_2 \rightarrow 2e+O_2^+$	$f(E/n)$	[143]
R6	$e+CH_4 \rightarrow 2e+CH_4^+$	$f(E/n)$	[144]
	Quenching of Excited Ar Atoms		
R7	$Ar^*+O_2 \rightarrow Ar+2O$	$2 \cdot 10^{-10}$	[68,145]
R8	$Ar^*+CH_4 \rightarrow Ar+CH_2+2H$	$3.25 \cdot 10^{-10}$	[68,145]
R9	$Ar^*+CH_4 \rightarrow Ar+CH+H+H_2$	$5.833 \cdot 10^{-11}$	[68,145]
R10	$Ar^*+CH_4 \rightarrow Ar+CH_3+H$	$5.833 \cdot 10^{-11}$	[68,145]
R11	$Ar^*+CH_4 \rightarrow Ar+CH_2+H_2$	$5.833 \cdot 10^{-11}$	[68,145]
	Charge Exchange		
R12	$CH_4^+ + O_2 \rightarrow CH_4 + O_2^+$	$5 \cdot 10^{-10}$	Estimate
R13	$Ar^+ + O_2 \rightarrow Ar + O_2^+$	$1 \cdot 10^{-10}$	[146]
R14	$Ar^+ + CH_4 \rightarrow Ar + CH_3^+ + H$	$1.056 \cdot 10^{-9}$	[146]
R15	$Ar^+ + CH_4 \rightarrow Ar + CH_2^+ + H_2$	$2.275 \cdot 10^{-10}$	[146]
	Electron–Ion Recombination		
R16	$e+O_2^+ \rightarrow O+O$	$2 \cdot 10^{-7} (300/T_e)$	[130]
R17	$e+CH_4^+ \rightarrow CH_3+H$	$1.75 \cdot 10^{-7} (300/T_e)^{0.5}$	[147,148]
R18	$e+CH_4^+ \rightarrow CH_2+2H$	$1.75 \cdot 10^{-7} (300/T_e)^{0.5}$	[147,148]
R19	$e+CH_3^+ \rightarrow CH_2+H$	$7.75 \cdot 10^{-8} (300/T_e)^{0.5}$	[149]
R20	$e+CH_3^+ \rightarrow CH+2H$	$2 \cdot 10^{-7} (300/T_e)^{0.4}$	[149]
R21	$e+CH_3^+ \rightarrow CH+H_2$	$1.95 \cdot 10^{-7} (300/T_e)^{0.5}$	[149]
R22	$e+CH_2^+ \rightarrow CH+H$	$1.60 \cdot 10^{-7} (300/T_e)^{0.6}$	[149]
R23	$e+CH_2^+ \rightarrow C+H_2$	$7.68 \cdot 10^{-8} (300/T_e)^{0.6}$	[149]
R24	$e+CH_2^+ \rightarrow C+2H$	$4.03 \cdot 10^{-7} (300/T_e)^{0.6}$	[149]

is noteworthy that the reduced electric field values after the discharge front do not exceed 150 Td, which allows the discharge to be modeled using a local electric field model.

The main initial voltage pulse in the discussed experiments was followed by several smaller reflected pulses. The generation of active particles during these subsequent discharge phases was modeled, highlighting a significant production of atoms, radicals, and excited neutral particles, while the generation of electrons and ions was considered minimal during this phase due to a significant reduction in the electric field and energy of electrons in these secondary pulses.

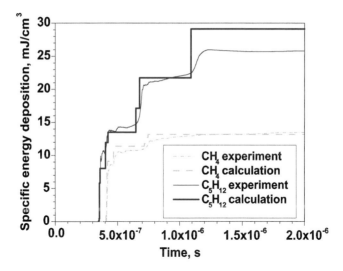

FIGURE 4.95 The evolution in time of the specific energy deposited in the Ar:O$_2$:CH$_4$ mixture at 1.1 atm and 1,530 K and in the Ar:O$_2$:C$_5$H$_{12}$ mixture at 0.52 atm and 1,390 K. Curves 1 correspond to calculation and curves 2 correspond to measurements.

The dynamics of active particle densities between and after voltage pulses was calculated by numerical solution of the balance equations, using rate coefficients for elementary processes obtained from the literature or approximated when necessary. In this zero electric field phase, charged particles were eliminated via dissociative electron–ion recombination reactions (e.g., reactions (R16)–(R24) in Table 4.27), yielding additional atoms and radicals. In addition, these particles were formed by charge exchange of certain positive ions with hydrocarbon molecules (e.g., reactions (R14) and (R15)) and by quenching of excited Ar atoms by oxygen and hydrocarbon molecules (e.g., reactions (R7)–(R11)). In our kinetic framework, charged and excited particles were eventually converted to unexcited, uncharged components in a ground state, leaving the gas mixture consisting of initial neutral species along with atoms and radicals generated during both the discharge and its afterglow. The calculated densities of atoms and radicals were then used as initial parameters in computational simulations to predict the dynamics of plasma-assisted ignition within the studied mixtures.

Figure 4.96 illustrates the temporal evolution of the densities of the predominant active species during the discharge afterglow in both the Ar:O$_2$:CH$_4$ and Ar:O$_2$:C$_5$H$_{12}$ mixtures, under the identical conditions presented in Figure 4.95. In particular, the variance in energy input between these mixtures is attributed to different gas densities: for CH$_4$, the gas number density is $n = 5.2 \cdot 10^{18}$ cm^{-3}, as opposed to $n = 2.7 \cdot 10^{18}$ cm^{-3} for C$_5$H$_{12}$.

Comparable analyses were also performed for several other hydrocarbon mixtures. In the scenarios studied, O atoms emerge as the main active species, generated primarily by electron impact dissociation of O$_2$ in the high electric field of the discharge. Conversely, the generation of atoms and radicals by electron impact dissociation of hydrocarbon molecules plays a minor role. De-excitation of Ar atoms contributes to an increased density of H atoms and radicals such as CH$_3$ and C$_5$H$_{11}$ during the initial stages of the discharge afterglow (lasting tens to hundreds of nanoseconds).

The potential influence of uncertainties in the electron impact dissociation cross sections for C$_3$H$_8$, C$_4$H$_{10}$, and C$_5$H$_{12}$ on the results of these simulations is expected to be minimal. To evaluate this premise, further simulations were performed in which the dissociation cross sections of these molecules were increased by an order of magnitude. Such changes resulted in the densities of H atoms and hydrocarbon molecule radicals increasing by a factor of 6–7, while the density of O atoms

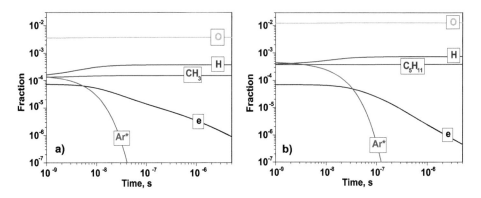

FIGURE 4.96 The evolution in time of the densities of dominant active species in the afterglow of the discharge in (a) the Ar:O_2:CH_4 mixture and (b) Ar:O_2:C_5H_{12} mixture. The curves correspond to the same conditions as those in Figure 4.95.

decreased by tens of percent. As a result, the cumulative density of atoms and radicals in mixtures containing C_3H_8, C_4H_{10}, and C_5H_{12} showed a marginal increase of about 10%. This variance has a negligible effect on the induction period for the mixtures studied; subsequent simulations indicated that at high temperatures, the ignition delay time is primarily a function of the overall density of atoms and radicals generated by the discharge, rather than the specific composition of the active species. For example, in simulations where CH_3 and H radicals were replaced by an equivalent number of O atoms within a CH_4:O_2:Ar mixture, the resulting deviation in ignition delay times was limited to tens of microseconds in contexts where the ignition delay itself spanned hundreds of microseconds.

4.14.3 Simulation of Autoignition and Plasma-Assisted Ignition. Comparison with the Experiments

Zero-dimensional simulations of both autoignition and plasma-assisted ignition at constant pressure were performed. For the modeling of autoignition, the mechanism proposed by Tan for mixtures containing CH_4 to C_3H_8 and the mechanism developed by Zhukov and Starikovskiy based on the Westbrook mechanism for mixtures containing C_4H_{10} and C_5H_{12} were used. These mechanisms were supplemented with reactions for the formation and quenching of the OH radical in the $A^2\Sigma$ state to facilitate a comparison between the computational predictions and the experimental results.

It was postulated that the nonequilibrium plasma of a pulsed gas discharge could be conceptualized as an instantaneous injection of dissociated species and radicals as previously described. A notable discrepancy was observed between the ignition delays for autoignition and plasma-assisted ignition, the latter induced by radical injection. This contrast is readily apparent in Figure 4.97, which shows two kinetic trajectories for OH radicals in a methane-oxygen mixture diluted with argon.

A zero-dimensional simulation of autoignition and plasma-assisted ignition by discharge at constant pressure was performed. Autoignition was modeled using the mechanism proposed by Tan [151] for mixtures with CH_4–C_3H_8 hydrocarbons, and the Westbrook mechanism [152] for mixtures containing C_4H_{10} and C_5H_{12}. These mechanisms were supplemented with reactions of formation and quenching of the OH radical in the excited $A^2\Sigma$ state to directly compare the numerical and experimental results.

Due to significantly different time scales, it has been suggested that the nonequilibrium plasma of a pulsed gas discharge can be considered as an instantaneous injection of dissociated species and radicals. This assumption indicates a significant difference in ignition delay times between autoignition and "plasma" ignition, i.e., ignition facilitated by radical injection. This difference is clearly

Mechanisms of Plasma-Chemical Reactions

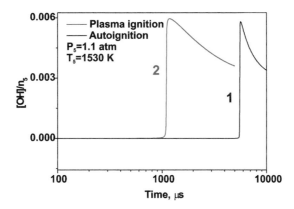

FIGURE 4.97 The evolution in time of the OH – density in the Ar:O$_2$:CH$_4$ mixture at 1.1 atm and 1,530 K. Curve 1 corresponds to autoignition and curve 2 corresponds to plasma-assisted ignition.

illustrated in Figure 4.97, where two kinetic curves for OH radicals are shown for a methane-oxygen mixture diluted with argon.

It should be noted that an exact agreement between the absolute values of the experimentally determined autoignition delay times and those obtained by numerical simulations for the methane-oxygen mixture has not been achieved; the calculated values are about twice as high as the measured ones.

Subsequent calculations show a good correlation between the experimentally measured and numerically calculated autoignition delay times, with the experimental results closely matching those reported by others. Figure 4.98 illustrates this correlation for mixtures containing propane and also highlights the agreement between the two methods of determining the ignition delay. In particular, it can be seen that the ignition delay times derived from OH emission at 306 nm are very similar to those derived from CH emission at 431 nm, and any discrepancies between these two measurements are small compared to the difference between autoignition and plasma-assisted ignition (Figure 4.99).

To elucidate the details of the kinetic mechanism, a sensitivity analysis of the kinetic scheme was performed. This analysis consisted of sequentially increasing the rate coefficients of each reaction by a factor of 1.5 and recalculating the induction delay time for each case. The sensitivity coefficients were determined by estimating $\left(\tau^i_{1.5} - \tau\right)/\tau$, where τ is the induction delay time calculated using the original rate coefficients and $\tau^i_{1.5}$ is the induction delay time calculated using the rate coefficient of the ith reaction increased by a factor 1.5. The results of this analysis for C1–C3 mixtures at a temperature of $T_5 = 1,670$ K are shown in Figure 4.100, with the top 17 reactions highlighted. It is clear from the analysis that certain reactions are critical for all mixtures (e.g., reactions (1), (4), or (8)), while others are particularly relevant only for mixtures containing methane (reaction (9)) or ethane and propane (reactions (11)–(17)). The variation of the importance of different reactions with increasing temperature is shown by yellow bars.

In addition, Figure 4.101 contrasts experimental observations with computational predictions for mixtures containing C_4H_{10} and C_5H_{12}. In particular, there is commendable agreement for both autoignition and plasma-assisted ignition facilitated by nanosecond discharges. These results, in conjunction with the experimental and computational data for other hydrocarbons, suggest that under high-temperature conditions the acceleration of ignition by nonequilibrium plasma of pulsed discharge at sufficiently high overvoltage can be attributed predominantly to the generation of ground-state radicals produced either by electron impact or by collisional quenching of electronically excited species.

Therefore, experiments and numerical simulations were performed to investigate the ignition of hydrocarbon-containing mixtures under the influence of pulsed nanosecond discharges.

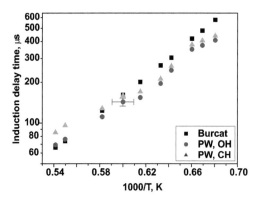

FIGURE 4.98 Comparison of autoignition delay time, experimentally determined from OH emission at 306 nm (PW, OH), experimentally determined from CH emission at 431 nm (PW, CH), and calculated in accordance with the formula given by Burcat [150].

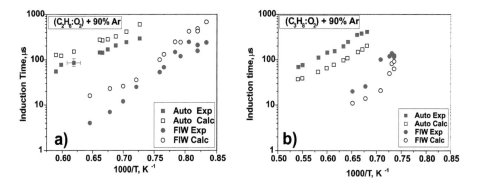

FIGURE 4.99 Autoignition (designated as "auto") and plasma-assisted ignition by nanosecond discharge in the form of fast ionization wave (designated as "FIW") for C_2H_6– and C_3H_8–containing mixtures. Comparison of experiments and calculations.

The experiments used a variety of stoichiometric mixtures, $C_nH_{2n+2}:O_2$ (10%) diluted with Ar (90%), containing hydrocarbons from CH_4 to C_5H_{12}. The temperature was ranged from 950 to 2,000 K, while the pressure ranged from 0.2 to 1.0 atm. For each experimental scenario, plasma-assisted ignition was contrasted with autoignition.

The numerical modeling included both the high-temperature kinetics of ignition and the generation of atoms and radicals by the nanosecond discharge at relatively high overvoltage. First, the energy distribution in the discharge was calculated based on experimental electric field measurements and energy input. Discharge afterglow modeling integrates reactions that mediated energy transfer from electronically excited particles to molecules of fuel and oxidizer, yielding atoms and radicals. High-temperature kinetics were then modeled, considering the atoms and radicals produced in the initial phase as an initial condition for ignition development.

The detailed comparison of experimental and numerical results yields a satisfactory agreement, confirming that, under the high-temperature conditions, the predominant mechanism of energy transfer is dissociation by electron impact along with the generation of electronically excited atoms and molecules. The additional generation of atoms and radicals, attributed to the quenching of electronically excited species in the immediate discharge afterglow, did not exceed a factor of 2–3. These generated species (O, H, OH, C_nH_{2n-2}, etc.) facilitate the spatially uniform ignition of gas mixtures at temperatures 500–200 K below the autoignition threshold.

Mechanisms of Plasma-Chemical Reactions

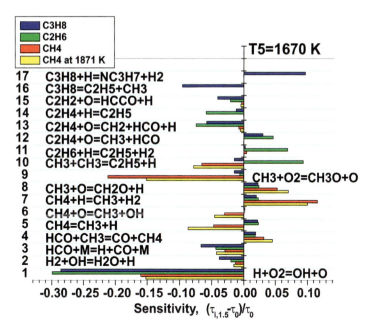

FIGURE 4.100 Comparison of sensitivity analysis of the kinetic scheme for CH_4 (red bars), C_2H_6 (green bars), and C_3H_8 (blue bars) – containing mixtures. Temperature $T_5 = 1,670$ K. For comparison, another temperature for methane-containing mixture (1,870 K) is given by yellow bars.

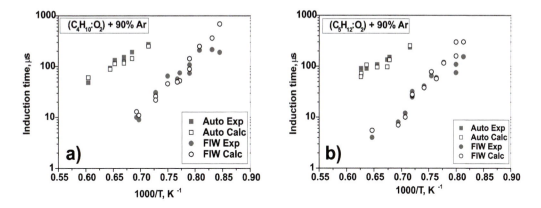

FIGURE 4.101 Autoignition (designated as "auto") and plasma-assisted ignition by nanosecond discharge in the form of fast ionization wave (designated as "FIW") for C_4H_{10}- and C_5H_{12}-containing mixtures. Comparison of experiments and calculations.

REFERENCES

[1] Zatsepin, D.V., S.M. Starikovskaia, and A. Yu. Starikovskiy, Hydrogen oxidation in the stoichiometric hydrogen-oxygen mixture in the fast ionization wave. *Chemical Physics Reports*, 2001. **20**(7): pp. 66–99.

[2] Pancheshnyi, S.V., S.M. Starikovskaya, and A.Yu. Starikovsky, Dynamics of population of electronic levels of molecular nitrogen and structure of fast ionization wave. *Plasma Physics*. 1998. **25**(4): p. 326.

[3] Starikovskaia, S.M., A.Yu. Starikovskiy, and D.V. Zatsepin, Hydrogen oxidation in a stoichiometric hydrogen-air mixtures in the fast ionization wave. *Combustion Theory and Modelling*, 2001. **5**(1): pp. 97–129.

[4] Anikin, N.B., S.V. Pancheshnyi, S.M. Starikovskaia, and A.Yu. Starikovskiy, Breakdown development at high overvoltage: electric field, electronic levels excitation and electron density. *Journal of Physics D: Applied Physics*, 1998. **31**: pp. 826–833.

[5] Alexandrov, N.L., and E.E. Son, Energy distribution and kinetic coefficients of electrons in gases in an electric field. In *Chemistry of Plasma*, Ed. B.M. Smirnov. 1980: Atomizdat, p. 35.

[6] Anikin, N.B., S.M. Starikovskaya, and A.Yu. Starikovsky, Dynamics of the profiles of charge density and longitudinal components of an electric field in a high-speed ionization wave. *Plasma Physics*. 1998. **24**(1): p. 9.

[7] Alexandrov, N.L., and E.E. Son, In *Chemistry of Plasma*, Ed. B.M. Smirnov. 1980: Atomizdat, p. 35.

[8] Itikawa, Y., Momentum-transfer cross sections for electron collisions with atoms and molecules: Revision and supplement, 1977. *Atomic Data and Nuclear Data Tables*, 1978. **21**(1): 69–75.

[9] Hake, R.D. and A.V. Phelps, Momentum-Transfer and Inelastic-Collision Cross Sections for Electrons in O_2, CO, and CO_2. *Physical Review*, 1967. **158**(1): p. 70.

[10] Hayashi, M., Electron collision cross sections for molecules determined from beam and swarm data, in *Swarm Studies and Inelastic Electron-Molecule Collisions*, Eds. L.C. Pitchford, B.V. McKoy, A. Chutjian, and S. Trajmar. 1987: Springer-Verlag.

[11] Rusanov, V.D. and A.A. Friedman. *Physics of Chemically Active Plasma*. 1984: Nauka, p. 415.

[12] Billing, G., Vibration-Vibration and Vibration-Translation Energy Transfer, Including Multiquantum Transitions in Atom-Diatom and Diatom-Diatom Collisions, in *Non-Equilibrium Vibrational Kinetics*, Ed. M. Capitelli, 1989: Mashinostroenie, p. 104.

[13] Ablekov, V.K., Yu.N. Denisov, F.N. Lyubchenko, and M. Mashinostroenie, *Reference Book on Gas-Dynamic Lasers*. 1982.

[14] Smith, K. and R.M. Thomson. *Computer Modeling of Gas Lasers*. 1978: Plenum Press.

[15] Zuev, A.P., S.A. Losev, A.I. Osipov, and A.M. Starik, Vibrational-translational energy exchange in collisions of triatomic molecules. *Chemical Physics*, 1992. **11**(1): pp. 4–34.

[16] Gardiner, W.C. Jr., Ed., *Combustion Chemistry*. 1984: Springer-Verlag.

[17] Erwin, D.A. and J.A. Kunc, *IEEE Transactions on Plasma Science*, 1983. **11**(4): p. 266.

[18] Galtsev, V.E., Demyanov A.V., Kochetov I.V., Pevgov V.G., Sharkov V.F. Boltzmann equation solver, Preprint IAE-3156. 1979.

[19] Gordeev, O.A. and D.V. Khmara, Software package for modeling of kinetic processes in a gas discharge plasma in the reduced field approximation, in IX Conference on Gas Discharge Physics, Ryazan. 1998: Abstracts, Chapter 2, p. 91.

[20] Schulz, G.J., Experiments on resonances in the elastic cross-section of electrons on rare-gas atoms. *Physical Review*, 1964. **135**: p. A938.

[21] Boness, M.J.W. and G.J. Schulz, Electron impact vibrational excitation of CO in the range 1–4 eV. *Physical Review*, 1973. **A8**: p. 2883.

[22] Cartwright, D.C., Electron impact excitation of the electronic states of N_2. I. Differential cross sections at incident energies from 10 to 50 eV. *Physical Review A*, 1977. **16**(3): p. 1013.

[23] Itikawa Y., Cross sections for collisions of electrons and photons with nitrogen molecules *journal of physical and chemical reference data*, *Journal of physical and chemical reference data*, 1986. **15**(3): p. 985–1010.

[24] Spence D. and P.D. Burrow, Resonant dissociation of N_2 by electron impact. *Journal of Physics B*, 1979. **12**: p. 179.

[25] Winters, H.F., Ionic adsorption and dissociation cross-section for nitrogen. *The Journal of Chemical Physics*, 1966. **44**: p. 1472.

[26] Zipf, E.C., On the dissociation of nitrogen by electron impact and by E.U.V. photo-absorption. Planetary and Space Science. *Planetary and Space Science*, 1978. **26**: p. 449–462.

[27] Rapp, D. and P. Englander-Golden, Total cross sections for ionization and attachment in gases by electron impact. I. Positive ionization. *The Journal of Chemical Physics*, 1965. **43**(5): p. 1464.

[28] Rapp, D., P. Englander-Golden, and D.D. Briglia, Cross sections for dissociative ionization of molecules by electron impact. *The Journal of Chemical Physics*, 1965. **42**: p. 4081.

[29] Onda, K., Rotational excitation of molecular nitrogen by electron impact. *Journal of the Physical Society of Japan*. 1985. **54**(12): p. 4544.

[30] Islamov, R.Sh., I.V. Kochetov, and V.G. Pevgov, Analysis of electron interaction processes with an oxygen molecule. FIAN preprint No 169. 1977: FIAN.

[31] Itikawa, Y., and A. Ichimura, Cross sections for collisions of electrons and photons with oxygen molecules. *Journal of Physical and Chemical Reference Data*, 1989. **18**(1): p. 23.

[32] Rapp, D., and D.D. Briglia, Total Cross sections for ionization and attachment in gases by electron impact. II. Negative-ion formation. *The Journal of Chemical Physics*, 1965. **43**(5): p. 1480.

[33] Melton, C.E., Cross sections and interpretation of dissociative attachment reactions producing OH⁻, O⁻, and H⁻ in H_2O. *The Journal of Chemical Physics*, 1972. **57**(10): p. 4218.

[34] Hayashi, M., Electron collision cross sections determined from beam and swarm data by Boltzmann analysis, in *Nonequilibrium Processes in Partially Ionized Gases*, Eds. M. Capitelli and J. N. Bardsley. 1990: Plenum Press.

[35] Khare, S.P. and W.J. Veath, Cross sections for the direct and dissociative ionization of NH_3, H_2O and H_2S by electron impact. *Journal of Physics B*, 1987. **20**: p. 2101.

[36] Kostinsky, A.Yu, A.A. Matveyev, V.P. Silakov, Kinetical processes in the non-equilibrium nitrogen oxygen plasma. Academy of science of USSR. General physics Institute. Plasma physics division, Preprint No. 87, IOF AS. 1990, 29 p.

[37] Silakov V.P., Non-equilibrium discharge in nitrogen oxygen mixtures. *Fizika Plasmy*. 1988. **14**: p.1209.

[38] Mitchell J. Atomic processes in electron-ion and ion-ion collisions, in: *Physics of Ion-Ion Collisions*, Eds. F. Brouillard, D. McGouen 1986: Plenum, p. 219.

[39] Yeremeitsev, I.G. and N.N. Pilyugin, Calculation of the nonequilibrium parameters of air at the surfaces of models and in the wakes behind them for the conditions of aeroballistic experiments. *Journal of Applied Mechanics and Technical Physics* 1986. **27**: p. 250–260.

[40] Eletsky A.V. and B.M. Smirnov, Dissociative recombination of electron and molecular ion. *Uspekhi Fizicheskikh Nauk*, 1982. **136**: p. 25.

[41] Biondi, M., et al., Recombination. *In the Book: Plasmas in Lasers*, Ed. J. Bekefi. 1982: p.145.

[42] Polak, L.S., et al., Plasma chemistry, *In Collected Works: Chemistry of Plasma*. Ed. L.S.Polak and Yu. A. Lebedev, Vol. 5. 1978: Cambridge Interscience Publ, p. 242.

[43] Virin, L.I., I.K.Larin, *Ion-Molecular Reactions in Gases*. 1979: Nauka.

[44] Alexandrov, N.L., Energy distribution and kinetic coefficients of electrons in gases in an electric field. *In Collected Works: Chemistry of Plasma*, Vol. 8. 1981: p. 90.

[45] Rodriguez, A.E., et al., An air breakdown kinetic model. *Journal of Applied Physics*, 1991. **70**(4): p. 2015.

[46] Losev, S.A., et al., Air ionization by laser plasma radiation. *Khimicheskaya Fizika*, 1987. **6**(12): p. 1677.

[47] Baulch, D.L., et al., Evaluated kinetic and photochemical data for atmospheric chemistry: supplement II. codata task group on gas phase chemical kinetics. *Journal of Physical and Chemical Reference Data*, 1984. **13**(2): p. 1259.

[48] Delcroix, J., et al., Metastable atoms and molecules in ionized gases. *In the Book: Plasmas in Lasers*, Ed. J. Bekefi. 1982: p. 176.

[49] Brunet, H., et al., Model for a glow discharge in flowing nitrogen. *Journal of Applied Physics*, 1985, **57**(5): p. 1574.

[50] Dvoryankin, A.N., et al., Mechanisms of electron relaxation in atomic-molecular media // Plasma Chemistry. *In Collected Works: Chemistry of Plasma*. 1987, Vol. 14: p. 102.

[51] Vasilieva, A.N., et al., Excitation of metastable states in oxygen-nitrogen plasma. *Plasma Physics*, 1989. **15**(2): p. 190.

[52] Pancheshnyi, S.V., S.M. Starikovskaya, and A.Yu. Starikovskiy, Measurements of rate constants of the $N_2(C^3\Pi_u)$ and $N_2^+(B^2\Sigma^+_u)$ deactivation by N_2, O_2, H_2, CO and H2O molecules in afterglow, *Chemical Physics Letters*, 1998. **294**: p. 523.

[53] Golubovsky, Y.B., et al., Excitation of the radiating states $C\ ^3\Pi_u$ and $C'\ ^3\Pi_u$ of the nitrogen molecule during binary collisions of $N_2\ (A\ ^3\Sigma_{u+})$ metastables. *Optics and Spectroscopy*, 1990. **69**(2): p. 322.

[54] Stott, I.P., et al., Laboratory studies of the mechanism of the oxygen airglow, *Proceedings of the Royal Society of London*, 1989. **424A**: p. 1.

[55] Loftus, A. and P.H. Krupenie, The spectrum of molecular nitrogen. *Journal of Physical and Chemical Reference Data*, 1977. **6**(1): p. 113.

[56] Ferreira, C.M., et al., Modeling of low-pressure microwave discharges in Ar, He, and O/sub 2/: similarity laws for the maintenance field and mean power transfer, *IEEE Transactions on Plasma Science*, 1991. **19**(2): p. 229.

[57] Pravilov, A.M., Photoprocesses in molecular gases, *In Collected Works: Chemistry of Plasma*. 1987, Vol. 14: p. 65.

[58] Capitelli, M., et al., *In Sb: Non-Equilibrium Vibrational Kinetics*. 1989: Mir: p. 360.

[59] Baulch, D.L., et al., Evaluated kinetic and photochemical data for atmospheric chemistry. *Journal of Physical and Chemical Reference Data*, 1980. **9**(2): p. 295.

[60] Nichipor, G.V. et al., Gas-phase radiation-chemical oxidation of naphthalene. *Chemical Physics*, 1994. **13**(4): pp. 69–76.

[61] Sadeghi, N. and P.W. Setser, Collisional coupling of $N_2(B^3\Pi_g)$ and $N_2(W^3\Delta_u)$ states studied by laser-induced fluorescence. *Chemical Physics Letters*, 1981. **77**(2): p. 304.

[62] Sadeghi, N. and P.W. Setser, Collisional coupling and relaxation of $N_2(B^3\Pi g)$ and $N_2(W^3\Delta u)$ vibrational levels in Ar and Ne. *The Journal of Chemical Physics*, 1983. **79**(6): p. 2710.

[63] Cahill P., Determination of the Lifetime of the $d^3\Pi_u$ State of H_2*. *Journal of the Optical Society of America*, 1969. **59**: p. 875.

[64] Kuznetsova, L.A., N.E., Kuzmenko, and Yu. Kuzyakov, *Probabilities of Optical Transitions of Two-Atomic Molecules*, Ed. R.V. Khokhlov. 1980: Nauka, Moscow.

[65] Johnson, C.E., Lifetime of the $c^3\Pi u$ metastable state of H2, D2, and HD. *Physical Review*, 1972. **A5**:p. 1026.

[66] Lewis, B.R., Experimentally-determined oscillator strengths for molecular hydrogen-II. The Lyman and Werner bands below 900 A, the B'-X and the D-X bands. *Journal of Quantitative Spectroscopy and Radiative Transfer*, 1974. **14**: p. 537.

[67] Hesser, J.E., Absolute transition probabilities in ultraviolet molecular spectra. *The Journal of Chemical Physics*, 1968. **48**: p. 2518.

[68] Smirnov, B.M., *Excited Atoms*. 1982: Energoizdat.

[69] Zuev, A.P. and A.Yu. Starikovskiy, High-Temperature Reactions Involving Nitrogen Oxides. Reactions $NO+O+M=NO_2+M$ and $NO_2+O=NO+O_2$. *Soviet Journal of Chemical Physics*, 1992. **10**(2): p. 273.

[70] Zuev, A.P. and A.Yu. Starikovskiy, Laser-Schlieren and IR measurements of VT- and VV-relaxation rates behind incident shock waves in NO_2/O_2 mixtures. *Soviet Journal of Chemical Physics*, 1991. **8**(8): p. 1822.

[71] Krivonosova, O.E., et al., AVOGADRO database, In *Sb. Chemistry of Plasma*. Ed B M Smirnov. 1987: Energoizdat, Vol. 14: p. 3.

[72] Tsang, W. and J. Herron, Chemical kinetic data base for propellant combustion I. Reactions involving NO, NO_2, HNO, HNO_2, HCN and N_2O. *Journal of Physical and Chemical Reference Data*, 1991. **20**(4): p. 609.

[73] Kalinenko, R.A., et al., *Plasma Chemistry-90*, Ed. L.S. Polak. 1990: INH, p. 41.

[74] Zuev, A.P. and A.Yu. Starikovskiy, Reactions in the N_2O-H_2 system at high temperatures. *Soviet Journal of Chemical Physics*, 1992. **10**(3): p. 520.

[75] Harradine, D.M., et al., Hydrogen/air combustion calculations - The chemical basis of efficiency in hypersonic flows. *AIAA Journal*, 1990. **28**(10): p. 1740.

[76] Zuev, A.P. and A.Yu. Starikovskiy, High-temperature reactions involving nitrogen Oxides. reactions $NO+O+M=NO_2+M$ and $NO_2+O=NO+O_2$. *Soviet Journal of Chemical Physics*, 1992. **10**(1): p. 80.

[77] Starikovskiy, A.Yu., Kinetics and mechanism of reactions in the N_2O-CO system at high temperatures. *Chemical Physics Reports*, 1995. **13**(8–9): p. 1422.

[78] Stuhl, F. and H. Niki, Measurements of rate constants for termolecular reactions of $O(^3P)$ with NO, O_2, CO, N_2, and CO_2 using a pulsed vacuum-uv photolysis—chemiluminescent method. *The Journal of Chemical Physics*, 1971. **55**: p. 3943.

[79] Zuev, A.P. and A.Yu. Starikovskiy, High-temperature reactions involving nitrogen oxides. Reactions of N_2O with atomic oxygen, *Soviet Journal of Chemical Physics*, 1992. **10**(2): p. 255.

[80] Starikovskiy, A.Yu., Kinetics and mechanism of reactions in the N_2O-CO system at high temperatures. *Chemical Physics Reports*, 1994. **13**(1): p. 151.

[81] Gardiner, W.C. Jr., Ed., *Chemistry of Combustion*. 1988: Mir: p. 217.

[82] Krinberg, I.A., *Kinetics of Electrons in the Ionosphere and Plasmasphere of the Earth*. 1978: Nauka.

[83] Jacquin D., et al., Kinetic study of low-pressure H_2 multipole discharge. *Plasma Chemistry and Plasma Processing*, 1989. **9**(2): p. 165.

[84] Zatsepin, D.V., S.M. Starikovskaia, and A.Yu. Starikovskiy. Non-thermal decomposition of N_2O in pulsed high current discharge. *Plasma Physics Reports*, 2003. **29**(6): pp. 517–527.

[85] Hayashi, M., JILA Atomic Collisions Data Center. 1987. CITATION 20397. Hayashi database. Available from: www.lxcat.net/Hayashi

[86] Starikovskaia, S.M., A.Yu. Starikovskiy, and D.V. Zatsepin, *14th International Symposium on Plasma Chemistry*, Vol. 2. 1999: Czech Republic, Prague, p. 1037.

[87] Capitelli, M., Ed. *Nonequilibrium Vibrational Kinetics*. 1989: Mir.

[88] Anikin, N.B., S.M. Starikovskaia, and A. Yu.Starikovskiy, Oxidation of saturated hydro- carbons under the effect of nanosecond pulsed space discharge. *Journal of Physics D: Applied Physics*, 2006. **39**: pp. 3244–3252.

[89] Asinovskii, E.I., L.M. Vasiluyak, A.V. Kirillin, and V.V. Markovets, Response of weakly ionized plasma to a high-voltage nanosecond pulse. *High Temperature*, 1975. **13**: p. 195.

[90] Anikin, N.B., S.M. Starikovskaia, and A.Yu. Starikovskiy, Uniform nanosecond gas breakdown of negative polarity: Initiation form electrode and propagation in molecular gases. *Journal of Physics D: Applied Physics*, 2001. **34**: p. 177.
[91] Anikin, N.B., S.M. Starikovskaia, and A.Yu. Starikovskiy, The development of fast ionization wave in the systems with different configurations of high-voltage electrodes. *High Temperature*, 1998. **36**: p. 992.
[92] Raizer, Yu. P., *Gas Discharge Physics*. 1991: Springer.
[93] Korolev, U.D. and G.A. Mesiats, *Field-Emission and Explosive Processes in Gas Discharge*. 1982: Nauka.
[94] Farley, D.R. and R.J. Cattolica, Collisional quenching and excitation cross-sections of the CO_2^+ $A^2\Pi(1\to 3,0,0)$ and $B^2\Sigma^+$ (0,0,0) excited states from electron-impact ionization. *Chemical Physics Letters*, 1997. **274**: p. 445.
[95] Kuznetsova, L.A, N.E. Kuzmenko, Yu.Ya. Kuziakov, and Yu.A. Plastinin, *Transition Probabilities of Diatomic Molecules*. 1980: Nauka.
[96] Pancheshnyi, S.V., S.M. Starikovskaia, and A.Yu. Starikovskiy, Population of nitrogen molecule electron states and structure of the fast ionization wave. *Journal of Physics D: Applied Physics*, 1999. **32**: pp. 2219–2227.
[97] Herzberg, G., *Electronic Spectra and Stricture of Mutiatomic Molecules*. 1969: Mir.
[98] Morgan, W.L. Kinema Software, J.-P. Boeuf, and L.C. Pitchford, *The Siglo Data base, CPAT and Kinema Software*, https://us.lxcat.net/data/set_databases.php.
[99] Bates, D.R., Super dissociative recombination. *Journal of Physics B: Atomic, Molecular and Optical Physics*, 1991: pp. 695–701.
[100] DeMore, W.B., C.J. Howard, S.P. Sander, A.R. Ravishankara, D.M. Golden, C.E. Kolb, R.F. Hampson, M.J. Molina, and M.J. Kurylo, *Chemical Kinetics and Photochemical Data for Use in Stratospheric Modeling Evaluation, N* 12. January 15, 1997, JPL Publication 97-4.
[101] Polak, L.S., D.I. Ovsyannikov, D.I. Slovetskii, and F.B. Vurzel, *Theoretical and Applied Plasmachemistry*. 1975: Nauka.
[102] Starikovskaia, S.M., E.N. Kukaev, A.Y. Kuksin, M.M. Nudnova, and A.Y. Starikovskiy, Combustion initiated by nonequilibrium plasma. *IEEE Transactions on Plasma Science*, 2008. **36**(4): pp. 904–905.
[103] Bozhenkov, S.A., S.M. Starikovskaia, and A.Yu. Starikovskiy. Nanosecond gas discharge ignition of H_2 and CH_4 containing mixtures. *Combustion and Flame*, 2003. **133**: pp. 133–146.
[104] Anikin, N.B., S.V. Pancheshnyi, S.M. Starikovskaia, and A.Yu. Starikovskiy, Break-down development at high overvoltage: electric field, electronic levels excitation and electron density. *Journal of Physics D: Applied Physics*, 1998. **31**: pp. 826–833.
[105] Zatsepin, D.V., Starikovskaia, S.M., and A.Yu. Starikovskiy. Non-thermal decomposition of N2O in pulsed high current discharge. *Plasma Physics Reports*, 2003. **29**(6): pp. 517–527.
[106] Starikovskaia, S.M., N.B. Anikin, S.V. Pancheshnyi, D.V. Zatsepin, A.Yu. Starikovskiy, Plasma decay in the afterglow of high-voltage nanosecond discharges in unsaturated and oxygenated hydrocarbons. *Plasma Sources Science and Technology*, 2001. **10**: p. 344.
[107] Aleksandrov, N.L. and E.E. Son, in *Chemistry of Plasma*, Ed., B.M. Smirnov. 1980: Atomizdat Press, Vol. 7, p. 35.
[108] Galtsev, V.E., Demyanov A.V., Kochetov I.V., Pevgov V.G., Sharkov V.F. Boltzmann equation solver. Preprint IAE-3156. 1979: Moscow.
[109] Rapp, D. and D.D. Briglia, Total cross sections for ionization and attachment in gases by electron impact. II. Negative-ion formation. *The Journal of Chemical Physics*, 1965. **43**(5): p. 1480.
[110] Rapp, D. and P. Englander-Golden, Total cross sections for ionization and attachment in gases by electron impact. I. Positive ionization. *The Journal of Chemical Physics*, 1965. **43**(5): p. 1464.
[111] Konnov, A.A., *Proceedings of the Combustion Institute*. 2000: The Combustion Institute, Pittsburg, Vol. 28, p. 317.
[112] Bhaskaran, K.A. and P. Roth, The shock tube as wave reactor for kinetic studies and material systems. *Progress in Energy and Combustion Science*, 2002. **28**: p. 151.
[113] Glushko, V.P., Ed., *Thermodynamical Properties of Individual Species*. 1978: Nauka.
[114] Vasilyak, L.M., S.V. Kostyuchenko, N.N. Kudryavtsev, and I.V. Filyugin, High–speed ionization wave s at an electric breakdown. *Physics – Uspekhi*, 1994. **163**: p. 263.
[115] Starikovskaia, S.M., N.B. Anikin, S.V. Pancheshnyi, and A.Yu. Starikovskiy, Selected research papers on spectroscopy of nonequilibrium plasma at elevated pressures. *Proc. SPIE*, 2002. **4460**: p. 63.
[116] Smith, G.P., et al. GRI-Mech 3.0. http://www.me.berkeley.edu/gri_mech/.

[117] Kosarev, I.N., N.L. Aleksandrov, S.V. Kindysheva, S.M. Starikovskaia, and A. Yu. Starikovskiy, Kinetics of ignition of saturated hydrocarbons by nonequilibrium plasma: CH_4-containing mixtures, *Combustion and Flame*, 2008. **154**: pp. 569–586.
[118] Semenov, N.N., *Nobel Lecture*, December 11, 1956, https://www.nobel.se/chemistry/laureates/.
[119] Eichenberger, D.A. and W.L. Roberts, Effect of unsteady stretch on spark-ignited flame kernel survival. *Combustion and Flame*, 1999. **118**: p. 469.
[120] Samano, E.C., W.E. Carr, M. Seidl, and B.S. Lee, An arc discharge hydrogen atom source. *Review of Scientific Instruments*, 1993. **64**(10): pp. 2746–2752.
[121] Starikovskaia, S.M., E.N. Kukaev, A.Yu. Kuksin, M.M. Nudnova, and A.Yu. Starikovskiy, Analysis of the spatial uniformity of the combustion of a gaseous mixture initiated by a nanosecond discharge. *Combustion and Flame*, 2004. **139**: pp. 177–187.
[122] Kosarev, I.N., S.V. Kindysheva, I.V. Kochetov, A.Yu. Starikovskiy, and N.L. Aleksandrov. Shock-tube study of dimethyl ether ignition by high-voltage nanosecond discharge. *Combustion and Flame*, 2019. **203**: pp. 72–82.
[123] Kosarev, I.N., S.O. Belov, S.V. Kindysheva, A.Y. Starikovskiy, and N.L. Aleksandrov. Inhibition of plasma-assisted ignition in hydrogen-oxygen mixtures by hydrocarbons. *Combustion and Flame*, 2018. **189**: pp. 163–172.
[124] Kosarev, I.N., S.V. Kindysheva, R.M. Momot, E.A. Plastinin, N.L. Aleksandrov, and A.Y. Starikovskiy. Comparative study of nonequilibrium plasma generation and plasma- assisted ignition for C_2-hydrocarbons. *Combustion and Flame*, 2016. **165**: pp. 259–271.
[125] Kosarev, I., S. Kindysheva, N. Aleksandrov, and A. Starikovskiy. Ignition in ethanol- containing mixtures after nanosecond discharge. *Combustion and Flame*, 2015. **162**(1): pp. 50–59.
[126] Aleksandrov, N.L., S.V. Kindysheva, E.N. Kukaev, S.M. Starikovskaya, and A. Yu. Starikovskiy. Simulation of the ignition of a methane-air mixture by a high-voltage nanosecond discharge. *Plasma Physics Reports*, 2009. **35**(10): pp. 867–882.
[127] Kosarev, I.N., N.L. Aleksandrov, S.V. Kindysheva, S.M. Starikovskaia, A. Yu. Starikovskiy, Kinetics of ignition of saturated hydrocarbons by nonequilibrium plasma: C_2H_6- to C_5H_{12}-containing mixtures, *Combustion and Flame*, 2009. **156**: pp. 221–233.
[128] Starikovskaia, S.M., On the energy distribution of high-voltage nanosecond pulsed-periodic discharge through internal degrees of freedom. O2 dissociation, *Plasma Physics Reports*, 1995. **21**: pp. 541–547.
[129] Krasnochub, A.V. and L.M. Vasilyak, Dependence of the energy deposition of a fast ionization wave on the impedance of a discharge gap. *Journal of Physics D: Applied Physics*, 2006. **34**: pp. 1678–1682.
[130] Kossyi, I.A., A.Yu. Kostinsky, A.A. Matveyev, and V.P. Silakov, Kinetic scheme of the non-equilibrium discharge in nitrogen-oxygen mixtures. *Plasma Sources Science and Technology*, 1992. **1**: pp. 207–220.
[131] Slovetskii, D.I., *Mechanisms of Chemical Reactions in Nonequilibrium Plasma*. 1980: Nauka, In Russian.
[132] Lukhovitskii, B.I., A.M. Starik, and N.S. Titova, Activation of chain processes in combustible mixtures by laser excitation of molecular vibrations of reactants. *Combustion, Explosion, and Shock Waves*, 2005. **41**: pp. 386–394.
[133] Starik, A.M. and N.S. Titova, Possibility of initiation of combustion of CH_4—O_2 (Air) mixtures with laser induced excitation of O_2 molecules. *Combustion, Explosion, and Shock Waves*, 2005. **40**: pp. 499–510.
[134] Starikovskaia, S.M., N.B. Anikin, I.N. Kosarev, N.A. Popov, and A.Yu. Starikovskiy, Analysis of ignition by nonequilibrium sources. Ignition of homological series of hydrocarbons by volume nanosecond discharge, in 44th AIAA Aerospace Sciences Meeting and Exhibit, 9–12 January 2006, Reno, Nevada, USA. 2006, AIAA-2006-616.
[135] Chintala, N., A. Bao, G. Lou, and I.V. Adamovich, Measurements of combustion efficiency in nonequilibrium RF plasma-ignited flows. *Combustion and Flame*, 2006. **144**(4): pp. 744–756.
[136] Anikin, N.B., S.M. Starikovskaia, and A.Yu. Starikovskiy, Study of the oxidation of alkanes in their mixtures with oxygen and air under the action of a pulsed volume nanosecond discharge. *Plasma Physics Reports*, 2004. **30**: pp. 1028–1042.
[137] von Elbe, G. and B. Lewis, Mechanism of the thermal reaction between hydrogen and oxygen. *The Journal of Chemical Physics*, 1942. **10**: pp. 366–393.
[138] Williams, S., S. Popovic, L. Vuskovic, C. Carter, L. Jacobson, S. Kuo, D. Bivolaru, S. Corera, M. Kahandawala, and S. Sidhu, Model and igniter development for plasma assisted combustion, in 42nd AIAA Aerospace Sciences Meeting and Exhibit, 5–8 January 2004, Reno, Nevada, USA. 2004. AIAA-2004-1012.

[139] Campbell, C.S. and F.N. Egolfopoulos, Kinetic paths to radical-induced ignition of methane/air mixtures. *Combustion Science and Technology*, 2005. **177**: pp. 2275–2298.

[140] K. Takita and Y. Ju, Effect of radical addition on extinction limits of H_2 and CH_4 flames, in *44th AIAA Aerospace Sciences Meeting and Exhibit*, 9–12 January 2006, Reno, Nevada, USA. 2006, AIAA-2006-1209.

[141] Lou, G., A. Bao, M. Nishihara, S. Keshav, Y.G. Utkin, and I.V. Adamovich, Ignition of premixed hydrocarbon–air flows by repetitively pulsed, nanosecond pulse duration introduction plasma, in *41st AIAA Aerospace Sciences Meeting and Exhibit*, 9–12 January 2003, Reno, Nevada, USA. 2003, AIAA-2006-1215.

[142] Kof, L.M. and A.Y. Starikovskiy, Oxygen–hydrogen mixtures ignition under the high– voltage ionization wave conditions at high temperatures, in *26th Int. Symposium on Combustion. Abstracts of Work–in–Progress Papers*. 1996: The Combustion Institute Publ., p. 406.

[143] Hagelaar, G.J.M. and L.C. Pitchford, Solving the Boltzmann equation to obtain electron transport coefficients and rate coefficients for fluid models. *Plasma Sources Science and Technology*, 2005. **14**(4): pp. 722–733. and https://www.bolsig.laplace.univ-tlse.fr/.

[144] Hayashi, M., Electron collision cross-sections determined from beam and swarm data by Boltzmann analysis. In *Swarm Studies and Inelastic Electron-Molecule Collisions*, Ed. L.C. Pitchford. 1987: Springer, pp. 167–187.

[145] Zhiglinskii, A.G., ed. *Handbook on Rate Coefficients for Elementary Processes with Atoms, Ions, Electrons and Photons*. 1994: St–Petersburg University Press.

[146] McEwan, M.J. and L.F. Phillips, *Chemistry of the Atmosphere*. 1975: Edward Arnold.

[147] Brian, J. and A. Mitchell. The dissociative recombination of molecular ions. *Physics Reports*, 1990. **186**(5): pp. 215–248.

[148] Bates, D.R., Products of dissociative recombination of polyatomic ions. *Astrophysical Journal*, 1986. **306**: pp. L45–L47.

[149] *The UMIST Database for Astrochemistry*, https://umistdatabase.uk/files/UDfA2024.pdf.

[150] Burcat, A., K. Scheller, and A. Lifshitz, Shock–tube investigation of comparative ignition delay times for C1—C5 alkanes. *Combustion and Flame*, 1971. **16**: p. 29.

[151] Tan, Y., P. Dagaut, M. Cathonnet, and J.C. Boettner, Oxidation and ignition of methane–propane and methane–ethane–propane mixtures: experiments and modelling. *Combustion Science and Technology*, 1994. **103**: p. 133.

[152] Curran, H.J., P. Gaffuri, W.J. Pitz, and C.K. Westbrook, A comprehensive modelling study of n-heptane oxidation. *Combustion and Flame*, 1998. **114**: pp. 149–177.

5 Plasma-Assisted Combustion Chemistry

5.1 ELEMENTARY REACTIONS IN COMBUSTION

5.1.1 Chain-Initiation, Propagation, and Termination Reactions in Combustion

Fuel oxidation and heat release in the combustion process are proceeded through radical production, multiplication, and recombination. As such, radicals such as H, O, OH, CH_3, and HO_2 play the most critical role in combustion. There are several key reaction processes that govern radical production and consumption. They are, respectively, the chain-initiation, branching, propagation, and termination reactions (Table 5.1).

Chain-initiation reaction: As shown in Table 5.1, the chain-initiation reaction is a reaction that reactants produce one or more radicals in the molecule collision processes. This reaction is very important in initiating the ignition and reaction processes because it produces the first group of radicals directly from reactants. However, as shown in Figure 5.1, the chain-initiation reactions typically have a very high activation energy ($E_{a,i}$) because the reactants (A and B) are stable molecules in the ground states. Therefore, this type of reaction is very slow and can be rate limiting in an ignition process. However, as shown later, plasma can provide new reaction pathways at low temperature via electrons and excited species to produce radicals rapidly by bypassing the rate-limiting slow chain-initiation reactions in combustion.

Chain-branching reaction: The chain-branching reaction is a reaction that one radical reacts with a reactant and produces two or more radicals. This reaction is like the nuclear fission reactions that magnify the radical pool to accelerate the reaction processes. However, the chain-branching reactions also have quite high activation energies ($E_{a,b}$), although they may be lower than that of the chain-initiation reactions. Therefore, the chain-branching reactions are the most important rate-limiting reactions (after initial radicals are produced) that dominate the rate of reaction processes and the burning rate of combustion. As such, understanding the key chain-branching reactions and knowing how to control them using plasma is critical in plasma-assisted combustion.

Chain-propagation reaction: The chain-propagation reaction is a reaction that one radical (e.g., H) reacts with another reactant and produces another radical (e.g., OH) in the combustion processes. Although this reaction is less important than the chain-branching reactions in affecting the rate of radical pool growth, it normally has lower activation energy and thus can strongly affect fuel oxidation, intermediate species production, and heat release rate.

TABLE 5.1
Schematic of Chain-Initiation, Branching, Propagation, and Termination Reactions, *A* and *B* Are the Reactants, *R* the Radicals, and *P* the Product, Respectively

$A + B = 2R$	Initiation
$A + R = 2R$	Branching
$A + R = R + P$	Propagation
$R + R = 2P + \text{heat}$	Termination

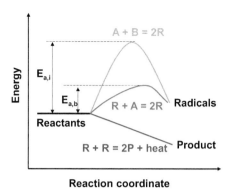

FIGURE 5.1 Schematic of activation energies and enthalpy changes of chain-initiation, branching, and termination reactions.

Chain-termination reaction: The chain-termination reaction is a reaction that two radicals react together or with a third molecule and produce a stable product with one or two less active radicals. Therefore, in this process, the radial number will decrease, and the reaction will slow down. Sometimes, we refer to it as radical quenching reaction. However, the chain-termination reaction is important to shift the chemical equilibrium toward the products and produce heat to raise the temperature to self-accelerate the reaction process.

Reaction rate and the Arrhenius law: For reaction to occur, as shown in Figure 5.1, the energy of the colliding reactants needs to be greater than or equal to the activation energy (E_a). The kinetic energy of a molecule is proportional to temperature. However, not every molecule has the same amount of energy. In equilibrium, the energy of molecules follows the Boltzmann distribution (Figure 1.25), that is, the number of molecules with energy, E, is proportional to $\exp(-E/R_0T)$. Thus, only a very small portion of the molecules have enough energy to induce a reaction after a collision. From statistical mechanics, by using the activation energy as the reaction threshold and the Boltzmann energy distribution, one can derive the *Arrhenius law* for the reaction rate, k, as a function of temperature, T, and the activation energy, E_a, as,

$$k = AT^n e^{-E_a/R_0 T} \tag{5.1}$$

where A is the pre-exponential factor, R_0 is the universal gas constant, and n is a constant, respectively. Therefore, to accelerate the reaction rate, one needs to reduce the activation energy by creating a new reaction pathway or raising the temperature.

5.1.2 Elementary Kinetics of Hydrogen

Hydrogen is a green and the simplest fuel. Moreover, the hydrogen combustion mechanism is the base mechanism for all combustion mechanisms of other fuels. Therefore, it is critical to understand the key chain-initiation, branching, propagation, and termination reactions in hydrogen combustion. A detailed hydrogen mechanism recently updated for high-pressure combustion [1] (HP-Mech) is shown in Table 5.2.

In the hydrogen reaction system, the major chain-initiation, branching, propagation, and termination reactions are as follows:

Chain-initiation reactions:

$$H_2 + O_2 = H + HO_2 \tag{R5.1a}$$

$$H_2 + O_2 = OH + OH \tag{R5.2}$$

TABLE 5.2
Detailed Reaction Kinetics of Hydrogen Combustion in HP-Mech [1]

Reactions	A (cm³/(mol·s·K))	n	E_a (cal/mol)
$H + O_2 = O + OH$	7.26E+14	−0.235	15928.7
$OH + OH = O + H_2O$	9.32E+03	2.564	−2603.7
$H + H + H_2 = H_2 + H_2$	1.02E+17	−0.6	0
$H + OH(+M) = H_2O(+M)$	2.51E+13	0.234	−114
LOW	4.50E+25	−3.064	1581.4
TROE	0.72	1.0E−30	1.0E+30
$H + O_2(+M) = HO_2(+M)$	1.03E+12	0.604	−241.1
LOW	1.74E+19	−1.23	0
TROE	0.495	1.0E−30	1.0E+30
$HO_2 + OH = O_2 + H_2O$	7.44E+12	0.055	−915.2
DUP	1.17E+23	−2.156	23,681
$HO_2 + HO_2 = H_2O_2 + O_2$	1.93E−02	4.12	−4,960
$HO_2 + HO_2 = OH + OH + O_2$	6.41E+17	−1.54	8,540
$H_2O_2 + H = H_2 + HO_2$	4.40E+01	3.45	712
$H_2O_2 + H = H_2O + OH$	3.35E+07	1.91	3,654
$H_2O_2(+M) = 2OH(+M)$	2.00E+12	0.9	48,749
LOW	2.49E+24	−2.3	48,749
TROE	0.43	1.0E−30	1.0E+30

LOW and TROE indicate the low- and high-pressure rate limiting, respectively; DUP indicates the reaction rate constant equals to the summation of those identical reactions; M indicates a third body.

Chain-branching reactions:

$$H + O_2 = OH + O \tag{R5.3}$$

$$O + H_2 = OH + H \tag{R5.4}$$

$$H_2O_2 = OH + OH \tag{R5.5}$$

Chain-propagation reactions:

$$OH + H_2 = H_2O + H \tag{R5.6}$$

$$HO_2 + H = OH + OH \tag{R5.7}$$

Chain-termination reactions:

$$H + O_2(+M) = HO_2(+M) \tag{R5.8}$$

$$H + OH(+M) = H_2O(+M) \tag{R5.9}$$

$$HO_2 + H = H_2 + O_2 \tag{R5.1b}$$

$$HO_2 + OH = H_2O + O_2 \tag{R5.10}$$

$$HO_2 + HO_2 = H_2O_2 + O_2 \tag{R5.11}$$

As discussed before, since the slowest chain-branching reaction (R5.3) is the rate-limiting reaction, from Figure 5.2 we can see that if there is one H radical in the reaction system, reaction R5.3 will produce one OH and one O. Since R5.6 is a very fast reaction, the OH produced from R5.3 will be immediately converted to a new H by R5.6. Then, because R5.4 is also faster than R5.3, the O produced in R5.3 will be converted to the second H and a new OH from R5.4. The resulting new OH will be converted to the third H via the fastest reaction R5.6 again. As a result, by adding up reactions R5.3 and R5.4 with twice of reaction R5.6, as seen in Figure 5.2, one H radial will produce 3H radicals in the hydrogen chain-branching and propagation system. Therefore, the slowest chain-branching reaction R5.3 is a dominant chain-branching reaction for H radical production at high temperature, especially in flames and high-temperature ignition (HTI):

High-temperature chain-branching reaction:

$$H + O_2 = OH + O \tag{R5.3}$$

Note that the H radical for the chain-branching reaction R5.3 can be suppressed via the following chain-termination reaction,

Chain-termination reaction:

$$H + O_2(+M) = HO_2(+M) \tag{R5.8}$$

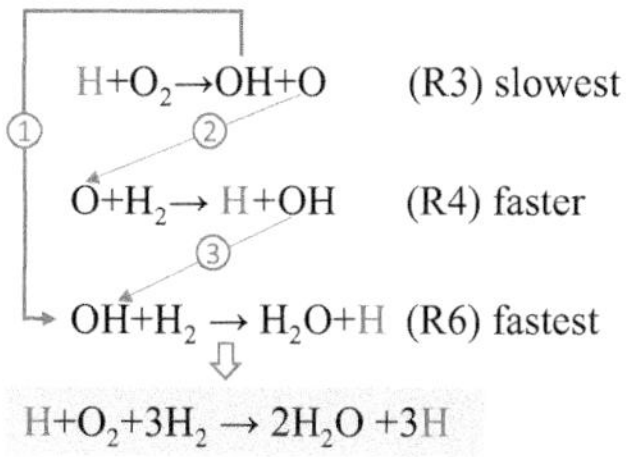

```
H+O₂ → OH+O       (R3) slowest
O+H₂ → H+OH       (R4) faster
OH+H₂ → H₂O+H     (R6) fastest
H+O₂+3H₂ → 2H₂O +3H
```

FIGURE 5.2 One H radical produces three H radicals via the slowest rate-limiting reaction of R5.3 and the faster reactions of R5.4 and R5.6 (R_3, R_4, and R_6 in this figure are referred to as R5.3, R5.4, and R5.6).

Figure 5.3 shows the comparison of the reaction rates between the chain-branching reaction R5.3 and the termination reaction R5.8 at different pressures with air, H_2O, and CO_2 as the third body (M), respectively. It is seen clearly that with the decrease of temperature, the termination reaction R5.8 is faster than the branching reaction R5.3. With the increase of pressure, the balance of these two competing reactions shift to higher temperature. At 25 atm, a typical gas turbine pressure, the temperature at which the two reaction rates become equal increases to 1,250 K. Moreover, if H_2O and CO_2 are used as the diluents, the increased third-body effect makes the termination reaction even faster and further raises the critical reaction temperature.

Therefore, the competition between the chain-branching reaction R5.3 and chain-termination reaction R5.8 will determine the combustion limits and is strongly pressure and diluent-dependent.

The hydrogen explosion limits:

Since reactions R5.4 and R5.6 are fast (Figure 5.2), the consumption rates of O and OH are faster than their production rates via R5.3. Therefore, the concentration of these two radicals (OH and O) will be in *quasi-steady state*, i.e., their net production rates are almost zero. With this assumption, the radical production via reactions R5.3, R5.4, R5.6, and R5.8 can be written as

$$\frac{dC_H}{dt} = -k_3 C_H C_{O_2} - k_8 C_H C_{O_2} C_M - k_4 C_O C_{H_2} - k_6 C_{OH} C_{H_2} \tag{5.2a}$$

$$\frac{dC_{OH}}{dt} = k_3 C_H C_{O_2} + k_4 C_O C_{H_2} - k_6 C_{OH} C_{H_2} \approx 0 \tag{5.2b}$$

$$\frac{dC_O}{dt} = k_3 C_H C_{O_2} - k_4 C_O C_{H_2} \approx 0 \tag{5.2c}$$

By submitting the OH and O concentrations from Eqs. 5.2b and 5.2c into Eq. 5.2a, we will have

$$\frac{dC_H}{C_H} = (2k_3 - k_8 C_M) C_{O_2} dt \tag{5.3}$$

The above equation indicates that there is a critical limit that define whether the hydrogen reaction system is explosive or non-explosive,

$$\frac{2k_3}{k_8 C_M} \begin{cases} > 1, & \text{explosive} \\ < 1, & \text{non-explosive} \end{cases} \tag{5.4}$$

This is the so-called *the second explosion limit* (Figure 5.4) [2], which is a result of the competition between the chain-branching and termination reactions, R5.3 and R5.8. The second explosion limit occurs at intermediate pressures and increases with the increase of temperature.

At low pressure, the reaction rate of R5.8 is very slow and negligible. As a result, the radical quenching is mainly via H radical diffusion and quenching on the wall. Therefore, the first explosion limit at low pressure (Figure 5.4) is governed by R5.3 and radical loss due to diffusion and increases with the decrease of temperature.

At high pressure, HO_2 radical production via R5.8 becomes more important (Figure 5.3). With the increase of HO_2 concentration, the reactions R5.5 and R5.7 become the new chain-branching reactions at high pressure. Therefore, the HO_2 chemistry leads to the third explosion limit at high pressure, which increases again with the decrease of the temperature (Figure 5.4). A sensitivity analysis of the burning rate of a hydrogen flame on the rates of elementary reactions is shown in Figure 5.5. It clearly shows the increased sensitivity of reaction R5.8 and R5.3 at high pressure.

Plasma-Assisted Combustion: Chemistry

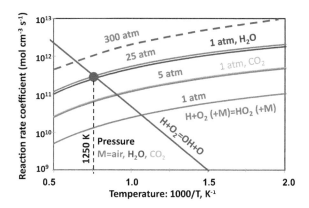

FIGURE 5.3 Reaction rate comparison between $H+O_2 = OH+O$ and $H+O_2(+M) = HO_2(+M)$ at different pressures and with air, H_2O, and CO_2 as the diluents, respectively.

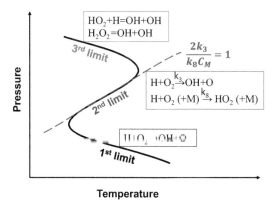

FIGURE 5.4 Schematic of the hydrogen explosion limits and the key reactions. The region on the right of the explosion limit curve is the explosive region (k_3 and k_8 are referred to as R5.3 and R5.8).

If the system temperature is further reduced (Figure 5.3), the H radical production via reaction R5.3 will be very slow and negligible. Therefore, the major chain-branching reaction at intermediate temperature (700–1,100 K) and high pressure will be:

Intermediate-temperature chain-branching reaction:

$$H_2O_2 \rightarrow OH + OH \tag{R5.5}$$

Therefore, the two chain-branching reactions of R5.3 and R5.5 play a dominant role in hydrogen combustion chemistry for H and HO_2 radical production at high and intermediate temperatures, respectively.

5.1.3 Elementary Kinetics of Methane

Methane is an important fuel for power, heating, carbon, and hydrogen production. Oxidation of methane at fuel lean conditions is very important to reduce emissions and methane slip. Unfortunately, unlike hydrogen, methane is much more difficult to oxidize than hydrogen because of the strong C–H bond in methane (~439 kJ/mole) and CH_3 (~463 kJ/mol) [3], resulting in slow CH_4 chain-initiation and branching channels.

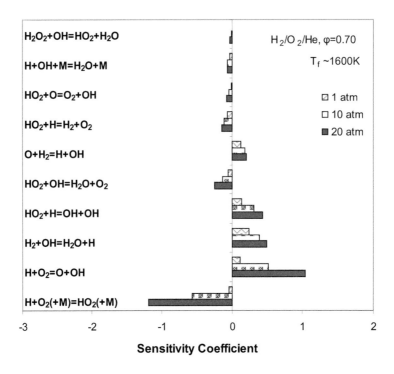

FIGURE 5.5 Reaction sensitivity of lean hydrogen combustion ($\varphi=0.7$) at 1, 10, and 20 atm, respectively.

The major chain-initiation, branching, propagation, and termination reactions of methane at high temperature are listed below. The first radical production in chain-initiation reaction is via R5.12. After this initiation reaction, CH_3 will be oxidized via chain-branching and propagation reactions R5.14 and R5.15 to form CH_3O as well as O or OH radicals. However, these reactions are slower than R5.3. Therefore, methane oxidation is slower than that of hydrogen. To proceed methane oxidation, it is critical to produce H radials to accelerate R5.3 branching reaction. CH_3O radicals will be converted to CH_2O and HCO via reactions R5.16 and R5.17, respectively. The H-abstraction reaction by radicals from CH_2O in reaction R5.17 is one of the major exothermic reactions in methane oxidation. In methane oxidation, reaction R5.18a is a critical reaction to produce H radicals for R5.3. However, R5.18b is a termination reaction to produce relatively inactive HO_2. Note that different from R5.8 in hydrogen mechanism, R5.18b is the major reaction of HO_2 production in methane oxidation. Therefore, like R5.3 and R5.8 for hydrogen, reactions R5.18a and R5.18b are the major competition reaction pairs for methane oxidation (Figure 5.6). As such, producing H and OH radicals by plasma can dramatically accelerate methane oxidation via R5.3. Depending on the temperature, CO will be oxidized to CO_2 and produce heat via reactions R5.19 and R5.20, respectively, at high and intermediate temperatures. Note that at high pressure, the pressure fall-off reaction R5.13 and the termination reactions R5.8 and R5.21 will play important roles in affecting methane oxidation.

Chain-initiation reactions:

$$CH_4 + O_2 = CH_3 + HO_2 \tag{R5.12}$$

$$CH_4(+M) = CH_3 + H(+M) \tag{R5.13}$$

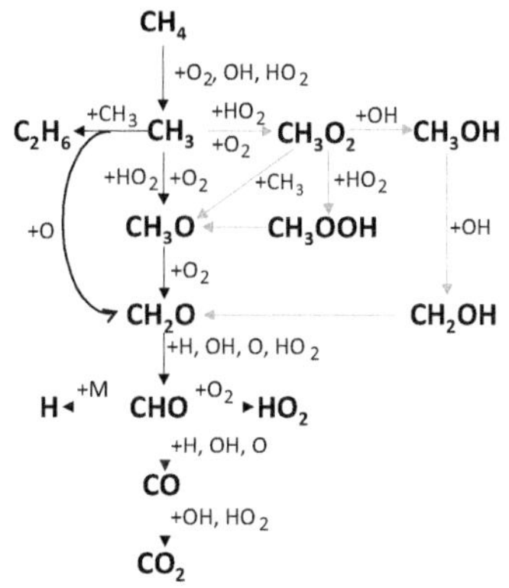

Plasma-Assisted Combustion: Chemistry

FIGURE 5.6 Schematic of methane oxidation at high (black arrow) and low (green arrow) temperatures, respectively. The dashed box indicates H radical production and termination competition reaction pare.

Chain-branching reactions:

$$H + O_2 = OH + O \qquad (R5.3)$$

$$CH_3 + O_2 = CH_3O + O \qquad (R5.14a)$$

Chain propagation reactions:

$$CH_3 + O_2 = CH_2O + OH \qquad (R5.14b)$$

$$CH_3 + HO_2 = CH_3O + OH \qquad (R5.15)$$

$$CH_3O + O_2 = CH_2O + HO_2 \qquad (R5.16)$$

$$CH_2O + X = HCO + XH \left(X = H, OH, O, HO_2\right) \quad \text{Exothermic} \qquad (R5.17)$$

$$HCO(+M) = CO + H(+M) \qquad (R5.18a)$$

$$HCO + O_2 = HO_2 + CO \qquad (R5.18b)$$

$$CO + HO_2 = CO_2 + OH \qquad (R5.19)$$

$$CO + OH = CO_2 + H \quad \text{Exothermic} \qquad (R5.20)$$

Chain-termination reactions:

$$H + O_2 (+M) = HO_2 (+M) \quad \text{(R5.8)}$$

$$CH_3 + CH_3 (+M) = C_2H_6 (+M) \quad \text{(R5.21)}$$

To summarize the discussions above, a schematic of the reaction pathways of methane oxidation at high and low temperatures is shown in Figure 5.6. Note that at low temperature and high pressure, production of methyldioxy radical, CH_3O_2, from CH_3 with O_2 and HO_2 becomes important. Therefore, new low-temperature reaction channels via $CH_3 \rightarrow CH_3O_2 \rightarrow CH_3O$ pathway and $CH_3 \rightarrow CH_3O_2 \rightarrow CH_3OOH \rightarrow CH_2O$ pathway will become important at low temperature and high pressure [4]. Understanding the rate-limiting reactions, the competing radical production reaction pairs, and the low and high-temperature/-pressure reaction pathways are important for plasma-assisted combustion. Since nonequilibrium plasma operates at low temperature, the low-temperature methane oxidation via CH_3O_2 will be accelerated.

5.1.4 Reaction Kinetics of Ammonia

Ammonia is likely to be an important hydrogen carrier for power generation and energy storage. However, ammonia is difficult to ignite and has a very low flame speed and high NO_x formation in combustion [5–7]. Therefore, it is necessary to understand ammonia combustion chemistry and to explore the possibility of plasma-enhanced ammonia combustion.

Ammonia combustion chemistry has been studied extensively since the 1980s [8,9]. To improve the burning properties of ammonia, more interest in ammonia combustion is to understand the combustion kinetics of ammonia in blended fuel mixtures with hydrogen, methane, large alkanes, and oxygenated fuels. Figure 5.7 shows a summary of the recent studies of ammonia combustion kinetics with hydrogen, alkanes, alcohols, and ethers. It is seen that most of the kinetic studies were conducted in high-temperature flames under low pressure (0.3–10 atm) [10–32] with hydrogen and hydrocarbon blends [6,7,33–50]. However, kinetic studies of ammonia oxidation with oxygenated fuels [51–61] are very scarce and many studies have been limited to low pressures. To data, only

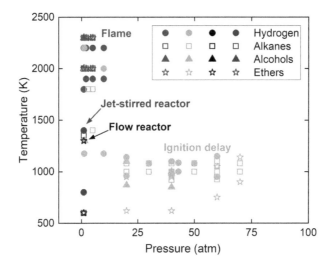

FIGURE 5.7 Pressure and temperature conditions of kinetic studies of NH_3 oxidation with different fuel blends.

the ignition delay time of ammonia with H_2, alcohols, and ethers are studied up to 70 atm [53–55,58,59,61] using a rapid compression machine (RCM) or shock tube [35]. To overcome this pressure limitation, Hashemi et al. [62,63] succeeded in species measurements at 100 atm for $CH_4/C_2H_6/C_3H_8/C_7H_{16}/NH_3$ oxidations [6,35,62–64]. Glarborg and Ju [35] also collaborated in extending these studies for high-pressure alkane/NH_3 oxidation using a new supercritical pressure jet-stirred reactor (SP-JSR), which can operate up to 250 atm with a well-defined temperature (±5 K) and engine relevant flow residence time (50–500 ms) [1,4,65–67]. The results showed a strong kinetic coupling between n-heptane and ammonia oxidation at low and intermediate temperature at high pressure.

The major chain-initiation, branching, propagation, and termination reactions of ammonia combustion are as follows [5,8,9,68,69],

Chain-initiation reactions:

$$NH_3(+M) = NH_2 + H(+M) \tag{R5.22}$$

$$NH_3 + O_2 = NH_2 + HO_2 \tag{R5.23}$$

Chain-branching reactions:

$$H + O_2 = OH + O \tag{R5.3}$$

$$NH_2 + NO = NNH + OH \tag{R5.24a}$$

Chain-propagation reactions:

$$NH_2 + O_2 = HNO + OH \tag{R5.25}$$

$$NH + O_2 = HNO + O \tag{R5.26}$$

$$NH_3 + X = NH_2 + XH \, (X = H, O, OH, HO_2) \tag{R5.27}$$

$$NH_2 + X = NH + XH \, (X = H, O, OH) \tag{R5.28}$$

$$NH_2 + O = HNO + H \tag{R5.29}$$

$$HNO(+M) = NO + H(+M) \tag{R5.30}$$

$$HNO + O_2 = HO_2 + NO \tag{R5.31}$$

$$NH + NO = N_2O + H \quad (R5.32a)$$

$$NH + NO = N_2 + OH \quad (R5.32b)$$

$$NH + OH = HNO + H \quad (R5.33)$$

$$NH + X = N + XH\,(X = H, O, OH) \quad (R5.34)$$

$$NH + O = NO + H \quad (R5.35)$$

$$N_2O + H = N_2 + OH \quad (R5.36)$$

$$N + NO = N_2 + O \quad (R5.37)$$

$$N + O_2 = NO + O \quad (R5.38)$$

$$NH_2 + HO_2 = H_2NO + OH \quad (R5.39a)$$

$$H_2NO + HO_2 = HNO + H_2O_2 \quad (R5.40)$$

$$NO + HO_2 = NO_2 + OH \quad (R5.41)$$

$$NH_2 + NO_2 = H_2NO + NO \quad (R5.42a)$$

Chain-termination reactions:

$$NH_2 + HO_2 = NH_3 + O_2 \quad (R5.39b)$$

$$NH_2 + HO_2 = HNO + H_2O \quad (R5.39c)$$

$$NH_2 + NO_2 = N_2O + H_2O \quad (R5.42b)$$

$$NH_2 + HNO = NH_3 + NO \quad (R5.43)$$

$$NH_2 + NO = N_2 + H_2O \tag{R5.24b}$$

$$NH_2 + NH_2(+M) = N_2H_4(+M) \tag{R5.44}$$

$$HNO + OH = NO + H_2O \tag{R5.45}$$

$$N + OH = NO + H \tag{R5.46}$$

$$H + O_2(+M) = HO_2(+M) \tag{R5.8}$$

The main chain-initiation reaction is reaction R5.23 which is very slow because of the strong N–H bond of NH_3. At high temperature, the ammonia oxidation is schematically shown in Figure 5.8a [9]. Equation R5.3 remains to be the key chain-branching reaction to produce H/O/OH radicals. NH_3 is then mainly consumed by H-abstraction reactions via R5.27 and R5.28 and from NH_2 and NH. Both of them are further oxidized by O, OH, and O_2 via reactions R5.24, R5.26, R5.29, R5.33, and R5.35 to form HNO and NO. NH can be further reduced to N with O and H radicals via reaction R5.34. In addition, NH_2 and NH reactions with NO will lead to NNH and N_2O via reactions R5.24a and R5.32b. Then, NO reactions with NH_2 and N can lead to NO reduction to N_2 from R5.24b and R5.37.

At lower temeprature and high pressure, due to the importance of HO_2 chemistry (R5.8), R5.39b and R5.39c start to compete with R5.39a for radical progagation and termination. Reactions R5.39–R5.42 will further couple NO, NO_2, and H_2NO production and consumption by HO_2, leading to a complicated coupling between NH_3 oxidation, HO_2 chemistry, H_2NO reactions, and NO_x chemistry at lower temperatures (Figure 5.8b). Unfortunately, this kinetic coupling at high pressure is not well understood.

Figure 5.9a shows the energy barriers of reactions R5.39, NH_2+HO_2 [69]. It is seen that the termination reaction R5.39b has the lowest energy barrier. Therefore, NH_3 oxidation at high pressure is very sensitive to the branching ratios of reaction R5.39. Unfortunately, although quantum chemistry calculations have been done for this reaction [69], few experimental measurements of this reaction have been conducted because of the difficulty in quantifying radial-radical reactions involving HO_2. With the introduction of oxygenated fuels and low-temperature plasma, more HO_2 will be formed and this reaction coupling will become even more important. As shown in the ammonia oxidation

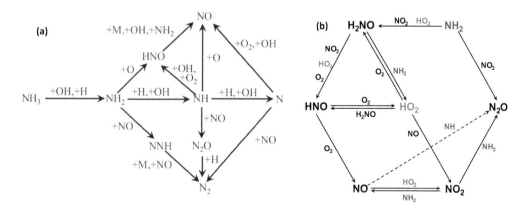

FIGURE 5.8 (a) Schematic of high-temperature NH_3 oxidation kinetics [9]. (b) Schematic of low-temperature and high-pressure kinetic coupling between HO_2 chemistry, ammonia oxidation, and NO_x chemistry.

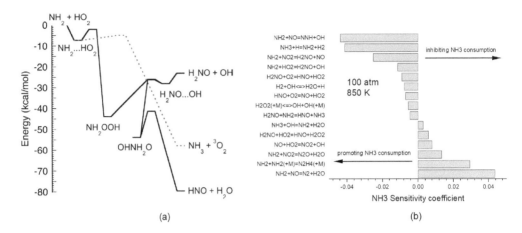

FIGURE 5.9 (a) Energy barriers of NH_2+HO_2 [69] reaction pathways. (b) Reaction sensitivity analysis of NH_3 concentration at 100 atm, 850 K, and fuel lean conditions.

sensitivity analysis of elementary reactions at high pressure and low temperature (Figure 5.9b), there is a strong kinetic coupling between ammonia oxidation via NH_x+HO_2 chemistry, NH_x+NO_x reaction channels, and NO_x chemistry as well as RO_2+NO and HO_2+NO reaction pathways [69–71].

5.1.5 Reaction Kinetics of Large Alkanes and Oxygenates at Different Temperatures

For large hydrocarbon fuels such as n-alkanes and oxygenated fuels such as alcohols [72,73], ethers [66,67,74,75], and esters [76,77], in addition to the low and intermediate temperature chain-branching reactions R5.48 and R5.50 (Table 5.3), there exist low-temperature chain-branching reactions via alkyl-peroxy radicals, which results in low-temperature chemistry (LTC) and cool flames below 900 K [78,79] (Figure 5.10). As shown in Figure 5.10, for a large alkane or oxygenate fuel, in addition to the HTI above 1,000 K, there exists a low-temperature ignition (LTI) between 500 and 900 K due to LTC. Corresponding to the LTI, there exists a cool flame at a temperature far below the hot flame (typically above the critical chain-branching temperature of R5.48 around 1,100 K). The existence of LTI and cool flame significantly changes the ignition delay time and the flammability limit of the combustion process and provides a great opportunity for nonequilibrium plasma to control low-temperature combustion [78,79].

Figure 5.11a shows the dependence of the ignition delay time on the initial temperature for a stoichiometric n-heptane/air mixture at 20 atm with and without 100 ppm OH addition (e.g., generated by plasma). The solid and dashed lines, respectively, represent the ignition delay times of HTI and LTI. It is clearly seen that radical addition by plasma in the initial mixture can dramatically reduce the ignition delay time of LTI to 100 μs. Thus, through chemical sensitization by plasma, LTI can be accelerated dramatically.

The schematic of temperature-dependent chain-branching reaction pathways for fuel oxidation is shown in Figure 5.12 [79–83] and the key elementary reactions for low and intermediate temperature are shown in Table 5.3. At a low temperature (below 700 K), LTC chain-branching pathway governs the rate of fuel oxidation. As shown in Figure 5.12, at LTC, fuel (RH) oxidation starts with an H-abstraction of fuel by radicals such as OH/HO_2 and forms a fuel radical (R). Then, the fuel radicals (R) are added by an O_2 molecule to form RO_2 or HO_2. The internal isomerization of RO_2 will lead to QOOH (hydrocarbon hydrogen peroxide). The subsequent oxygen addition to QOOH for O_2QOOH and its beta scissions produce multiple OH radicals (Figure 5.12). This LTC results in the first stage LTI and cool flame as shown in Figure 5.10. As such, the major cool flame chain-branching reaction pathway can be written as:

Plasma-Assisted Combustion: Chemistry

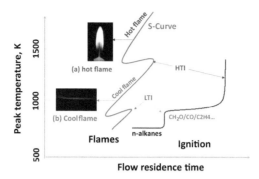

FIGURE 5.10 Schematic of the dependence of peak temperature for low-temperature ignition (LTI), high-temperature ignition (HTI), cool flame, and hot flame on flow residence time for a large n-alkane.

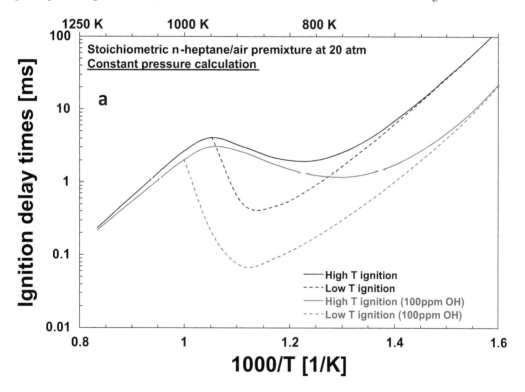

FIGURE 5.11 Dependence of the ignition delay time on the initial temperature for a stoichiometric *n*-heptane/air mixture at 20 atm with and without 100 ppm OH addition by plasma.

$$R \to RO_2 \to QOOH \to O_2QOOH \to OQ'O + 2OH \tag{R5.56}$$

At an intermediate temperature (700–1,050 K), as shown in Figure 5.12 and Table 5.3, the decompositions of RO_2, QOOH and O_2QOOH via $RO_2 \to R+O_2$ (R5.48), QOOH$\to$ HO_2+alkene or OH+cyclic ether (R5.50), and $O_2QOOH \to QOOH+O_2$ (R5.51) shut down the O_2QOOH chain-branching pathway and slow down the reactivity, leading to the negative temperature coefficient (NTC) effect [81,84,85]. As a result, in this NTC process, the concentrations of HO_2 and fuel radicals R increase. Therefore, HO_2 reactions with fuel radicals (R5.52) and partially oxidized intermediate species (R5.47 and R5.55) as well as the H_2O_2 decomposition reaction (R5.5) become critical to produce

TABLE 5.3
Low- and Intermediate-Temperature Chemistry

Fuel + HO_2 = R + H_2O_2	(R5.46)
Aldehyde + HO_2 = H_2O_2 + RCO	(R5.47)
RO_2 = R + O_2	(R5.48)
RO_2 → QOOH	(R5.49a)
RO_2 → aldehyde + HO_2	(R5.49b)
QOOH → cyclic ether + OH	(R5.50a)
QOOH → alkene + HO_2	(R5.50b)
O_2QOOH → QOOH + O_2	(R5.51)
R + HO_2 = RO + OH	(R5.52)
RO + O_2 = aldehyde + HO_2	(R5.53)
RCO + O_2 = aldehyde + CO + OH	(R5.54)
HO_2 + CH_2O = H_2O_2 + HCO	(R5.55)
HO_2 + HO_2 = H_2O_2 + O_2	(R5.11)
H_2O_2 = 2OH	(R5.5)

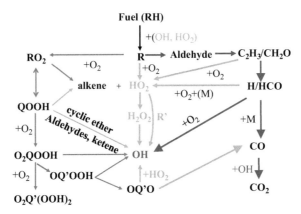

FIGURE 5.12 A schematic of the key reaction pathways at different temperatures (blue arrow: below 700 K; green arrow: 700–1,050 K; red: above 1,050 K).

OH radicals for chain-branching. Therefore, at the intermediate temperature, the HO_2 chemistry listed in Table 5.3 is the major chain-branching pathway for the second stage ITI (Figure 5.11) and warm flame formation [70,86–88]. At high pressure, recent studies have shown that HO_2 chemistry plays a critical role in fuel oxidation and the NTC effects [63,89]. As such, the major chain-branching reaction pathway for ITI and warm flame can be written as Ref. [79],

$$R' + HO_2 \rightarrow R'O + OH, \quad CH_2O \rightarrow HCO \rightarrow HO_2 \rightarrow H_2O_2 \rightarrow 2OH \quad (R5.57a)$$

$$R'CO + O_2 \rightarrow aldehyde + CO + OH, \quad CH_2O \rightarrow HCO \rightarrow HO_2 \rightarrow H_2O_2 \rightarrow 2OH \quad (R5.57b)$$

The above three sets of chain-branching reaction pathways in reactions R5.56, R5.57, and R5.3, respectively, represent the dominant radical production reactions at low, intermediate, and high temperatures. As will be discussed in the sections below, these three sets of temperature-dependent chain-branching reactions lead to three different kinds of flames: cool flame, warm flame, and hot flame.

5.2 CHEMICAL KINETICS IN PLASMA-ASSISTED COMBUSTION

5.2.1 NONEQUILIBRIUM PLASMA ENERGY TRANSFER AND ITS IMPACT ON COMBUSTION

As discussed in Chapters 3 and 4, plasma is a nonequilibrium energy transfer process. The heat, radicals, ions, and excited species produced in plasma will affect combustion chemistry and transport. Figure 5.13 schematically shows the nonequilibrium energy transfer processes as well as active species and fast and slow heating production in plasma. When an electron gains energy in an electric field, it will collide with neutral molecules and produce a second electron and a positive ion. When electrons and positive ions recombine, it will generate heat and emit photons. Both energetic electrons and photons will create ionization, excitation, and dissociations of neutral molecules and lead to the formation of ions, excited molecules, and radicals. This process occurs within a few nanoseconds. As shown in Figure 5.13, in addition to electrons, the resulting electronically excited molecules will also collide with neutral molecules to produce vibrational molecule excitations, radicals, and fast heating from nanoseconds to microseconds. After that molecules at higher vibrational energy states will be relaxed to lower states via vibrational-vibrational (V-V) and vibrational-translational (V-T) energy transfer and generate slow heating. This process occurs from a few microseconds to milliseconds depending on the excited molecules and pressure. If the rate of V-T energy transfer is faster than that of molecule excitation by electrons and photons, the plasma will be relaxed to equilibrium plasma. Otherwise, the plasma will be in nonequilibrium and the electron and vibrational energies are much higher than the rotational and translational energies of the neutral molecules. Therefore, plasma-assisted combustion is strongly affected by the active species production in plasma.

In fact, the active species production and nonequilibrium energy transfer in plasma are a function of electron energy distribution function (EEDF) or the reduced electric field (E/N). Figure 5.14 shows the energy transfer fractions from electrons to different energy states as a function of the reduced electric field for a $CH_4/H_2/He$ mixture. It is seen that when the E/N is below 50 Td, plasma delivers most energy

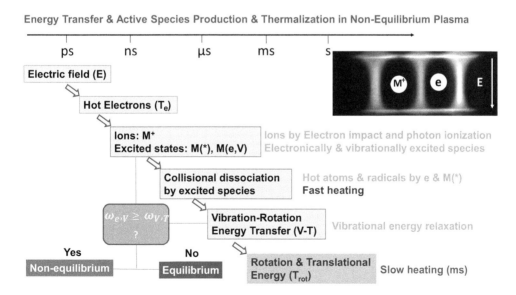

FIGURE 5.13 Timescales of nonequilibrium energy transfer, excitation, and relaxation in plasma and the production of ions, excited molecules, radicals, and heat via fast heating (via electrons and excited molecules) and slow heating processes (via V-V and V-T).

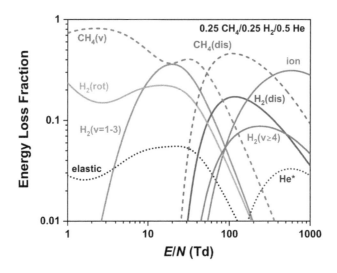

FIGURE 5.14 Computed energy loss fractions as a function of E/N in a methane/hydrogen/helium plasma.

to vibrational states of methane and hydrogen. However, when the E/N is increased above 50 Td, it is seen that CH_4 and H_2 dissociations and high-level H_2 vibrational excitation, H_2 ($v \geq 4$), will start to dominate. When the E/N is further increased above 100 Td, the ionizations of CH_4 and H_2 and the electronic excitations will dictate. Therefore, by controlling E/N, one can tune the active species production in plasma and control plasma chemistry for combustion and chemical manufacturing.

In the plasma-assisted combustion processes, the production of heat, chemically active species such as electrons, ions, vibrationally and electronically excited species, radicals, long-lifetime intermediate species, and fuel fragments as well as ionic wind and acoustic waves, and Coulomb and Lorentz forces, can affect the combustion process. Ju and Sun [90] summarized the three major pathways of the interactions between plasma and combustion, as shown in Figure 5.15. Plasma affects combustion via thermal, kinetic, and transport pathways. In the thermal enhancement pathway, the fast gas heating (due to rapid electronically excited state quenching) and slow gas heating (due to vibrationally excited state relaxation) from plasma increase temperature and accelerate chemical reactions and fuel oxidation according to the Arrhenius Law. In the kinetic enhancement pathway, high-energy electrons and photons are produced in plasma and lead to the production of active radicals (such as O, H, and OH) by direct electron impact and photo dissociation, ionization and recombination dissociation of ions (e.g., H_2^+ and O_2^+), and subsequent reactions involving electronically exited species (e.g. $O_2(a^1\Delta_g)$, N_2^* and $O(^1D)$) [91–94]. In addition, the long-lifetime

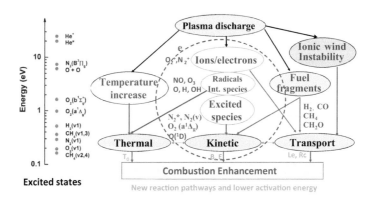

FIGURE 5.15 Schematic of major enhancement pathways of plasma-assisted combustion [90].

Plasma-Assisted Combustion: Chemistry

reactive (O_3) [95–97] and catalytic (NO) [70] intermediated species produced in the plasma also accelerate low-temperature fuel oxidation [78,79]. In the transport enhancement pathway, plasma dissociates the fuel molecules into fuel fragments, which changes the fuel diffusivity and therefore modifies the combustion process. In addition, the ionic wind as well as the acoustic waves, thermal expansion, and hydrodynamic instabilities produced by plasma can change the local flow velocity and increase the flow turbulence intensity and mixing.

In the sections below, we will discuss plasma chemistry and plasma-assisted combustion chemistry for hydrogen, methane, ammonia, and large hydrocarbons. In addition, the effects of a few key intermediate species produced in plasma such as ozone and NO_x on low-temperature fuel oxidation will be discussed.

5.2.2 Elementary Reactions of Plasma

Due to the electron-impact and photoionization processes in plasma (Figure 5.14), many chemically active species (Figure 5.15) are produced, and they will significantly affect plasma chemistry in combustion and manufacturing. These chemically active species need to be quantitatively quantified by using elementary reactions. Table 5.4 lists the major elementary reactions of electron-impact and photoionization processes in plasma. These processes include electron-impact excitation (electronically, vibrationally, and rotationally), electron-impact and photoionization to produce ions as well as new electrons, electron attachment and detachment, molecule dissociation by electrons and excited molecules, quenching of excited states, radiative and dissociative recombination, vibrational-vibrational energy transfer (V-V), and vibrational-translational (V-T) energy transfer. As shown in the representation reactions in the right column, these elementary processes govern the reactive energy transfer and chemically active species production.

TABLE 5.4
Elementary Reaction Processes via Electron Impact and Photon Ionization in Plasma

	Excitation of internal degrees of freedom	
$e + AB \rightarrow e + AB^*$	Electronic excitation	$e + N_2 \rightarrow e + N_2^*$
$e + AB \rightarrow e + AB(v)$	Vibrational excitation	$e + N_2 \rightarrow e + N_2(v)$
	Ionization	
$e + AB \rightarrow 2e + AB^+$	Ionization	$e + O_2 \rightarrow 2e + O_2^+$
$e + A^* \rightarrow 2e + A^+$	Multi-step ionization	$e + O(^1D) \rightarrow 2e + O^+$
$e + AB \rightarrow 2e + A + B^+$	Dissociative ionization	$e + N_2 \rightarrow 2e + N + N^+$
	Ionization	
$hv + AB \rightarrow AB^+ + e$	Photoionization	$hv + O_2 \rightarrow O_2^+ + e$
$nhv + AB \rightarrow AB^+ + e$	Multiphoton ionization	$nhv + O_2 \rightarrow O_2^+ + e$
	Attachment	
$e + AB \rightarrow A + B^-$	Dissociative attachment	$e + O_2 \rightarrow O + O^-$
$e + A + B \rightarrow A + B^-$	Three-body attachment	$e + O + O \rightarrow O + O^-$
	Detachment	
$A^- + B \rightarrow AB + e$	Detachment	$O + O^- \rightarrow O_2 + e$
	Dissociation	
$e + AB \rightarrow e + A + B$	Dissociation by electron	$e + H_2 \rightarrow e + H + H$

(Continued)

TABLE 5.4 (*Continued*)
Elementary Reaction Processes via Electron Impact and Photon Ionization in Plasma

$AB^* + CD \rightarrow AB + C + D$	Dissociation by excited molecules	$N_2^+ + H_2 \rightarrow N_2 + H + H$
	Quenching	
$e + A^* \rightarrow e + A$	De-excitation	$e + O(^1D) \rightarrow e + O$
	Recombination	
$e + e + A^+ \rightarrow e + A$	Three-body recombination	$e + e + O^+ \rightarrow e + O$
$e + AB^+ \rightarrow A + B$	Dissociative recombination	$e + O_2^+ \rightarrow O + O$
$e + A^+ \rightarrow A^* \rightarrow A + h\nu$	Radiative recombination	$e + O^+ \rightarrow O(^1D) \rightarrow O + h\nu$
	Vibrational energy transfer	
$AB(v=n) + CD(v=m) \rightarrow$	Vibrational-vibrational transfer (V-V)	$N_2(v=1) + N_2(v=1) \rightarrow$
$AB(v=n+1) + CD(v=m-1)$		$N_2(v=2) + N_2(v=0)$
$AB(v=n) + C \rightarrow AB(v=n-1) + C$	Vibrational-translational transfer (V-T)	$N_2(v=1) + O \rightarrow N_2(v=0) + O$

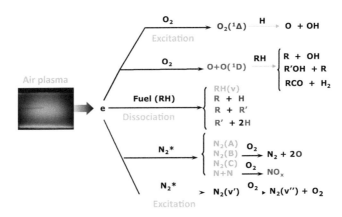

FIGURE 5.16 Schematic of chemically active species production via electron impact processes in an air/fuel mixture.

For example, in air plasma, the electron-impact excitation and dissociation of oxygen, nitrogen, and fuel (RH) molecules (Figure 5.16) will lead to many active species productions such as $O_2(a^1\Delta_g)$, $O(^1D)$, $N_2(A)$, $N_2(B)$, $N_2(C)$, $N_2(v)$, R, O, N, H, OH, and NO_x. Moreover, the reaction rates for this elementary process depend on the EEDF, thermal and non-thermal energy distributions of molecules, and the reaction cross-section areas. The production of active species will affect the chemical reactions in combustion and catalysis. Therefore, models for appropriate inclusion of these active species in plasma-assisted combustion and chemical manufacturing are needed.

In air plasma, in addition to the combustion species on the ground states, we also need to consider electronically excited states such as O_2^* (e.g., $O_2(a^1\Delta_g)$, $O_2(b^1\Sigma_g^+)$, $O_2(c^1\Sigma_u^-)$, $O_2(C^3\Delta_u)$, $O_2(A^3\Sigma_u^+)$); N_2^* (e.g., $N_2(A)$, $N_2(B)$, $N_2(a')$, $N_2(C)$), N^*(e.g., $N(^2D)$ and $N(^2P)$); O^* (e.g., $O(^1D)$ and $O(^1S)$); positive ions: O^+, O_2^+, O_4^+, N^+, N_2^+, N_3^+, N_4^+, NO^+, NO_2^+, N_2O^+; negative ions and charged particles: O^-, O_2^-, O_3^-, O_4^-, NO^-, NO_2^-, N_2O^-, e; vibrational states such as $O_2(v)$ and $N_2(v)$; and rotational states such as $O_2(rot)$ and $N_2(rot)$. All these species are chemically active species and their energy transfer and reactions with combustion species need to be appropriately considered in plasma-assisted combustion and chemical manufacturing. Note that depending on the chemical or combustion processes, not all these species and their associated reactions are equally important. Careful examination of the reaction pathways and

Plasma-Assisted Combustion: Chemistry

rate-limiting reactions for reactivity is needed. In the section below, we will provide a few examples of key elementary reactions in plasma-assisted combustion and fuel oxidation.

5.2.3 Elementary Reactions of Plasma-Assisted Hydrogen Combustion

Table 5.5 shows the key elementary reactions for radical production as well as fast and slow heating in the plasma-assisted hydrogen-air combustion system [90]. It includes the major electron-impact electronic and vibrational excitations, dissociation, and ionization, vibrational-translational (V-T) relaxation (slow heating), vibrational-vibrational (V-V) energy transfer, H abstraction of vibrationally excited species, dissociation by electronically excited species (fast heating), and electron-ion recombination.

Without plasma, as discussed in Section 5.1, the radicals (such as H and HO_2) are initially produced slowly by the chain-initiation reaction (R5.1). This process is very slow which makes hydrogen ignition delay time quite long even though it is a very reactive fuel. With plasma discharge, active radicals such as H, O, and OH will be produced in nanosecond timescale by electron-impact excitation and dissociation processes via reactions listed in Table 5.5. In addition, the V-T relaxation reactions (R5.14') generate slow heating between microsecond and millisecond. The dissociation reactions by excited species (R5.20'–R5.24') can not only produce radicals but also result in fast heating (via the extra energy of the excited states) from nanosecond to microsecond. Therefore, such rapid radical production and fast/slow heating in plasma will bypass the slow chain-initiation reaction (R5.1) and accelerate the chain-branching and propagation reactions (R5.3–R5.5) in hydrogen combustion, thus enhancing the combustion process.

Figure 5.17 [98–101] shows the comparison of the reaction rate of the key hydrogen chain-branching reaction (R5.3) at ground state with that when oxygen is electronically excited (R5.19'). It is seen that at low temperature (below 900 K), R5.19' is much faster than R5.3. Therefore, plasma-assisted chemical kinetics can dramatically accelerate ignition and combustion at low temperatures. Unfortunately, many reaction rates between the excited molecules with fuel molecules and intermediate species in combustion shown in Table 5.5 and Figure 5.16 remain unknown.

TABLE 5.5
Key Reactions in Plasma-Assisted $H_2/O_2/N_2$ Combustion System [90]

Electron Impact		Vibrational-Translational (V-T) Relaxation: Slow Heating	
$e + H_2 \rightarrow e + H_2(v=n)(n=1-3)$	(R5.1')	$N_2(v=n) + H_2 \rightarrow N_2(v=n-1) + H_2$	(R5.14')
$e + O_2 \rightarrow e + O_2\ (v=n)(n=1-4)$	(R5.2')	*Vibrational-vibrational (V-V) exchange*	
$e + N_2 \rightarrow e + N_2(v=n)(n=1-8)$	(R5.3')	$N_2(v=n) + N_2(v=m) \rightarrow N_2(v=n-1) + N_2(v=m+1)$	(R5.15')
$e + O_2 \rightarrow e + O_2(a^1\Delta_g)$	(R5.4')	*H-abstraction of vibrationally excited species*	
$e + N_2 \rightarrow e + N_2(A)/N_2(B)/N_2(a')/N_2(C)$	(R5.5')	$O + H_2(v=n) \rightarrow OH + H (n=1-3)$	(R5.16')
$e + H_2 \rightarrow e + H + H$	(R5.6')	$H_2(v=n) + OH \rightarrow H_2O + H (n=1-3)$	(R5.17')
$e + O_2 \rightarrow e + O + O$	(R5.7')	$H + O_2(v=n) \rightarrow O + OH (n=1-4)$	(R5.18')
$e + O_2 \rightarrow e + O + O(^1D)$	(R5.8')	*Dissociation by excited species: fast heating*	
$e + N_2 \rightarrow e + N + N(^2D)$	(R5.9')	$H + O_2(a^1\Delta_g) \rightarrow O + OH$ or HO_2	(R5.19')
$e + M \rightarrow e + e + M^+\ (M = H_2, O_2, N_2)$	(R5.10')	$O(^1D) + H_2 \rightarrow H + OH$	(R5.20')
Electron-ion recombination		$N_2(A)/N_2(B) + O_2 \rightarrow N_2 + O + O$	(R5.21')
$e + H_2^+ \rightarrow H + H$	(R5.11')	$N_2(a')/N_2(C) + O_2 \rightarrow N_2 + O + O(^1D)$	(R5.22')
$e + O_2^+ \rightarrow O + O/O(^1D)$	(R5.12')	$N_2(a') + H_2 \rightarrow N_2 + H + H$	(R5.23')
$e + N_2^+ \rightarrow N + N(^2D)$	(R5.13')	$N_2(C) + H_2 \rightarrow N_2 + H + H$	(R5.24')

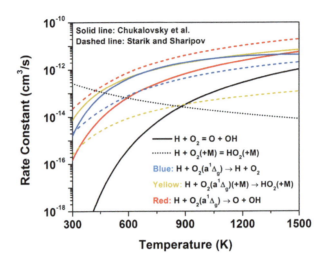

FIGURE 5.17 Reaction rate constants of different channels of $H + O_2(a^1\Delta_g) \to$ products from Chukalovsky et al. [102] and Sharipov and Starik et al. [100,101]. (Rate constant of the pressure dependence reaction $H + O_2(a^1\Delta_g)(+M) \to HO_2(+M)$ is presented at 1 atm as a second-order reaction.)

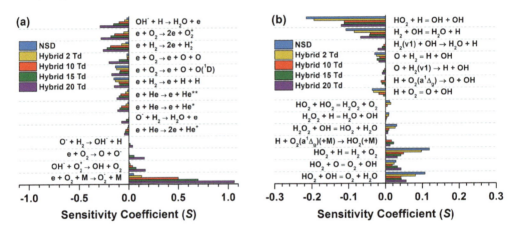

FIGURE 5.18 Sensitivity coefficients of the ignition delay time for NSD and hybrid discharge at different DC E/N values and 400 K, (a) electron impact reactions and (b) other reactions [99].

Figure 5.18 shows the sensitivity analysis of the ignition delay time of a $H_2/O_2/He$ mixture for NSD and hybrid discharge with different DC reduced electric field strengths at 400 K [99]. For electron impact reactions, Figure 5.18 shows that $e+O_2 \to e+O+O(^1D)$ (R5.8') and $e+H_2 \to e+H+H$ (R5.6') are the dominant reactions in initial radical production and ignition enhancement in the NSD-assisted ignition. This figure clearly indicates the key roles of O, $O(^1D)$ and H production by plasma in combustion. In addition, it is seen that electron concentration and excited helium atoms also affect the production of excited species and radicals via $He^* + O_2 \to e + He + O_2^+$, $He^* + H_2 \to e + He + H_2^+$ and $He^{**} + He \to He_2^+ + e$ as well as $e+O_2 \to e+O+O$. Figure 5.18 also shows that the OH production reaction $HO_2+H=OH+OH$ has the largest sensitivity and plays a key role in the low-temperature $H_2/O_2/He$ ignition. The major OH production is by $H_2(v1)$ and $O_2(a^1\Delta_g)$ via $O+H_2(v1) \to H+OH$, $H_2(v1)+OH \to H_2O+H$ and $H + O_2(a^1\Delta_g) \to O+OH$. This plasma kinetic enhancement of hydrogen combustion suggests that plasma is a promising tool to actively control combustion.

Plasma-Assisted Combustion: Chemistry

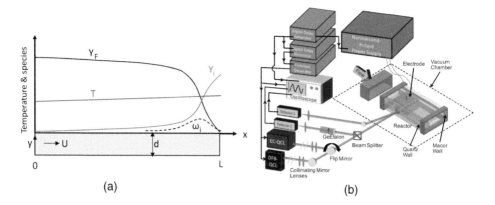

FIGURE 5.19 (a) Schematic of species and temperature distribution in a flow reactor. (b) Schematic of experimental setup for DBD flow reactor with TDLAS measurement system.

To obtain time-resolved species measurements and validated kinetic models, Lefkowitz et al. [103–105] developed a plasma reactor with *in situ* optical diagnostics (see Figures 9.11 and 9.12). This reactor (Figures 9.11 and 9.12) has been used to develop plasma-assisted combustion model for H_2 [99], CH_4 [105], C_2H_4 [104], C_5H_{12} [106], n-heptane [107], n-dodecane [108], and NH_3 [109]. Figure 5.19 provides a schematic of the experimental setup. The reactor is constructed primarily of quartz and Macor, with a stainless-steel inlet gas flow manifold and brackets. Each of the two 45 mm×45 mm stainless-steel electrodes is sandwiched between a quartz plate and a Macor plate which make up the top and bottom of the reactor, forming a plane-to-plane double DBD. There is a wedge-shaped calcium fluoride window in the sidewall to allow the mid IR laser beam to pass through the chamber wall and into the Herriott cell for laser absorption measurements. All experiments were conducted at a total pressure of 30–100 Torr and an initial temperature of 296 K.

To make kinetic measurement in a reactor, an important assumption is that the flow reaction system is either a time-evolving zero-dimensional or a steady-state one-dimensional plug flow reactor. In the formal case, the distribution of species and temperature in the reactor needs to be homogeneous, so that the species and energy governing equation can be written as

$$\frac{dY_i}{dt} = \frac{\omega_i}{\rho} \tag{5.5a}$$

$$\frac{dT}{dt} = -\sum_i \frac{\omega_i h_i}{\rho C_p}, \quad i = 1\ldots N \tag{5.5b}$$

Therefore, by measure the time history of *i*th species mole fraction, Y_i, and temperature, T, the reaction rate of ω_i can be validated.

In the steady-state one-dimensional plug flow reactor, two assumptions need to be valid: (1) the ratio of the diffusion time (d^2/D_i) to the flow residence time ($\tau_{res}=L/U$) is very small (*d* is the reactor's channel height) so that the diffusion in the direction vertical to the flow is fast so that the concentration and temperature gradient $dY_i/dy \approx 0$. (2) The diffusion flux in the flow direction is negligible, and (3) the reaction Damköhler number is close to unity

$$D_a = \frac{\tau_{res}}{\tau_r} = \frac{\tau_{res}}{\rho/\omega_i} \approx 1 \tag{5.6}$$

With the above three assumptions, the governing equations for energy and species in the coordinate of the flow direction normalized by the length of the reactor (*L*) in Figure 5.19a can be given as,

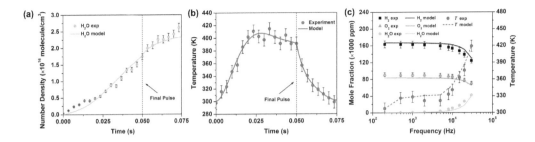

FIGURE 5.20 Time-dependent measured (a) H_2O number density and (b) temperature with fitted line with model prediction during and after the 1,500 pulse, 30 kHz nanosecond plasma in a burst mode; and (c) species concentration between steady-state measurement and model prediction as well as measured temperature in a continuous nanosecond plasma ranging from 200 Hz to 30 kHz for a 0.1667 H_2/0.0833 O_2/0.75 He mixture at 60 Torr [99].

$$\frac{dY_i}{dx} = \tau_{\text{res}} \frac{\omega_i}{\rho} \tag{5.7a}$$

$$\frac{dT}{dx} = -\tau_{\text{res}} \sum_i \frac{\omega_i h_i}{\rho C_p}, \quad i = 1\ldots N \tag{5.7b}$$

We will use the plasma flow reactor to study plasma chemistry. The purpose of the studies was to quantitatively measure time-resolved species production and temperature histories in a homogeneous plasma discharge and compare these results with kinetic modeling.

By using experiments in Figure 5.19 and the kinetic model in Table 5.5 by including helium, Mao et al. [99] studied the plasma-assisted low-temperature H_2/O_2/He combustion kinetics in a repetitively-pulsed nanosecond discharge (NSD). Figure 5.20a and b show the time-dependent measurement of H_2O number density and temperature measured *in situ* by mid-IR tunable diode laser absorption spectroscopy (TDLAS) during and after the 1,500 pulse, 30 kHz nanosecond plasma in a burst mode. Figure 5.20a shows that the H_2O number density increases with discharge pulses from room temperature due to the O and H radicals as well as electronically excited $O(^1D)$ produced by electron-impact dissociation reactions (R5.6'–R5.8'). With the progress of hydrogen oxidation, the temperature increases with plasma pulses as shown in Figure 5.20b. Note that the temperature peaks at the end of the discharge burst and decreases after that due to heat losses from the discharged mixture to the reactor walls. Figure 5.20c shows the species concentration measured *ex situ* by gas chromatograph (GC) measurements and temperature as a function of pulse repetition frequency. It is seen that the fuel consumption and temperature increase with the plasma frequency as more energy is deposited into the plasma with more pulses. In addition, the model prediction agrees well with the experimental data. Therefore, the development of experimentally validated kinetic models is critical for plasma modeling.

5.2.4 Kinetics of Plasma-Assisted CH_4 Combustion

Plasma-assisted methane oxidation plays a critical role in fuel flexible power generation, heating, and methane reforming. Previous studies of methane oxidation in plasma have been conducted using shock tubes [110–112], counterflow flames [113,114], and flow reactors [115–118]. In addition, there have been extensive studies of plasma fuel reforming using methane [119–127]. In shock tubes, Kosarev et al. [110] measured ignition delays after a fast ionization wave (FIW) in CH_4/O_2/Ar mixtures at initial temperatures from 1,230 to 1,719 K and pressures from 0.3 to 1.1 bar. It was shown that the ignition delay was shortened by a factor of 30 using a nanosecond pulsed plasma.

Lou et al. [115] studied the oxidation of methane-air mixtures in a NSD in a homogenous flow reactor at initial temperature of 290 K and pressure <100 Torr. The results showed that PAC could accelerate fuel oxidation. Further studies of plasma-assisted methane combustion [116–118] quantitatively measured the temperature, NO, O, and OH concentrations. Modeling of the PAC processes predicted the temperature and O-atom concentration, but different combustion kinetic models had mixed results in predicting the OH profile. In counterflow flames, Sun et al. [113,114] found that the extinction limits of a partially premixed $CH_4/O_2/Ar$ mixture could be extended by more than a factor of two when a nanosecond pulsed discharge was applied to the mixture at the exit of the fuel nozzle. Modeling efforts found that O and CH_2O concentrations were well predicted, but the concentration of H_2, CO, CO_2, and H_2O could not be matched by the model predictions. It was found that methane ignition was significantly accelerated by plasma at high oxygen concentrations.

However, few quantitative comparisons between measured and predicted time histories of species concentrations were made. Therefore, experimentally validated kinetic models for quantitatively predicting plasma-assisted methane combustion remain scarce. The model predictability, especially at low temperature, in terms of speciation data, is still limited, preventing a full assessment of the reaction pathways and computational design optimization of the reaction systems.

Recently, Lefkowitz et al. [105] measured plasma-assisted methane oxidation using the reactor in Figure 5.19 and conducted time-dependent and *in situ* measurements of reactants, products, and intermediate species as well as temperature. Figure 5.21 presents the mole fraction of CH_2O, a key low-temperature intermediate species of methane oxidation, as a function of time [105]. The model under-predicted the peak concentration of CH_2O by a factor of 5, indicating a major limitation in the model's predictive ability of this primary intermediate in low-temperature plasma-assisted methane oxidation. The primary formation pathways reported for CH_2O in methane PAC [110,128] are as follows:

$$CH_3 + O \rightarrow CH_2O + H \tag{R5.58}$$

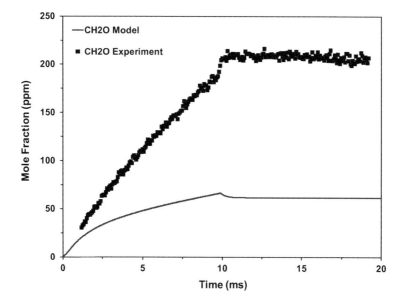

FIGURE 5.21 Formaldehyde measurement and model prediction during and after a 300 pulse burst plasma discharge at 30 kHz repetition rate and 8.76 kV peak voltage in a stoichiometric $CH_4/O_2/He$ mixture with 75% dilution [105].

$$O(^1D) + CH_4 \rightarrow CH_2O + H_2 \quad (R5.25a')$$

in which CH_3 is mainly produced via,

$$O(^1D) + CH_4 \rightarrow CH_3 + OH \quad (R5.25b')$$

$$e + CH_4 \rightarrow e + CH_3 + H \quad (R5.26a')$$

and O is formed from R5.7' and R5.8' by electron-impact oxygen dissociation. The discrepancy between experimental data and model prediction indicates that the reaction rates of Eqs. R5.25' and R5.26a' may not be properly modeled.

To collect a more complete set of species data, multiple species quantification was also made by using a gas chromatograph and temperature measurements were conducted by two-line laser absorption spectroscopy. Figure 5.22 presents measured and predicted species concentrations as a function of the NSD frequency. Figure 5.22a presents the concentrations of the reactants and H_2O. It is found that the fuel and oxygen consumption are predicted to be within 5% of the measured values. Therefore, the total electron collision rates were well modeled. The production of water is predicted within 20% of the measured value, which is in excellent agreement. Figure 5.22b presents the other major products: carbon monoxide, carbon dioxide, and hydrogen. The model captures the correct trends and relative concentrations of the three species, but under-predicts the absolute concentrations, particularly at the highest frequency conditions. The minor species are plotted in Figure 5.22c. Agreement is comparatively poor between the model and measurements of formaldehyde, methanol, ethane, ethylene, and acetylene. Like the time-dependent results, formaldehyde is under-predicted by approximately a factor of five, while methanol is over-predicted by an order of magnitude. In summary, the major trends of reactant consumption and major product species production are well captured by the kinetic model, indicating that the electron collision rates and dominant reaction pathways are well modeled, but the minor intermediate species modeling results are in significant disagreement, indicating that perhaps some secondary rates need further attention.

To understand which reaction pathways are of importance, Figure 5.23 presents the consumption pathways of methane for 30 kHz continuously pulsed plasma at steady-state temperature as predicted by the model. The major fuel consumption pathways are through electron collision reactions, reactions with $O(^1D)$, and reaction with OH.

Reactions R5.25a', R5.26a' and $CH_4 + OH = CH_3 + H_2O$ are the major reaction pathways (64%) for methane consumption. The electron-impact dissociative excitation reactions via R5.26a' and

$$e + CH_4 \rightarrow e + CH_2 + H_2 \quad (R5.26b')$$

$$e + CH_4 \rightarrow e + CH + H_2 + H \quad (R5.26c')$$

also account for 16% of CH_4 consumption, while electron-impact ionization and dissociative ionization reactions of $e + CH_4 \rightarrow 2e + CH_4^+$ and $e + CH_4 \rightarrow 2e + H + CH_3^+$ reactions account for 20%. The CH_4 dissociative excitation reactions of R5.26' lead to methyl radical (CH_3) and methylene radical (CH_2) formation, while the dissociative ionization reaction leads to methyl cation (CH_3^+). About 10% of CH_4 consumption leads to CH_2 radical production. At low temperature, the CH_2 radical is then oxidized by reactions with oxygen,

$$CH_2 + O_2 = CO + OH + H \quad (R5.59a)$$

$$= CO_2 + H + H \quad (R5.59b)$$

$$= CO_2 + H_2 \quad (R5.59c)$$

$$= CH_2O + O \quad (R5.59d)$$

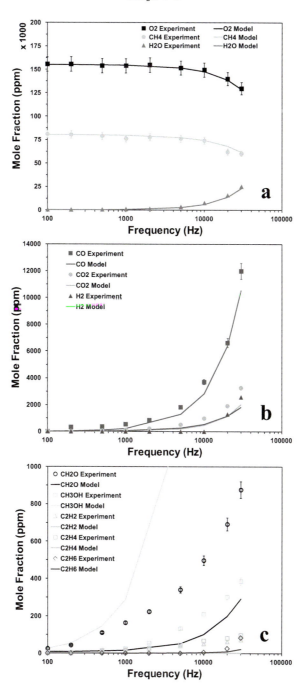

FIGURE 5.22 Species measurements and model predictions in a continuous plasma at 30 kHz repetition rate and 8.76 kV peak voltage in a stoichiometric $CH_4/O_2/He$ mixture with 75% dilution [105].

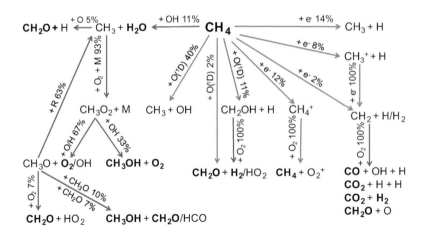

FIGURE 5.23 Path flux analysis of formaldehyde and methanol formation integrated over a single pulse period during continuous discharge at 30 kHz repetition frequency and steady-state temperature conditions. Bold species represent those which are measured in the experiment, red arrows refer to reactions from the combustion model, and blue arrows are from the plasma model. R represents any radical species.

resulting in the formation of CO, CO_2, and CH_2O. Reaction R5.59a accounts for 47% of carbon monoxide formation, and reactions R5.59b and R5.59c together account for 99% of carbon dioxide formation from CH_2. Reaction R5.59c accounts for 19% of hydrogen formation, while the dissociative excitation of methane (R5.26') accounts for another 35% of hydrogen production, thus accounting for a total of 54% of the hydrogen formation together. All three species are reasonably well predicted by the model.

In low-temperature plasma discharge, methyl radical is consumed by O and O_2 via,

$$CH_3 + O = CH_2O + H \tag{R5.60}$$

$$CH_3 + O_2 + M = CH_3O_2 + M \tag{R5.61}$$

The competition between these two reactions determines the production of methanol (Figure 5.6), as R5.61 leads to methyl peroxy radical (CH_3O_2) formation, which reacts with small radicals to form CH_3OH and CH_3O via reactions (Figure 5.6),

$$CH_3O_2 + OH \rightarrow CH_3OH + O_2 \tag{R5.62}$$

$$CH_3O_2 + H \rightarrow CH_3O + OH \tag{R5.63}$$

$$CH_3O_2 + CH_3 \rightarrow 2CH_3O \tag{R5.64}$$

$$CH_3O_2 + O \rightarrow CH_3O + O_2 \tag{R5.65}$$

Reaction R5.62 is responsible for 82% of methanol formation, while further reactions from CH_3O contribute to the remaining 18% of methanol formation, as presented in Figure 5.23.

To understand how the LTC in plasma-assisted combustion plays a critical role in model prediction, Figure 5.24 compares the experimental results and predictions by using two different models,

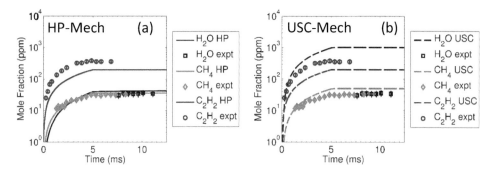

FIGURE 5.24 Measurements and predictions of C_2H_2, CH_4, and H_2O concentrations after 150 pulses at 30 kHz repetition rate for a mixture of $C_2H_4/O_2/Ar$: 6.25/18.75/75 by using (a) HP-Mech and (b) USC-Mech II [104].

HP-Mech [104] and USC-MECH [129], for plasma-assisted ethylene oxidation. The former includes the low-temperature reaction pathway discussed above and the latter only has high-temperature reaction pathways. It is clearly seen that the high-temperature model poorly predicts the formation of major products H_2O, C_2H_2, and CH_4.

Therefore, in plasma-assisted low-temperature methane oxidation, the reaction channels via CH_3O_2 and CH_3OH need to be appropriately included (Figure 5.6). In addition, some reaction channels for CH_2O may still be missing or inaccurate in the rate constants. Future studies are still needed.

5.2.5 Kinetics of Plasma-Assisted NH_3 Combustion and NO_x Emissions

Ammonia (NH_3) is considered to be an important hydrogen carrier for power generation [5,130,131] and distributed energy storage [132]. However, ammonia is difficult to ignite, has very low flame speeds, and high N_2O/NO_x emissions. The low reactivity of ammonia at low temperatures creates serious concerns of incomplete NH_3 combustion resulting from cold engine walls and poor fuel/air mixing. The unburnt ammonia emitted into the atmosphere will change the nitrogen cycle and contribute to smog formation due to the strong coupling of HO_x-NO_x-VOC–O_3 chemistry [133]. As such, enhancing low-temperature ammonia oxidation as well as controlling N_2O/NO_x emissions by using plasma is of paramount significance.

Recently, several research efforts have been devoted to plasma-assisted ammonia combustion. Studies in high-temperature ammonia flames [134–136] reported encouraging impacts of plasma on enhancing combustion and reduction in NO. However, the underlying nonequilibrium plasma kinetics for ammonia oxidation are not well understood. Several key technical questions remain to be answered: How will plasma chemistry affect ammonia oxidation and NO_x emissions? What are the major ammonia and NO_x reaction pathways in plasma? Unfortunately, few experimentally validated models exist.

More recently, Zhong et al. [137] conducted *in situ* diagnostics of nonequilibrium ammonia oxidation and developed an experimentally validated N_2O/NO_x chemistry in low-temperature plasma. The experimental setup is shown in Figure 5.25.

The plasma reactor is a rectangular flow reactor and is the same as that in Figure 5.19. The reactor is maintained at 30 Torr and the overall flow velocity was 0.3 m/s (the residence time is 0.15 s). The nanosecond (ns) voltage pulser operates repetitively in both burst and continuous modes. The burst mode was used to facilitate the time-dependent species production. The burst frequency was 0.2 Hz. Each burst had 400–600 pulses running at 30 kHz with a peak voltage of 11 kV and pulse duration of ~20 ns. TDLAS was used to scan two NH_3 lines at 1726.39 cm^{-1} and 1727.09 cm^{-1} for both temperature and concentration measurements. Two distributed feedback quantum cascade

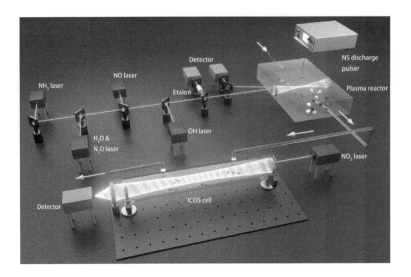

FIGURE 5.25 Experimental setup of ammonia oxidation in a plasma reactor with time-resolved laser diagnostics [137].

lasers (DFB QCLs) were used for NO and OH measurements at 1906.73 and 3568.41 cm^{-1}, respectively. An external-cavity QCL was also used for the detection of H$_2$O at 1338.55 cm^{-1} and N$_2$O at 1306.93 cm^{-1}. In addition, a sensitive off-axis integrated cavity output spectroscopy (ICOS) [93] was employed for NO$_2$ measurement at 6640.4 cm^{-1}.

A NH$_3$/O$_2$/N$_2$ plasma-combustion kinetic model was assembled and updated. The vibrationally excited species NH$_3$ (v=1–4), O$_2$ (v=1–4), N$_2$ (v=1–8); electronically excited species O$_2\left(a^1\Delta_g\right)$, O$_2\left(b^1\Sigma_g^+\right)$, O(^{1}D), O(^{1}S), N$_2$(A), N$_2$(B), N$_2$(a'), N$_2$(C), N(^{2}D); ions NH$^+$, NH$_2^+$, NH$_3^+$, NH$_4^+$, N$_2H^+$, O$_2^+$, O$_4^+$, N$_2^+$, N$_4^+$, H$_2^+$, H$_3^+$, H$_2O^+$, H$_3O^+$, O$^-$, O$_2^-$, O$_4^-$, OH$^-$; and electrons were included in the model. Both V-T and V-V energy transfers were considered. The reaction cross-section area data of electron impact reactions of NH$_3$, O$_2$, and N$_2$ were obtained from the database LXCat [138]. The NH$_3$ oxidation sub-mechanism was taken from Thorsen et al. [35] and the O$_3$ sub-mechanism of Zhao et al. [139] was added. This is the "Starting Model". Based on the experimental data, the electron impact cross sections of

$$e + NH_3 \rightarrow e + NH + H_2 \quad \text{(R5.27a')}$$

and the rate constants of N$_2$(a') and N$_2$(C) with NH$_3$ dissociation reactions,

$$N_2(a') + NH_3 \rightarrow N_2 + NH_2 + H \quad \text{(R5.28a')}$$

$$N_2(a') + NH_3 \rightarrow N_2 + NH + H_2 \quad \text{(R5.28b')}$$

$$N_2(C) + NH_3 \rightarrow N_2 + NH_2 + H \quad \text{(R5.29a')}$$

$$N_2(C) + NH_3 \rightarrow N_2 + NH + H_2 \quad \text{(R5.29b')}$$

were adjusted to fit the experimental data. In addition, several reaction rates involving NH and NH$_2$ reactions with NO and HO$_2$ were also changed within the uncertainty of these reactions to fit the

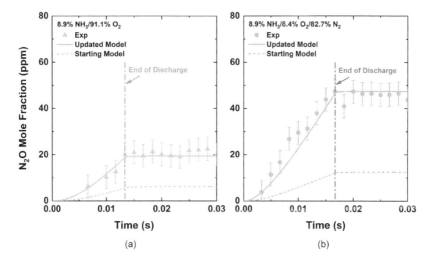

FIGURE 5.26 Nitrous oxide (N_2O) measurements (scatter) and model predictions (lines) during and after a train of plasma pulse burst at 30 kHz repetition rate: (a) without (400 pulses) and (b) with N_2 dilution (500 pulses). The vertical line indicates the end of the plasma discharge.

experimental measurements. This updated model is labeled as the "Updated Model" in the following comparison. The final kinetic model consists of 77 species and 894 reactions.

Figure 5.26 shows the comparisons of measured and computed time-dependent N_2O profiles. Significant amount of N_2O was formed at room temperature after hundreds of discharge pulses. Note that the Starting Model fails to predict N_2O formation by one order of magnitude.

Important reaction pathways for ammonia oxidation, NO, NO_2, and N_2O formation as well as NO reburning ($DeNO_x$) are shown in Figure 5.27. Like NH_3 combustion, the major oxidation sequence from NH_3 to NO in low-temperature plasma is,

$$NH_3 \rightarrow NH_2 \rightarrow HNO \rightarrow NO \tag{R5.66}$$

First, NH_3 is dissociated by electron impact via R5.27' and the collisions with electronically excited N_2(A, B, a', C) via R5.28' and R5.29'. In addition, R5.27 as well as the H-abstraction reactions from NH_3 by $O(^1D)$ and $N(^2D)$,

$$e + NH_3 \rightarrow e + NH_2 + H \tag{R5.27b'}$$

$$O(^1D) + NH_3 \rightarrow NH_2 + OH \tag{R5.30'}$$

$$N(^2D) + NH_3 \rightarrow NH + NH_2 \tag{R5.31'}$$

also contribute to NH_3 consumption. NH_2 is then mainly consumed by R5.29 to form HNO. Finally, HNO is converted to NO by reacting with H and O produced in plasma via $HNO+H=NO+H_2$ and $HNO+O=NO+OH$. Therefore, plasma-generated electrons and excited species accelerate NH_3 dissociation and NO formation via radical production of O/H/N (on the left in Figure 5.27). Note that most NO is formed from NH and NH_2, and only 2% of NO is formed directly through N atom via,

$$N(^2D) + O_2 \rightarrow NO + O(^1D) \tag{R5.32a'}$$

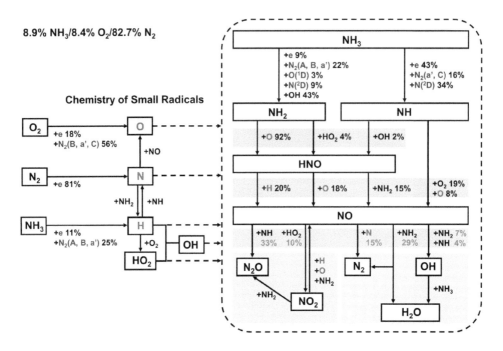

FIGURE 5.27 Path flux analysis of plasma-assisted ammonia oxidation and NO$_x$ formation. Species in purple indicate nonequilibrium species directly generated by plasma. The three blue shading in the right block indicates the major path flux for HNO formation, NO formation, and NO consumption. The orange shading in the right figure indicates NO$_x$ coupling. The green shading in the right figure indicates NH$_3$ oxidation. The percentages with black (green) color indicate the contributions accounting for the species total production (consumption).

$$N(^2D) + O_2 \rightarrow NO + O \tag{R5.32b'}$$

Figure 5.27 also shows that NO formed in plasma-assisted low-temperature ammonia combustion is either converted to N$_2$O, NO$_2$, or N$_2$ (DeNO$_x$ mechanism). One major channel of NO consumption is NO$_2$ by

$$NO + HO_2 = NO_2 + OH \tag{R5.41}$$

The other NO consumption channels are conversions to NNH, N$_2$, and N$_2$O by NH$_x$ through the thermal DeNO$_x$ mechanism,

$$NO + NH_2 = NNH + OH \tag{R5.24a}$$

$$NO + NH_2 = N_2 + H_2O \tag{R5.24b}$$

$$NO + NH = N_2O + H \tag{R5.32a}$$

$$NO + NH = N_2 + OH \tag{R5.32b}$$

This DeNO$_x$ mechanism plays a critical role in the development of NO$_x$ reduction strategies for low-temperature ammonia oxidation.

Plasma-Assisted Combustion: Chemistry

For N_2O formation in low-temperature plasma-assisted ammonia oxidation, in addition to R5.32, it is also formed from NO_2 via,

$$NH_2 + NO_2 = N_2O + H_2O \qquad \text{(R5.42b)}$$

Therefore, the NO_x formation in low-temperature plasma discharges follows a two-step mechanism (Figure 5.27). The first step is solely controlled by nonequilibrium plasma chemistry, where electron-impact ammonia dissociation and the collisional quenching of excited species provide amine radicals and $O/H/OH/HO_2$ radicals. The second step is the reactions between amine radicals and $O/H/OH/HO_2$ which lead to NO and NO_2 formation. N_2O is then formed from NO and NO_2 by the amine radicals produced in the plasma. Unfortunately, the rate constants for amine radical production via electronically excited nitrogen, nitrogen atom, and oxygen atom are not well determined. Future studies in these reactions are needed to improve the model predication.

5.2.6 Kinetics of Plasma-Assisted Large Hydrocarbon and Oxygenate Fuel Combustion

For hydrogen, methane, and ammonia plasma-assisted combustion, the major role of plasma is to generate radicals from electron-impact and excited molecules to accelerate the chain-initiation reaction R5.1 and then to promote the onset of the high-temperature chain-branching reaction R5.3. However, for large hydrocarbons or oxygenates, as discussed in Section 5.1.5, in addition to the high-temperature chain-branching reaction R5.3, the low-temperature and intermediate temperature chain-branching pathways via R5.56 and R5.57 are also important. These reactions have lower activation energy than R5.3. The question then becomes at what temperature plasma-assisted radical production can accelerate the low-temperature chain-branching reactions and at what electric field strength?

Figure 5.28a shows a comparison of the radical production reaction flux of the electron-impact atomic oxygen and $O(^1D)$ production (R5.8') as a function of the reduced electric field for reaction in comparison with that of important chain-branching and propagation reactions (R5.1–R5.5). Although the rate constant of the electron-impact reaction is much larger than that of the important radical chain-branching and initiation reactions, the reaction flux at a typical flame condition (the estimation condition of 1 atm, the electron number density of $5 \times 10^{11} cm^{-3}$, O and H at $4.9 \times 10^{15} cm^{-3}$, O_2 concentration at 19%, H_2 concentration of 0.5%, and H_2O_2 concentration of 500 ppm) [90] for the electron-impact reaction is only faster than that of the key combustion reactions at low temperatures ($T < 1,100$ K). Therefore, at low temperatures, plasma can effectively enhance combustion by accelerating the chain-initiation, branching, and propagation reactions

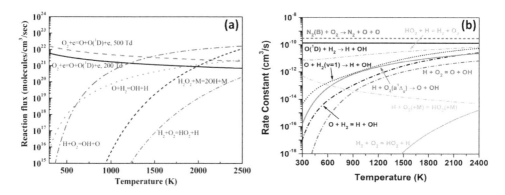

FIGURE 5.28 (a) Comparison of reaction flux between electron impact oxygen dissociation reaction and key chain-initiation and termination reactions of hydrogen. (b) Comparison of rate constants of oxygen and hydrogen dissociation reactions with key hydrogen elementary reactions [90].

R5.1–R5.5. However, when the temperature is higher than 1,100 K, the high-temperature combustion chain-branching reaction (R5.3) starts to dominate the radical production process. Note that for almost all the flames, their temperatures are higher than 1,100 K because of the rate-limiting reaction of R5.3. Therefore, the plasma effect on combustion at temperature above 1,100 K is more through the thermal effect (e.g., for flames) or bypassing the chain-initiation reaction (R5.1) (e.g., for ignition) than the kinetic enhancement.

Figure 5.28b shows the rate constants of key combustion reactions with H_2, N_2, O, and O_2 on the ground states in comparison to that of similar reactions at their vibrationally and electronically excited states as a function of temperature. It is seen that the rate constants of reactions involving vibrationally ($H_2(v=1)$) and electronically ($O_2\left(a^1\Delta_g\right)$, $O(^1D)$ and $N_2(B)$) excited species are several orders of magnitude higher than that of the combustion chain-initiation and chain-branching reactions, especially at low temperatures. Therefore, in plasma, in addition to the electron-impact reaction R5.8', the radical production reactions involving vibrationally and electronically excited molecules and ions (R5.14'–R5.21') will also help to kinetically accelerate fuel oxidation at low and intermediate temperatures.

To schematically show how plasma-assisted radical production in ignition of large hydrocarbons and oxygenated fuels, Figure 5.29 shows schematically the two-stage ignition, LTI and HTI, the plasma-assisted radical production, and the radical production via the major reaction pathways of combustion at low, intermediate, and high temperatures, respectively, via R5.56, R5.57, and R5.3. It is important to bear in mind that, at high temperature (above 1,100 K), when the temperature is above the critical temperature of the chain-branching reactions (R5.3), the major role of plasma is thermal enhancement or the chain-initiation enhancement (R5.1) because the branching rate of R5.3 is very high. As the temperature reduces, however, the kinetic effect of plasma chemistry on ignition plays an increasing role to accelerate ignition at low and intermediate temperatures to promote the chain-branching reactions R5.56 and R5.57. As discussed in Figure 5.11, radical production by plasma can accelerate LTI by one order of magnitude. Such significant enhancement is because at low temperature, the radical production by plasma is much faster than the rate-limiting chain-branching reactions R5.48–R5.51 at low temperature and R5.5 at intermediate temperature. As will be discussed in later sections, because of the significant plasma kinetic enhancement at low temperature, one can observe stable cool flames in laboratory even at atmospheric pressure.

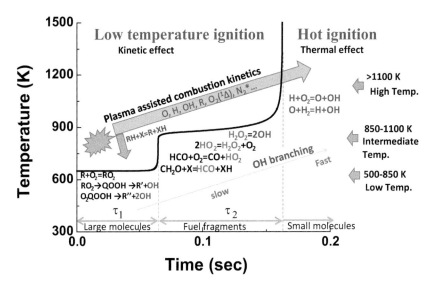

FIGURE 5.29 Schematic of kinetic and thermal enhancement pathways of plasma-assisted combustion for liquid fuels at high, intermediate, and low temperatures, respectively [90].

To develop experimentally validated kinetic models, plasma reactor experiments (Figure 5.19) were carried out for large n-alkanes (pentane, heptane, and dodecane) [106–108]. Here we will use n-dodecane as an example for the development of experimentally validated kinetic models and discuss what needs to be done in the future.

Both plasma-assisted pyrolysis and oxidation of n-dodecane were conducted using the same plasma reactor in Figure 5.19. In the pyrolysis case, 1% n-dodecane and 99% N_2 were used as the reactant mixture. In the oxidative case, 1% n-dodecane, 19% oxygen, and 80% N_2 were flown into the system. For *in situ* measurements of species-time profiles, a 24-pass Herriott multi-pass cell was utilized for mid-IR TDLAS measurements. An absorption line at $1{,}377\,cm^{-1}$ was chosen for n-dodecane with calibration. Time-resolved measurements of other species such as CH_4 at $1341.62\,cm^{-1}$, C_2H_2 at $1342.35\,cm^{-1}$, and H_2O at both $1338.55\,cm^{-1}$ and $1339.15\,cm^{-1}$ were also conducted using EC-QCL. CH_2O was quantified at $1726.79\,cm^{-1}$ using a DFB laser. For steady-state measurements, a gas chromatograph (GC) was used. A dual-modulation Faraday rotation spectroscopy (DM-FRS) system [140,141] was developed to detect steady-state nitric oxide (NO) concentration. The DM-FRS system targets $^{14}N^{16}O$ P(19/2)e doublet transition at $1842.946\,cm^{-1}$ (major isotope of NO). The pressure was at 80 Torr.

The numerical modeling is conducted by a zero-dimensional hybrid ZDPlasKin-CHEMKIN model [142]. The discharge voltage measured is used as input to calculate the E/N in the plasma. A plasma-assisted n-dodecane combustion mechanism was developed and validated. The mechanism consists of both plasma and combustion kinetic sub-mechanisms. The combustion sub-mechanism is reduced from Cai's model [143]. The C_0–C_2 sub-mechanism was updated by HP-Mech for accurate modeling of the LTC. The reactions of N and NO_x with fuel and fuel radicals, adapted from Ref. [144]. The plasma mechanism includes electronically excited species of O_2^*, O^*, N_2^*, and N^*, ions such as N_2^+ and O_2^+. The N_2 and O_2 rotationally and vibrationally excited species were also considered to provide slow gas heating. As the n-dodecane pyrolysis and oxidation experiments were conducted at the high diluted conditions and the cross sections of electron-impact n-dodecane dissociation reactions are not available, it was assumed that the fuel was mainly consumed by the electronically excited $N_2(A)$, $N_2(B)$, $N_2(a')$, $N_2(C)$) in pyrolysis and also by $O(^1D)$ and $O(^1S)$ in the oxidation case. Due to the fast collisional relaxation of $N_2(B)$ and $N_2(C)$ to $N_2(A)$ and $N_2(a')$ with N_2, the n-dodecane consumption by N_2^* in the pyrolysis can be simplified by $N_2(A)$ and $N_2(a')$. Therefore, the reactions of n-dodecane consumption via $N_2(A)$, $N_2(a')$, $O(^1D)$ and $O(^1S)$ are listed in Table 5.6. The branching ratios and reaction rates of these reactions were estimated based on fitting with experimental data and sensitivity analysis.

Figure 5.30a shows the comparison of measured and predicted time-dependent CH_4 and C_2H_2 number densities in the pyrolysis condition. Note that the measured temperature evolution was used in the modeling. These species-time histories and the data of the steady-state measurements

TABLE 5.6
Dissociation Reactions of n-Dodecane via $N_2(A)$, $N_2(a')$, $O(^1D)$ and $O(^1S)$ Collisions

Excited Species	Dissociation Reactions by Excited Molecules
$N_2(A)$ (6.17–7.8 eV)	$N_2(A)+C_{12}H_{26} \rightarrow N_2+C_{12}H_{25}+H\ N_2+C_{10}H_{21}+C_2H_5\ N_2+C_8H_{17}+C_3H_6+CH_3$ …
$N_2(a')$ (8.4–8.89 eV)	$N_2(a')+C_{12}H_{26} \rightarrow N_2+C_{10}H_{21}+C_2H_4+H\ N_2+C_{10}H_{21}+CH_3+CH_2$ …
$O(^1D)$	$O(^1D)+C_{12}H_{26} \rightarrow C_{12}H_{25}+OH\ C_{10}H_{21}+CH_3+CH_2O$ …
$O(^1S)$	$O(^1S)+C_{12}H_{26} \rightarrow C_{12}H_{25}+OH\ C_{10}H_{21}+CH_3+CH_2O$ …

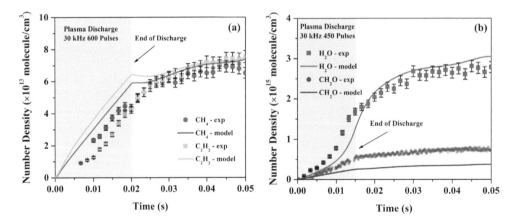

FIGURE 5.30 Time evolution of measured and predicted (a) CH_4 and C_2H_2 for the pyrolysis of 0.01 $C_{12}H_{26}/0.99N_2$ mixture and (b) H_2O and CH_2O for the oxidation of 0.01 $C_{12}H_{26}/0.19O_2/0.8N_2$ mixture.

by GC were used to determine the reaction channels and the branching ratios between $N_2(A, a')$ with fuel (Table 5.6). The developed model well-predicts the major pyrolysis species. Therefore, the time-dependent multispecies diagnostics provided a critical target to estimate the reaction branching ratios and rates in Table 5.6. The path flux analysis showed that n-dodecane was dissociated by $N_2(A)$ (61.4%) and $N_2(a')$ (24%) to produce small hydrocarbon directly or fuel radicals, as well as by H-abstraction reaction (14.6%) to produce H_2 and $C_{12}H_{25}$.

As shown in Figure 5.30b, the model also predicts the time profile of H_2O reasonably well. Up to 90% of H_2O was produced from the $C_{12}H_{26}+OH$ reaction. However, like the case of methane oxidation (Figure 5.21), there is an under-prediction for CH_2O, which may result from reactions directly converting hydroperoxy alkyl radicals such as O_2QOOH to CH_2O which were not included in this model.

Comparisons between the measured and predicted speciation in steady-state pyrolysis and oxidation cases of plasma-assisted n-dodecane are shown in Figure 5.31, respectively. In the pyrolysis case, the measured species matched relatively well with the simulation for different discharge frequencies. For oxidation, the primary oxidative products such as H_2O, CH_2O, CO, and CO_2 with high concentrations from the modeling matched reasonably well with the experimental measurements. However, small hydrocarbons were under-predicted. Note that in the oxidation case with a high oxygen concentration and atomic oxygen production, small hydrocarbons produced by the electron-impact dissociation or ionization reactions from n-dodecane may need to be considered appropriately. By estimating the rates of electron-impact reactions with n-dodecane using the similarity rule from pentane and including them in the model, it is seen that the updated model (Figure 5.31b) improved the prediction of the experimental data of small hydrocarbons production. This indicates that the accurate branching ratios and cross sections of electron impact reactions for large hydrocarbons are critical and dedicated measurements and quantum calculation for electron-impact dissociation and ionization reactions are required for future plasma-assisted kinetic studies.

The path flux analysis in Figure 5.32 shows that $C_{12}H_{26}$ consumption by $N_2(A, a')$ and $O(^1D,^1S)$ accounts for 3.5% and 6.1% of total fuel consumption, and contributes to the production of fuel radicals and CH_2O. N-dodecane consumption by OH and O was the dominant fuel consumption pathways, accounting for 65.3% and 23% of n-dodecane consumption, respectively. The O radical production in plasma promotes the H-abstraction reaction of n-dodecane to form $C_{12}H_{25}$ and OH. It is seen that the production of $C_{12}H_{25}$ leads to the formation of $C_{12}H_{25}O_2$ at low temperature via the addition of O_2 and subsequently to the formation $C_{12}H_{24}OOH$, which becomes a major channel of low-temperature chain-branching (R5.56).

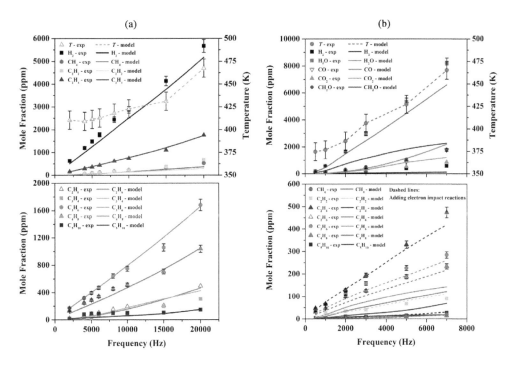

FIGURE 5.31 Comparisons between the measured and predicted species in steady state of plasma-assisted n-dodecane pyrolysis and oxidation cases. (a) Pyrolysis: 0.01 $C_{12}H_{26}$/0.99 N_2. (b) Oxidation: 0.01 $C_{12}H_{26}$/0.19 O_2/0.80 N_2.

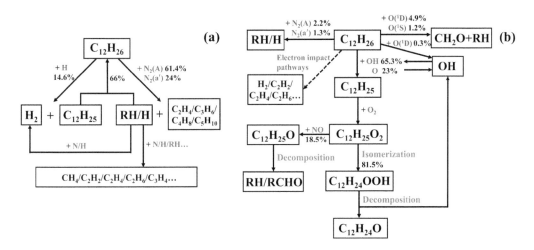

FIGURE 5.32 Path flux of fuel in the time-dependent simulations for (a) the pyrolysis case and (b) the oxidation case.

Therefore, the major plasma-combustion kinetic coupling in plasma-assisted combustion is through the production of small radicals and active intermediate species (Figure 5.33). In summary, for large hydrocarbons and oxygenated fuels, the direct impact dissociation by electron, excited molecules, and ions with fuel molecules in plasma discharge (R5.33'–R5.37') is important and needs to be appropriately considered. However, many cross-sections of electron impact dissociation

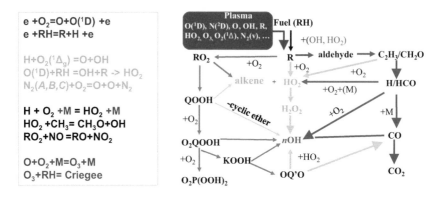

FIGURE 5.33 Schematic of plasma-assisted reaction pathways in affect chain-branching reactions at low, intermediate, and high temperatures, respectively.

TABLE 5.7
Available Electron-Impact Cross-Section Areas for Molecules in Atmosphere and Fuels

Molecules in Atmosphere	Saturated Hydrocarbon	Oxygenates	Isomers
N_2	H_2	CO	
O_2	CH_4	CH_3OH	
CO_2	C_2H_6	C_2H_5OH	
H_2O	C_3H_8	CH_3OCH_3	$i\text{-}C_3H_8$
O_3	C_4H_{10}		$i\text{-}C_4H_{10}$
Ar	C_5H_{12}		$i\text{-}C_5H_{12}$
N_2O			

reactions for large hydrocarbons and oxygenated fuels are not available. Table 5.7 shows the availability of electron-impact reaction cross-section areas for molecules in atmosphere, hydrocarbons, and oxygenates. It is seen that cross-section area data for many large hydrocarbons and oxygenates are still missing. Moreover, the reaction cross-section areas and rate constants for dissociation and oxidation of fuel molecules by excited molecules like N_2^*, O_2^*, O^*, N^* and nitrogen, oxygen, fuel ions, O_3, and NO_x are also not available.

$$e + RH \rightarrow e + R + H \quad (R5.33a')$$

$$e + RH \rightarrow e + R' + H + CH_3 \quad (R5.33b')$$

$$e + RH \rightarrow 2e + R^+ + H \quad (R5.33c')$$

$$N_2^+ + RH \rightarrow N_2 + R + H \quad (R5.34a')$$

$$N_2^+ + RH \rightarrow N_2 + R + R' \quad (R5.34b')$$

Plasma-Assisted Combustion: Chemistry

$$O_2^* + RH \rightarrow RO + R'O \quad \text{(R5.35a')}$$

$$O_2^* + RH \rightarrow R + R'O_2 \quad \text{(R5.35b')}$$

$$O^* + RH \rightarrow R + OH \quad \text{(R5.36a')}$$

$$O^* + RH \rightarrow R'O + R''H \quad \text{(R5.36b')}$$

$$N_2^+ + RH \rightarrow R^+ + R' + N_2 \quad \text{(R5.37')}$$

5.3 IMPACT OF O(^{1}D), NO$_x$, AND OZONE PRODUCTION IN PLASMA ON COMBUSTION KINETICS

5.3.1 O(^{1}D) Reactions with Saturated and Unsaturated Hydrocarbons and Oxygenates

The O(^{1}D)+fuel reactions are among the key reactions in plasma-assisted combustion, atmospheric chemistry, and chemical reforming. Because of the high internal energy of O(^{1}D), these reactions are very fast and have many possible reaction channels via H-abstraction and direct O-atom insertions. The resulting products have very different reactivities, thus affecting the reaction process significantly. Therefore, it is important to determine the branching ratios of O(^{1}D)+fuel reactions. With the recent progress in quantum chemistry calculations and *in situ* laser diagnostics, some progress has been made in quantifying O(^{1}D) reactions with hydrocarbons and oxygenated fuels [91,92,145–147]. In the sections below, we will summarize a few direct measurements and quantum chemistry calculations of O(^{1}D) reactions with CH_4, C_2H_2, alcohol, and ether, respectively.

5.3.1.1 O(^{1}D) + CH$_4$ Reactions

The reaction of O(^{1}D) with methane has several channels (Table 5.8). A recent quantum chemistry calculation at multireference configuration interaction MRCI+Q/CBS level with the zero-point energy correction obtained from CAS(10,10)/cc-pVDZ results show that the major reaction channels of O(^{1}D)+CH$_4$ are R5.25a'–R5.25f'. Equation R5.25b' is a dominant channel. The predicted branching ratios for OH and H production are, respectively, 0.73 and 0.18. Note that H radical is

TABLE 5.8
Reaction Channels of O(^{1}D)+CH$_4$ Reactions and Computed Branching Ratios [146]

O(^{1}D)+CH$_4$ → reactions	Branching Ratio	R5.25'
$H_2 + CH_2O$	0.0252	R5.25a'
$OH + CH_3$	0.7299	R5.25b'
$CH_2 + H_2O$	0.0648	R5.25c'
$H + CH_3O$	0.1178	R5.25d'
$H + CH_2OH$	0.0548	R5.25e'
$H + H + CH_2O$	0.0075	R5.25f'

more reactive than OH in many plasma-assisted combustion systems. As such, although reactions R5.25d' and R5.25f' have small product ratios but the production of H and other radicals render these reaction channels ineligible.

5.3.1.2 $O(^1D) + C_2H_2$ Reactions

For $O(^1D)$ reactions with unsaturated hydrocarbons, the rate constants and the branching ratios are not well known. Recently, by using highly selective mid-infrared Faraday rotation spectroscopy (FRS) (Figure 9.22) [140,141,148] with a digitally balanced detection scheme (Figure 9.23) to probe the time histories of $O(^1D)$, HO_2, OH, and H_2O in the reaction of $O(^1D)$ with acetylene, the branching ratios of $O(^1D) + C_2H_2$ reactions were determined in Table 5.9.

The experimental setup of the UV photolysis reactor integrated with FRS and LAS is shown in Figure 5.34 [92]. $O(^1D)$ was generated by UV photolysis of O_3 at 266 nm. To determine the important kinetic information, time-dependent measurements of OH, HO_2, C_2H_2, O_3, and H_2O were carried out by FRS and LAS in $C_2H_2/O_2/O_3/He$ mixtures. By comparing the kinetic model simulations with the time-dependent experimental measurements of O_3, OH, and HO_2, the important chain-branching ratios for the various channels of reaction between $O(^1D)$ and C_2H_2 were determined.

TABLE 5.9
Reaction Channels of $O(^1D) + C_2H_2$ Reactions and Computed Branching Ratios [92]

$O(^1D) + C_2H_2 \rightarrow$	Branching Ratio	R5.38'
C_2H_2O	0.56	R5.38a'
$HCCO + H$	0.22	R5.38b'
$CH_2 + CO$	0.22	R5.38c'
$C_2H + OH$	0	R5.38d'

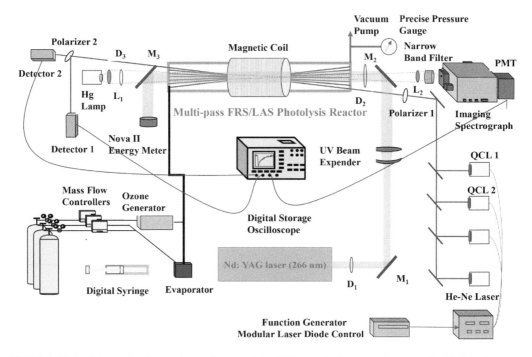

FIGURE 5.34 Schematic of experimental setup of the UV photolysis reactor integrated with Faraday rotation spectroscopy (FRS) and LAS [92].

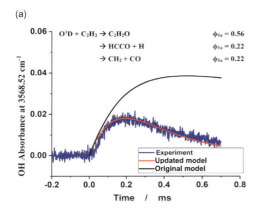 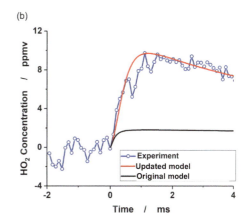

FIGURE 5.35 Comparison of the experimental data and simulation based on HP-Mech [92] with (updated model) and without (original model) the new rates of R38'. Experimental conditions: 60 Torr and 296 K in a gas mixture of $C_2H_2/O_2/O_3$/He. $[C_2H_2]=1.32\times10^{16}$, $[O_2]=4.94\times10^{15}$ and $[O_3]=2.95\times10^{14}$ molecules/cm. (a) The time-resolved measurement of OH radicals. (b) The time-resolved measurement of HO_2 radicals [92].

Figure 5.35a shows the measured and modeled OH radical production time history. There is a good agreement between simulation using an updated mechanism and the branching ratio of R5.38'. It is seen that right after the laser photolysis, OH was produced via $O(^1D)+C_2H_2$ reactions. By fitting the time history of the experimental data, the branching ratio of R5.38' for OH production was estimated. The derived branching ratios of channels R5.38a', R5.38b', and R5.38c' are 56%, 22%, and 22%, respectively. Time-resolved measurements of HO_2 formation from $O(^1D)+C_2H_2$ reactions are shown in Figure 5.35b. The reaction path flux analysis indicated that a missing reaction channel of

$$C_2H_2OH + O_2 \rightarrow HO_2 + HCCOH \quad (1\times10^{12}\ cm^3/mole\ s) \tag{R5.67}$$

needs to be added to HP-Mech. The updated model with inclusion of R5.67 significantly improved the prediction. As shown in Table 5.9, unlike saturated hydrocarbons, for $O(^1D)$ reactions with unsaturated hydrocarbons, the H-abstraction channel to form OH may not necessarily be the main channel.

5.3.1.3 $O(^1D)$ + alcohol Reactions

By using the same experimental setup in Figure 5.35, the $O(^1D)$ reactions with methanol and ethanol were also experimentally investigated. For example, Figure 5.36 shows the experimental measurement of OH production in $O(^1D)+CH_3OH$ reaction and the model prediction [91,149]. The measured branching ratios and the reaction rate of $O(^1D)+CH_3OH$ are shown in Table 5.10 and Figure 5.36. The present experimental data are consistent with the previous work of Huang et al. [149]. A similar experiment of ethanol was also conducted for $O(^1D)+C_2H_5OH$ [91]. The measured branching ratios for four major reaction pathways are listed in Table 5.10. It is seen that Reaction R5.40a' and R5.40d' are the major reaction channels. Therefore, both $O(^1D)+CH_3OH$ and $O(^1D)+C_2H_5OH$ reaction kinetics show that $O(^1D)$+fuel reactions are all multi-channeled and much more complicated than O+fuel reactions on the ground state. Care is needed in developing plasma-assisted combustion models for $O(^1D)$ reactions.

5.3.1.4 $O(^1D) + CH_3OCH_3$ Reactions

Recently Zhong et al. [150] experimentally and theoretically integrated the complex multichannel reaction dynamics and determined the total reaction rate and branching ratio of this reaction by using *in situ* laser spectroscopy and ab initio quantum chemistry theory. The computationally determined total reaction rate at 300 K and 30 Torr is $k=2.98\times10^{-10}$ cm^3/molecule s. The branching ratios

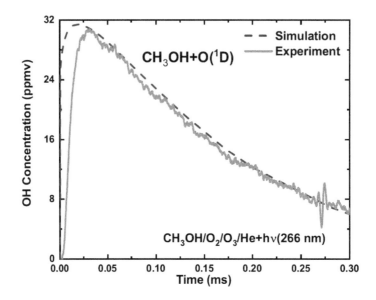

FIGURE 5.36 Comparison of measured and predicted OH production time history in $O(^1D)+CH_3OH$ reaction system at 296 K and 150 Torr. The total reaction rate is $k_{CH_3OH+O(^1D)} = (3.0 \pm 0.3) \times 10^{14}$ cm^3/mole s [91].

TABLE 5.10
Measured Branching Ratios of $O(^1D)$ Reactions with Methanol and Ethanol [91]

$O(^1D) + CH_3OH \rightarrow$	Branching Ratio	R5.39'
CH_2OH+OH	0.55	R5.39a'
$HOCHO+2H$	0.45	R5.39b'
$O(^1D) + C_2H_5OH \rightarrow$	Branching Ratio	R5.40'
$CH_3CHOH+OH$	0.46	R5.40a'
CH_2CH_2OH+OH	$0.12-x$	R5.40b'
CH_3CH_2O+OH	x	R5.40c'
CH_3O+CH_2OH	0.41	R5.40d'

TABLE 5.11
Measured Branching Ratios of $O(^1D)$ Reactions with Dimethyl Ether [150]

$O(^1D) + CH_3OCH_3 \rightarrow$	Branching Ratio 30 Torr	Branching Ratio 60 Torr	R5.41'
CH_3OCH_2+OH	0.03	0.06	R5.41a'
CH_3O+CH_3O	0.90	0.76	R5.41b'
CH_3O+CH_2OH	0.07	0.18	R5.41c'

for reactions R5.41' in Table 5.11 are, respectively, 0.03, 0.90, and 0.07. Clearly, R5.41b' to form two CH_3O molecules is the dominant one. At 60 Torr and 300 K, the predicted reaction rate increases to $k=4.41 \times 10^{-10}$ cm^3/molecule s and the experimentally fitted branching ratios become 0.06. 0.76, and 0.18, respectively. Therefore, the branching ratio is pressure-dependent.

5.3.2 Plasma-Produced NO_x and the Impact on Combustion Kinetics

Plasma discharge creates a lot of excited nitrogen and nitrogen atoms which lead to the formation of NO_x (e.g. R5.32'). The resulting NO_x has a significant impact on combustion kinetics [70,144,151–153]. In addition, the NO_x effect on combustion kinetics is temperature and fuel-dependent. For example, Figure 5.37 shows the measured NO production in plasma-assisted n-dodecane oxidation in the experiment of Figure 5.31. It is seen that for plasma discharge without fuel (n-dodecane), NO production monotonically increased with the discharge frequency (plasma power). However, with n-dodecane addition, NO production dramatically decreased. This result may give readers a misunderstanding that plasma-assisted combustion will decrease NO production. The real answer is that it may and may not.

To understand the impact of plasma-generated NO_x on combustion kinetics and the kinetic coupling between NO_x chemistry with combustion chemistry, Zhao et al. [70,144,153] conducted kinetic studies of NO_x addition on the oxidation of large hydrocarbons such as n-pentane and n-dodecane. The mutual oxidation of n-pentane/n-dodecane and NO_x (NO and NO_2) at 500–1,000 K were studied at fuel lean and rich conditions by using an atmospheric-pressure jet-stirred reactor (JSR). Multispecies measurements were conducted by using an electron-impact molecular beam mass spectrometer (EI-MBMS), a micro-gas chromatograph (μ-GC), and a mid-IR dual-modulation Faraday rotation spectrometer (DM-FRS). The results (Figure 5.38) show that at both lean and rich conditions, NO_x addition has different sensitization characteristics on fuel oxidation in three different temperature windows. For n-pentane, between 550 and 650 K (region 1 in Figure 5.38), NO addition inhibits low-temperature oxidation. With an increase of temperature to the NTC region (650–750 K) (region 2 in Figure 5.38), NO addition suppresses the NTC behavior at the fuel lean condition. In the intermediate and high-temperature region (750–1,000 K) (region 3 in Figure 5.38), fuel oxidation is accelerated with NO addition. Kinetic models predicted reasonably well the temperature-dependent NO/NO_2 sensitization effect on fuel oxidation [144,153]. The results also show that although NO_2 addition in n-pentane has similar effects to NO at many conditions due to fast NO and NO_2 interconversion at higher temperature, it affects low-temperature oxidation somewhat differently. When NO_2/NO interconversion is slow at low temperature, NO_2 is relatively inert while

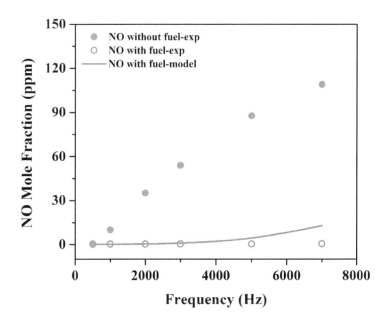

FIGURE 5.37 NO measurements with the increase of plasma pulse frequencies for normal air ($0.20O_2/0.80N_2$) and $0.01C_{12}H_{26}/0.19O_2/0.80N_2$ mixture [108].

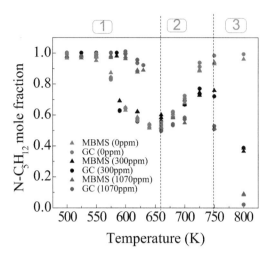

FIGURE 5.38 Temperature evolution of the mole fraction of n-pentane at the fuel lean conditions ($\varphi=0.5$) with different amounts of NO additions (0, 300, and 1,070 ppm) [144].

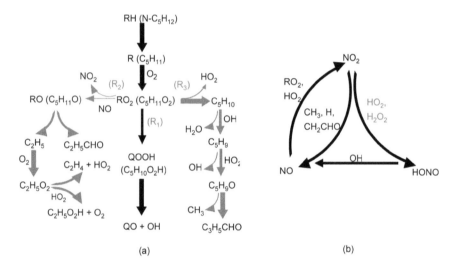

FIGURE 5.39 Kinetic coupling between NO_x chemistry at low (a) and intermediate temperature (b) fuel oxidation chemistry.

NO can strongly promote or inhibit fuel oxidation. In addition, NO addition delays the onset temperature of n-pentane low-temperature oxidation and strongly inhibits the NTC effect, while NO_2 has little effect on the onset temperature and weaker impact on NTC.

The kinetic coupling between NO_x chemistry, low-temperature combustion chemistry, and intermediate temperature HO_2 chemistry is schematically shown in Figure 5.39. In the low-temperature oxidation region, there are two reaction channels for NO to affect OH radical production, respectively, via RO_2 consumption (R5.68) and OH radical quenching (R5.69) to slow down the low-temperature reactivity,

$$RO_2 = QOOH \tag{R5.49a}$$

$$RO_2 = R' + HO_2 \tag{R5.49b}$$

$$RO_2 + NO = RO + NO_2 \tag{R5.68}$$

$$NO + OH + M = HONO + M \tag{R5.69}$$

Since R5.49a and R5.56 are the major low-temperature chain-branching reactions via RO_2, the RO_2 radical quenching via R5.68 will slow down OH radical production. Moreover, reaction R5.69 will further terminate OH radicals at low temperature. Therefore, the low-temperature oxidation (Figure 5.38) is inhibited and shifted to a higher temperature.

In the NTC region, the kinetic coupling between NO_x chemistry and HO_2 chemistry will accelerate radical production and suppress the NTC effects due to RO_2 and QOOH decomposition via R5.48 and R5.50, thus enhancing fuel oxidation (Figure 5.38). As shown in the reactions listed below, HO_2 starts to react with NO (R5.41) and CH_3 (R5.15) at elevated temperature to produce OH and NO_2. Then, HO_2 will react with NO_2 to form HONO and OH via reaction R5.70 and R5.69b. In addition, the resulting NO_2 from R5.41 will further react with CH_3 and CH_2O to form a catalytic reaction pathway for OH production (R5.41, R5.71, and R5.72). As such, one can see from Figure 5.38 that NO_x addition into n-heptane will significantly enhance the intermediate temperature fuel oxidation and suppress the NTC effect. Therefore, plasma-produced NO_x will dramatically impact fuel oxidation chemistry and such effects are temperature- and fuel-dependent. The plasma-produced NO_x coupling effect needs to be appropriately incorporated into plasma-assisted combustion chemistry when involving nitrogen containing species.

$$HO_2 + NO = OH + NO_2 \tag{R5.41}$$

$$NO_2 + HO_2 = HONO + O_2 \tag{R5.70}$$

$$HONO + M = OH + NO + M \tag{R5.69b}$$

$$CH_3 + HO_2 = CH_3O + OH \tag{R5.15}$$

$$NO_2 + CH_3 = CH_3O + NO \tag{R5.71}$$

$$CH_2O + NO_2 = HCO + HONO \tag{R5.72}$$

5.3.3 PLASMA-PRODUCED, OZONE-ASSISTED COMBUSTION KINETICS

Ozone is one of the key species produced in nonequilibrium plasma involving oxygen. In addition, it is a long-lived species in dry air at temperatures below 400 K (approximately 1,500 min at room temperature). It can be efficiently and economically produced at high pressure using DBD discharge in industry [154]. It is also an important species in atmospheric chemistry [155]. The interest in ozone effects on combustion can be traced back to the 1950s [156]. Since then, ozone effects on chemistry [96,97,139,157–161], ignition [162,163], hot flames [139,156–158,163–168], cool flames and warm flames [78,79,158,166,169–172], detonation [95,165], and engines [162,163,173] have been extensively explored. The studies showed that ozone can promote extreme low-temperature oxidation [96,174–176], ignition [162,163], cool flames [157,158,167], and detonation [95,165].

There are several ways (Table 5.12) that ozone will affect combustion kinetics at different temperatures and fuel molecule structures.

The first pathway of ozone impact on combustion is producing radicals via thermal decomposition or H-abstraction (R5.73–R5.74). Zhao et al. [139] measured ozone decomposition in a flow reactor with flow residence time approximately at 0.5 s under different temperatures. Figure 5.40 shows that ozone started to decompose around 410 K and the decomposition ratio reached 50% at 500 K. Therefore, ozone can produce O radicals via thermal decomposition at a temperature where most fuels do not have any reactivity. As such, the radical production via ozone decomposition at

TABLE 5.12

Low-Temperature Ozone Termination and Radical Production and Propagation Reactions

Ozone Reactions	Impact on Combustion	
Radical Production Reactions		
$O_3 + M = O_2 + O + M$	Producing O radicals below 500 K	R5.73
$O + RH = R + OH$	OH production from saturated hydrocarbons/oxygenates	R5.74
Radical Propagation Reactions		
$O_3 + H = O_2 + OH$	Changing chain-branching reaction rate	R5.75
$O_3 + OH = O_2 + HO_2$	Reducing low-temperature reactivity	R5.76
$O_3 + HO_2 = 2O_2 + OH$	Changing low-temperature reactivity	R5.77
Ozonolysis Reactions with Unsaturated Hydrocarbons and Oxygenates		
O_3 + alkene → POZ → Criegee		
$O_3 + C_2H_4 = CH_2O + CH_2OO^*$	Rapid radical production at low temperature	R5.78
Quenching Reactions		
$O_3 + O = O_2 + O_2$	Losing chemical reactivity for heating	R5.79
$NO + O_3 = NO_2 + O_2$	Losing chemical reactivity but modifying NO_x chemistry	R5.80
$SO_2 + O_3 = SO_3 + O_2$	Loss of chemical reactivity but enabling SO_2 capture	R5.81

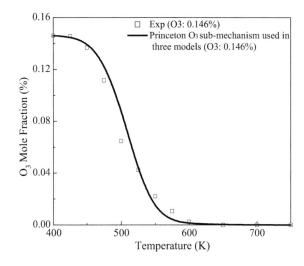

FIGURE 5.40 Measured and predicted O_3 mole fraction profiles with varied temperatures in a flow reactor O_3 decomposition experiment with flow residence time around 0.5 s [139].

extremely low temperature can initiate low-temperature fuel oxidation at a temperature outside the conventional fuel LTC window.

Recent experiments by Zhao et al. [139] and Rousso et al. [175] studied the ozone-initiated low-temperature oxidation of dimethyl ether and methyl hexanoate, respectively. Both studies reported that ozone addition into these oxygenated fuels induced low-temperature fuel oxidation at temperatures below 500 K. Figure 5.41 shows the ozone depletion in the absence of other reactants and the modeled atomic O concentration as well as alkyl hydroperoxides distribution of ozone-assisted methyl hexanoate (MHX) oxidation in a JSR [175]. It is clear from this figure that about 80% of O_3 were decomposed and consumed at the extreme low-temperature combustion (ELTC) region centered at 500 K. Simultaneously, the model predicted that the O concentration rose due to the thermal decomposition (R5.73). At the same time, as indicated by methyl hydroperoxide CH_4O_2 and methyl-formate $C_2H_4O_2$ concentrations, LTC occurred in the ELTC region. In the non-ozone case, the dashed red line shows zero CH_4O_2 signal. Therefore, the fuel oxidation at the ELTC region was induced by ozone-sensitized LTC, which promotes radical production via R5.74. Quantum chemistry simulation showed that the reaction rate of methyl hexanoate with O (R5.74) is considerably higher than that of dimethyl ether. Once the fuel radical was formed by R5.73 and R5.74, the LTC chemistry pathway (R5.56) would be enhanced and resulting in the new reaction zone at ELTC. The resulting temperature-dependent profiles of the hydroperoxide species (including the keto-hydroperoxide) were measured and shown in Figure 5.42. It is seen that the hydroperoxide concentrations in the ELTC region are higher than that in the LTC region. This may be because temperatures were not sufficiently high enough for the keto-hydroperoxide to dissociate as in the normal LTC region.

The second reaction pathway of ozone radical propagation reactions in Table 5.12 may only occur during *in situ* plasma discharge because ozone does not exist at high temperature at which H and OH are created. With *in situ* plasma discharge which creates high concentrations (10–100 ppm level) of $H/OH/HO_2$ and 0.1%–10% of ozone at room temperature, these radical propagation reactions will occur. This reaction pathway needs to be included in plasma-assisted combustion and fuel reforming.

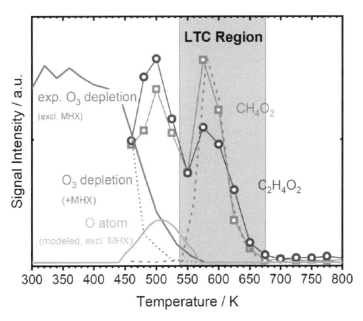

FIGURE 5.41 Temperature profiles of ozone depletion (solid green line), ozone depletion in the presence of MHX (dotted green line), modeled O-atom concentration (O_3 only in the absence of the reactant MHX) and observed double-peak structures of methyl hydroperoxide CH_4O_2 and $C_2H_4O_2$ profiles. The dashed line for CH_4O_2 represents data from the non-ozone case.

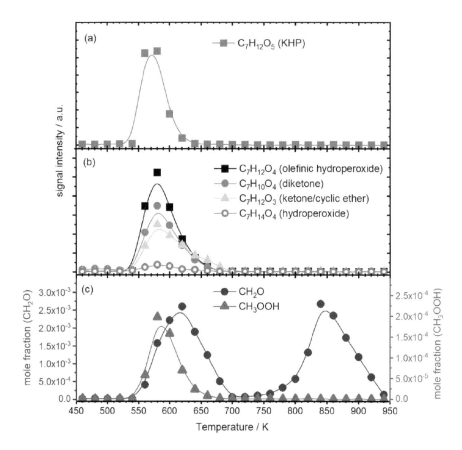

FIGURE 5.42 Temperature dependence of panel (a) keto-hydroperoxides (KHPs), (b) olefinic hydroperoxides, diketones, ketones/cyclic ethers, and hydroperoxides, and (c) methyl hydroperoxide and formaldehyde follow the trends typical for LTC behavior. This non-ozone data was taken using electron ionization (EI).

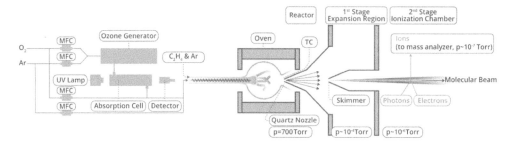

FIGURE 5.43 Experimental Setup for the JSR. The ozone detection cell is just upstream of the JSR. The stages of the MBMS sampling from the JSR are shown on the right.

The third reaction pathway is the ozonolysis reaction [174,177] in which the unsaturated hydrocarbon bonds are cleaved with ozone to form reactive products such as the criegee intermediate. Recent reviews of the chemistry of criegee intermediates can be found in Refs. [178,179], Rousso et al. [174], and Sun et al. [176] studied the low-temperature reaction kinetics of ozone reaction with ethylene. Figure 5.43 shows the experimental apparatus for the study of ozone-assisted ethylene oxidation in a heated JSR [174].

For ethylene, the ozonolysis reaction pathway is schematically shown in Figure 5.44 following the studies of Refs. [180,181]. The ozonolysis initially forms a chemically activated primary ozonide

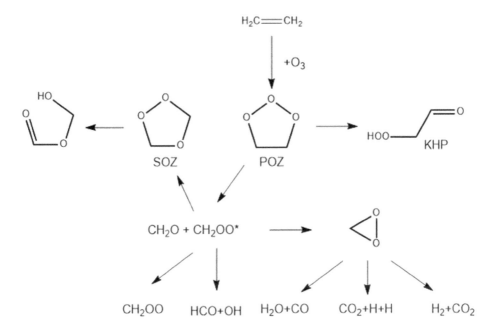

FIGURE 5.44 Ethylene ozonolysis reaction pathway with major intermediate species.

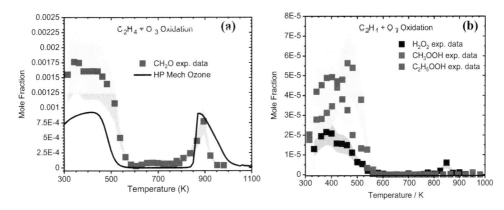

FIGURE 5.45 Species dependence on temperature measured and predicted in ozone-assisted reaction with ethylene in a JSR. (a) CH_2O mole fraction, (b) measurements of the H_2O_2, CH_3OOH, and C_2H_5OOH concentrations.

(POZ) and then quickly dissociates in the formation of formaldehyde and CH_2OO^*, the simplest criegee intermediate, which subsequently stabilizes or further decomposes to radicals.

The ozonolysis reaction between ozone and ethylene occurs at room temperature. Rate coefficients for many of these reactions have been experimentally determined or calculated between 293 and 300 K and 1 atm, with branching fractions of ~37% (CI stabilization), 15% (HCO+OH), 30% (H_2O+CO), <1% (CO_2+H+H), and 18% (H_2+CO_2) [180]. However, experimental studies of ozonolysis reactions at elevated temperatures have not been well investigated.

Figure 5.45 shows the measured and simulated temperature dependence of intermediate species formation in ozone-ethylene reactions in a JSR (Figure 5.43). It is clearly seen that CH_2O and peroxide species were formed between room temperature and 575 K due to the ozonolysis reactions. Above 575 K, ozone will decompose so quickly that the ozonolysis reaction is suppressed and only the ozone decomposition reaction proceeds which led to very low reactivity, because ethylene does not have LTC under atmospheric conditions. It is seen that only very low reactivity was observed until the temperature reaches close to 900 K at which the HO_2 chemistry starts to oxidize ethylene and form CH_2O and other peroxide species.

Therefore, in this chapter, we summarized the major reaction pathways of several important fuels including H_2, CH_4, ammonia, and large hydrocarbons and oxygenated fuels. We discussed the rate-limiting chain-initiation, branching, propagation, and termination reactions of these fuel oxidations. Then, we introduced the key elementary reactions in plasma and production of active radicals. The key chain-branching reactions, respectively, at low, intermediate and high temperatures by R5.56, R5.5, and R5.3 are discussed. We demonstrated that plasma-generated key active species and intermediate species as well as products such as NO_x and ozone can promote or inhibit low-temperature combustion. The kinetic impacts of plasma-generated species on combustion strongly depend on temperature and fuel molecule structures. More future research is needed to develop predictive plasma chemistry for combustion and chemical reforming.

In Chapter 6, we will discuss how the plasma-enhanced new reaction pathways, LTC, and ELTC will impact ignition, flame speeds, burning limit, flame regimes, the minimum ignition energy, and detonation.

REFERENCES

1. Wang, Z., et al., Methanol oxidation up to 100 atm in a supercritical pressure jet-stirred reactor. *Proceedings of the Combustion Institute*, 2022. **39**: pp. 1–9.
2. Law, C.K., *Combustion Physics*. 2010: Cambridge University Press.
3. Ruscic, B., Active thermochemical tables: Sequential bond dissociation enthalpies of methane, ethane, and methanol and the related thermochemistry. *The Journal of Physical Chemistry A*, 2015. **119**(28): pp. 7810–7837.
4. Zhao, H., et al., Studies of low and intermediate temperature oxidation of propane up to 100 atm in a supercritical-pressure Jet-Stirred reactor. *Proceedings of the Combustion Institute*, 2023. **39**: pp. 2715–2723.
5. Kobayashi, H., et al., Science and technology of ammonia combustion. *Proceedings of the Combustion Institute*, 2019. **37**(1): pp. 109–133.
6. Song, Y., et al., Ammonia oxidation at high pressure and intermediate temperatures. *Fuel*, 2016. **181**: pp. 358–365.
7. Sullivan, N., et al., Ammonia conversion and NO_x formation in laminar coflowing nonpremixed methane-air flames. *Combustion and Flame*, 2002. **131**(3): pp. 285–298.
8. Glarborg, P., et al., Modeling nitrogen chemistry in combustion. *Progress in Energy and Combustion Science*, 2018. **67**: pp. 31–68.
9. Miller, J.A., et al., Kinetic modeling of the oxidation of ammonia in flames. *Combustion Science and Technology*, 1983. **34**(1–6): pp. 149–176.
10. Li, J., et al., Study on using hydrogen and ammonia as fuels: Combustion characteristics and NO_x formation. *International Journal of Energy Research*, 2014. **38**(9): pp. 1214–1223.
11. Ichikawa, A., et al., Laminar burning velocity and Markstein length of ammonia/hydrogen/air premixed flames at elevated pressures. *International Journal of Hydrogen Energy*, 2015. **40**(30): pp. 9570–9578.
12. Han, X., et al., Experimental and kinetic modeling study of laminar burning velocities of NH_3/air, NH_3/H_2/air, NH_3/CO/air and NH_3/CH_4/air premixed flames. *Combustion and Flame*, 2019. **206**: pp. 214–226.
13. Wang, S., et al., Experimental study and kinetic analysis of the laminar burning velocity of NH_3/syngas/air, NH_3/CO/air and NH_3/H_2/air premixed flames at elevated pressures. *Combustion and Flame*, 2020. **221**: pp. 270–287.
14. Mei, B., et al., Enhancement of ammonia combustion with partial fuel cracking strategy: Laminar flame propagation and kinetic modeling investigation of NH_3/H_2/N_2/air mixtures up to 10 atm. *Combustion and Flame*, 2021. **231**: p. 111472.
15. Lee, J.H., et al., Studies on properties of laminar premixed hydrogen-added ammonia/air flames for hydrogen production. *International Journal of Hydrogen Energy*, 2010. **35**(3): pp. 1054–1064.
16. Lee, J.H., S.I. Lee, and O.C. Kwon, Effects of ammonia substitution on hydrogen/air flame propagation and emissions. *International Journal of Hydrogen Energy*, 2010. **35**(20): pp. 11332–11341.

17. Kumar, P. and T.R. Meyer, Experimental and modeling study of chemical-kinetics mechanisms for H_2–NH_3–air mixtures in laminar premixed jet flames. *Fuel*, 2013. **108**: pp. 166–176.
18. Lesmana, H., et al., Experimental and kinetic modelling studies of laminar flame speed in mixtures of partially dissociated NH_3 in air. *Fuel*, 2020. **278**: p. 118428.
19. Lhuillier, C., et al., Experimental investigation on laminar burning velocities of ammonia/hydrogen/air mixtures at elevated temperatures. *Fuel*, 2020. **263**: p. 116653.
20. Lesmana, H., et al., Experimental and kinetic modelling studies of flammability limits of partially dissociated NH_3 and air mixtures. *Proceedings of the Combustion Institute,* 2021. **38**(2): pp. 2023–2030.
21. Osipova, K.N., O.P. Korobeinichev, and A.G. Shmakov, Chemical structure and laminar burning velocity of atmospheric pressure premixed ammonia/hydrogen flames. *International Journal of Hydrogen Energy*, 2021. **46**(80): pp. 39942–39954.
22. Shrestha, K.P., et al., An experimental and modeling study of ammonia with enriched oxygen content and ammonia/hydrogen laminar flame speed at elevated pressure and temperature. *Proceedings of the Combustion Institute,* 2021. **38**(2): pp. 2163–2174.
23. Wang, N., et al., Laminar burning characteristics of ammonia/hydrogen/air mixtures with laser ignition. *International Journal of Hydrogen Energy*, 2021. **46**(62): pp. 31879–31893.
24. Alfazazi, A., et al., Counterflow flame extinction of ammonia and its blends with hydrogen and C1-C3 hydrocarbons. *Applications in Energy and Combustion Science*, 2022. **12**: p. 100099.
25. Gotama, G.J., et al., Measurement of the laminar burning velocity and kinetics study of the importance of the hydrogen recovery mechanism of ammonia/hydrogen/air premixed flames. *Combustion and Flame*, 2022. **236**: p. 111753.
26. Han, X., et al., Uniqueness and similarity in flame propagation of pre-dissociated NH_3+air and NH_3+H_2+air mixtures: An experimental and modelling study. *Fuel*, 2022. **327**: p. 125159.
27. Hayakawa, A., et al., Experimental and numerical study of product gas and N_2O emission characteristics of ammonia/hydrogen/air premixed laminar flames stabilized in a stagnation flow. *Proceedings of the Combustion Institute,* 2022. **39**: pp. 1–9.
28. Jin, B.-Z., et al., Experimental and numerical study of the laminar burning velocity of NH_3/H_2/air premixed flames at elevated pressure and temperature. *International Journal of Hydrogen Energy*, 2022. **47**(85): pp. 36046–36057.
29. Berwal, P., and Kumar, S., Laminar burning velocity measurement of CH_4/H_2/NH_3-air premixed flames at high mixture temperatures. *Fuel*, 2023. **331**: p. 125809.
30. Chen, X., et al., Experimental and chemical kinetic study on the flame propagation characteristics of ammonia/hydrogen/air mixtures. *Fuel*, 2023. **334**: p. 126509.
31. Wang, Z., et al., Experimental and numerical study on laminar premixed NH_3/H_2/O_2/air flames. *International Journal of Hydrogen Energy*, 2023. **48**: pp. 14885–14895.
32. Zhou, S., et al., An experimental and kinetic modeling study on NH_3/air, NH_3/H_2/air, NH_3/CO/air, and NH_3/CH_4/air premixed laminar flames at elevated temperature. *Combustion and Flame*, 2023. **248**: p. 112536.
33. Zhang, X., et al., Combustion chemistry of ammonia/hydrogen mixtures: Jet-stirred reactor measurements and comprehensive kinetic modeling. *Combustion and Flame*, 2021. **234**: p. 111653.
34. Osipova, K.N., et al., Ammonia and ammonia/hydrogen blends oxidation in a jet-stirred reactor: Experimental and numerical study. *Fuel*, 2022. **310**: p. 122202.
35. Thorsen, L.S., et al., High pressure oxidation of NH_3/n-heptane mixtures. *Combustion and Flame*, 2023. **254**, p. 112785.
36. Dai, L., et al., Experimental and numerical analysis of the autoignition behavior of NH_3 and NH_3/H_2 mixtures at high pressure. *Combustion and Flame*, 2020. **215**: pp. 134–144.
37. Rocha, R.C., et al., Combustion of NH_3/CH_4/air and NH_3/H_2/air mixtures in a porous burner: Experiments and kinetic modeling. *Energy & Fuels*, 2019. **33**(12): pp. 12767–12780.
38. Puechberty, D. and M.J. Cottereau, Nitric oxide formation in an ammonia-doped methane-oxygen low pressure flame. *Combustion and Flame*, 1983. **51**: pp. 299–311.
39. Rosier, B., et al., Carbon monoxide concentrations and temperature measurements in a low pressure CH_4–O_2–NH_3 flame. *Applied Optics*, 1988. **27**(2): pp. 360–364.
40. Garo, A., C. Hilaire, and D. Puechberty, Experimental study of methane-oxygen flames doped with nitrogen oxide or ammonia. Comparison with modeling. *Combustion Science and Technology*, 1992. **86**(1–6): pp. 87–103.
41. Rahinov, I., A. Goldman, and S. Cheskis, Absorption spectroscopy diagnostics of amidogen in ammonia-doped methane/air flames. *Combustion and Flame*, 2006. **145**(1): pp. 105–116.

42. Tian, Z., et al., An experimental and kinetic modeling study of premixed $NH_3/CH_4/O_2$/Ar flames at low pressure. *Combustion and Flame*, 2009. **156**(7): pp. 1413–1426.
43. Li, B., et al., Measurements of NO concentration in NH_3-doped CH_4+air flames using saturated laser-induced fluorescence and probe sampling. *Combustion and Flame*, 2013. **160**(1): pp. 40–46.
44. Lamoureux, N., et al., Measurements and modelling of nitrogen species in $CH_4/O_2/N_2$ flames doped with NO, NH_3, or NH_{3+NO}. *Combustion and Flame*, 2017. **176**: pp. 48–59.
45. Brackmann, C., et al., Formation of NO and NH in NH_3-doped $CH_4 + N_2 + O_2$ flame: Experiments and modelling. *Combustion and Flame*, 2018. **194**: pp. 278–284.
46. Ramos, C.F., et al., Experimental and kinetic modelling investigation on NO, CO and NH_3 emissions from NH_3/CH_4/air premixed flames. *Fuel*, 2019. **254**: pp. 115693.
47. Colson, S., et al., Experimental and numerical study of NH_3/CH_4 counterflow premixed and non-premixed flames for various NH_3 mixing ratios. *Combustion Science and Technology*, 2021. **193**(16): pp. 2872–2889.
48. Shi, H., et al., Experimental study and mechanism analysis of the NO_x emissions in the NH_3 MILD combustion by a novel burner. *Fuel*, 2022. **310**: p. 122417.
49. Montgomery, M.J., et al., Effect of ammonia addition on suppressing soot formation in methane co-flow diffusion flames. *Proceedings of the Combustion Institute*, 2021. **38**(2): pp. 2497–2505.
50. Bell, J.B., et al., Detailed modeling and laser-induced fluorescence imaging of nitric oxide in a NH_3-seeded non-premixed methane/air flame. *Proceedings of the Combustion Institute*, 2002. **29**(2): pp. 2195–2202.
51. Wang, Z., et al., Experimental and kinetic study on the laminar burning velocities of NH_3 mixing with CH_3OH and C_2H_5OH in premixed flames. *Combustion and Flame*, 2021. **229**: p. 111392.
52. Ronan, P., et al., Laminar flame speed of ethanol/ammonia blends: An experimental and kinetic study. *Fuel Communications*, 2022. **10**: p. 100052.
53. Li, M., et al., An experimental and modeling study on auto-ignition kinetics of ammonia/methanol mixtures at intermediate temperature and high pressure. *Combustion and Flame*, 2022. **242**: p. 112160.
54. Li, X., et al., Effect of methanol blending on the high-temperature auto-ignition of ammonia: An experimental and modeling study. *Fuel*, 2023. **339**, 126911: pp. 1–11.
55. Li, M., et al., Experimental and kinetic modeling study on auto-ignition properties of ammonia/ethanol blends at intermediate temperatures and high pressures. *Proceedings of the Combustion Institute*, 2022. **39**: pp. 1–9.
56. Xiao, H. and H. Li, Experimental and kinetic modeling study of the laminar burning velocity of NH_3/DME/air premixed flames. *Combustion and Flame*, 2022. **245**: pp. 112372.
57. Chen, J. and X. Gou, Experimental and kinetic study on the extinction characteristics of ammonia-dimethyl ether diffusion flame. *Fuel*, 2023. **334**: p. 126743.
58. Issayev, G., et al., Ignition delay time and laminar flame speed measurements of ammonia blended with dimethyl ether: A promising low carbon fuel blend. *Renewable Energy*, 2022. **181**: pp. 1353–1370.
59. Dai, L., et al., Ignition delay times of NH_3/DME blends at high pressure and low DME fraction: RCM experiments and simulations. *Combustion and Flame*, 2021. **227**: pp. 120–134.
60. Murakami, Y., et al., Effects of mixture composition on oxidation and reactivity of DME/NH_3/air mixtures examined by a micro flow reactor with a controlled temperature profile. *Combustion and Flame*, 2022. **238**: p. 111911.
61. Issayev, G., et al., Combustion behavior of ammonia blended with diethyl ether. *Proceedings of the Combustion Institute,* 2021. **38**(1): pp. 499–506.
62. Hashemi, H., et al., High-pressure oxidation of ethane. *Combustion and Flame*, 2017. **182**: pp. 150–166.
63. Hashemi, H., et al., High-pressure oxidation of propane. *Proceedings of the Combustion Institute*, 2019. **37**(1): pp. 461–468.
64. Payne, M.C., et al., Iterative minimization techniques for ab initio total-energy calculations: Molecular dynamics and conjugate gradients. *Reviews of Modern Physics*, 1992. **64**(4): p. 1045.
65. Zhao, H., et al., Studies of high-pressure n-butane oxidation with CO_2 dilution up to 100 atm using a supercritical-pressure jet-stirred reactor. *Proceedings of Combustion Insittute*, 2021. **38**: pp. 279–287.
66. Yan, C., et al., Low-and intermediate-temperature oxidation of dimethyl ether up to 100 atm in a supercritical pressure jet-stirred reactor. *Combustion and Flame*, 2022. **243**: p. 112059.
67. Wang, Z., et al., Study of low-and intermediate-temperature oxidation kinetics of diethyl ether in a supercritical pressure jet-stirred reactor. *The Journal of Physical Chemistry A*, 2023, **127**(2): pp. 506–516.
68. Klippenstein, S.J., et al., The role of NNH in NO formation and control. *Combustion and Flame*, 2011. **158**(4): pp. 774–789.

69. Klippenstein, S.J. and P. Glarborg, Theoretical kinetics predictions for NH_2+HO_2. *Combustion and Flame*, 2022. **236**: p. 111787.
70. Zhou, M., et al., Kinetic effects of NO addition on n-dodecane cool and warm diffusion flames. *Proceedings of the Combustion Institute*, 2021. **38**(2): pp. 2351–2360.
71. Glarborg, P., et al., On the rate constant for NH_2+HO_2 and third-body collision efficiencies for NH_2+H (+M) and NH_2+NH_2 (+M). *The Journal of Physical Chemistry A*, 2021. **125**(7): pp. 1505–1516.
72. Sarathy, S.M., et al., Alcohol combustion chemistry. *Progress in Energy and Combustion Science*, 2014. **44**: pp. 40–102.
73. Wang, Z., et al., Study on cool flame radical index and oxygen concentration dependence of oxygenated fuels. *Combustion and Flame*, 2023, **257**, 112493: pp. 1–7.
74. Wang, Z., et al., Kinetics and extinction of non-premixed cool and warm flames of dimethyl ether at elevated pressure. *Proceedings of Combustion Insittute*, 2022. **39**: pp. 1–8.
75. Wang, Y., et al., Ignition of dimethyl ether/air mixtures by hot particles: Impact of low temperature chemical reactions. *Proceedings of the Combustion Institute*, 2021. **38**: pp. 2459–2466.
76. Felsmann, D., et al., Contributions to improving small ester combustion chemistry: Theory, model and experiments. *Proceedings of the Combustion Institute*, 2017. **36**(1): pp. 543–551.
77. Diévart, P., et al., A comparative study of the chemical kinetic characteristics of small methyl esters in diffusion flame extinction. *Proceedings of the Combustion Institute*, 2013. **34**(1): pp. 821–829.
78. Ju, Y., Understanding cool flames and warm flames. *Proceedings of the Combustion Institute*, 2021. **38**(1): pp. 83–119.
79. Ju, Y., et al., Dynamics of cool flames. *Progress in Energy and Combustion Science*, 2019. **75**(100787): p. 39.
80. Curran, H.J., et al., A comprehensive modeling study of n-heptane oxidation. *Combustion and Flame*, 1998. **114**(1–2): pp. 149–177.
81. Westbrook, C.K., et al., A detailed chemical kinetic reaction mechanism for n-alkane hydrocarbons from n-octane to n-hexadecane. *Combustion and Flame*, 2009. **156**(1): pp. 181–199.
82. Harper, M.R., et al., Comprehensive reaction mechanism for n-butanol pyrolysis and combustion *Combustion and Flame*, 2011. **158**(1): pp. 16–41.
83. Herbinet, O., W.J. Pitz, and C.K. Westbrook, Detailed chemical kinetic mechanism for the oxidation of biodiesel fuels blend surrogate. *Combustion and Flame*, 2010. **157**(5): pp. 893–908.
84. Curran, H.J., Developing detailed chemical kinetic mechanisms for fuel combustion. *Proceedings of the Combustion Institute*, 2019. **37**(1): pp. 57–81.
85. Metcalfe, W.K., et al., A hierarchical and comparative kinetic modeling study of C1−C2 hydrocarbon and oxygenated fuels. *International Journal of Chemical Kinetics*, 2013. **45**(10): pp. 638–675.
86. Yehia, O.R., C.B. Reuter, and Y. Ju, Low-temperature multistage warm diffusion flames. *Combustion and Flame*, 2018. **195**: pp. 63–74.
87. Yehia, O.R., C.B. Reuter, and Y. Ju, On the chemical characteristics and dynamics of n-alkane low-temperature multistage warm diffusion flames. *Proceedings of the Combustion Institute*, 2019. **37**(2): pp. 1717–1724.
88. Farouk, T., D. Dietrich, and F.L. Dryer, Three stage cool flame droplet burning behavior of n-alkane droplets at elevated pressure conditions: Hot, warm and cool flame. *Proceedings of the Combustion Institute*, 2019. **37**: pp. 3353–3361.
89. Skodje, R.T., et al., Theoretical validation of chemical kinetic mechanisms: Combustion of methanol. *The Journal of Physical Chemistry A*, 2010. **114**(32): pp. 8286–8301.
90. Ju, Y. and W. Sun, Plasma assisted combustion: Dynamics and chemistry. *Progress in Energy and Combustion Science*, 2015. **48**: pp. 21–83.
91. Zhong, H., et al., Kinetic studies of excited singlet oxygen atom O (1D) reactions with ethanol. *International Journal of Chemical Kinetics*, 2021. **53**(6): pp. 688–701.
92. Yan, C., et al., The kinetic study of excited singlet oxygen atom O(1D) reactions with acetylene. *Combustion and Flame*, 2020. **212**: pp. 135–141.
93. Ombrello, T., et al., Flame propagation enhancement by plasma excitation of oxygen. Part II: Effects of O_2 (a 1 Δ g). *Combustion and Flame*, 2010. **157**(10): pp. 1916–1928.
94. Ombrello, T., et al., Flame propagation enhancement by plasma excitation of oxygen. Part I: Effects of O_3. *Combustion and flame*, 2010. **157**(10): pp. 1906–1915.
95. Sepulveda, J., et al., Kinetic enhancement of microchannel detonation transition by ozone addition to acetylene mixtures. *AIAA Journal*, 2019. **57**(2): pp. 476–481.

96. Sun, W., et al., The effect of ozone addition on combustion: Kinetics and dynamics. *Progress in Energy and Combustion Science*, 2019. **73**: pp. 1–25.
97. Rousso, A.C., et al., Low-temperature oxidation of ethylene by ozone in a jet-stirred reactor. *The Journal of Physical Chemistry A*, 2018. **122**(43): pp. 8674–8685.
98. Mao, X., et al., Numerical modeling of ignition enhancement of CH_4/O_2/He mixtures using a hybrid repetitive nanosecond and DC discharge. *Proceedings of the Combustion Institute*, 2019. **37**: pp. 5545–5552.
99. Mao, X., et al., Effects of controlled non-equilibrium excitation on H_2/O_2/He ignition using a hybrid repetitive nanosecond and DC discharge. *Combustion and Flame*, 2019. **206**: pp. 522–535.
100. Alexander, S. and S. Alexander, Theoretical analysis of reaction kinetics with singlet oxygen molecules. *Physical Chemistry Chemical Physics (PCCP)*, 2011. **13**(36): pp. 16424–16436.
101. Sharipov, A.S. and A.M. Starik, Analysis of the reaction and quenching channels in a $H+O_2(a1\Delta g)$ system. *Physica Scripta*, 2013. **88**(5): p. 058305.
102. Chukalovsky, A.A., et al., Reaction of hydrogen atoms with singlet delta oxygen ($O_2(a1\Delta g)$). Is everything completely clear? *Journal of Physics D Applied Physics*, 2016. **49**(48): p. 485202.
103. Ju, Y., et al., Plasma assisted low temperature combustion. *Plasma Chemistry and Plasma Processing*, 2016. **36**(1): pp. 85–105.
104. Lefkowitz, J.K., et al., In situ species diagnostics and kinetic study of plasma activated ethylene dissociation and oxidation in a low temperature flow reactor. *Proceedings of the Combustion Institute*, 2015. **35**(3): pp. 3505–3512.
105. Lefkowitz, J.K., et al., Species and temperature measurements of methane oxidation in a nanosecond repetitively pulsed discharge. *Philosophical Transactions of the Royal Society A*, 2015. **373**(2048): p. 20140333.
106. Rousso, A., et al., Kinetic studies and mechanism development of plasma assisted pentane combustion. *Proceedings of the Combustion Institute*, 2019. **37**(4): pp. 5595–5603.
107. Rousso, A., et al., Low temperature oxidation and pyrolysis of n-heptane in nanosecond-pulsed plasma discharges. *Proceedings of the Combustion Institute*, 2017. **36**: pp. 4105–4112.
108. Zhong, H., et al., Kinetic studies of plasma assisted n-dodecane/O_2/N_2 pyrolysis and oxidation in a nanosecond-pulsed discharge. *Proceedings of Combustion Insittute*, 2021. **38**: pp. 6521–6531.
109. Zhong, H., et al. Kinetic studies of low-temperature ammonia oxidation in a nanosecond repetitively-pulsed discharge. In *AIAA SCITECH Forum, National Harbor*, Maryland, Jan. 23-27, 2023.
110. Kosarev, I.N., et al., Kinetics of ignition of saturated hydrocarbons by nonequilibrium plasma: CH_4-containing mixtures. *Combustion and Flame*, 2008. **154**(3): pp. 569–586.
111. Aleksandrov, N.L., et al., Mechanism of ignition by non-equilibrium plasma. *Proceedings of the Combustion Institute*, 2009. **32**(1): pp. 205–212.
112. Starikovskaya, S.M., et al., Ignition with low-temperature plasma: Kinetic mechanism and experimental verification. *High Energy Chemistry*, 2009. **43**(3): pp. 213–218.
113. Sun, W., et al., Kinetic effects of non-equilibrium plasma-assisted methane oxidation on diffusion flame extinction limits. *Combustion and Flame*, 2012. **159**(1): pp. 221–229.
114. Sun, W., et al., Direct ignition and S-curve transition by in situ nano-second pulsed discharge in methane/oxygen/helium counterflow flame. *Proceedings of the Combustion Institute*, 2013. **34**(1): pp. 847–855.
115. Lou, G., et al., Ignition of premixed hydrocarbon–air flows by repetitively pulsed, nanosecond pulse duration plasma. *Proceedings of the Combustion Institute*, 2007. **31**(2): pp. 3327–3334.
116. Uddi, M., et al., Nitric oxide density measurements in air and air/fuel nanosecond pulse discharges by laser induced fluorescence. *Journal of Physics D: Applied Physics*, 2009. **42**(7): p. 075205.
117. Uddi, M., et al., Atomic oxygen measurements in air and air/fuel nanosecond pulse discharges by two photon laser induced fluorescence. *Proceedings of the Combustion Institute*, 2009. **32**(1): pp. 929–936.
118. Yin, Z., et al., Measurements of temperature and hydroxyl radical generation/decay in lean fuel–air mixtures excited by a repetitively pulsed nanosecond discharge. *Combustion and Flame*, 2013. **160**(9): pp. 1594–1608.
119. Suib, S.L. and R.P. Zerger, A direct, continuous, low-power catalytic conversion of methane to higher hydrocarbons via microwave plasmas. *Journal of Catalysis*, 1993. **139**: pp. 381–391.
120. Lesueur, H., A. Czernichowski, and J. Chapelle, Electrically assisted partial oxidation of methane. *International Journal of Hydrogen Energy*, 1994. **19**(2): pp. 139–144.
121. Okumoto, M., et al., Nonthermal plasma approach in direct methanol synthesis from CH_4. *IEEE Transactions on Industry Applications*, 1998. **34**(5): pp. 940–944.
122. Bromberga, L., et al., Plasma catalytic reforming of methane. *International Journal of Hydrogen Energy*, 1999. **24**: pp. 1131–1137.

123. Lee, D.H., et al., Optimization scheme of a rotating gliding arc reactor for partial oxidation of methane. *Proceedings of the Combustion Institute,* 2007. **31**(2): pp. 3343–3351.
124. Hwang, N., Y.-H. Song, and M.S. Cha, Efficient use of CO_2 reforming of methane with an arc-jet plasma. *IEEE Transactions on Plasma Science,* 2010. **38**(12): pp. 3291–3299.
125. Zhang, X. and M.S. Cha, Electron-induced dry reforming of methane in a temperature-controlled dielectric barrier discharge reactor. *Journal of Physics D: Applied Physics,* 2013. **46**(41): p. 415205.
126. Zhang, X. and M.S. Cha, Partial oxidation of methane in a temperature-controlled dielectric barrier discharge reactor. *Proceedings of the Combustion Institute,* 2015. **35**: pp. 3447–3454.
127. Jasiński, M., M. Dors, and J. Mizeraczyk, Production of hydrogen via methane reforming using atmospheric pressure microwave plasma. *Journal of Power Sources,* 2008. **181**(1): pp. 41–45.
128. Sun, W., et al., Effects of non-equilibrium plasma discharge on counterflow diffusion flame extinction. *Proceedings of the Combustion Institute,* 2011. **33**(2): pp. 3211–3218.
129. Wang, H., et al., USC mech version II. High-temperature combustion reaction model of H_2/CO/C1-C4 compounds, 2007.
130. MacFarlane, D.R., et al., A roadmap to the ammonia economy. *Joule,* 2020. **4**(6): pp. 1186–1205.
131. Valera-Medina, A., et al., Ammonia for power. *Progress in Energy and Combustion Science,* 2018. **69**: pp. 63–102.
132. Afif, A., et al., Ammonia-fed fuel cells: A comprehensive review. *Renewable and Sustainable Energy Reviews,* 2016. **60**: pp. 822–835.
133. Pai, S.J., C.L. Heald, and J.G. Murphy, Exploring the global importance of atmospheric ammonia oxidation. *ACS Earth and Space Chemistry,* 2021. **5**(7): pp. 1674–1685.
134. Tang, Y., et al., Flammability enhancement of swirling ammonia/air combustion using AC powered gliding arc discharges. *Fuel,* 2022. **313**: p. 122674.
135. Choe, J., et al., Plasma assisted ammonia combustion: Simultaneous NO_x reduction and flame enhancement. *Combustion and Flame,* 2021. **228**: pp. 430–432.
136. Kim, G.T., et al., Effects of non-thermal plasma on turbulent premixed flames of ammonia/air in a swirl combustor. *Fuel,* 2022. **323**. p. 124227.
137. Zhong, H., et al., Understanding non-equilibrium N_2O/NO_x chemistry in plasma-assisted low-temperature NH_3 oxidation. *Combustion and Flame,* 2023. **256**: p. 112948.
138. Pancheshnyi, S., et al., The LXCat project: Electron scattering cross sections and swarm parameters for low temperature plasma modeling. *Chemical Physics,* 2012. **398**: pp. 148–153.
139. Zhao, H., X. Yang, and Y. Ju, Kinetic studies of ozone assisted low temperature oxidation of dimethyl ether in a flow reactor using molecular-beam mass spectrometry. *Combustion and Flame,* 2016. **173**: pp. 187–194.
140. Zhong, H., et al., Kinetic study of reaction $C_2H_5 + HO_2$ in a photolysis reactor with time-resolved Faraday rotation spectroscopy. *Proceedings of the Combustion Institute,* 2021. **38**: pp. 871–880.
141. Teng, C.C., et al., Time-resolved HO_2 detection with Faraday rotation spectroscopy in a photolysis reactor. *Optics Express,* 2021. **29**(2): pp. 2769–2779.
142. Kee, R. J., Grear, J. F., Smooke, M. D. & Miller, J. A. 1985 Sandia Rep. SAND85-8240.
143. Cai, L., et al., Optimized reaction mechanism rate rules for ignition of normal alkanes. *Combustion and Flame,* 2016. **173**: pp. 468–482.
144. Zhao, H., et al., Studies of low temperature oxidation of n-pentane with nitric oxide addition in a jet stirred reactor. *Combustion and Flame,* 2018. **197**: pp. 78–87.
145. Lin, J., et al., Multiple dynamical pathways in the O (1 D)+ CH_4 reaction: A comprehensive crossed beam study. *The Journal of Chemical Physics,* 2000. **113**(13): pp. 5287–5301.
146. Yu, H.G. and J.T. Muckerman, MRCI calculations of the lowest potential energy surface for CH_3OH and direct ab initio dynamics simulations of the O (1D)+ CH_4 reaction. *The Journal of Physical Chemistry A,* 2004. **108**(41): pp. 8615–8623.
147. Luntz, A., Chemical dynamics of the reactions of O (1 D 2) with saturated hydrocarbons. *The Journal of Chemical Physics,* 1980. **73**(3): pp. 1143–1152.
148. Brumfield, B., et al., Dual modulation Faraday rotation spectroscopy of HO 2 in a flow reactor. *Optics Letters,* 2014. **39**(7): pp. 1783–1786.
149. Huang, C.-K., et al., Dynamics of the reactions of O (1D) with CD_3OH and CH_3OD studied with time-resolved Fourier-transform IR spectroscopy. *The Journal of Chemical Physics,* 2012. **137**(16): p. 164307.
150. Zhong, H., et al., Direct kinetic measurements and theoretical predictions of singlet oxygen atom reaction with dimethyl ether. *The Journal of Physical Chemistry Letters,* 2024. **15**, pp. 6158–6165.

151. Zhong, H., et al., Plasma thermal-chemical instability of low-temperature dimethyl ether oxidation in a nanosecond-pulsed dielectric barrier discharge. *Plasma Sources Science and Technology*, 2022. **31**(11): p. 114003.
152. Am Ano, T. and F.L. Dryer. Effect of dimethyl ether, NOx, and ethane on CH_4 oxidation: High pressure, intermediate-temperature experiments and modeling. in *Symposium (International) on Combustion*, 1998. Elsevier.
153. Zhao, H., et al., Experimental and modeling study of the mutual oxidation of n-pentane and nitrogen dioxide at low and high temperatures in a jet stirred reactor. *Energy*, 2018. **165**: pp. 727–738.
154. Eliasson, B., M. Hirth, and U. Kogelschatz, Ozone synthesis from oxygen in dielectric barrier discharges. *Journal of Physics D: Applied Physics*, 1987. **20**(11): p. 1421.
155. Johnston, H.S., Atmospheric ozone. *Annual Review of Physical Chemistry*, 1992. **43**(1): pp. 1–31.
156. Sandri, R., On the decomposition flame of liquid ozone-oxygen mixtures in a tube. *Combustion and Flame*, 1958. **2**(4): pp. 348–352.
157. Hajilou, M., et al., Experimental and numerical characterization of freely propagating ozone-activated dimethyl ether cool flames. *Combustion and Flame*, 2017. **176**: pp. 326–333.
158. Won, S.H., et al., Self-sustaining n-heptane cool diffusion flames activated by ozone. *Proceedings of the Combustion Institute*, 2015. **35**(1): pp. 881–888.
159. Halter, F., P. Higelin, and P. Dagaut, Experimental and detailed kinetic modeling study of the effect of ozone on the combustion of methane. *Energy & Fuels*, 2011. **25**(7): pp. 2909–2916.
160. Hippler, H., R. Rahn, and J. Troe, Temperature and pressure dependence of ozone formation rates in the range 1–1000 bar and 90–370 K. *The Journal of Chemical Physics*, 1990. **93**(9): pp. 6560–6569.
161. Patrick, R. and D.M. Golden, Kinetics of the reactions of amidogen radicals with ozone and molecular oxygen. *The Journal of Physical Chemistry*, 1984. **88**(3): pp. 491–495.
162. Masurier, J.-B., et al., Ozone applied to the homogeneous charge compression ignition engine to control alcohol fuels combustion. *Applied Energy*, 2015. **160**: pp. 566–580.
163. Tachibana, T., et al., Effect of ozone on combustion of compression ignition engines. *Combustion and Flame*, 1991. **85**(3–4): pp. 515–519.
164. Chen, C., et al., Experimental and kinetic modeling study of laminar burning velocity enhancement by ozone additive in $NH_{3+O2+N2}$ and $NH_{3+CH4}/C_2H_6/C_3H_{8+air}$ flames. *Proceedings of the Combustion Institute*, 2023. **39**: pp. 4237–4246.
165. Crane, J., et al., Isolating the effect of induction length on detonation structure: Hydrogen–oxygen detonation promoted by ozone. *Combustion and Flame*, 2019. **200**: pp. 44–52.
166. Alfazazi, A., et al., Cool diffusion flames of butane isomers activated by ozone in the counterflow. *Combustion and Flame*, 2018. **191**: pp. 175–186.
167. Alam, F.E., et al., Ozone assisted cool flame combustion of sub-millimeter sized n-alkane droplets at atmospheric and higher pressure. *Combustion and Flame*, 2018. **195**: pp. 220–231.
168. Warnatz, J., Calculation of the structure of laminar flat flames I: Flame velocity of freely propagating ozone decomposition flames. *Berichte der Bunsengesellschaft für physikalische Chemie*, 1978. **82**(2): pp. 193–200.
169. Lee, M., et al., Experimental observation and numerical simulation of wall-stabilized premixed cool flames. *Proceedings of the Combustion Institute*, 2019. **37**: pp. 1749–1756.
170. Reuter, C.B., et al., Study of the low-temperature reactivity of large n-alkanes through cool diffusion flame extinction. *Combustion and Flame*, 2017. **179**: pp. 23–32.
171. Reuter, C.B., S.H. Won, and Y. Ju, Experimental study of the dynamics and structure of self-sustaining premixed cool flames using a counterflow burner. *Combustion and Flame*, 2016. **166**: pp. 125–132.
172. Won, S.H., et al., A new cool flame: Establishment and studies of dynamics and kinetics. *52nd Aerospace Sciences Meeting*, (doi: 10.2514/6.2014-0818), in AIAA paper-2014-0818, 2014.
173. Foucher, F., et al., Influence of ozone on the combustion of n-heptane in a HCCI engine. *Proceedings of the Combustion Institute*, 2013. **34**(2): pp. 3005–3012.
174. Rousso, A.C., et al., Identification of the Criegee intermediate reaction network in ethylene ozonolysis: Impact on energy conversion strategies and atmospheric chemistry. *Physical Chemistry Chemical Physics*, 2019. **21**(14): pp. 7341–7357.
175. Rousso, A.C., et al., Extreme low-temperature combustion chemistry: Ozone-initiated oxidation of methyl hexanoate. *The Journal of Physical Chemistry A*, 2020. **124**(48): pp. 9897–9914.
176. Wu, B., et al., Dynamics of laminar ethylene lifted flame with ozone addition. *Proceedings of the Combustion Institute*, 2021. **38**(4): pp. 6773–6780.

177. Gutbrod, R., et al., Formation of OH radicals in the gas phase ozonolysis of alkenes: The unexpected role of carbonyl oxides. *Chemical Physics Letters*, 1996. **252**(3–4): pp. 221–229.
178. Osborn, D.L. and C.A. Taatjes, The physical chemistry of Criegee intermediates in the gas phase. *International Reviews in Physical Chemistry*, 2015. **34**(3): pp. 309–360.
179. Taatjes, C.A., et al., Direct observation of the gas-phase Criegee intermediate (CH_2OO). *Journal of the American Chemical Society*, 2008. **130**(36): pp. 11883–11885.
180. Anglada, J.M., R. Crehuet, and J.M. Bofill, The ozonolysis of ethylene: A theoretical study of the gas-phase reaction mechanism. *Chemistry: A European Journal*, 1999. **5**(6): pp. 1809–1822.
181. Criegee, R., Mechanism of ozonolysis. *Angewandte Chemie International Edition in English*, 1975. **14**(11): pp. 745–752.

6 Plasma-Assisted Combustion Dynamics

6.1 PLASMA-ASSISTED IGNITION

6.1.1 Ignition and Ignition Delay Time

Autoignition is a process where the combustion mixture auto-ignites with an exponential increase of temperature, heat release rate, fuel consumption, product formation, and chemiluminescence. The ignition process is governed by the chain-initiation, branching, and termination reactions with a positive feedback of chemical heat release on reactions. Therefore, during ignition, there is a strong positive coupling between the chemistry and chemical heat release.

One of the key parameters to characterize ignition is the ignition delay time. Ignition delay time is a critical combustion property for compression ignition engines, gas turbine, and detonation engines as well as fire safety. Ignition delay time is often measured either in a shock tube or in rapid combustion machine (RCM). Figure 6.1 shows a schematic of the shock dynamics in shock tube ignition measurement and an experimental result of the ignition delay time measured for real fuels [1]. As shown in Figure 6.1a, when a diaphragm that separates a high-pressure driver gas from a combustion mixture at location x_0 breaks, the gas expansion from the high-pressure driver gas side to the lower pressure combustion mixture side generates a shock wave propagating to the right into the combustible mixture. At the same time, an expansion fan is generated and propagates to the left in the driver gas. In addition, there is a contact surface separating the combustible mixture and driver gas, moving to the right. After the shock wave hits and reflects from the right end wall of the shock tube at $x = L$ and $t = 0$, it creates a stationary shocked mixture (no flow) in zone 5 with an instant rise of temperature (T_5) and pressure (p_5). Under the high-temperature and high-pressure environment of T_5 and p_5, the chain-initiation, branching, and termination reactions (Chapter 5) will occur in the shocked mixture of zone 5. After some delay of $\Delta t = t_{ig}$, the chain-branching reactions accelerate so fast that they lead to an exponential increase in temperature and pressure as well as an increase in chemiluminescence as autoignition (Figure 6.1b). This time delay is called

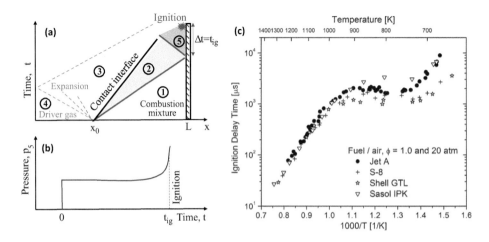

FIGURE 6.1 (a–c) Comparison of stoichiometric jet fuel/air ignition delay times at 20 atm for jet-A and three Fischer-Tropsch fuels [1].

the autoignition delay time. By recording the pressure and chemiluminescence of the mixture and ensuring that the pressure rise before the ignition event in the shock tube is negligible (Figure 6.1b), the ignition delay time of a mixture under the initial condition of shocked pressure and temperature can be measured. Figure 6.1c [1] shows the measured ignition delay time of jet-A and three Fischer-Tropsch alternative jet fuels at 20 atm at the stoichiometric condition. It is seen that at high temperature, the ignition delay times for all studied fuels are almost the same. This is because at high temperature, the chain-initiation reactions are fast, and it is the chain-branching reaction (R5.3) that dominates the ignition for all tested fuels. However, at intermediate temperatures, there exists a negative temperature coefficient (NTC) region, where the ignition delay time increases with the increase of temperature. In this region, the ignition delay time is now a function of fuel molecule structure, thus is strongly fuel-dependent. As the temperature further decreases, the disparities between the ignition delay time of different fuels further increase because reactions in this region are governed by low-temperature chemistry (LTC), which is more sensitive to fuel molecule structures. Therefore, the measurements of ignition delay time are important to characterize the reactivity of fuels and the effectiveness of plasma-assisted combustion as well as to develop an experimentally validated kinetic model.

Because of the limitation in available time of p_5 and T_5 in shock tube experiments due to the shock interaction with the moving contact surface and the reflected expansive waves, shock tube measurements of ignition delay time are often limited to a short ignition delay time from microseconds to a few milliseconds [2]. To measure longer ignition delay time at lower temperatures, an RCM [3] is often used. Figure 6.2 shows [3] a typical pressure time history in an RCM. Initially, the mixture is compressed by the motion of a piston and its temperature and pressure rise simultaneously. At $t = 0$, the piston stops and the compressed high-temperature and high-pressure mixture will auto-ignite, allowing the measurements of the first-stage ignition (LTI) and the second stage ignition (HTI). Therefore, if heat losses to the wall and near wall flow motion are appropriately considered, RCM provides a complimentary method to shock tube to quantify both LTI and HTI at elevated temperatures and pressures with a longer ignition timescale to study LTC and ITC kinetics. Note that in both shock tube and RCM, appropriate modeling of the preignition pressure rise, heat loss, and flow motion is important to provide good kinetic data to assess fuel chemistry and the effect of plasma-assisted ignition [4].

The ignition delay time can be simply analyzed using one-step chemistry from fuel (F) to product (P) with a chemical heat release rate, Q,

$$F \rightarrow P + Q, \omega = B\rho Y_F e^{-E_a/RT} \qquad (6.1)$$

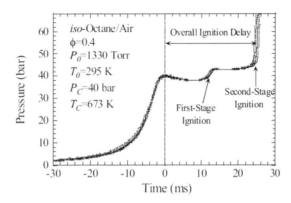

FIGURE 6.2 Autoignition of an iso-octane/air mixture under conditions of two-stage ignition. Conditions: $\phi = 0.4$, $p_0 = 1330$ Torr, $T_0 = 295$ K, $p_c = 30$ bar, and $T_c = 673$ K [3].

where ω is the reaction rate, B the rate coefficient, ρ the density, Y_F the fuel mass fraction, E_a the activation energy, R the universal gas constant, and T the temperature.

The governing equation for fuel mass fraction (Y_F) and temperature (T) can be given as

$$\rho \frac{dY_F}{dt} = -B\rho Y_F e^{-\frac{E_a}{RT}} \tag{6.2a}$$

$$\rho C_p \frac{dT}{dt} = QB\rho Y_F e^{-\frac{E_a}{RT}} \tag{6.2b}$$

With the initial condition of

$$t = 0: T = T_0, \quad Y_F = Y_{F0} \tag{6.3}$$

By introducing the normalized parameters of,

$$\theta \equiv \frac{T}{T_0}, \quad y = \frac{Y_F}{Y_{F0}}, \quad q = \frac{QY_{F0}}{C_p T_0}, \quad \beta = \frac{E_a}{RT_0}, \quad \tau = \frac{t}{t_0}, \quad \frac{1}{\tau_0} = \beta B e^{-\frac{E_a}{RT_0}} \tag{6.4}$$

Equations 6.1 and 6.2 give the following nondimensionalized relations,

$$y = q^{-1}(1 + q - \theta) \tag{6.5}$$

and

$$\frac{d\theta}{d\tau} = \frac{1}{\beta}(1 + q - \theta)e^{\frac{\beta(\theta-1)}{\theta}} \tag{6.6a}$$

$$\theta(\tau = 0) = 1 \tag{6.6b}$$

If one assumes the activation energy (E_a and β) is very large, i.e., $\beta \to \infty$, then we can expand the temperature from the initial temperature using a small perturbation of $1/\beta$ as,

$$\theta - 1 \equiv \vartheta/\beta \to O(1/\beta) \tag{6.7}$$

Substituting Eq. 6.7 into Eq. 6.6 and by keeping the terms of $O(1/\beta)$ and neglecting the terms of $O(1/\beta^2)$, we can have a perturbed equation of temperature change from the initial value as,

$$\frac{d\vartheta}{d\tau} = qe^{\vartheta}, \quad \vartheta(0) = 0 \tag{6.8}$$

The above equation has a solution of

$$\vartheta = -\ln(1 - q\tau) \tag{6.9}$$

If we define that ignition occurs at $\vartheta \to \infty$, then we have,

$$\tau_{ig} = 1/q \tag{6.10}$$

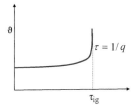

FIGURE 6.3 Ignition curve of large activation analysis (temperature dependence on time).

Therefore, the dimensional ignition delay time becomes,

$$t_{ig} = \tau_{ig}\tau_0 = q^{-1}(\beta Be^{-E/RT_0})^{-1} = \frac{RC_pT_0^2}{BQY_{F0}E_a}e^{E_a/RT_0} \quad (6.11)$$

From the result of Eq. 6.11, it is clearly seen that plasma can enhance ignition and shorten the ignition delay time in the following ways: (1) increase the temperature T_0 via fast heating and slow heating in plasma, (2) increase the reaction rates of chain-initiation and chain-propagation reactions (increase of B) by creating initial radicals, and (3) creating new reaction pathways to lower the activation energy (smaller E_a) of the key chain-branching reactions.

6.1.2 Plasma-Assisted Ignition

Many experiments and computations have been done to examine how plasma enhances ignition. Figure 6.4a [5] and b [6], respectively, show the results of measured and calculated ignition delay time as a function of temperature for hydrogen and methane mixtures with and without a nanosecond discharge in a reflected shock tube. For hydrogen mixtures, it is seen that nanosecond plasma discharge shortened the ignition delay time by 5–10 times. In addition, with the increase in temperature, the ignition enhancement becomes greater. This is likely because the H and O radicals generated from electron-impact reactions R5.6'–R5.8' and collisional dissociation by electronically excited oxygen atom, $O(^1D)$, excited nitrogen, N_2^*, and Ar* (R5.20'–R5.24') can directly accelerate the chain-branching reaction R5.3 at higher temperatures. Figure 6.4b shows that for methane mixtures, the ignition enhancement by plasma was even stronger than that of hydrogen. This is because in methane oxidation CH_3 reactions with O_2 and HO_2 to produce CH_3O and OH radicals

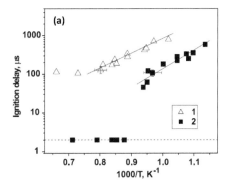
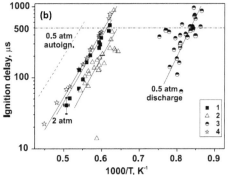

FIGURE 6.4 (a) Measured ignition delay time as a function of gas temperature for mixture $H_2/O_2/N_2/Ar$ = 6/3/11/80 with pressure about 20 Torr. 1 (triangle): autoignition and 2 (solid square): ignition by nanosecond discharge at voltage of 160 kV [5]. (b) Ignition delay time as a function of temperature for mixture of $CH_4/O_2/N_2/Ar$ = 1/4/15/80. 1 (solid square): 2 atm, autoignition; 2 (triangle): 2 atm, ignition with discharge; 3 (circle): 0.5 atm, ignition with discharge; dashed line: 0.5 atm, autoignition (calculated) and 4 (star): 2 atm, autoignition (calculated) [6].

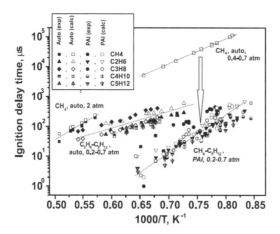

FIGURE 6.5 The ignition delay time for autoignition with and without plasma discharge in hydrocarbon-containing mixtures as a function of temperature. Closed points correspond to measurements, and open points correspond to calculations. Solid lines are approximations of the data. Vertical arrow indicates the displacement of the ignition delay time for methane-containing mixture under the action of the discharge. PAI: plasma-assisted ignition [7].

via R5.14–R5.15 are very slow. Plasma can accelerate radical production and directly accelerate the chain-branching reaction R5.3 and the H radical production via R5.18.

Plasma-assisted ignition for larger hydrocarbons was also studied in a shock tube for stoichiometric $C_nH_{2n+2}/O_2/Ar$ mixtures with 90% Ar dilution [7]. Figure 6.5 shows that nanosecond plasma also significantly accelerated the ignition of C_2–C_5 n-alkanes. Note that the ignition enhancement of higher alkanes is smaller compared to methane because of the slow oxidation kinetics of methane itself. In addition, it also shows that the ignition delay time of larger n-alkanes at high temperature is similar. Moreover, like hydrogen, the plasma enhancement effect is greater when the shocked mixture temperature is higher because plasma chemistry bypasses the rate-limiting chain-initiation reaction of hydrogen combustion (R5.1) to produce radicals with less quenching at higher temperature. Also, at higher temperatures, the E/N is higher. Plasma can generate more energetic electrons and modify the molecular excitations and plasma energy deposition and transfer.

The fast radical production by plasma not only can enhance ignition but also extend the ignition limit (Figure 5.4). An earlier study by Gorchakov and Lavrov [8,9] suggested that the hydrogen explosion limits of H_2/O_2 mixture under the action of a plasma discharge can be extended to even lower temperatures (Figure 6.6). The results revealed that a higher discharge current led to broader explosion limits. Figure 6.6 shows that at the discharge current of 2 A, the second explosion limit (Figure 5.4) can be extended by more than 100 K. To explain this phenomenon, we can add an electron-impact oxygen dissociation reaction into the kinetic analysis of the second explosion limit in Eq. 5.3,

$$e + O_2 \rightarrow e + O + O \quad \text{(R5.7')}$$

Then, the growth rate of H radical concentration in Eq. 5.3 will be modified to,

$$\frac{dC_H}{C_H} = \left[(2k_3 - k_8 C_M)C_{O2} + 4k_{7'}\frac{C_e C_{O2}}{C_H}\right]dt \quad (6.12)$$

In a plasma discharge, the electron number density, C_e, mainly depends on the discharge current. Before ignition, in a plasma discharge C_e is typically much greater than the number density H atom, C_H, and $k_{7'}$ is much greater than k_3 (Figures 5.27 and 5.28). Therefore, the appearance of the second

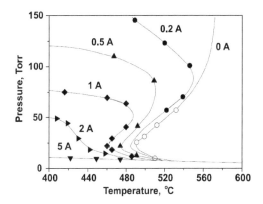

FIGURE 6.6 Extension of explosion limits of H_2/O_2 mixture under the action of a plasma discharge. Different curves correspond to different currents in a primary electric circuit [8,9].

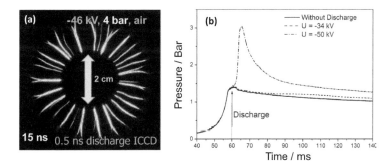

FIGURE 6.7 (a) SDBD discharge system, (b) Effect of the voltage pulse amplitude on the pressure profile during plasma-assisted ignition experiments for stoichiometric n-heptane/air mixtures at 1.5 bar and $T_C = 650$ K [10,11].

term on the right of Eq. 6.12 due to plasma chemistry clearly shows that the explosion limit will be significantly extended to lower temperature due to the radical production via the electron-impact oxygen dissociation reaction of R5.7'. Of course, other radical production reactions in plasma will also extend the explosion limit like R5.7'.

To further explore plasma-assisted ignition at elevated pressure and lower temperatures for higher hydrocarbons, experiments using a nanosecond surface dielectric barrier discharge (SDBD) in an RCM with a specially designed two-dimensional electrode (Figure 6.7) [10,11]. Figure 6.7a shows the streamers of SDBD at 4 bar in air. As will be shown later, ignition initiation is strongly dependent on the initial ignition kernel size. The large two-dimensional multi-channel streamers help to increase the initial ignition kernel size and accelerate ignition. As shown in Figure 6.7b for the n-heptane case [11], with no plasma discharge or with a lower discharge voltage, ignition did not occur. However, when the discharge voltage was increased to 50 kV, ignition occurred right after the discharge, indicating that the SDBD streamers at high voltages were strong enough to initiate the ignition. More experiments were conducted for methane and n-butane-containing mixtures. These results suggested that the production of atomic oxygen via reactions such as R5.7'–R5.8' and R5.21'–R5.22' in plasma discharge enhanced the ignition chemistry by increasing the radical pool.

The effect of plasma discharge in air stream on the minimum autoignition temperature was also quantified by using a rotating gliding arc in a counterflow flame [12,13]. As shown in Figure 6.8a, in a counterflow flame, air or H_2-blended air was issued from the bottom burner and discharged by a rotating gliding arc [12]. A fuel mixture such as hydrogen and methane diluted with nitrogen was

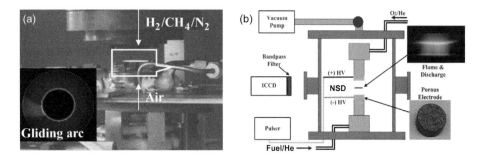

FIGURE 6.8 (a) Direct image of a counterflow burner and the rotating gliding arc stabilized by magnet. (b) Schematic and image of plasma-assisted ignition and flame using a nanosecond discharge.

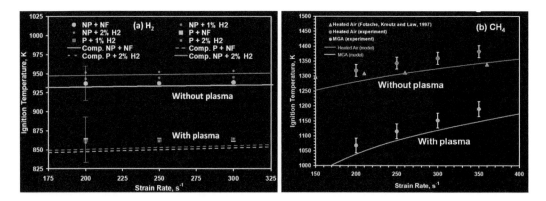

FIGURE 6.9 Counterflow diffusion flame autoignition temperatures as a function of strain rate for air and H_2-blended air with and without gliding arc activation: (a) hydrogen and (b) methane.

issued from the top burner. The air temperature was raised by using an electrical heater and monitored by a thermal couple. Ignition temperature was measured when an autoignition occurred at the minimum air temperature at a given flow velocity gradient (the strain rate). Due to the heat loss from the flow, a higher strain rate resulted in a higher autoignition temperature.

Figure 6.9a and b show the measured autoignition temperature for H_2 and CH_4 with and without a gliding arc at different flow strain rates [12]. For the hydrogen case, with gliding arc activation in air, the autoignition temperature was decreased by approximately 100 K. Also, because hydrogen chemistry is fast, the autoignition temperature is not sensitive to the strain rate in the measured flow conditions. For methane, however, because of its slow oxidation kinetics, the autoignition temperature is very high without plasma (above 1300 K). In addition, the autoignition temperature is sensitive to the flow strain rate. With gliding arc discharge in air, the autoignition temperature decreased by more than 200 K for methane. The results showed that plasma enhancement of autoignition in this experiment was mostly because of the plasma-produced NO_x. Modeling showed that NO_x reactions with HO_2 accelerated the slow radical production reactions from CH_3 via the kinetic pathway discussed in Chapter 5 such as,

$$NO + HO_2 = NO_2 + OH \tag{R5.41}$$

$$NO_2 + CH_3 = CH_3O + NO \tag{R5.72}$$

6.1.3 PLASMA-ASSISTED IGNITION AND EXTINCTION

For many combustion applications, ignition, flame stabilization, and blow-off are important combustion characteristics, especially for scramjet engine applications. As shown in Figure 6.10a, in the conventional ignition and flame extinction events, the ignition and extinction are governed by the so-called S-curve. Ignition and extinction transition is a hysteresis process in which the ignition timescale is longer than extinction. Therefore, once flame extinction occurs one needs to reduce the flow speed to reignite the mixture. The question is then whether plasma can enhance the ignition so much that the extinction limit disappears, and the S-curve will be stretched out (Figure 6.10a).

To demonstrate this possibility, the ignition and extinction S-curve of methane in a diffusion reaction system was studied by measuring the OH* emission intensity and temperature in a nanosecond plasma-assisted counterflow flame (Figure 6.8b) [14]. During the experiments, the flow strain rate (400 1/s), oxygen mole fraction, X_O, and the plasma discharge frequency ($f = 24$ kHz) were held constant, while the CH_4 mole fraction, X_F, was varied. The measured relationship between OH* emission intensity as well as reaction zone peak temperature as a function of X_F at two different oxygen mole fractions, $X_O = 0.34$ and 0.62 are shown in Figure 6.10b and c. The temperatures of the reaction zone were measured by the Rayleigh scattering method. In Figure 6.10a, it is seen that an increase of X_F resulted in an abrupt increase in OH* intensity, indicating the occurrence of ignition. After ignition, a diffusion flame was formed. When the X_F of the flame was decreased to 0.20, an abrupt decrease in the OH* intensity was observed and the visible flame emission disappeared, indicating the occurrence of flame extinction. The hysteresis of OH* emission intensity between ignition and extinction represents the S-curve in Figure 6.10a. However, when X_O was increased to 0.62, the ignition and extinction limits merged at $X_F = 0.09$, resulting in a stretched ignition and extinction S-curve in Figure 6.10c. The temperature measurements also demonstrated a similar monotonic increase in the local maximum temperatures. The fully stretched ignition and extinction S-curve is the outcome that the plasma-generated reactive species accelerated the ignition so much that the ignition time becomes shorter than the flame burning time. This means that the extinction limit did not exist, and flame stabilization was fully dictated by the ignition limit. Kinetic analysis showed that atomic oxygen production from electron-impact reaction R5.7' and R5.8' played the dominant role in producing O radicals to accelerate ignition chemistry.

To observe plasma-enhanced ignition with LTC and ignition transition to cool flames (CFs), Sun et al. [15] conducted experiments of non-equilibrium plasma-activated LTC on the ignition and extinction of dimethyl ether (DME)/O_2/He mixtures in a counterflow burner with *in situ* nanosecond pulsed discharge at 72 Torr [15]. Like the high-temperature ignition of methane in Figure 6.10, a low-temperature ignition S-curve was observed by recording the OH and CH_2O

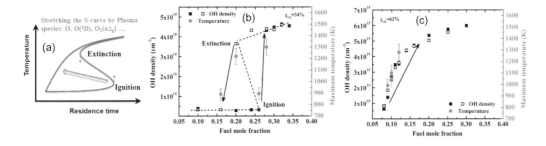

FIGURE 6.10 (a) Schematic of plasma-enhanced transition from the conventional ignition S-curve to a stretched S-curve without an extinction limit. (b) plasma-assisted methane ignition S-curve with oxygen mole fraction of 34% at $T_0 = 650$ K, $T_f = 600$ K, He/O_2 = 0.66:0.34, $P = 72$ Torr, $f = 24$ kHz, $a = 400$ 1/s; and (c) plasma-assisted methane ignition stretched S-curve with oxygen mole fraction of 62% at He/O_2 = 0.38:0.62, $P = 72$ Torr, $f = 24$ kHz, $a = 240$ 1/s [14].

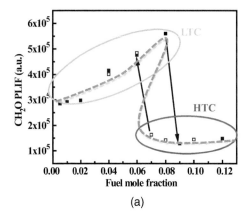 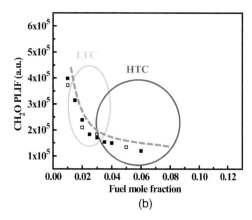

FIGURE 6.11 Relationship between CH_2O PLIF and fuel mole fraction $P = 72$ Torr, $a = 250$ 1/s (symbols: experiments, lines: modeling). (a) $X_O = 0.4$, $f = 24$ kHz, (b) $X_O = 0.6$, $f = 34$ kHz, $a = 250$ 1/s [15].

PLIFs (Figure 6.11a). The relationship between CH_2O PLIF signal intensity and fuel mole fraction at the fuel side nozzle exit, X_F, with a repetitive plasma discharge is shown in Figure 6.11a at $f = 24$ kHz and $X_O = 0.4$. It is seen that with the increase of the fuel mole fraction (X_F), the CH_2O PLIF signal intensity increased. This linear increase of the CH_2O PLIF signal before ignition indicates the occurrence of low-temperature DME oxidation. When the DME mole fraction at the fuel side is larger than 9%, ignition occurs with a sharp decrease in the CH_2O PLIF signal intensity (CH_2O is consumed by OH). After the ignition, the CH_2O PLIF signal intensity became insensitive to the change in DME mole fraction. On the other hand, when the DME mole fraction was decreased slowly to 7%, extinction occurred along with a rapid increase in CH_2O PLIF signal intensity.

To understand whether plasma can enhance LTC so fast that the *S*-curve in Figure 6.11a can also be stretched out without an extinction limit, the discharge frequency was increased to 34 kHz to further increase the O radical production. As shown in Figure 6.11b, a monotonically stretched *S*-curve without hysteresis between ignition and extinction was observed. Therefore, this experiment confirmed that plasma can significantly enhance LTC and autoignition to enhance flame stabilization at a very short flow residence time. This observation can be generalized to other larger hydrocarbon fuels (such as *n*-heptane, jet fuel) with low-temperature reactivity.

6.1.4 Control of Non-equilibrium Plasma for Ignition Enhancement

The above research shows that atomic oxygen and other radical production by electrons as well as electronically and vibrationally excited molecules help to enhance ignition and combustion. As shown in Figure 5.14, the excited molecules and electron-impact oxygen dissociation are strongly affected by electron energy or electric field (*E/N*). It would be interesting to understand whether one can control plasma properties (*E/N*) to manipulate electronically and vibrationally excited molecules as well as electron-impact dissociation reactions to maximize the impact on plasma-assisted combustion.

Recently, Mao et al. [16] conducted numerical modeling of controlled non-equilibrium excitation of H_2/O_2/He mixtures and examined the effect of *E/N* on the ignition delay time by using a hybrid repetitive nanosecond (NSD) and DC discharge at atmospheric pressure. The schematic of the hybrid repetitive nanosecond and DC discharge is shown in Figure 6.12a. A short pulse of high *E/N* was created by the nanosecond discharge to produce very high electron temperatures (Figure 1.21) and to initiate the electron-impact electronic excitations and dissociations (Figure 5.14). Then, a long period of a low DC electric field was introduced to maintain the selective excitation of vibrationally excited species and electronically excited $O_2(a^1\Delta_g)$ with a low *E/N* (Figure 6.12a).

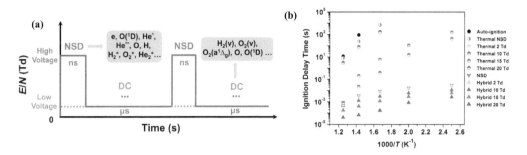

FIGURE 6.12 (a) Schematic of hybrid nanosecond discharge and DC discharge. (b) Ignition delay times for autoignition, thermal ignition, NSD-assisted ignition and hybrid discharge-assisted ignition with different temperatures at 1 atm [16].

A zero-dimensional model incorporating the plasma kinetics solver ZDPlasKin [17] and the combustion chemical kinetics solver CHEMKIN [18] was used to calculate the time evolution of species densities and temperature in a plasma-assisted combustion system. The rate constants of electron-impact reactions were solved by the Boltzmann equation solver BOSLIG+ [19] incorporated in ZDPlasKin. The detailed descriptions of the hybrid nanosecond and DC discharge and the hybrid ZDPlasKin-CHEMKIN model were described in Ref. [16].

Figure 6.12b shows the comparison of the ignition delay times for autoignition, thermal ignition, NSD-assisted ignition, and hybrid discharge-assisted ignition with different initial temperatures at 1 atm. Note that the thermal ignition in this study was to increase the temperature by adding the energy of plasma-generated species (enthalpy changes). It is seen clearly the effective enhancement of NSD and hybrid NSD/DC discharge in promoting $H_2/O_2/He$ ignition, especially at low temperatures. Moreover, compared with the NSD, the addition of a DC electric field in the hybrid discharge further shortened the ignition delay time. For autoignition and thermal ignition below 800 K, ignition did not occur in most conditions. This was because the chain-initiation and chain-branching reactions were too slow at low temperatures to produce active radicals. It is interesting to note that the energy deposited by plasma was the largest at hybrid discharge of 10 Td, however, the ignition enhancement was the most effective at hybrid 20 Td. Therefore, chemical kinetic enhancement by hybrid plasma is more important than thermal enhancement. It was shown that besides the excitation of $O_2(v)$, $H_2(v)$ and $O_2(a^1\Delta_g)$, a considerable percentage of plasma energy went to the dissociations of O_2 and H_2 for radical production and ignition enhancement.

To understand the reaction kinetics, Figure 6.13 shows the comparisons of the sensitivity coefficients of the ignition delay time for NSD and hybrid discharge with different DC reduced electric field strengths at 400 K. For electron-impact reactions, Figure 6.13a shows that electron-impact reactions $e + H_2 \rightarrow e + H + H$ (R5.6') and $e + O_2 \rightarrow e + O + O(^1D)$ (R5.8') are the dominant reactions in enhancing ignition in the NSD case, indicating the key roles of O, $O(^1D)$ and H produced under a high E/N. In the hybrid plasma-assisted ignition, however, it was shown that reactions corresponding to electron production reactions show negative sensitivities. In addition, reactions $e + He \rightarrow e + He^*$ and $e + He \rightarrow e + He^{**}$ also have negative sensitivities. Interestingly, it shows that the OH production via $HO_2 + H = OH + OH$ (R5.7) has the largest sensitivity, indicating the key role of H and HO_2 production in the low-temperature $H_2/O_2/He$ ignition by reaction channels other than electron-impact electronic excitations. Due to the fact that $H_2(v = 1)$ and $O_2(a^1\Delta_g)$ were produced efficiently in the hybrid discharge, reactions $O + H_2(v = 1) \rightarrow H + OH$ (R5.16'), $H_2(v = 1) + OH \rightarrow H_2O + H$ (R5.17') and $H + O_2(a^1\Delta_g) \rightarrow O + OH$ (R5.19') show a positive sensitivity (promotive effect) on ignition enhancement. At the same time, Figure 5.14 shows more plasma energy goes to the dissociation of O_2 and H_2 above 10 Td. The production of O, H, and OH with higher E/N further accelerates these corresponding reactions. The addition reaction $H + O_2(a^1\Delta_g)(+M) \rightarrow HO_2(+M)$ also shows an inhibitive effect on ignition enhancement because it quenches the active radicals of $O_2(a^1\Delta_g)$.

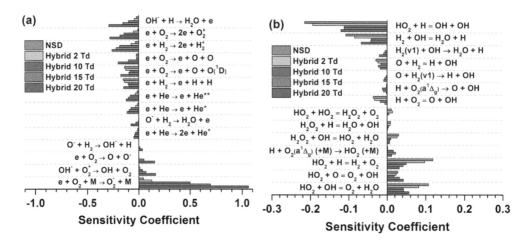

FIGURE 6.13 (a and b) Sensitivity coefficients of the ignition delay time for NSD and hybrid discharge at different DC E/N values and 400 K [16].

Therefore, vibrational excitation $H_2(v = 1)$ and $O_2(v = 1–4)$ as well as electronic excitation of $O_2(a^1\Delta_g)$ at lower E/N in a hyrid plasma during the DC discharge period contribute more to the prodcution O and OH, and thus further promote plasma ignition enhancement. The results of hybrid discharge-assisted H_2/O_2 ignition reveal that an optimized ignition enhancement can be achieved when both excited species and radicals are produced efficiently in a hybrid plasma. The modeling provided useful insight into the plasma-combustion model development and the development of controlled plasma discharge to achieve efficient ignition with optimized non-equilibrium excitation of reactants.

6.2 PLASMA-ASSISTED FLAME PROPAGATION

6.2.1 Flame Propagation Speed

For a premixed mixture, a propagating or a stabilized flame will be formed after ignition. Flame is the most important process to convert the chemical energy in fuel to heat in an industrial combustion system such as internal combustion engines, gas turbines, detonation engines, and rockets. A flame is an exothermic, luminescent, diffusion-ignition front that consumes fuel and produces heat via a chain-branching reaction process. Therefore, it involves diffusion, ignition, and exothermic reactions. Figure 6.14a shows a schematic of a planar premixed flame structure propagating in a quiescent mixture. It is seen that the heat is diffused from the high-temperature flame zone to the unburned premixture. The fuel and oxygen are diffused into a thin reaction zone with a thickness of δ_f, and then auto-ignited and consumed in the reaction zone. As a result, this self-sustaining diffusion and reaction front will propagate into the unburned mixture with a laminar flame speed, S_L. Therefore, the laminar flame speed will depend on chemistry, convection, and diffusion.

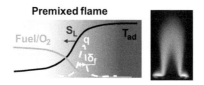

FIGURE 6.14A Schematic of a premixed flame structure.

If we use the same one-step reaction in Eq. 6.1 and assume the flame propagation is in steady state, we can write the fuel and energy conservation equations as,

$$\rho u \frac{dY_F}{dx} = \rho D \frac{d^2 Y_F}{dx^2} - M_F B \rho Y_F e^{-\frac{E_a}{RT}} \qquad (6.13a)$$

$$\rho u C_p \frac{dT}{dx} = \lambda \frac{d^2 T}{dx^2} + Q B \rho Y_F e^{-\frac{E_a}{RT}} \qquad (6.13b)$$

with boundary conditions,

$$T(-\infty) = T_0, \quad Y_F(-\infty) = Y_{F0}$$

$$T(\infty) = T_{ad}, \quad Y_F(\infty) = 0$$

Adding Eq. 6.13a divided by M_F to Eq. 6.13b by Q and then integrating the resulting equation from $-\infty$ to ∞, we can obtain the adiabatic flame temperature as,

$$T_{ad} = T_0 + \frac{Q Y_{F0}}{C_p M_F} \qquad (6.14)$$

Also, by integrating the same equation from x to the end of the flame front δ_f^+ in the reaction-diffusion zone and neglecting the convection terms in Eq. 6.13, we also have

$$\frac{d}{dx}\left(\frac{\lambda}{q} T + \frac{\rho D}{M_F} Y\right)_x^{\delta_f^+} = 0 \qquad (6.15)$$

Substituting Eq. 6.14 into 6.15, we obtain

$$\frac{T - T_{ad}}{T_{ad} - T_0} = -\frac{1}{Le} \frac{Y_F}{Y_{F0}}, \quad Le \equiv \frac{\lambda}{\rho C_p}/D \qquad (6.16)$$

Here Le is the Lewis number, a ratio between the thermal diffusivity to the mass diffusivity.

In the reaction-diffusion zone, the mass conservation equation of fuel Eq. 6.13a reduces to,

$$\frac{d^2 Y_F}{dx^2} = -\frac{M_F}{\rho D} \omega, \quad \omega \equiv B \rho Y_F e^{-\frac{E_a}{RT}} \qquad (6.17a)$$

The above equation can be rewritten as

$$d\left[\frac{dY_F}{dx}\right]^2 = -2 \frac{M_F}{\rho D} \omega dY_F \qquad (6.17b)$$

By using normalized parameters of

$$\alpha \equiv \frac{T_{ad} - T_0}{T_{ad}}, \quad \beta \equiv \frac{T_{ad} - T_0}{T_{ad}} \frac{E_a}{RT_{ad}}, \quad \varsigma \equiv \frac{\beta}{Le} \frac{Y_F}{Y_{F0}} \qquad (6.18)$$

And integrating Eq. 6.17b across the flame,

$$\int_{\delta_f^-}^{\delta_f^+} d\left[\frac{dY_F}{dx}\right]^2 = -2\frac{M_F}{\rho D}\int_{Y_F^-}^{Y_F^+} \omega\, dY_F \quad (6.19)$$

Since we know

$$\frac{dY_F}{dx}\bigg|_{\delta_f^+} = 0 \text{ and } Y_F\big|_{\delta_f^+} = 0 \quad (6.20)$$

By further assuming constant density and using ς to replace Y_F, Eq. 6.19 becomes

$$\left(\frac{dY_F}{dx}\bigg|_{\delta_f^-}\right)^2 = -2\frac{\rho BM_F}{\rho D}\left(\frac{Y_{F0}\text{Le}}{\beta}\right)^2 e^{-E_a/RT_{ad}}\int_{-\infty}^{0}\varsigma e^{-\varsigma}\,d\varsigma = -2\frac{\rho BM_F}{\rho D}\left(\frac{Y_{F0}\text{Le}}{\beta}\right)^2 e^{-E_a/RT_{ad}} \quad (6.21)$$

This is the boundary condition of the fuel mass fraction gradient upstream of the flame front.

Integrating Eq. 6.13a of the convection-diffusion region from negative infinity to the upstream of the flame front and neglecting the reaction term, we have

$$\int_{-\infty}^{\delta_f^-} \rho u\frac{dY_F}{dx}dx = \int_{-\infty}^{\delta_f^-} \rho D\frac{d^2Y_F}{dx^2}dx \quad (6.22)$$

Using the boundary conditions of

$$Y_F\big|_{\delta_f^-} = 0 \text{ and } \frac{dY_F}{dx}\bigg|_{-\infty} = 0 \quad (6.23)$$

Equation 6.22 becomes,

$$\rho u Y_{F0} = \rho D\frac{dY_F}{dx}\bigg|_{\delta_f^-} \quad (6.24)$$

Substituting Eq. 6.21 to Eq. 6.24, we obtain the laminar flame speed relation of the mass burning rate as

$$(\rho S_L)^2 \equiv (\rho u)^2 = 2\text{Le}\frac{\rho BM_F \lambda}{\beta^2 C_p}e^{-E_a/RT_{ad}} \quad (6.25)$$

Although the above result of flame speed is from the one-step and single reactant reaction model (6.1), similar results for one-step and two reactants with an arbitrary reaction order of fuel and oxidizer can be derived in a similar way [20]. It is seen from Eq. 6.25 that the laminar flame speed depends on, (1) adiabatic flame temperature (T_{ad}), (2) the activation energy (E_a), (3) the reaction rate (B), (4) the Lewis number (Le), and the flow field. What are the roles of plasma in affecting flame speeds? Plasma can break down the fuel molecules to smaller ones to reduce the fuel Lewis number. In addition, like the ignition case, plasma can increase the reaction rate, reduce the activation energy, and raise the adiabatic flame temperature with both chemistry effect and thermal effect. Note that as we discussed before, flame chemistry at high temperature can produce radicals very fast by itself. Therefore, the role of plasma is more about to increase the thermal heating to increase the flame speed [21].

Many experiments have been carried out to measure the laminar flame speed for various fuels [22]. Figure 6.14b shows an experiment of the measured and simulated laminar flame speed of methane as a function of equivalence ratio at different pressures. Figure 6.14b shows that the laminar flame speed peaks near the rich side of the stoichiometric condition ($\phi = 1$) but decreases with the increase of pressure (as seen from Eq. 6.25). This is true for other hydrocarbon fuels. For hydrogen

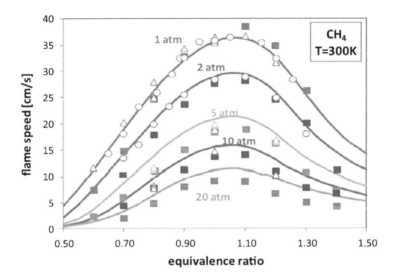

FIGURE 6.14B Comparison measured and predicted methane/air flame speed as a function of equivalence ration at different pressures [22].

mixtures, however, the flame speed initially increases with and then decreases after reaching a maximum with the pressure [23]. This is happening because of the unique pressure dependence of the HO_2 chemistry of hydrogen discussed in Chapter 5 (Section 5.1.2).

6.2.2 Effect of Electric Field and Plasma on Flame Propagation Speed

There have been numerous attempts to use electric field and plasma to enhance flame propagation speeds [21,24,25]. Today, it remains in debate (to some extent) whether electric field and plasma can enhance flame propagation speed non-thermally or kinetically significantly.

It is well known that flame itself is a weak plasma [26,27] with electron and ion density in the order of $10^{17}/m^3$. Therefore, when a low electric field (without causing breakdown) is imposed on a flame it results in two outcomes. Firstly, electrons will gain velocity and increase temperature. The collisions between electrons and neutral molecules will transfer the electron energy to molecules and lead to gas heating. The electron energy exchange with neutral gas via collision is,

$$\sigma E^2 = \frac{3}{2}\alpha v_e n_e k_B (T_e - T) \tag{6.26}$$

where $\sigma = (n_e e^2/m_e v_e)$ is the plasma conductivity, v_e the electron collision frequency with molecules, m_e the electron mass, E the electric field strength, α the fractional loss of energy per collision, k_B the Boltzmann constant, T_e the electron temperature, and T the gas temperature. Secondly, the ions in the flame or created by electron ionization will be accelerated by the electric field and induce a flow motion: the ionic wind. The gas heating due to electron collisions will increase the flame speed thermally. The ionic wind will change the flow velocity in the combustion zone and modify the flame shape, temperature distribution, fuel/air mixing, and soot formation. The question becomes whether the active species production such as ions and radicals due to electron-impact ionization and dissociation can be strong enough to enhance flame speed kinetically. As discussed in Chapter 5 (Sections 5.1.2–5.1.4), if the flame temperature is above 1,200 K, which is true for almost all hot flames (HFs), the active species production by plasma is not comparable with that by the combustion chain-branching reactions in flames.

Early studies of the effect of electric field on flames by Bradley and Nasser [28] showed that the ionic wind dramatically increased the flame stability and the increase of electron temperature also increased the flame temperature. Various studies of electric field effect on flames using DC, AC, and high-frequency (MHz) electric fields as well as microwaves (GHz) for combustion enhancement were conducted [27,29–32]. Unfortunately, earlier results in the 1970s were not very conclusive as to whether an electric field can increase flame speed kinetically because of the complication induced by the coupling between the ionic wind and the gas heating and kinetic effect. To remove the effect of ionic wind, high-frequency microwaves were later used to examine the impact of electric field effect on flame speed. The experiment by Maclatchy et al. [27] used a microwave at a frequency of 2.7 GHz for atmospheric propane-air flame at an electric field strength of 200 V/cm. The results showed a non-discernible increase in flame speed even though the electron temperature was significantly increased. By using microwave and further increasing the electric field strength close to the breakdown limit (~2,000 V/cm) of the burned gas, an enhancement of flame speed up to 20% was observed for ethylene-air flames only near the lean limit [33]. It was concluded that flame speed enhancement by an electric field was only possible for very lean mixtures at near breakdown electric fields.

A recent experiment [34,35] on microwave combustion enhancement using 2.45 GHz high power microwave input (up to 3.4 KW) and a stagnation flame in a high-quality-factor resonator (Figure 6.15) successfully demonstrated that the flame speed of the methane–air mixture at equivalence ratio of 0.7 was increased by approximately 35% at an input power of 800 W (Figure 6.15). Similar increases of flame speeds of lean propane- and ethylene-air flames were also observed. Note that the absorbed power by the flame is below 10% of the input microwave power.

To understand the mechanism of the observed flame speed enhancement by microwave, a numerical simulation was conducted for microwave-assisted laminar flame propagation [36]. The distribution of the accumulated fraction of electron Joule heating (q_e) in comparison to the total chemical heat release from combustion (q_{rt}) in a planar propagation flame,

$$\frac{q_{e0-x}}{q_{rt}} = \frac{\int_0^x \dot{q}_e \, dx}{\int_0^\infty \dot{q}_r \, dx} \tag{6.27}$$

is shown in Figure 6.16 for a stoichiometric methane–air flame at an electric filed strength of $E = 900$ V/cm. It is seen that at the flame downstream location of $x = 0.8$ cm, the accumulated fraction

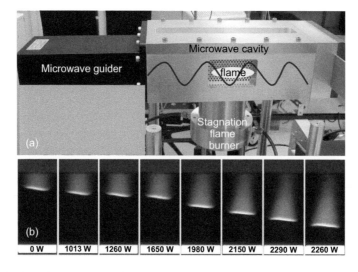

FIGURE 6.15 (a) Experimental setup of microwave combustion enhancement using 2.45 GHz wave input (up to 3.4 kW) for a stagnation flame in a high-quality-factor microwave resonator, (b) images of flame front moving upstream (downward) with the increase of microwave power [34].

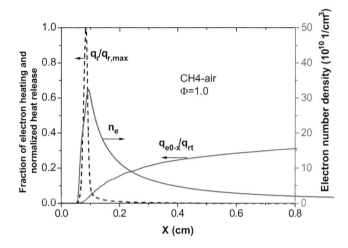

FIGURE 6.16 Distributions of the accumulated fraction of electron heating in total chemical heat release, the chemical heat release, and the electron number density at $E = 900$ V/cm.

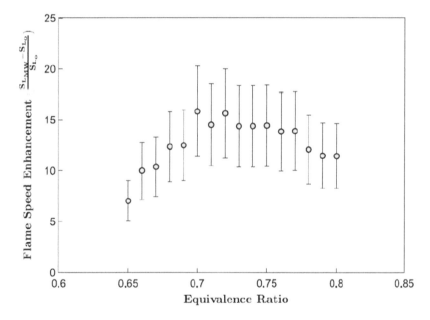

FIGURE 6.17 Variation in flame speed enhancement with equivalence ratio in a premixed CH_4/air flame [35].

of electron heating is about 16% of the total chemical heat release. However, the energy absorbed before the ending of the flame zone is less than 5%, which means that less than 30% of the microwave energy absorbed energy was directly added to the unburned mixture and the flame zone to raise the flame temperature and speed and that much of the absorbed microwave energy was only to heat the burned gas. The reason is that the electron number density is very low in front of the flame and most electrons created in the flame only exist in the burned region (Figure 6.16).

Stockman et al. [35] examined the microwave enhancement on flame speed for different fuel lean methane/air mixtures using the experimental setup in Figure 6.15 and the results are shown in Figure 6.17. A peak increase in flame speed of 16% occurred at the equivalence ratio of 0.72. For leaner mixtures, the flame ionization level was low so the microwave heating effect was smaller.

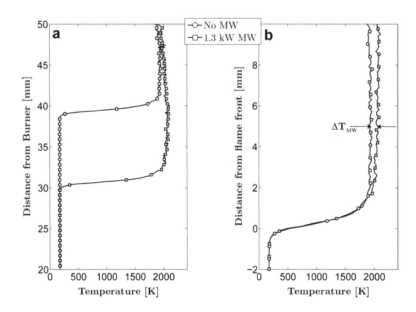

FIGURE 6.18 Measured temperature profiles demonstrating the change in temperature when microwave energy is coupled to enhance the flame speed of $\phi = 0.75$, CH_4/air flame. (a) Absolute coordinates showing enhanced flame position and equilibration of temperature. (b) Flame-shifted coordinate system displaying ~125 K increase in temperature just out of the flame front [35].

At higher equivalence ratios, the flame speed and heat release rate were greater so that the microwave heating effect was reduced. The microwave heating effect was further investigated by using filtered Rayleigh scattering for temperature measurements across the flame along the centerline of the stagnation flow. A measured temperature profile of the flame at an equivalence ratio of 0.75 is shown in Figure 6.18 where the parameter, ΔT_{MW}, corresponds to the temperature increase associated with the application of the microwaves. It is seen that, like the prediction in Figure 6.16, there was no observable increase in the temperature of the cold unburned reactant gases of the microwave-assisted flame. Therefore, the flame speed enhancement did not arise from the microwave heating of the unburned mixture. However, there was a significant increase in temperature throughout the post-flame zone. The temperature of microwave-enhanced flames was higher than that without microwave. This clearly indicates that microwave heating effect occurred mainly in the burned gas region. The measured increases in a post-flame gas temperature of 100–200 K are indicative of the mechanism of significant flame speed enhancement by the thermal effect. Diagnostics with OH PLIF also showed an increase in peak OH-number density and suggested that the enhanced flame speed came mainly from microwave heating of the burned gas where electron number density was higher (Figure 6.16).

Therefore, summarizing the results in Figures 6.16 and 6.18, one can conclude that the electric field or microwave effect on flame propagation speed before breakdown is mainly from the thermal heating effect. There are two types of electric field induced Joule heating which increase the flame temperature, one is the direct heating in the unburned flame region and flame zone (upstream) and the other is the conductive heating in the burned gas region due to the temperature gradient created by the microwave. By using large activation energy analysis and assume that the microwave heating effect is small, the normalized flame speed enhancement by microwave heating to the adiabatic flame speed can be given as

$$m^2 \ln(m^2) \equiv q^- + q^+, \, m = (\rho S_L)/(\rho S_L)_{ad} \quad (6.28)$$

Plasma-Assisted Combustion: Dynamics

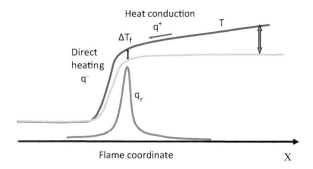

FIGURE 6.19 Schematic of temperature distribution with and without microwave heating effect in a flame. q^-: microwave heating in the unburned zone. q^+: conductive heat transfer from microwave heating in the burned zone to the flame.

Therefore, when q^- and/or q^+ are greater than zero, the normalized mass burning rate, m, will be greater than unity, indicating an increase of flame speed.

More experimental studies on plasma-assisted flame stabilization and propagation were conducted by using plasma breakdown [24,37,38]. Many experiments reported enhanced flame stabilization, lean burn limits, and heat release rate with plasma breakdown. However, it was not clearly demonstrated whether the plasma-enhanced flame stabilization effect was due to thermal effect or kinetic effect. To answer this question, Wolk et al. [38] investigated the enhancement of laminar flame development using microwave-assisted spark ignition for methane–air mixtures at a range of initial pressures and equivalence ratios in a constant volume combustion chamber. Microwave enhancement was evaluated based on two several parameters at two different flame propagation stages using the flame development time (time for the initial 0%–10% of total net heat release of flame kernel) and the flame rise time (time for 10%–90% of total net heat release after the initial flame kernel stage). The results showed that compared to a capacitive spark ignition (SI), microwave-assisted SI extended the lean and rich ignition limits at all pressures investigated. As shown in Figure 6.20a, the addition of microwave discharge reduced the initial flame development time and increased the flame kernel size for all equivalence ratios tested and resulted in increases in the spatial flame speed for sufficiently lean flames. It concluded that flame enhancement might be caused by a non-thermal chemical kinetic enhancement from energy deposition to free electrons in

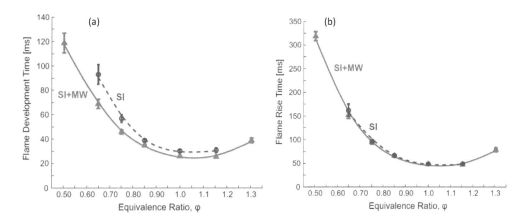

FIGURE 6.20 (a) Flame development time for the SI and SI + MW modes as a function of equivalence ratio at an initial pressure of 1.08 bar and 300 K. (b) Flame rise time for the SI and SI + MW modes as a function of equivalence ratio at an initial pressure of 1.08 bar and 300 K. The flame rising time is unchanged between the two modes [38].

the flame front and induced flame wrinkling from excitation of flame (plasma) instability. However, it was also shown that the enhancement of flame development time by microwaves diminished as the initial pressure of the mixture was increased above 3 atm. Moreover, as shown in Figure 6.20b, no enhancement was observed for the flame rise time. These results implied that the non-thermal effects of microwave were negligible at high pressure for the flame kernel development and there was no effect on flame speed after a normal flame structure was developed. Therefore, it remains not conclusive whether the observed increase of the initial flame kernel size was due to non-thermal kinetic effect or due to plasma heating or instability (plasma instability).

The question becomes why so many experiments have reported that flame stabilization was enhanced, but few data showed that the observed enhancement of flame stabilization was caused by the kinetic effects. Is it possible that plasma can enhance flame speeds via kinetic effect? The answer is yes, for example, the case of ignition kernel development [39], ignition-assisted flame [40], and plasma-enhanced CFs [41,42]. The question really is to what extent and in what conditions.

6.2.3 Ignition-Assisted Flame Propagation

As discussed in Chapter 5, plasma can significantly enhance autoignition kinetically. In turbulent combustion, detonation, mild combustion, cool flames, and warm flames. The unburned mixtures are preheated by turbulent mixing, shock compression, and burned gas preheating. Therefore, the ignition delay time, t_{ig}, of the unburned mixture is shortened significantly with the increase of the mixture initial temperature. If a plasma discharge is added to these flames, then the autoignition delay time (t_{ig}) will be reduced so much that it becomes comparable to the characteristic combustion time in the flame zone ($t_f = \delta_f/S_L$). As a result, one of the following criteria of the ignition Damköhler number for ignition-assisted flame may be satisfied,

$$\text{Da}_{ig} = \frac{t_{res}}{t_{ig}(T_0, p, \phi)} \sim O(1) \tag{6.29a}$$

$$\text{Da}_{fig} = \frac{\delta_f / S_L}{t_{ig}} \sim O(1) \tag{6.29b}$$

The schematic of the flame structure of an autoignition-assisted flame is shown in Figure 6.21a. There are four different zones in the structure of ignition-assisted flame propagation [43]. In region I, this is a convection-autoignition zone where the mixture with a high ignition Damköhler number undergoes pre-flame fuel oxidation and autoignition processes and the temperature increases from T_0 to T_p. In region II, this is a thin convection-diffusion zone where more heat is diffused upstream and further raises the temperature to autoignition temperature (T_i) and accelerates fuel oxidation. Region III is the reaction-diffusion zone where fuel is burned and most of the heat is generated and raises the flame temperature to T_f. Region IV is the after flame burned gas zone.

For simplicity, we assume that autoignition in region I only modifies the mixture temperature and that other changes such as compositions are negligible. We also assume the constant thermal dynamic properties and a one-step global reaction with the activation temperature, T_a (E_a/R). The energy equation based on the flame front coordinate in the convection-autoignition zone I can be written as:

$$\frac{DT}{Dt} = u\frac{dT}{dx} = \dot{\omega} = AT^b \exp\left(-\frac{T_a}{T}\right), \quad x_0 < x < x_p \tag{6.30a}$$

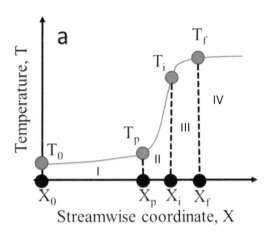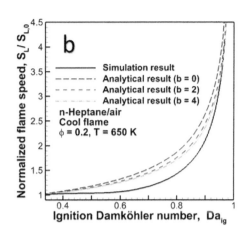

FIGURE 6.21 (a) Illustration of flame structure for the theoretical model of autoignition-assisted flame. (b) the comparison for the simulation result and theoretical analysis result for the autoignition-assisted lean cool flame at $T_0 = 650$ K, $P = 1$ atm, $\phi = 0.2$, $S_{L,Da_{ig}=0} = 27.2$ cm/s and $T_a = 1871$ K [43].

Where $\dfrac{D}{Dt}$ represents material derivative, A is the reduced pre-factor of the reaction rate including the specific heat and density, and b is a constant. Equation (6.30a) can be rewritten in Lagrangian coordinate fixed on a certain mixture pocket:

$$\frac{dT}{dt} = AT^b \exp\left(-\frac{T_a}{T}\right),\ 0 < t < t_p \tag{6.30b}$$

where t_p is the residence time for the mixture flowing through convection-autoignition zone (x_0, x_p). Integrating Eq. 6.30b from $T_0|_{t=0}$ to $T_p|_{t=t_p}$, we have

$$\int_{T_0}^{T_p} A^{-1} T^{-b} \exp\left(\frac{T_a}{T}\right) dT = t_p \tag{6.31a}$$

Equation (6.30b) can also be used to estimate the ignition delay time t_{ig} under constant pressure:

$$t_{ig} \approx \int_{T_0}^{T_i} A^{-1} T^{-b} \exp\left(\frac{T_a}{T}\right) dT \tag{6.31b}$$

Here T_i is the ignition temperature. By substituting Eqs. 6.31a and 6.31b into Eq. 6.29, we can have:

$$\text{Da}_{ig} \int_{T_0}^{T_i} A^{-1} T^{-b} \exp\left(\frac{T_a}{T}\right) dT = \int_{T_0}^{T_p} A^{-1} T^{-b} \exp\left(\frac{T_a}{T}\right) dT \tag{6.32}$$

An explicit solution of the autoignition preheating temperature, T_p, can be estimated directly from Eq. 6.32 by using a one-step reaction model. For example, if $b = 2$, a simple explicit expression of T_p is given as:

$$T_p = T_a / \ln\left[(1 - \text{Da}_{ig}) e^{T_a/T_0} + \text{Da}_{ig} e^{T_a/T_i}\right] \tag{6.33}$$

Furthermore, by using the expression for flame speed from the classical laminar flame theory using two reaction zone approximation [44,45] (a simplified form of Eq. 6.25), we can have:

$$S_L = \sqrt{\frac{\lambda}{c_p} \frac{T_f - T_i}{T_i - T_p} \frac{\dot{\omega}}{\rho}} \qquad (6.34)$$

In the limit of large activation energy ($T_i \approx T_f$), we can approximate:

$$\frac{T_i - T_0}{T_i - T_p} \approx \frac{T_f - T_0}{T_f - T_p} \qquad (6.35a)$$

$$\mathrm{Da_{ig}} e^{T_a/T_i} \approx \mathrm{Da_{ig}} e^{T_a/T_f} \qquad (6.35b)$$

Therefore, the laminar flame speed for autoignition-assisted flame becomes,

$$S_{L,\mathrm{Da_{ig}}} = \sqrt{\frac{\lambda}{c_p} \frac{T_f - T_i}{T_i - T_p} \frac{\dot{\omega}}{\rho}} = \sqrt{\frac{\lambda}{c_p} \frac{\dot{\omega}}{\rho} \frac{T_f - T_i}{T_i - T_0}} \sqrt{\frac{T_i - T_0}{T_i - T_p}} = S_{L,\mathrm{Da_{ig}}=0} \sqrt{\frac{T_f - T_0}{T_f - T_p}} \qquad (6.36)$$

As a result, an analytical solution of ignition-assisted flame speed can be achieved by combining Eqs. 6.33 and 6.36 directly,

$$S_L = f\left(\mathrm{Da_{ig}}, b, S_{L,\mathrm{Da_{ig}}=0}, T_a\right) \qquad (6.37)$$

Here the right-hand side of Eq. 6.37 is an explicit function of the ignition Damköhler number, $\mathrm{Da_{ig}}$, while $S_{L,\mathrm{Da_{ig}}=0}$ is the laminar flame speed at zero ignition Damköhler number (Eq. 6.25).

To demonstrate the impact of autoignition on the flame propagation speed, the laminar flame speeds and structures of near-limit autoignition-assisted n-heptane/air cool flames at different ignition Damköhler numbers, elevated temperatures and pressures were studied computationally and analytically over a broad range of equivalence ratios. Figure 6.21b shows that Eq. 6.36 predicts an exponential increase of the flame propagation speed with the ignition Damköhler number. Figure 6.21 also shows the comparison between the simulated and the analytical results with $b = 0$, 2, and 4, respectively, for flame speeds. Both the numerical and analytic results show that the normalized cool flame speed increases nonlinearly with the ignition Damköhler number and asymptotically goes to infinity as $\mathrm{Da_{ig}} \to 1$. The discrepancy between the analytical result and numerical result may be caused by the change of species and activation energy in the process of autoignition.

The significant increase of flame speed due to enhanced ignition in Figure 6.22 suggests that if plasma can kinetically shorten the ignition delay time so much that the criterion of ignition-assisted flame propagation (Eq. 6.29) is met, then plasma can kinetically enhance flame speed. This result may provide the explanation to the experimentally observed flame acceleration by microwave (Figure 6.20a) at the initial ignition kernel stage and enhanced flame stabilization in a recirculation zone [24,37]. In addition, it supports the experimental results in Figure 6.10 in which when ignition is enhanced by plasma the flame extinction limit disappears (exponential increase of the flame speed due to autoignition in Figure 6.21b). Therefore, to enhance flame stabilization and speed kinetically by using plasma, one needs to design the plasma discharge under the condition that the autoignition delay time is accelerated and becomes comparable to the flame burning time (Eq. 6.29). Otherwise, mostly the thermal effect will be observed.

6.2.4 Extinction and Flammability Limits of Stretched Flames

In the above sections, we discussed two things: the thermal effect and the kinetic effects of plasma chemistry on ignition and flame speed. We only used planar flames as examples. However, as discussed in Figure 5.15, in addition to the thermal effect, plasma can also create transport effect in

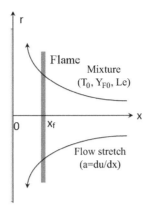

FIGURE 6.22 Schematic of a premixed counterflow flame (right nozzle only) with a flow stretch rate, a, and mixture Lewis number, Le. The stagnation plane is located at $x = 0$.

affecting flames by changing the Lewis number effect with fuel fragmentation or reforming and the flow stretch via ionic wind or flow turbulization. Therefore, it is necessary to understand how the transport process will affect flame speed and flame burning limit.

To understand the transport effect on flames, it is convenient to use a premixed counterflow flame (Figure 6.22) to examine the effect of the fuel Lewis number (Le) and flow stretch rate (a) as well as the activation energy (E_a) on the dynamics of the flame [46,47]. The premixture with an initial temperature of T_0, fuel Lewis number of Le, and fuel mass fraction of Y_{F0} is impinged toward each other in the opposite direction from two nozzles. A stagnation plane is formed at $x = 0$ (Figure 6.22) and the flow stretch is defined by the gradient of the velocity in the x direction, $a = du/dx$. By assuming the constant properties, steady-state flow, and a one-step thin-flame reaction model (delta function), the normalized governing equations of temperature and fuel mass fraction in Figure 6.22 with a flame at $x = x_f$ can be written as [46],

$$-\rho C_p ax \frac{dT}{dx} = \lambda \frac{d^2 T}{dx^2} - Q_r (T - T_0) + Q Y_{F0} B e^{-E_a/2RT_f} \delta(x - x_f) \quad (6.38a)$$

$$-\rho ax \frac{dY_F}{dx} = \rho D \frac{d^2 Y_F}{dx^2} - Y_{F0} B e^{-E_a/2RT_f} \delta(x - x_f) \quad (6.38b)$$

with boundary conditions of

$$x \to \infty, \quad T = T_0, \quad Y_F = Y_{F0}; \quad x = 0, \quad \frac{dT}{dx} = 0, \quad \frac{dY_F}{dx} = 0$$

where x is the streamwise coordinate, Q_r the heat loss from thermal radiation or heat gain by plasma, Q the chemical heat release rate, λ and D the thermal conductivity and mass diffusivity, respectively, a the stretch rate, T_f the flame temperature, and δ the Dirac delta function, respectively,

$$Q_r = 4K_p \sigma \left(T_{ad}^3 - T_0^3\right), \quad a = \frac{du}{dx}, \quad T_{ad} = T_0 + \frac{QY_{F,0}}{C_p}, \quad \rho S_{L,ad} = B e^{-E_a/2RT_{ad}} \quad (6.39)$$

where K_p and σ are the Planck absorption coefficient and the Stefan–Boltzmann constant, respectively. $S_{L,ad}$ is the adiabatic flame speed and B the reaction rate coefficient.

By using the normalized variables,

$$X = \frac{x}{x_{\text{ref}}}, \quad \theta \equiv \frac{T-T_0}{T_{\text{ad}}-T_0}, \quad y_F = \frac{Y_F}{Y_{F0}}, \quad q = \frac{QY_{F0}}{C_p T_0}, \quad x_{\text{ref}} = \frac{\lambda/\rho C_p}{S_{L,\text{ad}}}, \quad t_{\text{ref}} = \frac{x_{\text{ref}}}{S_{L,\text{ad}}}$$

and

$$\eta = X\sqrt{\frac{at_{\text{ref}}}{2}}, \quad \text{Le} = \frac{\lambda/\rho C_p}{D}, \quad H = \frac{\beta 4\sigma K_p T_{\text{ad}}^3}{\rho C_p S_{L,\text{ad}}} \frac{\lambda}{\rho C_p S_{L,\text{ad}}}, \quad \beta = \frac{E_a}{RT_0}\frac{T_{\text{ad}}-T_0}{T_{\text{ad}}}$$

Here η is the stretch normalized coordinate, Le the Lewis number, H the normalized heat loss or gain, and β the reduced activation energy.

By using the above nondimensionalized variables, Eq. 6.38 can be rewritten as

$$2\eta\frac{d\theta}{d\eta} + \frac{d^2\theta}{d\eta^2} - \frac{2}{a}\frac{H\theta}{\beta} + \sqrt{\frac{2}{a}}e^{\beta(\theta_f-1)/2}\delta(\eta-\eta_f) = 0 \quad (6.40a)$$

$$2\eta\frac{dy_F}{d\eta} + \frac{1}{\text{Le}}\frac{d^2 y_F}{d\eta^2} - \sqrt{\frac{2}{a}}e^{\beta(\theta_f-1)/2}\delta(\eta-\eta_f) = 0 \quad (6.40b)$$

Note here a is normalized by t_{ref}. In the large activation energy limit ($\beta \to \infty$), variables in above equation can be expanded by introducing small perturbation variables,

$$\theta = \theta^0 + \frac{\theta^1}{\beta} + \cdots, \quad y_F = y_F^0 + \frac{y_F^1}{\beta} + \cdots, \quad p = \theta^1 + y_F^1 + \cdots, \quad \text{Le} = 1 + l/\beta + \cdots \quad (6.41)$$

By substituting Eq. 6.41 into Eq. 6.40, solutions of the leading order variables of θ^0 and y_F^0 are obtained as,

$$\theta^0 = \frac{\int_{-\infty}^{\eta} e^{-t^2} dt}{\int_{-\infty}^{\eta_f} e^{-t^2} dt}, \quad y_F^0 = 1 - \int_{-\infty}^{\eta} e^{-t^2} dt \Big/ \int_{-\infty}^{\eta_f} e^{-t^2} dt \quad \text{for } \infty \geq \eta \geq \eta_f \quad (6.42a)$$

$$\theta^0 = 1, \quad y_F^0 = 0 \quad \text{for } \eta_f \geq \eta \geq 0 \quad (6.42b)$$

By integrating the energy equation in Eq. 6.40a across the flame in the reaction-diffusion zone, we can obtain a jump condition across the flame zone as,

$$\left[\frac{d\theta}{d\eta}\right]_-^+ = \sqrt{\frac{2}{a}}e^{\theta_f^1/2} \quad (6.43)$$

Similarly, by adding Eq. 6.40a to Eq. 60.40b and integrating it across the reaction-diffusion zone, we can obtain another jump condition,

$$\left[\frac{dp}{d\eta} - l\frac{dy_F^0}{d\eta}\right]_-^+ = 0 \quad (6.44)$$

Substituting the expansion relations in Eq. 6.41 to the governing equation of Eq. 6.40 and only keeping the first-order perturbations of temperature and fuel concentration, we can obtain a small perturbation equation,

$$2\eta \frac{dp}{d\eta} + \frac{d^2 p}{d\eta^2} - l\frac{d^2 Y_F^0}{d\eta^2} - \frac{2}{a}\frac{HT^0}{\beta} = 0 \qquad (6.45)$$

with boundary condition of,

$$p = 0 \text{ at } \eta = 0, \text{ and } \infty \qquad (6.46)$$

The above equation has an analytical solution that gives the normalized flame mass burning rate, m, and the flame location, η_f, as a function of stretch rate, a, heat loss or gain (H), and the reduced Lewis number (l).

$$m = \eta_f \sqrt{2a} \qquad (6.47)$$

$$\frac{-\sqrt{a}}{\sqrt{2}\eta_f g_1} = e^{p_f/2} \qquad (6.48)$$

where the flame enthalpy change, p_f, is given as,

$$p_f = l\left(-\frac{1}{2} - \eta_f^2 - \frac{1}{g_1}\right) + \frac{2H^+}{a}\eta_f^2 g_1 g_2 + \frac{H^-}{a}\left(-g_3 + I_1/g_1\right) \qquad (6.49a)$$

$$g_1 = \int_\infty^1 e^{(1-n^2)\eta_f^2} dn, \quad g_2 = \int_0^1 e^{(n^2-1)\eta_f^2} dn, \quad g_3 = \int_\infty^1 \frac{1 - e^{(1-n^2)\eta_f^2}}{n^2 - 1} dn \qquad (6.49b)$$

$$I_1 = \int_\infty^1 e^{(1-n^2)\eta_f^2} \int_\infty^1 \frac{1 - e^{(1-k^2)n^2\eta_f^2}}{k^2 - 1} dk\, dn \qquad (6.49c)$$

where m is the normalized flame speed [46].

In Eq. 6.49, the first term on the right-hand side represents the Lewis number effect, that is, if l is negative (Le < 1), the flame speed will increase. The second and third terms represent the reduction (increase) of flame temperature caused by the radiation heat loss (plasma energy addition) in the burned or unburned region, respectively.

In the limit of $a \to 0$ and $\eta_f \to \infty$, the solution of Eq. 6.47-6.49 reduces to the freely propagating flame, we have,

$$-2\eta_f^2 g_1 = 1$$

Then, Eq. 6.49 gives the flame speed of a freely propagating flame as a function of heat loss or heat addition,

$$m^2 \ln m^2 = -\left(H^- + H^+\right) \qquad (6.28)$$

which recovers the same result in Eq. 6.28.

Therefore, plasma can affect the flame speed and burning limits of stretch flames by varying the activation energy (E_a or β), reducing the Lewis number (Le or l), heat addition (H), and the flow motion (a).

To demonstrate the flame dynamics of a stretched counterflow flame with heat loss, we can use Eq. 6.48 to examine the dependence of flame location, flame speed, and flame temperature as a function of stretch rate at different equivalence ratios. To mimic the CH_4-air combustion, we choose $\rho = 1.0$ kg/m²s, $Q = 4.8 \times 10^7$ J/kg, $C_p = 1,360$ J/kg K, $E_a = 1.25 \times 10^5$ J/mol K, $\lambda = 0.076$ J/mK s, Le = 0.9, $T_0 = 300$ K and the reaction frequency constant $B = 10$ kg/m²s [46]. This choice together with the heat loss H results in adiabatic burning velocities and temperatures at standard flammability limit ($Y_F = 0.0295$, the lean limit of the one-dimensional planar propagating flame) with 3.5 cm/s and 1,340 K and at stochiometric condition ($Y_F = 0.057$) with 37.6 cm/s and 2,300 K, respectively [46].

Figure 6.23 shows a typical dependence of flame location as a function of normalized stretch rate (a) at fuel concentrations, respectively, below ($Y_F = 0.029$), equal to, and above ($Y_F = 0.0296$ and 0.03) the flammability limit ($Y_F = 0.0295$) of the freely propagating planar premixed flame in this study. For $Y_F = 0.029$, it is seen that in a stretched counterflow flame with Le < 1, a flame can exist in a broad range of stretch rates even for a mixture with fuel concentration below the lean flammability limit. It is also seen that when the stretch rate is high, there is a stretch extinction limit at which the flame is quenched near the stagnation plane ($x \to 0$) by the high flow rate (not enough time for combustion). As the stretch rate decreases, the flame will move upstream and increase its temperature and the volume of burned gas region. However, as the stretch rate decreases to $a = 1$, the flame location from the stagnation plane and temperature both reach the maximum. A further decrease of stretch rate will lead to an increase of the thermal radiation (proportional to time and the volume of the burned gas) and a decrease of flame temperature and flame location. As a result, at $a = 0.5$, radiation extinction happens. Therefore, for a stretched flame with Le < 1, a sub-flammability limit flame exists in a range of the stretch rates bounded by the radiation extinction limit and the stretch extinction limit [48–51].

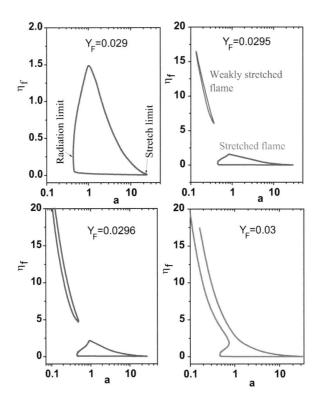

FIGURE 6.23 Flame location as a function of stretch rate at fuel concentrations, respectively, equal to, and above the flammability limit ($Y_F = 0.0295$) of the corresponding planar freely propagating flame. The stretched flame and weakly stretched flame have two extinction limits: radiation extinction limit and stretch extinction limit [46].

As the fuel concentration increases close to the flammability limit ($Y_F = 0.0295$), in Figure 6.23 there exist two flame branches, a weakly stretch flame standing far away from the stagnation plane and the stretched flame existing near the stagnation plane. Note that it is the weakly stretched flame (WSF) that connects to the flammability limit of the freely propagating flame at zero stretch rate.

As the fuel concentration further increases to above the flammability limit, at $Y_F = 0.03$, it is seen that the WSF branch merges with the stretched flame branch. Only under this condition, the stretched flame will reduce to a planar freely propagating flame at zero stretch. Therefore, appropriate understanding of the near-limit stretched flame dynamics is critical to understand the phenomena observed in experiments and plasma-assisted near-limit combustion.

To observe these near-limit stretched flames experimentally, Maruta et al. conducted microgravity experiments using lean methane ($Le < 1$)) and propane ($Le > 1$) air mixtures [51,52]. Figure 6.24 shows the results of the microgravity experiment and numerical modeling of these two fuel mixtures, respectively. For methane, the simulated planar freely propagating flame was at $\phi = 0.485$. However, experiments showed that the stretch flame existed even at $\phi = 0.47$ and was bounded by two extinction limits, a stretch extinction limit at a higher stretch rate and a radiation extinction limit at lower stretch rate. These observations were consistent with the prediction of the theory in Figure 6.23. However, it is interesting to note that the extinction limit of stretched counterflow flames cannot be simply extrapolated to the flammability limit of the freely propagating flame for methane/air mixture ($Le < 1$). Moreover, the flammability limit measured at microgravity at zero stretch rate using spherical flames for methane is different from the experimentally measured extinction limits of counterflow flames. However, for lean propane/air mixtures ($Le > 1$), no sublimit stretched flames were observed and only the stretched extinction limit was observed. Moreover, the measured extinction limit can be linearly extrapolated to the results of 1D planar flame and the spherical flame experiments at zero stretch rate.

To answer the questions above and establish the relationship between stretched flame extinction limits and the flammability limit of the planar free-propagating flames at different Lewis numbers, Ju et al. conducted numerical simulations for methane and propane mixtures with helium dilution to change the Lewis number [48,49,51,54]. By summarizing the analytical results in Figure 6.23, experimental data in Figure 6.24, and the simulation results in Refs. [48,49,51,54], a schematic of the flammable regions and burning limits of three different flame regimes, WSF, normally stretched flame

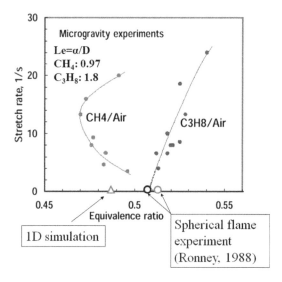

FIGURE 6.24 Measured extinction limits of methane and propane/air mixtures in microgravity counterflow flames [52] in comparison with 1D simulations and the experimental data of spherical flames [53].

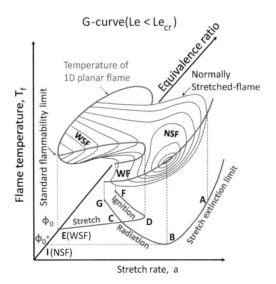

FIGURE 6.25 Schematic of the flammable regions and burning limits of various flame regimes of stretched flames in the coordinates of equivalence ratio, stretch rate, and flame temperature for Lewis number less than a critical value (Le < Le_{cr} > 1). WSF: weakly stretched flame, WF: weak flame, and NSF: normally stretched flame.

(NSF), and weak flame (WF), of stretched flames in relation to the 1D planar unstretched flames in the coordinates of equivalence ratio, stretch rate, and flame temperature is shown in Figure 6.25.

For 1D unstretched freely propagating flames ($a = 0$), Figure 6.25 shows that the flame temperature peaks near the stoichiometric condition and there are two flammability limits, one on the fuel rich side and the other on the fuel lean side at $E(\phi_0)$. In the discussion below, we only focus on fuel lean mixtures. Similar insights can be gained for the fuel rich flames.

For fuel lean stretched flames, note that there are three different flame regimes: (1). WSFs at equivalence ratio slightly above ϕ_0, (2): NSF at equivalence ratios below and above ϕ_0, and WF at equivalence ratios above ϕ_0. It is seen that below ϕ_0, only the NSF exists and is bounded by the radiation extinction limit (BC) and the stretch extinction limit (AB) (as discussed in Figure 6.23). Therefore, the NSF burning limit is ϕ_{0*}, which is lower than ϕ_0 for Le < Le_{cr} (the critical Lewis number). As shown in Figure 6.25, in this case, neither the extension of the radiation limit of BC nor the stretch limit AB to zero stretch leads to the flammability limit of ϕ_0. This explains well the experimental data in Figure 6.24. When the equivalence ratio is above ϕ_0, a new WSF branch appears at low stretch rates and the flammable region of NSF is extended to lower and higher stretch rates because of the increase of heat release rate in higher fuel concentration. Note that the extinction limit of WSF (CE) is caused by the flow stretch and thus is a stretch extinction limit too. As the equivalence ratio further increases, the WSF branch and the NSF branch merge at location D, resulting in a WF region between the reignition limit at FD and the radiation extinction limit at CG. Therefore, an increase and decrease of stretch rate will cause the WF to reignite at FD and become an NSF with reduced heat loss and radiative extinction at CG with increased radiative heat loss. Therefore, the WF is an outcome of thermal radiation. Figure 6.25 clearly shows that only the extension of the weakly stretch flame extinction limit (DCE) to zero stretch rate can lead to the flammability limit of the 1D planar unstretched flame, ϕ_0. This also explains well the simulated 1D flame data in Figure 6.24.

Therefore, if one projects the extinction limits in Figure 6.25 to the plane of equivalence ratio and stretch rate, the flammable region (extinction limits) on this phase plane is a G-curve [48]. This G-curve defines all the limits of the three different stretched flames and their relation to the flammability limit, ϕ_0. As discussed in Refs. [49,51], due to the effect of thermal radiation, the critical

Lewis number Le_{cr} is greater than unity. Moreover, the stronger the radiation heat loss is, the larger the Le_{cr} will be. When the Lewis number is greater than Le_{cr}, the sublimit NSF flame island in Figure 6.25 disappears. As a result, the G-curve becomes a K-curve, and the stretch extinction limits of stretched flames can be extrapolated to zero stretch to obtain the flammability limit ϕ_0, like the case the fuel lean propane/air flames shown in Figure 6.24.

Therefore, understanding the flame dynamics of stretch flames and their relationship to the planar unstretched flames and the flammability limit is important to interpret the data of plasma-assisted combustion for near-limit flames, because plasma changes the heating, the Lewis number, and the stretch rate of the flow.

6.2.5 Plasma-Assisted Flame Propagation and Stabilization by Ozone and Singlet Oxygen

6.2.5.1 Ozone-Assisted Hot Flame Propagation

Ozone and singlet oxygen, $O_2(a^1\Delta_g)$, produced in plasma have long lifetime and strong kinetic effects in combustion (Tables 5.12 and 5.5). There have been many studies to understand whether plasma can enhance flame speeds kinetically by generating active species such as ozone and singlet oxygen.

Early studies of ozone-assisted flames were reported by Lewis and von Elbe [55] and Wilde [56]. These early studies showed that ozone decomposition to atomic oxygen enhanced flame speeds significantly. Streng and Grosse later [57] measured the flame speed of combustion/decomposition of O_3 in O_3/O_2 mixtures in a tube and Bunsen flames (Figure 6.26). It was shown that the flame speed was linearly dependent on the mole fraction of O_3. Nomaguchi et al. [58] measured the flame speed of methane and methanol in ozonized air. The results showed that 5,000 ppm O_3 addition in methane stoichiometric mixture increased the burning velocity by 5% and the enhancement was stronger at fuel lean conditions. Ombrello et al. [59, 60] conducted experiments of ozone and singlet oxygen-assisted lifted flames to understand whether plasma-generated ozone and singlet oxygen can change the lifted flame speed and height of ethylene and propane in air. Later, Wang et al. [61] measured the effect of O_3 on methane/air flame speeds by using a heat flux. More recently, Sun et al. conducted experiments of ozone-assisted ethylene lifted flames [62]. These studies significantly

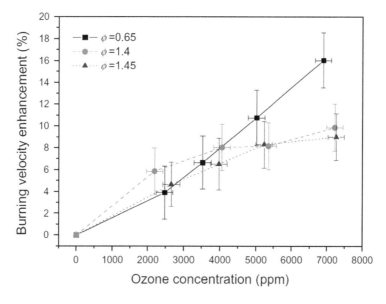

FIGURE 6.26 Measured burning velocity enhancement of methane/air flames as a function of O_3 concentration at different equivalence ratio [61].

advanced the understanding of the thermal and kinetic effects of ozone on flame propagation and stabilization.

In the study of O_3-enhanced flame propagation of methane/air mixture, the experiments by Wang et al. [61] were conducted by using a heat-flux burner. It was shown that the addition of O_3 in the methane/air mixtures enhanced the flame burning velocity through both kinetic and thermal effects. Around 16% enhancement was observed on the lean side ($\phi = 0.65$) but only 9.8% and 9.0% on the rich side with 7,000 ppm ozone addition in the oxidizer. A linear relationship between the enhancement and increase of O_3 concentration at $\phi = 0.65$ was observed. It was found that the amounts of O radicals produced by ozone decomposition in the preheated zone initiating the chain-propagating and branching reactions. The kinetic effects were the main contributor to the burning velocity increase as compared to the thermal effects.

The ozone effects on flame speeds were further studied for larger alkanes and alkenes by Gao et al. [63] for C_2H_4 and C_3H_8. In this study, experiments were carried out by using a high-pressure Bunsen flame facility. (Figure 6.27). Axisymmetric Bunsen flames were produced with a laminar flow nozzle installed in a pressure chamber with optical access for flame imaging. The flow velocity at the nozzle exit velocity was uniform. A sintered plate surrounding the nozzle exit produced a near-stoichiometric, flat, CH_4/air pilot flame to anchor the main flame for flame speed measurements. Flame chemiluminescence images were acquired using an intensified charge-coupled device (ICCD) camera. An image intensity gradient-based edge detection algorithm was used to determine the flame front and the surface area. The laminar flame speed was calculated by dividing the volumetric flow rate of the reactants by the flame surface area.

The experimental results of 6,334 ppm ozone addition on C_3H_8 flame speed are shown in Figure 6.28. It is seen that ozone addition increased the flame speed of propane by 7.6% at the stoichiometric condition. The simulations matched the experiments at all equivalence ratios reasonably well. In addition, the effect of ozone addition on flame propagation was also studied at higher pressures. A significant increase of flame speed enhancement at elevated pressure was observed by comparing it to atmospheric pressure. Experiments showed that the enhancement in the stoichiometric CH_4/air flame speed for 6,334 ppm ozone addition was from 7.7% at 1 atm to 11% at 2.5 atm. Simulations showed that two pressure effects on O_3 reactions accounted for this increased enhancement at high pressure. First, a pressure increase promoted ozone decomposition in the preheat zone.

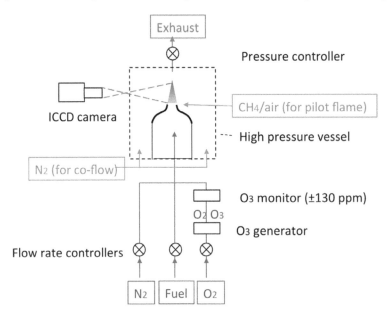

FIGURE 6.27 Schematic of high-pressure Bunsen flame facility [63].

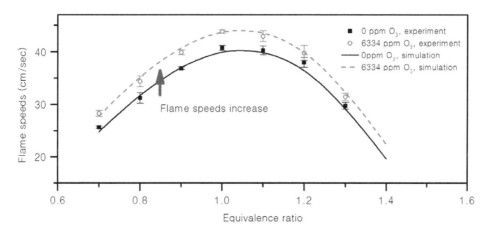

FIGURE 6.28 Effect of O_3 addition on C_3H_8 flame speed at atmospheric pressure and 300 K [63].

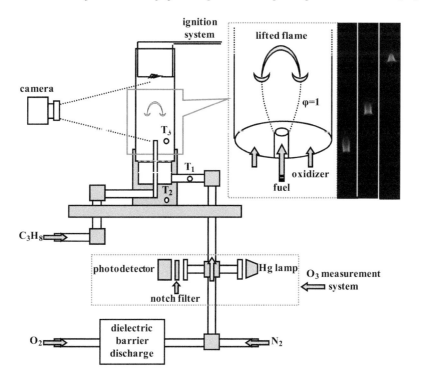

FIGURE 6.29 Experimental setup schematic of the lifted flame burner integrated with a dielectric barrier plasma discharge device and O_3 absorption measurement system.

Second, the reaction $O_3 + H = OH + O_2$ was suppressed due to reduced diffusivity of H atoms from the flame zone, thus increasing the ozone effect. It was also shown that the ozone enhancement increased linearly with ozone concentration with a slightly stronger enhancement effect on leaner mixtures (Figure 6.28).

The thermal and kinetic effects of O_3 on flame propagation and stabilization were investigated experimentally and numerically by using $C_3H_8/O_2/N_2$ laminar lifted flames [59,60]. In this experiment, ozone was produced by a dielectric barrier plasma discharge and was isolated and measured quantitatively by using absorption spectroscopy. The effect of ozone addition in oxidizer stream on the flame liftoff height (flame propagation velocity) was examined. A schematic of the experimental platform

TABLE 6.1
Important Reactions and Their Reaction Rate Constants with Wall and Oxygen

Reaction	Reaction Constant [cm³/molecule/s]
$O + O_2 + M \rightarrow O_3 + M$	1.0×10^{-33}
$O(^1D) + O_2 \rightarrow O + O_2$	3.84×10^{-11}
$O_2(v) + O_2 \rightarrow O_2 + O_2$	1.0×10^{-14}
$O_2(b^1\Sigma_g) + O_2 \rightarrow O_2(a^1\Delta_g) + O_2$	3.6×10^{-17}
$O_2(b^1\Sigma_g) + O_2 \rightarrow O_2 + O_2$	4.0×10^{-18}
$O_2(a^1\Delta_g) + O_2 \rightarrow O_2 + O_2$	2.0×10^{-18}

is shown in Figure 6.29. The lifted flame burner consisted of a central fuel jet with an inner diameter of 0.254 mm that was in a 90 mm inner diameter fused silica tube to contain the co-flow of oxidizer. The gases used in the experiments were C_3H_8 and O_2 and N_2 mixed for the oxidizer. The undiluted ultra-high purity O_2 was passed through a dielectric barrier discharge device for ozone generation and was then merged with the N_2 stream to be introduced to the lifted flame burner. The dielectric barrier discharge produced multiple oxygen-containing species including O, O_3, $O_2(v)$, $O(^1D)$, $O(^1S)$, $O_2(a^1\Delta_g)$, $O_2(b^1\Sigma_g)$, etc. To ensure that O_3 was the only species present in the flow when merged with the N_2 stream, a sufficient residence time was given to quench all plasma-produced species other than O_3. Table 6.1 lists the quenching rates of some common plasma-produced oxygen species. The atomic oxygen rapidly recombined with O_2 to produce O_3 at 101.3 kPa and 300 K and had a lifetime of approximately 20 milliseconds, which was much shorter than the flow residence time over 100 ms.

The high-velocity fuel jet (3.5–10 m/s) and low-velocity oxidizer co-flow (0.049 m/s) created a flow field with a stoichiometric contour where the premixed flame head of a lifted flame was located (top right insert in Figure 6.29). The lifted flame has a premixed flame head anchored on the stoichiometric contour, followed by a diffusion flame tail. The lifted flame is stabilized at a location where the flame speed and the local flow velocity are balanced. Therefore, if the flame speed increases, the lifted flame will move upstream toward the burner exit and the flame liftoff height will decrease to re-establish a local velocity balance at a higher flow velocity.

The O_3 produced by the dielectric barrier discharge was measured using a one-pass, line-of-sight absorption cell in the flow downstream of where the O_2 and N_2 streams merged. A mercury light provided ultraviolet light at the wavelength of 253.7 nm where O_3 has a peak absorption cross-section of 1.137×10^{-17} cm² (at 300 K) in the Hartley band [64]. Therefore, the change in the transmittance of the cell with the plasma on and off could be used to determine the O_3 concentration through the Beer-Lambert law,

$$N_{ozone} = \frac{-\ln\left(\frac{I}{I_0}\right)}{\sigma_{ozone} L} \tag{6.50}$$

where N_{ozone} is the absolute number density of the absorbing species, O_3; I the intensity of light with the presence of O_3; I_0 the intensity of light without the presence of O_3; σ_{ozone} the absorption cross-section of O_3 at the excitation wavelength of 253.7 nm; and L the path length in the absorption cell (12.48 cm).

Noticeable flame speed enhancement by O_3 was observed by comparing flame stabilization locations with and without O_3 production. The results are shown in Figure 6.30 as a function of fuel mixture fraction gradient. It is shown clearly that there is an enhancement of the lifted flame speed with increasing O_3 concentration. Interestingly, the enhancement of lifted flame speed increases with increasing fuel mixture fraction gradient for the same concentration of O_3. This can be explained

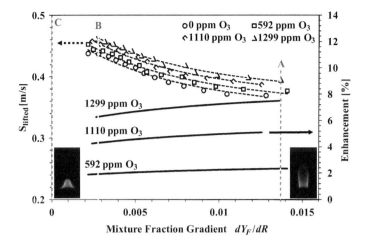

FIGURE 6.30 Plot of the lifted flame speed, S_{lifted}, and percent enhancement of S_{lifted} as a function of mixture fraction gradient at 101.3 kPa with and without O_3 addition. The inset pictures of the lifted flames show the differences in the flame front near the fuel nozzle (right) and far from the fuel nozzle (left) [60].

reasonably with a coupling effect between kinetic enhancement and changes to the flame front curvature (flame stretch) leading to a hydrodynamic enhancement by considering the unique characteristics of the triple flame structure of laminar lifted flames. Experiments at atmospheric pressures showed an 8% enhancement in the flame propagation speed for 1,299 ppm of O_3 addition to the O_2/N_2 oxidizer. Numerical simulations showed that the O_3 decomposition early in the preheat zone of the flame produced O which rapidly reacted with C_3H_8 to abstract an H and produced OH. The subsequent reaction of OH with fuel provided chemical heat release and H_2O at lower temperatures to enhance the flame propagation speed. It was also shown that the kinetic effect on flame propagation enhancement by O_3 reaching the preheat zone of the flame for early oxidation of fuel via O is much greater than that by the thermal effect from the energy contained within O_3.

6.2.5.2 Ozone-Assisted Low-Temperature Flame Propagation and Ozonolysis Effect

To study the effect of ozone addition on the liftoff flame propagation for unsaturated hydrocarbons, the effect of ozone addition on ethylene (C_2H_4) non-premixed jet flames (similar to Figure 6.29) was investigated [62,65]. The oxidizer co-flow composition was 11.5% O_2 + 88.5% N_2 with the velocity of 0.016 m/s. Different from the results of saturated hydrocarbons (e.g., propane in Figure 6.30), it was observed that with O_3 addition, the flame liftoff height could either increase or decrease, depending on the initial value of the flame liftoff height and jet velocity before O_3 was added. Figure 6.31a shows the liftoff heights of flames with different amounts of O_3 in the co-flow. It is interesting to note that with the same O_3 in the co-flow, there are opposite trends on the change of flame liftoff height. The liftoff height decreased with O_3 addition if the jet velocity was small, while it increased if the value of the jet velocity was relatively large. At the jet velocity of 3.68 m/s, for example, the addition of O_3 higher than 150 ppm in the oxidizer co-flow resulted in flame blow-out. The effect of fuel jet velocity at constant O_3 on the flame liftoff height was also studied and shown in Figure 6.31b. It is seen that the curves with and without O_3 addition crossed at fuel jet velocity near 3.7 m/s. Therefore, the effect of O_3 addition on the lifted ethylene flame was a function of the fuel jet velocity. At low jet velocities, O_3 addition decreased the flame liftoff height, while at higher jet velocities, O_3 addition increased the liftoff height. Therefore, there were two competing mechanisms controlling the flame dynamics.

To understand the kinetic mechanism, CH_2O PLIF measurements were conducted for lifted flames with and without O_3 addition. Figure 6.32 shows overlaid images of CH_2O PLIF (blue) and broadband chemiluminescence (red) collected by the same ICCD camera at fuel jet velocity of

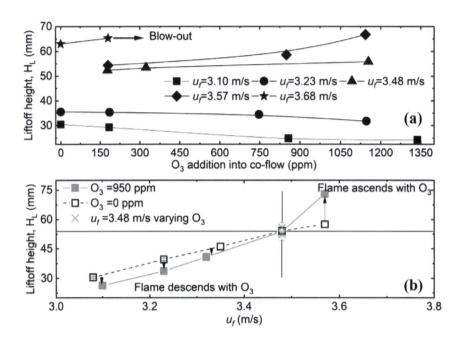

FIGURE 6.31 Relationship of liftoff height with (a) increasing O_3 addition in co-flow at different fuel jet velocities and (b) varying fuel jet velocity with a constant O_3 addition in co-flow [62].

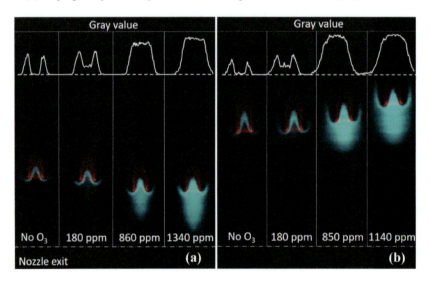

FIGURE 6.32 CH_2O PLIF (blue) and broadband chemiluminescence (red) of lifted flames with different O_3 concentrations at fuel jet velocity of (a) 3.10 m/s and (b) 3.57 m/s [62].

(a) 3.10 m/s and (b) 3.57 m/s, corresponding to cases of increasing liftoff height and decreasing liftoff height with fuel jet velocity, respectively. The gray value of CH_2O PLIF was sampled at the horizontal level of the triple points of each lifted flame, as illustrated in Figure 6.32 as well. For both cases with no O_3 addition, Figure 6.32 CH_2O was detected only at the location of the triple flames. Once O_3 was added to the co-flow, CH_2O started to appear upstream of the flame. With the increase of O_3, CH_2O fluorescence also increased. In addition, it was confirmed that CH_2O formed immediately near the nozzle exit where C_2H_4 and O_3 mixed in the shear layer. This rapid formation of CH_2O

at low temperature indicated that the ozonolysis reaction (Figure 5.44) occurred immediately when C_2H_4 and O_3 mixed at room temperature.

Such ozonolysis reaction upstream of the flame increased the propagation speed of the triple flame and enhanced the axial jet velocity and mixing along the stoichiometric contour, therefore, creating competing processes for establishing the flame liftoff height. The increase of flame speed led to the decrease of the liftoff height with the addition of ozone. On the other hand, the increase of jet velocity and mixing of fuel products with oxidizer pushed flame downstream and diluted the mixture in front of the triple flame, leading to an increase of the liftoff height with the increase of ozone. Moreover, the initial flame liftoff height determines the residence time for the ozonolysis reaction progress and which process dominates and the change of the liftoff height with O_3 addition.

6.2.5.3 Ozone-Assisted Cool Flame Propagation

To study ozone-assisted low-temperature flame propagation, an experimental study of lean premixed DME CF was studied by Hajilou et al. [66] by using a laminar flat flame Hencken burner [67] at low pressure (7.3 kPa). In a low-pressure Hencken burner, the fuel and oxidizer diffusively mix fast at the exit of the fuel and oxidizer microtube nozzles due to the increased diffusivity (Figure 6.33a) [67], allowing a stabilization of a premixed flame.

Figure 6.33a shows the sequence of images of the DME CFs at $\phi = 0.6$ over a range of flow rates. The horizontal white line denotes the burner surface. By determining the balance of flow velocity and flame speed at different flow rates, the ozone-assisted CF propagation speeds from $\phi = 0.4$ to 1.4 were measured (Figure 6.33b). It is seen that ozone-assisted DME/O_2 flame speeds were as high as 48 cm/s and decreased with the increase of equivalence ratio. This was the first measurement of ozone-assisted low-pressure premixed CF speed for a DME/O_2 mixtures. Note that without ozone, the CF speed of DME/O_2 mixtures was too low to be measured. Therefore, as discussed in Chapter 5, plasma-generated ozone can dramatically enhance low-temperature combustion kinetics non-thermally and increase low-temperature flame speed.

6.2.5.4 $O_2(a^1\Delta_g)$ Assisted Cool Flame Propagation

Like ozone, $O_2(a^1\Delta_g)$ is another long lifetime species created in plasma and may have a significant impact on combustion chemistry because it creates a much faster reaction channel via reaction R5.19' for radical production than the key chain-branching reaction of R5.3 in combustion.

$$H + O_2(a^1\Delta_g) \rightarrow O + OH \tag{R5.19'}$$

$$H + O_2 = O + OH \tag{R5.3}$$

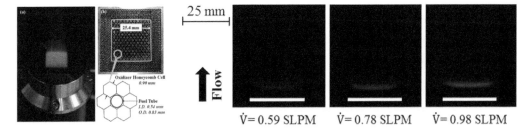

FIGURE 6.33A Left: Image of a Hencken burner under operation [67] and image and schematic representation of burner the nozzles with microtubes; Right: Liftoff heights for a $DME/O_2/O_3$ cool flame over a range of reactant flow rates ($\phi = 0.6$, $p = 7.3$ kPa). Burner-stabilized and freely propagating flame modes are located respectively at lower and higher flow rates [66].

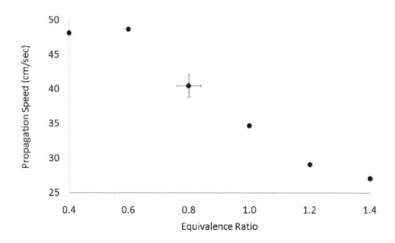

FIGURE 6.33B Propagation speeds, derived from cool flame liftoff height analyses, for DME/O_2/O_3 cool flames at $\phi = 0.4$–1.4 and $P = 7.3$ kPa [66].

However, few experimental studies have been conducted to understand the impact of $O_2(a^1\Delta_g)$ on flame propagation. The effect of plasma-generated $O_2(a^1\Delta_g)$ on the flame propagation of C_2H_4 lifted flames was also conducted by Ombrello et al. [59] using the same flame liftoff experimental setup in Figure 6.29 at reduced pressures (3.61 kPa and 6.73 kPa). The $O_2(a^1\Delta_g)$ was produced in a microwave discharge plasma and was isolated from O and O_3 by NO addition to the plasma afterglow in a flow residence time on the order of one second. The concentrations of $O_2(a^1\Delta_g)$ and O_3 were measured quantitatively through absorption by sensitive off-axis integrated cavity output spectroscopy (ICOS) and one-pass line-of-sight absorption, respectively.

In the experiment, the ICOS cell was installed downstream of the plasma and 82.5 cm long. The flow residence time in the cell was approximately 1.5 s. The effective path length was greater than 78 km due to multiple passes and provided accurate measurements down to 10^{14} molecules/cm^3. In the oxygen plasma, active species such as O, O_3, $O_2(a^1\Delta_g)$, $O_2(b^1\Sigma_g)$, and metastable O and Ar were produced. The key elementary reactions related to active species production and quenching in plasma are given in Table 6.2. With the one-second-long residence time, the only species that were not quenched were O_3 and $O_2(a^1\Delta_g)$. To isolate $O_2(a^1\Delta_g)$ from O_3 to observe the individual effect of $O_2(a^1\Delta_g)$ on flame propagation. To remove O_3, NO was added to the flow downstream of the plasma. The isolation of $O_2(a^1\Delta_g)$ with NO addition relies upon the reaction of NO with O_3 via

$$NO + O_3 = NO_2 + O_2 \tag{R5.81}$$

TABLE 6.2
Rate Constants of Production and Quenching Reactions of Plasma-Produced Oxygen Species [59]

Reaction	Reaction Constant [cm^3/molecule/s]
$O + O_2 + M \rightarrow O_3 + M$	1.0×10^{-33} = (HP limit 6.02×10^{-10})
$O(^1D) + O_2 \rightarrow O + O_2$	3.84×10^{-11}
$O_2(v) + O_2 \rightarrow O_2 + O_2$	1.0×10^{-14}
$O_2(b^1\Sigma_g) + O_2 \rightarrow O_2(a^1\Delta_g) + O_2$	3.6×10^{-17}
$O_2(b^1\Sigma_g) + O_2 \rightarrow O_2 + O_2$	4.0×10^{-18}
$O_2(a^1\Delta_g) + O_2 \rightarrow O_2 + O_2$	2.0×10^{-18}
$O_2(a^1\Delta_g) + Ar \rightarrow O_2 + Ar$	1.0×10^{-20}

Reaction R5.81 is three-orders of magnitude faster than NO reaction with $O_2(a^1\Delta_g)$, as shown in Table 6.3. Furthermore, the subsequent reaction of NO_2 with $O_2(a^1\Delta_g)$ is slow in comparison to reaction R5.81. Therefore, NO acted as a catalyst to fully eliminate the O_3 so that the individual effects of $O_2(a^1\Delta_g)$ were examined.

The effect of $O_2(a^1\Delta_g)$ on the flame liftoff height is shown in Figure 6.34. It is seen that approximately ten times the amount of $O_2(a^1\Delta_g)$ (approximately 5,500 ppm) was needed to achieve the same enhancement as O_3 (approximately 500 ppm) in Figure 6.30. The observed effect of $O_2(a^1\Delta_g)$ on flame speed enhancement was much lower than expected because the rate constant of reaction R5.19' is two orders of magnitude higher than the key chain-branching reaction R5.3.

The C_2H_4 laminar and lifted flame speed enhancement by $O_2(a^1\Delta_g)$ was also computed [59]. Numerical simulation showed that the $O_2(a^1\Delta_g)$ effect on flame speed enhancement was much stronger than the experimentally observed. The reason could be the collisional quenching of $O_2(a^1\Delta_g)$ by C_2H_4, reducing the kinetic effect of R5.19' on flame propagation. Note that this was the first isolated experimental result of the enhancement of an unsaturated hydrocarbon fueled flame by $O_2(a^1\Delta_g)$. The results showed clearly that the kinetic effect of $O_2(a^1\Delta_g)$ on flame speeds was lower than that predicted by using R5.19'. Therefore, $O_2(a^1\Delta_g)$ quenching via $C_mH_n + O_2(a^1\Delta_g)$ reactions is needed to appropriately model kinetic enhancement of $O_2(a^1\Delta_g)$ in plasma-assisted flame propagation. More kinetic experiments of $O_2(a^1\Delta_g)$ with fuels at elevated temperature are needed.

TABLE 6.3
Reaction Rate Constants of $O_2(a^1\Delta_g)$ and O_3 with NO and NO_2 [59]

Reaction	Reaction Constant (cm³/molecule/s)
$O_2(a^1\Delta_g) + NO \rightarrow O_2 + NO$	4.48×10^{-17}
$O_2(a^1\Delta_g) + NO \rightarrow O + NO_2$	4.88×10^{-18}
$O_2(a^1\Delta_g) + NO_2 \rightarrow O_2 + NO_2$	5.00×10^{-18}
$O_3 + NO \rightarrow O_2 + NO_2$	1.80×10^{-14}
$O_3 + NO_2 \rightarrow O_2 + O_2 + NO$	1.00×10^{-18}

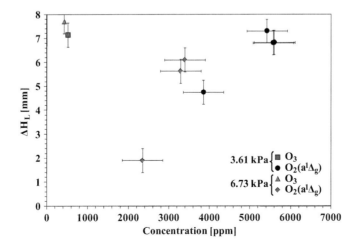

FIGURE 6.34 Plot of experimental results of C_2H_4 flame liftoff change with $O_2(a^1\Delta_g)$ and O_3 concentration for a plasma power of 80 Watts [59].

In summary, ozone has both kinetic and thermal effects in enhancing HF propagation. However, to reach about 10% of flame speed enhancement it requires thousands of ppms of ozone addition. On the other hand, ozone has a significant kinetic effect on low-temperature flames via ozonolysis or ozone-enhanced LTC. Singlet oxygen also has kinetic effect in enhancement flame propagation; however, the enhancement effect is much smaller compared to the predicted results. As such, the quenching effect of singlet oxygen by hydrocarbons and radicals at elevated temperature needs to be appropriately considered.

6.3 MINIMUM IGNITION ENERGY

6.3.1 The Minimum Ignition Energy and the Critical Radius

Ignition failure at fuel lean conditions or at high altitude remains to be a big challenge for internal combustion engines [68–70] and gas turbine relight [71]. Many experimental [72–75] and theoretical and computational studies [39,72,76,77] have been carried out to understand the mechanism of the minimum ignition energy (MIE) for successful flame initiation with spark or non-equilibrium plasma discharge. As shown in Figure 6.35, for a fuel mixture above or below the flammability limit (Figure 6.25), depending on the energy of the spark, there are four outcomes: (1) a successful flame initiation (ignition energy $Q \geq$ MIE and $\phi > \phi_0$), (2) a self-extinguishing flame ($Q \geq$ MIE and $\phi_0^* < \phi < \phi_0$, the sublimit stretched flame in Figure 6.25), (3) a stationary flame ball ($Q \geq$ MIE, Le $\ll 1$, and $\phi_0^* < \phi < \phi_0$) [78], and a quenching ignition kernel ($Q <$ MIE).

Recent theories and experiments [39,75,77] have clearly shown that MIE is governed by a critical radius ($R_{f,cr}$) below which an initial ignition kernel will quench and flame initiation will fail (Figure 6.35a). The theory [39] also showed that the critical flame radius and the MIE increase rapidly with the increase of the Lewis number (or the size of fuel molecules). Moreover, the theory in Refs. [39,72] also showed that the critical flame initiation radius is strongly affected by the fuel activation energy. Unfortunately, few experimental data are available to confirm the validity of the theory until recently.

The existence of a critical radius for successful flame initiation has been observed in experiments for hydrogen, methane, and other small hydrocarbon flames [72,73,77,79,80] in spherical chambers. Direct measurements of the critical radius as a function of fuel equivalence ratio for n-heptane and higher hydrocarbons were made by Kim et al. [75,81]. The experimental results showed that the critical radius of fuel/air mixtures increased significantly with the decrease of equivalence ratio [81]. The impact of the critical radius on flame initiation was also observed in a high-speed flow by using repetitive non-equilibrium plasma discharge [82,83]. More recently, a numerical simulation

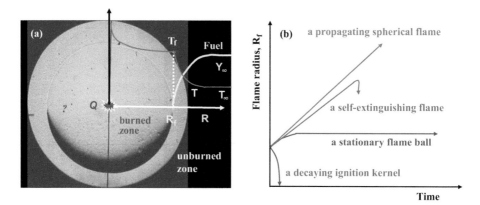

FIGURE 6.35 (a) Image of flame initiation by a spark with energy Q and the schematic of flame structure in the unburned and burned zone. (b) Four outcomes of spark ignition in a flammable mixture depending on the ignition energy, equivalence ratio, and mixture Lewis number.

was carried out for non-equilibrium repetitive discharge and demonstrated that high-frequency inter-pulse discharge coupling can increase the ignition kernel size and enable successful flame initiation in a reactive flow [84].

To understand the physics of MIE by a plasma energy deposition, analytical and computation studies were done by Chen and coworkers [39, 72, 76, 77]. In the attached coordinate moving with the flame front (Figure 6.35), the flame structure can be considered as in quasi-steady [39, 85]. By assuming constant thermal properties and a one-step irreversible reaction, the non-dimensional conservation equations for energy and fuel mass fraction are given as [39],

$$-S_L \frac{d\theta}{dr} = \frac{1}{r^2}\frac{d}{dr}\left(r^2 \frac{d\theta}{dr}\right) + \omega \tag{6.51}$$

$$-S_L \frac{dy_F}{dr} = \frac{1}{\text{Le}}\frac{1}{r^2}\frac{d}{dr}\left(r^2 \frac{dy_F}{dr}\right) - \omega \tag{6.52}$$

where S_L is the flame propagating speed normalized by the adiabatic planar flame speed, $S_{L,\text{ad}}$; r, the spatial coordinate normalized by the planar flame thickness, $\delta_f = \alpha/S_{L,\text{ad}}$; α the thermal diffusivity; Le, the Lewis number; θ, the normalized temperature, $\theta = (T - T_\infty)/(T_{\text{ad}} - T_\infty)$; y_F, the normalized fuel mass fraction normalized by Y_∞ (Figure 6.35), and $T_{\text{ad}} = T_\infty + Y_\infty q/C_P$ is the adiabatic flame temperature of the corresponding planar flame, with q and C_p being the reaction heat release per unit mass of fuel and the specific heat capacity at constant pressure, respectively. For simplicity, the effect of radiative heat loss studied in Ref. [39] is not included here. For those who are interested in self-extinguishing flames, please read [39].

In the limit of large activation energy, chemical reaction occurs only within a very thin zone of high temperature and the reaction term can be replaced by a Delta function with jump conditions at the flame front [77]

$$\omega = \exp\left[\frac{Z}{2}\frac{\theta_f - 1}{\sigma + (1-\sigma)\theta_f}\right] \cdot \delta(r - R_f) \tag{6.53}$$

where Z is the Zeldovich number; and $\sigma = T_\infty/T_{\text{ad}}$, the density ratio. By integrating the conservation equations around the flame front, $r = R$, the following jump relationships across the flame could be obtained [86],

$$\left.\frac{d\theta}{dr}\right|_{R^-} - \left.\frac{d\theta}{dr}\right|_{R^+} = \frac{1}{\text{Le}}\left(\left.\frac{d\theta}{dr}\right|_{R^+} - \left.\frac{d\theta}{dr}\right|_{R^-}\right) = \exp\left[\frac{Z}{2}\frac{\theta_f - 1}{\sigma + (1-\sigma)\theta_f}\right] \tag{6.54}$$

The boundary conditions for temperature and fuel mass fraction can be given as

$$r = 0, \quad r^2 \partial \theta / \partial r = -Q, \quad y_F = 0 \tag{6.55}$$

$$r = R_f, \quad \theta = \theta_f, \quad y_F = 0 \tag{6.56}$$

$$r = \infty, \quad \theta = 0, \quad y_F = 1 \tag{6.57}$$

where $Q = q/\left[4\pi\lambda\delta_f(T_{\text{ad}} - T_\infty)\right]$ is the normalized ignition power deposited in the center.

The above equations can be solved analytically [39]. For a plasma energy flux deposited in the center of quiescent mixture ($Q > 0$), the following relationship [39] between normalized flame speed S_L, flame radius R_f, flame temperature θ_f, and energy deposition flux Q is obtained as

$$\frac{\theta_f R_f^{-2} e^{-S_L R_f}}{\int_{R_f}^{\infty} \tau^{-2} e^{-S_L \tau} d\tau} - Q \cdot R_f^{-2} e^{S_L R_f} = \frac{1}{Le} \frac{R_f^{-2} e^{-S_L Le R_f}}{\int_{R_f}^{\infty} \tau^{-2} e^{-S_L Le \tau} d\tau} = \exp\left[\frac{Z}{2} \frac{\theta_f - 1}{\sigma + (1-\sigma)\theta_f}\right] \quad (6.58)$$

In the limit of zero propagating flame speed for the flame ball case ($S_L = 0$), Eq. 6.58 reduces to the following solution for adiabatic Zeldovich flame ball radius (R_Z) with energy deposition at the center ($r = 0$),

$$\ln(R_Z Le) + \frac{Z}{2} \frac{Le^{-1} + Q/R_Z - 1}{\sigma + (1-\sigma)\left(Le^{-1} + q/R_Z\right)} = 0 \quad (6.59)$$

In the limit of a large flame radius ($R >> 1$), the detailed model reduces to the simplified form,

$$\left(S_L + \frac{2}{R_f}\right)\ln\left(S_L + \frac{2}{R_f}\right) = \frac{Z}{R_f}\left(\frac{1}{Le} - 1\right) - \frac{2}{R_f}\left(\frac{1}{Le} - 1\right) \quad (6.60)$$

which governs the relation shift between the flame speed (S_L), flame propagation velocity (V_f), and the flame stretch rate,

$$a = \frac{2V_f}{R_f} = \frac{2S_L}{\sigma R_f} \quad (6.61a)$$

Equation 6.60 is similar to the theory presented in Ref. [87] but has a new curvature term on the right-hand side of Eq. 6.60, which was not considered in Ref. [87] because of the assumption of $Le \rightarrow 1$. Note, for most mixtures, the Zeldovich number is in the range of 5~15 and the deviation of Lewis number from unity is not small, i.e., $Le^{-1} \sim O(1)$. As a result, the first curvature term in Eq. 6.60 cannot be neglected. For very WSFs, the stretched flame speed is nearly the adiabatic unstretched flame speed (i.e. $S_L = 1 + \varepsilon$ with $\varepsilon \ll 1$). Then, Eq. 6.60 reduces to the linear function of the stretch rate [45,88].

By solving Eq. 6.59 numerically, the relation between flame propagating speed, flame radius, flame temperature, flame ball size (Zeldovich radius), and the existence of different flame regimes for different Lewis numbers and ignition powers can be obtained [39].

Figure 6.36 shows the analytical results of Eq. 6.58. The normalized flame speed is plotted as a function of flame radius normalized by flame thickness at different energy deposition powers (Q) for a mixture with $Le = 2$. For $Q = 0$, only a single flame branch exists at large flame radius ($R_f > 10$). It means that a flame cannot exist at a smaller radius without the support of external ignition power. When the ignition energy is small ($Q = 0.5$), there appears a new flame kernel branch (left branch) at a small flame radius. However, the flame speed decreases rapidly as the flame kernel grows due to the curvature effect and decay of the deposited energy; and flame fails to propagate. When the energy deposition Q is increased to 0.97 (the MIE), it is seen that the flame speed decreases first as it propagates outwardly, then reaches the minimum flame speed at a critical flame radius ($R_C = R_C^- = R_C^+$). After that with the increase of flame kernel size, the flame speed increases and reaches unity when it becomes close to a planar flame at a very large radius, that is, a successful ignition. Figure 6.36 also shows that for any energy deposition power greater than the MIE of $Q = 0.97$, ignition will always be successful. Therefore, it can be concluded that there is a critical flame initiation radius R_C that determines the MIE which is required to drive the initial flame kernel size reaching

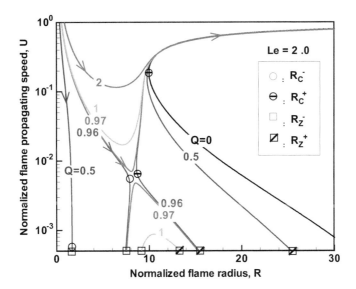

FIGURE 6.36 Normalized flame propagating speed as a function of flame radius at different ignition power for mixtures with Le = 2 [77].

R_C. Therefore, even for a flammable mixture, to initiate a successful ignition, we need ignition energy greater than the MIE.

Note that R_C is normalized by flame thickness. Therefore, at lower pressure or for fuel lean mixtures, the dimensional critical radius will increase significantly. That is why it is difficult to ignite a mixture at lower pressure or at fuel lean conditions. Moreover, because R_C is in the order of flame thickness, therefore, the flame stretch at the critical flame radius,

$$a = \frac{2S_L}{\sigma R_C} \tag{6.61b}$$

is very large. As a result, the mixture's Lewis number will play a significant role in affecting the critical radius.

To see how the critical flame radius correlated with the Lewis number, the minimum ignition power, Q_{min}, and the cube of critical flame radius, R_C^3, for mixtures with different activation energies (Z = 10, 13), Figure 6.37 plots the dependence of the MIE as a function of R_C^3. It is clearly seen that the minimum ignition power changes almost linearly with the cube of critical flame radius at different Lewis numbers, i.e., $Q_{min} \sim R_C^3$. In addition, Figure 6.37 shows that Q_{min} and R_C^3 increase linearly with the Lewis number. Therefore, for a mixture with large Le, the MIE will need to be dramatically increased, making it difficult for ignition of fuel lean gasoline mixtures and jet fuels, especially at lower pressure or flying high altitude.

Many experiments were carried out to measure the critical flame radius [72,73,75,77,81] for H_2, propane, large alkanes and aromatics. These experiments confirmed the prediction of the theory [39,72,76,77] and provided important data of the critical radius for different fuels with different molecular structures and Lewis numbers at different pressures and equivalence ratios. To show the experimental observation of the critical radius and its dependence on mixture equivalence ratio, Figure 6.38 shows the results of n-decane/air mixture at 400 K and atmospheric conditions under fuel lean and rich conditions [75]. It is interesting to note that for fuel rich mixtures ($\phi = 1.4$), since Le < 1, after the SI with an energy greater than the MIE, the ignition kernel grew, and the flame speed decreased with the decrease of stretch rate (Eq. 6.58). As the flame radius becomes very large, the flame speed linearly decreases with the decrease of the stretch rate (Eq. 6.61a). The flame speed dependence on stretch rate can be extrapolated linearly to the adiabatic flame speed at zero stretch

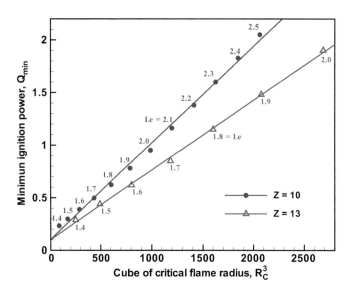

FIGURE 6.37 Minimum ignition power and cube of critical flame radius for mixtures with different Lewis numbers and Zeldovich numbers [77].

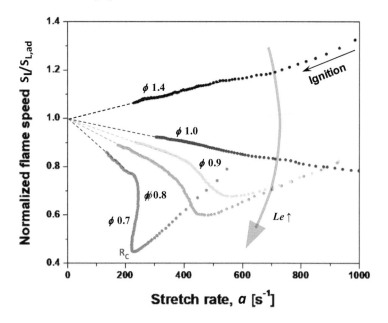

FIGURE 6.38 Measured and extrapolated flame propagation speed after ignition at different equivalence ratios for n-decane/air mixtures [75].

rate (Eq. 6.60). This linear extrapolation of flame speed also confirmed the relationship between the WSF extinction limits and the flammability limit of the 1D planer freely propagating flames in Figures 6.24 and 25 at $Le < Le_{cr}$.

For lean n-decane/air mixtures, the Lewis becomes larger than unity. For $Le > 1$, the strong curvature (stretch) of the flame at the small flame radius (initial ignition kernel) significantly reduced the flame speed (Eq. 6.58). In addition, the weakening of the effect of external energy deposition on the flame speed resulted in a rapid decrease of flame speed as the flame kernel grew (Figure 6.39). For $\phi = 0.7$, for example, it is seen that there existed a critical flame radius, R_C, at which flame

Plasma-Assisted Combustion: Dynamics 443

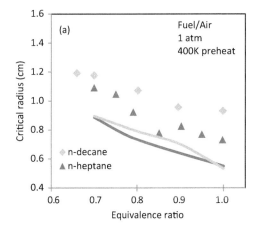
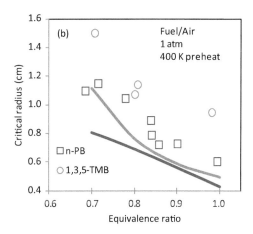

FIGURE 6.39 Critical radius R_c as a function of equivalence ratio at $p =1$ atm, (a) for n-$C_{10}H_{22}$/air and n-C_7H_{16}/air, and (b) n-propyl-benzene and 1,3,5 trimethyl-benzene/air [81] (Symbols: experimental measurement; Lines: numerical calculation).

speed became minimum (Figure 6.36). After that flame speed rapidly increased and then linearly approached the adiabatic flame speed as the flame radius became very large. The experimental observation confirmed the theoretical prediction of Figure 6.36 and Eq. 6.58. In addition, it is noted that the leaner the mixture is, the larger the critical radius becomes because of the increase of flame thickness.

Figure 6.39 shows the measured critical radius of n-heptane, n-decane, n-propyl-benenzene, and 1,3,5 trimethyl-benzene/air mixtures at 1 atm and 400 K as a function of equivalence ratio. It is seen that the critical radius increased quickly as the equivalence ratio decreased for all measured n-alkanes and aromatics. The critical radius of aromatic fuels is larger than that of n-alkanes at comparable molecular weight because of their high activation energies. Moreover, the critical radius was greater than 1 cm at fuel lean conditions, which explained why it is difficult to ignite lean fuel mixtures, especially at lower pressures. Some plasma-assisted ignition technologies need to be developed to overcome the increase of the critical ignition radius for fuel lean combustion.

6.3.2 Impact of Plasma Properties on the Minimum Ignition Energy and Flame Initiation

To overcome the critical radius for ignition and reduce the MIE in combustors, many new techniques to modify plasma ignition systems have been developed. Such techniques include using radio frequency spark [89], laser-induced spark [69,80,90,91], high-frequency microwave [38,68,92], nanosecond repetitively pulsed plasma [82,83,93] and a SDBD discharge system (Figure 6.7). All the above studies confirmed that the spark kernel volume affected the MIE dramatically, and the increased spark kernel volume could extend the ignition limit and improve the operating stability of engines. Meanwhile, an experimental study [94] showed that under the same operating condition, the increase of discharge energy had limited effect on enhancing the initial flame growth and extending the ignition limits, while the multiple spark strategy had significant influence on the ignition process. Besides, by using the repetitive nanosecond pulsed discharge, Lefkowitz and co-workers [82,83,93] experimentally investigated the impact of the initial ignition radius and the plasma discharge frequency on flame initiation in a high-speed turbulent flow. They observed that multiple overlapping pulse discharges enhanced the ignition probability with constant discharge energy. In addition, Zhao et al. and Lin et al. [95,96] compared the ignition behaviors of multi-channel nanosecond discharge with single spark. The results demonstrated that compared to spark and single-channel nanosecond discharge, multi-channel nanosecond discharge could generate a much larger ignition kernel with stronger flame wrinkling, and thus had a higher ignition probability. These studies have confirmed that overcoming the critical radius (Eq. 6.58) using a multi-spark system and high-frequency repetitive discharge to create a large ignition kernel side is important.

To quantitatively understand the role of the spark kernel volume in affecting combustion ignition at different equivalence ratios and pressure with fixed total ignition energy by using multi-channel sparks, Zhao et al. and Lin et al. [95, 96] developed a novel multi-channel spark system to increase the spark kernel volume while maintaining constant discharge energy. Figure 6.40a shows the comparison of ignition failure and success with a single-channel spark system and a cascading multi-channel spark system for a lean n-pentane/air mixture at the same ignition energy. It is seen that a multi-channel cascading spark created 3 cascading sparks and formed a larger ignition kernel greater than the critical radius, thus leading to a successful ignition at which a single spark failed. Figure 6.40b shows the ignition probability dependence on the equivalence ratio for the n-pentane/air mixture at 0.5 atm with the spark gap distance of 2.5 mm. It is seen that the multi-channel cascading spark system significantly extended the ignition probability to lower equivalence ratios. To demonstrate the effectiveness of multi-channel spark system on ignition at different pressures, Lin et al. [96] conducted ignition experiments for propane/air mixtures. Figure 6.41 shows the comparison of ignition for four different ignition methods, spark, single-channel nanosecond discharge (SND) (gap distance: 1 mm), SND (gap distance: 4 mm) and MND, at reduced pressures. It is observed that the ignition probability of SND (4 mm) with a larger gap (larger ignition volume) was higher than that of SND (1 mm) with a smaller gap (smaller ignition volume) at the same pressure. The ignition probability of SND with both 4 mm gap and 1 mm gap decreased rapidly as the pressure decreased. However, the MND ignition system was able to extend the ignition probability to 0.3 atm, suggesting a superior performance of a larger ignition kernel ignitor. Therefore, for a successful ignition and flame initiation, it is not just about energy greater than the MIE, it is more important to have an ignition kernel size greater than the critical radius under fuel lean and low-pressure conditions.

To understand how flow affects the minimum ignition power (MIP) or MIE, a parametric exploration of the dynamics of ignition using nanosecond pulsed high-frequency discharges (NPHFD) in flowing mixtures of methane and air was conducted by Lefkowitz and co-workers [83,93] to determine the "inter-pulse coupling" (overlap of discharge to increase the ignition kernel size) effect of the number of burst of high-frequency discharges (N) in the pulse repetition frequency (PRF) range between 1 and 300 kHz. The impacts of PRF, number of pulses, equivalence ratio, discharge gap distance, and flow velocity are quantified in terms of ignition probability and minimum ignition power, and Schlieren images of ignition kernel development were examined. Three regimes of

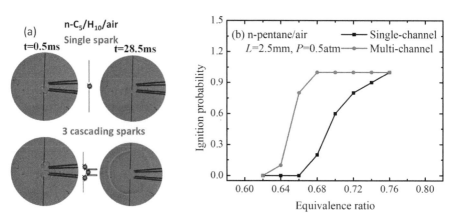

FIGURE 6.40 (a) Schematic of single spark and multi-channel cascading spark ignition for lean n-pentane/air mixture flames, (b) Ignition probability as a function of equivalence ratio for the n-pentane/air mixture at $L = 2.5$ mm: $P = 0.5$ atm [95].

Plasma-Assisted Combustion: Dynamics

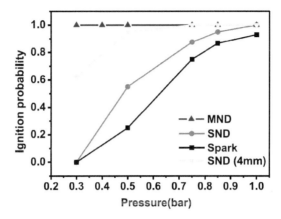

FIGURE 6.41 Comparison of ignition probability of four ignition methods: Spark (61.6 mJ), single-channel nanosecond discharge (SND) (1 mm, 15 kHz, 75 pulses, 43.15 mJ), SND (4 mm, 15 kHz, 75 pulses, 46.62 mJ) and multi-channel nanosecond discharge MND (15 kHz, 75 pulses, 48.46 mJ) for stoichiometric propane/air mixture [96].

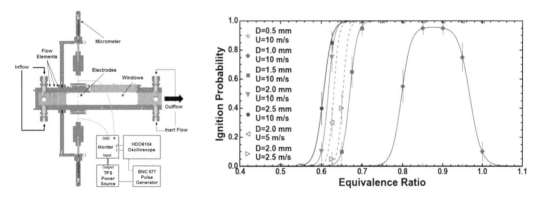

FIGURE 6.42 Schematic of the experimental design and the probability of a single discharge pulse igniting a flowing methane/air mixture as a function of equivalence ratio, flow velocity, and gap distance. Unless otherwise noted, $U = 10$ m/s and $D = 2.0$ mm [83].

inter-pulse coupling were found for different values of PRF: fully coupled, partially coupled, and decoupled.

The experimental system [83] included a small tunnel with well-defined in-flow conditions for plasma ignition experiments (Figure 6.42). The tunnel consists of a constant area cross-section with dimensions of 3.81 cm by 3.81 cm. There are two 1.6 mm diameter tungsten electrodes in a pin-to-pin geometry with a variable gap distance (D). Bulk flow velocities tested in the tunnel were in the range of 2.5–10 m/s and were deemed to be within the turbulent regime. Static pressure in the tunnel was maintained at approximately 100 kPa for all experiments. High-frame-rate Schlieren was used to capture the global ignition kernel development. For plasma ignition, controllable parameters were the magnitude of the peak voltage, the discharge frequency, and the number of pulses. The system produced approximately 10 ns FWHM pulses at a maximum frequency of 300 kHz for up to 500 pulses. Figure 6.42 presents the probability of a single pulse igniting methane–air mixtures as a function of equivalence ratio and for different gap distances and flow velocities. Ignition probability increases with equivalence ratio. Moreover, ignition was not possible at small spark gap distance of $D = 0.5$ mm for any equivalence ratio. An increasing probability in broader equivalence ratios was observed with the increase of the spark gap distance through 2.5 mm. The results confirmed that the theoretical prediction of the critical radius in Eq. 6.58 and experimental ignition data in a quiescent flow can be applied to plasma-assisted ignition in a turbulent flow.

The increase of ignition gap is often limited by the plasma voltage and source. Therefore, in a reactive flow one can change the plasma discharge frequency to create a large ignition volume to facilitate successful ignition. The effect of inter-pulse coupling of plasma discharge on ignition was also examined by Lefkowitz et al. [83,93] by varying the plasma frequency and the number of plasma burst (N) at the same discharge gap distance (2 mm) (Figure 6.43). The trends of ignition probability with inter-pulse time or plasma burst number are non-monotonic, with a clear minimum in the ignition probability at intermediate inter-pulse times, and higher probability at both shorter and longer inter-pulse times. For all values of N, the ignition probability was 1 for discharge period of 2×10^{-5} s, and decreased with increasing inter-pulse time until reaching a minimum around 2×10^{-4} s. At discharge period greater than 3×10^{-4} s the ignition probability again increased but remained lower. It was shown that for each set of conditions, the fully coupled regime (short inter-pulse time) was in the inter-pulse time range when ignition probability was 1, the partially coupled regime was in the range when ignition probability began to decrease after the fully coupled regime, but before it reached a constant probability at long inter-pulse times. The decoupled regime was at inter-pulse times in which ignition kernels did not interact, which results in a constant ignition probability. For increasing N, the inter-pulse time at which the partially coupled regime transitions into the fully coupled regime shifted to longer times, and the ignition probability in the partially coupled regime also increased because of the increase of energy deposition (Figure 6.43). In the decoupled regime, the ignition probability increased linearly with the number of pulses, with an ignition probability approximately equal to the product of N and the single pulse probability. High-speed laser diagnostics [93] showed that fully coupled regime occurs for the highest PRF and led merging of complete ignition of the kernels and the highest ignition probability. The partially coupled regime occurred for intermediate PRF and exhibited only local ignition of portions of the kernel and had the lowest ignition probability. The decoupled regime occurs for the lowest PRF and exhibited isolated and non-interacting ignition events with ignition probability being a linear function of the number of pulses.

To understand the inter-pulse coupling and provide guidance of plasma design optimization, Mao et al. [84,97] computationally examined the effect of inter-pulse coupling on the MIE of a reactive flow. Figure 6.44 shows the schematic of the plasma ignition geometry. An axisymmetric

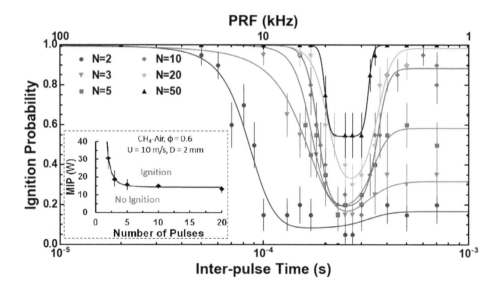

FIGURE 6.43 Probability of ignition in methane/air as a function of inter-pulse time for various number of pulses, at $\phi = 0.6$, $U = 10$ m/s, and $D = 2$ mm. The inlay presents the minimum ignition power as a function of pulse number [83].

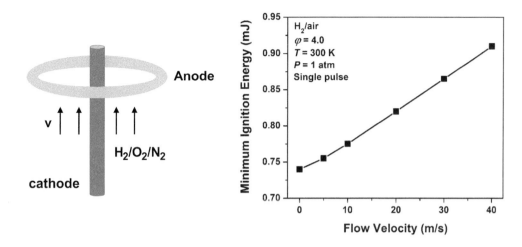

FIGURE 6.44 Plasma discharge geometry and the minimum ignition energy dependence on the flow velocity [84].

nanosecond plasma ignitor was installed in a lean hydrogen/air mixture (0.627 H_2/0.078 O_2/0.295 N_2-air and $\phi = 4.0$) at 1 atm and 300 K with a flow velocity varying from 0 to 50 m/s. The plasma had a peak voltage of 9 kV, rise time of 2 ns, and gap distance from the center electrode to the cathode ring of 1.5 mm. The study investigated the effects of non-equilibrium nanosecond plasma discharge PRF, pulse number, and flow velocity on the critical ignition volume, MIE, and chemistry. Numerical simulations were done using a multi scale adaptive reduced chemistry solver for plasma-assisted combustion (MARCS-PAC) [97].

Figure 6.44 shows the dependence of the MIE on the flow velocities. The MIE at different velocities was obtained at the minimum discharge energy for successful ignition and flame propagation. The results show that the MIE increases monotonically with the flow velocity due to the increase of convective heat loss. Note that this result is different from that in Figure 6.42, probably because of the heat loss to the electrodes at low flow speeds in the experiments.

To study the effects of inter-pulse coupling on ignition, two sequential pulses with the same discharge energy but different time differences with a flow velocity of 20 m/s were compared. The discharge energy in each pulse was set as $Q = 0.6$ mJ, which was below the MIE of 0.82 mJ (Figure 6.44). Figure 6.45 shows the time evolutions of mole fractions of OH (left) and temperature (right) at PRF = 8, 15, 18 and 50 kHz, respectively. Images in the first row show the OH distribution and temperature profiles right after the second pulse with the inter-pulse delay of 125, 66.7, 55.6, and 20 μs. At 8 kHz, the first ignition kernel moved downstream far enough due to flow motion before the second discharge pulse. Two isolated ignition kernels were generated at $t = 150$ μs and propagated without any coupling. This case is the decoupled regime of inter-pulse coupling discussed above, at which ignition failed. At 15 kHz, the results of OH concentration at $t = 66.7$ μs showed that a part of the discharge volume generated at the second pulse overlapped with the previous ignition kernel. This is in the partially coupled regime and the ignition kernels started to interact with each other. However, the resulting ignition kernel was still smaller than the critical ignition volume and thus ignition kernel quenched downstream. When the frequency is greater than 18 kHz, the second ignition kernel is coupled fully with the first. This is the fully coupled regime that led to successful ignition and flame propagation. The computational results supported the experimental observation in Figure 6.43 and concluded that for successful ignition, the discharge pulse must overlap with the developing ignition kernel in the inter-electrode region to achieve an ignition kernel size greater than the critical radius.

Similar to experimental results in Figure 6.43, the effects of pulse number N on ignition at PRF = 200 kHz in the fully coupled regime were also computationally examined. The total discharge

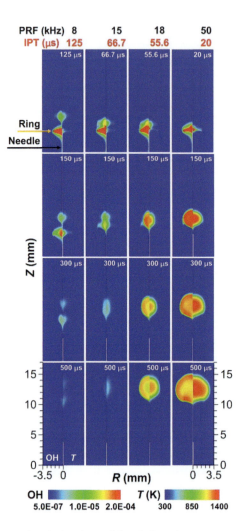

FIGURE 6.45 Time evolutions of mole fractions of OH (left) and temperature (right) in 20 m/s flowing H_2/air mixtures at various pulse repetition frequency (PRF)/inter-pulse time (IPT) conditions with 2 pulses, respectively, at decoupled (8 kHz), partially coupled (15 kHz) and fully coupled (18–50 kHz) regimes. The yellow arrow indicates the location of the ring electrode. The vertical white region at the bottom indicates the needle electrode [84].

energy was maintained as Q_{total} = 0.75 mJ. The energy was split evenly into each pulse of the N total pulses. Figure 6.46 shows the time evolutions of ignition kernel volume in 20 m/s flowing H_2/air mixtures with different N. It is seen that there was a non-monotonic dependence of ignition kernel volume on N. At first, the ignition kernel volume increased with N. The maximum ignition enhancement was achieved when N = 3. However, a further increase of N decreased the ignition kernel volume. This is because the increase of pulse number N also decreased the discharge energy deposited in each pulse for a fixed total energy. The ignition kernel volume significantly decreased at N = 10 due to the low discharge energy of 0.075 mJ in each pulse. This result is also consistent with the experimental result in Figure 6.43. As such, there exists an optimal pulse number at which the energy deposition was most effective for ignition kernel development.

In addition to plasma discharge gap and frequency, the electrode geometry and discharge voltage waveform also affect ignition process. A computational work by Mao et al. [97] was conducted to investigate the effects of non-equilibrium excitation and electrode geometries on H_2/air ignition in

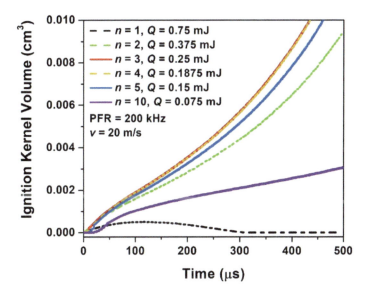

FIGURE 6.46 Time evolutions of ignition kernel volume in 20 m/s flowing H_2/air mixtures with different pulse numbers at PRF = 200 kHz [84].

a nanosecond plasma discharge. The results showed that the production of electronically excited species ($N_2(A)$, $N_2(B)$, $N_2(a')$, $N_2(C)$, and $O(^1D)$) and ground-state atoms H and O at high reduced electric field significantly promoted the ignition kernel development. Vibrational excitation of nitrogen and reactants was less efficient than the electronic excitation and dissociation. Fast gas heating mechanism via excited molecules played a major role in temperature increase during nanosecond discharge phase and early afterglow. In addition, it was shown that the plasma-assisted ignition was significantly affected by the electrode geometry including electrode shape, diameter, and gap size because of the redistribution of the electric field and the plasma density. A larger discharge volume with active species production is critical to the ignition enhancement in addition to the electric field.

In summary, co-optimization of discharge gap geometry with plasma voltage waveform and PRF can promote plasma chemistry and increase ignition kernel size to overcome the critical ignition radius and allow reaching the maximum ignition enhancement with the same discharge energy.

6.4 PLASMA-ASSISTED HOT AND COOL DIFFUSION FLAMES

6.4.1 Extinction Limits of Hot Diffusion Flames

Many combustion systems are non-premixed flames. To understand the dynamics of hot diffusion flames, it is necessary to understand their burning limits. The diffusion flame regimes and extinction limits are governed by the Damköhler number, Da. As shown in the schematic in Figure 6.47, with the decrease of Da, there is not enough flow residence time for combustion to be completed, leading to a decrease of the flame temperature from the adiabatic flame temperature, T_{ad}, and the stretch extinction will occur at Da_L. When thermal radiation is considered, the flame radiation will also reduce the flame temperature as the flow residence time increases, leading to the radiation extinction limit [48,98,99] at Da_H. Together with ignition (point E) and reignition limit (point B), Figure 6.47 represents the ignition and extinction S-curve of non-adiabatic diffusion flames.

Linan [100] developed ignition and extinction theory of adiabatic (non-radiative) counterflow diffusion flames. After that, many researchers conducted analytical studies of non-adiabatic diffusion flame extinction [98,99,101] and studied the relationship between the stretched extinction and radiation extinction limits. Maruta et al. [50] conducted the first experiments of measuring the

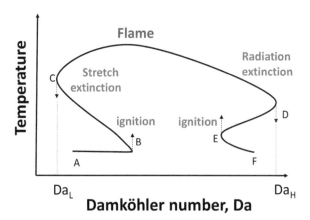

FIGURE 6.47 Schematic of the dependence of diffusion flame temperature on the Damköhler number as well as the ignition and extinction limits.

radiation extinction limit in a counterflow flame in microgravity. Such extinction limits were also reported in droplet combustion much earlier [102–104]. Today the physics and dynamics of both stretch and radiation extinction of hot diffusion flames are well understood.

The dynamics of diffusion flame extinction can be well explained using a counterflow diffusion flame (Figure 6.48) in a pair of opposed jets. By assuming constant properties, steady-state flow, and a one-step reaction model from fuel (F) and oxidizer (O) to product (P).

$$v_F F + v_O O = v_P P, \quad \bar{\omega} = B\rho^2 Y_F Y_O e^{-E_a/RT} \tag{6.62}$$

The normalized governing equations of temperature (θ) and mass fraction (y) in a counterflow diffusion flame (Figure 6.48) can be given as [99],

$$\eta \frac{d\theta}{d\eta} + \frac{d^2\theta}{d\eta^2} = q + \omega \tag{6.63a}$$

$$\eta \frac{dy_F}{d\eta} + \frac{1}{\text{Le}_F} \frac{d^2 y_F}{d\eta^2} = -\omega \tag{6.63b}$$

$$\eta \frac{dy_O}{d\eta} + \frac{1}{\text{Le}_O} \frac{d^2 y_O}{d\eta^2} = -\omega \tag{6.63c}$$

with boundary conditions of

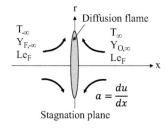

FIGURE 6.48 Schematic of a premixed counterflow flame (right nozzle only) with a flow stretch rate, a, and mixture Lewis number, Le. The stagnation plane is located at $x = 0$.

$$\eta \to -\infty, \quad \theta = \theta_{-\infty}, \quad y_F = Y_{F,-\infty}, \quad y_O = 0;$$

$$\eta \to \infty, \quad \theta = \theta_\infty, \quad y_F = 0, \quad y_O = Y_{O,\infty}(\nu_F W_F / \nu_O W_O)$$

where nondimensionalized reaction rate ω becomes

$$\omega = \text{Da}\, y_F y_O e^{-\theta_a/\theta}, \quad \text{Da} = \frac{B\rho \nu_O W_O / \nu_F W_F}{n W_F a} \tag{6.64}$$

Here a stretch rate, n the geometry factor of the flow (1 for planar and 2 for axisymmetric flow), W the molecular mass, ν the stoichiometric coefficient, and

$$\eta = \frac{x}{\sqrt{\lambda/\rho C_p a n}}, \quad \theta = \frac{T}{Q/C_p}, \quad y_i = \frac{Y_i}{\nu_i W_i / \nu_F W_F}, \quad \theta_a = \frac{E_a C_p}{RQ}, \quad q \approx \frac{4 K_p \sigma Q^3 \theta^4}{C_p^4 \rho a n} \tag{6.65}$$

Equation (6.63) can be solved in three zones: the convection-diffusion zone without reaction and radiation, the convection-diffusion-radiation zone, and the thin reaction-diffusion zone without convection and radiation. In the convection-diffusion zone, Eq. 6.63 in the region of $\eta < \eta_f$ has the solution of,

$$\theta = \theta_{-\infty} + \frac{\theta_f - \theta_{-\infty}}{\int_{-\infty}^{\eta_f} e^{-t^2/2}\, dt} \int_{-\infty}^{\eta} e^{-t^2/2}\, dt$$

$$y_F = y_{F,-\infty} - \frac{y_{F,-\infty}}{\int_{-\infty}^{\eta_f} e^{-\text{Le}_F t^2/2}\, dt} \int_{-\infty}^{\eta} e^{-\text{Le}_F t^2/2}\, dt \tag{6.66}$$

$$y_O = 0$$

For $\eta > \eta_f$, we have

$$\theta = \theta_\infty + \frac{\theta_f - \theta_\infty}{\int_{\eta_f}^{\infty} e^{-t^2/2}\, dt} \int_{\eta}^{\infty} e^{-t^2/2}\, dt$$

$$y_F = 0 \tag{6.67}$$

$$y_O = y_{O,\infty} - \frac{y_{O,\infty}}{\int_{\eta_f}^{\infty} e^{-\text{Le}_O t^2/2}\, dt} \int_{\eta}^{\infty} e^{-\text{Le}_O t^2/2}\, dt$$

In the convection-diffusion-radiation zone, the temperature equation in Eq. 6.63 can be rewritten as,

$$\frac{d^2\theta}{dz^2} = q e^{\eta^2}, \quad z \equiv \int_{-\infty}^{\eta} e^{-t^2/2}\, dt \tag{6.68}$$

By introducing a small thermal radiation parameter, $\varepsilon \ll \varepsilon_r \ll 1$, and assuming the radiation heat loss and the temperature perturbation in the radiation zone can be approximately written in the form of,

$$q = q_f e^{\frac{\theta - \theta_f}{\varepsilon_r \theta_f}}, \quad \theta = \theta_f + \varepsilon_r \vartheta_r \tag{6.69}$$

where q_f is the radiation heat loss at the flame temperature and ϑ_r the deviation of temperature from the flame temperature due to radiation.

By further expanding the coordinate of z using ε_r, in a radiation coordinate of ξ_r,

$$\xi_r = \frac{z - z_f}{\varepsilon_r} \tag{6.70}$$

Equation 6.68 can be rewritten as,

$$\frac{d^2 \vartheta_r}{d\xi_r^2} = \varepsilon_r \, q_f e^{(\eta_f^2 + \vartheta_r / \theta_f)} \tag{6.71}$$

Integration of this equation, respectively, from upstream of fuel ($-\infty$) and oxidizer (∞) to left ($-$) and positive ($+$) flame fronts of the flame and applying the boundary condition for temperature perturbation gradients give the temperature gradient resulted from thermal radiation,

$$\left[\left(\frac{d\vartheta_r}{d\xi_r} \right)^2 \right]_{-\infty}^{-} = 2 \, \varepsilon_r \, q_f \theta_f e^{\eta_f^2} \quad \text{and} \quad \left[\left(\frac{d\vartheta_r}{d\xi_r} \right)^2 \right]_{+}^{\infty} = -2 \, \varepsilon_r \, q_f \theta_f e^{\eta_f^2} \tag{6.72}$$

In the reaction-diffusion zone with flame thickness of $O(\varepsilon)$, by ignoring the radiation and convection, the governing equations become,

$$\frac{d^2 \theta}{d\eta^2} = \omega \tag{6.73a}$$

$$\frac{1}{\text{Le}_F} \frac{d^2 y_F}{d\eta^2} = -\omega \tag{6.73b}$$

$$\frac{1}{\text{Le}_O} \frac{d^2 y_O}{d\eta^2} = -\omega \tag{6.73c}$$

Adding Eq. 6.73a–c, respectively, and integrating the resulting equations across the flame, we will obtain the jump conditions of the flame fronts,

$$\left[\frac{d\theta}{d\eta} + \frac{1}{\text{Le}_F} \frac{dy_F}{d\eta} \right]_{-}^{+} = 0 \quad \text{and} \quad \left[\frac{1}{\text{Le}_O} \frac{dy_O}{d\eta} - \frac{1}{\text{Le}_F} \frac{dy_F}{d\eta} \right]_{-}^{+} = 0 \tag{6.74}$$

Substituting the solutions of Eqs. 6.66 and 6.67 in the convection-diffusion zone to the jump conditions of the reaction zones in Eq. 6.74, we have,

$$\left[\frac{d\theta}{d\eta} \right]_{-}^{+} = -\frac{y_{F,-\infty}}{\text{Le}_F \int_{-\infty}^{\eta_f} e^{-\text{Le}_F t^2 / 2} dt} e^{-\text{Le}_F \eta_f^2 / 2} \tag{6.75a}$$

$$\frac{y_{F,-\infty}}{\text{Le}_F \int_{-\infty}^{\eta_f} e^{-\text{Le}_F t^2/2} dt} e^{-\text{Le}_F \eta_f^2/2} = \frac{y_{O,\infty}}{\text{Le}_O \int_{\eta_f}^{\infty} e^{-\text{Le}_O t^2/2} dt} e^{-\text{Le}_O \eta_f^2/2} \qquad (6.75b)$$

Now by using the temperature solutions of Eqs. 6.66 and 6.67 for θ in the convection-diffusion zone to the temperature gradient $(d\vartheta_r/d\xi_r)$ in the convection-diffusion-radiation zone at $\xi_r \to \infty$ and $-\infty$, and then using the jump condition of Eq. 6.72, we can obtain the temperature gradient $(d\vartheta_r/d\xi_r)$ in the convection-diffusion-radiation zone at the flame fronts, of $\xi_r \to 0^+$ and 0^-,

$$\left[\frac{d\vartheta_r}{d\xi_r}\right]_{-} = F_F(\eta_f, \theta_f, \theta_{-\infty}) \equiv \sqrt{2\epsilon_r\, q_f \theta_f e^{\eta_f^2} + \left(\frac{\theta_f - \theta_{-\infty}}{\int_{-\infty}^{\eta_f} e^{-t^2/2} dt}\right)^2} \qquad (6.76a)$$

$$\left[\frac{d\vartheta_r}{d\xi_r}\right]_{+} = F_O(\eta_f, \theta_f, \theta_{\infty}) \equiv \sqrt{2\epsilon_r\, q_f \theta_f e^{\eta_f^2} + \left(\frac{\theta_f - \theta_{\infty}}{\int_{\eta_f}^{\infty} e^{-t^2/2} dt}\right)^2} \qquad (6.76b)$$

Substituting the above two temperature gradient boundary conditions to the jump conditions of the temperature gradients at the flame fronts of the reaction-diffusion zone in Eq. 6.75, we have,

$$F_F(\eta_f, \theta_f, \theta_{-\infty}) + F_F(\eta_f, \theta_f, \theta_{\infty}) = -\frac{y_{F,-\infty}}{\text{Le}_F \int_{-\infty}^{\eta_f} e^{-\text{Le}_F t^2/2} dt} e^{(1-\text{Le}_F)\eta_f^2/2} \qquad (6.77)$$

This above equation (Eq. 6.77) gives the relationship between flame temperature and location as a function of radiation, diffusion, and reaction. Note Eq. 6.75 gives the relationship between flame, reactant concentrations, and diffusion. These two equations provide the solutions of the diffusion flame structure.

To further analyze the flame extinction, we need to analyze the structure of the reaction zone and its response to the heat loss and stretch rate. By introducing a small perturbation ($\varepsilon = \theta_f^2/\theta_a$) of temperature and reactant concentration from its adiabatic equilibrium values,

$$\theta = \theta_f - \epsilon\, \vartheta, \quad y_F = \epsilon\, \varphi_F, \quad y_O = \epsilon\, \varphi_O \qquad (6.78)$$

and introducing an expanded flame coordinate,

$$\xi \equiv \frac{\eta - \eta_f}{\varepsilon A}, \quad A = \frac{\text{Le}_F \int_{-\infty}^{\eta_f} e^{-\text{Le}_F t^2/2} dt}{y_{F,-\infty} e^{-\text{Le}_F \eta_f^2/2}} \qquad (6.79)$$

substituting the new variables of Eqs. 6.78 and 6.79 into Eq. 6.73, and applying the temperature boundary conditions in Eq. 6.76 on both sides of the flame fronts, we can obtain two new boundary conditions for the gradient of perturbed flame temperature, ϑ, in the reaction zone,

$$\left.\frac{d\vartheta}{d\xi}\right|_{-\infty} = -m \text{ and } \left.\frac{d\vartheta}{d\xi}\right|_{\infty} = 1 - m \qquad (6.80)$$

where

$$m = F_F(\eta_f, \theta_f, \theta_{-\infty}) \frac{\text{Le}_F \int_{-\infty}^{\eta_f} e^{-\text{Le}_F t^2/2} dt}{y_{F,-\infty} e^{(1-\text{Le}_F)\eta_f^2/2}} \quad (6.81)$$

represents the ratio of heat loss from the flame to the fuel side to the heat release due to chemical reaction.

By substituting the fuel and oxidizer conservation equations into the energy equations and representing their relationships as a function of temperature with approximation of Le→1 in the reaction zone, we can obtain the perturbed energy equation in the form of,

$$\frac{d^2 \vartheta}{d\xi^2} = \Lambda(\vartheta + m\xi)[\vartheta + (m-1)\xi]e^{-\vartheta} \quad (6.82)$$

where Λ is the reduced Damköhler number,

$$\Lambda = \varepsilon^3 \text{Da} \text{Le}_O \text{Le}_F A^2 e^{-\theta_a/\theta_f} \quad (6.83)$$

The above equation with the boundary conditions in Eq. 6.80 has an analytic solution of the reduced extinction Damköhler number which can be approximately represented by [99,100],

$$\Lambda_e \approx \frac{ec}{2}(1 - 2c + 1.04c^2 + 0.44c^3), \quad c = \min(m, 1-m) \quad (6.84)$$

Extinction will occur if $\Lambda \leq \Lambda_e$.

By using the parameters of $\theta_a = 5$, $v_O W_O / v_F W_F = 0.9$, $K_p = 1$ m^{-1}, $n = 2$, $C_p = 1400$ J/kg, $T_{-\infty} = T_\infty = 300$ K, $\rho = 0.8$ kg/m^3, $\varepsilon_r = 0.25$, $B = 2.4 \times 10^{10}$ m^3/mol/s, and $W_F = 0.016$ kg/mol to mimic methane/air mixtures, the extinction strain rate (from Eq. 6.84) as function of fuel mass fraction was obtained [99] in Figure 6.49a for different fuel and oxidizer Lewis numbers. Once again, like the premixed flames in Figure 6.24, this is a C-shaped extinction curve with the upper branch defining the stretch extinction limit and the lower branch identifying the radiation extinction limit. The merging point of the two branches defines the flammability limit in terms of the fuel concentration. The experimental confirmation by Maruta and coworkers of two different extinction modes of counterflow diffusion flames was conducted in microgravity by using methane/air mixtures (Figure 6.49b) [50].

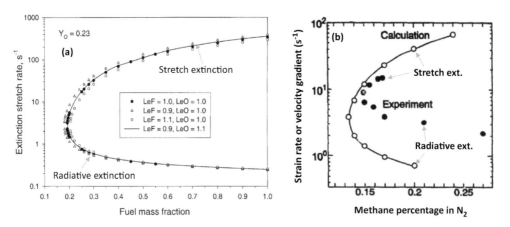

FIGURE 6.49 Stretch and extinction limits as a function of fuel concentration in counterflow diffusion flames. (a) Analytical results [99]. (b) Microgravity experimental results [50].

Plasma has been used to extend the extinction limit of hot diffusion flames. However, the conclusion whether plasma can significantly enhance the extinction limit of a hot diffusion flame kinetically was still in debate. It is convenient to introduce plasma in the air stream to assist flame stabilization. In the early 2000s, it was not clear whether a plasma discharge in an oxidizer stream will generate non-thermal kinetic effects. Ombrello et al. [12,13] used a gliding arc to study the flame extinction limit. Zhu et al. conducted various laser diagnostics of gliding arc structures, dynamics, and OH radical production [105] (Figure 6.50). The gliding arc discharge has both thermal and non-thermal plasma properties. It garners the benefits of high temperatures (1,000–3,000 K) while still having relatively high electron number density and high electron temperature (>1 eV). Thus, it can create both thermal enhancement effect and the kinetic enhancement effect (Figure 6.9) by providing excited species and radicals (Figure 6.50b). There are two types of gliding arc, one is convective or naturally convective gliding arc (Figure 6.50a and b). The other is magnetic stabilized gliding arc (Figure 6.50c). Many previous applications used the former [105,106]. However, because of the turbulent effect induced by this type of gliding arcs, it is difficult to understand the thermal and non-thermal effects of gliding arcs to flame. On the other hand, the magnetic stabilized gliding arc can create a well-defined discharge in a flow, thus benefiting the understanding of the kinetic mechanism of the gliding arc effect.

By using the gliding arc-assisted counterflow diffusion flame setup (Figure 6.8a), the extinction limits of nitrogen-diluted methane–air diffusion flames with and without gliding arc activation on the oxidizer side were measured. As shown in Figure 6.51a, with only 78 Watts of average plasma power input, there was a 220% increase in the extinction strain rate. The power input was less than 6% of the flame power from the fuel stream in the mixture. The results showed that by using a magnetic gliding arc discharge, there was significant enhancement of flame stabilization by having higher strain rates at extinction.

To understand the kinetic effect by gliding arc discharge in the oxidizer stream on the extinction limit, temperature and OH concentration measurements were performed by using Rayleigh scattering thermometry and OH PLIF. The experimental results were compared to numerical modeling by only considering the thermal energy addition of the gliding arc to examine whether only the thermal effect was able to explain the observed extinction limit extension. The air temperature was found experimentally from the Rayleigh scattering and given as the only input to the numerical simulations. Figure 6.51b shows the measured OH-number density change with the increase of plasma powers and the comparison of measured and simulated OH distributions in the flame with plasma power of 78 W at strain rate of 218 s^{-1} using two different kinetic mechanisms (C_1 and GRI 3.0)

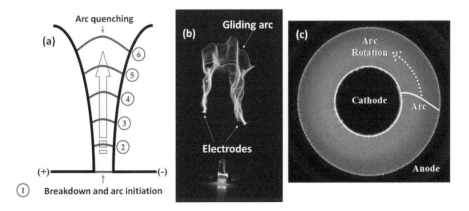

FIGURE 6.50 (a) Schematic of gliding arc initiation, gliding, and quenching by natural convection; (b) OH high-speed images of gliding arc dynamics [105]; and (c) magnetic stabilized gliding arc [12,13].

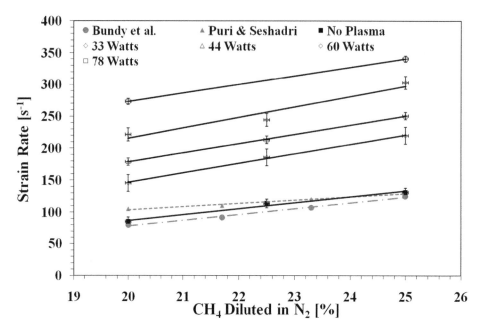

FIGURE 6.51A Effects of plasma power addition on the strain rates at extinction for different levels of N_2 dilution [13].

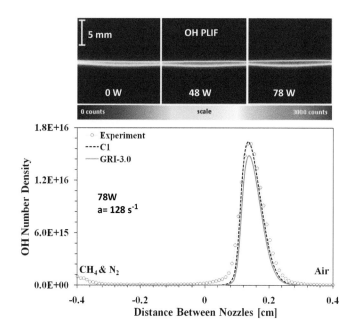

FIGURE 6.51B Comparison of the OH profiles between the two nozzles of the counterflow burner measured through PLIF, with comparison to numerical simulations using only elevated boundary temperatures [12]

[107,108]. Note that the experimental profiles were broader than the computational profiles, but overall, the agreement was within the uncertainty of the measurement, ±15%. The slight broadening of the experimental profile might be caused by the thickness of the laser sheet (250 μm). Nonetheless, the close agreement between the measurements and modeling of OH peak concentration suggested again that the effect of the magnetic gliding arc on the flame was predominately thermal.

To further understand whether an *in situ* plasma discharge can enhance the diffusion flame extinction limits of methane/air mixtures, Sun et al. [109] used a nanosecond plasma discharge in a counterflow burner with direct measurements of atomic oxygen production near the limit of methane diffusion flames at low pressures (Figure 6.8b). The O atom was measured by using TALIF (see Section 9.7 in Chapter 9) [110]. The results (Figure 6.52a) showed the dependence of the measured atomic oxygen concentration on the PRF. It is seen that atomic oxygen concentration increased with the increase of the pulse repetition rate. At plasma frequency of 40 kHz, the measured atomic oxygen concentration was approximately 1.1×10^{16} cm^{-3} (5,000 ppm). The results clearly indicate that non-equilibrium plasma can generate significant amount of O atoms in the oxidizer upstream.

Figure 6.52b shows the comparison of the measured extinction limits with the nanosecond plasma discharge and the electrical heater to the same temperatures as well as the simulation results with (dashed line) and without (solid line with symbols) atomic oxygen production for repetition rates of $f = 20$ kHz at 8,000 Pa. The experimental results showed that for the repetition rate of 5 kHz, the enhancement by the nanosecond discharge and the electrical heater at the same oxidizer temperature of 398 K were approximately equal. This result again implied that the enhancement at a low pulse frequency was predominately by the thermal effect. However, when the repetition rate was increased to 20 kHz, it was seen that the extinction limit by nanosecond plasma was about 10% higher than that by the electrical heater at the same oxidizer temperature of 528 K, indicating that the large amount of atomic oxygen production by plasma could kinetically enhance the extinction limit by promoting the chain-branching reactions to some extent. In addition, the modeling well reproduced the extinction limit without plasma discharge. The results also showed that at 20 kHz, the predicted kinetic enhancement with 1,200 ppm atomic oxygen addition was only 2.5%, much below the kinetic enhancement observed in the experiment. The above comparison demonstrated that non-equilibrium *in situ* plasma discharge with significant amount of O atoms and other excited species production can enhance flame stabilization kinetically. However, there is a crossover temperature between 900 K and 1,000 K, above (below) which the kinetic enhancement by atomic oxygen production was governed by chain-branching (chain termination) reaction (R5.3). The results gave a good explanation to the controversial conclusions made in different experimental conditions for plasma-assisted flame speed and extinction enhancement and suggested that in order to achieve a significant kinetic enhancement effect from atomic oxygen production on flame stabilization, the plasma discharge or gas temperature needs to be above the critical crossover temperature so that the radical quenching is slow and that plasma repetition rate needs to be high enough to produce sufficient atomic oxygen comparable or greater than the level in the flames.

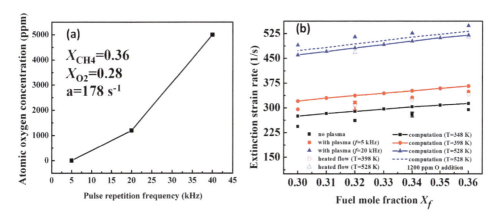

FIGURE 6.52 (a) Dependence of average atomic oxygen concentration at the burner exit on the pulse repetition frequency (b) and comparisons of extinction strain rates between experimental results with plasma activation and an electrically heated flow as well as numerical simulations [109].

6.4.2 Extinction Limits of Cool Diffusion Flames

In hot diffusion flames, we have shown that the kinetic effect of plasma on flame extinction and stabilization is small, and the dominant effects are thermal and transport effect. In this section, we will discuss the plasma kinetic effect on diffusion CFs.

The effect of plasma discharge on CFs was first observed in nanosecond discharge on low-pressure DME diffusion flames [15]. In this experiment, an *in situ* repetitive NSD integrated into counterflow flame was developed by placing porous electrodes at the ends of the burner nozzles (Figure 6.8b). The experiment demonstrated for the first time the phenomenon of plasma-activated low-temperature combustion at low pressure by using the *in situ* non-equilibrium plasma discharge. The experiment was also extended to n-heptane diffusion CF. It was found that plasma-assisted diffusion CF occurred at 72 Torr with a flow residence time shorter than the conversional LTC timescale by several orders of magnitude. It was also shown that plasma discharge not only enhanced ignition but also extended extinction limits. The results revealed that the radical production from plasma accelerated low-temperature ignition was much greater than the high-temperature ignition. Moreover, this study demonstrated that the global reactivity of LTC at low pressure was enhanced so greatly that no low-temperature flame extinction limit was observed with the decrease of fuel concentration (Figure 6.11a). In addition, a reignition limit from CF to HF was observed with the increase of fuel concentration (Figures 6.11 and 6.53).

To understand the chemistry of plasma-assisted diffusion CFs, Won et al. [111,112] used ozone produced by a DBD discharge to enhance the LTC and stabilize an *n*-heptane/N_2/O_2 diffusion CF in a counterflow burner (Figure 6.54a–c). The ozone mole fraction in oxygen varied between 0% and 10%. Figures 6.54a and b show the comparison between an ozone-assisted CF and a HF at the same boundary conditions (7% of *n*-heptane in nitrogen at a stretch rate of 100 s^{-1}). It is seen in Figure 6.54a that with ozone addition, a stable CF was formed without soot formation. On the other hand, the HF has a strong yellow soot emission. The cool diffusion flame temperature was below 800 K and the corresponding hot diffusion flame was around 2,000 K.

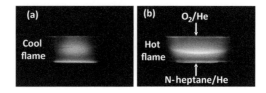

FIGURE 6.53 Direct image of *n*-heptane/O_2/He diffusion (a) cool flame and (b) hot flame sustained by an *in situ* nanosecond repetitive plasma discharge at 72 Torr. Fuel and oxygen mole fractions at fuel and oxidizer stream: $X_F = 0.1$ and $X_{O2} = 0.3$. (Provided by Dr. Tomoya Wada, 2014, Princeton.)

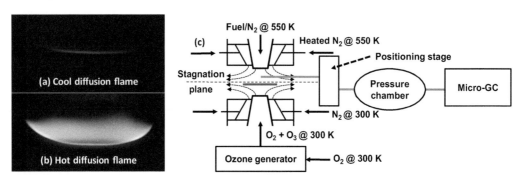

FIGURE 6.54 Direct photos of atmospheric (a) cool and (b) hot *n*-heptane/O_2–O_3 diffusion flames, observed at the identical flow condition of $X_F = 0.07$ and $a = 100$ s^{-1}. (c) Experimental setup for establishing self-sustaining counterflow cool diffusion flames [111,112].

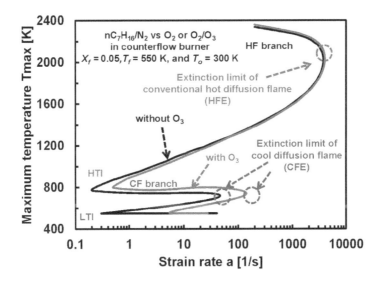

FIGURE 6.55 Calculated S-curves of n-heptane/O_2–O_3 diffusion flames in counterflow configuration for $X_F = 0.05$, with and without 3% ozone addition at the oxidizer side [112].

The ozone-enhanced LTC was presented already in Chapter 5. To elucidate the kinetic effect of ozone on cool and hot diffusion flames, numerical calculations were performed to investigate the ignition-extinction S-curve [111,112]. Figure 6.55 shows the calculated variation of the maximum flame temperature (T_{max}) as a function of the stretch rate (a) for steady-state n-heptane diffusion flames for fuel mole fraction at $X_F = 0.05$ with and without 3% ozone addition at the oxidizer stream. Numerical calculations showed clearly the two stable flame branches, a HF and a CF. The temperature on the HF (> 2,000 K) was much higher than that of the CF (< 800 K). Note that 3% ozone addition only slightly extended the HF extinction (HFE) limit from 3,700 s^{-1} to 3950 s^{-1}, indicating again a small kinetic effect. However, for the CF, 3% ozone addition significantly extended the CF extinction limit from 46 s^{-1} without ozone to 141 s^{-1}, indicating a strong kinetic effect. In addition, with ozone addition, the low-temperature ignition (LTI) limit was so greatly extended that it was even much higher than the high-temperature ignition (HTI) limit. As such, the modeling showed that both CF and LTI were dramatically enhanced by ozone addition. Note that the kinetic effect of ozone addition on CFs was also confirmed for different large alkanes and oxygenated fuels [113–115].

Recently, the ozone-assisted cool diffusion flames were also studied in elevated pressure for alkanes and oxygenated fuels [116–119]. Moreover, the plasma NO$_x$ effect on CF was also studied [118,120]. Figure 6.56a–c depicts the non-premixed DME CF images without/with ozone addition in the oxidizer at 1 and 3 atm. It can be noted that a CF could not be observed without ozone addition for fuel mole fraction of 0.3 and strain rate of 100 s^{-1} at 1 atm (case a), while it can exist with 1% ozone addition at the same condition (case b). Moreover, as pressure increases to 3 atm, the CF existed even without ozone addition (case c). Moreover, ozone addition can also promote CF transition to a new warm flame at a higher flame temperature: warm flame (Figure 6.56d). Therefore, plasma-generated ozone or other active species can enhance the CF chemistry and enable CF formation at lower fuel concentration and higher flow rate and promote CF reignition to warm flame or HF.

6.4.3 Extinction Limits of Warm Flames

While the above experiments showed that plasma can enhance CF formation [121,122], perhaps one of the more significant findings of plasma-assisted cool diffusion flames [103,104,111,112,123]

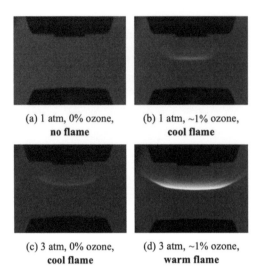

FIGURE 6.56 Non-premixed DME cool (a–c) and warm (d) flame images with fuel mole fraction of 0.3 and at 1 and 3 atm strain rate of 100 s^{-1} [116].

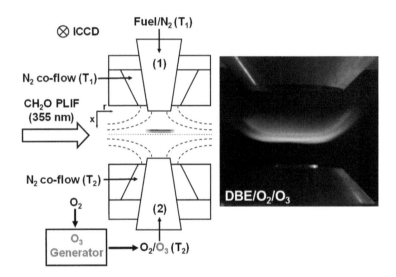

FIGURE 6.57 Experimental apparatus (left) and visual image (right) of multistage warm diffusion flame of dibutyl ether/oxygen/ozone (X_{O3} = 0.0409) at a strain rate of 55 s^{-1} [124,125].

was the experimental observation of two-stage warm diffusion flames [124–127]. By using ozone addition, Yehia et al. [124,125] conducted experiments using non-premixed dibutyl ether (DBE)/oxygen flames in a counterflow geometry (Figure 6.57) and observed experimental evidence of a self-sustaining low-temperature multistage warm diffusion flame, existing between the CF and HF, at atmospheric pressure. Figure 6.57 (right) shows an unfiltered visual image of a two-stage dibutyl ether warm diffusion flame. The warm flame has two luminescent reaction zones, a weaker CF reaction zone on the fuel side (top burner) and a stronger intermediate-temperature reaction zone on the oxidizer side (bottom burner).

Figure 6.58 shows the comparison of key species and chemical heat release rates of dibutyl ether/oxygen warm flames without and with ozone addition. It is clearly seen that warm flames have

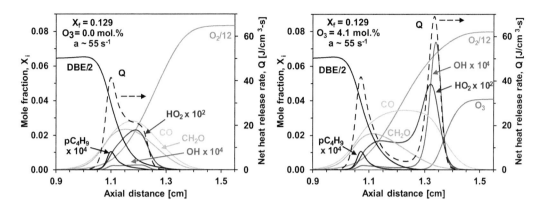

FIGURE 6.58 Heat release rate and key species in warm diffusion flames without (left) and with (right) ozone-doped ($X_{O3} = 0.0409$) oxidizer [124,125].

two-stage reaction zones with a CF on the fuel side and an intermediate-temperature flame on the oxidizer side. In the first-stage CF zone, CH_2O, pC_4H_9, and CO are the major products from the RO_2 LTC (Table 5.3 and Figure 5.12). In the second-stage reaction zone, significant HO_2, OH, CO and chemical heat release are formed via the intermediate-temperature chain-branching pathway. Specifically, OH radicals were mainly produced from $CH_2CHO + O_2 = OH + CO + CH_2O$, $C_2H_5 + HO_2 = OH + C_2H_5O$, $CH_3 + HO_2 = OH + CH_3O$, and $CH_2O + O = HCO + OH$. This is exactly the intermediate-temperature chain-branching pathway (Table 5.3). It is noticed that ozone addition significantly enhanced the intermediate-temperature flame zone of the warm flame.

In addition, with ozone sensitization, Yehia et al. [125] demonstrated experimentally that the two-stage warm diffusion flames are also possible for *n*-alkanes such as *n*-heptane, *n*-decane, and *n*-dodecane in an atmospheric counterflow burner. Furthermore, at high pressure, as shown in Figure 6.56d, ozone addition (1%) enhanced the intermediate-temperature fuel oxidation chemistry, and a warm flame was successfully observed at 3 atm.

The kinetic effects of plasma-generated NO on the flame dynamics and burning limits of cool and warm diffusion flames were investigated experimentally and computationally by using *n*-dodecane in a counterflow system. As discussed in Chapter 5, NO plays distinctive roles in cool and warm flame chemistry due to their different reaction pathway sensitivities to and interactions with NO. Figure 6.59 shows the effect of NO on the temperatures of cool and warm *n*-dodecane diffusion flames with and without 300 ppm NO addition as a function of flow stretch rate at 8 atm. It is clearly seen that a warm flame extinguishes to a CF from the extinction transition point (WCE) with the temperature decreasing or can jump to the HF through the ignition transition point (WHI) with the

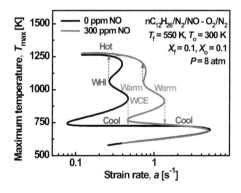

FIGURE 6.59 Calculated *S*-curves depicting different flame regimes with and without NO additions (300 ppm) for $X_O = 0.1$, $X_F = 0.1$ and 8 atm [118].

increase of temperature. Moreover, the strain rates of both WCE and WHI increase with NO addition, indicating that NO delays the warm flame extinction to CF and promotes the transition to HF. Both experiments and modeling showed that NO addition significantly inhibited the CF extinction limits but enhanced warm flames. The warm flame extinction transition to CF was delayed by NO addition while the warm flame ignition transition to HF was promoted.

6.5 PLASMA-ASSISTED CONTROL OF DETONATION AND DEFLAGRATION TO DETONATION TRANSITION

6.5.1 Detonation

Recently, there is increasing interest in the development of pressure gain combustion engines or rotating detonation engines [128–130]. Compared to conventional constant pressure gas turbine engines which operate in the Brayton cycle, pressure gain detonation engines operate at constant volume cycles. As shown in Figure 6.60, detonation engines will produce additional work and thus are more efficient than conventional gas turbine engines. However, to operate for fuel lean air mixtures, detonation initiation and instability remain very challenging. Therefore, plasma provides a promising technique to enhance ignition and can potentially accelerate deflagration to detonation transition (DDT) and enhance detonation wave stability [131–133]. However, it is not well understood how plasma kinetically enhances detonation transition.

Different from a premixed flame with a reaction-diffusion wave front, a detonation wave is an exothermic convection-reaction front (Figure 6.61). When the heat release in a reaction zone of a propagating wave (Figure 6.61) is so fast that it cannot be dissipated quickly enough by thermal expansion via acoustic waves, a significant pressure rise via acoustic compression will occur in the reaction zone. When the acoustic compression becomes very strong, it can lead to a shock wave and

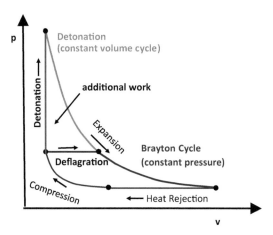

FIGURE 6.60 Schematic of comparison between the constant pressure combustion cycle and the constant volume detonation cycle in the pressure-volume diagram.

FIGURE 6.61 Schematic of the thermal dynamic property changes across a reaction wave front.

shock coupled autoignition, i.e., a detonation wave. The speed of a detonation wave depends on the acoustic wave speed and heat release rate of the reactants as well as thermodynamic properties.

If one stands on the reaction wave front in Figure 6.61 and assumes the wave propagation is steady state, the conservation laws of mass, momentum, and energy as well as the equation of state across the reaction zone can be given as,

$$\rho_1 u_1 = \rho_2 u_2 \tag{6.85}$$

$$p_1 + \rho_1 u_1^2 = p_2 + \rho_2 u_2^2 \tag{6.86}$$

$$C_p T_1 + \frac{1}{2} u_1^2 + q = C_p T_2 + \frac{1}{2} u_2^2 \tag{6.87}$$

$$p_1 = \rho_1 R T_1, \; p_2 = \rho_2 R T_2 \tag{6.88}$$

where q is the volumetric chemical heat release.

Through some algebraic manipulations of Eqs. 6.85–6.88 [45,134,135], the following two new equations for pressure and density changes after the reaction front can be derived,

$$\frac{\gamma}{\gamma-1}\left(\frac{p_2}{\rho_2} - \frac{p_1}{\rho_1}\right) - \frac{1}{2}(p_2 - p_1)\left(\frac{1}{\rho_1} + \frac{1}{\rho_2}\right) = q \tag{6.89}$$

$$\gamma M_1^2 = \left(\frac{p_2}{p_1} - 1\right) / \left[1 - \frac{\rho_1}{\rho_2}\right] \tag{6.90}$$

where γ is the ratio of specific heat and M is the Mach number.

Equation 6.89 represents the Hugoniot relation with chemical heat release of q, whose solution can be represented by a hyperbola curve (Figure 6.62). Equation 6.90 is the Rayleigh line starting from the upstream condition. The gradient of the Rayleigh line indicates the upstream Mach number. The intersections of the Hugoniot curve and Rayleigh line are the solutions to the governing equations. It is seen that there are two solution branches in terms of the Mach number. The first solution with a significant pressure and density increase on the upper left branch is detonation. Detonation is a supersonic shock induced ignition and combustion wave ($M_1 > 1$). The second

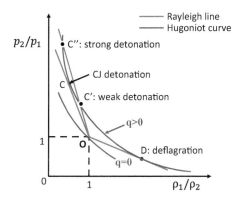

FIGURE 6.62 Schematic illustrating the Rankine-Hugoniot solutions.

solution is a subsonic combustion wave ($M_1 < 1$), and the corresponding intersection is on the lower right of the diagram. This solution is the reaction-diffusion deflagration wave, i.e., a flame. For the deflagration wave, the pressure rise across the reaction zone is very small. The solution of the deflagration wave has been discussed in Section 6.3 in detail.

On the detonation branch, for a given detonation speed (the same Rayleigh line), there are two possible solutions on points C' and C'', respectively, representing the weak and strong (overdriven) detonations. Both the strong and weak detonation waves are not self-sustainable or stable, the only stable steady-state solution is the minimum detonation speed at which the Rayleigh line is tangential to the Hugoniot curve. This unique detonation state is called the Chapman–Jouguet (CJ) detonation at which the detonation speed is the sonic speed of the burned gas ($M_2 = 1$). Based on Eqs. 6.85–6.88, the CJ detonation speed can be derived as,

$$M_1^2 = M_{CJ}^2 = 1 + \frac{\gamma^2 - 1}{\gamma} q \left\{ 1 + \left[1 + \frac{2\gamma}{(\gamma^2 - 1)q} \right] \right\}^{1/2} \qquad (6.91)$$

The one-dimensional, quasi-steady-state structure of CJ detonation was studied by Zel'dovich, von Neumann, and Doring (ZND) independently [134,135]. The ZND model is shown schematically in Figure 6.63. The detonation structure includes a leading shock wave, an ignition induction zone, and a combustion zone. The supersonic shock wave travels at the Mach number of M_{CJ}. After the shock wave, the pressure, temperature, and density increase, but the velocity decreases and the flow is subsonic. After the ignition delay time, t_{ig}, with the induction length of x_{ig}, ignition occurs, leading to an increase in temperature and velocity. The combustion heat release produces compression waves which propagate upstream in the subsonic flow and support the leading shock wave. At the end of the combustion zone, the flow Mach number reaches unity so that the burned gas is choked and the ZND detonation structure is self-sustained and isolated from any perturbations downstream.

A key question in detonation is how to initiate detonation and how DDT occurs.

6.5.2 Deflagration to Detonation Transition

There have been many studies in the mechanisms of DDT [134,136–139]. Several mechanisms such as the hot spot theory, pressure gradient (hydraulic resistance) theory, and turbulent flame acceleration hypothesis have been proposed (Figure 6.64).

Hot spot and gradient mechanism: The "hot spot" or "temperature gradient" mechanism for ignition-to-detonation transition was first proposed by Zeldovich [137] (Figure 6.64). In this mechanism, a temperature gradient in space will result in a gradient of ignition delay time, leading to a

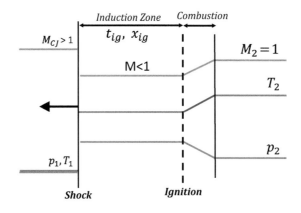

FIGURE 6.63 Schematic of the ZND detonation structure

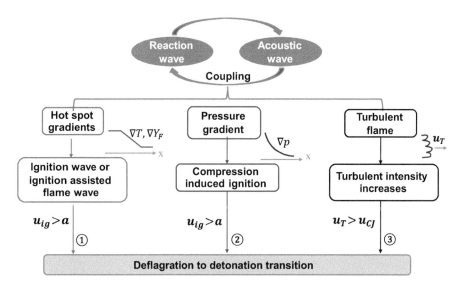

FIGURE 6.64 Proposed different mechanisms for DDT.

propagating autoignition wave from high temperature to low temperature in the hot spot. Therefore, there is a critical temperature gradient for ignition and acoustic wave coupling when the ignition wave speed (dt_{ig}/dT) equals to the sonic speed (a),

$$\nabla T_{\text{cri}} = \left(\frac{dT}{dx}\right)_{\text{cri}} = a^{-1}\left(\frac{\partial t_{ig}}{\partial T}\right)^{-1} \tag{6.92}$$

It was shown that there are, respectively, three different ignition modes—spontaneous ignition, detonation, and deflagration—corresponding, respectively, to small, moderate, and large ξ,

$$\xi = \frac{\nabla T}{\nabla T_{\text{cri}}} \tag{6.93}$$

- $\xi \ll 1$, spontaneous ignition (ignition speed $\gg$ sonic speed)
- $\xi \sim 1$, detonation (ignition speed $\sim$ sonic speed)
- $\xi \gg 1$, deflagaration (ignition speed $\ll$ sonic speed)

Bradley and coworkers [138,140,141] performed numerical studies using the hot spot model with both one-step and detailed chemistry, and obtained a detonation diagram by using the normalized critical temperature gradient, ξ, and the normalized chemical reaction time, ε,

$$\varepsilon = \frac{\delta}{a\tau_e} \tag{6.94}$$

where δ is the hot spot size, and τ_e is the chemical heat release time or excitation time. The results showed that both the ignition wave speed and the heat release time relative to the acoustic wave time are important.

Recently, Chen and coworkers [142] extended the modeling the ignition-to-detonation transition near the low-temperature NTC region of n-heptane/air mixtures by using a cold spot (negative temperature gradient). The results showed that like a hot spot, the cold spot can also generate detonation due to the NTC effect on the ignition delay time.

To understand the effect of concentration gradient on DDT formation, Sun et al. [143] studied the DDT of stratified n-heptane and toluene/air mixtures with a given concentration gradient and uniform temperature. It was shown that a concentration gradient can result in strong pressure rise and supersonic ignition waves. To further understand the concentration gradient and its impact on DDT, Zhang et al. [144] and Qi et al. [145] simulated the effects of the combined effect of concentration and temperature gradient on DDT. By introducing both temperature and concentration gradients, the critical gradients of temperature and concentration for ignition and acoustic wave coupling in Eq. 6.92 were extended to [144],

$$\nabla(T,\phi)_{cri} = a^{-1}\left(\frac{\partial t_{ig}(T,\phi)}{\partial T}\right)^{-1} = a^{-1}\left(\frac{\partial t_{ig}}{\partial T}\frac{dT}{dx} + \frac{\partial t_{ig}}{\partial T}\frac{d\phi}{dx}\right)^{-1} \quad (6.95)$$

The normalized coupled temperature and concentration gradient can then be approximated linearly as [146]:

$$\xi_{T\phi} = \xi_T + \xi_\phi \quad (6.96)$$

Figure 6.65a shows the computed DDT boundaries of heptane/air mixtures with a combination of temperature and concentration gradients. The results show that both concentration gradient and temperature gradient can induce DDT. Moreover, the results show that the temperature gradient can strongly couple with the concentration gradient to affect DDT. A unified critical gradient criterion involving both temperature and concentration gradients was proposed and tested (Eq. 6.96). A critical ignition-to-detonation initiation diagram with concentration gradients and heat release time is plotted in Figure 6.65a [145]. It is seen that the concentration gradient acts like the temperature gradient (hot or cold spot) in affecting DDT. In addition, Figure 6.65b shows the DDT peninsula using the unified temperature and concentration gradient as well as the normalized chemical reaction time [147]. It is seen that the normalized gradient of hot spot and the normalized chemical reaction time can describe well the DDT onset boundaries.

Pressure gradient and hydraulic resistance mechanism: When a flame propagates in a narrow channel, a channel with obstacles, a porous media, or a dusty mixture, the wall friction, or the drag from the obstacles, porous materials, and particles can induce a pressure gradient (Figure 6.64). If the pressure gradient is so large, thermodynamic compression can generate a temperature gradient and trigger autoignition in this pressure gradient region. Sivashisky [136, 148] as well as Ju and Law [149] studied analytically the hydraulic drag effect on DDT and multiple combustion regimes. It was shown that pressure gradient generated by hydraulic drag can promote coupling between acoustic compression and autoignition, leading to DDT.

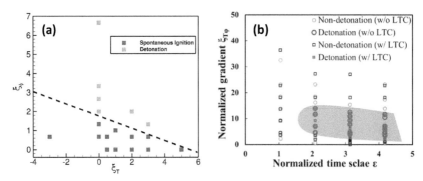

FIGURE 6.65 (a) ignition-to-detonation transition boundary with both temperature and concentration gradients for n-heptane/air mixtures at 900 K and 40 atm [144]. (b) Detonation peninsula by two non-dimensional parameters: normalized gradient $\xi_{T\phi}$ or $\xi_{T\phi}^*$ and normalized time scale ε [147].

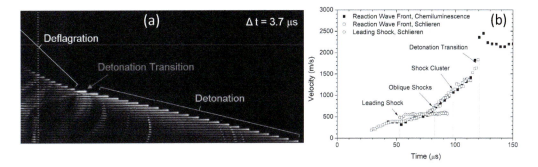

FIGURE 6.66 (a) DDT of stoichiometric C_2H_4/O_2 mixture at ambient temperature and pressure in 1 mm channel (b) velocity time history of DDT in a microchannel [150].

Recently, many experimental and numerical simulations [132,150–153] have also demonstrated that flame acceleration due to the viscous boundary layer effect in a microchannel can lead to DDT. Figure 6.66 shows the DDT development of the flame front and shock wave formation processes of stoichiometric ethylene and oxygen mixture in a 1×1 mm^2 channel by Chan and Wu [150]. Several different flame acceleration stages were identified prior to DDT. The formation of leading shock waves, oblique shocks, and shock clusters was observed due to acoustic compression and interaction with the boundary layer. It was observed that a local explosion (autoignition) drove the reaction front into detonation. Therefore, like the hot spot mechanism, the pressure gradient or hydraulic drag mechanism also requires an autoignition process before detonation occurs.

Turbulent flame mechanism: Some researchers also proposed turbulent flame acceleration to DDT model. In this hypothesis, the turbulent flame speed, S_T, is increased by the turbulence and reaches to the flame speed at which the burned gas is choked (sonic speed) [139,154],

$$S_T = S_{CJ} = a_2 \rho_2 / \rho_1 \qquad (6.97)$$

By using a turbulent shock tube facility, DDT in a highly turbulent reactive flow was explored by observing the runaway acceleration with a pressure buildup that led to a turbulence-induced DDT. The flame dynamics and the associated reacting flow field were characterized using simultaneous high-speed particle image velocimetry, OH chemiluminescence, pressure sensor measurements, and Schlieren imaging. It was observed that the locally measured turbulent flame speed was greater than that of a Chapman–Jouguet deflagration speed, S_{CJ}, before detonation occurred. The observed flame regimes varied from slow deflagrations to detonations as shown in Figure 6.67. There were four flame velocity regimes in the laboratory reference frame: slow quasi-isobaric deflagration, fast compressible deflagration, shock-flame complex, and detonation. Three critical boundaries between these regimes are shown in Figure 6.67: (1) sonic velocity of the reactants (a_1), (2) sonic velocity of the products (a_2), and (3) CJ detonation velocity (u_{CJ}). As shown in Figures 6.67a and b, fast deflagration wave that surpassed the sonic velocity of the unreacted mixture (a_1) began to generate compression waves. Such compression waves appeared as lighter color structures in the Schlieren image (Figure 6.67). These compression waves gained strength as the flame accelerated, leading to the formation of a secondary shock between the flame and the initially transmitted shock. When the flame speed reached the isobaric product sonic velocity (a_2), there was a tight acoustic coupling of the flame and the compressed region. Shortly after that, the secondary shock caught up with the leading shock and merged with it (Figure 6.67c) causing its significant amplification. The distance between the flame and this global shock decreased (Figure 6.67d), and the shock-flame complex propagates supersonically in the laboratory frame of reference. At the same time, the turbulent flame displacement speed, S_T, relative to the highly compressed reactants behind the global shock remained subsonic. This coupled structure traveled together to accelerate, eventually transitioning

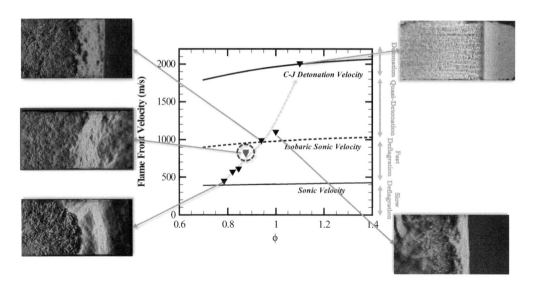

FIGURE 6.67 Propagation regimes of highly turbulent fast flames from deflagration (a)–(d) to detonation (e). Red highlighted frame indicates critical CJ deflagration regime for onset of runaway flame [154].

to a detonation. The resulting detonation shown in Figure 6.67e formed in the overdriven state and quickly relaxed to speed u_{CJ}. It was concluded that the critical stage for turbulent flame DDT was the formation of pressure waves and shocks within the turbulent flame brush, which occurred when the flame speed approached and exceeded that of a CJ deflagration (S_{CJ}), thus forming a sonic plane in products and a choking flow. Different from the pressure gradient mechanism and the hot spot mechanism, in these studies [139,154] no ignition event was reported or discussed between the chocking product flow and detonation. It is not clear how a turbulent flame brush transferred to a detonation wave without shock induced autoignition. More exploration in autoignition events before detonation in turbulent flame to DDT studies is needed.

6.5.3 Control of Detonation Using Plasma-Enhanced Combustion Chemistry

The above studies showed that detonation is strongly affected by the coupling between autoignition with shockwave and acoustic compression. As discussed in combustion and plasma chemistry in Chapter 5, plasma can significantly accelerate autoignition, especially at low temperatures. Therefore, it would be interesting to know whether one can control DDT by using plasma to enhance ignition chemistry at low and intermediate temperature to enhance ignition-shock wave coupling.

Many studies of plasma effect on DDT were conducted [155]. The first experiment to explore plasma effect on DDT in a microchannel was carried out by Sepulveda et al. and Vorenkamp et al. [131–133, 156] by using ozone to sensitize the ignition chemistry. As shown in Figure 6.68 (left), with plasma discharge, the active species production in plasma can accelerate the chain-initiation and branching reactions, thus shortening the ignition delay time and the induction zone. The shortening of the induction zone will strengthen the coupling between the leading shockwave and combustion, thus enhancing DDT. Figure 6.68 (right) shows the comparison between the ignition delay time of a lean acetylene/oxygen mixture with and without 1% ozone addition. It is seen that not only ozone addition shortened the ignition delay time by one-order of magnitude at intermediate temperatures but also it changed the gradient of ignition delay time to temperature. Therefore, plasma discharge will also change the critical temperature or concentration gradient of detonation transition (Eq. 6.95), thus modifying the DDT process.

Plasma-Assisted Combustion: Dynamics

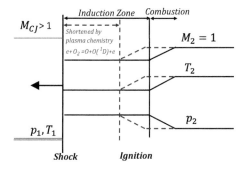
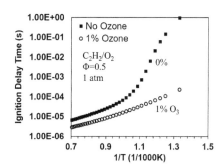

FIGURE 6.68 Left: schematic of plasma chemistry effect on the shortening of induction length and shock-ignition coupling of a detonation wave. Right: comparison of predicted ignition delay times for acetylene/O_2 mixtures at $\phi = 0.5$ with and without 1% ozone addition at atmospheric pressure.

The experiment of Sepulveda et al. [132,133] examined the chemical kinetic effect of ozone addition on the onset of DDT for C_2H_2/O_2 mixtures at fuel lean conditions in a square microchannel (1 mm²). It was found that the addition of ozone significantly reduced the DDT onset time by up to 77.5%, whereas it only slightly increased CJ velocities. In addition, the 1% ozone addition extended the DDT limit from equivalence ratio of 0.3 to 0.2. It was also shown that ozone addition had a much larger effect on DDT time than the increase of equivalence ratio. The results clearly revealed that, for accelerating DDT, the kinetic effect via ozone addition was much greater than the thermal effect. The results provided a good insight to develop non-equilibrium plasma to control detonation.

Since ozone decomposes quickly at temperatures above 400 K, it is of great interest to see how *in situ* plasma discharge will affect DDT. Vorenkamp et al. developed a microchannel DDT reactor with a nanosecond dielectric barrier discharge (ns-DBD) [131,156] (Figure 6.69). They examined the kinetic enhancement by repetitive ns-DBD on the DDT process of fuel lean DME, oxygen, and argon premixtures. As shown in Figure 6.70, the ns-DBD plasma microchannel has a rectangular cross-section of 1 × 4 mm² and a length of 600 mm. A dielectric barrier discharge was formed across the channel between copper electrodes isolated from the discharge gap by a Kapton film of 0.089 mm thickness. The distance between the top and bottom electrodes was 1 mm, the width of the electrodes and the channel was 4 mm, and the length of the electrodes was 500 mm. An SI system was installed at a distance of 50 mm from the edge of the DBD electrodes to initiate a deflagration wave in the channel. The channel windows were fused silica to allow optical diagnostics. The plasma discharge was controlled by a signal generator that initiated a sequence of 10 kV, 10 kHz, 200 ns pulses ahead of ignition. A variable number of pulses were applied, with the final pulse 100 μs prior to SI. To conduct time-resolved measurements of temperature and species, high-speed imaging, one-dimensional, two-beam, femtosecond/picosecond, coherent anti-Stokes Raman scattering (CARS), as well as CH_2O PLIF were conducted.

Figure 6.69 (bottom) shows the comparison of flame front time histories of a DME/oxygen mixture without and with 15 and 5,000 preignition plasma discharge pulses, respectively. It is clearly seen that without plasma discharge, the flame front accelerated with time and DDT occurred with a sudden increase of chemiluminescence. It is also seen that with 15 pulses of plasma discharge, DDT was dramatically accelerated. However, with 5,000 preignition plasma discharge pulses, although the initial chemiluminescence was increased, the occurrence of DDT was surprisingly delayed.

Figure 6.70 shows the history of reaction front velocities without and with 15 and 5,000 preignition plasma discharge pulses. Without plasma discharge, two distinct velocity peaks occurred, one around 70 μs and the other around 115 μs. This is unique for DME, which has LTC, and was not observed for C_2H_2 DDT experiments [132]. The first peak was suspected to be caused by the LTC of DME and the heat loss near the wall. The second peak was due to the high-temperature autoignition and DDT where a steep overdriven velocity profile was observed before the wave speed equilibrated close to

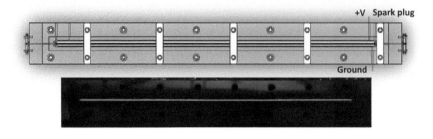

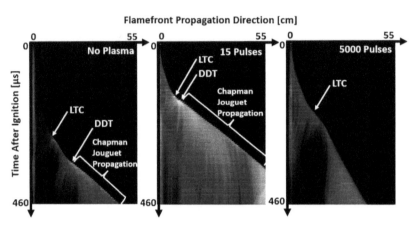

FIGURE 6.69 Top: shematic design and a direct image of nanosecond plasma discharge in a microchannel for DDT study, Bottom: comparison of flame front time histories of a DME/oxygen mixture without and with 15 and 5,000 preignition plasma discharge pulses.

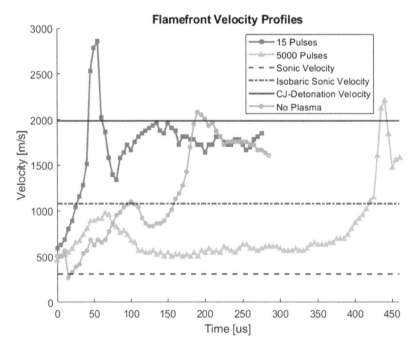

FIGURE 6.70 Flame-front velocity as a function of time for the control case, no plasma pulses, as well as the cases resulting in the greatest reduction and increase in DDT onset distance and time, 15 pulses and 5,000 pulses, respectively [131,156].

Chapman–Jouguet velocity. When 15 ns-DBD pulses were applied, the experiments exhibited the highest degree of DDT enhancement. It is seen that the distance between the two peaks was dramatically shortened, indicating an accelerated transition from LTI to high-temperature ignition. However, when 5000 plasma pulses were applied to the mixture prior to ignition, two velocity peaks were once again observed, and the second peak was greatly delayed. The experiment and simulation showed that this delay at a large number of plasma pulses was caused by the over consumption of the fuel by the preignition plasma discharges and thus the reduction of the chemical heat release in the deflagration process. In summary, the results showed that plasma discharge nonlinearly affected the onset time and distance of DDT. A small number of plasma discharge pulses prior to ignition resulted in reduced DDT onset time and distance by 60% and 40%, respectively, when compared to the results without plasma. The results also showed that a substantial increase of plasma discharge pulses resulted in an extended DDT onset time and distance of 224% and 94%, respectively. Therefore, plasma discharge can accelerate DDT. However, at the same time, consumption of fuel with excessive plasma discharge enhanced fuel oxidation can reduce the mixture fuel concentration and heat release rate (Eq. 6.94), leading to a delayed or failed DDT process.

In Chapter 5, we have shown that plasma can significantly enhance low- and intermediate-temperature chemistry. Since DME has strong LTC, it is necessary to demonstrate that plasma discharge and the resulting accelerated LTC affect the DDT process. From the low and intermediate chain-branching reaction pathways in R5.57 of Chapter 5, we know that one of the major indicators of the LTC is the formation of CH_2O. To measure the time history of CH_2O in plasma-assisted DDT, Vorenkamp et al. [156] conducted *in situ* laser diagnostics and computational modeling to examine the influence of the ns discharge on the LTC of DME and DDT using formaldehyde (CH_2O) laser-induced fluorescence (LIF) and detailed kinetic analysis. In the experimental study, a fixed quantity of discharge pulses, 500, was applied prior to the SI in the DDT channel (Figure 6.69) to observe the plasma enhancement of LTC via CH_2O measurements. The CH_2O formation in the region associated with plasma-enhanced DME low-temperature oxidation is shown in Figure 6.71. In Figure 6.71a, a case without plasma discharge, one can see strong chemiluminescence in the flame zone but only a very weak CH_2O LIF signal near the flame front in this region. This is because near the flame front, CH_2O was formed as an intermediate species in the preheating zone via CH_3O + OH/O reactions). In Figure 6.71b, with plasma, we instead see a very strong CH_2O LIF signal in a broad plasma region ahead of the flame front spanning nearly 2 cm. This region of the flow was a

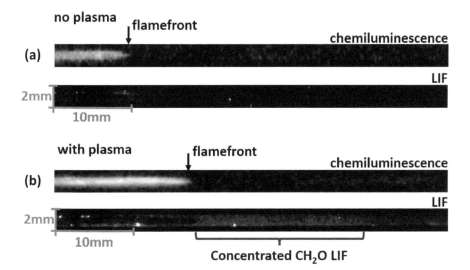

FIGURE 6.71 Chemiluminescence and corresponding CH_2O LIF images for combustion experiments for cases both (a) without and (b) with plasma applied prior to ignition [156].

shock compression zone, suggesting that there was a coupling between plasma-enhanced LTC and the shock clusters. Kinetic modeling showed that the formation of CH_2O at shock compressed flow temperature was via reaction,

$$CH_2OCH_2O_2H = 2CH_2O + OH \qquad (R5.50)$$

Therefore, the experiments not only showed that a ns discharge accelerated LTC of DME in a microchannel but also accelerated the detonation transition via the coupling of shock compression and plasma-enhanced LTC.

Recently, more experiments on the effect of plasma discharge on detonation cell size were conducted. Cherif et al. [157] studied the effect of a volumetric nanosecond discharge on detonation cell size in CH_4:O_2:Ar and CH_4:O_2:Ar:H_2 mixtures, at 180 mbar and 120 mbar, respectively. It was shown that the detonation cell size was reduced by a factor of 1.5–3, while passing through the region of the discharge. Similar experiment was also conducted for hydrogen/oxygen mixtures by Crane et al. by using ozone [133]. The observed reduction of detonation cell size indicates the decrease of the ignition delay time (Figure 6.68), thus the enhancement of ignition and shock wave coupling.

In summary, in this chapter, we summarized the theory of ignition, flame propagation, ignition-assisted flame propagation, extinction, and detonation initiation. We have shown that plasma can dramatically enhance ignition and ignition-assisted flame propagation. We also have shown that the plasma effect on HF propagation and extinction is dominated by thermal effect. Extinction can be dramatically extended by plasma via shortening the ignition time and rendering a smooth transition from ignition to flame without an extinction limit. Moreover, it is shown that plasma can dramatically enhance the flame speed and extend extinction limit of low-temperature flames via plasma-enhanced LTC and ozonolysis. Plasma enabled the observation of diffusion and premixed cool flames at low pressure and a high-speed flow. Finally, we showed that plasma can dramatically accelerate DDT by strengthening the coupling between autoignition and shockwaves and accelerating the LTC. With the understanding of the fundamentals of plasma-assisted ignition, flame propagation and detonation, we will discuss the plasma effects on various combustion application systems in Chapter 7.

REFERENCES

1. Wang, H. and M.A. Oehlschlaeger, Autoignition studies of conventional and Fischer–Tropsch jet fuels. *Fuel*, 2012. **98**: pp. 249–258.
2. Vasu, S.S., D.F. Davidson, and R.K. Hanson, Jet fuel ignition delay times: Shock tube experiments over wide conditions and surrogate model predictions. *Combustion and Flame*, 2008. **152**(1–2): pp. 125–143.
3. Sung, C.-J. and H.J. Curran, Using rapid compression machines for chemical kinetics studies. *Progress in Energy and Combustion Science*, 2014. **44**: pp. 1–18.
4. Shen, H.-P.S. and M.A. Oehlschlaeger, The autoignition of C_8H_{10} aromatics at moderate temperatures and elevated pressures. *Combustion and Flame*, 2009. **156**(5): pp. 1053–1062.
5. Bozhenkov, S., S. Starikovskaia, and A.Y. Starikovskiia, Nanosecond gas discharge ignition of H_2– and CH_4– containing mixtures. *Combustion and Flame*, 2003. **133**(1–2): pp. 133–146.
6. Starikovskaia, S., et al., Analysis of the spatial uniformity of the combustion of a gaseous mixture initiated by a nanosecond discharge. *Combustion and Flame*, 2004. **139**(3): pp. 177–187.
7. Aleksandrov, N.L., et al., Mechanism of ignition by non-equilibrium plasma. *Proceedings of the Combustion Institute*, 2009. **32**(1): pp. 205–212.
8. Starikovskaia, S., Plasma assisted ignition and combustion. *Journal of Physics D: Applied Physics*, 2006. **39**(16): p. R265.
9. Gorchakov, G. and F. Lavrov, Influence of electric discharge on the region of spontaneous ignition in the mixture $2H_2$–O_2. *Acta Physicochim. URSS*, 1934. **1**: pp. 139–144.
10. Boumehdi, M.A., et al., Ignition of methane-and n-butane-containing mixtures at high pressures by pulsed nanosecond discharge. *Combustion and Flame*, 2015. **162**(4): pp. 1336–1349.

11. Vanhove, G., et al., A comparative experimental kinetic study of spontaneous and plasma-assisted cool flames in a rapid compression machine. *Proceedings of the Combustion Institute*, 2017. **36**(3): pp. 4137–4143.
12. Ombrello, T., Y. Ju, and A. Fridman, Kinetic ignition enhancement of diffusion flames by nonequilibrium magnetic gliding arc plasma. *AIAA Journal*, 2008. **46**(10): pp. 2424–2433.
13. Ombrello, T., et al., Combustion enhancement via stabilized piecewise nonequilibrium gliding arc plasma discharge. *AIAA Journal*, 2006. **44**(1): pp. 142–150.
14. Sun, W., et al., Direct ignition and S-curve transition by in situ nano-second pulsed discharge in methane/oxygen/helium counterflow flame. *Proceedings of the Combustion Institute*, 2013. **34**(1): pp. 847–855.
15. Sun, W., S.H. Won, and Y. Ju, In situ plasma activated low temperature chemistry and the S-curve transition in DME/oxygen/helium mixture. *Combustion and Flame*, 2014. **161**(8): pp. 2054–2063.
16. Mao, X., et al., Numerical modeling of ignition enhancement of CH_4/O_2/He mixtures using a hybrid repetitive nanosecond and DC discharge. *Proceedings of the Combustion Institute*, 2019. **37**: pp. 5545–5552.
17. S. Pancheshnyi, B.E., G.J.M. Hagelaar, L.C. Pitchford, Computer code ZDPlasKin, http://www.zdplaskin.laplace.univ-tlse.fr/ (University of Toulouse, LAPLACE, CNRS-UPS-INP, Toulouse, France, 2008).
18. Lutz, A.E., R.J. Kee, and J.A. Miller, SENKIN: a FORTRAN Program for Predicting Homogenous Gas Phase Chemical Kinetics with Sensitivity Analysis, 1988.
19. Hagelaar, G.J.M. and L.C. Pitchford, Solving the Boltzmann equation to obtain electron transport coefficients and rate coefficients for fluid models. *Plasma Sources Science & Technology*, 2005. **14**(4): pp. 722–733.
20. Mitani, T., Propagation velocities of two-reactant flames. *Combustion Science and Technology*, 1980. **21**(3–4): pp. 175–177.
21. Ju, Y. and W. Sun, Plasma assisted combustion: Dynamics and chemistry. *Progress in Energy and Combustion Science*, 2015. **48**: pp. 21–83.
22. Ranzi, E., et al., Hierarchical and comparative kinetic modeling of laminar flame speeds of hydrocarbon and oxygenated fuels. *Progress in Energy and Combustion Science*, 2012. **38**(4): pp. 468–501.
23. Burke, M.P., F.L. Dryer, and Y. Ju, Assessment of kinetic modeling for lean $H_2/CH_4/O_2$/diluent flames at high pressures. *Proceedings of the Combustion Institute*, 2011. **33**(1): p. 905–912.
24. Lacoste, D.A., Flames with plasmas. *Proceedings of the Combustion Institute*, 2022. **39**(4): pp. 5405–5428.
25. Starikovskiy, A. and N. Aleksandrov, Plasma-assisted ignition and combustion. *Progress in Energy and Combustion Science*, 2013. **39**(1): p. 61–110.
26. Calcote, H. Ion and electron profiles in flames. in *Symposium (International) on* Combustion, 1963. 9: pp. 622–637.
27. Maclatchy, C., R. Clements, and P. Smy, An experimental investigation of the effect of microwave radiation on a propane-air flame. *Combustion and Flame*, 1982. **45**: pp. 161–169.
28. Bradley, D. and S. Nasser, Electrical coronas and burner flame stability. *Combustion and Flame*, 1984. **55**(1): pp. 53–58.
29. Jaggers, H. and A. Von Engel, The effect of electric fields on the burning velocity of various flames. *Combustion and Flame*, 1971. **16**(3): pp. 275–285.
30. Ward, M., Potential uses of microwaves to increase internal combustion engine efficiency and reduce exhaust pollutants. *Journal of Microwave Power*, 1977. **12**(3): pp. 187–198.
31. Won, S.H., et al., Effect of electric fields on the propagation speed of tribrachial flames in coflow jets. *Combustion and Flame*, 2008. **152**(4): pp. 496–506.
32. Lee, S.M., et al., Effect of electric fields on the liftoff of nonpremixed turbulent jet flames. *IEEE Transactions on Plasma Science*, 2005. **33**(5): pp. 1703–1709.
33. Clements, R., R. Smith, and P. Smy, Enhancement of flame speed by intense microwave radiation. *Combustion Science and Technology*, 1981. **26**(1–2): pp. 77–81.
34. Zaidi, S., et al. Measurements of hydrocarbon flame speed enhancement in high-Q microwave cavity. in *44th AIAA Aerospace Sciences Meeting and Exhibit*. Reno, Nevada, Jan. 9-12 2006.
35. Stockman, E.S., et al., Measurements of combustion properties in a microwave enhanced flame. *Combustion and Flame*, 2009. **156**(7): pp. 1453–1461.
36. Ju, Y., et al. Numerical study of the effect of microwave discharge on the premixed methane-air flame. in *40th AIAA/ASME/SAE/ASEE Joint Propulsion Conference and Exhibit*. Fort Lauderdale, Florida, July 11-14, 2004.
37. Pilla, G., et al., Stabilization of a turbulent premixed flame using a nanosecond repetitively pulsed plasma. *IEEE Transactions on Plasma Science*, 2006. **34**(6): pp. 2471–2477.

38. Wolk, B., et al., Enhancement of flame development by microwave-assisted spark ignition in constant volume combustion chamber. *Combustion and Flame*, 2013. **160**(7): pp. 1225–1234.
39. Chen, Z. and Y. Ju, Theoretical analysis of the evolution from ignition kernel to flame ball and planar flame. *Combustion Theory and Modelling*, 2007. **11**(3): pp. 427–453.
40. Zhang, T. and Y. Ju. Propagation Speeds and Kinetic Analysis of Premixed Heptane/Air Cool and Warm Flames at Large Ignition Damköhler Numbers. in *11th U. S. National Combustion Meeting*. 2019, Pasadena, California.
41. Ju, Y., Understanding cool flames and warm flames. *Proceedings of the Combustion Institute*, 2021. **38**(1): pp. 83–119.
42. Ju, Y., et al., Dynamics of cool flames. *Progress in Energy and Combustion Science*, 2019. **75**(100787): p. 39.
43. Zhang, T. and Y. Ju, Structures and propagation speeds of autoignition-assisted premixed n-heptane/air cool and warm flames at elevated temperatures and pressures. *Combustion and Flame*, 2020. **211**: p. 8–17.
44. Glassman, I., R.A. Yetter, and N.G. Glumac, *Combustion*. 2014: Academic press.
45. Law, C.K., *Combustion Physics*. 2010: Cambridge university press.
46. Ju, Y. and S. Minaev, Dynamics and flammability limit of stretched premixed flames stabilized by a hot wall. *Proceedings of the Combustion Institute*, 2002. **29**(1): p. 949–956.
47. Buckmaster, J., The effects of radiation on stretched flames. *Combustion Theory and Modelling*, 1997. **1**(1): p. 1.
48. Ju, Y., et al., On the extinction limit and flammability limit of non-adiabatic stretched methane–air premixed flames. *Journal of Fluid Mechanics*, 1997. **342**: pp. 315–334.
49. Ju, Y., et al., Effects of the Lewis number and radiative heat loss on the bifurcation and extinction of CH_4/O_2-N_2-He flames. *Journal of Fluid Mechanics*, 1999. **379**: pp. 165–190.
50. Maruta, K., et al., Extinction of low-stretched diffusion flame in microgravity. *Combustion and Flame*, 1998. **112**(1): pp. 181–187.
51. Maruta, K., et al., Lewis number effect on extinction characteristics of radiative counterflow CH_4–O_2–N_2–He flames. *Proceedings of Symposium (International) on Combustion*, 1998. **27**: pp. 2611–2617.
52. Maruta, K., et al., Experimental study on methane-air premixed flame extinction at small stretch rates in microgravity. *Proceedings of Symposium (International) on Combustion*, 1996. **26**: pp. 1283–1289.
53. Ronney, P.D., Effect of chemistry and transport properties on near-limit flames at microgravity. *Combustion Science and Technology*, 1988. **59**(1–3): pp. 123–141.
54. Ju, Y., K. Maruta, and T. Niioka, Combustion limits. *Applied Mechanics Reviews*, 2001. **54**(3): pp. 257–277.
55. Lewis, B. and G. von Elbe, On the theory of flame propagation. *The Journal of Chemical Physics*, 1934. **2**(8): pp. 537–546.
56. Wilde, K.A., Boundary-value solutions of the one-dimensional laminar flame propagation equations. *Combustion and Flame*, 1972. **18**(1): pp. 43–52.
57. Streng, A. and A. Grosse, The pure ozone to oxygen flame 1. *Journal of the American Chemical Society*, 1957. **79**(6): pp. 1517–1518.
58. Nomaguchi, T. and S. Koda. Spark ignition of methane and methanol in ozonized air. *Proceedings of Symposium (International) on Combustion*, 1989. **22**: pp. 1677–1682.
59. Ombrello, T., et al., Flame propagation enhancement by plasma excitation of oxygen. Part II: Effects of O_2 ($a^1\Delta_g$). *Combustion and Flame*, 2010. **157**(10): pp. 1916–1928.
60. Ombrello, T., et al., Flame propagation enhancement by plasma excitation of oxygen. Part I: Effects of O3. *Combustion and Flame*, 2010. **157**(10): pp. 1906–1915.
61. Wang, Z., et al., Investigation of combustion enhancement by ozone additive in CH4/air flames using direct laminar burning velocity measurements and kinetic simulations. *Combustion and Flame*, 2012. **159**(1): pp. 120–129.
62. Wu, B., et al., Dynamics of laminar ethylene lifted flame with ozone addition. *Proceedings of the Combustion Institute*, 2021. **38**(4): pp. 6773–6780.
63. Gao, X., et al., The effect of ozone addition on laminar flame speed. *Combustion and Flame*, 2015. **162**(10): pp. 3914–3924.
64. Malicet, J., et al., Ozone UV Spectroscopy.2. Absorption cross-sections and temperature-dependence. *Journal of Atmospheric Chemistry*, 1995. **21**(3): pp. 263–273.
65. Sun, W., et al., The effect of ozone addition on combustion: Kinetics and dynamics. *Progress in Energy and Combustion Science*, 2019. **73**: pp. 1–25.

66. Hajilou, M., et al., Experimental and numerical characterization of freely propagating ozone-activated dimethyl ether cool flames. *Combustion and Flame*, 2017. **176**: pp. 326–333.
67. Ombrello, T., C. Carter, and V. Katta, Burner platform for sub-atmospheric pressure flame studies. *Combustion and Flame*, 2012. **159**(7): pp. 2363–2373.
68. Lefkowitz, J.K., et al., *A Study of Plasma-Assisted Ignition in a Small Internal Combustion Engine*. AIAA paper-2012-1133, 2012.
69. Lee, T.-W., V. Jain, and S. Kozola, Measurements of minimum ignition energy by using laser sparks for hydrocarbon fuels in air: propane, dodecane, and jet-A fuel. *Combustion and Flame*, 2001. **125**(4): pp. 1320–1328.
70. Ju, Y. and K. Maruta, Microscale combustion: Technology development and fundamental research. *Progress in Energy and Combustion Science*, 2011. **37**(6): pp. 669–715.
71. Colket, M.B., et al., *Fall Technical Meeting: Eastern States Section of the Combustion Institute*. 2011: Combustion Institute.
72. Chen, Z., M. Burke, and Y. Ju, Effects of Lewis number and ignition energy on the determination of laminar flame speed using propagating spherical flames. *Proceedings of the Combustion Institute*, 2009. **32**(1): p. 1253–1260.
73. Kelley, A.P., G. Jomaas, and C.K. Law, Critical radius for sustained propagation of spark-ignited spherical flames. *Combustion and Flame*, 2009. **156**(5): pp. 1006–1013.
74. Bradley, D., M. Lawes, and M.S. Mansour, Explosion bomb measurements of ethanol–air laminar gaseous flame characteristics at pressures up to 1.4 MPa. *Combustion and Flame*, 2009. **156**(7): pp. 1462–1470.
75. Kim, H.H., et al., Measurements of the Critical Initiation Radius and Unsteady Propagation of n-Decane/Air Premixed Flames. *Proceedings of the Combustion Institute*, 2013. **34**: pp. 929–936.
76. Chen, Z., *Studies on the Initiation, Propagation, and Extinction of Premixed Flames*. 2008: Princeton University.
77. Chen, Z., M.P. Burke, and Y. Ju, On the critical flame radius and minimum ignition energy for spherical flame initiation. *Proceedings of the Combustion Institute*, 2011. **33**(1), pp. 1219–1226.
78. Ronney, P.D. Understanding combustion processes through microgravity research. *Proceedings of Symposium (International) on Combustion*, 1998. **27**: pp. 2485–2506.
79. Blanc, M., et al., Ignition of explosive gas mixtures by electric sparks. I. Minimum ignition energies and quenching distances of mixtures of methane, oxygen, and inert gases. *The Journal of Chemical Physics*, 1947. **15**(11): pp. 798–802.
80. Beduneau, J.-L., et al., Measurements of minimum ignition energy in premixed laminar methane/air flow by using laser induced spark. *Combustion and Flame*, 2003. **132**(4): pp. 653–665.
81. Santner, J.S., S.H. Won, and Y. Ju, Chemistry and transport effects on critical flame initiation radius for alkanes and aromatic fuels. *Proceedings of the Combustion Institute*, 2016. **36**(1): pp. 1457–1465.
82. Lefkowitz, J.K., et al., Schlieren imaging and pulsed detonation engine testing of ignition by a nanosecond repetitively pulsed discharge. *Combustion and Flame*, 2015. **162**(6): pp. 2496–2507.
83. Lefkowitz, J.K. and T. Ombrello, An exploration of inter-pulse coupling in nanosecond pulsed high frequency discharge ignition. *Combustion and Flame*, 2017. **180**: pp. 136–147.
84. Mao, X., et al., Effects of inter-pulse coupling on nanosecond pulsed high frequency discharge ignition in a flowing mixture. *Proceedings of the Combustion Institute*, 2023. **39**(4): pp. 5457–5464.
85. He, L.T., Critical conditions for spherical flame initiation in mixtures with high Lewis numbers. *Combustion Theory and Modelling*, 2000. **4**(2): pp. 159–172.
86. Joulin, G. and P. Clavin, Linear-stability analysis of non-adiabatic flames - diffusional-thermal model. *Combustion and Flame*, 1979. **35**(2): pp. 139–153.
87. Frankel, M.L. and G. Sivashinsky, On effects due to thermal expansion and Lewis number in spherical flame propagation. *Combustion Science and Technology*, 1983. **31**(3–4): pp. 131–138.
88. Clavin, P., Dynamic behavior of premixed flame fronts in laminar and turbulent flows. *Progress in Energy and Combustion Science*, 1985. **11**(1): pp. 1–59.
89. Mariani, A. and F. Foucher, Radio frequency spark plug: An ignition system for modern internal combustion engines. *Applied Energy*, 2014. **122**: pp. 151–161.
90. Lyon, E., et al., Multi-point laser spark generation for internal combustion engines using a spatial light modulator. *Journal of Physics D: Applied Physics*, 2014. **47**(47): pp. 475501.
91. Bladh, H., et al., Flame propagation visualization in a spark-ignition engine using laser-induced fluorescence of cool-flame species. *Measurement Science and Technology*, 2005. **16**(5): pp. 1083.
92. Wang, Q., et al., Visual features of microwave ignition of methane-air mixture in a constant volume cylinder. *Applied Physics Letters*, 2013, **103**, 204104.

93. Lefkowitz, J.K., et al., Elevated OH production from NPHFD and its effect on ignition. *Proceedings of the Combustion Institute*, 2021. **38**(4): pp. 6671–6678.
94. Yu, S., M. Wang, and M. Zheng, *Distributed Electrical Discharge to Improve the Ignition of Premixed Quiescent and Turbulent Mixtures*. 2016, SAE Technical Paper.
95. Zhao, H., et al., Studies of multi-channel spark ignition of lean n-pentane/air mixtures in a spherical chamber. *Combustion and Flame*, 2020. **212**: pp. 337–344.
96. Lin, B.-x., et al., Multi-channel nanosecond discharge plasma ignition of premixed propane/air under normal and sub-atmospheric pressures. *Combustion and Flame*, 2017. **182**: pp. 102–113.
97. Mao, X., et al., Modeling of the effects of non-equilibrium excitation and electrode geometry on H_2/air ignition in a nanosecond plasma discharge. *Combustion and Flame*, 2022. **240**: p. 112046.
98. Chao, B., C. Law, and J. T'ien. Structure and extinction of diffusion flames with flame radiation. *Proceedings of Symposium (International) on Combustion*, 1991. **23**: pp. 523–531.
99. Liu, F., et al., Asymptotic analysis of radiative extinction in counterflow diffusion flames of nonunity Lewis numbers. *Combustion and Flame*, 2000. **121**(1–2): pp. 275–287.
100. Linan, A., The asymptotic structure of counterflow diffusion flames for large activation energies. *Acta Astronautica*, 1974. **1**(7–8): pp. 1007–1039.
101. Tien, J.S., Diffusion flame extinction at small stretch rates: the mechanism of radiative loss. *Combustion and Flame*, 1986. **65**(1): p. 31–34.
102. Nayagam, V., et al., Microgravity n-heptane droplet combustion in oxygen-helium mixtures at atmospheric pressure. *AIAA Journal*, 1998. **36**(8): pp. 1369–1378.
103. Nayagam, V., et al., Can cool flames support quasi-steady alkane droplet burning? *Combustion and Flame*, 2012. **159**(12): pp. 3583–3588.
104. Dietrich, D.L., et al., Droplet combustion experiments aboard the International Space Station. *Microgravity Science and Technology*, 2014. **26**(2): p. 65–76.
105. Zhu, J., et al., Dynamics, OH distributions and UV emission of a gliding arc at various flow-rates investigated by optical measurements. *Journal of Physics D: Applied Physics*, 2014. **47**(29): p. 295203.
106. Fridman, A., et al., Gliding arc gas discharge. *Progress in Energy and Combustion Science*, 1999. **25**(2): pp. 211–231.
107. Kee, R., et al., *CHEMKIN Collection, Release 3.6, Reaction Design*. Inc., San Diego, CA, www.chemkin.com. 2000.
108. Smith, G.P., et al., *GRI-Mech* 3.0, https://www.me.berkeley.edu/gri_mech, 1999. **51**: p. 55.
109. Sun, W., et al., Effects of non-equilibrium plasma discharge on counterflow diffusion flame extinction. *Proceedings of the Combustion Institute*, 2011. **33**(2): pp. 3211–3218.
110. Uddi, M., et al., Atomic oxygen measurements in air and air/fuel nanosecond pulse discharges by two photon laser induced fluorescence. *Proceedings of the Combustion Institute*, 2009. **32**(1): pp. 929–936.
111. Won, S.H., et al., A new cool flame: establishment and studies of dynamics and kinetics. in *52nd Aerospace Sciences Meeting*, (doi: 10.2514/6.2014-0818), in AIAA paper-2014-0818. 2014.
112. Won, S.H., et al., Self-sustaining n-heptane cool diffusion flames activated by ozone. *Proceedings of the Combustion Institute*, 2015. **35**(1): pp. 881–888.
113. Reuter, C.B., S.H. Won, and Y. Ju, Experimental study of the dynamics and structure of self-sustaining premixed cool flames using a counterflow burner. *Combustion and Flame*, 2016. **166**: pp. 125–132.
114. Reuter, C.B., S.H. Won, and Y. Ju, Flame structure and ignition limit of partially premixed cool flames in a counterflow burner. *Proceedings of the Combustion Institute*, 2017. **36**: pp. 1513–1522.
115. Reuter, C.B., et al., Study of the low-temperature reactivity of large n-alkanes through cool diffusion flame extinction. *Combustion and Flame*, 2017. **179**: pp. 23–32.
116. Wang, Z., et al., Kinetics and extinction of non-premixed cool and warm flames of dimethyl ether at elevated pressure. *Proceedings of Combustion Insittute*, 2022. **39**: pp. 1–8.
117. Wang, Z., et al., Study on cool flame radical index and oxygen concentration dependence of oxygenated fuels. *Combustion and Flame*, 2022. **257**: p. 112493.
118. Zhou, M., et al., Kinetic effects of NO addition on n-dodecane cool and warm diffusion flames. *Proceedings of the Combustion Institute*, 2021. **38**(2): pp. 2351–2360.
119. Wang, Z., et al., Pressure effects on reactivity and extinction of n-dodecane diffusion cool flame. *Combustion and Flame*, 2023. **254**: p. 112829.
120. Murakami, Y., et al., Studies of autoignition-assisted nonpremixed cool flames. *Proceedings of the Combustion Institute*, 2021. **38**(2): pp. 2333–2340.
121. Ju, Y., E. Lin, and C.B. Reuter. The effect of radiation on the dynamics of near limit cool flames and hot flames. in *55th AIAA Aerospace Sciences Meeting*. 2017, Dallas.

122. Lin, E., C.B. Reuter, and Y. Ju, Dynamics and burning limits of near-limit hot, mild, and cool diffusion flames of dimethyl ether at elevated pressures. *Proceedings of the Combustion Institute*, 2019. **37**(2): pp. 1791–1798.
123. Deng, S., et al., NTC-affected ignition and low-temperature flames in nonpremixed DME/air counterflow. *Combustion and Flame*, 2014. **161**(8): pp. 1993–1997.
124. Yehia, O.R., C.B. Reuter, and Y. Ju, Low-temperature multistage warm diffusion flames. *Combustion and Flame*, 2018. **195**: pp. 63–74.
125. Yehia, O.R., C.B. Reuter, and Y. Ju, On the chemical characteristics and dynamics of n-alkane low-temperature multistage warm diffusion flames. *Proceedings of the Combustion Institute*, 2019. **37**(2): pp. 1717–1724.
126. Farouk, T., D. Dietrich, and F.L. Dryer. Three stage quasi-steady droplet burning behavior of n-alkane droplets at elevated pressure conditions: Hot, warm and cool flame combustion. in *Eastern States Section of the Combustion Institute*. March 4–7, 2018, State College, Pennsylvania
127. Farouk, T., D. Dietrich, and F.L. Dryer, Three stage cool flame droplet burning behavior of n-alkane droplets at elevated pressure conditions: Hot, warm and cool flame. *Proceedings of the Combustion Institute*, 2019. **37**: pp. 3353–3361.
128. Yokoo, R., et al., Experimental study of internal flow structures in cylindrical rotating detonation engines. *Proceedings of the Combustion Institute*, 2021. **38**(3): pp. 3759–3768.
129. Schwer, D. and K. Kailasanath, Numerical investigation of the physics of rotating-detonation-engines. *Proceedings of the Combustion Institute*, 2011. **33**(2): pp. 2195–2202.
130. Schauer, F., J. Stutrud, and R. Bradley. Detonation initiation studies and performance results for pulsed detonation engine applications. in *39th Aerospace Sciences Meeting and Exhibit*. Reno, Nevada, Jan. 8-11, 2001.
131. Vorenkamp, M., et al., Plasma-assisted deflagration to detonation transition in a microchannel with hybrid fs/ps coherent anti-stokes Raman scattering measurements. *Proceedings of the Combustion Institute*, 2023. **39**(4): pp. 5561–5569.
132. Sepulveda, J., et al., Kinetic enhancement of microchannel detonation transition by ozone addition to acetylene mixtures. *AIAA Journal*, 2019. **57**(2): pp. 476–481.
133. Crane, J., et al., Isolating the effect of induction length on detonation structure: Hydrogen–oxygen detonation promoted by ozone. *Combustion and Flame*, 2019. **200**: pp. 44–52.
134. Lee, J.H., *The Detonation Phenomenon*. 2008.
135. Zel'dovich, I.A., et al., *Mathematical Theory of Combustion and Explosions*. Springer US, 1985.
136. Sivashinsky, G.I., Some developments in premixed combustion modeling. *Proceedings of the Combustion Institute*, 2002. **29**(2): pp. 1737–1761.
137. Zeldovich, Y.B., Regime classification of an exothermic reaction with nonuniform initial conditions. *Combustion and Flame*, 1980. **39**(2): pp. 211–214.
138. Gu, X., D. Emerson, and D. Bradley, Modes of reaction front propagation from hot spots. *Combustion and Flame*, 2003. **133**(1): pp. 63–74.
139. Poludnenko, A.Y., et al., A unified mechanism for unconfined deflagration-to-detonation transition in terrestrial chemical systems and type Ia supernovae. *Science*, 2019. **366**(6465): p. eaau7365.
140. Bradley, D. and G. Kalghatgi, Influence of autoignition delay time characteristics of different fuels on pressure waves and knock in reciprocating engines. *Combustion and Flame*, 2009. **156**(12): pp. 2307–2318.
141. Bates, L., et al., Engine hot spots: Modes of auto-ignition and reaction propagation. *Combustion and Flame*, 2016. **166**: pp. 80–85.
142. Dai, P., et al., Numerical experiments on reaction front propagation in n-heptane/air mixture with temperature gradient. *Proceedings of the Combustion Institute*, 2015. **35**(3): pp. 3045–3052.
143. Sun, W., et al., Multi-scale modeling of dynamics and ignition to flame transitions of high pressure stratified n-heptane/toluene mixtures. *Proceedings of the Combustion Institute*, 2015. **35**(1): pp. 1049–1056.
144. Zhang, T., W. Sun, and Y. Ju, Multi-scale modeling of detonation formation with concentration and temperature gradients in n-heptane/air mixtures. *Proceedings of the Combustion Institute*, 2017. **36**(1): pp. 1539–1547.
145. Qi, C., et al., Different modes of reaction front propagation in n-heptane/air mixture with concentration non-uniformity. *Proceedings of the Combustion Institute*, 2017. **36**(3): p. 3633–3641.
146. Zhang, T., et al., Effects of thermal and fuel stratifications and turbulence transport on knocking formation for dimethyl ether/air mixtures in *37th International Symposium on Combustion*, Dublin, Ireland. 2018.

147. Zhang, T. et al., Effects of low temperature chemistry and turbulence transport on knocking formation for stratified dimethyl ether/air mixtures. *Combustion and Flame*, 2019. **200**: p. 342–353.
148. Brailovsky, I. and G.I. Sivashinsky, Hydraulic resistance as a mechanism for deflagration-to-detonation transition. *Combustion and Flame*, 2000. **122**(4): pp. 492–499.
149. Ju, Y. and C.K. Law, Propagation and quenching of detonation waves in particle laden mixtures. *Combustion and Flame*, 2002. **129**(4): pp. 356–364.
150. Chan, H. and M. Wu. Stages of flame acceleration and detonation transition in a thin channel filled with stoichiometric ethylene/oxygen mixture. in *26th International Colloquium on the Dynamics of Explosions and Reactive Systems,* Boston, MA. 2017.
151. Wu, M.H., et al., Flame acceleration and the transition to detonation of stoichiometric ethylene/oxygen in microscale tubes. *Proceedings of the Combustion Institute*, 2007. **31**(2): pp. 2429–2436.
152. Houim, R.W., A. Ozgen, and E.S. Oran, The role of spontaneous waves in the deflagration-to-detonation transition in submillimetre channels. *Combustion Theory and Modelling*, 2016. **20**(6): pp. 1068–1087.
153. Du, N., et al., Flame propagation in millimeter-scale tubes for lean ethylene–oxygen mixtures. *AIAA Journal*, 2020. **58**(3): pp. 1337–1347.
154. Chambers, J., et al., Spontaneous runaway of fast turbulent flames for turbulence-induced deflagration-to-detonation transition. *Physics of Fluids*, 2022. **34,** 015114.
155. Starikovskaia, S., D.A. Lacoste, and G. Colonna, Non-equilibrium plasma for ignition and combustion enhancement. *The European Physical Journal D*, 2021. **75**(8): p. 231.
156. Vorenkamp, M., et al., Effect of plasma-enhanced low-temperature chemistry on deflagration-to-detonation transition in a microchannel. *AIAA Journal*, 2023. **61**(11): pp. 1–7.
157. Cherif, M.A., et al., Effect of non-equilibrium plasma on decreasing the detonation cell size. *Combustion and Flame*, 2020. **217**: pp. 1–3.

7 Plasma-Assisted Combustion Applications

With detailed understanding of how plasma affects fundamental combustion properties of ignition, the minimum ignition energy, flame propagation, extinction, ignition-assisted flames, and detonation, various plasmas have been widely developed and used for numerous applications in propulsion, power generation, transportation, and emission control. For example, nanosecond pulsed plasma discharge, dielectric barrier discharge (DBD), gliding arc, corona, plasma torch, microwave (MW), radio frequency (RF) discharge, and lasers have been tested and applied in supersonic combustion, detonation engines, gas turbine engines, internal combustion engines, industrial burners, and chemical reactors to control ignition, flame holding, flame stability, lean burn, detonation, emissions. In this chapter, a summary of major research efforts, progress, and challenges in these areas will be highlighted and discussed.

7.1 SUPERSONIC COMBUSTION, SCRAMJET ENGINES, AND DETONATION ENGINES

The major challenges in supersonic ramjet engines (Figure 7.1) for hypersonic propulsion are fuel/air mixing, ignition, and flame stabilization [1–7]. At a Mach number below 6, the stagnation temperature of the flow is not very high and the autoignition time of a jet fuel (t_{ig} ~ 1–2 ms) is relatively long compared to the flow residence time in the engine (t_{res} ~ 0.5 ms) [8]. In addition, due to the flight Mach number change and there is a concern that the flame stabilization may fail if the combustion time (t_c) or the timescale of the key chain-branching reactions is shorter than the flow residence time. Therefore, without ignition enhancement, both the ignition Damköhler number (Da_{ig}) and combustion Damköhler number (Da_c) are less than unity, rendering ignition and flame stabilization difficult.

$$Da_{ig} = \frac{t_{res}}{t_{ig}(T_0, p, \phi)} \sim O(1) \tag{6.29a}$$

$$Da_c = \frac{\tau_{res}}{\tau_c} < 1 \tag{7.1}$$

Another challenge of supersonic combustion at lower Mach number ($M < 6$) is the engine unstart caused by the combustion-induced shock trains and flow choking by forming a forward propagating normal shock at the inlet or isolator (Figure 7.1). This happens if the combustion heat release is too fast, the rapid heat release rate may cause a huge pressure rise and the increase of the intensity of the oblique shock waves in the isolator. Therefore, it is necessary to shorten the ignition delay time and the combustion heat release rate to avoid both flame blowoff and engine unstart. As discussed in Chapters 5 and 6, because the fast energy transfer from electrons to molecules in plasma (in a few nanoseconds) compared to the flow residence time in a scramjet engine, plasma provides a great opportunity to shorten the ignition delay time and extend extinction limit either thermally or kinetically in a supersonic combustor.

Complex Fluid Dynamic and Combustion Interaction

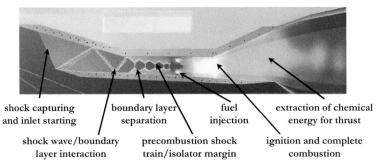

FIGURE 7.1 Schematic of complex fluid dynamics and combustion interaction in a scramjet engine. Courtesy from Timothy Ombrello.

7.1.1 Plasma Torch and Arc Jet

The earliest attempt to use plasma to enhance ignition in a supersonic flow was conducted by Kimura et al. using a thermal plasma torch [9,10]. Arc discharge can provide high elevated temperature (up to 10,000 K) and power to thermally enhance combustion. Over the last two decades, a large number of studies have been carried out to control ignition in a supersonic flow by using different plasma torches and gliding arc systems [4,10–19]. For example, Takita and coworkers [11,14,15] developed a plasma igniter consisting of one and two plasma torches with different plasma feedstocks (e.g., H_2/N_2/O_2) for fuels such as H_2, CH_4, and C_2H_4. The experiment was conducted using an intermittent suction-type wind tunnel. The plasma torch was tested in a supersonic flow of Mach number (M) 2.3. Figure 7.2 [20,21] shows the successful H_2 ignition by a plasma torch in a $M=2.3$ supersonic flow and the effect of total plasma heat and feedstock enthalpy addition on combustion-induced shock wave. It is seen that with the increase of H_2 mole fraction and plasma power, supersonic combustion was thermally enhanced and became stronger, and the upstream shock train induced by combustion heat release was strengthened. To understand the kinetic enhancement by plasma-generated radicals and NO_x catalytic effect on ignition (Chapter 6), the effect of different plasma feedstocks using nitrogen containing H_2/N_2/O_2 mixtures for the twin plasma torches was also studied [15]. Numerical analysis showed that the production of a small amount of NO_x by plasma drastically reduced ignition delay time of H_2 and hydrocarbon fuels at a relatively low initial temperature. It was shown that plasma produced NO_x was more effective than atomic O radicals for ignition enhancement of CH_4. Ignition tests by a N_2/O_2 plasma torch at $M=1.7$ showed that the ignitability of the plasma torch was affected by the composition of the feedstock and that pure O_2 was not the optimum feedstock for ignition with downstream fuel injection.

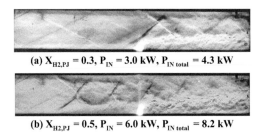

(a) $X_{H2,PJ} = 0.3$, $P_{IN} = 3.0$ kW, $P_{IN\,total} = 4.3$ kW

(b) $X_{H2,PJ} = 0.5$, $P_{IN} = 6.0$ kW, $P_{IN\,total} = 8.2$ kW

FIGURE 7.2 H_2 ignition by plasma torch in a $M=2.3$ flow and the effect of total heat addition on precombustion shock wave, X_{H2} is hydrogen mole fraction in H_2/N_2 plasma torch; P_{in}: plasma torch electric power, $P_{in\,total}$: total heat addition (P_{in}+H_2 enthalpy flux) [20].

Plasma-Assisted Combustion: Applications 481

To further explore the kinetic effect by plasma, Matsubara et al. [22,23] extended the plasma torch ignition enhancement by combining it with a DBD in a supersonic flow. As shown in Figure 7.3a, H_2 was injected upstream of a back-step flame holder. The non-equilibrium DBD discharge produced active radicals (Chapter 5) in a large volume in the back-step recirculation zone. The thermal N_2 arc plasma jet was issued from a flat wall at the downstream of the back-step flame holder to stabilize the flame. Figure 7.3a shows that non-equilibrium DBD plasma was successfully generated in a supersonic flow at $M = 2.0$ with maximum power consumption of about 8 W. Figure 7.3b

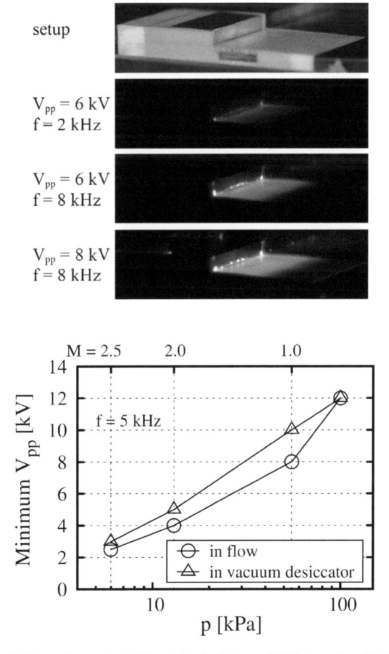

FIGURE 7.3 (a) Direct photographs of dielectric barrier discharge (DBD) plasma in a $M = 2.0$ supersonic flow [22]. (b) Comparison of the minimum V_{pp} to initiate DBD under different flow conditions [22].

shows the dependence of the minimum V_{pp} on the flow Mach number and the comparison with the results of the same DBD discharge in a low-pressure desiccators without a flow. It was found that the minimum peak-to-peak amplitude of the applied sinusoidal voltage of DBD, V_{pp}, to initiate DBD decreased with the increase in the flow Mach number (M) due to the decreased static pressure (about 55, 13, and 6 kPa in $M = 1.0$, 2.0, and 2.5, respectively). Ignition tests were conducted in an $M = 2.0$ supersonic flow. It was confirmed that ignition occurred by only operating the DBD discharge. On the other hand, the wall pressure measurements showed that with a high plasma torch power at $P_{in} = 4.05$ kW, strong combustion occurred, but no combustion enhancement was observed by adding the DBD. However, for relatively weak and unstable combustion at a lower plasma torch power, significant enhancement was obtained when the fuel was injected upstream of the DBD device. It was shown that the wall pressure distribution of plasma torch at $P_{in} = 2.4$ kW with DBD operation was almost equivalent to that at $P_{in} = 3.8$ kW without DBD operation, indicating a significant drop of total electrical energy consumption with combined plasma discharge. This experiment suggested that plasma thermal effect is needed to stabilize a flame in a high-speed flow and that non-equilibrium plasma can kinetically enhance flame stabilization near its stabilization (or blowoff) limit when there is sufficient thermal effect from the plasma torch.

More recently, to enable ignition-assisted flame stabilization (Chapter 6), the ethylene flame dynamics in a model scramjet was investigated using a pulsed-arc-heated hypersonic wind tunnel simulating supersonic/hypersonic flight conditions (Mach number 4.5, 6, and 9 flows of up to 3,500 K stagnation temperature). The air flow was heated by an arc discharge powered by a 260 kW DC power supply delivering arc currents up to 630 A. A partially-premixed ethylene flame was auto-ignited with Mach number 4.5, 6, and 9 freestream flows in a region downstream of a C_2H_4 fuel jet where the high enthalpy flow was compressed and decelerated by incident/reflected shockwaves and development of boundary layers. It was shown that the inlet unstart, flame propagation, and anchoring location were significantly affected by the flow conditions and mixture equivalence ratio (autoignition delay time). In the case of Mach number 4.5 freestream flows, the results showed that the inlet unstart phenomenon was observed by excessive heat release from ethylene combustion reactions over a range of fuel concentration (overall equivalence ratio = 1.5–2.3). In addition, it was shown that autoignition-assisted flame stabilization dictated the flame anchoring location. Figure 7.4 shows the dependence of ignition-assisted flame anchoring location as a function of equivalence ratio of ethylene-air mixtures. Because of the arc jet activation, plasma-generated radicals and heating will shorten the ignition delay time. Moreover, with the increase of the equivalence ratio, the ignition delay time decreases. Therefore, as seen in Figure 7.4, with the increase of equivalence ratio, the flame front moved upstream. Even when the equivalence ratio passed the stoichiometric ratio, at which the flame speed maximized, the flame front (ignition front) continued to propagate upstream. This is a clear indication that plasma-assisted autoignition dominated the flame stabilization in a model scramjet engine when the total temperature (the ignition Damköhler number, Da_{ig}) was very high because the autoignition-assisted flame speed can increase exponentially with the increase with Da_{ig}.

7.1.2 Gliding Arc

Gliding arc has the merits of both thermal effect and kinetic effect on ignition and flame speed enhancement. Gliding arc has also been widely examined for supersonic combustion to promote kinetic enhancement of combustion with lower energy cost than arc discharge. Leonov et al. [17] developed an experimental platform of plasma-assisted supersonic combustion by integrating multi-electrode quasi DC discharge (gliding arc) behind a back-step cavity in H_2 and C_2H_4 with air mixtures (Figure 7.5). The discharge dynamics in flow behind the back-step and in the cavity was explored. It was found that the discharge effect on the flow structure in the cavity and behind the back-step lied in an intensive turbulization of gas in the interaction area and simultaneously slight increase of the separation zone volume. The direct photographs of the multi-electrode discharge

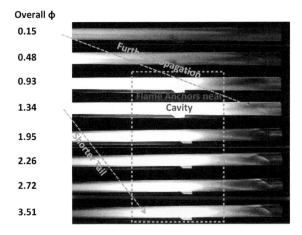

FIGURE 7.4 The effect of ethylene equivalence ratio on arc jet-assisted ignition and flame anchoring location in a Mach number 4.5 air free stream with stagnation temperature and pressure of 2,500 K and 1 atm, respectively [24,25]. Courtesy of Hyungrok Do.

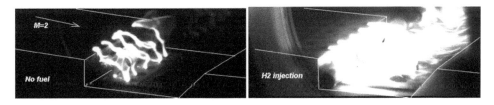

FIGURE 7.5 Left: discharge without fuel injection. Right: discharge interaction with H_2 injection [17].

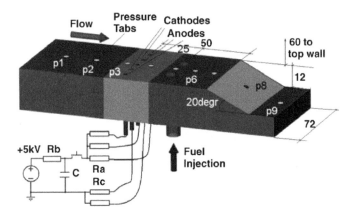

FIGURE 7.6 Experimental schematic of the combustor bottom wall and the test arrangement [27].

and associated combustion are shown in Figure 7.5. With H_2 injection, combustion took place in the cavity as well as in the shear layer. For C_2H_4, combustion was only detected in the cavity. The authors proposed three mechanisms for high-speed combustion control: plasma-induced ignition, plasma-intensified mixing, and flame holding by plasma generation.

Leonov et al. [12,26,27] further conducted experiments on plasma-induced H_2 and C_2H_4 ignition and flame holding by near-surface electrical discharges in a Mach number 2 flow. The modified experimental setup is shown in Figure 7.6. The fuel injectors were located downstream of the plasma generator at a distance that was less than the plasma filament length. The injectors were arranged in line with the electrodes for the most intensive air-plasma interaction with the fuel jet. Fuel injection was started prior to the discharge initiation and was switched off after completion of the discharge. The discharge appeared in the form of oscillating plasma filaments. The individual filaments were blown down due to the main flow at a velocity that was a bit higher than the core value. It was found that the flame could only exist in the presence of gliding arc. The authors proposed a two-zone model of plasma-induced ignition to explain the experimental data. In zone 1, the low-temperature fuel oxidation was induced by plasma discharge with relatively small heat release. In zone 2, combustion was completed with high heat release. Plasma enhanced the low-temperature fuel oxidation inside zone 1 owing to the generation of high amounts of active species. The lengths of zone 1 in the tests were measured by Schlieren technique in the range from 50 to 150 mm, which corresponded to the induction time range from 0.1 to 0.3 ms. However, the transition mechanism of plasma-assisted ignition to strong combustion in zone 2 was not well understood. It might be plasma-enhanced autoignition-assisted flame propagation. More detailed flame regime diagnostics needs to be conducted to understand the combustion regime.

More recently, a multi-channel gliding arc (MCGA) plasma installed on the sidewall surface of a scramjet combustor was employed to enhance combustion near the flame blowout limit in a C_2H_4-fueled and cavity-based model scramjet combustor [28] (Figure 7.7). The results showed that the MCGA plasma demonstrated an excellent flame stabilization ability. The flame blowout limit of the scramjet combustor in the presence of the MCGA plasma was extended by approximately 29%. Two plasma-enhanced combustion processes, including the plasma enhancement mode and the re-ignition mode, were identified to play a vital role in extending the flame blowout limit. The

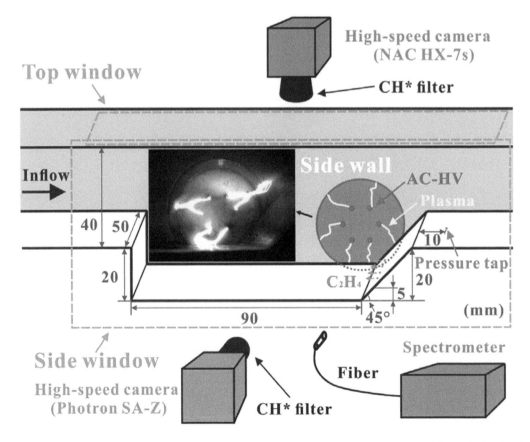

FIGURE 7.7 Schematic of the experimental setup and optical diagnostic system, along with a typical MCGA photo with an exposure time of ½,500s [28].

two plasma-enhanced combustion processes showed that the spark-type discharge generated by the MCGA could ignite the local fuel-rich mixture and increase the flame blowout limit. It was shown that local ignition was dominated by the continuous accumulation of the massive flame kernels with the local fuel-rich mixture, as well as more heat and species exchanges. This result is consistent with the finding by Do et al. (Figure 7.4) [24]. High-temperature plasma can enable ignition-assisted flames to extend the blowout limit. The underlying physics discussed in Chapter 6 related to plasma-assisted ignition enhancement can stretch out the ignition-extinction S-curve.

7.1.3 Nanosecond Discharge

Nanosecond discharge can provide very high electron temperature (10–20 eV) to effectively excite, dissociate, and ionize reactant mixtures to enhance ignition and combustion. Therefore, if the mixture temperature is close to autoignition temperature, it can dramatically accelerate ignition by increasing the chain-reactions (Chapter 5). In 2010, Do et al. [29] integrated a nanosecond pulsed discharge in wall cavity-stabilized supersonic combustion. They investigated nanosecond plasma discharge-assisted ignition of H_2 and C_2H_4 in a supersonic crossflow. The schematic of the experimental design mimicking a supersonic combustor in a scramjet engine is shown in Figure 7.8. The discharge consists of a rectangular aluminum plate with inject nozzle and electrodes, and a ceramic plate embedded for electrical insulation of the pulsed discharge electrodes. The electrodes were 2% thoriated tungsten rods with 1 mm diameter. Non-equilibrium plasma was used to generate active species at a lower temperature. The nanosecond discharge was produced by repetitive pulses of 15 kV peak voltage, 20 ns pulse width, and 50 kHz repetition rate between the two electrodes. Fuel jets at sonic condition were impinged into the free stream air at Mach number between 1.7 and 3.0. Flame stabilization was found to be improved by the nanosecond plasma discharge.

Figure 7.9 provides a comparison of autoignition cavity-stabilized flames and nanosecond plasma-assisted cavity-stabilized flames at two different hydrogen flow conditions. The images in the presence of the plasma were taken 1 µs after the discharge pulse and approximately 200 µs after the start of the test time. The plasma was turned on before the beginning of the test. Here, J_n (Figure 7.9) is the cross jet momentum ratio, the ratio of the normal component of the hydrogen jet momentum to the free stream momentum: $J_n = 0.5(\rho u^2)_{jet}/(\rho u^2)_\infty$. The factor of 0.5 is the projection factor of the momentum in the vertical direction from the 30° injection relative to the free stream. The plasma-enhanced cavity-stabilized flames in Figure 7.9b and d showed that the flame was more extended into the free stream flow from the cavity. The results showed that the nanosecond pulsed plasma in the cavity served as a source of both heat and radical production, shortening the ignition delay time of the well-mixed flammable hydrogen/air mixture. In the case with $M = 2.0$ and stagnation enthalpy ~2.0 MJ/kg, ignition of the H_2 jet was observed within the test time about 500 µs only with plasma enhancement. A simple model was employed to simulate the experimental results. The authors concluded that the reduction of the ignition delays was due to H and O production through electron impact dissociation of H_2 and O_2 in the nanosecond pulsed plasma discharge.

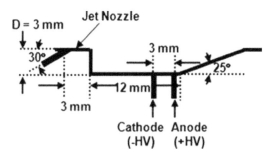

FIGURE 7.8 Schematic of the cavity model [29].

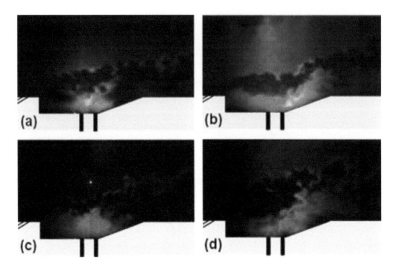

FIGURE 7.9 OH PLIF images of a cavity flame in supersonic flows of two different enthalpies: (a) without the plasma and (b) with plasma at $M = 2.9$, $J_n \sim 4$ of H_2 jet, and (c) without plasma and (d) with plasma at $M = 2.6$ $J_n \sim 3.5$ of H_2 jet [29].

To reduce the pressure loss and eliminate the coupling between cavity and plasma discharge, Do et al. [29] further conducted the flame-holding experiments with plasma discharge in a supersonic crossflow without a cavity. Subsonic (upstream) and sonic (downstream) fuel (H_2 and C_2H_4) jets were issued into a pure oxygen free stream of $M = 1.7$–2.4. The configuration of dual fuel jet injection with plasma in-between enabled the coupling between the plasma energy and the plasma activation of the upstream subsonic fuel/oxidizer mixture that served as a pilot flame to enhance the ignition of the main supersonic flame downstream. Radical production by this pilot flame can be several orders of magnitude more than that by the discharge alone. The observed flame stabilization enhancement was achieved by both the thermal and kinetic effects for flame holding in a supersonic flow.

7.1.4 Microwave and RF Discharge

Microwave and RF discharges provide not only high electron temperature but also electrodeless discharge at high pressures. Esakov et al. [30] studied microwave-assisted combustion of C_3H_8/air in a supersonic flow. The experimental setup is shown in Figure 7.10. The steady-state airstream was formed at the outlet of the nozzle with $M = 2$ and a static pressure of 100 Torr, static temperature of 150 K, and stream velocity of 490 m/s. The free stream pressure and temperature at stagnation conditions were 550 Torr and 300 K, respectively. To create a local electric field that was larger than the critical breakdown field to initiate the discharge, a passive electromagnetic vibrator was used, and the discharge was initiated in the base of the electromagnetic vibrator, which was immersed in the cold supersonic airflow. A linearly polarized beam of electromagnetic radiation with a wavelength of 12.5 cm, power of 1.5 kW, and a typical transverse size of 9 cm was introduced perpendicular to the screen surface. A C_3H_8/air mixture was injected through internal tubes in the pylon and vibrator. The stream velocity on the stream axis was approximately 200 m/s. The experiments demonstrated that the use of a deeply undercritical microwave discharge effectively enhanced combustion and increased the fuel combustion efficiency of C_3H_8/air mixture in a cold supersonic flow. Stable operation of a hot combustion torch in a supersonic stream in steady state was also achieved.

Plasma-Assisted Combustion: Applications

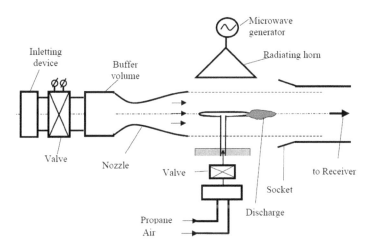

FIGURE 7.10 Schematic of experimental setup [30].

7.1.5 LASER IGNITION AND ELECTRON BEAM

Plasmas generated by lasers via thermal and photon ionization or electron beam via charging and electron impact energy transfer have also been examined in supersonic combustion. Brieschenk et al. [31] reported experimental investigation on laser-induced plasma ignition in a model scramjet engine for the first time. The experiments were conducted in the T-ADFA free piston shock tunnel using a flow condition with a total specific enthalpy of 2.7 MJ/kg and a free stream velocity of 2,075 m/s. The scramjet model features a rectangular duct with a $9°$ compression ramp, followed by a constant-area combustor. H_2 is injected through four 2 mm diameter holes located on the compression ramp of the model, which were distributed across the compression ramp 120 mm downstream of the leading edge. The laser-induced plasma was formed by a Q-switched ruby laser with pulse energies of 750 mJ (only about 54% of the laser energy was absorbed). The location of the plasma was in the shear layers 1.7 mm downstream of the injectors. The temporal evolution of the laser-induced plasma ignition region was monitored using the OH Planar Laser-Induced Fluorescence (PLIF) technique. Hydroxyl radical (OH) formation was successfully observed. Unfortunately, the subsequent ignition and flame propagation events were not reported, possibly because of insufficient laser power or small ignition volume. As discussed in the concept of the minimum ignition energy, both the ignition energy and the critical ignition volume are important for successful ignition.

To increase the ignition volume, the concept of using a multipoint plasma ignition scheme with electron beams or high-voltage pulses was examined to reduce the ignition power budget in a supersonic flow by orders of magnitude [32]. The authors explored an approach where the fuel was injected into the core flow as a liquid jet, and low-power electron beams were used to electrostatically charge the droplets and thus to control droplet breakup, atomization, and mixing, as well as ignition. With a subcritical microwave field applied to the injection region, local enhancement of electric field strength at the surface of droplets, due to both the electric charge and the field increase near the polarized dielectric body, together with seed electrons and ions produced by the electron beam, would create subcritical microwave discharges and thus initiate combustion in multiple spots to enlarge the ignition volume. The multi-spot ignition would help in spreading the flame across the combustor, especially for fuel lean (in average) mixtures.

In summary, plasma activation can enhance ignition and flame stabilization in a high-speed flow through both thermal and kinetic effects. However, due to strong turbulent mixing (heat/radical loss) and short flow residence time (requiring a very short ignition delay time and fast heat release rate), combustion enhancement using nonthermal plasma alone seems to be very challenging. A combination of both thermal and nonthermal plasmas such as the gliding arc/microwave and plasma jet/DBD discharge might be more effective for high-speed combustion and propulsion applications. As such, the kinetic mechanisms to optimize the combination of thermal and nonthermal plasmas need to be further understood. Furthermore, the introduction of plasma discharge may significantly modify the flow field. The effect of flow perturbation by plasma in affecting ignition and flame stabilization in a high-speed flow also needs to be addressed. Moreover, recent studies show that plasma-assisted combustion can also affect plasma dynamics and instability [33–36]. Therefore, appropriate consideration of plasma-combustion coupling in both kinetics and dynamics is important for plasma-assisted combustion in a high-speed flow.

7.2 DETONATION ENGINES

Improvement of energy conversion efficiency of power generation through combustion is important in reducing carbon emissions. Existing electricity generation and propulsion systems such as gas turbine engines use a constant pressure Brayton cycle, which limits their thermal efficiency. Advanced pressure-gain power generation using a constant volume cycle, such as a rotating detonation engine (RDE) or a pulse detonation engine (PDE), has the great potential to increase efficiency by up to 30% [37]. However, fuel/air mixing and detonation control remain a big challenge for air-breathing or lean-burn RDEs [38–41] as well as PDEs [42–44]. Plasma can generate fast and slow heating as well as active species to accelerate ignition. Therefore, plasma-assisted combustion provides a great opportunity to accelerate deflagration-to-detonation transition (DDT) as well as promises stronger shock-flame coupling to increase detonation stability which would ultimately make such advanced engines feasible [6,45–50].

For initiation of DDT, Starikovskiy et al. [45,50] proposed using a distributed, non-equilibrium gas-discharge plasma for gas excitation and reduction of the chemical induction time. It was shown that synchronization of the ignition of different parts of the gas using gas-discharge excitation led to a sufficient reduction in the DDT length and time. An experimental demonstration of a longitudinal, high-voltage, nanosecond discharge was used for detonation initiation in a smooth-wall tube [51]. Three different modes of combustion: flame front propagation, deflagration, transient detonation, and Chapman–Jouguet (CJ) detonation, were observed.

An experimental demonstration of the application of a longitudinal, high-voltage, nanosecond gas discharge for initiation of detonation was performed [51]. A pulsed, nanosecond discharge initiated detonation at a DDT length of up to 130 mm in a tube with a diameter of 140 mm. The detonation was initiated by an energy of 70 mJ in different C_3H_8 and C_6H_{14} mixtures at an initial pressure between 0.15 and 1 atm (Figure 7.11). In the first setup, a distributed, non-equilibrium,

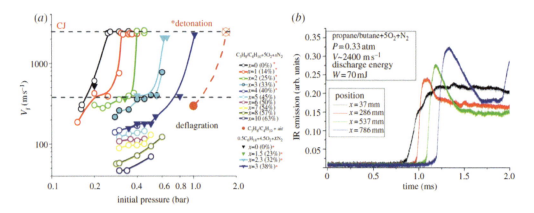

FIGURE 7.11 (a) Velocity of the flamefront at 412 mm from the discharge chamber in $C_3H_8/C_4H_{10}+5O_2+xN_2$ and $0.5C_6H_{14}+5O_2+xN_2$ mixtures. (b) Dynamics of DDT in propane–butane/oxygen/nitrogen mixture. Discharge energy $W=70$ mJ [51].

nanosecond discharge was used for mixture excitation and ignition. In the second setup, a localized, microsecond, pulsed spark discharge with a stored energy of 14 J was implemented. The electrical parameters of the discharges, ignition delay time, flame front, and shock-wave velocities were measured in the experiments. Essentially, a higher efficiency of the nanosecond discharge in comparison to the microsecond one as a detonation initiator was demonstrated. The DDT was observed at three tube diameters away from the discharge chamber in different $C_3H_8/C_4H_{10}+5O_2+xN_2$ and $0.5C_6H_{14}+5O_2+xN_2$ mixtures with nitrogen concentrations up to 63% and 38%, respectively. The energy input in these cases did not exceed 3 J, and the DDT time was less than 1 ms. The microsecond spark discharge ignited the mixture efficiently, but the transition length and time increased significantly [51].

A review of multi-institutional collaborations on plasma-assisted PDEs was made by Cathey et al. in [42]. It was demonstrated that at high flow rates where spark-initiated flames were normally extinguished, the transient plasma was able to ignite and effectively create a detonation wave in the PDE experiment. A significant reduction (factor of 4) in ignition delay time was also shown in a PDE experiment for C_2H_4/air mixtures as in Figure 7.12 together with a direct photograph of the PDE setup. Tests were conducted at Air Force Research Laboratory (ARFL) with H_2/air and aviation gasoline/air mixture. A reduction in the ignition delay by factor of 2 was obtained for both H_2/air and aviation gasoline/air mixture. The transient plasma was able to reliably ignite aviation gasoline/air mixture at equivalence ratios of nearly 0.65, whereas the baseline's lower limit was 0.71. The transient plasma ignition of C_2H_4/O_2 mixture resulted in ignition delay reduction by nearly one order of magnitude (factor 9) in the PDE setup at Stanford University.

Recently, Lefkowitz et al. [52] studied the effect of high-frequency nanosecond pulsed discharges on PDEs as shown in Figure 7.13a. By comparing the ignition delay times as well as high-speed imaging of the ignition kernel growth with different igniters (igniter powered by nanosecond pulsed power and conventional multi-spark igniter), they found a significant decrease in the ignition time in the PDE for a variety of fuels and equivalence ratios (ϕ). As demonstrated in Figure 7.13b, with the same amount of total energy input, higher frequency discharges can initiate flame propagation. Figure 7.13c shows the difference between the nanosecond pulsed plasma igniter and the multiple spark discharge (MSD) igniter. With roughly the same amount of total energy consumption, the MSD ignition kernel eventually extinguished, while the plasma ignited kernel went on to become a self-propagating flame. In addition, ignition at both the leaner and the richer conditions was achieved with the nanosecond pulsed igniter.

More recently, Lacoste and coworkers used nanosecond repetitively pulsed (NRP) discharges to enhance the DDT in a propagating flame [53]. With a pin-ring electrode geometry (Figure 7.14) located in the middle of a detonation tube, reliable DDT of stoichiometric hydrogen-air flames could

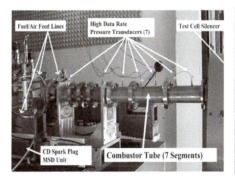
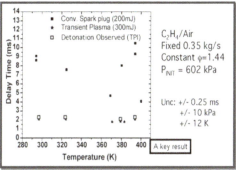

FIGURE 7.12 Left: A valve less PDE setup at the Naval Postgraduate School. This type of architecture requires a booster and its anticipated applications are missiles or rockets. Right: Comparison of ignition delays for C_2H_4/air mixture using spark plug and transient plasma igniter [42].

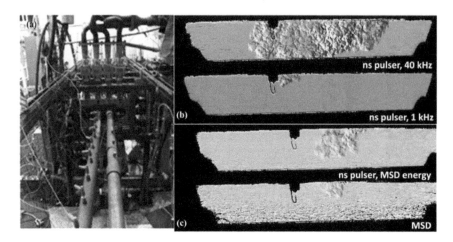

FIGURE 7.13 (a) PDE engine facility at the Air Force Research Lab at Wright-Patterson Air Force Base, (b) Schlieren imaging of nanosecond pulsed discharge igniter in CH_4/air mixture, $\phi=1$, (c) Schlieren imaging of nanosecond pulsed discharge igniter in CH_4/air mixture, $\phi=0.8$ [52].

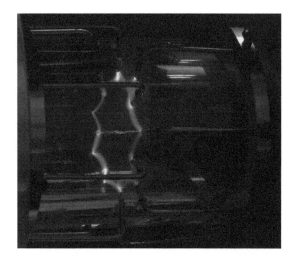

FIGURE 7.14 Photograph of NRP discharges in air, obtained for a gap distance of 9. 1 mm, a penetration depth of $d=1$ mm, and an applied voltage of $V=15$ kV. The exposure time is 2 s [53].

be achieved for a plasma power of less than 0.14% of the thermal power of the flame. Time-of-flight measurements were used in combination with energy deposition measurements and high-speed OH∗-chemiluminescence imagery to investigate the flame acceleration process. The velocities obtained from the time-of-flight data using the ionization probes are shown in Figure 7.15 with NRP discharges (red triangles) and without (black circles). By applying NRP discharges, the flame is accelerated to the point where it transitions to detonation in all cases. In the measurement runs in which no plasma was applied, the flame did not transition to detonation before the I7th probe, although a few runs may transition between the I7th and I8th probes. For the cases with plasma, propagation velocities in excess of CJ velocity are observed in 50% of the cases between I3th and I4th probes, while for the other 50%, the flame velocities are just above 1,000 m/s. This difference suggested that two differing modes of DDT took place during plasma actuation. For stoichiometric hydrogen-air flames, successful transition to detonation was achieved by applying a burst of 110 pulses at 100 kHz, with energies as low as 10 mJ per pulse. The essential role of shock-flame interaction was established as being the main mechanism for flame acceleration when the discharges were located near the wall.

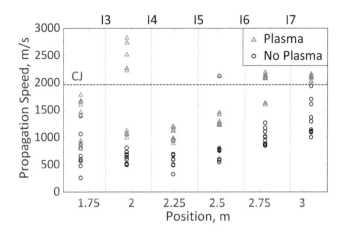

FIGURE 7.15 Time-of-flight measurements for a configuration with an applied voltage of 21 kV, a penetration depth of 1 mm, and a gap distance of 9.1 mm. The blue dashed line corresponds to CJ velocity. Location of the I3–I7 probes is noted [53].

To understand the mechanism of plasma-assisted detonation initiation, studies of the changes of DDT time with high-speed laser diagnostics [46–48] and detonation cell size [54] by plasma discharge and ozone addition were recently conducted. Ju and coworkers [46–48] revealed that ozone or plasma discharge can produce active radicals such as O and excited species that can dramatically reduce the ignition delay time and accelerate shock-ignition coupling as well as DDT. Moreover, the results of high-speed CH_2O PLIF and shadow graph imaging showed that plasma or ozone can enhance low-temperature chemistry and promote low-temperature fuel oxidation and DDT (Figures 6.68–6.70). Furthermore, it was shown that plasma-enhanced low-temperature fuel oxidation can also delay DDT with excessive fuel consumption of precombustion mixture. A recent study by Thawko et al. [55] (Figure 7.16) showed that with ozone addition in a dimethyl ether/oxygen mixture, the O atom production by precursor shock heating (Figure 7.16-I) induced an autoignition immediately after the precursor shock and triggered rapid coupling between ignition and precursor shock and accelerated DDT via direct ignition-shock coupling (Figure 7.16-II–VI).

Also, the effect of a volumetric nanosecond discharge on detonation cell size was demonstrated experimentally in a detonation tube test rig by Cherif et al. [54] (Figure 7.17). The experiments were performed in $CH_4:O_2:Ar = 1:2:2$ and $CH_4:O_2:Ar:H_2 = 3:7:8:2$ mixtures at 180 and 120 mbar, respectively, and ambient temperature. The plasma was generated by two consecutive pulses of 50 and 32 kV amplitude on the high-voltage electrode and 25 ns pulse duration. The analysis of the detonation cell size with and without plasma generation was performed via sooted-plate technique. The results showed that detonation cell size was reduced by a factor of 1.5–3, while passing through the region of the plasma discharge.

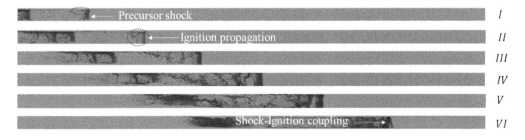

FIGURE 7.16 Shadowgraph images of the DDT process with 1.2% ozone addition. (I) autoignition immediately behind the precursor shock, (II–V) spontaneous ignition propagation (VI) ignition-shock coupling [55].

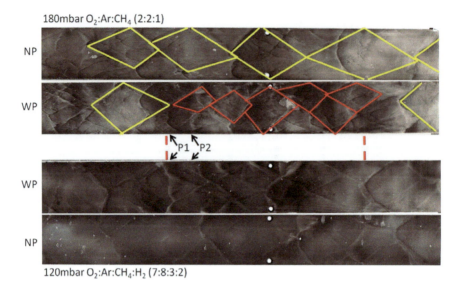

FIGURE 7.17 Soot prints with No Plasma (NP) and With Plasma (WP). Typical detonation cells are shown by red and yellow diamonds. The DWF/plasma interaction zone is marked by red dashed lines [54].

Therefore, from the above experiments, it is clearly seen that plasma-enhanced combustion chemistry can accelerate low-temperature and high-temperature ignition, modify the gradient of temperature and reactivity (Zel'dovich DDT mechanism), strengthen or advance shock-ignition and shock-flame interaction, and reduce detonation cell size, leading to the acceleration of DDT. In addition, plasma can also rapidly accelerate the fuel oxidation in the predetonation mixture and delay DDT. As such, plasma provides a promising technology to control DDT and detonation stability for detonation engines. The challenge is how to develop a desired plasma discharge for applications in detonation engines at elevated pressures.

7.3 EXTENSION OF BLOWOFF AND LEAN-BURN LIMITS

Lean blowout limit, flame stabilization, and instability are the key issues of gas turbine engines. Plasma has also been used as a new technology to increase flame stability and achieve ultra-lean combustion. Serbin et al. [56] concluded that a gas turbine combustor with piloted flame stabilization by non-equilibrium plasma provided better performance, wider turndown ratios, and lower emissions of carbon and nitrogen oxides. Moeck et al. [57] studied the effect of nanosecond pulsed discharge on combustion instabilities. The nanosecond pulsed discharges had 10 ns pulse width at 10–80 kHz pulse repetition frequency (f) and 12 kV amplitude. The discharge was coupled at the nozzle exit of a swirl-stabilized combustor as shown in Figure 7.18. A central pin electrode was installed at the swirler outlet to the combustion chamber that served as the cathode. With this electrode arrangement, nanosecond pulsed discharge filaments were generated between the central pin electrode and the loop-electrode in a disk-shaped area. The effect of the discharges on the mean flame illustrated based on averaged CH* chemiluminescence images is shown in Figure 7.19. The images were taken at an equivalence ratio $\phi=0.62$ and a thermal power of 50 kW. The Reynolds number based on the bulk velocity in the burner passage, the diameter thereof, and the viscosity in the unburned gases was 37,000. The mean flame shape without plasma discharges was compared to the case with discharges at pulse repetition frequencies between 5 and 25 kHz, which corresponded to an electrical power between 30 and 150 W. Without plasma, the flame was lifted off and stabilized in the shear layers associated with the recirculation zone (Figure 7.19a). As the discharges were activated and the pulse repetition frequency was increased, the flame moved successively further upstream and eventually stabilized in the burner passage (Figure 7.19b–d).

Plasma-Assisted Combustion: Applications 493

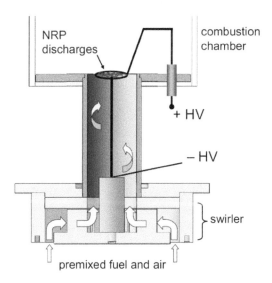

FIGURE 7.18 Installation of the electrodes in the swirl-stabilized burner [57].

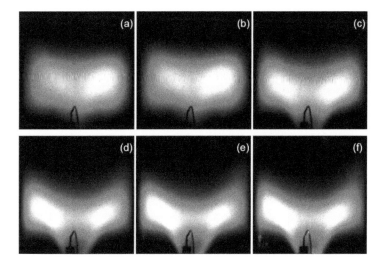

FIGURE 7.19 Averaged CH* chemiluminescence intensity for the case without plasma discharges (a), and with f = 5, 10, 15, 20, and 25 kHz (b–d); equivalence ratio ϕ = 0.62, thermal power 50 kW [57].

The effect of steady pulsed discharges on combustion dynamics was further investigated. It was found that the discharges had a strong effect on the pressure pulsations associated with thermoacoustic dynamics. With the consumption of less than 1% of the total power of the flame, the nanosecond discharge can significantly reduce the oscillation amplitude of the acoustic pressure. However, at some conditions, plasma forcing also destabilized a nominally stable combustion system. Since the discharges affected both the flame shape and the position, they had an influence on the dynamic response of the flame to acoustic perturbations.

Later, Lacoste et al. [58] modified the experimental rig and studied the effects of nanosecond pulsed discharges on the dynamics and instability of a swirl-stabilized lean premixed CH_4 and propane/air flame stabilization. Barbosa et al. [59] studied the effects of a nanosecond pulsed discharge plasma on the flame stability domain of a swirling burner in a premixed lean C_3H_8/air mixture. A two-stage swirled injector was connected to a rectangular combustion chamber with optical access ports. Both stages were supplied with C_3H_8 and air. Both the swirlers were oriented in the same direction to ensure a strong co-rotating swirling motion. The primary and the secondary stage C_3H_8/air mixtures mixed upon entering the combustion chamber. The peak power of the combustor was 52 kW with exit velocity 40 m/s. The combustion chamber had a square cross-section of 100×100 mm² and a length of 500 mm. All studies were performed at ambient pressure and temperature. A direct photograph of the experimental setup and flame stabilization is shown in Figure 7.20. The discharge was produced using a pulse generator with 10 ns pulse width. The voltage amplitude was 30 kV at a repetitive frequency $f = 30$ kHz which corresponded to approximately 350 W power.

The operating regimes of the burner were determined as a function of the global equivalence ratio, ϕ_g using CH* chemiluminescence. Figure 7.21 shows the evolution of the flames when the fuel flow rate injected through the primary stage was decreased. It is clearly showing that without plasma, the flame extinction was observed when the equivalence ratios reached $\phi_g = 0.4$, whereas using plasma the flame stabilization limits were reduced to $\phi_g = 0.11$. Thus, the plasma discharge significantly

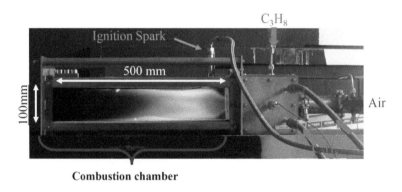

FIGURE 7.20 Direct photograph of the combustion chamber during operation [59].

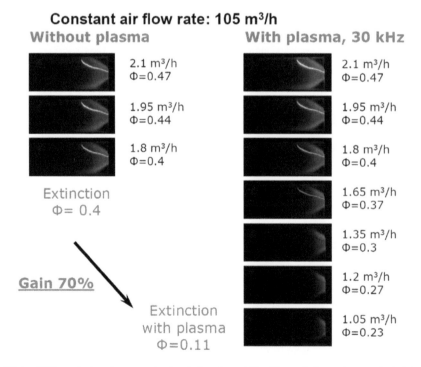

FIGURE 7.21 CH* emission images obtained for $Q_{air} = 105$ m³/h and for primary fuel injection rate $Q_{C3H8} < 2.1$ m³/h or $\phi_g < 0.4$. The repetition rate of the discharge was 30 kHz [59].

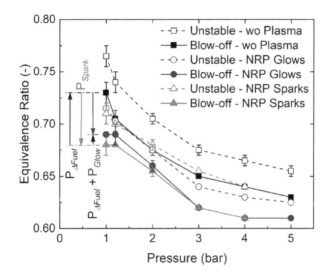

FIGURE 7.22 Stability and lean blowoff limits of a methane-air swirl flame with and without NRP discharges as a function of pressure. Error bars represent standard deviation in reproducibility of results [60].

extended the global flame stabilization limit. By using a small fraction (~1%) of the flame power, extended stability domain of a laboratory scale gas turbine combustor was demonstrated.

Recently, Di Sabatino and Lacoste examined the impact of different discharges such as NRP glow discharge and NRP spark on flame lean stability at pressures above atmospheric pressure [60]. The results showed that at pressures higher than 2 bar, NRP glow discharges (with a higher kinetic effect and lower thermal effect) were more efficient than NRP spark discharges (with a strong thermal effect but lower kinetic effect) in extending the lean stability and blowoff limits of methane-air swirl flames, as shown in Figure 7.22. The ratio of plasma power to flame thermal power of the flame was kept constant for all pressures at 0.32% for NRP glow discharges and 0.7% for NRP spark. It was shown that at 4 bar the use of NRP glow discharges obtained with an applied voltage of 13 kV and a pulse repetition frequency of 30 kHz, could save up to 650 W of thermal power while maintaining a 16 kW flame. For the same conditions, NRP spark discharges could save 600 W only. The authors explained that the less efficient effect of the NRP spark discharges at elevated pressures could be due to the generation of strong shock waves close to the flame stabilization area. In addition, modeling results showed that the equilibrium energy transfer between excited states and rotational states for fast and slow heating also affected the NRP glow effect at different pressures.

Therefore, the above plasma-assisted combustion experiments have clearly shown that non-equilibrium plasma can enhance flame stabilization and extend the lean-burn limit and increase flame stability significantly via both kinetic enhancement and thermal effects.

7.4 GAS TURBINE ENGINES AND THERMAL ACOUSTIC INSTABILITY CONTROL

Plasma-assisted combustion via active species production and ionic wind can enable rapid control and enhancement of combustion and heat release as well as flow field and flow-flame interaction. Therefore, plasma has also been used to control combustion thermal acoustic instability. Lacoste et al. [58] used the experimental rig in Figure 7.23 and studied the effects of nanosecond pulsed discharges on the dynamics of a swirl-stabilized lean premixed CH_4/air flame at 1 atm with thermal power 4 kW. The discharge pulses were 8 kV in amplitude and 10 ns in duration at $f = 30$ kHz, which was equivalent to 40 W electric power. The swirl number was 0.53. The flame image is shown in Figure 7.23 (right). The velocity and CH* chemiluminescence signals were used to determine the

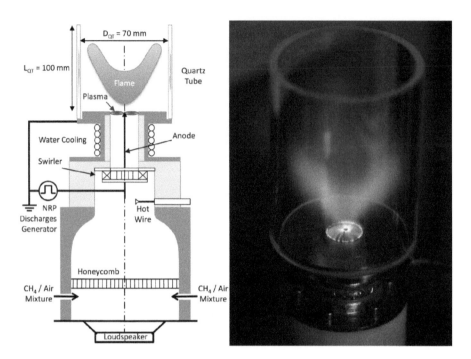

FIGURE 7.23 Left: schematic of burner equipped with a loudspeaker and a nanosecond pulsed discharge generator; Right: photographic representation of the burner with a lean premixed CH_4/air swirl-stabilized flame (blue) and discharges (purple) [58].

flame transfer function. The flame response to acoustic perturbations of the incoming flow was investigated in a range of frequencies from 16 to 512 Hz at equivalence ratio $\phi = 0.7$. It was found that in almost the entire range of excitation frequencies, the plasma discharge affected the gain of the flame transfer function. The phase was also affected by the discharges, but only at high frequencies. In addition, the study showed the flame instability could be effectively mitigated by the application of nanosecond pulsed discharges, reducing the velocity fluctuation amplitude by an order of magnitude. The authors explained that this phenomenon was mainly caused by the thermal effect from plasma. The ultra-fast heating of the flow by plasma was followed by the expansion of gas and generation of shock waves. This aerodynamical impact on transport affected flow velocity and altered vortex structure and therefore changed the flame transfer function.

Lacoste et al. [61] investigated the flame responses of methane-air mixtures to forcing by acoustic waves, AC electric fields, and NRP glow discharges by using an axisymmetric burner with a nozzle made from a quartz tube (Figure 7.24). Three different flame geometries were studied: conical, M-shaped and V-shaped flames. In the experiments, a central stainless steel rod was used as a cathode for the electric field and plasma excitations. The acoustic forcing was obtained with a loudspeaker located at the bottom part of the burner. For forcing by AC electric fields, a metallic grid was placed above the rod and connected to an AC power supply. Plasma forcing was obtained by applying high-voltage pulses of 10-ns duration applied at 10 kHz, between the rod and an annular stainless steel ring, placed at the outlet of the quartz tube. The chemiluminescence of CH^* was used to determine the heat release rate fluctuations. The gain of the transfer function of a flame subjected to velocity perturbations u was defined as

$$F_{ac}(\omega) = \frac{\hat{Q}(\omega)/\overline{Q}}{\hat{u}(\omega)/\overline{u}} \qquad (7.2)$$

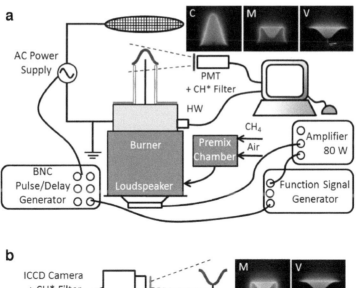

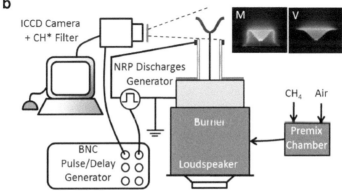

FIGURE 7.24 Experimental setup used for (a) acoustic and AC electric field forcing with three flame geometries: conical (C), M-shaped (M), and V-shaped (V) flames and (b) plasma forcing by NRP glow discharges with two flame geometries: M-shaped (M) and V-shaped (V) flames [61].

the ratio of the relative heat release rate $(\dot{Q})$ fluctuation and the relative velocity fluctuation in the frequency domain, where ^ denotes the Fourier transform of a variable. In this experiment, plasma forcing was found to be up to five times higher than that of acoustic forcing.

Figure 7.25 presents the transfer functions obtained for plasma forcing for the C-, V-, and M-flames with an averaged plasma power of about 1.8% of the thermal power of the flame. The gain maxima were obtained for a reduced frequency of about 5, but the amplitude of the response of the V-flame was two times higher than the M-flame. In addition, even though the phases were both nearly linear, their slopes were very different. Thus, in the case of plasma forcing by NRP glow discharges, the shape of the flame strongly influenced the response of the combustion front. In summary, this experiment demonstrated that in the range of 4–450 Hz, plasma and electric field forcing significantly affected the heat release rate of the C-, V-, and M-flames. In addition, it was found that in the case of forcing by acoustic waves and plasma generated by NRP glow discharges, the geometry of the flame played a key role in the response of the combustion, while the flame shape did not greatly affect the response of the combustion to AC electric field forcing. For plasma forcing, a local increase in the burning velocity close to the rod, inducing wrinkling of the flame front, explained the results pf plasma forcing effect on the transfer function.

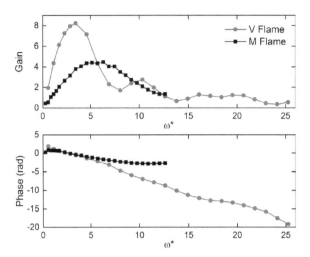

FIGURE 7.25 C-flame, V-flame and M-flame transfer functions to nonthermal plasma generated by NRP glow discharges at 10 kHz (forcing amplitude 1.8%) [61].

7.5 INTERNAL COMBUSTION ENGINE APPLICATIONS

Plasma-assisted combustion has also been applied for in internal combustion engines such as gasoline and diesel engines [62–79] (Suckewever, A., *personal communication*. 2010). In spark ignition (SI) engines, as we discussed in previous chapters, the development of the spark ignition kernel size strongly influences the lean-burn limit and emission characteristics. A larger ignition kernel size can extend the lean-burn limit for large hydrocarbon fuels [75–77]. In order to improve the ignition of SI engines, different plasmas such as microwave [62,63], NSD [74], and gliding arc have been used to replace or integrate with a conventional spark plug. A schematic of the integration of different plasmas with a conventional spark plug and a comparison of spark sizes are shown in Figures 7.26 and 7.27 [6], respectively. Figure 7.27 shows that using a gliding arc (Suckewever, A.,

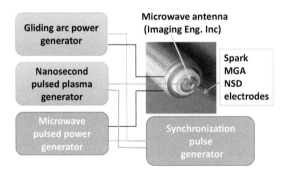

FIGURE 7.26 Concept of plasma-assisted spark ignition using NSD, gliding arc, and microwave [6].

FIGURE 7.27 Comparison of ignition using spark plug (left), microwave (middle), gliding arc (right). Photos were provided by Knite Inc. and Imagineering Inc. [6].

personal communication. 2010) and microwave discharge [62] at the same pressure of 1 MPa the ignition kernel sizes are much greater than that of the conventional spark plug.

Ikeda and collaborators developed and tested a microwave-integrated spark igniter [62–64] that can be applied to existing engine systems without any engine modification. As shown in Figure 7.26, a microwave discharge at a frequency of 2.45 GHz was generated by a magnetron into a non-resistor spark plug which had the same geometry as a conventional spark plug. The two electrodes of the spark plug also acted as a microwave antenna. The microwave system operated in burst mode, in which a series of pulses of 2.45 GHz microwave energy were delivered to the spark plug starting before and ending after the standard spark. The microwave energy was then absorbed by the electrons generated by the spark and raised the electrons to higher temperature. These electrons collided with the gas molecules and created excited species and active radicals. In an earlier work [78], the rotational temperature and N_2 density were measured by rotational Raman scattering in the region of the microwave discharge. It was shown that at a microwave power less than the ignition energy of a spark plug, the rotational temperature was raised to a maximum of 1,500 K in a nitrogen/helium mixture by microwave.

Figure 7.28 shows direct photographs of the C3H8 flames with SI and microwave plasma ignition with the equivalence ratio of $\phi=1$ [79]. In the case of using a normal spark plug, spark discharge was observed at −14.2° after top dead center (ATDC). The flame kernel was formed at −7.0° ATDC, and then the flame propagated at −0.2° ATDC. With the microwave-enhanced spark plug, intense light from the plasma was observed shortly after SI. Then, a large flame kernel was formed by the microwave energy addition at 0.2° ATDC.

The microwave-enhanced spark plug was further tested using a 499.6 cc, four-stroke-single cylinder gasoline research engine at a constant indicated mean effective pressure (IMEP) of 275 kPa and 2,000 rpm [63,79]. The effects of the microwave-enhanced spark plug on the lean limit, fuel consumption, and exhaust emissions were evaluated. The results are shown in Figure 7.29. With a spark plug igniter, at air/fuel mass ratio (A/F) of 17, the cyclic variation coefficient of IMEP (COV_{IMEP}) increased dramatically, indicating close to the lean limit of the engine. In contrast, with the microwave-enhanced spark plug, the COV remained essentially constant until A/F=22. Moreover, MW also reduced the fuel consumption (FC) and CO emissions for A/F greater than 18.

The pressure dependence of the effectiveness of the microwave addition on extending the lean and rich ignitability limits was also investigated [64]. Figure 7.30 compares the ignition limits versus initial pressure with and without microwave addition for initial pressures ranging from 1.08 to 7.22 bar at 300 K. The ignition limits were both extended at the lean and rich limits with microwave use. The experiments also showed that for methane/air mixtures at $\phi=0.65$, 0.75, and 1.0, the

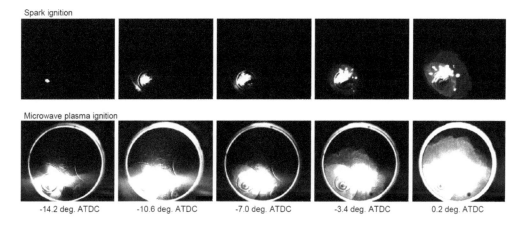

FIGURE 7.28 Comparison of C_3H_8 flame images in a compression-expansion engine using conventional spark plug and microwave-enhanced spark plug, $\phi=1$, initial pressure 600 kPa, initial engine speed 600 rpm [79].

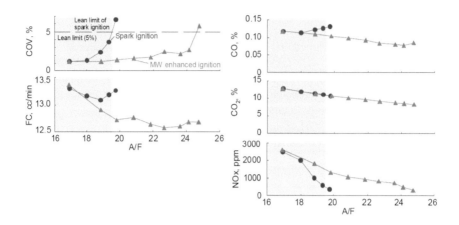

FIGURE 7.29 The effect of spark ignition and microwave-enhanced spark ignition on COV_{IMEP}, fuel consumption, and exhaust emission [79].

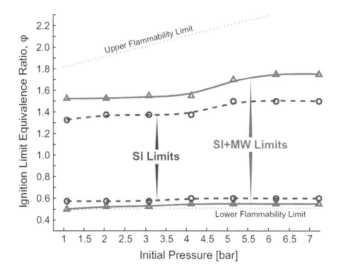

FIGURE 7.30 Lean and rich ignition limits for SI and SI+MW ignitions for a range of initial pressures at 300 K [64].

reduction of flame initiation time by microwaves decreased with the increase in initial pressure for all equivalence ratios and the enhancement effect became negligible when the initial pressure was greater than 3 bar.

The MW-assisted ignition system was also tested for small engines which are affected more by the SI energy and radical losses to the engine wall due to the increased surface/volume ratio. A single cylinder, 34 cc 4-stroke gasoline engine (Fuji Imvac Model BF-34EI) is shown in Figure 7.31 [63]. It was found that the microwave-enhanced SI produced a larger ignition kernel and led to an overall faster ignition and increase of the peak pressure with about 750 mJ MW energy output. By comparison, only less than 50 mJ output energy was obtained by the standard capacitive spark plug. As shown in Figure 7.31, the experimental results of COV_{imep} showed that the lean-burn limit was extended by 20%–30% in terms of the air/fuel ratio. The results suggested that the microwave-enhanced spark plug is an effective way to improve lean burn for micro- and mesoscale engines.

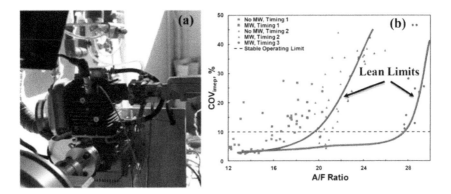

FIGURE 7.31 (a) Direct photograph of plasma-assisted 34 cc Fuji engine test setup and (b) the comparison of limits of stable engine operating conditions with and without microwave (MW) discharge at 2000 rpm [63].

Using the same small engine as Lefkowitz et al. [63], Thelen at al. [80] investigated the effect of a RF plasma ignition system. The RF power generator pulses at 20 kHz and the output pulses are a burst of alternating high (30 kV) and low (0 V) voltage square waves with 25 μs width. The maximum pulse burst duration was 1.2 ms with the RF system providing maximum amount of energy of 400 mJ to the discharge. The engine was tested at multiple engine speeds and air fuel ratios. The results also demonstrated an improvement in combustion stability at all operating conditions and an extension of the engine lean flammability limit. It was reported that the power output of the small engine was also higher with the RF plasma ignition system for stoichiometric and leaner air fuel ratios, and the hydrocarbon emissions were reduced.

The USC (University of Southern California) group developed various transient plasma igniters (TPIs) using corona and nanosecond discharge [73,74] for gasoline engines. It was found that by using a TPI, the ignition delay time can be improved. When a 20 ns pulse was used, stable lean combustion was realized at conditions not ignitable by a spark plug. It was also found that the peak pressure relative to SI increased by 20% using a 20 ns pulse TPI. The authors inferred that during the initial phase of ignition of a hydrocarbon fuel (RH), the chain initiation reaction $RH+O_2=R+HO_2$ was replaced by fuel and O_2 dissociation reactions, $RH+e \rightarrow R+H+e$ and $O_2+e \rightarrow 2O+e$ to accelerate the radical production. Shiraishi et al. [74] further investigated ignition characteristics in the lean combustion region and the relationship between the pulse width and ignition characteristics. The transient plasma ignition used in this study operated at 80 or 25 ns pulse width with a fixed input energy of 60 mJ. The test fuel was a 90 research octane number (RON) primary reference fuel (a mixture of 90% iC_8H_{18} (iso-octane) and 10% nC_7H_{16} (n-heptane) in volume). With the increase of the maximum voltage of the streamer discharge at the same pressure condition, an increase in E/N by roughly 30%–40% was observed. The results indicated that a shortened pulse width produced faster combustion.

The ignition mechanism of transient plasma ignition was investigated by Cathey et al. [81] in a quiescent, stoichiometric CH_4/air mixture by measuring the OH production in a large cylindrical chamber with the anode rod located in the center. The transient plasma was generated using a 70 ns FWHM, 60 kV, 800 mJ pulse. The total time for avalanche development, avalanche to streamer development and streamer propagation between the electrodes was typically 100–300 ns. Using a short high-voltage pulse (<100 ns) in this experiment, energy was effectively coupled into a non-equilibrium plasma without a transition from the streamer to the arc discharge. The number density of OH was measured using PLIF. It was found that OH was produced throughout the chamber volume after the discharge and decreased below detectability. The results showed that the production of OH by the streamers in the bulk of the volume was not the cause for ignition because of the low energy release. The authors suggested that a high voltage to produce OH throughout the chamber was not, apparently, critical for ignition. The authors further explained that it was possible due to an electrically enhanced surface catalytic reaction at the anode surface and/or local joule heating of the electrode. Different from the MW experiment, the transient plasma ignition experiments indicated that the thermal effect played an important role in the enhancement of ignition and that the OH production by streamer affected the flame propagation.

Laser as another high-temperature ignition source for internal combustion engine has been tried for gasoline and natural gas engines for a number of years by Toyota, Ford, Caterpillar, and Cummins engines [31,70,71]. It has been reported that laser ignition was able to ignite leaner mixtures with NO_x reduction. The fundamental mechanisms for laser ignition can be summarized as (1) Thermal ignition in which molecules absorb photon energy and increase temperature; (2) photochemical ignition in which molecules dissociate or are ionized after absorbing multiple photon energy; (3) resonant breakdown in which dissociation and ionization of target molecules or atoms by the resonant multi-photon-ionization process; and (4) non-resonant thermal breakdown occurs when the focus laser power is sufficiently high to influence the gas molecules and initiate the electrical breakdown of the gas. By far, non-resonant breakdown has been the most frequently adopted ignition mode to initiate combustion primarily because of its freedom in selecting the laser wavelength and power. The advantage of laser ignition is that it provides remote and multipoint ignition and avoids heat loss to the electrode. Figure 7.32 shows a compact prototype laser igniter which is a passively Q-switched Nd:YAG/$^{Cr4+}$:YAG giant pulse emitting micro-laser with three beam output [82]. Pavel et al. [82] reported that at a 5 Hz repetition rate, each line delivered laser pulses with approximately 2.4 mJ energy and 2.8 MW peak power within the pulse duration. Recently, Princeton Optronics developed Vertical Cavity Surface Emitting Laser and increased the efficiency (up to 63.4%) and energy output to 7 mJ per pulse with a repetition rate of 20 Hz at pulse duration of 100 μs [83]. Although many engine tests have demonstrated the ability of laser ignition to ignite ultra-lean fuel-air mixtures and mixtures with high Exhaust Gas Recirculation (EGR), the laser stability at a high temperature (e.g., 200°C–400°C), vibration, and soot formation are still challenging problems.

Ozone (O_3), which can be efficiently produced by plasma and has a strong low-temperature kinetic enhancement effect [84] and modified flame speed [85–87], has also been used for engine combustion control. The effect of O_3 on the combustion of n-heptane (nC_7H_{16}) in a Homogeneous Charge Compression Ignition (HCCI) engine was studied by Foucher et al. [88]. Experiments were performed in a single cylinder diesel engine fueled with n-heptane at a constant equivalence ratio of 0.3, intake temperature of 300 K, and engine speed of 1,500 rpm. The intake manifold was seeded by ozone produced by a DBD

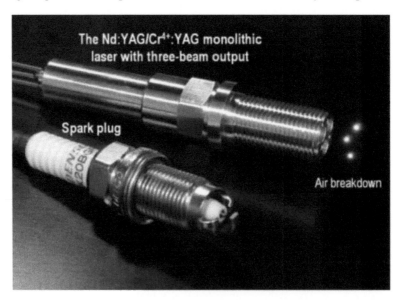

FIGURE 7.32 Direct photograph of a prototype laser igniter showing breakdown in air at multiple points [82].

reactor. Experimental results showed that low ozone concentrations (<50 ppm) had an important impact on the phasing of the cool and main flame in the engine. It was found that the low-temperature ignition and main flame ignition timing decreased significantly when ozone concentration increased from 0 to 20 ppm. However, it was also reported that the influence of O_3 on flame speed became less important above 20 ppm although the ignition time monotonically decreased. If the O_3 concentration increased from 0 to 54 ppm at the intake of the HCCI engine, it advanced the main flame phasing by 7 CAD (Crank Angle Degree). Comparatively, a variation of 0%–60% of EGR retards the phasing of the main flame by 7 CAD. A variation of 60 K of the intake gases mixture temperature affected 3 CAD and a variation of 49 ppm of NO had an effect of 1.5 CAD in this engine at the same condition.

In summary, many engine experiments have shown that non-equilibrium plasma such as microwave, RF, corona, gliding arc, and ozone, can enhance engine performance. However, since most of the practical engines work at high pressure, there is a great need to understand the kinetic properties of plasma and to develop a large volume plasma discharge at high pressure to enhance combustion at engine conditions.

7.6 EMISSION CONTROL

Plasma-assisted combustion can lower the flame temperature, extend burning limits, increase fuel/air mixing, and produce intermediate radicals such as NH_x, OH, H, and CH_x via energetic electrons and excited species. Many studies have been conducted to use non-equilibrium plasma for emission control such as NO_x [89–95], SO_x [96], unburned hydrocarbons (UHCs) [97], and soot [98,99]. In this section, we will only briefly highlight the results of NO_x and soot emission control by plasma, ozone, and electric field. We will also focus on ammonia and small hydrocarbons because of recent interest in hydrogen and e-fuels. Readers can find other applications in the literature.

7.6.1 NO$_x$ Emission Control

NO_x formation has different mechanisms such as the Zel'dovich thermal mechanism, prompt mechanism, NNH mechanism, fuel NO_x mechanism [100–104], which is sensitive to flame temperature, fuel, fuel/air mixture ratio, flow residence time, and radical formation. The reduction of NO_x by plasma can be achieved by changing flame temperature, combustion composition, burning time, flame geometry, and radical production by plasma. In fuel-rich combustion or fuel NO_x formation, there are several NO_x reburning mechanisms [101,102,105] to reduce NO_x, for example,

$$CH_i + NO \rightarrow HCN + H_iO \tag{R7.1}$$

$$HCN + O \rightarrow NCO+H \rightarrow NH+CO \tag{R7.2}$$

$$NH + H \rightarrow N + H_2 \tag{R7.3}$$

$$N + NO \rightarrow N_2 + O \tag{R7.4}$$

$$NH_i + NO \rightarrow N_2 + H_iO \tag{R7.5}$$

$$NH_2 + NO \rightarrow NNH + OH \tag{R7.6}$$

In combustion of hydrocarbon fuels, plasma-assisted combustion may reduce NO emissions if the above reburning mechanism is activated. It may also increase NO_x emissions with fast O atom and N atom production by plasma discharge. Kim et al. [94] studied nitric oxide emission control in an ultra-short pulse repetitive discharge stabilized premixed methane/air flame. It was shown that the flame exhibited a dual-layer structure. A pre-flame was observed to form at lower equivalence ratio and contributed to partially consuming NO via the NO_x reburning mechanism. In the absence of ignition of either this pre-flame, the production of NO correlated well with the concentration of excited molecular nitrogen and the power deposition by the plasma. Following the ignition of the main flame at higher equivalence ratio, the NO was seen to rise abruptly. Therefore, NO increased in plasma-assisted main flames. The increase of NO production in a laminar premixed methane/air flame by non-equilibrium plasma discharge was possibly caused by the excited metastable molecular nitrogen. A later study by Lacoste et al. [58] also showed that NO concentration in the burned gases only depended on the power of the plasma per unit mass of gas. The results were in good agreement with the observation by Kim et al. [94]. The dominant chemical mechanism for NO_x production in plasma discharges in air-hydrocarbon mixtures was fast heating and atomic oxygen production. Further investigation is necessary to understand how to optimize plasma-assisted combustion to lower NO emissions in combustion of hydrocarbon fuels.

Ozone was also used to oxidize NO and subsequently remove NO_2, NO_3, and N_2O_5 with a wet flue gas desulfurization system for SO_2 [95]. Want et al. [95] used PLIF of NO and NO_2 to investigate the reaction structures between a turbulent O_3 jet and a laminar coflow of simulated flue gas (containing 200 ppm NO) in co-axial tubes. The results showed that about 62% of NO was oxidized at $15d$ (d, jet orifice diameter) by a 30 m/s O_3 jet with an influence width of about $6d$ in radius. Therefore, plasma-generated ozone is an effective way to oxidize NO in flue gas and then to be removed by using a wet flue gas desulfurization system.

For plasma-assisted ammonia combustion, there are some encouraging findings for NO_x reduction [89–92]. Choe and Sun [89,90] investigated the effect of nanosecond pulsed non-equilibrium plasma on ammonia combustion in a gas turbine combustor using NH_{2*} chemiluminescence and OH PLIF. It was shown that plasma could improve the stabilization of ammonia/air flame and extend the attached flame regime to a lower equivalence ratio. It was also shown that with the increase of discharge voltage or discharge power, both NH_{2*} chemiluminescence intensity and OH PLIF signal intensity increased. This observation gave an explanation of the improved stabilization of ammonia flames and might also shed light on the mechanism of NO_x reduction.

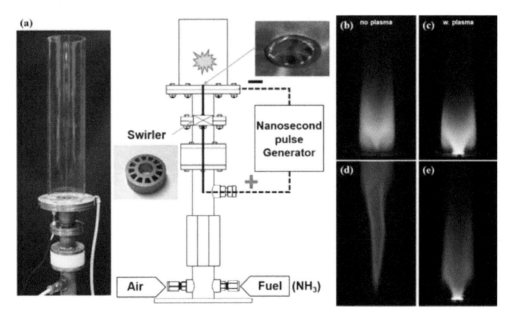

FIGURE 7.33 (a) Schematic of the experimental setup, and direct photographs (ISO100, F/10, 2 s) of flames (b) no plasma, $\phi=0.94$ (c) with plasma, $\phi=0.94$, (d) no plasma, $\phi=0.71$ (e) with plasma, $\phi=0.71$ [89].

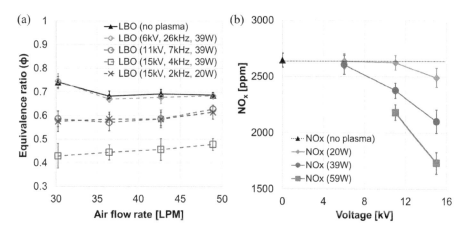

FIGURE 7.34 (a) Lean blowoff limits of ammonia/air flames with and without plasma, (b) NO_x emissions without and with plasma ($\phi=0.94$) [89].

The experimental setup [89] of the premixed ammonia/air swirling flame is shown in Figure 7.33. Non-equilibrium plasma was generated by a nanosecond high-voltage pulse generator between the center copper electrode (diameter of 5 mm) and outer ring (diameter of 19 mm) at the nozzle exit. Figure 7.33b–e shows direct photographs of ammonia/air flames without and with plasma (discharge voltage $V=11$ kV, pulse repetition frequency $f=7$ kHz, 39 W). The ranges of voltage and frequency used in the study were 6–15 kV and 2–26 kHz, respectively. Figure 7.34a shows the lean blowoff limits (LBO) of ammonia/air flames with and without plasma. Without plasma, an initially attached, stable flame Figure 7.33b was lifted (Figure 7.33d) from the dump plane with the increase of air flow rate at a fixed fuel flow rate. When the equivalence ratio (ϕ) was decreased below the values indicated by the black line in Figure 7.34a, blowoff occurred. With plasma activation, flames were always stabilized near the burner exit (Figure 7.34c and e). With plasma at $V=11$ kV, $f=7$ kHz (discharge power 39 W, corresponding to 1.9% of thermal load at $\phi=0.94$), the lean blowout limit was significantly extended. The discharge power was proportional to the increase of both reactant temperature and number of reactive species produced by the plasma. However, 39 W thermal power only increased the average reactant temperature by 6 K, which had very limited effect on flame blowout limit. Therefore, the effect of plasma on the extended flame stabilization limit might be primarily kinetic in nature. The measured NO_x concentrations are shown in Figure 7.34b. Without plasma, ammonia/air combustion produced approximately 2,645 ppm NO_x at $\phi=0.94$ by estimation. With plasma, NO_x emissions are significantly reduced. The corresponding adiabatic flame temperature was 2,009 K, assuming 59 W plasma power was purely thermal effect. It is seen that NO_x concentration further decreased with the increase of both discharge power and voltage. NO_x concentration decreased as the discharge voltage of plasma was increased at constant discharge power. If the discharge voltage was fixed, NO_x concentration decreased with the increase of discharge power (via increased plasma frequency). The tendency of NO_x formation is opposite to that for methane/air flames discussed above. Therefore, likely that plasma-generated NH_i radicals may contributed to the NO_x reduction via reactions such as (R7.5). This result demonstrates that plasma is a promising technology to reduce NO_x emissions from fuel NO_x.

7.6.2 Soot Emission Control

Soot emissions are strongly affected by fuel/air premixing, temperature, and OH and H radicals. Plasma not only can change temperature and radical production in combustion but also creates ionic wind to affect fuel/air mixing, leading to lower soot emissions. Cha et al. [98,99] studied DBD and electric field effect on soot emissions in a counterflow diffusion flame.

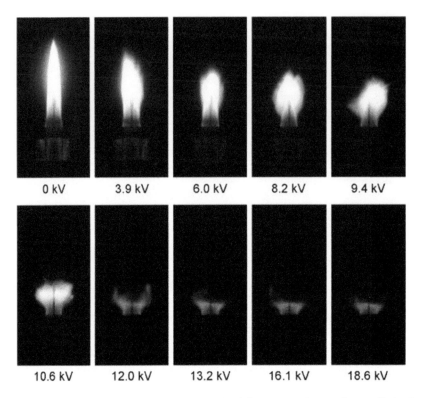

FIGURE 7.35 Flame images taken by digital camera with 0.5 s exposure by varying applied voltage [98].

As shown in Figure 7.35, the apparatus consisted of a coflow burner and flow controllers, a nonthermal plasma-generating system, and a visualization setup. The burner had an inner fuel nozzle made of quartz tube with 8 mm i.d. and 10 mm o.d., respectively. The fuel was chemically pure grade (>99.5%) propane and compressed air was used as an oxidizer. The plasma reactor had wire-cylinder-type electrodes with AC power supply operated at 400 Hz. The effect of streamers on flame behavior revealed that the flame length was significantly decreased as the applied voltage increased. The yellow luminosity by the radiation of soot particles was also significantly diminished. Figure 7.35 [98] shows the direct photographs of the variation of flame shape with increasing applied voltage to the DBD reactor. Three distinguished features were observed with increasing voltage: (1) the decrease in flame length, (2) the decrease in yellow luminosity, and (3) the change of flame shape from near conical to crown shape. The flame length, defined from the visual luminosity at the center axis, decreased with the increase of voltage. It was shown that the formation of polycyclic aromatic hydrocarbons (PAHs) and soot was influenced appreciably by the nonthermal plasma, while the flame temperature and the concentration of major species were not influenced much with the plasma generation. The results demonstrated that the application of nonthermal plasmas can effectively suppress PAH and soot formation in the flames with a very low power consumption. The mechanism is mostly because of the plasma-enhanced fuel/air mixing.

A later study by Park et al. using DC electric field in a counterflow ethylene diffusion flame [99] to study electric field affected soot formation and oxidation. It was shown that the flames subjected to the negative electric fields moved toward the fuel nozzle because of an ionic wind. In addition, the yellow luminosity from soot significantly decreased, indicating changes in the sooting characteristics. Planar laser-induced incandescence and fluorescence techniques were used to visualize the soot, PAHs, and OH radicals. The results showed that soot particles were reduced significantly by applying negative voltages. Modification of the ionic wind-driven flow field reduced soot particles by affecting the flow residence time, the temperature field, and radical distributions, leading to the reduction of soot formation and growth.

REFERENCES

1. Choubey, G., et al., Recent advances in cavity-based scramjet engine-a brief review. *International Journal of Hydrogen Energy*, 2019. **44**(26): pp. 13895–13909.
2. Miller, J.D., et al., Investigation of transient ignition processes in a model scramjet pilot cavity using simultaneous 100kHz formaldehyde planar laser-induced fluorescence and CH* chemiluminescence imaging. *Proceedings of the Combustion Institute*, 2017. **36**(2): pp. 2865–2872.
3. Segal, C., *The Scramjet Engine: Processes and Characteristics*. Vol. 25. 2009: Cambridge University Press.
4. Jacobsen, L.S., et al., Plasma-assisted ignition in scramjets. *Journal of Propulsion and Power*, 2008. **24**(4): pp. 641–654.
5. Colket, M.B. and L.J. Spadaccini, Scramjet fuels autoignition study. *Journal of Propulsion and Power*, 2001. **17**(2): pp. 315–323.
6. Ju, Y. and W. Sun, Plasma assisted combustion: Dynamics and chemistry. *Progress in Energy and Combustion Science*, 2015. **48**: pp. 21–83.
7. Starikovskiy, A. and N. Aleksandrov, Plasma-assisted ignition and combustion. *Progress in Energy and Combustion Science*, 2013. **39**(1): pp. 61–110.
8. Dooley, S., et al., The experimental evaluation of a methodology for surrogate fuel formulation to emulate gas phase combustion kinetic phenomena. *Combustion and Flame*, 2012. **159**(4): pp. 1444–1466.
9. Kimura, M. and Y. Itikawa, *Advances in Atomic, Molecular, and Optical Physics: Electron Collisions with Molecules in Gases: Applications to Plasma Diagnostics and Modeling*. 2000: Elsevier.
10. Kimura, I., H. Aoki, and M. Kato, The use of a plasma jet for flame stabilization and promotion of combustion in supersonic air flows. *Combustion and Flame*, 1981. **42**: pp. 297–305.
11. Takita, K., et al., Ignition characteristics of plasma torch for hydrogen jet in an airstream. *Journal of Propulsion and Power*, 2000. **16**(2): pp. 227–233.
12. Leonov, S.B., et al., Plasma-assisted combustion of gaseous fuel in supersonic duct. *IEEE Transactions on Plasma Science*, 2006. **34**(6): pp. 2514–2525.
13. Barbi, E., et al., Operating characteristics of a hydrogen-argon plasma torch for supersonic combustion applications. *Journal of Propulsion and Power*, 1989. **5**(2): pp. 129–133.
14. Takita, K., Ignition and flame-holding by oxygen, nitrogen and argon plasma torches in supersonic airflow. *Combustion and Flame*, 2002. **128**(3): pp. 301–313.
15. Takita, K., et al., Ignition enhancement by addition of NO and NO_2 from a N_2/O_2 plasma torch in a supersonic flow. *Proceedings of the Combustion Institute*, 2007. **31**(2): pp. 2489–2496.
16. Wagner, T.C., et al., Plasma torch igniter for scramjets. *Journal of Propulsion and Power*, 1989. **5**(5): pp. 548–554.
17. Leonov, S.B. and D.A. Yarantsev, Plasma-induced ignition and plasma-assisted combustion in high-speed flow. *Plasma Sources Science and Technology*, 2006. **16**(1): p. 132.
18. Feng, R., et al., Multi-channel gliding arc plasma-assisted ignition in a kerosene-fueled model scramjet engine. *Aerospace Science and Technology*, 2022. **126**: p. 107606.
19. Leonov, S.B., et al., Modes of plasma-stabilized combustion in cavity-based M = 2 configuration. *Experimental Thermal and Fluid Science*, 2021. **124**: p. 110355.
20. Masuya, G., et al., Effects of airstream Mach number on H/N plasma igniter. *Journal of Propulsion and Power*, 2002. **18**(3): pp. 679–685.
21. Takita, K., et al., Ignition and flame-holding of H_2 and CH_4 in high temperature airflow by a plasma torch. *Combustion and Flame*, 2003. **132**(4): pp. 679–689.
22. Matsubara, Y., K. Takita, and G. Masuya, Combustion enhancement in a supersonic flow by simultaneous operation of DBD and plasma jet. *Proceedings of the Combustion Institute*, 2013. **34**(2): pp. 3287–3294.
23. Takita, K., et al., A novel design of a plasma jet torch igniter in a scramjet combustor. *Proceedings of the Combustion Institute*, 2005. **30**(2): pp. 2843–2849.
24. Liu, Q., et al., Ethylene flame dynamics and inlet unstart in a model scramjet. *Journal of Propulsion and Power*, 2014. **30**(6): pp. 1577–1585.

25. Do, H., et al., Ethylene flame dynamics in an arc-heated hypersonic wind tunnel, in *51st AIAA Aerospace Sciences Meeting Including the New Horizons Forum and Aerospace Exposition*. Jan. 7-10, Grapevine, Texas, 2013.
26. Leonov, S., et al., Plasma-assisted ignition and flameholding in high-speed flow, in *44th AIAA Aerospace Sciences Meeting and Exhibit*. Jan. 9-12, Reno, Nevada, 2006.
27. Leonov, S., D. Yarantsev, and C. Carter, Experiments on electrically controlled flameholding on a plane wall in a supersonic airflow. *Journal of Propulsion and Power*, 2009. **25**(2): pp. 289–294.
28. Tian, Y., et al., Enhancement of blowout limit in a Mach 2.92 cavity-based scramjet combustor by a gliding arc discharge. *Proceedings of the Combustion Institute*, 2023. **39**(4): pp. 5697–5705.
29. Do, H., M.A. Cappelli, and M.G. Mungal, Plasma assisted cavity flame ignition in supersonic flows. *Combustion and Flame*, 2010. **157**(9): pp. 1783–1794.
30. Esakov, I., et al., Efficiency of propane-air mixture combustion assisted by deeply undercritical MW discharge in cold high-speed airflow, in *44th AIAA Aerospace Sciences Meeting and Exhibit*. Jan. 9-12, Reno, Nevada, 2006.
31. Brieschenk, S., S. O'Byrne, and H. Kleine, Laser-induced plasma ignition studies in a model scramjet engine. *Combustion and Flame*, 2013. **160**(1): pp. 145–148.
32. Macheret, S.O., M.N. Shneider, and R.B. Miles, Plasma-assisted fuel atomization and multipoint ignition for scramjet engines. *Journal of Propulsion and Power*, 2020. **36**(3): pp. 357–362.
33. Zhong, H., et al., Plasma thermal-chemical instability of low-temperature dimethyl ether oxidation in a nanosecond-pulsed dielectric barrier discharge. *Plasma Sources Science and Technology*, 2022. **31**(11): p. 114003.
34. Zhong, H., et al., Dynamics and chemical mode analysis of plasma thermal-chemical instability. *Plasma Sources Science and Technology*, 2021. **30**: p. 035002.
35. Rousso, A.C., et al., Time and space resolved diagnostics for plasma thermal-chemical instability of fuel oxidation in nanosecond plasma discharges. *Plasma Sources Science and Technology*, 2020. **29**(10): p. 105012.
36. Zhong, H., et al., Thermal-chemical instability of weakly ionized plasma in a reactive flow. *Journal of Physics D: Applied Physics*, 2019. **52**(48): p. 484001.
37. National Academies of Sciences, Engineering, and Medicine, *Advanced Technologies for Gas Turbines*. 2020: National Academies Press.
38. Journell, C.L., et al., High-speed diagnostics in a natural gas–air rotating detonation engine. *Journal of Propulsion and Power*, 2020. **36**(4): pp. 498–507.
39. Stechmann, D.P., et al., Role of ignition delay in rotating detonation engine performance and operability. *Journal of Propulsion and Power*, 2019. **35**(1): pp. 125–140.
40. Yokoo, R., et al., Experimental study of internal flow structures in cylindrical rotating detonation engines. *Proceedings of the Combustion Institute*, 2021. **38**(3): pp. 3759–3768.
41. Schwer, D. and K. Kailasanath, Numerical investigation of the physics of rotating-detonation-engines. *Proceedings of the Combustion Institute*, 2011. **33**(2): pp. 2195–2202.
42. Cathey, C., et al., Transient plasma ignition for delay reduction in pulse detonation engines, in *45th AIAA Aerospace Sciences Meeting and Exhibit*. Jan. 8-11, Reno, Nevada, 2007.
43. Schauer, F., J. Stutrud, and R. Bradley, Detonation initiation studies and performance results for pulsed detonation engine applications, in *39th Aerospace Sciences Meeting and Exhibit*. Jan. 8-11, Reno, Nevada, 2001.
44. Kailasanath, K., Recent developments in the research on pulse detonation engines. *AIAA Journal*, 2003. **41**(2): pp. 145–159.
45. Starikovskiy, A., N. Aleksandrov, and A. Rakitin, Plasma-assisted ignition and deflagration-to-detonation transition. *Philosophical Transactions of the Royal Society A: Mathematical, Physical and Engineering Sciences*, 2012. **370**(1960): pp. 740–773.
46. Vorenkamp, M., et al., Effect of plasma-enhanced low-temperature chemistry on deflagration-to-detonation transition in a microchannel. *AIAA Journal*, 2023. **61**: pp. 1–7.
47. Vorenkamp, M., et al., Plasma-assisted deflagration to detonation transition in a microchannel with hybrid fs/ps coherent anti-stokes Raman scattering measurements, *Proceedings of the Combustion Institute*, 2023. **39** (4): pp. 5561–5569.
48. Sepulveda, J., et al., Kinetic enhancement of microchannel detonation transition by ozone addition to acetylene mixtures. *AIAA Journal*, 2019. **57**(2): pp. 476–481.
49. Lacoste, D.A., Flames with plasmas. *Proceedings of the Combustion Institute*, 2022. 39: pp. 5405–5428.

50. Starikovskiy, A., Deflagration-to-detonation control by non-equilibrium gas discharges and its applications for pulsed detonation engine, in *39th AIAA/ASME/SAE/ASEE Joint Propulsion Conference and Exhibit*. July 20-23, Huntsville, Alabama, 2003.
51. Zhukov, V. and A. Starikovskiy, Deflagration-to-detonation control by non-equilibrium gas discharges and its applications for pulsed detonation engine, in *43rd AIAA Aerospace Sciences Meeting and Exhibi*. Jan. 10-13, Reno, Nevada, 2005.
52. Lefkowitz, J.K., et al., The effects of repetitively pulsed nanosecond discharges on ignition time in a pulsed detonation engine, in *49th AIAA/ASME/SAE/ASEE Joint Propulsion Conference*. July 14-17, San Jose, CA, 2013.
53. Gray, J.A. and D.A. Lacoste, Effect of the plasma location on the deflagration-to-detonation transition of a hydrogen–air flame enhanced by nanosecond repetitively pulsed discharges. *Proceedings of the Combustion Institute*, 2021. **38**(3): pp. 3463–3472.
54. Cherif, M.A., et al., Effect of non-equilibrium plasma on decreasing the detonation cell size. *Combustion and Flame*, 2020. **217**: pp. 1–3.
55. Thawko, A., et al., Accelerated ignition-shock coupling and deflagration to detonation transition by kinetic enhancement of ozone in a dimethyl ether mixture, in *40th International Symposium on Combustion*. 2024, Combustion Institute: Milan, Italy.
56. Serbin, S., et al., Improvement of the gas turbine plasma assisted combustor characteristics, in *49th AIAA Aerospace Sciences Meeting Including the New Horizons Forum and Aerospace Exposition*. Jan. 4-7, 2011, Orlando, Florida, 2011.
57. Moeck, J., et al., Control of combustion dynamics in a swirl-stabilized combustor with nanosecond repetitively pulsed discharges, in *51st AIAA Aerospace Sciences Meeting including the New Horizons Forum and Aerospace Exposition*. 7-10 January 2013, Grapevine, Texas, USA, 2013.
58. Lacoste, D., et al., Effect of nanosecond repetitively pulsed discharges on the dynamics of a swirl-stabilized lean premixed flame. *Journal of Engineering for Gas Turbines and Power*, 2013. **135**(10): p. 101501.
59. Barbosa, S., et al., Influence of a repetitively pulsed plasma on the flame stability domain of a lab-scale gas turbine combustor, in *Fourth European Combustion Meeting*, Vienna University of Technology, 14-17 April, 2009.
60. Di Sabatino, F. and D.A. Lacoste, Enhancement of the lean stability and blow-off limits of methane-air swirl flames at elevated pressures by nanosecond repetitively pulsed discharges. *Journal of Physics D: Applied Physics*, 2020. **53**(35): p. 355201.
61. Lacoste, D.A., et al., Transfer functions of laminar premixed flames subjected to forcing by acoustic waves, AC electric fields, and non-thermal plasma discharges. *Proceedings of the Combustion Institute*, 2017. **36**(3): pp. 4183–4192.
62. Ikeda, Y., A. Moon, and M. Kaneko, Development of microwave-enhanced spark-induced breakdown spectroscopy. *Applied Optics*, 2010. **49**(13): pp. C95–C100.
63. Lefkowitz, J.K., et al., A study of plasma-assisted ignition in a small internal combustion engine. AIAA paper-2012-1133, 2012.
64. Wolk, B., et al., Enhancement of flame development by microwave-assisted spark ignition in constant volume combustion chamber. *Combustion and Flame*, 2013. **160**(7): pp. 1225–1234.
65. Wang, Q., et al., Visual features of microwave ignition of methane-air mixture in a constant volume cylinder. *Applied Physics Letters*, 2013. **103**(20): p. 204104.
66. Maly, R., Spark ignition: Its physics and effect on the internal combustion engine. *Fuel Economy: In Road Vehicles Powered by Spark Ignition Engines*, Hilliard, J.C., (ed.), 1984: Springer, pp. 91–148.
67. Mariani, A. and F. Foucher, Radio frequency spark plug: An ignition system for modern internal combustion engines. *Applied Energy*, 2014. **122**: pp. 151–161.
68. Dale, J., P. Smy, and R. Clements, Laser ignited internal combustion engine—An experimental study. *SAE Transactions*, 1978: pp. 1539–1548.
69. Ma, J.X., D.R. Alexander, and D.E. Poulain, Laser spark ignition and combustion characteristics of methane-air mixtures. *Combustion and Flame*, 1998. **112**(4): pp. 492–506.
70. Herdin, G.N., et al., Laser ignition: A new concept to use and increase the potentials of gas engines, in ASME *Internal Combustion Engine Division Fall Technical Conference*. Sept. 11–14, Ottawa, Ontario, 2005.
71. Morsy, M.H., Review and recent developments of laser ignition for internal combustion engines applications. *Renewable and Sustainable Energy Reviews*, 2012. **16**(7): pp. 4849–4875.
72. Cathey, C.D., et al., Nanosecond plasma ignition for improved performance of an internal combustion engine. *IEEE Transactions on Plasma Science*, 2007. **35**(6): pp. 1664–1668.

73. Wang, F., et al., Transient plasma ignition of quiescent and flowing air/fuel mixtures. *IEEE Transactions on Plasma Science*, 2005. **33**(2): pp. 844–849.
74. Shiraishi, T., T. Urushihara, and M. Gundersen, A trial of ignition innovation of gasoline engine by nanosecond pulsed low temperature plasma ignition. *Journal of Physics D: Applied Physics*, 2009. **42**(13): p. 135208.
75. Chen, Z., M. Burke, and Y. Ju, Effects of Lewis number and ignition energy on the determination of laminar flame speed using propagating spherical flames. *Proceedings of the Combustion Institute*, 2009. **32**(1): pp. 1253–1260.
76. Santner, J.S., S.H. Won, and Y. Ju, Chemistry and transport effects on critical flame initiation radius for alkanes and aromatic fuels. *Proceedings of the Combustion Institute*, 2016. **36**(1): pp. 1457–1465.
77. Kim, H.H., et al., Measurements of the critical initiation radius and unsteady propagation of n-decane/air premixed flames. *Proceedings of the Combustion Institute*, 2013. **34**: pp. 929–936.
78. ElSabbagh, M., et al., Measurements of rotational temperature and density of molecular nitrogen in spark-plug assisted atmospheric-pressure microwave discharges by rotational Raman scattering. *Japanese Journal of Applied Physics*, 2011. **50**(7R): p. 076101.
79. Ikeda, Y., A. Nishiyama, and M. Kaneko, Microwave enhanced ignition process for fuel mixture at elevated pressure of 1MPa, in *47th AIAA Aerospace Sciences Meeting Including the New Horizons Forum and Aerospace Exposition*. Jan. 05-08, 2009, Orlando, Florida, 2009.
80. Thelen, B.C., et al., A study of an energetically enhanced plasma ignition system for internal combustion engines. *IEEE Transactions on Plasma Science*, 2013. **41**(12): pp. 3223–3232.
81. Cathey, C., et al., OH production by transient plasma and mechanism of flame ignition and propagation in quiescent methane–air mixtures. *Combustion and Flame*, 2008. **154**(4): pp. 715–727.
82. Pavel, N., M. Tsunekane, and T. Taira, Composite, all-ceramics, high-peak power Nd: YAG/Cr 4+: YAG monolithic micro-laser with multiple-beam output for engine ignition. *Optics Express*, 2011. **19**(10): pp. 9378–9384.
83. Seurin, J.-F., et al., High-brightness pump sources using 2D VCSEL arrays, in *Vertical-Cavity Surface-Emitting Lasers XIV*. 2010: SPIE.
84. Won, S.H., et al., Self-sustaining n-heptane cool diffusion flames activated by ozone. *Proceedings of the Combustion Institute*, 2015. **35**(1): pp. 881–888.
85. Ombrello, T., et al., Flame propagation enhancement by plasma excitation of oxygen. Part II: Effects of O_2 (a 1 Δ g). *Combustion and Flame*, 2010. **157**(10): pp. 1916–1928.
86. Ombrello, T., et al., Flame propagation enhancement by plasma excitation of oxygen. Part I: Effects of O_3. *Combustion and flame*, 2010. **157**(10): pp. 1906–1915.
87. Ehn, A., et al., Plasma assisted combustion: Effects of O_3 on large scale turbulent combustion studied with laser diagnostics and large Eddy Simulations. *Proceedings of the Combustion Institute*, 2015. **35**(3): pp. 3487–3495.
88. Foucher, F., et al., Influence of ozone on the combustion of n-heptane in a HCCI engine. *Proceedings of the Combustion Institute*, 2013. **34**(2): pp. 3005–3012.
89. Choe, J., et al., Plasma assisted ammonia combustion: Simultaneous NOx reduction and flame enhancement. *Combustion and Flame*, 2021. **228**: pp. 430–432.
90. Choe, J. and W. Sun, Experimental investigation of non-equilibrium plasma-assisted ammonia flames using nh2* chemiluminescence and oh planar laser-induced fluorescence. *Proceedings of the Combustion Institute*, 2023. **39**(4): pp. 5439–5446.
91. Zhong, H., et al., Kinetic studies of low-temperature ammonia oxidation in a nanosecond repetitively-pulsed discharge, in *AIAA SCITECH 2023 Forum*. 2023.
92. Radwan, A.M. and M.C. Paul, Plasma assisted NH_3 combustion and NOx reduction technologies: Principles, challenges and prospective. *International Journal of Hydrogen Energy*, 2024. **52**: pp. 819–833.
93. Lacoste, D.A., et al., Effect of plasma discharges on nitric oxide emissions in a premixed flame. *Journal of Propulsion and Power*, 2013. **29**(3): pp. 748–751.
94. Kim, W., et al., Investigation of NO production and flame structure in plasma enhanced premixed combustion. *Proceedings of the Combustion Institute*, 2007. **31**(2): pp. 3319–3326.
95. Wang, Z., et al., Investigation of flue-gas treatment with O_3 injection using NO and NO_2 planar laser-induced fluorescence. *Fuel*, 2010. **89**(9): pp. 2346–2352.
96. Sun, W., et al., Non-thermal plasma remediation of SO_2/NO using a dielectric-barrier discharge. *Journal of Applied Physics*, 1996. **79**(7): pp. 3438–3444.
97. Kirkpatrick, M.J., et al., Plasma assisted heterogeneous catalytic oxidation of carbon monoxide and unburned hydrocarbons: Laboratory-scale investigations. *Applied Catalysis B: Environmental*, 2011. **106**(1–2): pp. 160–166.

98. Cha, M.S., et al., Soot suppression by nonthermal plasma in coflow jet diffusion flames using a dielectric barrier discharge. *Combustion and Flame*, 2005. **141**(4): pp. 438–447.
99. Park, D.G., et al., Soot reduction under DC electric fields in counterflow non-premixed laminar ethylene flames. *Combustion Science and Technology*, 2014. **186**(4–5): pp. 644–656.
100. Glarborg, P., et al., Modeling nitrogen chemistry in combustion. *Progress in Energy and Combustion Science*, 2018. **67**: pp. 31–68.
101. Miller, J.A. and C.T. Bowman, Mechanism and modeling of nitrogen chemistry in combustion. *Progress in Energy and Combustion Science*, 1989. **15**(4): pp. 287–338.
102. Glarborg, P., J.A. Miller, and R.J. Kee, Kinetic modeling and sensitivity analysis of nitrogen oxide formation in well-stirred reactors. *Combustion and Flame*, 1986. **65**(2): pp. 177–202.
103. Miller, J.A., et al., Kinetic modeling of the oxidation of ammonia in flames. *Combustion Science and Technology*, 1983. **34**(1–6): pp. 149–176.
104. Guo, H., et al., The effect of hydrogen addition on flammability limit and NOx emission in ultra-lean counterflow CH_4/air premixed flames. *Proceedings of the Combustion Institute*, 2005. **30**(1): pp. 303–311.
105. Smoot, L.D., S. Hill, and H. Xu, NOx control through reburning. *Progress in Energy and Combustion Science*, 1998. **24**(5): pp. 385–408.

8 Electrified Non-Equilibrium Chemical Manufacturing

With rapid transition to electric vehicles (EVs) and decarbonization in transportation, carbon emissions from chemical manufacturing remain high (Figure 1.13). Therefore, there is a great need to decarbonize the sectors in chemicals and materials manufacturing using renewable electricity. Unfortunately, today's chemicals and materials manufacturing still highly rely on fossil fuels and equilibrium chemical processes such as the Haber–Bosch (HB) for ammonia [1] and steam-reforming for syngas and hydrogen [2,3]. The fossil fuel-based chemical processes produce enormous carbon emissions. Moreover, the yield and energy efficiency using an equilibrium chemical process are governed by the thermodynamic equilibrium and chemical equilibrium at a given pressure, temperature, and composition [4,5]. Electrified chemical manufacturing such as electrochemical, photochemical, and plasma processing using electricity or electron energy for non-equilibrium chemical manufacturing so that it can achieve higher yield, higher selectivity, higher energy efficiency, and net zero carbon emissions. Unfortunately, electrified chemical manufacturing is still a young research field, and many of the kinetic mechanisms in electrified non-equilibrium chemical manufacturing are not well understood. In this chapter, we will introduce two different non-equilibrium chemical manufacturing processes, one using transient variation of thermodynamic properties and the other using non-equilibrium plasma. The former remains close to thermodynamic equilibrium but is kinetic non-equilibrium. The latter is both thermodynamic non-equilibrium in energy states and chemical non-equilibrium in reactions.

8.1 ELECTRIFIED TRANSIENT NON-EQUILIBRIUM CHEMICAL MANUFACTURING

8.1.1 Chemical Equilibrium

For a chemical reaction with reactants A and B and products of E and F,

$$aA + bB + \ldots = eE + fF + \cdots \tag{R8.1}$$

where a, b, e, and f, are, respectively, the stoichiometric coefficient coefficients. At chemical equilibrium, the system entropy reaches the maximum and the Gibbs energy of the mixture (per unit mole), G_{mix}, at a given temperature and pressure becomes the minimum, that is,

$$dG_{\text{mix}} = \sum_{i=1,n} g_i(T) dn_i = \sum_{i=1,n} \left(\bar{g}_i^0(T) + R_0 T \ln\left(x_i p / 1\,\text{atm}\right) \right) dn_i = \sum_{i=1,n} \mu_i dn_i = 0 \tag{8.1}$$

where the superscript "0" represents standard pressure conditions (1 atm) and

$$\mu_i \equiv \left(\frac{\partial G}{\partial n_i} \right)_{n_j, T, p, j \neq i} \tag{8.2}$$

is defined as the chemical potential of species "i".

Under constant pressure and temperature, the chemical equilibrium described in Eqs. 8.1 and 8.2 can be rewritten as

$$-a\left(\bar{g}_a^0(T) + R_0 T \ln(x_i p/1\,\text{atm})\right) - b\left(\bar{g}_b^0(T) + R_0 T \ln(x_i p/1\,\text{atm})\right) - \cdots$$
$$+ e\left(\bar{g}_e^0(T) + R_0 T \ln(x_i p/1\,\text{atm})\right) + f\left(\bar{g}_f^0(T) + R_0 T \ln(x_i p/1\,\text{atm})\right) + \cdots = 0 \tag{8.3}$$

Regrouping the above equation by separating temperature from pressure and mole fractions, we have

$$e\left(\bar{g}_{e,T}^0(T)\right) + f\left(\bar{g}_{f,T}^0(T)\right) + \cdots - a\left(\bar{g}_{a,T}^0(T)\right) - b\left(\bar{g}_{b,T}^0(T)\right)$$
$$= -R_0 T \ln\left[\frac{(x_e p)^e (x_f p)^f \cdots}{(x_a p)^a (x_b p)^b \cdots}\right] = -R_0 T \ln\left[\frac{(p_e)^e (p_f)^f \cdots}{(p_a)^a (p_b)^b \cdots}\right] \tag{8.4}$$

Note that in Eq. 8.4, the pressure, p, now is normalized by the standard reference pressure ($p_{\text{ref}} = 1$ atm), so that the units of p in this equation are atm. x_i is the mole fraction and p_i is the partial pressure of the ith species in atm. By defining the left-hand side of Eq. 8.4 as the standard state Gibbs free energy change of the reaction R8.1, ΔG_T^0, we have

$$e\left(\bar{g}_{e,T}^0(T)\right) + f\left(\bar{g}_{f,T}^0(T)\right) + \cdots - a\left(\bar{g}_{a,T}^0(T)\right) - b\left(\bar{g}_{b,T}^0(T)\right) \equiv \Delta G_T^0 \tag{8.5}$$

Similarly, by defining the right-hand side of Eq. 8.4, which is only the function of partial pressure or species mole fraction, as the equilibrium constant of the reaction, K_p, we have

$$\ln\left[\frac{(x_e p)^e (x_f p)^f \cdots}{(x_a p)^a (x_b p)^b \cdots}\right] = \ln\left[\frac{(p_e)^e (p_f)^f \cdots}{(p_a)^a (p_b)^b \cdots}\right] \equiv \ln[K_p] \tag{8.6}$$

Then, Eq. 8.4, which governs the relation between temperature and species concentrations at chemical equilibrium, becomes

$$K_p(T) = \frac{(p_e)^e (p_f)^f \cdots}{(p_a)^a (p_b)^b \cdots} = \exp\left[-\frac{\Delta G_T^0}{R_0 T}\right] \tag{8.7}$$

Therefore, in chemical synthesis, the equilibrium constant in Eq. 8.4 determines the yield and selectivity of the products at a given temperature, pressure, and reactant equivalence ratio.

By using the relationship between the Gibbs energy, enthalpy, and entropy of the reaction system at the standard condition, we have

$$\Delta G_T^0 = \Delta H^0 - T\Delta S^0 \tag{8.8}$$

Thus, the equilibrium constant is a function of the changes in enthalpy and entropy.

$$K_p(T) = \exp\left[-\frac{\Delta H^0}{R_0 T}\right]\exp\left[\frac{\Delta S^0}{R_0}\right] \tag{8.9}$$

For small changes in T near equilibrium, ΔH^0 and ΔS^0 are relatively constant, so we have the relation between the equilibrium constant and the change in the enthalpy of the reaction,

$$\frac{d}{dT}\ln K_p(T) = -\frac{\Delta H^0}{R_0 T^2} \tag{8.10}$$

This is the Van't Hoff equation. By integrating Eq. 9.10 from temperature T_1 to T_2, the Van't Hoff equation becomes

$$\ln\left(\frac{K_p(T_2)}{K_p(T_1)}\right) = -\frac{\Delta H^0}{R_0}\left(\frac{1}{T_2} - \frac{1}{T_1}\right) \quad (8.11)$$

The Van't Hoff equation now tells which direction the reaction will proceed from equilibrium when the system temperature is varied. We can see that if $T_2 > T_1$, when ΔH^0 is negative (exothermic reaction), $K_p(T_2)$ will decrease relative to $K_p(T_1)$. Therefore, for an exothermic reaction, the equilibrium will shift toward reactants if the system temperature is increased. On the other hand, if the reaction is endothermic, the equilibrium will shift toward products as the temperature is increased.

Therefore, the chemical equilibrium represented by Eqs. 8.7 and 8.10 not only govern the yield and selectivity of the product of a reaction system but also dictate the reaction direction if the system pressure or temperature is varied. Therefore, in order to increase the yield and selectivity, one has to break or shift the chemical equilibrium. This will be the focus of the discussion of non-equilibrium manufacturing in the following sections because electrification produces a great opportunity to change the temperature and species of a reaction system faster than the thermodynamic and chemical equilibrium.

8.1.2 Concept of Non-Equilibrium Chemical Manufacturing

As discussed above, equilibrium chemical synthesis limits the yield and selectivity of product at a given temperature and pressure. For example, for high-temperature methane pyrolysis, Figure 8.1 shows that at low temperature (below 800 K), the equilibrium states produce low yield of hydrogen, and the synthesis rate is very slow. On the other hand, at high temperature, although the reaction rate is high, the chemical equilibrium leads to increase of gaseous carbon formation with lower hydrogen yield and selectivity. At intermediate temperatures (e.g., around 1,500 K), chemical equilibrium produces high hydrogen and high solid carbon. If one wants to produce high yield of hydrogen and ethylene selectively, then chemical equilibrium synthesis may not be a good option.

To overcome the limit of yield and selectivity of thermal equilibrium synthesis, one needs to either dynamically change the thermodynamic state in a time scale so short that the reaction is frozen before chemical equilibrium is established (Figure 8.2) or to modify the thermodynamic states of the reactant molecules (Figures 1.24 and 1.25) so that new reaction pathways with lower activation energy are available.

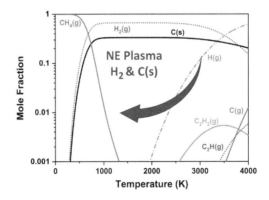

FIGURE 8.1 Mole fraction of methane pyrolysis in chemical equilibrium and schematic of non-equilibrium plasma-assisted methane pyrolysis at reduced temperature.

Electrified Non-Equilibrium Chemical Manufacturing

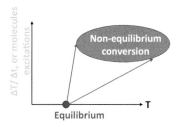

FIGURE 8.2 Schematic of the yield of equilibrium chemical synthesis and non-equilibrium synthesis in the phase diagram of temperature and transient thermodynamic states as well as excited states of molecules.

In Figure 8.2, it is seen that at chemical equilibrium, the product yield and selectivity are governed by the equilibrium constant in Eq. 8.7, which is a function of pressure and temperature. To break the chemical equilibrium and increase the yield and selectivity, one can vary the temperature as a function of time before chemical equilibrium is reached or change the thermodynamic equilibrium using excited vibrational and electronic states to create new non-equilibrium reaction pathways (Figures 1.24 and 1.25). Since electrons can be manipulated at a timescale (e.g., ns) much shorter than fossil fuel combustion and that of typical chemical synthesis, electrified non-equilibrium chemical synthesis provides a great potential to increase the energy conversion efficiency and selectivity of chemical synthesis. However, many of the reaction kinetics of non-equilibrium chemical synthesis are poorly known. In the following sections, we will highlight a few examples and opportunities of electrified non-equilibrium chemical synthesis using transient Joule heating and low-temperature plasmas.

8.1.3 Non-Equilibrium Chemical Manufacturing by Pulsed Electrical Joule Heating

8.1.3.1 Methane Pyrolysis

Based on the concept of non-equilibrium chemical synthesis shown in Figure 8.2, recently, a novel approach using programmable heating and quenching (PHQ) for efficient thermochemical synthesis of valuable chemicals from methane pyrolysis and ammonia from hydrogen and nitrogen was developed [4]. Different from conventional constant temperature chemical equilibrium approaches (e.g., 1,273 K in Figure 8.3), this transient Joule heating method allows for rapid switching between low and high temperatures (e.g., between 650 and 2,000 K in Figure 8.3b) in just a few milliseconds to achieve non-equilibrium thermochemical synthesis. When this PHQ method is applied for CH_4 pyrolysis, Figure 8.3c shows much higher selectivity (>75% vs. <35%) to value-added C_2 products at comparable CH_4 conversions (about 13%) in comparison with the conventional constant temperature thermal equilibrium CH_4 pyrolysis. The C_2 product selectivity by the PHQ technique even outperformed most literature reports of CH_4 pyrolysis using optimized catalysts by continuous heating (Figure 8.3d). This result clearly shows that by breaking the chemical equilibrium (Figure 8.2), one can not only increase the yield but also increase the selectivity. The PHQ technique provides a new platform for non-equilibrium chemical synthesis via transient Joule heating.

8.1.3.2 Ammonia Synthesis

The PHQ process was also applied to catalytic ammonia (NH_3) synthesis from N_2 and H_2. In conventional chemical equilibrium synthesis, NH_3 synthesis often suffers from problems such as poor catalyst stability at high synthesis temperature or a low reaction rate at low synthesis temperature [6]. The PHQ process provides a good solution to resolve these conflicts. PHQ can use transient pulsed high-temperature heating to activate and dissociate nitrogen molecules and therefore a high NH_3 production rate with a rapid quenching to ensure good catalyst stability (Figure 8.4a). Ru nanoparticles (supported on a carbon felt heater) were selected as a model catalyst for comparison. Using a typical

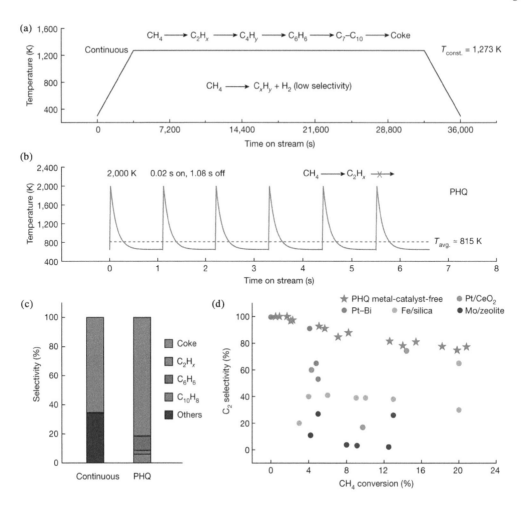

FIGURE 8.3 Comparison between our PHQ method and conventional continuous heating using a CH_4 pyrolysis model reaction. (a) Schematic of a typical temperature profile by continuous heating. Continuous heating creates a variety of products owing to the lack of temporal tunability over the temperature profile and thus the resulting reaction pathways. (b) Schematic of an estimated temperature profile by the PHQ method. PHQ selectively produces value-added C_2 products. The high temperature ensures high conversion, whereas the transient heating duration enables high selectivity. (c) Comparison of CH_4 pyrolysis by PHQ and continuous heating reported in the literature. At comparable CH_4 conversions (about 13%), continuous heating of the non-catalytic CH_4 pyrolysis reaction results in coke as the major product (grey) with other minor products (maroon), whereas CH_4 pyrolysis by PHQ showcases its >75% selectivity to C_2 products (red). (d) Metal-catalyst-free PHQ technique (red stars) even outperforms most literature reports of CH_4 pyrolysis reactions with optimized catalysts conducted by continuous heating in terms of the C_2 product selectivity at a wide range of CH_4 conversions [4].

heating and quenching program (0.11 s heating, 0.99 s quenching; T_{high} of 1,400 K; Figure 8.4b), the PHQ operation showed a stable performance that lasted for around 20 h with an NH_3 synthesis rate of about 7,000 μmol/gR_u h, after which the activity started to decay (Figure 8.4c). In comparison, we measured the NH_3 synthesis rates by continuous heating at T_{high} (1,400 K), T_{low} (about 700 K), and $T_{avg.}$ (about 900 K). In comparisons (Figure 8.4c), continuous heating at T_{high} showed good activity that was comparable with PHQ but only lasted for about 2 h. Meanwhile, continuous heating at $T_{avg.}$ showed much worse catalyst activity, albeit with a stable NH_3 synthesis rate. Lastly, continuous heating at T_{low} showed an almost zero NH_3 synthesis rate owing to the poor N_2 activation under low temperature. It was shown that the Ru nanoparticles retained their original size and distribution

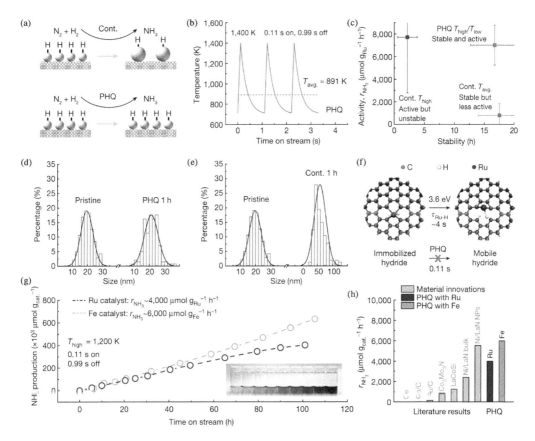

FIGURE 8.4 NH_3 synthesis by PHQ under ambient pressure: (a) Catalyst morphological evolution by means of the metal hydride intermediate during NH_3 synthesis by continuous heating and PHQ. (b) An estimated temperature profile of the PHQ process used in (c). (c) Comparison of the activity and stability of NH_3 synthesis by PHQ (0.11 s on, 0.99 s off; T_{high} of 1,400 K, T_{low} of around 700 K and $T_{avg.}$ of around 900 K) (red) and by continuous heating at 1,400 K (corresponding to T_{high} of PHQ) (blue) and 900 K (corresponding to $T_{avg.}$ of PHQ) (cyan). Error bars denote standard deviation with the value of $n \geq 3$. (d) Ru catalyst size and distribution after PHQ for 1 h (0.11 s on, 0.99 s off; T_{high} of 1,400 K). (e) Ru catalyst size and distribution after continuous heating for 1 h at 1,400 K (corresponding to T_{high} of PHQ). (f) DFT modeling for the activation barrier and timescale of Ru hydride (Ru-H) migration at 1,300 K. (g) NH_3 production as a function of the time on stream by PHQ (0.11 s on, 0.99 s off; T_{high} of 1,200 K) using non-optimized Ru and Fe catalysts. The error range for production rate was found to be ±3%. The inset shows the testing solutions for NH_3 quantification by the Berthelot method. (h) Comparison of NH_3 between the PHQ method using non-optimized Ru and Fe catalysts and literature reports with material innovations. NPs, nanoparticles [4].

after PHQ for 1 h (Figure 8.4d) but were severely sintered after continuous heating at T_{high} for the same duration (Figure 8.4e). Therefore, once again, the experiments revealed that non-equilibrium chemical synthesis offers the merits of increasing the yield and stabilizing catalysis in chemical synthesis.

8.1.3.3 Plastic Pyrolysis and Recycling

Non-equilibrium electrified chemical processing can also be applied to waste materials recycling to increase the yield and selectivity. Here we highlight a recent collaborative work by Dong et al. using the PHQ reactor to depolymerize plastics for chemicals without any catalysts [7].

In the PHQ depolymerization process, as shown in Figure 8.5a, a bilayer of porous carbon felt was placed in contact above a reservoir of solid plastic reactant. A pulsed electrical Joule heating

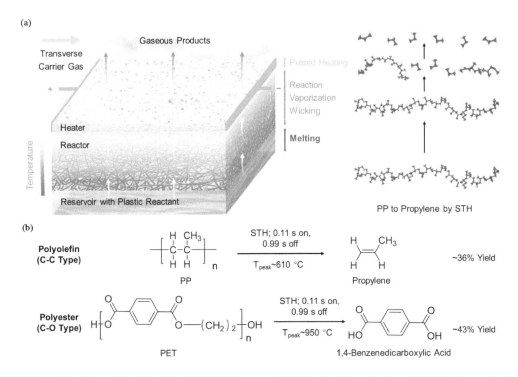

FIGURE 8.5 Schematic of the electrified PHQ approach for the depolymerization of polyolefins and polyesters. (a) Schematic demonstrating the continuous process in the bilayer configuration and the molecular transformation from polymer to monomer using PP as a model plastic species. Pulsed electrical heating of the top carbon heater layer creates a temperature gradient from the top to the bottom of the carbon reactor layer, which first melts the plastic reactant in the reservoir. The plastic melt is then wicked by capillary forces through the bottom carbon reactor layer (composed of thicker, non-electrified carbon felt), entering regions of higher temperature while climbing. After reaching certain threshold temperatures, the polymer chains begin to break and gradually form smaller polymer chains, oligomers, and monomers. The volatile species diffuse through the porous scaffold, which allows them to experience more heating pulses to reach a higher depolymerization degree before being transported away from the heating zone by the transverse inert carrier gas (e.g., argon). Meanwhile, the transient heating timescale keeps the reaction in the far-from-equilibrium regime by suppressing secondary reactions of the formed monomers. (b) The depolymerization reactions of PP (as a model polyolefin) and PET (as a model polyester) to their monomers, with yields of ~36% and ~43%, respectively, using electrified spatiotemporal heating (STH) with a typical pulsed heating program of 0.11 s power on and 0.99 s power off. T_{peak} refers to the peak temperature within the bilayer structure during the pulsed electrical heating [7].

was applied to the top carbon heater to pyrolyze the precursors from the decomposed plastic polymers. The heat also generated a spatial temperature gradient in the porous carbon felt to melt the plastic in the reservoir and wick upward the melted polymer liquid. Depolymerization occurs as the liquid polymer experiences higher temperatures when moving upward. With the increase of temperature, decomposed volatile gas phase species with carbon numbers below 20 (e.g., at ~350°C) entered the top carbon heater layer as the decomposed plastic precursor (Figure 8.5a). In the top heating layer, pulsed heating with high peak temperatures (e.g., up to ~610°C) and fast heating rates (e.g., >10^3°C/s) were designed to only break the weakest C–C or C–O bonds that connect the monomer fragments, while shortening the transient heating timescales (e.g., 0.11 s) to prevent the reaction system from approaching its chemical equilibrium that would otherwise result in the formation of undesired secondary products (e.g., aromatics) from the monomers via dehydrogenation, C–C bond coupling and aromatization. The continuous melting, wicking, vaporization, and pulsed reaction process within the bilayer porous carbon felt structure extended the non-equilibrium depolymerization process toward high monomer yield. The process of using PHQ to depolymerize polypropylene (PP) and polyethylene terephthalate (PET) is schematically shown in Figure 8.5b.

PP holds one of the largest market shares among all synthetic polymers [8]. The depolymerization of PP to its monomer propylene (C_3H_6) has proven challenging due to PP's high ceiling temperature. As a result, the competition among various decomposition pathways during PP pyrolysis under continuous heating often leads to a variety of products of different carbon numbers [9–16]. The results of depolymerization using the PHQ process for PP as a model plastic are shown in Figure 8.6 [7]. Since the PHQ approach operates in the far-from-equilibrium regime and thus provides an opportunity for higher monomer selectivity and yield. As seen in Figure 8.6a, PHQ method achieved a high C_3H_6 monomer yield of ~36%. In addition to the high monomer yield, the PHQ system also exhibited good reusability over multiple heating cycles with consistently high C_3H_6 monomer yields of ~36% (Figure 8.6b). Note that, unlike continuous heating, the electrified non-equilibrium process can be controlled by varying the input power and the on/off timescale, enabling us to tune the heating duration, frequency, and heating pulse temperature, as well as the temperature distribution in the reactor layer to optimize the depolymerization performance. Despite being non-optimized, the 36% C_3H_6 yield by the catalyst-free PHQ process was already much higher compared to the reported value in the literature via catalyst-free pyrolysis using continuous heating at a slow heating rate (~10% with 0.33°C/s using furnace heating in a non-isothermal mode (Figure 8.6d) as well as compared to continuous Joule heating at a fast heating. The yield by the PHQ approach was among the highest compared to the literature results even using optimized catalysts [9–16]. These results indicate that the catalyst-free, far-from-equilibrium electrified PHQ approach enables a unique depolymerization process and superior monomer yield compared to conventional thermochemical methods.

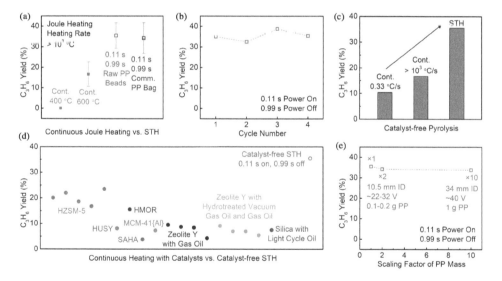

FIGURE 8.6 The depolymerization performance of the PHQ process using PP as a model plastic. (a) The C_3H_6 monomer yield for continuous Joule heating (400°C and 610°C) and PHQ experiments (with PP beads and commercial PP bag). (b) The C_3H_6 monomer yield as a function of PHQ operation cycle. (c) Comparison of the C_3H_6 monomer yield among our work using PHQ under far-from-equilibrium conditions, a representative literature report using catalyst-free pyrolysis by continuous heating at a low heating rate (0.33°C/s) under near-equilibrium conditions,[30] and our control experiment using continuous Joule heating at a high heating rate (>10³°C/s, same as in the PHQ process). (d) Comparison of the C_3H_6 monomer yield between our work using PHQ under far-from-equilibrium conditions and literature reports using different catalysts (or with different reaction media) by continuous heating under near-equilibrium conditions [9–16]. (e) The C_3H_6 monomer yield as a function of reaction scales. The scaling factor is based on the PP mass feed, with the initial scale of 0.1 g defined as 1. All PHQ processes used a program of 0.11 on, 0.99 s off, and T_{peak}~610°C. Error bars denote standard deviation with the value of $n \geq 3$ [7].

In addition to depolymerizing PP as a representative polyolefin, PHQ was also applied to depolymerize PET as a representative polyester. Polyesters have been widely used as plastics but also as fabrics whose recycling holds great importance for environmental and economic considerations [8]. Although polyesters can be depolymerized via hydrolysis or glycolysis, this process is time-consuming and sometimes environmentally unfriendly [17]. While thermochemical approaches can be more efficient and scalable, conventional methods typically suffer from low yield and poor selectivity [18–21]. To depolymerize PET, the PHQ system applied a higher pulse peak temperature compared to PP, given PET's higher melting point (~250°C vs. ~150°C for PP). As a non-optimized, proof-of-concept demonstration, a power on (heating) duration of 0.11 s at ~50 V and a power off (cooling) duration of 0.99 s were applied to the top porous carbon heater layer to generate a pulse peak temperature of ~950°C. The reported monomer yield (i.e., relative abundance of 1,4-benzenedicarboxylic acid) was up to ~43% along with ~6% monomer-related product (i.e., relative abundance of 1,4-benzenedicarboxylic acid, 1-ethenyl ester). The total of ~49% monomer-related yield was the highest compared to the literature of conventional thermochemical methods under near-equilibrium conditions [18–21]. Similarly, when the PHQ process was applied to a commercial PET plastic bottle with 0.11 s power on and 0.99 s power off, ~45% total yield of 1,4-benzenedicarboxylic acid was obtained with 1,4-benzenedicarboxylic acid and 1-ethenyl ester, which was also close to the ~49% total yield of the monomer-like species from the PET pellets under the same conditions.

These results of depolymerization of PP and PET using the PHQ technique once again demonstrated that non-equilibrium Joule heating can potentially be used for more efficient and selective polymer pyrolysis and recycling with renewable electricity in comparison to conventional near-equilibrium thermal pyrolysis techniques.

8.1.3.4 Control of Reactor Residence Time Distribution in Non-Equilibrium Chemical Synthesis

As seen above, non-equilibrium chemical synthesis needs accurate control of the synthesis and quenching time. Unfortunately, in a practical reactor, there is always a distribution of reaction time. This broadened distribution of reaction time in a reactor will significantly affect the yield and selectivity of chemical synthesis.

Fluidized bed (FB) reactors are often used to study the pyrolysis of different feedstocks [22]. In these reactors, solid particles are moved along the streamlines of the flow. Each streamline in the reactor may have a different flow residence time. Given the complex nature of the pyrolysis process depending on the temperature, pressure, and flow residence time, the residence time distribution (RTD) of the reactor will factors affect the efficiency and selectivity of the products. Mastral et al. [23] explored the effect of the average residence time on product distribution of high-density polyethylene using FB reactor experiments. They found that pyrolysis temperatures affected the product distribution and the gas compositions.

To understand the impact of reactor flow RTD on non-equilibrium product selectivity, Lele and Ju [24] modeled non-equilibrium polypropylene pyrolysis using reactive molecular dynamic simulations with 6 different RTDs. Figure 8.7a–c shows the three different idealized RTDs used in the study: iRTD-a a delta function, iRTD-b a normal distribution, and iRTD-c a uniform distribution. For comparison, the timescales for all the iRTDs were normalized with respect to the longest mass residence time in each iRTD modeling. The residence time scales are represented by average residence time and the iRTDs have the same average residence time. For practical understanding, three engineering FB reactor RTDs selected from the literature [25–27] were also considered (see Figure 8.7d–f). RTD-a and RTD-c have similar average residence time of around 0.43 normalized time units, but RTD-c has a uniform distribution whereas RTD-a distribution has two peaks. RTD-c is closest to iRTD-c in terms of mass residence time being equally distributed over a very large range of time. RTD-b on the other hand has a single peak in the distribution with slightly lower average residence time of 0.35.

Electrified Non-Equilibrium Chemical Manufacturing

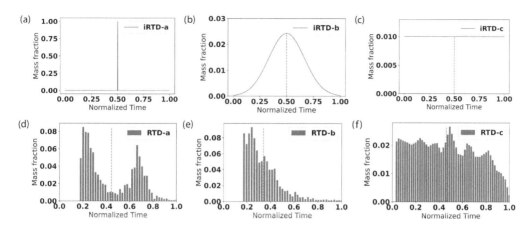

FIGURE 8.7 Normalized mass residence time distributions (RTD) used in this work. (a–c) idealized and (d–f) derived from literature [25–27]. The dotted red line shows the average residence time.

Figure 8.8 shows the relative product concentrations of C_3H_6, C_2H_4, and CH_4 with all three iRTDs. The product mass fractions are obtained with respect to the maximum number density of the given product of each case. It is clearly seen that iRTD-a is the best iRTD which has the best selectivity because all the reactants spend the same amount of time in the reactor of iRTD-a. In addition, the results of iRTD-a also show a strong sensitivity to the iRTD timescale. The uniform iRTD in iRTD-c shows the worst performance in terms of product selectivity. For example, the amount of C_2H_4 from a reactor with iRTD-c could be 30% less than a reactor with iRTD-a. This implies that not only a narrow RTD but also an accurate RTD control is important for efficient and selective non-equilibrium chemical manufacturing. Although the three iRTDs and product selectivity represent idealized cases, they clearly demonstrate the effect of RTD on selectivity.

The results of three different FB reactors with different RTDs in Figure 8.7 also showed that in case of C_3H_6, RTD-b gives the best product selectivity (narrower distribution), which is around 10% lower than the best C_3H_6 selectivity offered by iRTD-a. In comparison to RTD-b, the C_3H_6 selectivity in RTD-a decreases by 15%. However, in the case of C_2H_4 and CH_4, RTD-c gives the best product yield among the three different RTDs. This dependence of peak concentration on the shape of RTDs demonstrates the product selectivity is also dependent on the timescale of the product formation (chemical timescale) in comparison to the average residence time (Figures 8.7 and 8.8). For example, C_3H_6 production has a shorter timescale than C_2H_4 production. Hence, C_3H_6 product selectivity changes significantly with RTD.

In summary, the yield and selectivity of non-equilibrium chemical synthesis are not only sensitive to the RTD of the reactor but also to the ratio of the chemical timescale to the flow residence time (Damköhler number). Therefore, experimentally validated chemical kinetic models or reactive molecular dynamics simulation potentials need to be developed to optimize non-equilibrium chemical reactors.

8.2 KINETICS OF CATALYSIS REACTIONS

In many non-equilibrium chemical syntheses (e.g., Figure 8.3), not only gas phase reactions are involved but also surface reactions are important. Especially for plasma-assisted catalysis, the interaction of gas phase excited molecules and radicals with surface plays a critical role in affecting the reaction mechanisms and conversion pathways. Therefore, it is important to understand the key reaction mechanisms of surface reactions. Several major surface catalysis mechanisms have been described in previous books and lectures [28–31]. Here, to facilitate the discussions of plasma catalysis, only a brief summary of the key surface reaction mechanisms will be made.

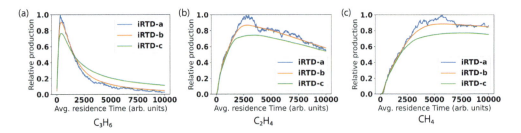

FIGURE 8.8 Relative product mass fractions at different average residence times during polypropylene pyrolysis using normalized ideal mass residence time distribution profiles: (a) C_3H_8, (b) C_2H_4, and (c) CH_4. [24].

8.2.1 Binding Energy and Its Impact on the Activation Energy of Catalytic Reactions

Heterogeneous catalysis is a great enabler of gas phase reactions that have high activation energies and are energetically challenging to complete. As shown in Figure 8.9a, the non-catalytic gas phase reaction has a very high activation energy to reach the transition state. However, once the reactants adsorb on the surface, the catalyst surface lowers the reaction barriers due to the surface binding energy and new reaction pathways on the surface. One very important example of such a reaction is the HB [1] process. The HB process is used to synthesize ammonia (NH_3) at elevated temperatures (~700 K) and pressures (~200 atm) using transition metal catalysts such as Fe or Ru to reduce the activation energy. The global reaction for NH_3 synthesis can be expressed as:

$$N_2(g) + 3H_2(g) = 2NH_3(g) \quad \Delta H = -92.4 \text{ kJ/mol} \tag{R8.2}$$

N_2 molecule is formed by a triple bond between two N atoms and has a very high bond dissociation enthalpy of 945 kJ/mol [32]. The catalyst surface facilitates the easier breaking of the $N \equiv N$ by lowering the barrier to 60–115 kJ/mol for Fe and Ru catalysts [33]. The energy barrier to $N \equiv N$ is lowered as N_2 binds to the catalyst surface.

Heterogeneous catalytic processes involve several intermediate reactions taking place with the aid of the catalyst surface. Surface binding energy or adsorption energy of reactants as well as intermediate species to the catalyst surface are key to determining the effectiveness of a catalyst. The binding energy or adsorption energy of an adsorbate on a catalyst surface is typically defined as:

$$E_{Ads} = E_{Surface+Adsorbate} - E_{Surface} - E_{Adsorbate} \tag{8.12}$$

The overall rate of the catalysis process is affected or reduced by the binding energies of chemical species involved in these reactions (Figure 8.9a). These binding energies are dependent on the catalyst element, the atomic arrangement on the catalyst surface (surface termination of the catalyst) as well as the thermodynamic conditions. More importantly, it has been shown empirically for a wide variety of reactions on different catalytic surfaces that these adsorption energies are correlated to each other by linear relationships. Not only that but the activation energies of catalytic reactions involved in a given catalytic process have also been shown to be linearly related with binding energies (Brønsted–Evans–Polanyi (BEP) relations) [34]. This means that the activity of a catalyst for a given catalytic process can be defined based on just the binding energies of certain species involved in the process (Figure 8.9a). Hence, only a handful of reaction parameters are required to tune the performance of a catalyst.

A classic example to demonstrate the importance of binding energies on the catalytic reaction rate is NH_3 synthesis. Figure 8.9b shows the relationship between NH_3 synthesis rate in terms of the turnover frequency (TOF) on different catalyst surfaces and thermodynamic conditions as a

Electrified Non-Equilibrium Chemical Manufacturing 523

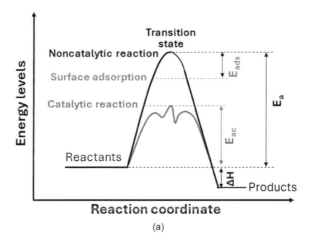

FIGURE 8.9A Schematic of activation energies (E_a) of a typical non-catalytic and catalytic reaction in relation with surface adsorption energy (E_{ads}). A catalyst helps reduce the energy barrier of a catalytic chemical reaction (E_{ac}) by allowing reactant species to bind to the surface and enabling surface reactions.

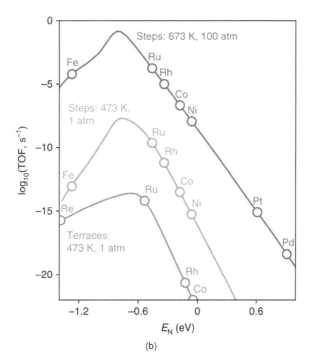

FIGURE 8.9B NH_3 synthesis rates (TOF) on different surface cites and operating conditions plotted as a function of dissociative adsorption energy of N_2.

function of dissociative adsorption energy of N_2 on different catalysts. Here we are using a different form of binding energy, dissociative adsorption energy, which is defined as

$$2E_N = 2E_{N+\text{Surface}} - 2E_{\text{Surface}} - E_{N_2} \tag{8.13}$$

E_N here represents the adsorption or binding of atomic N in reference to molecular N_2 instead of atomic N. All the intermediate reactions involved in the catalytic synthesis of NH_3 as well as their barriers have empirically shown to have a linear relationship with E_N. Hence, the entirety of the NH_3 catalytic synthesis process can be represented by just E_N. As shown in Figure 8.9b [35], NH_3 synthesis requires an optimal E_N to maximize the NH_3 synthesis rate. Fe, which has the strongest binding affinity for N_2 among the catalysts shown in the Figure 8-9b is the most effective catalyst along with Ru. Please note that smaller values (negative x-axis direction) represent stronger binding between N and the surface. Any catalysts with E_N stronger than that of Fe or weaker than that of Ru would show worse catalytic performance for NH_3 synthesis. This type of relationship between the catalysts and their performance is known as the "volcano curve" given the hill-like nature of the curve.

Similar relationships can also be established for the activation of other catalytic processes involving CO, NO, and O_2 dissociation as well.

8.2.2 Adsorption Dissociation Reactions

8.2.2.1 Unimolecular Surface Adsorption

A molecule A adsorbed by the surface can be described by an unimolecular surface reaction shown in Figure 8.10. Here "*" represents the surface site and A* denotes A molecule adsorbed on the surface. The unimolecular surface adsorption reaction can be written as

$$A + * \overset{k_A}{\leftrightarrow} A^* \tag{R8.3}$$

where k_A is the net reaction rate from reactant to the product. The forward and backward reaction rates are, respectively, k_{Af} and k_{Ab}. There are many practical reactions in chemical synthesis in this class. For example, in catalytic ammonia synthesis, the reaction

$$N_2 + * = N_2^* \tag{R8.4}$$

is one of the most rate-limiting reactions.

Therefore, the change of the surface site coverage of θ_{A^*} by A* can be given as

$$\frac{d\theta_{A^*}}{dt} = C_A k_{Af}\left(1 - \theta_{A^*}\right) - k_{Ab}\theta_{A^*} = 0 \tag{8.14}$$

If one assumes A* is in quasi-steady state, i.e.,

$$\frac{d\theta_{A^*}}{dt} = 0 \tag{8.15}$$

Then, from Eqs. 8.14 and 8.15, we have

$$\theta_{A^*} = \frac{K_A C_A}{1 + K_A C_A}, \quad K_A = \frac{k_{Af}}{k_{Ab}} \tag{8.16}$$

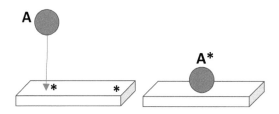

FIGURE 8.10 Schematic of unimolecular adsorption in surface catalytic reactions.

where K represents the equilibrium constant of reaction R8.3. Therefore, the open site coverage, on the surface will be,

$$\theta_* = 1 - \theta_{A^*} = \frac{1}{1 + K_A C_A} \tag{8.17}$$

Therefore, the surface coverage of θ_{A^*} by A* depends on reaction rate (equilibrium constant), K_A and molecular concentration, C_A. At the low pressure limit, θ_{A^*} is linearly proportional to the product of equilibrium constant and the concentration of A molecule. On the other hand, in the high-pressure limit, θ_{A^*} is close to unity. Therefore, the surface adsorption is strongly dependent on pressure or reactant concentration as well as the number of surface sites.

8.2.2.2 Unimolecular Surface Adsorption and Dissociation

If molecule A is a diatomic or polyatomic species, the adsorbed molecule on the surface may be quickly catalyzed and dissociated to atomic species or radicals on the surface (Figure 8.11). For example,

$$N_{2^+} * + * = 2N^* \tag{R8.5}$$

Then the diatomic molecular surface adsorption and dissociation reaction can be written as,

$$A_2 + 2* \overset{k_A}{\leftrightarrow} 2A^* \tag{R8.6}$$

By using the same assumption of quasi-steady state for A*, we have

$$\frac{d\theta_{A^*}}{dt} = C_{A_2} k_{Af} \left(1 - \theta_{A^*}\right)^2 - k_{Ab} \theta_{A^*}^2 = 0 \tag{8.18}$$

and the relation between the surface coverage of A* and the equilibrium constant K_A and fuel concentration C_{A2} can be obtained

$$\theta_{A^*} = \frac{\sqrt{K_A C_{A2}}}{1 + \sqrt{K_A C_{A2}}}, \quad K_A = \frac{k_{Af}}{k_{Ab}} \tag{8.19}$$

The open surface site coverage will become

$$\theta_* = 1 - \theta_{A^*} = \frac{1}{1 + \sqrt{K_A C_{A2}}} \tag{8.20}$$

From Eqs. 8.19 and 8.20, now at low pressure, the surface site coverage is proportional to square root of the equilibrium constant and the concentration of diatomic molecules, which is different from the unimolecular adsorption only.

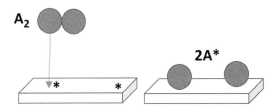

FIGURE 8.11 Schematic of diatomic molecule adsorption and dissociation on the catalytic surface.

8.2.2.3 Multispecies Surface Adsorption

Sometimes, catalytic reactions involve multiple species adsorptions and dissociations. As shown in Figure 8.12, both A and B molecules adsorb on the surface and form A* and B*, respectively. Depending on their concentrations and reaction rates, they will compete for the open surface site and ultimately the catalytic reaction rates.

The reactions can be written as

$$A + * \overset{k_A}{\leftrightarrow} A^* \tag{R8.3}$$

$$B + * \overset{k_B}{\leftrightarrow} B^* \tag{R8.7}$$

and the quasi-steady state relations for surface coverage by A and B molecules become

$$\frac{d\theta_{A^*}}{dt} = C_A k_{Af} \theta^* - k_{Ab} \theta_{A^*} = 0 \tag{8.21a}$$

$$\frac{d\theta_{B^*}}{dt} = C_B k_{Bf} \theta^* - k_{Bb} \theta_{B^*} = 0 \tag{8.21b}$$

Then, the surface coverage of A* and B* as a function of the equilibrium constant and gas molecule concentration can be obtained

$$\theta_{A^*} = K_A C_A \theta^* \quad \text{and} \quad \theta_{B^*} = K_B C_B \theta^* \tag{8.22}$$

Note that from surface site conservation, we have

$$\theta_{A^*} + \theta_{B^*} + \theta^* = 1 \tag{8.23}$$

The relationship between the surface coverages and open site via the competition between bimolecular adsorption reactions can be given as

$$\theta^* = \frac{1}{1 + K_A C_A + K_B C_B} \tag{8.24a}$$

$$\theta_{A^*} = \frac{K_A C_A}{1 + K_A C_A + K_B C_B} \tag{8.24b}$$

$$\theta_{B^*} = \frac{K_B C_B}{1 + K_A C_A + K_B C_B} \tag{8.24c}$$

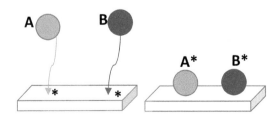

FIGURE 8.12 Schematic of bimolecular adsorption on the catalytic surface.

Electrified Non-Equilibrium Chemical Manufacturing

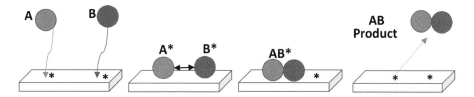

FIGURE 8.13 Schematic of Langmuir–Hinshelwood mechanism: bimolecular adsorption reactions.

Therefore, the surface coverage by A and B molecules now are coupled by the adsorption reactions of both molecules. The relationship between the surface coverage of one molecule and its concentration will depend on the other molecule's concentration and reaction rate. This surface site competition between A and B molecules will ultimately affect the catalytic reaction rate.

8.2.3 Langmuir–Hinshelwood Mechanism

The Langmuir–Hinshelwood (L–H) mechanism is one of the major mechanisms of catalytic surface reactions. L–H model assumes a bimolecular surface reaction of two adjacent molecules on the surface sites (Figure 8.13), which are both adsorbed via the biomolecular adsorption process shown in Figure 8.12. The L–H reaction mechanism can be described as

$$A + * \overset{k_A}{\leftrightarrow} A^* \tag{R8.3}$$

$$B + * \overset{k_B}{\leftrightarrow} B^* \tag{R8.7}$$

$$A^* + B^* \overset{k_*}{\leftrightarrow} AB^* + * \tag{R8.8}$$

$$AB^* \overset{fast}{\leftrightarrow} AB + * \tag{R8.9}$$

From Eqs. 8.21 assuming that AB* desorption rate is very fast, we can have [28,30],

$$\theta_* = 1 - (\theta_{A^*} + \theta_{B^*}) = \frac{1}{1 + K_A C_A + K_B C_B} \tag{8.25a}$$

$$\omega_{AB} = k_* \theta_{A^*} \theta_{B^*} = n_* \frac{k_* K_A K_B C_A C_B}{(1 + K_A C_A + K_B C_B)^2} \tag{8.25b}$$

where n^* is the number of catalyst site.

Therefore, the L–H mechanism shows that the catalytic reaction rate is strongly dependent on the partial pressure of the bimolecular reactants. The optimum catalytic reaction rate for AB production is a function of the ratio of the partial pressure of C_A/C_B and the ratio of adsorption reactions K_A/K_B.

8.2.4 Eley–Rideal Mechanism

The Eley–Rideal (E–R) mechanism is another major mechanism for catalytic reactions. E–R model assumes only one of the reactant molecules adsorbs on the surface (A*) and the other reactant (B) reacts with it directly from the gas phase to form reactant complex (AB*) and then to the product (AB) (Figure 8.14). The E–R reaction mechanism can be described as

$$A + * \overset{k_A}{\leftrightarrow} A^* \tag{R8.3}$$

$$A^* + B \overset{k_*}{\leftrightarrow} AB^* \tag{R8.10}$$

$$AB^* \overset{fast}{\leftrightarrow} AB + * \tag{R8.9}$$

A typical example of an E–R reaction in the ammonia synthesis is

$$H + * \leftrightarrow H^* \tag{R8.11}$$

$$H^* + N \leftrightarrow NH^* \tag{R8.12}$$

Therefore, the catalytic reaction rate becomes

$$\theta_{A^*} = \frac{K_A C_A}{1 + K_A C_A} \tag{8.16}$$

$$\omega_{AB} = n_* \frac{k_* K_A C_A C_B}{1 + K_A C_A} \tag{8.26}$$

Again, at low pressure, the reaction rate is proportional to the product of the reactants A and B. On the other hand, at high pressure or high concentration of A, it is only proportional to the concentration of reactant B.

8.2.5 Mars-Van Krevelen Mechanism

The Mars-Van Krevelen (MVK) mechanism is another type of surface reaction (Figure 8.15). It involves an atomic vacancy or crystal defect on the surface of the materials (*), surface adsorption of one reactant molecule to the vacancy, and the following reaction of the surface adsorbed reactant with a gas phase molecule. Note that here the surface atomic vacancy (*) is a crystal defect and different from the surface site of "*" in the L-H and E-R mechanisms. The mechanism can be described as

$$A + (*) \overset{k_A}{\leftrightarrow} A(*) \tag{R8.13}$$

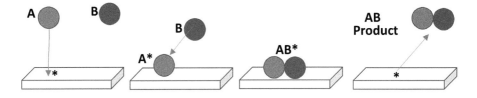

FIGURE 8.14 Schematic of Eley–Rideal mechanism.

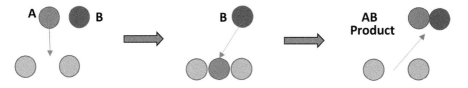

FIGURE 8.15 Schematic of Mars-Van Krevelen mechanism involving a vacancy from the defects of a surface.

$$A(*) + B \overset{k*}{\leftrightarrow} AB(*) \tag{R8.14}$$

$$AB(*) \overset{fast}{\leftrightarrow} AB + (*) \tag{R8.15}$$

A typical example of the MVK reaction scheme in ammonia synthesis on the surface of titanium nitride (TiN) is through the nitrogen vacancy

$$N + T_i(*) = T_i N(*) \tag{R8.16}$$

$$T_i N(*) + H = T_i(*) + NH \tag{R8.17}$$

Note that both E–R and MVK mechanisms play a more important role in plasma-assisted catalysis. This is because plasma can generate reactive radicals which facilitate the surface reactions via E–R and MVK mechanisms. In addition, recent studies of mechanocatalytic reactions of ammonia synthesis also showed that the MVK mechanism is the dominant reaction pathway under mechanically activated conditions on a TiN surface [36].

8.3 PLASMA CATALYSIS FOR NON-EQUILIBRIUM CHEMICAL SYNTHESIS

8.3.1 Ammonia Synthesis

Ammonia has attracted significant attention as a hydrogen carrier for power generation and transportation because of its high energy density, easy storage, and transportation. However, the current ammonia production using the thermal catalytic HB process [1],

$$N_2(g) + 3H_2(g) = 2NH_3(g) \quad \Delta H = -92.4 \text{ kJ/mol} \tag{R8.2}$$

requires high temperature (~500°C) for N_2 activation and high pressure (~200 atm) to increase the reactivity and shift the chemical equilibrium. The estimated energy intensity of ammonia is 39.3 GJ/ton. If H_2 is produced from methane reforming, 1.9 tons of carbon dioxide (CO_2) will be generated for producing one ton of NH_3. Therefore, the HB process is very energy and CO_2 intensive [4,37–39]. Moreover, this process is limited to large-scale plants to be economically viable. Therefore, it is necessary to develop a new electrified ammonia synthesis method using renewable resources for a distributed, efficient, and environmentally friendly NH_3 manufacturing.

In the HB process, ammonia is catalytically formed from N_2 and H_2 using Fe_3O_4 catalyst. A free energy diagram for thermal catalytic ammonia synthesis over Ru catalyst via the L–H mechanism (R8.18) is shown in Figure 8.16 [40]. It is seen that in this process, both H_2 and N_2 undergo dissociative adsorption (R8.6) onto the surface of the catalyst via reactions R8.18.1 and R8.18.2. The N_2 adsorption dissociation reaction has the highest activation energy close to 2 eV depending on the temperature. Thus, this reaction is the rate-limiting reaction in the L–H mechanism and requires a high temperature to activate. However, with the increase of temperature, Figure 8.16 shows that the activation energy of R8.18.1 increases due to the increase of binding energy. Therefore, there is a need to develop a plasma-assisted adsorption dissociation process to lower the activation energy of nitrogen adsorption dissociation. Also, from chemical equilibrium and the definition of equilibrium constant (Eq. 8.6), it is clear that the HB process favors high pressure. High pressure can not only reduce the activation energy of R8.18.1 but also increase the exothermicity of the H-B process. This energy diagram clearly demonstrates the need for high temperature and high pressure for the H-B process and also reveals the need to develop a novel method to reduce the adsorption dissociation energy of R8.18.1 or to create a new reaction pathway.

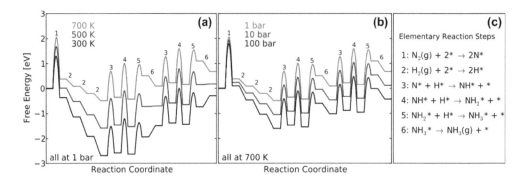

FIGURE 8.16 Free energy diagram for thermal catalytic ammonia synthesis over Ru at different temperatures and pressures: (a) effect of temperature, (b) effect of pressure, and (c) key catalytic reactions [40].

$$N_2(g) + 2^* \rightarrow 2N^* \tag{R8.18.1}$$

$$H_2(g) + 2^* \rightarrow 2H^* \tag{R8.18.2}$$

$$N^* + H^* \rightarrow NH^* + * \tag{R8.18.3}$$

$$NH^* + H^* \rightarrow NH_2^* + * \tag{R8.18.4}$$

$$NH_2^* + H^* \rightarrow NH_3^* + * \tag{R8.18.5}$$

$$NH_3^* \rightarrow NH_3(g) + * \tag{R8.18.6}$$

In plasma-assisted ammonia synthesis, vibrational nitrogen $N_2(v)$ and N radicals can be produced by electron impact excitation and dissociation via

$$e + N_2 \rightarrow e + N_2(v) \tag{R8.19}$$

$$e + N_2 \rightarrow e + 2N \tag{R8.20}$$

Therefore, plasma-assisted ammonia synthesis will create new reaction pathways such as

$$N_2(v) + 2^* \rightarrow 2N^* \tag{R8.21}$$

$$N + H^* \rightarrow NH^* \tag{R8.12}$$

$$N + * \rightarrow N^* \tag{R8.22}$$

Therefore, R8.21 reaction with vibrational excited nitrogen will significantly reduce the activation energy (Figure 8.17) and reaction R8.12 enhances the E–R reaction pathway.

As shown in Figure 8.17, compared to thermal catalysis, non-equilibrium plasma excited states and surface catalysis can both reduce the activation energy of the rate limiting nitrogen adsorption dissociation reaction and enhance plasma catalysis for ammonia production.

To examine how much vibrational excitation of nitrogen can enhance the catalytic reaction rate, Mehta et al. computationally studied the impact of plasma-induced vibrational excitations of N_2 on the decrease dissociation barriers and the enhancement of the catalytic reaction rate of R8.21. As shown in Figure 8.18, it predicted that plasma-induced vibrational excitation of nitrogen could increase the reaction rate. The comparison between thermal catalysis and plasma-assisted catalysis

Electrified Non-Equilibrium Chemical Manufacturing

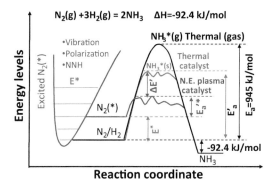

FIGURE 8.17 Schematic of activation energy reduction of thermal reactions for ammonia synthesis by using plasma-generated excited states and intermediate species and catalysis.

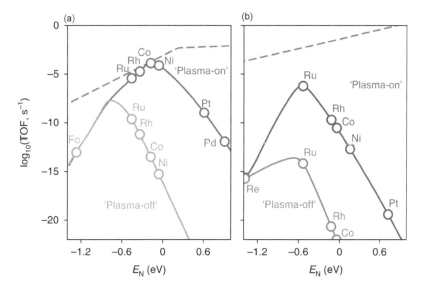

FIGURE 8.18 Rate enhancements with plasma-induced N_2 vibrational excitation. (a,b) Comparison of N_2 vibrational-distribution-weighted (plasma-on) and thermal (plasma-off) ammonia synthesis rates on step (a) and terrace (b) sites. Reaction conditions: 1 atm, T_{gas} = 473 K, T_{vib} = 3,000 K, conversion = 1%. The dashed lines are the maximum possible rates for the hydrogenation reactions [35].

is shown in Figure 8.18 for step and terrace sites. It is seen that for both types of sites, the optimal plasma-on reaction rates are several orders of magnitude higher than those for thermal catalysis. For step sites (Figure 8.18a), the peak of the plasma-on volcano curve shows a distinct shift toward weaker (to the right) nitrogen binding energies, predicting Co and Ni to be near the volcano peak. Comparing the low-temperature and -pressure plasma-on volcano curve with the high-temperature and -pressure HB curve, it was shown that rates on the optimal plasma-on catalyst are comparable to those on the optimal thermal catalyst at 673 K and 100 atm Ru steps. The reported significant plasma enhancement stimulated many research works to verify the predicted results.

To experimentally verify the effect of plasma excited states on ammonia synthesis, recently Yang et al. [41] studied plasma-catalytic ammonia synthesis in a NSD and a 'hybrid' ns pulse/RF discharge in plane-to-plane geometry with a preheated H_2–N_2 mixture and Ni/γ-Al_2O_3 or Co/γ-Al_2O_3 catalyst. In the 'single-stage' process, where a ns pulse discharge in the H_2–N_2 mixture was sustained

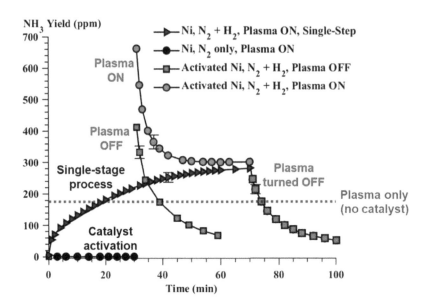

FIGURE 8.19 Comparison of the time-resolved ammonia yield in a 10% H_2–N_2 mixture at $T = 573$ K, $P = 190$ Torr, and total flow rate of 100 sccm over nickel/alumina catalyst, for a single-stage process (plasma is on continuously) and a two-stage process (catalyst activation by a 1 kHz ns pulse discharge in pure N_2, followed by the activated catalyst exposure to the H_2–N_2 flow, with and without a 1 kHz ns pulse discharge). Dashed line shows the steady-state yield in the discharge in the empty reactor.

continuously. In the 'two-stage' process, where the catalyst is first activated by the ns pulse discharge sustained in nitrogen, and then the activated catalyst is exposed to the H_2–N_2 flow, with or without the discharge. The results are shown in Figure 8.19. The first set of data is the plasma-catalytic process in H_2–N_2 with Ni catalyst in a *Single-stage* process (10% H_2–N_2 mixture at $P = 190$ Torr and $T = 573$ K, excited by a 1 kHz ns pulse discharge in the presence of the Ni/γ-Al_2O_3 catalyst). The time for the ammonia yield to reach the quasi-steady state was much longer compared to the empty reactor, approximately 5 min. In the second experiment of a *Two-stage with plasma OFF* process, the Ni/γ-Al_2O_3 catalyst in the reactor was first 'activated' by operating a 1 kHz ns pulse discharge in pure N_2 flow for 30 min and then exposed to the 10% H_2–N_2 flow while the discharge was turned off. In this case, the ammonia yield exhibited a strong overshoot, exceeding the quasi-steady-state value measured in the single-stage process, and then decreased by more than a factor of 6 over the next 30 min. In the third experiment of a *Two-stage with plasma ON* process, the catalyst was 'activated' for 30 min in the same way by a 1 kHz discharge in pure N_2, and then exposed to a 10% H_2–N_2 flow while the discharge was kept on. In this case, the initial NH_3 overshoot when the nitrogen flow was switched to H_2–N_2 was even higher, after which the yield decreased significantly over the next 40 min and approached the quasi-steady-state yield of the single-stage process.

The results demonstrate that the ammonia yield in the plasma-catalytic reactor is controlled by the N atom production in the plasma and accumulation on the catalyst surface, which reacted with H atoms thermally dissociated on the catalyst or generated in the plasma. The results also show that the plasma-catalytic ammonia yield was significantly higher compared to that in the ns pulse discharge without the catalyst. An additional series of measurements was made with a sub-breakdown RF waveform overlapped with the ns pulse discharge train, to enhance the vibrational excitation of nitrogen. The ammonia yield measured with the RF waveform added was approximately 20% higher compared to that at the baseline ns pulse discharge conditions, both with and without the catalyst. This effect was weaker compared to that of the catalyst activation by N atoms and was much weaker than the predicted enhancement in Figure 8.18. Therefore, the experiment suggests that

Electrified Non-Equilibrium Chemical Manufacturing

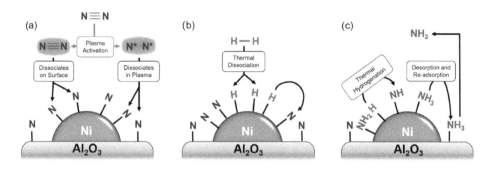

FIGURE 8.20 Proposed reaction pathways for (a) nitrogen activation and adsorption on a Ni particle, (b) nitrogen hydrogenation and hydrogen uptake due to thermal H_2 exposure, and (c) thermal desorption of ammonia from the catalyst and support surfaces [42].

electron impact nitrogen dissociation via R20 is more important than vibrational excitation via R19. Moreover, R8.22 plays a more important role in both the L–H mechanism and the E–R mechanism.

To understand the surface kinetics involving N atom surface adsorption, Barboun et al. [42] reported inelastic neutron scattering (INS) observations of alumina-supported Ni particles after two-stage treatment with N_2 and H_2 plasmas, sequentially. INS experiments revealed the presence of NH_x species and hydrides on Ni sites after exposure to the two-stage sequential N_2 and H_2 plasma treatments. Figure 8.20 shows the proposed pathways for ammonia synthesis over plasma-exposed Ni/γ-Al_2O_3. Non-thermal plasma activates nitrogen (e.g., vibrationally, electronically, or via dissociation pathways via R8.19–R8.22) to facilitate the adsorption of N atom on the catalyst surface (possibly both the metal surface and the alumina support) as shown in Figure 8.20a. Following nitrogen adsorption, hydrogen atoms in the second stage of plasma are introduced to hydrogenate the surface-bound N species and form NH_x on the surface via the E–R mechanism. Finally, ammonia desorbs from the catalyst surface to gas phase. These results directly implicate plasma excitation or dissociation of dinitrogen in the generation of surface-bound nitrogen atoms that participate in further hydrogenation reactions driven either thermally or by H_2 plasma. The *in situ* diagnostic results are consistent with the experimental observation shown in Figure 8.19.

Plasma not only produces N atoms in gas phase but also produces NH and NNH radicals. It is of interest to understand how plasma-generated NH and NNH intermediate species affect ammonia synthesis. Zhao et al. [43] examined the impact of NH, N_2H_2, and NNH produced in plasma on ammonia synthesis using a coaxial dielectric barrier discharge (DBD) plasma reactor without packing and with porous γ-Al_2O_3, 5 wt% Ru/γ-Al_2O_3, or 5 wt% Co/γ-Al_2O_3 catalyst particles (Figure 8.21a). Gas phase species were monitored *in situ* using an electron impact molecular-beam mass spectrometer (EI-MBMS). In this study, MBMS measurements were made while the DBD reactor was operated either empty with no packing or packed with porous γ-Al_2O_3 particles or 5 wt% Ru/γ-Al_2O_3 or Co/γ-Al_2O_3 catalyst particles, respectively. Figure 8.21b shows the NH_3 concentrations measured by MBMS for various cases. It is seen that the NH_3 concentration was below the detection limit of ~20 ppm in this study when the plasma was off, but its concentration greatly increased when the plasma was turned on. Packing the plasma zone with γ-Al_2O_3 support beads led to a doubling of the NH_3 concentration from that observed in the empty tube plasma reactor. Further improvement of the NH_3 yield by ~20% was obtained using a 5 wt % Co/γ-Al_2O_3 catalyst. Surprisingly, the 5 wt% Ru/γ-Al_2O_3 catalyst gave nearly the same NH_3 yield as the γ-Al_2O_3 support. These observations suggested NNH and N_2H_2 formation in gas phase plasma discharge is another pathway for ammonia synthesis or interfacial surface reactions. More future studies of the role of intermediate species production by plasma in ammonia synthesis are needed. In addition, the interaction of active species produced in plasma with catalysts and support materials is poorly understood. Moreover, the effect of surface charge on reactions such as R8.13-18 needs to be understood.

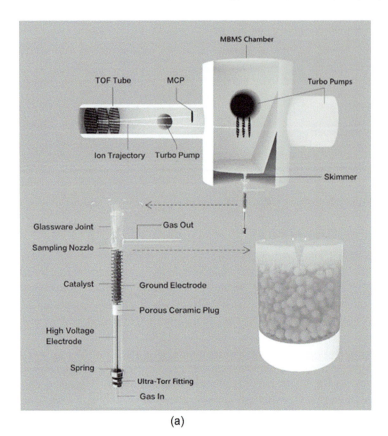

(a)

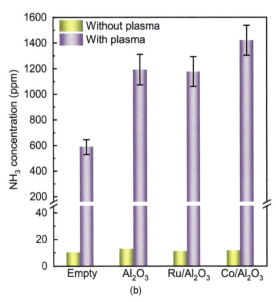

(b)

FIGURE 8.21 (a) Schematic of the molecular-beam mass spectrometer (MBMS) and the custom interface for *in situ* sampling of gas phase species in the plasma-assisted catalytic reactor. (b) NH_3 concentrations in the gas phase detected by MBMS during DBD plasma-assisted NH_3 synthesis. Data were obtained for an empty reactor and for the reactor loaded with porous γ-Al_2O_3, 5 wt% Ru/γ-Al_2O_3, or 5 wt% Co/γ-Al_2O_3 particles. Reaction conditions: 1 atm, 533 K, total flow rate 353 sccm, feed gas 20% N_2/10% H_2/70% He, catalyst particle mass 1.2 g.

8.3.2 Nitric Oxide for Nitrogen Fixation

Plasma-assisted nitrogen fixation is an important way of nitrogen fixation in nature by lightning. The earlier Birkland–Eyde (BE) process uses an arc plasma at high temperature for the synthesis of nitric oxide [44]. It has a great advantage compared to the HB process without using expensive H_2 and can also be powered by renewable electricity. However, to date, the energy efficiency of the BE process still cannot compete with the HB process. It has been shown that the theoretical energy consumption of arc plasma for NO synthesis via the BE process is 0.86 MJ/mol of NO under hypothetical conditions of 20–30 atm and 3,000–3,500 K [44]. On the other hand, non-thermal plasma has a theoretical limit of energy consumption of ~0.2 MJ/mol [45], which is lower than the HB process. Therefore, there is an increasing interest in developing efficient plasma-assisted nitric oxide synthesis methods.

It is well known that thermal NO formation in a nitrogen/oxygen mixture is through the Zeldovich mechanism,

$$N_2 + O = NO + N, \quad E_a = 318 \text{ kJ/mol} \quad \text{(R8.23)}$$

$$N + O_2 = NO + O, \quad E_a = 26 \text{ kJ/mol} \quad \text{(R8.24)}$$

Reaction R8.23 has a very high activation energy (318 kJ/mol) [46]. Figure 8.22 shows the thermal NO_x production as a function of gas temperature in a 0.5 N_2/0.5 O_2 mixture at atmospheric pressure. The thermal NO_x (NO/NO_2) concentration is negligible at low temperatures because of the high activation energy of R8.23. At higher temperatures, NO formation increases. Figure 8.22 shows that the peak NO concentration at chemical equilibrium is located around 3,500 K, which accounts for 7%–8% NO formed in the mixture. However, a further temperature increase results in a decrease of NO concentration due to the decomposition of NO to atomic species.

With plasma activation, several new reaction pathways via electron impact dissociation and vibrational nitrogen, $N_2(v)$, will replace the rate limiting reaction R8.23 in the Zeldovich mechanism to accelerate low-temperature NO formation.

$$e + N_2 \rightarrow e + N + N(^2D) \quad \text{(R8.25)}$$

$$N_2(A) + O \rightarrow NO + N \quad \text{(R8.26)}$$

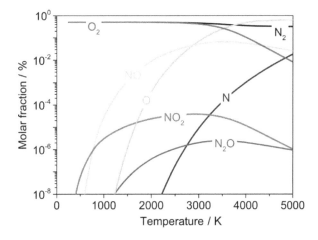

FIGURE 8.22 The NO_x production as a function of gas temperature in a 0.5 N_2/0.5 O_2 mixture at atmospheric pressure [47] (calculated by chemical equilibrium).

$$N_2(v) + O \rightarrow NO + N \quad (v = 1, 26.9 \text{ kJ/mol}; v = 2, 53.8 \text{ kJ/mol}; v = 10, 261 \text{ kJ/mol}) \quad \text{(R8.27)}$$

$$O_2(v) + N \rightarrow NO + N \quad \text{(R8.28)}$$

At a high reduced electric field (E/N), for example in a nanosecond discharge, at 9.8 eV, the N_2 is directly dissociated by electrons and is activated to generate electronically excited nitrogen, such as $N_2(A)$, respectively, via R8.25 and R8.26 even at room temperature. The resulting N and N(^{2}D) subsequently react with O_2 and vibrationally excited $O_2(v)$ to form NO and O through R8.24 and R8.28. At a low reduced electric field such as the gliding arc plasma (0.6–4.0 eV), 50%–90% of the plasma electron energy transfers to vibrational excitation of $N_2(v)$. Moreover, higher vibrational levels, for example $N_2(v = 5$–$10)$, are produced via vibrational-vibrational energy exchange from the low vibrational levels. The high vibrational levels of $N_2(v)$ can effectively overcome the high reaction energy barrier of Zeldovich mechanism (318 kJ/mol in R8.23). Then, N_2 is mainly dissociated by the reaction between $N_2(v)$ and O to generate NO and N via R8.27. This reaction chain is further accelerated by plasma-generated O and $O_2(v)$. The non-thermal plasma-assisted Zeldovich mechanism is critical to reduce energy conversion efficiency to NO.

To demonstrate that the energy efficiency of NO formation using a gliding arc is much better than that for the thermal process, Figure 8.23 [47] shows the calculated and measured energy efficiency of a gliding arc, with and without considering the energy cost of gas preparation, were around 0.5%–1.7%, whereas the thermal energy efficiency calculated for the same specific energy input (SEI) of 1.4 kJ /L was only about 0.2%. This comparison clearly demonstrated the non-equilibrium pathways of the gliding arc for NO_x synthesis, contributing to energy-efficient NO conversion.

Recently, Jardali et al. [48] used a rotating gliding arc in the air for direct NO_x production. The efficiency of NO_x production was explored in a wide range of feed gas conditions for both rotating and steady gliding arc. It was reported that in the case of steady gliding arc, up to 5.5% NO_x concentrations were achieved which were 1.7 times higher than the maximum concentration obtained by the rotating arc mode. The energy consumption was 2.5 MJ/mol, the lowest value achieved by atmospheric pressure plasma reactors. Computer modeling suggested that the combined thermal and vibrationally promoted Zeldovich mechanisms were responsible for the increased NO formation [48]. This study demonstrated experimentally gliding arc assisted NO formation from the air can be a promising technology for nitrogen fixation or an alternative technology to the HB process.

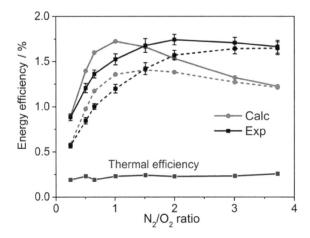

FIGURE 8.23 Experimental and calculated energy efficiency of NO_x formation as a function of the N_2/O_2 ratio in the mixture, for a gas flow rate of 2 L/min and a SEI of 1.4 kJ/L (or 0.35 eV/molec), and comparison with the thermal energy efficiency at the same SEI value. The solid and dashed (black and red) lines indicate without and with the energy cost of air separation, respectively [47].

8.3.3 Hydrogen Production from Methane Pyrolysis

Hydrogen is an ideal green energy carrier. Low-cost and energy-efficient production are critical to the transition of hydrogen energy. As shown in Figure 8.24 [49], there are four ways to produce hydrogen with different carbon emission levels, the green hydrogen from water electrolysis, the turquoise hydrogen and solid carbon production from methane/natural gas pyrolysis, the blue hydrogen from steam methane reforming (SMR) and carbon capture, and the grey hydrogen from methane SMR.

Table 1.5 lists the energy cost and carbon emission levels of each process. SMR is the current state-of-the-art SMR process. However, it produces 4 moles of H_2 with 1 mole of CO_2 [50]. While H_2 generation from water electrolysis is ideal in the long term, it is currently not cost- or energy-competitive [51]. However, the energy cost per kg H_2 production from the turquoise hydrogen (methane pyrolysis) is much less than that of water electrolysis per mole H_2 without CO_2. The relatively low cost and abundance of natural gas and renewable electricity in the U.S. make turquoise hydrogen an appealing source for H_2 with renewable electricity.

Plasma powered by renewable electricity is therefore a promising technology for turquoise hydrogen production with high-value carbon materials such as graphene and carbon nanotubes (CNTs) and higher efficiency compared with SMR.

Various plasmas have been applied to hydrogen production from methane pyrolysis. For example, atmospheric pressure non-thermal plasmas such as DBD [52,53], corona discharge [54], and glow discharge [55] were investigated. The relatively low electron number density and low gas temperature in these weakly ionized plasmas make it difficult to achieve a high conversion efficiency [56,57]. Gliding arc, which has high electron number density and gas temperature, has been shown to have advantages over other low-temperature plasmas and produce hydrogen and carbon more energy efficiently [56–58]. Zhang et al. [57] (Figure 8.25) conducted a chemical kinetics study on methane pyrolysis for hydrogen production in a N_2 rotating gliding arc (RGA) in comparison with experimental data. The results reveal that methane conversion ratio was above 90% at lower methane concentrations, but both selectivity and conversion ratio decreased with the increase of methane concentration. It is shown that the electrons and excited nitrogen species (mainly $N_2(A)$) played a dominant role in the initial dissociation of CH_4.

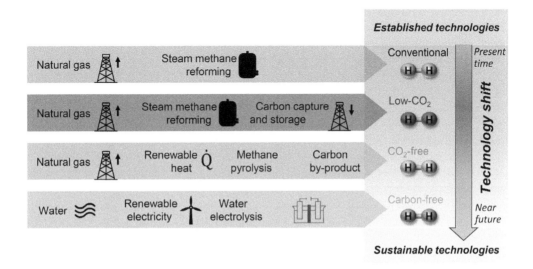

FIGURE 8.24 Common differentiation of hydrogen production pathways into conventional, low-CO_2, C_{O2}-free, and carbon-free production routes and frequently associated colors (grey, blue, turquoise, and green) [49].

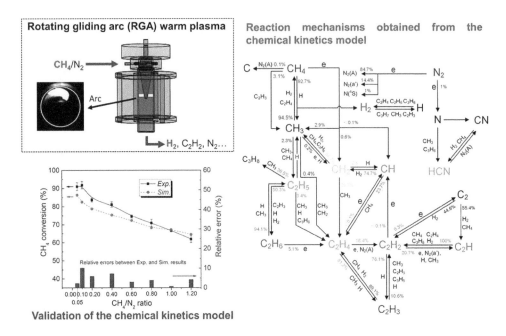

FIGURE 8.25 Schematic of experimental setup of a rotational gliding arc for methane pyrolysis in nitrogen, methane conversion ratio, and kinetic pathways analysis [57].

$$e + CH_4 \rightarrow e + CH_3 + H \quad (R8.29)$$

$$N_2(A) + CH_4 \rightarrow CH_3 + H + N_2 \quad (R8.30)$$

The H abstraction reaction via $CH_4 + H = CH_3 + H_2$ was the major contributor to both the conversion of CH_4 and H_2 production at high temperatures (above 1,200 K). In addition, C_2 hydrocarbons were formed via $C_2H_6 \rightarrow C_2H_4 \rightarrow C_2H_2$ pathway.

Since the temperature distribution in a gliding arc is highly non-uniform due to the thin arc structure. To explore the plasma temperature uniformity on methane pyrolysis, high-temperature microwave plasma torch (Figure 8.26) was also used to produce hydrogen and carbon from methane in comparison to gliding arc between 1,000 and 3,000 K [58].

The experiment showed that acetylene and hydrogen were the main products of methane pyrolysis by microwave torch. Figure 8.27 shows the selectivity toward hydrogen [58]. With increasing specific energy input and increasing methane molar ratio, the selectivity toward hydrogen increased to above 87% while the selectivity toward acetylene dropped to 0.4. At low methane mole fraction, the selectivity toward hydrogen decreased to 60%–70% and the selectivity of acetylene was greater than 0.75.

The comparison revealed considerable differences between the performance of both gliding arc and microwave discharges. It was found that gliding arc had higher methane conversions and a higher energy efficiency to produce hydrogen compared to the microwave discharge at similar operating parameters. However, the product compositions differed significantly. The gliding arc had higher selectivity toward acetylene but lower selectivity toward hydrogen. Moreover, the carbon produced with gliding arc consisted of graphitic flakes, while the carbon produced with the microwave consisted of smaller spherical particles of amorphous carbon. The results suggested that the larger hot center of the microwave increases the conversion of hydrogen and agglomeration and annealing of carbon particles.

Electrified Non-Equilibrium Chemical Manufacturing

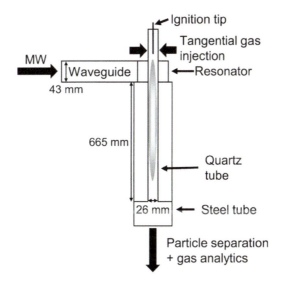

FIGURE 8.26 Schematic of microwave discharge for methane pyrolysis [58].

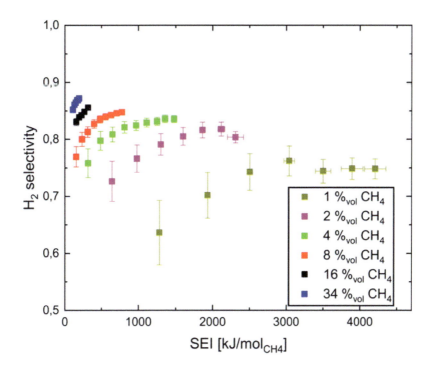

FIGURE 8.27 Selectivity toward hydrogen in the MPT as a function of the specific energy input (SEI) at 60 slm total volumetric flow rate [58].

8.3.4 CO_2 Reduction

The conversion of CO_2 into value-added chemicals and e-fuels by renewable electricity is one of the most attractive ways to decarbonize transportation and chemicals. The high electron energy and non-equilibrium excitation of CO_2 in plasma are very effective to break the bond of CO_2 to produce CO. Moreover, plasma-assisted catalytic reforming of CO_2 with H_2O, H_2, and CH_4 also provides viable pathways to produce syngas, e-fuels, and chemicals. There have been increasing research interests in plasma-assisted CO_2 reduction. In this section, we will provide a brief review of the recent research progress and findings.

8.3.4.1 CO_2 Dissociation

Plasma-assisted splitting of CO_2 to produce syngas as a feedstock for chemicals and e-fuels is one important way for long-term electrical energy storage in chemical bonds. CO_2 bond is very strong (carbon-oxygen bonds: 783 kJ/mol) and the reaction is endothermic.

$$CO_2 \rightarrow CO + 1/2 O_2, \quad \Delta H = 283 \text{ kJ/mol} \tag{R8.31}$$

As shown in the energy diagram of CO_2 electronic and vibrational energy levels in Figure 8.28, it is obvious that electron impact dissociation and vibrational excitation of CO_2 via reactions

$$e + CO_2 \rightarrow CO + O + e \tag{R8.32a}$$

$$e + CO_2 \rightarrow CO + O(^1D) + e \tag{R8.32b}$$

$$e + CO_2 \rightarrow CO_2(v) + e \tag{R8.33}$$

$$CO_2(v) + O \rightarrow CO + O_2 \tag{R8.34a}$$

$$CO_2 + O(^1D) \rightarrow CO + O_2 \tag{R8.34b}$$

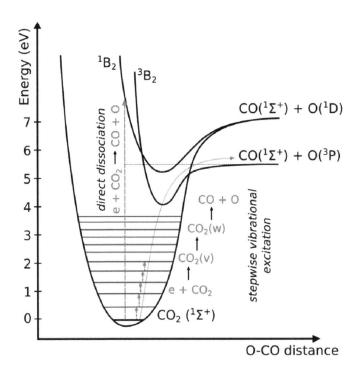

FIGURE 8.28 Schematic diagram of some CO_2 electronic and vibrational levels, illustrating that much more energy is needed for direct electronic excitation–dissociation than for stepwise vibrational excitation [64].

$$CO_2(m) + CO_2(n) \rightarrow CO_2(m+1) + CO_2(n-1) \tag{R8.35}$$

can potentially reduce the thermal energy barrier of CO_2 dissociation and increase energy conversion efficiency and selectivity.

Recently, many studies have been conducted for CO_2 dissociation using non-equilibrium plasmas. For example, DBD reactors with different catalysts [59] were used to examine the energy efficiency and CO_2 conversion. MW and RF discharges were used to examine and compare the conversion efficiency [60]. Nanosecond discharge was also adopted to examine CO_2 conversion efficiency at high pressure [61]. The results showed that DBD plasma can operate at low temperatures but are not energy-efficient. On the other hand, MW and RF showed high conversion efficiency, but these improvements were limited to lower pressures. To overcome these difficulties, gliding arcs were used to operate at atmospheric pressure and to achieve high efficiency at the same time. Most gliding arc experiments reported a maximum energy efficiency of 40%–50% or higher [62,63]. Unfortunately, because of the thin channel of gliding arc, the CO_2 conversion is relatively low in comparison to microwave discharge.

Figure 8.29 [61] shows, respectively, the energy efficiency and conversion ratio of various plasmas and their dependence on the reduced electric field E/N. The results show that low pressure MW plasma (50–200 Torr) had energy efficiency as high as 80%, but the conversion ratio was limited to 30%. Other low-pressure plasmas such as radio frequency (RF) [60] and DC glow discharges showed conversion efficiency <15%. The gliding arcs at atmospheric pressure, however, showed capable of reaching energy efficiency of 60%.

Recently, nanosecond repetitively pulsed (NRP) discharges in a high-pressure batch reactor [61] were used for CO_2 splitting up to 12 atm. The results showed that the extent of CO_2 conversion was almost linearly dependent on the specific energy invested. A conversion rate as high as 14% was reported with an energy efficiency of 23% (Figure 8.30). The computational modeling revealed

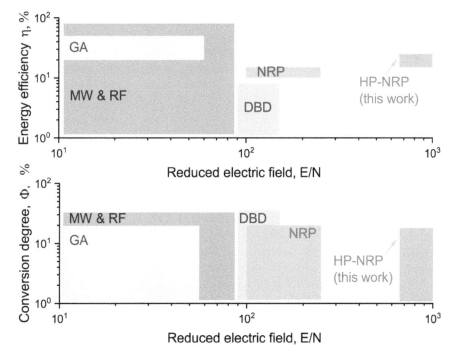

FIGURE 8.29 Variation of energy efficiency η and conversion degree Φ with respect to the reduced electric field E/N in different types of plasma discharges [61].

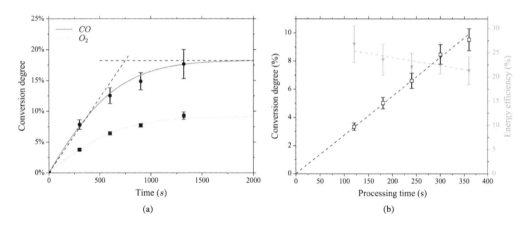

FIGURE 8.30 Effects of processing time on the CO_2 conversion degree and energy efficiency at experimental conditions as pressure $p = 5$ bar, initial temperature $T_0 = 300$ K, initial $XCO_2 = 1$, discharge frequency $f = 30$ kHz. (a) CO_2 conversion degree within 1,500 s. Solid circles: measurements. Lines: exponential trend for CO (blue) and O_2 (yellow). Average deposited energy per pulse $E = 700$ μJ. (b) CO_2 conversion degree Φ (black) and energy efficiency η (blue) within 400 s. Solid circles: measurements. Lines: linear trend. Average deposited energy per pulse $E = 800$ μJ.

the presence of three-stage kinetics (R8.32–R8.35) between NRP pulses, controlled by electron impact CO_2 dissociation, vibrational relaxation, and neutral elementary kinetics. This work concluded that different from the kinetics in gliding arc and MW, for high-pressure NRP CO_2 conversion, the electronic excitation is more significant than the vibrational mechanism. More future studies in the role of electronic excitation and vibrational excitation of CO_2 in plasma-assisted CO_2 splitting are needed.

8.3.4.2 CO_2/CH_4 Dry Reforming

Electrified CO_2 dry reforming of methane (DRM) to syngas and liquid fuels (E-fuels) is another key method for energy storage, green transportation and chemical manufacturing, and decarbonization [65–68]. Both CO_2 and methane are stable molecules. As such, the DRM to syngas and liquid fuels

$$CO_2 + CH_4 = 2CO + 2H_2, \quad \Delta H = 247 \text{ kJ/mol} \quad \text{(R8.36)}$$

$$CO_2 + CH_4 = CH_3COOH, \quad \Delta H = 71 \text{ kJ/mol} \quad \text{(R8.37)}$$

are not only endothermic but also need to be carried out at high temperatures (900–1,200 K) with a catalyst such as Ni, Co, Fe, and precious metals. Figure 8.31 [69] shows the predicted theoretical thermal conversion efficiency and energy efficiency as a function of temperature. It is seen that almost complete conversion can be achieved around 1,500 K with an energy efficiency of 60%. However, the maximum energy efficiency of 70% occurs near 1,000 K with a conversion of efficiency of 83%. Unfortunately, at low temperatures, the reaction rates are very slow. On the other hand, at high temperatures, the soot and carbon formation become problematic to deactivate the catalyst and further reduce energy efficiency. Due to these challenges, to date, DRM has not been well developed on an industrial scale. Therefore, novel methods such as plasma catalysis need to be developed to lower the synthesis temperature and improve conversation efficiency and energy efficiency.

There have been many efforts to use various plasma discharges with and without catalysts for CO_2 dry reforming [69]. By using DBD discharge in a FB reactor and a packed-bed (PB) reactor, Nozaki and coworkers [66,67] showed that dry reforming was promoted dramatically in the

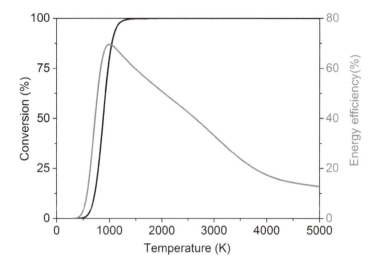

FIGURE 8.31 Calculated theoretical thermal conversion (left axis) and corresponding energy efficiency (right axis) as a function of temperature for the dry reforming of methane [65].

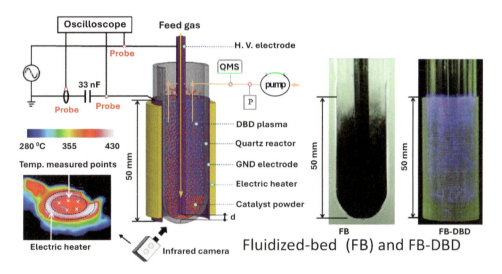

FIGURE 8.32 Schematic diagram of fluidized bed (FB) and FB-DBD and experimental system. Inset thermograph corresponds to the hybrid reaction with catalyst temperature of 404°C [67].

fluidized bed reactor. In addition to syngas production, minor amounts of large hydrocarbons such as C_2H_6, C_2H_4, C_2H_2, and C_3H_8 were also observed. Figure 8.32 shows the schematic diagram of plasma-assisted FB-DBD reactor in Ref. [67]. The quartz reactor had a dimension of 20 mm i.d. × 23 mm o.d. and the high-voltage (HV) electrode was coaxially placed in the quartz reactor with 7 mm gas gap in the radial direction. The La-Ni/Al$_2$O$_3$ catalyst particles (La-3wt%, Ni-11wt%) were used for the DRM study.

Figure 8.33a–c [67] shows feed gas conversion and H_2 selectivity in FB- and PB-DBD. The dotted line represents the chemical equilibrium at 10 kPa and $CH_4/CO_2 = 0.8$. The hollow symbol indicates PB-DBD. Blank experiments (without catalysts) were conducted in FB and PB reactors and data at 500°C, as shown in Figure 8.32d. The results show that feed gas conversion without DBD was well below the thermal equilibrium regardless of the residence time. In contrast, feed gas

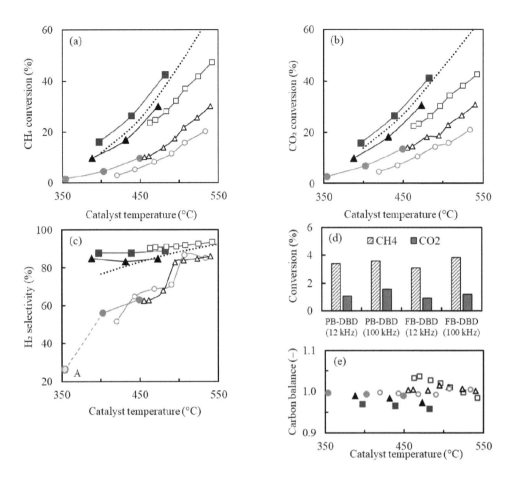

FIGURE 8.33 DMR performance in FB-DBD (●: thermal, ▲: 12 kHz, ■: 100 kHz) and PB-DBD (○: thermal, Δ: 12 kHz, □: 100 kHz). (a) CH_4 conversion, (b) CO_2 conversion, (c) H_2 selectivity, (d) CH_4 and CO_2 conversion in blank reactors at 500°C (without catalyst), (e) carbon balance [67].

conversion increased by FB-DBD at fixed residence time, which was even more drastic than that of PB-DBD. It was concluded that reaction promotion in FB-DBD was possible by the enhanced coupling between plasma-generated reactive species (most likely vibrationally excited species) and fluidized powdered catalysts. It was noted that although CH_4 and CO_2 conversion in 100 kHz FB-DBD exceeded the thermal equilibrium, it was not conclusive because deviation from the thermal equilibrium was small. The particle interaction with plasma and heat and mass transfer augmentation in the fluidized bed reactor are critically important.

A study of microwave discharge in DRM for syngas was also conducted [70]. The experiments reported CH_4 and CO_2 conversions of 71% and 69%, respectively, but the energy efficiency was low. DRM studies were also conducted by using gliding arcs [71,72]. The results showed that gliding arc resulted in similar conversion to DBD, but the energy efficiency was significantly higher. A summary of recent progress of plasma-assisted DRM was by Snoeckx and Bogaerts [69] for energy cost as a function of conversion, and results are shown in Figure 1.27. Like CO_2 decomposition, DBD and other weakly ionized plasma-assisted dry reforming do not allow the energy-efficient conversion of CO_2 and CH_4 into syngas, partly because of the low electron number density. FB-DBD may increase the energy conversion efficiency is still lower than the gliding arc. Gliding arc has demonstrated the best results of up to 40% conversion and higher energy efficiency than thermal

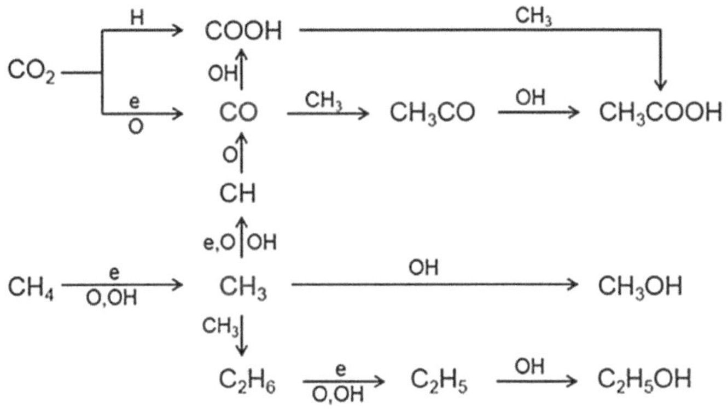

Electrified Non-Equilibrium Chemical Manufacturing

FIGURE 8.34 Possible reaction pathways for the formation of CH_3COOH, CH_3OH, and C_2H_5OH in the direct reforming of CH_4 and CO_2 with DBD [68].

equilibrium processes. By further raising the temperature using spark and arc discharges, conversions up to 85%–95% have already been achieved.

Recently, CO_2 DRM for oxygenates and hydrocarbon synthesis was also attempted. Wang et al. [68] developed a DBD plasma process (Figure 1.28) for the direct activation of carbon dioxide and methane into oxygenated liquid fuels and chemicals (e.g., acetic acid, methanol, ethanol, acetone, and formaldehyde) at room temperature and atmospheric pressure with a water electrode. The reaction pathways from DRM to acetic acid and alcohols are schematically shown in Figure 8.34. CO reacts with a CH_3 radical produced by electron impact dissociation via reaction

$$CO + CH_3 \rightarrow CH_3CO \quad (R8.38)$$

to form an acetyl radical (CH_3CO) with a low energy barrier of 28.77 kJ/mol. Then, the recombination of CH_3CO with OH

$$CH_3CO + OH \rightarrow CH_3COOH \quad (R8.39)$$

to produce acetic acid with no energy barrier. The selectivity to acetic acid increased initially and then decreased with the CH_4/CO_2 ratio with the optimal acetic acid formation at a CH_4/CO_2 ratio of 1:1 (Figure 8.35). It was observed that decreasing the CH_4/CO_2 molar ratio led to the decrease of the generation of CH_3 radicals with an increase of OH formation. In addition, as shown in Figure 8.35, direct coupling of CH_3 and carboxyl radicals (COOH) also forms acetic acid.

DRM can also lead to alcohol formation. As shown in Figure 8.34, CH_3 and C_2H_5 reactions with OH via

$$OH + CH_3 \rightarrow CH_3OH \quad (R8.40)$$

$$OH + C_2H_5 \rightarrow C_2H_5OH \quad (R8.41)$$

will respectively form methanol and ethanol. As shown in Figure 8.35, decreasing the CH_4/CO_2 molar ratio increased OH and the formation of CH_3OH. A peak formation of CH_3OH occurred at a CH_4/CO_2 molar ratio of 1:1. By contrast, the formation of C_2H_5OH decreased continuously as the CH_4/CO_2 molar ratio decreased. These results suggested that the production of CH_3OH depended on both CH_3 and OH radicals while the formation of C_2H_5OH was more sensitive to CH_3 radicals in the plasma reaction to form C_2H_5. The results showed total selectivity of oxygenates of 50%–60% and a 40.2% selectivity to acetic acid. In addition, Figure 8.35 also shows that plasma-assisted DRM can

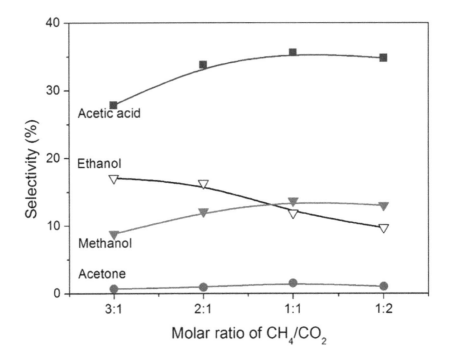

FIGURE 8.35 Effect of the CH_4/CO_2 molar ratio on the selectivity for oxygenates without a catalyst (total flow rate 40 mL/min, discharge power 10 W) [68].

also lead to formation of other oxygenates such as ethers. The CH_4/CO_2 molar ratio and the type of catalyst and plasma properties can be used to manipulate the production of different oxygenates.

Scapinello et al. [73] further investigated the catalytic effect of the electrode surface (Cu and Ni) on the conversion of CH_4 and CO_2 to $C_nH_{2n+1}COOH$ and CH_2O_2 at different discharge powers in a DBD reactor. It was shown that CO_2 hydrogenation on the metal surface was enhanced when increasing the discharge power and that the selectivity of the end products was related to the CO_2/CH_4 molar ratio. Although these experiments showed that CO_2 conversion with methane is a promising process for the synthesis of oxygenates, more research is required to improve the selectivity and conversion efficiency of these value-added products.

8.3.4.3 CO_2 Conversion with H_2O

Electrified syngas and oxygenated fuel production from CO_2 and H_2O from reactions of

$$CO_2 + H_2O \rightarrow CO + H_2 + O_2, \quad \Delta H = 525 \text{ kJ/mol} \quad \text{(R8.42)}$$

$$CO_2 + 2H_2O \rightarrow CH_3OH + 3/2O_2, \quad \Delta H = 676 \text{ kJ/mol} \quad \text{(R8.43)}$$

is also attractive because of carbon utilization and energy storage as well as the abundance of CO_2 and H_2O. However, these reactions are much more endothermic than DRM reactions and require high temperatures with catalysts.

Chen et al. investigated the conversion of carbon dioxide and water vapor into value-added chemicals (R8.42) in a surface-wave microwave discharge at 915 MHz in a pulse regime [74]. The results showed that syngas with a ratio close to unity was produced when CO_2/H_2O ratio in the gas mixture was 50:50. The optimum SEI for H_2 and CO formation for this mixture ratio was 1.6 eV/mol. Figure 8.36 shows the formation rates of H_2 and CO as a function of the specific energy input (SEI). Figure 8.36a reveals that the formation rate of H_2 for different H_2O/CO_2 ratios follows a similar

Electrified Non-Equilibrium Chemical Manufacturing

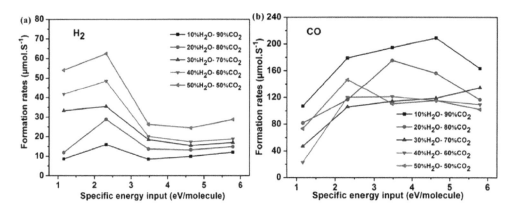

FIGURE 8.36 Formation rates of H_2 (a) and CO (b) as a function of SEI for different H_2O/CO_2 ratios in the gas mixture. The total flow rate is fixed at 6 slm [74].

trend. With the increase of SEI, the H_2 formation rate initially increased and reached a maximum at 2.3 eV. A further increase of the SEI led to a decrease of H_2. In addition, the production of H_2 increased with increasing the H_2O content in the feed mixture. Figure 8.36b shows that the CO formation rates were higher at high CO_2 content and maximized at SEI between 2 and 3 eV/molecule. The results suggested that syngas with a high H_2/CO molar ratio could be generated when the proportion of H_2O was 50 vol% and at SEI of about 2 eV/molecule. The study concluded that in order to increase the energy efficiency and the conversion rate of CO_2 and water to syngas or other products, a catalyst is needed in plasma synthesis. Microwave discharge was also tried for CO_2 + H_2O reforming [75]. The production of oxygenated species such as methanol, oxalic acid, and H_2O_2 was observed.

DBD plasmas [76] in a packed-bed reactor with $BaTiO_3$ pellets were used. The results showed CO_2 conversion of 12.3% and yields of 12.4% H_2, 11.8% CO and 2.8% O_2. Mahammadunnisa et al. [77] later used a Ni/g-Al_2O_3 catalyst for the same reaction. They obtained the conversion to syngas ratio of 24%–36% for the partially reduced catalyst. In addition, the formation of CH_3OH and other compounds such as C_2H_2 was also detected. In addition, gliding arc was also used for the same reaction [62]. However, the energy conversion efficiency was low. It was shown that even a small amount of water addition reduced the CO_2 conversion, possibly due to the vibrational energy losses to water molecules or because of the decrease of electron number density. Therefore, more kinetic understanding of energy transfer between CO_2 and H_2O excited states in non-equilibrium plasma-assisted CO_2 and water conversion to valued chemicals is needed.

8.3.4.4 CO_2 + H_2 Hydrogenation

With the rapid development of green hydrogen production, CO_2 hydrogenation reactions with hydrogen have attracted significant interest in producing CO or syngas, methane, alcohols, and e-fuels. In these reactions

$$CO_2 + H_2 \rightarrow CO + H_2O, \quad \Delta H = 37.2 \text{ kJ/mol} \quad \text{(R8.44)}$$

$$CO_2 + 4H_2 \rightarrow CH_4 + 2H_2O, \quad \Delta H = -165 \text{ kJ/mol} \quad \text{(R8.45)}$$

$$CO_2 + 3H_2 \rightarrow CH_3OH + H_2O, \quad \Delta H = -41 \text{ kJ/mol} \quad \text{(R8.46)}$$

hydrogen can provide H radicals to accelerate CO_2 reduction to CO and OH. In addition, the formation of OH radicals can also provide reaction pathways for the formation of alcohols and other oxygenates.

Plasma provides a promising technique to excite and dissociate CO_2 and H_2. Without catalysts, the major products in plasma-assisted hydrogenation of CO_2 are CO and H_2O. With catalysis, methane, methanol, and other oxygenates can be formed at low temperatures. Below we discuss, respectively, some recent research findings, respectively, from CO_2 hydrogenation to CO, methane, and oxygenated fuels.

8.3.4.4.1 CO_2 Hydrogenation to CO

Maya [78] employed a microwave discharge in a gaseous mixture of CO_2/H_2 for syngas production via R8.44. The results showed that the major products were CO and H_2O although a small amount of formic acid was observed. Reduction of CO2 with hydrogen in a non-equilibrium microwave plasma reactor was also examined by de la Fuente et al. [79]. The main products were also CO and H_2O. The reported CO_2 conversion ratio was 65%. Recently, to understand the non-equilibrium mechanism of plasma catalysis of CO_2 hydrogenation for CO production, Kim et al. [80] investigated in Pd_2Ga/SiO_2 alloy catalysts in comparison to thermal catalysts with *in situ* laser absorption spectroscopy. It was shown that both thermal and plasma catalysis had close to 100% CO selectivity. It also demonstrated that plasma catalysis increased CO_2 conversion by more than 2-fold and broke the thermodynamic equilibrium limitation.

The experimental setup of Kim et al. [80] is shown in Figure 8.37 using a fluidized bed dielectric barrier discharge (FB-DBD) reactor (Figure 8.32) was used for CO_2 hydrogenation over Pd_2Ga/SiO_2. The *in situ* IR characterization of gaseous and surface adsorbed species in catalytic CO_2 hydrogenation under plasma was conducted using a DBD flow-type IR cell under a pressure of 10 kPa (Figure 8.37). Stainless-steel (1 mm o.d.) electrodes wrapped with a quartz sheath were inserted

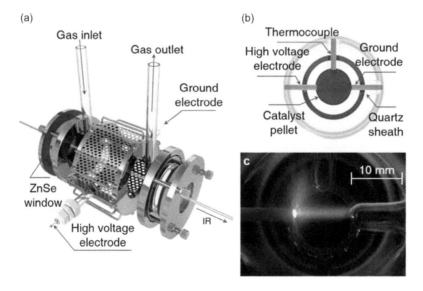

FIGURE 8.37 Design for the *in situ* DBD-TIR cell. (a) Schematic of the *in situ* TIR cell capable of identifying surface species with and without DBD irradiation characterized in this work (b) and (c) cross-sectional schematic (b) and a corresponding photograph (c) during DBD irradiation under CO_2 (10 mL/min) + H_2 (30 mL/min) diluted in Ar (150 mL/min). The high-voltage electrode is on the left exposing the metallic tip, while the grounded electrode is fully covered by glass sheath. The catalyst pellet is located behind the luminescence discharge channel which is separated by 5 mm to avoid direct contact of plasma channel to the catalyst surface [80].

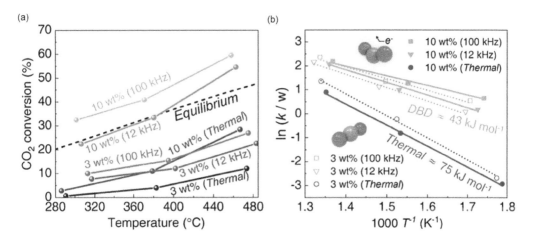

FIGURE 8.38 Kinetic study of CO_2 hydrogenation over the 10 and 3 wt% Pd_2Ga/SiO_2 with the FB-DBD reactor. (a) CO_2 conversion under thermal and DBD (12 and 100 kHz) conditions. (b) Arrhenius plot. Total flow rate = 200 mL/min (STP); H_2/CO_2 = 3; WHSV = 3,000 cm^3/g h (STP); pressure = 15 kPa; and SEI = 2.28 eV/molecules [80].

inside the reactor to form a DBD discharge. The gap between point-to-point electrodes was 10 mm. The applied voltage and frequency for all catalyst tests were 2 kV and 19 kHz, respectively, and kept constant. The catalyst was powdered (25–30 mg) and placed evenly in a disc kit in the form of a pellet (10 mm o.d., 1 mm thickness). The catalyst pellet was placed in a glass holder and inserted into the 5 mm downstream from the pin-to-pin discharge zone.

Figure 8.38 [80] shows the CO_2 conversion results over Pd_2Ga/SiO_2 with different Pd_2Ga loadings in SiO_2 (10 and 3 wt%) under thermal and two DBD conditions at 12 and 100 kHz, respectively. It is seen in Figure 8.38a that CO_2 conversion increased with catalyst temperature. In the case of catalysts, the reaction activity was significantly improved compared to thermal only conditions. The CO selectivity was close to 100% for all conditions. Moreover, CO_2 conversion for 10 wt% Pd_2Ga/SiO_2 under 100 kHz DBD conditions was higher than the thermodynamic equilibrium limitation. The study examined the role of three different vibrational modes of CO_2, bending and symmetric and asymmetric stretching. It was concluded that the bending-mode CO_2 contributed additional CO_2 conversion, yielding higher CO_2 conversion than that of the ground-state CO_2. The corresponding Arrhenius plot is shown in Figure 8.38b. The extracted activation energy for the DBD conditions (43 kJ/mol) was clearly smaller than that of the thermal equilibrium conditions (75 kJ/mol). The reduction of activation energy implies that the reaction pathways or the rate-determining step for the two catalysts were changed by the vibrationally excited molecules. *In situ* IR diagnostics and density functional theory (DFT) modeling of the reaction mechanism showed that the fast formation and decomposition of m-HCOO for CO_2 hydrogenation via vibrationally excited CO_2 and H_2 activation over Pd_2Ga/SiO_2 alloy catalysts was the primary mechanism for the enhanced conversion. This study suggested the high designability of CO_2 hydrogenation plasma catalysts toward value-added chemical synthesis.

8.3.4.4.2 CO_2 Hydrogenation to Methane
Methanation of CO_2 with hydrogen (R8.45) was another important research direction of CO_2 hydrogenation. CO_2 methanation in a DBD packed Ni/zeolite catalyst pellets was investigated by Jwa et al. in the 1990s [81]. More than 95% conversion of CO_2 was observed. It was inferred that the rate limiting reaction was adsorbed CO dissociation to carbon and oxygen atom on the catalyst surface and that the reactive species generated in the plasma reactor can speed up the rate-determining-step of the catalytic hydrogenation. Nizio et al. [82] conducted hybrid plasma-catalytic methanation

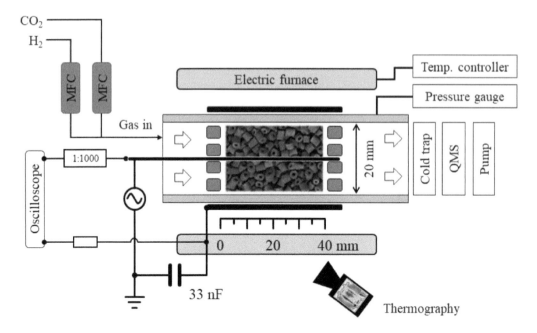

FIGURE 8.39 Experimental setup for non-thermal plasma catalysis of CO_2 methanation over multi-metallic Ru-based catalysts [83].

of CO_2 at low temperature and atmospheric conditions over ceria zirconia-supported Ni catalysts (Ni-$Ce_xZr_{1-x}O_2$). The experiments showed a promising result that CO_2 conversion was as high as 80% with 100% selectivity toward methane at 90 °C.

More recently, Nozaki and coworkers [83] investigated the non-thermal plasma effect on CO_2 methanation over Ru-based multi-metallic catalysts in a packed-bed DBD reactor at reduced pressure. The experimental setup is shown in Figure 8.39. The catalyst pallets were fully packed in the discharge region (40 mm in length) of a packed-bed dielectric barrier discharge (PB-DBD) reactor which was a quartz tube with an inner diameter of 20 mm. The high-voltage electrode (3 mm diameter) was laid at the tube center and the ground electrode was placed outside the tube. Plasma was generated in the packed-bed region by applying a HV power source (12 and 100 kHz). Discharge power was kept at 30 W. Ru-modified La-Ni/Al_2O_3 catalyst was used in this study. The Ni, La, and Ru loading were 11, 3, and 1 wt%, respectively. Another catalyst of Ni-modified Ru/Al_2O_3 was employed as the control catalyst.

Figure 8.40 [83] shows the comparison of the catalytic performance of Ru(La-Ni)/Al_2O_3 and Ni(Ru)/Al_2O_3 at 30 kPa. Equilibrium CO_2 conversion at 30 and 100 kPa is also shown for comparison. The onset temperature was 200°C and not influenced by the type of catalysts and the application of plasma. The CO_2 conversion reached the maximum limit of 75% in all four examined cases. The CO_2 conversion at 30 kPa in plasma-assisted catalytic conversion was obviously lower than that of 100 kPa. However, the equilibrium conversion was not influenced very much by pressure. This was caused by the reduction of the net reaction time as pressure was decreased. The results showed one noticeable superiority of Ru(La-Ni)/Al_2O_3 over Ni(Ru)/Al_2O_3 was that CO_2 conversion developed at lower temperature and reached the maximum limit at around 300°C. Reaction promotion by DBD was observed clearly when CO_2 conversion was much lower than the equilibrium. *In situ* DRIFTS study provided the understanding of the possible reaction pathway of CO_2 methanation. La provided CO_2 adsorption sites and helped carbonate formation which was the key initiation step of methanation reaction. When DBD plasma was applied, higher catalytic activities at low temperatures were confirmed because vibrationally excited CO_2 contributed to the formation of

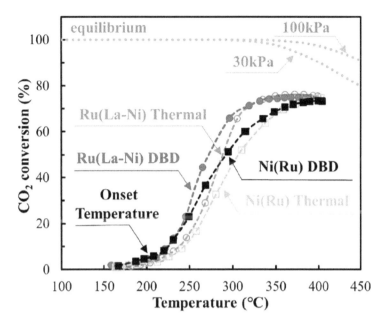

FIGURE 8.40 Comparison of catalytic performance on Ni(Ru)/Al$_2$O$_3$ and Ru(La-Ni)/Al$_2$O$_3$ at 30 kPa [83].

carbonate. Moreover, Ru promoted hydrogen spillover that hydrogenates CO$_2$-derived surface species. Therefore, the combination of La and Ru in the catalysts was essential to promote the hybrid reaction with DBD.

8.3.4.4.3 CO$_2$ Hydrogenation to Oxygenates

Direct production of methanol and oxygenate synthesis from CO$_2$ hydrogenation is of interest for e-fuels. CH$_3$OH is an important chemical feedstock for fuels and chemicals. However, direct production of methanol from CO$_2$ hydrogenation using thermal catalysis often requires high temperatures and high pressure (usually 30–300 atm) to activate CO$_2$ and shift chemical equilibrium, thus is very energy intensive. Non-equilibrium plasma offers a very promising technique for direct CO$_2$ hydrogenation to methanol under low temperatures and ambient pressure. Eliasson et al. [84] studied the hydrogenation of CO$_2$ to methanol in a DBD reactor with CuO/ZnO/Al$_2$O$_3$ catalyst. The methanol yield at high pressure (8 atm) increased to 40% and the selectivity was around 10%.

Recently, Wang et al. [85] reported a plasma-catalytic process for the direct CO$_2$ hydrogenation to methanol at room temperature and ambient pressure. A DBD reactor using a water ground electrode was used to shift the chemical equilibrium to low temperature. A methanol yield of 11.3% and selectivity of 53.7% were achieved with Cu/γ-Al$_2$O$_3$ catalyst in the plasma. They also found that the reactor structure greatly affected the production of methanol. Figure 8.41 [85] shows the effects of H$_2$/CO$_2$ molar ratio and catalyst on methanol yield. Figure 8.41a and b reveals that the concentration and yield of CH$_3$OH, as well as the conversion of CO$_2$ were affected significantly by the H$_2$/CO$_2$ molar ratio and especially by the catalysts. Increasing the H$_2$/CO$_2$ molar ratio from 1:1 to 3:1 increased the concentration and yield of CH$_3$OH from 1.7 to 3.7 mmol/L and 6.0%–7.2%, respectively. However, Figure 8.41c demonstrated that the corresponding selectivity of CH$_3$OH only varied slightly in the range of 50%–60% while the selectivity of CO decreased from 40.0% to 30.0% with the increase of the H$_2$/CO$_2$ molar ratio.

The study [85] also evaluated Cu/γ-Al$_2$O$_3$ and Pt/γ-Al$_2$O$_3$ catalysts in the plasma-catalytic CO$_2$ hydrogenation to methanol. Figure 8.41 shows that Cu and Pt catalysts significantly enhanced the concentration and yield of methanol in plasma catalysis but also increased the formation of

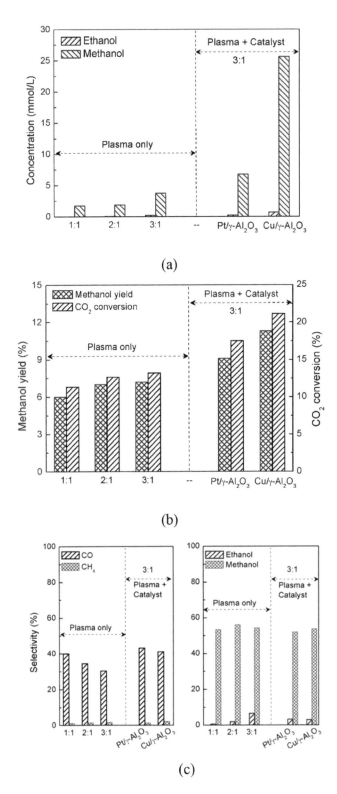

FIGURE 8.41 Effect of H_2/CO_2 molar ratio and catalysts on the reaction performance of the plasma hydrogenation process: (a) concentration of oxygenates; (b) methanol yield and CO_2 conversion; (c) selectivity of gas products and oxygenates [85].

byproduct CO. Moreover, Cu/γ-Al$_2$O$_3$ catalyst exhibited much better activity toward CH$_3$OH formation compared to the Pt/γ-Al$_2$O$_3$ catalyst. At the H$_2$/CO$_2$ molar ratio of 3:1, compared to the plasma hydrogenation without a catalyst, placing the Cu/γ-Al$_2$O$_3$ catalyst in the plasma discharge significantly increased the CH$_3$OH concentration by a factor of ~7 (from 3.7 to 25.6 mmol/L) and enhanced the CH$_3$OH yield (from 7.2% to 11.3%) and CO$_2$ conversion (from 13.2% to 21.2%), while maintaining the CH$_3$OH selectivity at a similar level of around 54.0%. The study also showed that the energy efficiency of methanol production in the plasma CO$_2$ hydrogenation was also strongly dependent on the structure of the DBD reactors.

In summary, the studies of plasma-assisted CO$_2$ hydrogenation to methanol have demonstrated that the coupling of plasma with catalysts is critical to enable selective formation of methanol at low temperature and atmospheric pressure. The increase of methanol yield with plasma-catalyst coupling indicates that plasma catalysis provides a unique electrified process for the conversion of low value feedstock (CO$_2$) to commodity liquid fuels and platform chemicals (e.g., methanol).

8.4 PLASMA-ASSISTED CHEMICAL LOOPING

Chemical-looping combustion (CLC) is an emerging method of power and chemical generation while also producing high concentrations of CO$_2$ in the reactor exhaust gas. CCS requires concentrated amounts of CO$_2$ for efficient carbon capture and storage. Compared to conventional hydrocarbon-based power generation and chemical systems, CLC systems are able to yield considerably higher CO$_2$ concentrations as there is no N$_2$ present to dilute the exhaust gases [86–88]. As shown in Figure 8.42, in CLC, there are two separate reactors: an oxidation reactor and a reduction reactor. These two reactors are connected via the transport of fluidized metal oxide and reduced metal particles between them. These metal particles are often oxides of transition metals such as CuO, NiO, and Fe$_2$O$_3$ on support materials such as Al$_2$O$_3$, and MgO [86], and are referred to as 'oxygen carriers' for transferring oxygen atoms from the oxidation reactor to the reduction reactor. The process first starts with the fluidized particles fully oxidizing in the oxidation reactor through reaction with air at elevated temperatures. This is then followed by the transfer of the oxidized particles to the reduction reactor where the fuel is introduced. The net result is the transfer of bonded oxygen atoms in metal oxides without the transfer of N$_2$. Therefore, the fuel will react with oxygen in metal

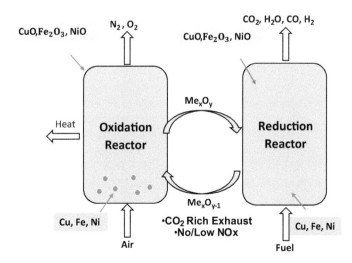

FIGURE 8.42 Schematic of chemical looping using an oxidization reactor and a reduction reactor with metal particles.

TABLE 8.1
Comparison of Heat of Reaction of Different Metals in Oxidation and Reduction

Metal	Fe	Cu	Ni	Mn	Co
Heats of Reaction for the Net Reaction $C + O_2 \rightarrow CO_2$					
Oxidation (kJ/mol)	−549	−315	−326	−639	−446
Reduction (kJ/mol)	155	−79	−68	246	52

oxides only without any nitrogen, resulting in the desired CO_2 rich exhaust or syngas that is free from N_2. This overall process often needs to operate at temperatures <1,000°C due to the temperature constraints of the system.

The problems arising with CLC operating at high temperature are the exothermicity in both reduction and oxidation processes as well as the costs and survivability of oxygen-carrier. Table 8.1 compares the exothermicity of different metals in oxidation and reduction. It is clear that both Cu and Ni have large exothermicity in both oxidation and reaction. Fe is an abundant metal. Cobalt is a good catalyst for hydrocarbons. Therefore, transition metals like Ni, Cu, Fe, and Co are the best choices for CLC. Among them, Ni-based oxygen carriers were the most popular choice as NiO is very reactive with methane, partly due to the fact that reduced Ni is also a methane reforming catalyst [89]. A thermodynamic analysis between NiO and methane found that conversion to CO_2 and H_2O was between 97.7% and 99.8% in the temperature range 700°C–1,200°C [90]. However, the risk of carbon deposits forming on NiO can be problematic. As such, potential carbon formation in NiO lead to CuO becoming more popular for investigation in recent years. CuO-based oxygen carriers have the benefit of reacting very exothermically with methane, and can even be reduced directly with solid carbon particles [91]. In addition, Cu was found to not have the issue with carbon deposits that NiO has, as Cu is not a methane reforming catalyst. While kinetically CuO trumps many other materials, questions remain around the lifetime of a Cu-based particle, related to the low melting temperature of metallic copper, which could result in sintering or agglomeration. Fe-based materials suffer from reduced reactivity compared to NiO and CuO and must be operated at higher temperatures [92], but has the benefit of low cost and being non-toxic/ environmentally friendly [93].

While the study of kinetics for CLC is often done with pure or high concentrations of metal oxide particles, pure metal oxide particles on their own are often not considered as oxygen carriers, as sintering results in a rapid loss of reactivity after only a few cycles [94,95]. To avoid long-term sintering or agglomeration, the metal oxide particles are often impregnated on support materials, which reduces the overall concentration of metal, adds mechanical support, and reduces direct contact between metal/metal oxide particles. However, attention must be paid to the choice of support. A comparative study on the effect of different support materials found that NiO impregnated on Al_2O_3 had reduced reactivity compared to NiO impregnated on $MgAl_2O_4$ or $CaAl_2O_4$, as it was found that NiO could directly react with the support material to from $NiAl_2O_4$ [96]. The most commonly studied support materials include SiO_2, MgO, and Al_2O_3 [97,98]. While various support materials can help delay agglomeration or sintering, they don't solve a fundamental issue facing the metal particles. The issue is that the temperature in the CLC environment is often too high for long-term survivability of the particles, regardless of the support they are impregnated on.

Also, the melting point of the metal or metal oxide particles presents a temperature limit for CLC processes. If this temperature limit is crossed, control over the size and distribution of the particles and particle morphology will be lost. The Tamman temperature, an empirical temperature of the material which often approximated as half the melting temperature, is the temperature at which

Electrified Non-Equilibrium Chemical Manufacturing

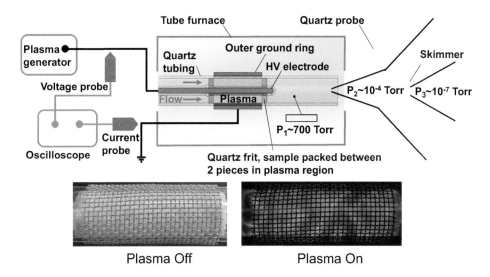

FIGURE 8.43 Schematic of experimental setup of DBD-assisted CLC and images of CLC with and without plasma.

motion of the ions in the crystalline lattice of the solid is sufficient for sintering and agglomeration to occur between particles in significant amounts [99]. As a result, the CLC process that operates below the melting point of the materials yet above the Tamman temperature will result in long-term agglomeration and eventual deactivation of the particles involved. Therefore, for the survivability of the particles, a low temperature is preferred. However, at lower temperatures reaction becomes limited. This is the fundamental problem related to the use of metal oxide particles in CLC: at high temperatures, there is significant reaction but the particles agglomerate, whereas at low temperature there is no agglomeration, but the reaction is limited. One potential technology that would allow for increased reaction at low temperatures is non-equilibrium plasma.

Plasma-assisted low-temperature CLC has great potential to reduce the temperature and accelerate fuel oxidation on the surface of metal oxides. Monazam et al. [100] studied the oxidation of CH_4 by CuO. Goldstein and Mitchell [101] described the oxidation of CuO with CO. However, a mechanistic understanding of the plasma-assisted gas-solid reactions in CLC is still missing. To understand the mechanism of plasma-assisted CLC, Burger et al. [87] studied DBD plasma-assisted methane oxidation on CuO in CLC by using time-resolved species measurements using EI-MBMS.

Experimental setup of plasma-assisted CLC is shown in Figure 8.43 [87]. The fixed-bed, cylindrical, double DBD plasma reactor was composed of a 17 mm inner diameter and 419 mm in length quartz tube. An inner quartz tube that is 6.35 mm in diameter contains the inner HV electrode. The plasma output power was adjusted from 10 to 300 W with frequency control from 20 to 70 kHz. Quartz wool containing CuO particles is placed inside the plasma zone and reactant gases were flown through the CuO particles.

Figure 8.44 shows the normalized CO_2 signals without and with plasma of 50 sccm of 10% methane reacting with 1 g of CuO at various temperatures. For the non-plasma conditions (left), no CO_2 was detected at 300°C and only minor amounts produced at 400°C. There was a sharp increase in CO_2 for 500°C and 600°C. Three existed distinct reaction stages in CLC, i.e., a gas phase transport limited reaction stage, a surface site limited kinetic stage, and a solid-phase oxygen ion diffusion stage, were observed. At 700°C, there was a noticeable increase of CO_2 production beyond 1,000 s (the last reaction stage), indicating the increase of oxygen (vacancy) diffusion from the core of the particle caused by the higher temperature. With the presence of plasma, it is seen that there was a significant change in the CO_2 profiles. At 300°C with plasma, the CO_2 signal was non-zero and

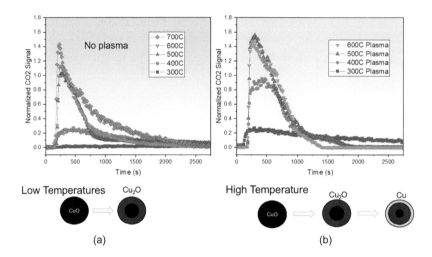

FIGURE 8.44 (a) CO_2 Signal from measurement of 50 sccm of 10% CH_4 reacting with 1 g of CuO at various temperatures without plasma. (b) Time-dependent species measurement of 50 sccm of 10% Methane reacting with 1 g of CuO at 600 °C with plasma [87].

approximately equivalent to the CO_2 production in the non-plasma condition at 400°C. At 400°C with plasma, a moderate amount of CO_2 was produced, and CO_2 production extended beyond 1,000 s while the non-plasma case only existed up to 750 s. One explanation for the broadening of the surface reaction limited kinetic stage in the presence of plasma involved the plasma-enhanced reduction of CuO and Cu_2O. With Cu_2O, another stable oxidation state occurs. As CuO was reduced, the reduction followed the following steps:

$$CuO \rightarrow Cu_2O \rightarrow Cu \qquad (R8.47)$$

At lower temperatures without plasma, CuO was not fully reduced to Cu by CH_4 due to high energy barrier. However, with the plasma present, the excited state of $CH_4(*)$ and the radical species such as CH_3 and H had lower energy barriers to react with O^{2-} ions from the particle surface. This resulted in a greater conversion of CuO to Cu and thus an extension of CuO reduction in time. By 500°C with plasma, a significant increase in the amount of CO_2 was observed, with the peak CO_2 signal even being larger than that of 700°C under non-plasma conditions. Therefore, plasma discharge had an effective benefit equivalent to a 200+°C reduction in reactor temperature.

More recently, the chemical reaction network of low-temperature plasma-assisted oxidation of methane (CH_4) and ethylene (C_2H_4) with nickel oxide (NiO) was investigated in temperature ranges from 300°C to 700°C [102]. Significant enhancement of methane oxidation was observed with plasma between 400°C and 500°C, where no oxidation was observed under non-plasma conditions. The effect of the oxidation of CH_4 by NiO without and with plasma on is shown in Figure 8.45. In the absence of plasma, NiO was unable to achieve methane oxidation at low temperatures, with minimal CO_2 being detected until the reactor was brought up to 600°C. However, with the presence of plasma, notable conversion to CO_2 was detected at temperatures as low as 400°C, a significant reduction from the non-plasma conditions. At a reactor temperature of 700°C, the differences between the plasma and non-plasma cases were smaller than at lower temperatures. This is because at higher temperature, NiO can oxidize CH_4 without the presence of plasma. Thus, the largest benefit that the plasma can contribute to the oxidation process on NiO occurred at the lower temperatures of 400°C–500°C. DFT-based reactive molecular dynamics simulations (ReaxFF) simulations were also preformed to create an experimentally validated oxidation reaction schematic which included

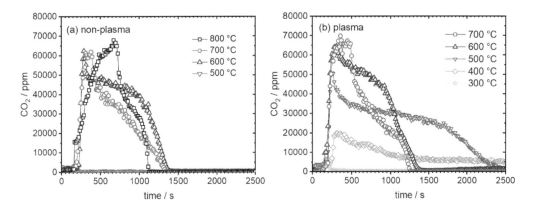

FIGURE 8.45 CO_2 time profiles from oxidation of 50 sccm of 10% methane with 1 g NiO at various temperatures, with and without the presence of plasma [102]. (a) Non-plasma and (b) plasma.

both CH_4 and C_2H_4. The simulations were able to largely predict the experimentally observed species, but incorrectly predicted intermediate oxygenated species such as CH_2O, CH_3OH, and C_2H_6O, which were only observable in the presence of plasma. Therefore, although these experimental studies demonstrated that plasma can dramatically reduce CLC temperature and accelerate oxidation, more detailed plasma-assisted CLC mechanism studies are needed.

8.5 PLASMA FOR MATERIALS SYNTHESIS AND WASTE MATERIALS RECYCLING

Non-equilibrium chemistry and a broad range of temperatures in plasma provide unprecedented opportunities in electrified materials manufacturing and waste resource recycling. Today, chemical processing in producing materials like cement, steel, plastics, and electrode materials remains heavily reliant on fossil energy. In addition, there are increasing waste materials such as plastics and used batteries which have created serious environmental concerns. For example, prolific plastic use has resulted in a surge of waste plastics and associated environmental concerns. To date, there exist ~8.3 billion metric tons of plastics globally [4], among which ~75% are wastes that can dramatically harm the environment if not properly treated. However, waste plastics offer alternative sources of high-value chemical products [103]. Existing steady-state plastic thermal pyrolysis-gasification via catalytic reforming at high temperature (~800°C) is not only carbon and energy intensive but also causes carbonaceous coke deposition on the catalyst due to chemical equilibrium. Thus, as discussed in Section 8.1.2, there is a need to develop a novel electrified non-equilibrium process for plastic pyrolysis using low-temperature plasma catalysis to increase selectivity and efficiency and suppress coke formation. Similarly, by 2025, it is estimated that the global lithium-ion battery market will reach 10 million tons of cathode materials. Therefore, by 2035 these large amounts of used or aged cathode materials need to be recycled. However, plasma-assisted materials manufacturing (except for semiconducting materials) is still a young research area. In the section below, we will highlight the potential of plasma-assisted recycling of battery electrode materials and manufacturing of extreme materials only. More commercial applications in plasma-assisted manufacturing and recycling will be developed in the next decades.

8.5.1 Plasma-Assisted Waste Resources Recycling and Upcycling of Battery Electrode Materials

With the rapid increase of EVs and lithium-ion batteries and the rise of electrode materials such as lithium, nickel, and cobalt [104], these aged lithium-ion batteries will need to be recycled. However, the current battery recycling technology heavily relies on the pyrolysis and mechanical-hydro-metallurgical processes and requires a series of thermal treatment and chemical leaching to break down the cathode materials into atomic form by using H_2SO_4. As such, these recycling processes are very energy and chemicals intensive, and environmentally unfriendly. Therefore, there is an urgent need to develop a novel method using combined electric heating and plasma that directly upgrades the aged low energy density cathode materials without breaking down the materials structure and producing wastewater [105].

Researchers at Princeton NuEnergy Inc. [105] have developed and patented a novel plasma-assisted recycling method for aged cathode powders by removing fluorine (f) in the form of metal fluorides (MF_x) in the electrode metal oxide materials (MO_x: example $Ni_{1-x-y}Mn_xCo_yO_2$) and regenerating the original cathode material (Figure 8.46). The novelty of this direct recycling method is the active species such as O and O(^{1}D) production from plasma to remove F (from fluorine-based electrolyte) formed on the surface of used batteries. In addition, as shown in Figure 8.46, in the plasma-assisted recycling and regenerative process, low energy capacity aged $Ni_{0.4}Mn_{0.3}Co_3O_2$ can be upgraded to high energy capacity $Ni_{0.8}Mn_{0.1}Co_{0.1}O_2$ (NCM811) electrode materials. As such, this plasma-aided direct recycling method can be a game-changer because it does not require any acid treatment or produce wastewater.

To demonstrate the potential of plasma-assisted electrode materials recycling, preliminary experiments of plasma-assisted battery electrode material $LiCoO_2$ (LCO) recycling were conducted. Figure 8.47 shows the comparison of X-Ray diffraction analysis (XRD) spectra, capacity, and cycling between aged, commercial, and recycled electrode materials. Figure 8.47 shows clearly that the recycled LCO has a sharp XRD spectra indicating an excellent crystalline layered structure. In addition, the recycled LCO shows the same capacity performance in charging and discharge.

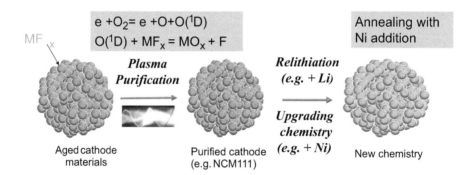

FIGURE 8.46 Schematic of plasma-assisted lithium-ion battery cathode materials recycling and to remove fluorine ion on the surface of used electrode materials and upgrade the NCM111 to NCM811 [105].

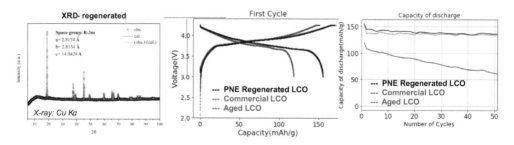

FIGURE 8.47 Comparison between aged, commercial, and regenerated lithium nickel oxide (LCO). (Courtesy to Princeton NuEnergy (PNE).)

Furthermore, the capacity cycling test also demonstrated that the recycled materials have comparable cycling performance to the commercial materials and are much better than the aged one. These preliminary data demonstrate that non-equilibrium plasma may play a critical role in recycling of aged batteries and other waste materials.

8.5.2 Plasma-Assisted Materials Synthesis

Electrified materials manufacturing is important to reduce CO_2 emissions in industrial processing [106] such as cement, steel, glasses, electrochemical materials, and new materials. As discussed above, the increased reactivity of plasma provides a great advantage in materials synthesis and may create new reaction pathways using non-equilibrium chemical processes. Moreover, the high temperature of plasma also provides a great opportunity to manufacture novel and extreme materials. Here we show an example of a newly developed fiber stabilized ultrahigh-temperature and stable plasma (USP) for extreme high-temperature materials synthesis.

It is well known that plasma can be used for the synthesis of a wide range of materials [107]. However, the limited volume, instability, and non-uniformity of plasmas have made it challenging to scalably manufacture bulk, high-temperature materials. In a recent study by Xie et al. [107], a new plasma platform consisting of a pair of carbon-fiber-tip enhanced electrodes that enable the generation of a uniform, ultrahigh-temperature and stable plasma (USP, up to 8,000 K) at atmospheric pressure via a combination of vertically oriented long and short carbon fibers was developed and used for extreme materials synthesis.

The uniform, ultrahigh-temperature (up to 8,000 K), stable plasma (USP) setup, and test results are shown in Figure 8.48. USP consists of a pair of carbon tip-enhanced electrodes (Figure 8.48a and b). The electrodes are composed of a high density (~$10^5/cm^2$) of short, vertically oriented carbon fibers (~10 μm diameter), as well as some long carbon fibers that extend into the gap between the two electrodes and form contact. Under an applied voltage, Joule heating is intensified at defective regions or contact points of the long carbon fibers, where the resistance is highest, reaching an ultrahigh temperature until the fibers break, creating very small micron sized gaps. The locally enhanced electric field at these newly formed fiber tips promotes secondary electron emission, resulting in a spark discharge across the narrow inter-fiber gaps that helps initiation of the plasma at a record-low breakdown voltage. Meanwhile, the densely spaced short carbon fibers, as shown in the scanning electron microscopy (SEM) image of Figure 8.48c, produce tip-enhanced electric fields that merge across the surface of the electrodes (Figure 8.48d), accelerating the Townsend breakdown to arc transition, expanding the plasma size and volume, and increasing the plasma uniformity. Therefore, a continuous, volumetric plasma featuring a highly controllable temperature of between 3,000 and 8,000 K as well as a uniform temperature distribution (Figure 8.48e) is established.

This continuous, volumetric, uniform, and stable ultrahigh-temperature plasma was employed for the synthesis of various high-temperature materials [107] (Figure 8.49). For example, USP was used to synthesize and sinter Hf(C,N) (Figure 8.49a and b), an ultrahigh-temperature ceramic that is very challenging to synthesize due to its record high melting point (>4,000 K). The cross-sectional SEM images of the HfC/HfN pellet before (Figure 8.49c) and after (Figure 8.49d) the USP treatment (5,150 K for 10 s) demonstrated the successful sintering of the powder precursor mixture. After the USP heating, XRD confirmed that a predominantly Hf(C,N) single phase of the rock salt crystal structure was successfully achieved (Figure 8.49e). USP was also employed to synthesize a tungsten-based refractory alloy directly from metallic elemental powders (e.g., $W_{1.5}Nb_{0.5}Ti$) to form a dense alloy with more uniform distribution of the elements than the compositional inhomogeneity observed when using conventional arc melting. USP was also demonstrated to generate high-value carbon materials, such as CNTs, simply by heating carbon black without any catalysts.

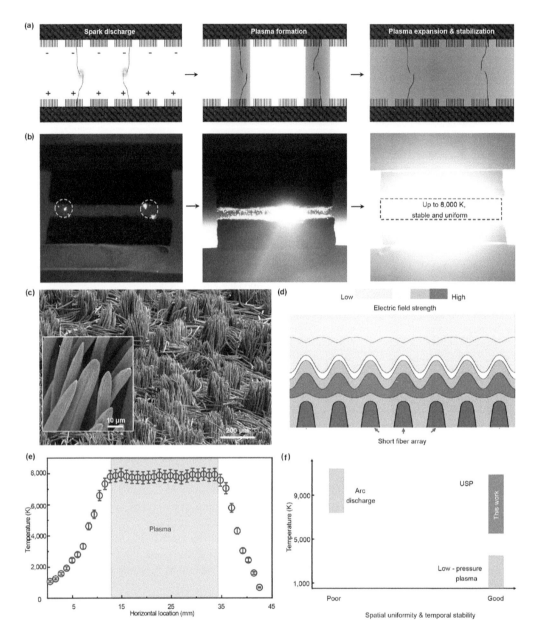

FIGURE 8.48 A uniform, ultrahigh-temperature stable plasma (USP) at atmospheric pressure enabled by a novel electrode design featuring long carbon fibers and a densely spaced array of numerous, small-diameter carbon fiber tips. (a) Schematic and (b) corresponding photographs of the volumetric plasma generation via the carbon fiber tip electrodes. (c) SEM image of the short fiber tip array on the surface of the carbon felt electrode. The inset is a zoomed-in image showing the fibers. (d) Schematic of the short carbon fibers, showing the tip-enhanced electric field distribution when the plasma discharge breakdown takes place, in which the carbon tip array spreads and strengthens the local electric field intensity across the electrode surface. (e) The temperature profile of the plasma along the carbon-felt electrodes, which can reach temperatures as high as ~8,000 K with excellent spatial uniformity. Error bars show s.d.; $n = 50$. (f) Temperature and uniformity comparison of USP with other plasma techniques using similar level of current input, showing the advantage of USP in generating a steady, large-area atmospheric plasma with high temperatures (up to ~8,000 K) [107].

Electrified Non-Equilibrium Chemical Manufacturing 561

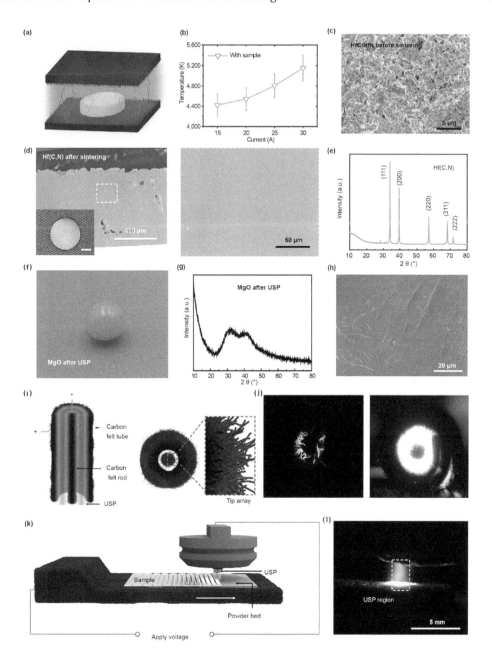

FIGURE 8.49 Application of USP for the synthesis of various high-temperature materials. (a) Schematic demonstrating the use of USP to synthesize and sinter a ceramic pellet. (b) The temperature of the plasma center with the HfC/HfN pellet at an input current of 15–30 A, which is consistent with the plasma temperature without samples present (Fig.8.48f). Error bars show s.d.; $n = 50$. (c) Cross-sectional SEM image of the HfC/HfN precursor pellet. (d) Cross-sectional SEM image of the resulting Hf(C,N) after USP sintering. The inset photo shows the Hf(C,N) pellet after plasma sintering (scale bar: 5 mm). (e) XRD pattern of the synthesized Hf(C,N). (f) Photo of the amorphous MgO synthesized from its crystalline powder via USP rapid heating and quenching. (g) XRD pattern of the MgO achieved via the USP method. (h) Cross-sectional SEM image of the amorphous MgO phase showing no obvious grain boundaries. (i) Schematic of a coaxial carbon felt rod-tube USP setup, which can be used to form (j) a long, volumetric plasma channel. (k) Schematic of a focused USP beam that can be used in a powder bed fusion/sintering process. (l) The plasma beam can be focused into a filament with a column radius of ~1 mm [107].

Therefore, the above example demonstrated that plasma provides an unprecedented opportunity to enable the processing of a wide range of materials and synthesis of novel extreme materials to decarbonize the current industrial fossil fuel-based materials manufacturing processes.

REFERENCES

1. Smith, C., A.K. Hill, and L. Torrente-Murciano, Current and future role of Haber–Bosch ammonia in a carbon-free energy landscape. *Energy & Environmental Science*, 2020. **13**(2): pp. 331–344.
2. Rostrup-Nielsen, J.R., J. Sehested, and J.K. Nørskov, Hydrogen and synthesis gas by steam-and CO_2 reforming. *Advances in Catalysis*, 2002. **47**: pp. 65–139.
3. Spath, P.L. and M.K. Mann, *Life cycle assessment of hydrogen production via natural gas steam reforming*. 2000, National Renewable Energy Lab (NREL), Golden, CO (United States).
4. Dong, Q., et al., Programmable heating and quenching for efficient thermochemical synthesis. *Nature*, 2022. **605**(7910): pp. 470–476.
5. Denbigh, K.G., *The Principles of Chemical Equilibrium: With Applications in Chemistry and Chemical Engineering*. 1981: Cambridge University Press.
6. Saadatjou, N., A. Jafari, and S. Sahebdelfar, Ruthenium nanocatalysts for ammonia synthesis: A review. *Chemical Engineering Communications*, 2015. **202**(4): pp. 420–448.
7. Dong, Q., et al., Depolymerization of plastics by means of electrified spatiotemporal heating. *Nature*, 2023. **616**(7957): pp. 488–494.
8. Ellis, L.D., et al., Chemical and biological catalysis for plastics recycling and upcycling. *Nature Catalysis*, 2021. **4**: pp. 539–556.
9. Kots, P.A., et al., Polypropylene plastic waste conversion to lubricants over Ru/TiO_2 catalysts. *ACS Catalysis*, 2021. **11**: pp. 8104–8115.
10. Lin, Y.-H. and P.N. Sharratt, Catalytic conversion of polyolefins to chemicals and fuels over various cracking catalysts. *Energy & Fuels*, 1998. **12**: pp. 767–774.
11. Scotp, D.S., et al., Production of liquid fuels from waste plastics. *The Canadian Journal of Chemical Engineering*, 1999. **77**: pp. 1021–1027.
12. Shabtai, J., X. Xiao, and W. Zmierczak, Depolymerization-liquefaction of plastics and rubbers. 1. Polyethylene, polypropylene, and polybutadiene. *Energy & Fuels*, 1997. **11**: pp. 76–87.
13. Lovás, P., et al., Catalytic cracking of heavy fractions from the pyrolysis of waste HDPE and PP. *Fuel*, 2017. **203**: pp. 244–252.
14. Encinar, J.M. and J.F. González, Pyrolysis of synthetic polymers and plastic wastes. Kinetic study. *Fuel Processing Technology*, 2008. **89**: pp. 678–686.
15. Rorrer, J.E., et al., Hydrogenolysis of polypropylene and mixed polyolefin plastic waste over Ru/C to produce liquid alkanes. *ACS Sustainable Chemistry & Engineering*, 2021. **9**: pp. 11661–11666.
16. Arandes, J.M., et al., Valorization by thermal cracking over silica of polyolefins dissolved in LCO. *Fuel Processing Technology*, 2004. **85**: pp. 125–140.
17. Lu, H., et al., Machine learning-aided engineering of hydrolases for PET depolymerization. *Nature*, 2022. **604**: pp. 662–667.
18. Dimitrov, N., et al., Analysis of recycled PET bottles products by pyrolysis-gas chromatography. *Polymer Degradation and Stability*, 2013. **98**: pp. 972–979.
19. Yoshioka, T., et al., Pyrolysis of poly(ethylene terephthalate) in a fluidised bed plant. *Polymer Degradation and Stability*, 2004. **86**: pp. 499–504.
20. Jia, H., et al., Catalytic fast pyrolysis of poly (ethylene terephthalate) (PET) with zeolite and nickel chloride. *Polymers*, 2020. **12**: p. 705.
21. Diaz-Silvarrey, L.S., A. McMahon, and A.N. Phan, Benzoic acid recovery via waste poly(ethylene terephthalate) (PET) catalytic pyrolysis using sulphated zirconia catalyst. *Journal of Analytical and Applied Pyrolysis*, 2018. **134**: pp. 621–631.
22. Wang, Z., et al., Co-pyrolysis of waste plastic and solid biomass for synergistic production of biofuels and chemicals: A review. *Progress in Energy and Combustion Science*, 2021. **84**: p. 100899.
23. Mastral, F., et al., Pyrolysis of high-density polyethylene in a fluidised bed reactor. Influence of the temperature and residence time. *Journal of Analytical and Applied Pyrolysis*, 2002. **63**(1): pp. 1–15.
24. Lele, A.D. and Y. Ju, Assessment of the impact of reactor residence time distribution on non-equilibrium product selectivity of polypropylene pyrolysis using reactive molecular dynamics simulations. *Fuel*, 2023. **338**: p. 127328.

25. Yang, S., X. Liu, and S. Wang, CFD simulation of air-blown coal gasification in a fluidized bed reactor with continuous feedstock. *Energy Conversion and Management*, 2020. **213**: p. 112774.
26. Lu, L., et al., Multiscale CFD simulation of biomass fast pyrolysis with a machine learning derived intra-particle model and detailed pyrolysis kinetics. *Chemical Engineering Journal*, 2022. **431**: p. 133853.
27. Hu, C., et al., Computational fluid dynamics/discrete element method investigation on the biomass fast pyrolysis: The influences of shrinkage patterns and operating parameters. *Industrial & Engineering Chemistry Research*, 2018. **58**(3): pp. 1404–1416.
28. Cesario, M.R. and D.A. de Macedo, *Heterogeneous Catalysis: Materials and Applications*. 2022: Elsevier.
29. Ross, J.R., *Heterogeneous Catalysis: Fundamentals and Applications*. 2011: Elsevier.
30. Lefferts, L., Catalysis expertise for understanding plasma-catalysis. *2023: Heraeus Seminar*, Bad Honnef, April 2023.
31. Zhu, X., et al., Chemical looping beyond combustion: A perspective. *Energy & Environmental Science*, 2020. **13**(3): pp. 772–804.
32. Wang, P., et al., Bond dissociation energy of N_2 measured by state-to-state resolved threshold fragment yield spectra. *The Journal of Chemical Physics*, 2024. **160** (1), 014304.
33. Rouwenhorst, K.H. and L. Lefferts, On the mechanism for the plasma-activated N_2 dissociation on Ru surfaces. *Journal of Physics D: Applied Physics*, 2021. **54**(39): p. 393002.
34. Bligaard, T., et al., The Brønsted–Evans–Polanyi relation and the volcano curve in heterogeneous catalysis. *Journal of Catalysis*, 2004. **224**(1): pp. 206–217.
35. Mehta, P., et al., Overcoming ammonia synthesis scaling relations with plasma-enabled catalysis. *Nature Catalysis*, 2018. **1**(4): p. 269.
36. Tricker, A.W., et al., Mechanocatalytic ammonia synthesis over TiN in transient microenvironments. *ACS Energy Letters*, 2020. **5**(11): pp. 3362–3367.
37. Kobayashi, H., et al., Science and technology of ammonia combustion. *Proceedings of the Combustion Institute*, 2019. **37**(1): pp. 109–133.
38. Gorky, F., et al., Plasma ammonia synthesis over mesoporous silica SBA-15. *Journal of Physics D: Applied Physics*, 2021. **54**(26): p. 264003.
39. Erisman, J.W., et al., How a century of ammonia synthesis changed the world. *Nature Geoscience*, 2008. **1**(10): p. 636.
40. Vojvodic, A., et al., Exploring the limits: A low-pressure, low-temperature Haber–Bosch process. *Chemical Physics Letters*, 2014. **598**: pp. 108–112.
41. Yang, X., C. Richards, and I.V. Adamovich, Ammonia generation in Ns pulse and Ns pulse/RF discharges over a catalytic surface. Plasma Sources Science and Technology, 2023.
42. Barboun, P.M., et al., Inelastic neutron scattering observation of plasma-promoted nitrogen reduction intermediates on Ni/γ-Al_2O_3. *ACS Energy Letters*, 2021. **6**(6): pp. 2048–2053.
43. Zhao, H., et al., In situ identification of NNH and N_2H_2 by using molecular-beam mass spectrometry in plasma-assisted catalysis for NH_3 synthesis. *ACS Energy Letters*, 2021. **7**: pp. 53–58.
44. Li, S., et al., Recent progress of plasma-assisted nitrogen fixation research: A review. *Processes*, 2018. **6**(12): p. 248.
45. Rusanov, V.D., A. Fridman, and G.V.E. Sholin, The physics of a chemically active plasma with nonequilibrium vibrational excitation of molecules. *Soviet Physics Uspekhi*, 1981. **24**(6): p. 447.
46. Zhong, H., et al. Kinetic studies of low-temperature ammonia oxidation in a nanosecond repetitively-pulsed discharge. in *AIAA SCITECH 2023 Forum*, Jan. 23-27, 2023, National Harbor, Maryland, 2023.
47. Wang, W., et al., Nitrogen fixation by gliding arc plasma: Better insight by chemical kinetics modelling. *ChemSusChem*, 2017. **10**(10): pp. 2145–2157.
48. Jardali, F., et al., NO_x production in a rotating gliding arc plasma: Potential avenue for sustainable nitrogen fixation. *Green Chemistry*, 2021. **23**(4): pp. 1748–1757.
49. Hermesmann, M. and T. Müller, Green, turquoise, blue, or grey? Environmentally friendly hydrogen production in transforming energy systems. *Progress in Energy and Combustion Science*, 2022. **90**: p. 100996.
50. Chase, M.W. and N.I.S. Organization, *NIST-JANAF Thermochemical Tables*, Vol. 9. 1998: American Chemical Society.
51. Diab, J., et al., Why turquoise hydrogen will be a game changer for the energy transition. *International Journal of Hydrogen Energy*, 2022. **47**(61): pp. 25831–25848.

52. Zheng, X., et al., Silica-coated LaNiO₃ nanoparticles for non-thermal plasma assisted dry reforming of methane: Experimental and kinetic studies. *Chemical Engineering Journal*, 2015. **265**: pp. 147–156.
53. Chen, Q., et al., Pyrolysis and oxidation of methane in a RF plasma reactor. *Plasma Chemistry and Plasma Processing*, 2017. **37**(6): pp. 1551–1571.
54. Ravari, F., et al., Kinetic model study of dry reforming of methane using cold plasma. *Physical Chemistry Research*, 2017. **5**(2): pp. 395–408.
55. Li, D., et al., CO_2 reforming of CH_4 by atmospheric pressure glow discharge plasma: A high conversion ability. *International Journal of Hydrogen Energy*, 2009. **34**(1): pp. 308–313.
56. Garduño, M., et al., Hydrogen production from methane conversion in a gliding arc. *Journal of Renewable and Sustainable Energy*, 2012. **4**(2), 021202.
57. Zhang, H., et al., Plasma activation of methane for hydrogen production in a N_2 rotating gliding arc warm plasma: A chemical kinetics study. *Chemical Engineering Journal*, 2018. **345**: pp. 67–78.
58. Kreuznacht, S., et al., Comparison of the performance of a microwave plasma torch and a gliding arc plasma for hydrogen production via methane pyrolysis. *Plasma Processes and Polymers*, 2023. **20**(1): p. 2200132.
59. Ozkan, A., A. Bogaerts, and F. Reniers, Routes to increase the conversion and the energy efficiency in the splitting of CO_2 by a dielectric barrier discharge. *Journal of Physics D: Applied Physics*, 2017. **50**(8): p. 084004.
60. Spencer, L. and A. Gallimore, CO_2 dissociation in an atmospheric pressure plasma/catalyst system: A study of efficiency. *Plasma Sources Science and Technology*, 2012. **22**(1): p. 015019.
61. Yong, T., et al., High-pressure CO_2 dissociation with nanosecond pulsed discharges. *Plasma Sources Science and Technology*, 2023. **32**(11): p. 115012.
62. Nunnally, T., et al., Dissociation of CO_2 in a low current gliding arc plasmatron. *Journal of Physics D: Applied Physics*, 2011. **44**(27): p. 274009.
63. Li, L., et al., Magnetically enhanced gliding arc discharge for CO_2 activation. *Journal of CO_2 Utilization*, 2020. **35**: pp. 28–37.
64. Bogaerts, A., et al., Plasma-based conversion of CO_2: Current status and future challenges. *Faraday Discussions*, 2015. **183**: pp. 217–232.
65. Uytdenhouwen, Y., et al., How process parameters and packing materials tune chemical equilibrium and kinetics in plasma-based CO_2 conversion. *Chemical Engineering Journal*, 2019. **372**: pp. 1253–1264.
66. King, B., et al., Comprehensive process and environmental impact analysis of integrated DBD plasma steam methane reforming. *Fuel*, 2021. **304**: p. 121328.
67. Chen, X., et al., CH_4 dry reforming in fluidized-bed plasma reactor enabling enhanced plasma-catalyst coupling. *Journal of CO_2 Utilization*, 2021. **54**: p. 101771.
68. Wang, L., et al., One-step reforming of CO_2 and CH_4 into high-value liquid chemicals and fuels at room temperature by plasma-driven catalysis. *Angewandte Chemie*, 2017. **129**(44): pp. 13867–13871.
69. Snoeckx, R. and A. Bogaerts, Plasma technology: A novel solution for CO_2 conversion? *Chemical Society Reviews*, 2017. **46**(19): pp. 5805–5863.
70. Zhang, J., et al., Study on the conversion of CH_4 and CO_2 using a pulsed microwave plasma under atmospheric pressure. *Acta Chimica Sinica*, 2002. **60**(11): p. 1973.
71. Martin-del-Campo, J., S. Coulombe, and J. Kopyscinski, Influence of operating parameters on plasma-assisted dry reforming of methane in a rotating gliding arc reactor. *Plasma Chemistry and Plasma Processing*, 2020. **40**(4): pp. 857–881.
72. Chun, Y.N., et al., Hydrogen-rich gas production from biogas reforming using plasmatron. *Energy & Fuels*, 2008. **22**(1): pp. 123–127.
73. Scapinello, M., L.M. Martini, and P. Tosi, CO_2 hydrogenation by CH_4 in a dielectric barrier discharge: Catalytic effects of nickel and copper. *Plasma Processes and Polymers*, 2014. **11**(7): pp. 624–628.
74. Chen, G., et al., Simultaneous dissociation of CO_2 and H_2O to syngas in a surface-wave microwave discharge. *International Journal of Hydrogen Energy*, 2015. **40**(9): pp. 3789–3796.
75. Ihara, T., M. Kiboku, and Y. Iriyama, Plasma reduction of CO_2 with H_2O for the formation of organic compounds. *Bulletin of the Chemical Society of Japan*, 1994. **67**(1): pp. 312–314.
76. Futamura, S. and H. Kabashima, Synthesis gas production from CO_2 and H_2O with nonthermal plasma, *Studies in Surface Science and Catalysis*. **153**, pp. 119–124, 2004.
77. Mahammadunnisa, S., et al., CO_2 reduction to syngas and carbon nanofibres by plasma-assisted in situ decomposition of water. *International Journal of Greenhouse Gas Control*, 2013. **16**: pp. 361–363.
78. Maya, L., Plasma-assisted reduction of carbon dioxide in the gas phase. *Journal of Vacuum Science & Technology A: Vacuum, Surfaces, and Films*, 2000. **18**(1): pp. 285–287.

79. de la Fuente, J.F., et al., Reduction of CO_2 with hydrogen in a non-equilibrium microwave plasma reactor. *International Journal of Hydrogen Energy*, 2016. **41**(46): pp. 21067–21077.
80. Kim, D.-Y., et al., Cooperative catalysis of vibrationally excited CO_2 and alloy catalyst breaks the thermodynamic equilibrium limitation. *Journal of the American Chemical Society*, 2022. **144**(31): pp. 14140–14149.
81. Jwa, E., et al., Plasma-assisted catalytic methanation of CO and CO_2 over Ni–zeolite catalysts. *Fuel Processing Technology*, 2013. **108**: pp. 89–93.
82. Nizio, M., et al., Hybrid plasma-catalytic methanation of CO_2 at low temperature over ceria zirconia supported Ni catalysts. *International Journal of Hydrogen Energy*, 2016. **41**(27): pp. 11584–11592.
83. Zhan, C., et al., Nonthermal plasma catalysis of CO_2 methanation over multi-metallic Ru based catalysts. *International Journal of Plasma Environmental Science and Technology*, 2022. **16**: p. e03006.
84. Eliasson, B., et al., Hydrogenation of carbon dioxide to methanol with a discharge-activated catalyst. *Industrial & Engineering Chemistry Research*, 1998. **37**(8): pp. 3350–3357.
85. Wang, L., et al., Atmospheric pressure and room temperature synthesis of methanol through plasma-catalytic hydrogenation of CO_2. *ACS Catalysis*, 2018. **8**(1): pp. 90–100.
86. Adanez, J., et al., Progress in chemical-looping combustion and reforming technologies. *Progress in Energy and Combustion Science*, 2012. **38**(2): pp. 215–282.
87. Burger, C.M., et al., Plasma-assisted chemical-looping combustion: Mechanistic insights into low temperature methane oxidation with CuO. *39th International Symposium on Combustion*, 2022. **39**: pp. 1–8.
88. Burger, C.M., et al., Experimental and computational investigations of ethane and ethylene kinetics with copper oxide particles for chemical looping combustion. *Proceedings of the Combustion Institute*, 2021. **38**(4): pp. 5249–5257.
89. Lyngfelt, A., et al., 11,000 h of chemical-looping combustion operation: Where are we and where do we want to go? *International Journal of Greenhouse Gas Control*, 2019. **88**: pp. 38–56.
90. Mattisson, T., M. Johansson, and A. Lyngfelt, The use of NiO as an oxygen carrier in chemical-looping combustion. *Fuel*, 2006. **85**(5): pp. 736–747.
91. Zhu, W., et al., Mechanistic study of chemical looping reactions between solid carbon fuels and CuO. *Combustion and Flame*, 2022. **244**: p. 112216.
92. Abad, A., et al., Mapping of the range of operational conditions for Cu-, Fe-, and Ni-based oxygen carriers in chemical-looping combustion. *Chemical Engineering Science*, 2007. **62**(1): pp. 533–549.
93. Dou, B., et al., Solid sorbents for in-situ CO_2 removal during sorption-enhanced steam reforming process: A review. *Renewable and Sustainable Energy Reviews*, 2016. **53**: pp. 536–546.
94. Ishida, M. and H. Jin, A novel chemical-looping combustor without NO_x formation. *Industrial & Engineering Chemistry Research*, 1996. **35**(7): pp. 2469–2472.
95. de Diego, L.F., et al., Impregnated CuO/Al_2O_3 oxygen carriers for chemical-looping combustion: Avoiding fluidized bed agglomeration. *Energy & Fuels*, 2005. **19**(5): pp. 1850–1856.
96. Gayán, P., et al., Effect of support on reactivity and selectivity of Ni-based oxygen carriers for chemical-looping combustion. *Fuel*, 2008. **87**(12): pp. 2641–2650.
97. San Pio, M.A., et al., Kinetics of CuO/SiO_2 and CuO/Al_2O_3 oxygen carriers for chemical looping combustion. *Chemical Engineering Science*, 2018. **175**: pp. 56–71.
98. Rydén, M., et al., NiO supported on $Mg-ZrO_2$ as oxygen carrier for chemical-looping combustion and chemical-looping reforming. *Energy & Environmental Science*, 2009. **2**(9): pp. 970–981.
99. Ertl, G., H. Knözinger, and J. Weitkamp, *Preparation of Solid Catalysts*. 2008: John Wiley & Sons.
100. Monazam, E.R., et al., Kinetics of the reduction of CuO/bentonite by methane (CH_4) during chemical looping combustion. *Energy & Fuels*, 2012. **26**(5): pp. 2779–2785.
101. Goldstein, E.A. and R.E. Mitchell, Chemical kinetics of copper oxide reduction with carbon monoxide. *Proceedings of the Combustion Institute*, 2011. **33**(2): pp. 2803–2810.
102. Burger, C.M., et al., Plasma-assisted chemical-looping combustion: Low-temperature methane and ethylene oxidation with nickel oxide. *The Journal of Physical Chemistry A*, 2023. **127**(3): pp. 789–798.
103. Wilk, V. and H. Hofbauer, Conversion of mixed plastic wastes in a dual fluidized bed steam gasifier. *Fuel*, 2013. **107**: pp. 787–799.
104. Rothermel, S., et al., Graphite recycling from spent lithium-ion batteries. *ChemSusChem*, 2016. **9**(24): pp. 3473–3484.
105. Koel, B., et al., Recycling lithium-ion batteries, 2021.
106. Spreitzer, D. and J. Schenk, Reduction of iron oxides with hydrogen: A review. *Steel Research International*, 2019. **90**(10): p. 1900108.
107. Xie, H., et al., A stable atmospheric-pressure plasma for extreme-temperature synthesis. *Nature*, 2023. **623**(7989): pp. 964–971.

9 Plasma Diagnostics

Plasma dynamics and chemistry have a broad range of timescales from picoseconds to milliseconds. In addition, it involves nonequilibrium energy distribution, electrons, ions, electronically and vibrationally excited states, radicals, intermediate species, reactants and products, as well as surface charges and species. To understand plasma physics and chemistry, it is essential to conduct time and space resolved, quantitative detection of nonequilibrium temperature distributions, electron energy and number density, electric field, and species concentrations. There are enormous publications and review articles on this subject. The focus of this chapter is to be placed on the most recent progress in gas phase plasma properties and chemistry, especially on optical emission spectroscopy, laser absorption spectroscopy (LAS), Faraday rotational spectroscopy, Raman and Thompson scattering, femtosecond and picosecond (fs/ps) coherent anti-Stokes Raman scattering (CARS) spectroscopy, and electric field-induced second harmonic generation methods.

9.1 OPTICAL EMISSION SPECTROSCOPY

Optical emission spectroscopy (OES) for plasma diagnostics is based on the detection and collection of emitted light photons from rotationally, vibrationally, and electronically excited states in plasmas (Figure 9.1). OES serves as a crucial diagnostic tool in plasma and reactive flows. OES has been widely used for monitoring/detecting excited species and characterizing basic parameters such as gas temperature, vibrational temperature, electron temperature, and electron number density. OES is critical for probing the energy transfer process, species time evolution, and elementary chemical kinetics. The flexibility and versatility make OES an indispensable tool, contributing significantly to both scientific research and industrial processes. The following section introduces the basic concept and implementation and provides several examples of OES. Readers are referred to reviews [1–3] for more details.

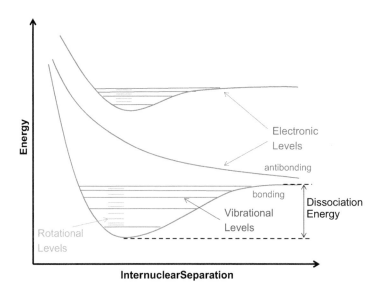

FIGURE 9.1 Energy diagram of a molecule showing bonding and antibonding states. Courtesy from Joseph Leftkowitz.

9.1.1 Optical Emission Spectroscopy Method

Gaseous atoms and molecules in the plasma are excited through inelastic collisions by free electrons, typically with electron energies above a threshold of ~10 eV, which is the "high-energy tail" of the non-Boltzmann electron energy distribution function (EEDF). The excitation leads to radiative states of gaseous species, with a short shelf-time on the order of nanoseconds or shorter. This ultra-short time scale ensures relaxation through radiation rather than other pathways, such as collisional quenching.

Different atoms and molecules radiate in specific wavelengths corresponding to potential energy differences between upper and lower states. As a result, OES usually exhibits in separated spectral lines. In some cases, the lower state is not stable but dissociates. Then, the molecule will relax by continuum radiation [4]. The radiation intensity depends on multiple factors, so OES is primarily qualitative. Semi-quantitative analysis is possible by intentionally adding a known concentration of another atom or molecule, termed actinometry or titration [5]. As the spectral response at a specific wavelength is far from being constant, calibration with a standard source is recommended at any attempt of OES.

OES using standard spectrometers usually cover the visible, near-infrared, and ultraviolet ranges (200–1,200 nm); thus, the detected photon energies are typically in the range from 1 to 6 eV. For radiations with higher energies (below 200 nm), for example, atomic transitions from excited Ar, He, O, N, and H, to the ground state, VUV spectrometers (wavelength 100–200 nm) with more complicated experimental setups [6] are required.

OES measurements are straightforward to set up and perform. No direct optical access is required for OES. A window or flange, through which an optical fiber can collect light, is sufficient. Even if the window or flange is not on-axis with the preferred line of sight, the measurement can be performed with lenses or mirrors. In addition, with a sufficient signal-to-noise ratio, the measurements are fast: the integration time is in the order of milliseconds. One example of the experimental setup is shown in Figure 9.2.

In an OES experiment, it is critical to collect a spectrum of a light source of a monoatomic substance that belongs to a known element and calibrate the lines and identify the line shift and

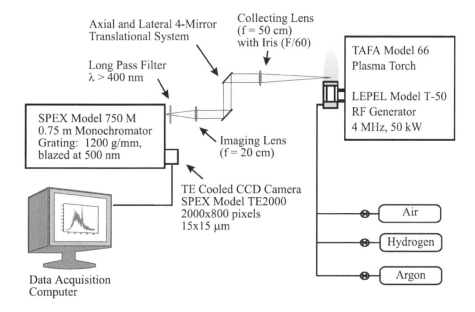

FIGURE 9.2 Typical experimental setup for OES [3].

broadening using the spectral database. Low-temperature nonequilibrium plasmas exhibit emission at specific wavelengths only. These line spectra need to be interpreted by so-called Collisional–Radiative Models (CRMs) [7]. CRMs describe the spectral line intensities as a function of plasma parameters such as electron density and electron temperature. They can theoretically be set up for any kind of discharge gas if its atomic structure is precisely known, and the required input data are available. Molecular gases are significantly more complex in modeling due to their additional vibrational and rotational states. Rare gases, like argon, are often used for fundamental investigations as well as for industrial etching processes. Reactive gases, such as CO_2 [8,9] and hydrocarbons [10], are less explored in CRMs. More recent advances in the CRM development are provided in [11].

9.1.2 Applications of OES in Plasma

Gas temperature measurements: In local thermodynamic equilibrium (LTE) plasmas, a single temperature characterizes the energy of all degrees of freedom (vibrational, rotational, and electronic). This single temperature can be determined from the absolute intensity of any atomic or molecular spectra, or from Boltzmann plots of vibrational or rotational population distributions. One such example is shown in Figure 9.3. The measured vibrational, rotational, and electronic temperature profiles are within experimental uncertainty.

In nonequilibrium plasmas, because of departures from equilibrium in the distribution of internal modes, the gas temperature is then best inferred from the intensity distribution of rotational lines. Various transitions of O_2, N_2, N_2^+, NO (dry air), and OH (humid air) can be used, depending on the level of the emission intensity. For example, in low-temperature humid air plasmas, the emission spectrum of the OH A–X(0, 0) transition around 308 nm provides a particularly convenient way to measure the rotational temperature [12,13]. The rotational temperature can be obtained by fitting the entire band, or more simply, from the relative intensities of two groups of rotational lines corresponding to the R and P branches of the OH A–X (0, 0) vibrational band. These branches form distinct peaks at about 307 and 309 nm, respectively. The potential difficulties are associated with self-absorption, non-Boltzmann rotational population distribution, and interferences with other species present in the plasma. Another emission band is the N_2 C–B system (Figure 9.4) or the second positive system of nitrogen [3,14]. The second positive system of nitrogen is often responsible for the dominant emission in air or N_2-containing plasmas. Its different vibrational branches range between 280 and 500 nm with the head of the (0,0) vibrational transition at 337 nm (Figure 9.4). Most often, the (0,0) transition or the (0,2) transition at 380 nm is used to obtain the rotational temperature. At even

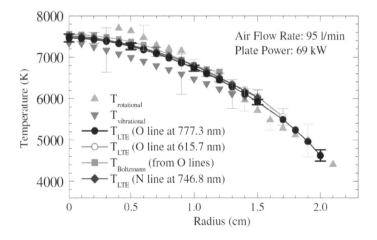

FIGURE 9.3 Measured electronic vibrational, and rotational temperature profiles in LTE air [3].

Plasma Diagnostics

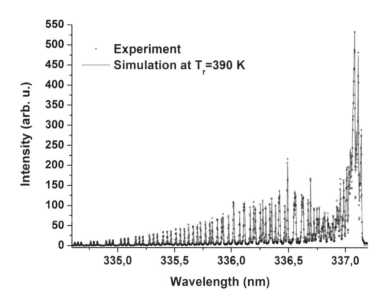

FIGURE 9.4 High resolution (4 pm, FWHM) rotational spectrum of the N_2(C-B;0-0) recorded in a RF He-H_2O diffusive glow discharge at atmospheric pressure [2].

higher temperatures or higher electric fields, many molecular transitions appear in the spectrum, and an accurate spectroscopic model is required to extract individual lines of a particular system.

Despite the extensive use of rotational spectra for temperature measurements, one thing to keep in mind is that it is not automatically guaranteed that in nonequilibrium plasmas the density distribution in the monitored rotational manifold of the diatomic molecule is in equilibrium with the translational (gas) temperature. Production mechanism of the species in the excited state, the thermalization process related to the rotational energy transfer (RET), can affect the validity of the OES gas temperature measurements. Readers are referred to [3] for more details.

Electron temperature measurements: Under nonequilibrium plasma conditions, emission intensities from two well-studied rare gas (e.g., Ar and helium) levels of differing energy can be used to obtain electron energy [15]. Often the two levels are of the neutral and the ion of the same atomic species. The ratio of these emission lines depends on the electron temperature (T_e) in a way that can be predicted. In practice, uncertainties in the excitation mechanism and in the cross-sections for electron impact excitation, combined with complications arising from radiation trapping, make it difficult to obtain accurate results from this method.

Another method is to detect emission from multiple rare gas mixtures that are added to a plasma at such low levels that they cause minimal perturbation to the plasma. Furthermore, their partial pressures are sufficiently low for their emissions to be under optically thin conditions. This trace rare gases OES method has been used to measure the electron energy in O_2, and fluorocarbon-containing inductively coupled plasma (ICP) [16]. This method involves a complete treatment of the complex excitation mechanism and relies on the use of many emission lines to minimize the errors introduced by the uncertainty in the electron impact cross-sections.

Electron number density measurements: In plasmas with high electron number densities (higher than $\sim 5 \times 10^{13}$ cm^{-3}), spatially and temporally resolved electron number density can be inferred from OES, specifically, the line shape of the Balmer α and β transition of atomic hydrogen or Ar Stark broadening [17]. The hydrogen Balmer α and β transition technique requires a small amount (typically 1% or 2%) of hydrogen, which may either come from dissociated water vapor in humid air or from premixed H_2 in the air stream. The lineshape of the H_α and H_β transition is determined by Lorentzian (Stark, van der Waals, resonance, natural) and Gaussian (Doppler, instrumental) broadening mechanisms that result in a Voigt profile (Figure 9.5). Among those broadening mechanisms, only the Stark broadening depends explicitly on the electron density. Stark broadening results from Coulomb interactions between the radiator (the hydrogen atom for the H_α/H_β transition) and the charged particles present in the plasma. Both ions and electrons induce Stark broadening, but electrons are responsible for the major part because of their higher relative velocities [3]. The addition of hydrogen to plasma may change plasma properties. Therefore, Ar Stark broadening and optical emission line ratio are also used for the measurement of electron number density. By using a combination of Ar Stark broadening and the optical emission line-ratio method, the temporal evolution of electron density in a nanosecond-pulsed argon microplasma was measured

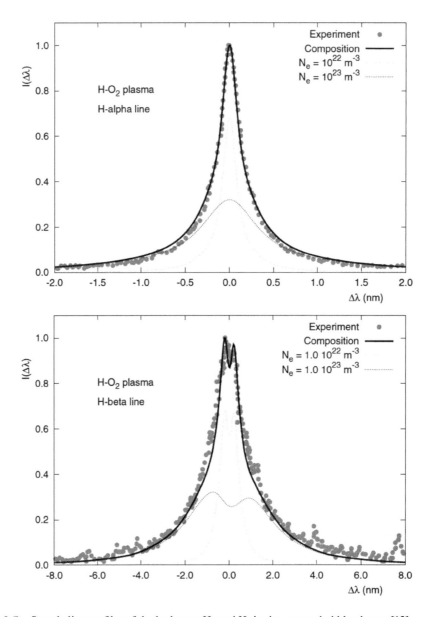

FIGURE 9.5 Sample line profiles of the hydrogen H_α and H_β in the oxygen bubble plasma [13].

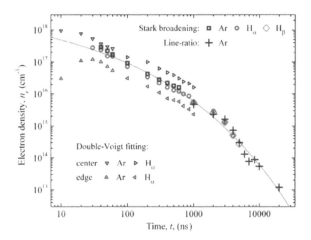

FIGURE 9.6 Experimentally measured electron densities in a high-pressure nanosecond-pulsed microplasma by using the Ar Stark broadening method and line-ratio methods (Ar/He = 700/30 Torr, discharge current lasts for about 100 ns, pulse period 1 ms) [17].

using [17] (Figure 9.6). The uncertainties of the Stark broadening method (using Ar 2p–1s lines) and the line-ratio method were also examined. It was shown that the Stark broadening method was valid when the electron number density $n_e > 10^{16}\,\text{cm}^{-3}$. When $n_e < 10^{16}\,\text{cm}^{-3}$, it was shown that the Ar line-ratio method was valid. More details about how to extract the electron number density from the H_α/H_β transition and the Stark broadening method can be found in [3,17].

Application of OES for probing plasma kinetics involving excited states: OES has also been applied to provide self-consistent experimental datasets in well-defined discharge and gas conditions for plasma kinetic studies. For example, one study [14] integrated the OES measurements for the spectra of the first and second positive systems of nitrogen to investigate the ultrafast heating mechanism in the nanosecond repetitively pulsed air discharge. The temporal evolution of the gas temperature measured by OES showed that a temperature increase of about 900 K in 20 ns was observed, corresponding to an ultrafast heating rate of 4.5×10^{10} K/s. The absolute density of $N_2(B)$, $N_2(C)$ from OES further confirmed that the ultrafast dissociation of molecular oxygen is mainly due to the quenching reaction of $N_2(B)$ and, to a much lesser extent, of $N_2(C)$ by oxygen. OES measurements show that the ultrafast mechanism creates excited electronic states of nitrogen molecules, which then dissociate molecular oxygen through quenching reactions, with the remaining energy dissipated as fast gas heating.

In another study [13], OES was used to characterize plasmas in and in contact with liquids, in which a strong emission of OH(A–X) and of hydrogen lines is demonstrated. For 600 ns pulsed discharges in He, Ar, and O_2 bubbles in tap water, OES shows that the gas temperature of the He plasma remains low (400 K) while the temperature in Ar and O_2 reaches 3,000 and 5,000 K. The electron densities obtained with Stark broadening of the H_β (and H_α) lines are in the range $(1–4.7) \times 10^{21}\,\text{m}^{-3}$ for He and 10^{22}–$10^{23}\,\text{m}^{-3}$ for O_2 and Ar. Assuming dissociative recombination as the primary loss mechanism, the estimates from the time constants of the emission decay of OH(A–X) yield electron density values of similar magnitudes with the OES measurements.

OES measurements are also used for studying kinetics in plasma-assisted combustion. In one recent work [18], the plasma discharge is created in a mixture of burnt gases from CH_4-air mixtures. OES provided temporally (2 ns) and spatially (0.5 mm) resolved evolutions of the temperatures and concentrations of $N_2(B)$, $N_2(C)$, $N_2^+(B)$, OH(A), NH(A), and CN(B) in the discharge, which further provides the insight into the plasma chemistry such as CN formation. The ratio of the densities of $N_2(B)$ and $N_2(C)$ in the burnt gas ranges between 10 and 40, which is close to what was found in preheated air and shows that these burnt gas discharges shared similar plasma chemistry.

In summary, OES provides non-intrusive, real-time, and cost-effective approaches for probing key parameters in low-temperature plasmas. However, quantitative species diagnostics in nonequilibrium plasma using OES remains challenging. The combination of OES with other plasma diagnostics can further unravel the underlying plasma physics and chemistry.

9.2 LASER ABSORPTION SPECTROSCOPY

9.2.1 Introduction

This section gives a brief description of LAS theory and methods. For more in-depth discussions, the reader is directed to the following references [19–24].

Every atom or molecule will absorb electromagnetic radiation at particular, discrete wavelengths corresponding to electronic, vibrational, and/or rotational energy level transitions (Figure 9.1). The total internal energy, E, of a molecule can be described by:

$$E_{\text{internal}} = E_{\text{electronic}} + E_{\text{vibrational}} + E_{\text{rotational}} \tag{9.1}$$

When an atom or molecule undergoes a transition by absorbing or emitting a photon, or by collisions with other species in the gas, the energy contained one or more of these modes will change. For many small molecules, the absorption/emission lines have been tabulated in the HITRAN database [25]. The energy difference between the upper and the lower energy states can be defined using Planck's Law:

$$\Delta E = E_{\text{upper}}(E') - E_{\text{lower}}(E'') = h\nu = hc\tilde{\nu} = hc/\lambda \tag{9.2}$$

where $h = 6.626 \times 10^{-34}$ J-s is Planck's constant and ν [s^{-1}] is the frequency of the photon, c is the speed of light in a vacuum, $c = 3 \times 10^{10}$ cm/s, $\tilde{\nu}$ is the wavenumber (cm^{-1}), and λ is the wavelength of light in cm.

A general diagram of potential wells, which depicts the energy levels of a molecule as a function of the internuclear distance, is shown in Figure 9.1. Since different electronically excited states alter the shape of the molecule, each electronic state has different vibrational and rotational energy levels.

The vibrational states of a molecule correspond to internuclear motion, which can be characterized by stretching or bending motions. For a transition to occur, the dipole moment of the molecule, which the product of the magnitude of charges and the separation between them, must be changed.

$$\vec{\mu} = \sum_i q_i \vec{r}_i \tag{9.3}$$

Where $\vec{\mu}$ is the dipole moment, q_i is the charge of particle i and $\vec{r}_i$ is the position vector of the particle defined from the center of mass. In a diatomic molecule with two identical atoms (such as O_2 and N_2), vibrational motion does not cause the dipole moment to move since the center of charge remains fixed due to molecular symmetry. The same is true of rotational transitions, since the center of charge is not changing in space as the molecule rotates. Thus, only heteronuclear diatomic molecules exhibit rotational and vibrational transitions.

The selection rules for transitions between excited states are governed by the Schrödinger equation. Detailed discussions of the spectra for different molecules can be found in the classic books by Herzberg [19]. In general, the structure of rovibrational spectra is separated into groups of transitions (or "lines") corresponding to specific vibrational transitions. These groups are called bands. Each band has many individual rotational lines within the upper and lower vibrational state. Figure 9.7 presents the absorption spectra of a few selected molecules.

As can be seen in Figure 9.7, many rovibrational absorption bands exist between $\tilde{v} = 1{,}000$–$8{,}000\,\text{cm}^{-1}$. Different regions of the spectrum have different line strengths. For diatomic molecules, there is only one very strong band (i.e. NO, CO, OH). This is called the "fundamental" band and is for transitions in which the vibrational quantum number "v" is changed from 0 to 1. At approximately twice the wavenumber of the strong band will appear a weaker band, which represents transitions in which the vibrational quantum number is changed by ± 2. This is called the "first overtone". There will be progressively weaker bands at every multiple of the fundamental, which are also called overtone bands. Polyatomic species have more complex spectra involving multiple vibrational modes, and thus multiple fundamental bands. Any combination of these bands and the overtones thereof will be part of the spectra.

Figure 9.8 presents the absorption spectra of CO for the fundamental stretch mode. The right-hand side branch (higher energy) is the "R branch" and the left-hand side branch (lower energy) is the "P

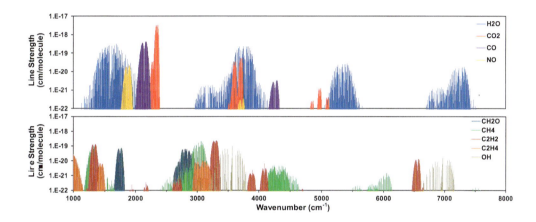

FIGURE 9.7 Spectra of selective species in the near to mid-IR [24].

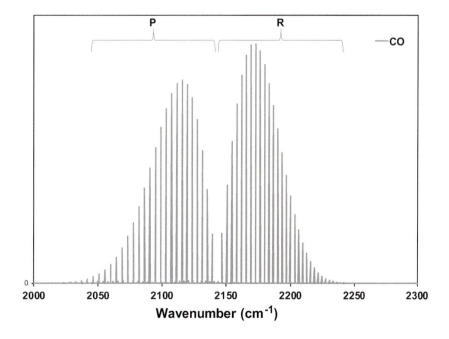

FIGURE 9.8 Absorption features in the fundamental CO band [24].

branch". Each line in the fundamental band represents a $\Delta v = \pm 1$ vibrational number change and a ± 1 change in the rotational quantum number, $\Delta J = \pm 1$. $\Delta J = -1$ represents the P branch and $\Delta J = +1$ is in the R branch of $v = 0 \rightarrow 1$. Polyatomic molecules (and some diatomic molecules) also allow for $\Delta J = 0$ transitions. This results in another branch, called the "Q-branch" in the center of the P and R branches [19].

The strength of a particular line is determined by the population in the lower state by the Boltzmann distribution (in equilibrium):

$$F_i = \frac{n_i}{n} = \frac{g_i \exp\left(-\frac{\varepsilon_i}{k_B T}\right)}{Q} \tag{9.4}$$

where n_i [cm^{-3}] is the number density of species in the state of i, g_i is the degeneracy of state i (or number of states with the same energy), ε_i [J] is the energy of the state i, k_B is the Boltzmann constant (1.38×10^{-23} J/K), T is temperature, and Q is the partition function,

$$Q = Q_{elec} Q_{vib} Q_{rot} = \sum_i g_i \exp\left(-\frac{\varepsilon_i}{k_B T}\right) \tag{9.5}$$

Therefore, the line strengths are dependent on temperature and the degeneracy of the state.

9.2.2 The Beer-Lambert Law

LAS is generally performed with a laser that can be tuned in terms of its wavelength. The light is transmitted through the sample (e.g. plasma) to a detector, as shown in Figure 9.9.

LAS is to determine the spectral absorbance by the molecules in the sample. If we assume the spectral emission from the sample is much lower than the transmitted light from the laser, we then have,

$$\alpha_v = 1 - \tau_v \tag{9.6}$$

where α is the spectral absorptivity and τ indicates the spectral transmissivity at laser frequency v. Then the change of the laser intensity after passing a small distance becomes,

$$\frac{dI_v}{I_v} = -\alpha_v = -k_v dx \tag{9.7}$$

where k_v [cm^{-1}] is the spectral absorption coefficient of the sampled molecules. By integrating Eq. 9.7 a path length, L [cm], we obtain the Beer-Lambert Law (without emission),

$$\frac{I_v}{I_{v,0}} = \exp(-k_v L) \tag{9.8}$$

where $I_{v,0}$ is the initial intensity of light (Figure 9.9).

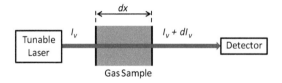

FIGURE 9.9 Light transmission through a gas [24].

The spectral absorption coefficient at a specific spectral line depends on the shape of the line. Let us use a generic line shape factor $\phi(v)$ [s], which is normalized by the integrated absorption coefficient to describe the shape,

$$\phi(v) = \frac{k_v}{\int_{\text{line}} k_v dv} \tag{9.9}$$

We can now redefine the Einstein coefficients in a portion of frequency space $v \rightarrow v+dv$ by multiplying each coefficient by the lineshape function at that wavelength. We will also redefine our equation for spectral absorption and emission in a differential wavelength region.

Note that there is induced emission, spontaneous emission, and induced absorption. If one assumes the spontaneous emission is negligible, we can find the integrated absorption or the "line strength", S_{12} [cm^{-1}s^{-1}],

$$S_{12} = \int_{\text{line}} k_v dv = \frac{hv}{c} n_1 B_{12} \left[1 - \exp\left(\frac{hv_0}{k_B T}\right)\right] = \frac{\lambda^2}{8\pi} n_1 A_{21} \frac{g_2}{g_1} \left[1 - \exp\left(\frac{hv_0}{k_B T}\right)\right] \tag{9.10}$$

where A_{21} [Hz] is the Einstein coefficient for spontaneous emission and B_{12} [cm^3 Hz] is the Einstein coefficient for stimulated absorption. Therefore, the line strength is a function of the incident light frequency, number density of the absorber, temperature, and the Einstein coefficient. The Beer-Lambert law can be written in terms of the line strength,

$$\frac{I_v}{I_{v,0}} = \exp\left[-S_{12}\phi(v)L\right] \tag{9.11}$$

The line strength in Eq. 9.11 is dependent on temperature. By using a reference temperature of T_0, its temperature dependence can be given as,

$$S_{12}(T) = S_{12}(T_0)\frac{Q(T_0)}{Q(T)}\exp\left[-\frac{hcE''}{k_B}\left(\frac{1}{T}-\frac{1}{T_0}\right)\right]\frac{\left[1-\exp\left(-\frac{hv_0}{k_B T}\right)\right]}{\left[1-\exp\left(-\frac{hv_0}{k_B T_0}\right)\right]} \tag{9.12}$$

where E'' is the energy of the lower state of the transition in cm^{-1}. The partition function, Q, can be approximated by a polynomial function of temperature.

$$Q(T) = a + bT + cT^2 + dT^3 \tag{9.13}$$

where the coefficients a, b, c, and d are tabulated in the HITRAN database as well as the lower state energy levels and integrated line strengths. If the temperature is known, Eqs. 9.11 and 9.12 can be used to measure the species number density, n. In the case one needs to measure the temperature by using two absorption lines of 1 and 2, we can use the ratio of 9–12 in the two different absorption lines,

$$\frac{S_{i,1}(T)}{S_{i,2}(T)} = \frac{S_{i,1}(T_0)}{S_{i,2}(T_0)} \times \exp\left[-\frac{hc(E_1''-E_2'')}{k}\left(\frac{1}{T}-\frac{1}{T_0}\right)\right] \times \frac{\left[1-\exp\left(-\frac{hv_{0,1}}{k_B T}\right)\right]}{\left[1-\exp\left(-\frac{hv_{0,2}}{k_B T}\right)\right]} \times \frac{\left[1-\exp\left(-\frac{hv_{0,1}}{k_B T_0}\right)\right]^{-1}}{\left[1-\exp\left(-\frac{hv_{0,2}}{k_B T_0}\right)\right]^{-1}} \tag{9.14}$$

If the two absorption lines are nearby in frequency space (i.e. $v_{0,1} \cong v_{0,2}$), the temperature can be approximately given by

$$T = \frac{\dfrac{hc(E_2'' - E_1'')}{k}}{\ln\left[\dfrac{S_{i,1}(T)}{S_{i,2}(T)}\right] + \ln\left[\dfrac{S_{i,2}(T_0)}{S_{i,1}(T_0)}\right] + \dfrac{hc}{k}\dfrac{(E_2'' - E_1'')}{T_0}} \tag{9.15}$$

This is often referred to as the "two-line" method of temperature measurement [26].

Light emissions and line shape are affected by the Heisenberg uncertainty principle as well as molecule collisions and motion. At elevated pressures and temperature, the increased collisions and random motion of molecules will broaden the spectral line. The ones of particular concern in probing rovibrational lines are natural broadening (uncertainty principle), collisional (or pressure) broadening, and Doppler broadening (Boltzmann velocity distribution). Both natural and collisional broadening take on Lorentzian lineshapes, while Doppler broadening takes on a Gaussian lineshape. The convolution of these profiles results in the Voigt profile as follows:

$$\phi_V(v) = \frac{2}{\Delta v_D}\sqrt{\frac{\ln 2}{\pi}}\frac{\Delta v_C/2}{\pi}\int_{-\infty}^{\infty}\left\{\exp\left[-\left(\frac{\sqrt{\ln 2}}{\frac{\Delta v_D}{2}}(u)\right)^2\right]\right\}\left[\frac{1}{(v - v_0 - u)^2 + (\Delta v_C/2)^2}\right]du \tag{9.16}$$

where u is the shifted convolution variable $(v - v_0)$. Δv_D and Δv_C are, respectively, the Full width at half maximum (FWHM) of the Doppler broadening and collisional broadening:

$$\Delta v_D = \frac{v_0}{c}\sqrt{\frac{8k_B T \ln 2}{m}} \tag{9.17}$$

$$\Delta v_C = \sum_i P_i\, 2\gamma_i, \quad \gamma_{Ai} = \gamma_i^{300}(300/T)^n \tag{9.18}$$

where m is the mass of molecules and γ_A [cm^{-1}/atm] is the collisional half-width for species i, defined for each collision of the absorbing species with all other particles in the gas, including itself. Both of them are temperature-dependent.

Figure 9.10 shows a comparison of the normalized Doppler, collisional, and Voigt broadened profiles with the same FWHM of all three lineshapes. It is seen that the Voigt profile contains

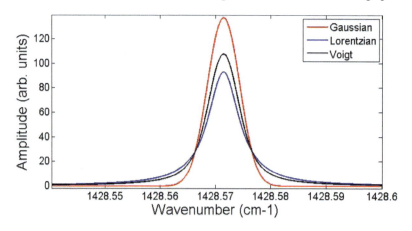

FIGURE 9.10 Comparison of Lorentzian, Gaussian, and Voigt lineshapes with equivalent FWHM [24].

Plasma Diagnostics

characteristics of both the Gaussian and Lorentzian lineshapes, with the broad wings of the Lorenztian and the wide peak of the Gaussian.

9.2.3 Experimental Method of LAS and Applications

Recently, continuous wave External-Cavity mode hop-free quantum cascade lasers (EC-QCL) and distributed feedback quantum cascade lasers (DFB-QCL) have been widely used for *in situ* species and temperature measurements in combustion and plasma. The EC-QCL has a line width resolution of less than 0.001 cm^{-1} and a wide tunability in mid-infrared wavelength covering the fundamental rovibrational absorption bands of species of interest in ethylene oxidation such as H_2O, CO, CH_2O, CH_4, and C_2H_2 [24,27–30]. For example, the absorption lines of H_2O for temperature and concentration measurements can be located at 1,338.55 and 1,339.15 cm^{-1}. The lines of CH_4 for temperature and concentration measurements are at 1,343.56, 1,343.63, and 1,341.32 cm^{-1}. The line for C_2H_2 is at 1,342.35 cm^{-1} and the line for CH_2O quantification with the DFB-QCL is at 1,726.79 cm^{-1}.

Figure 9.11 shows the experimental setup of *in situ* LAS with a Herriott cell for nonequilibrium plasma diagnostics. The reactor is constructed of quartz and Macor with stainless-steel brackets to hold the walls in place. One sidewall is quartz, which allows observation into the cell for laser alignment. Each of the two 45 mm×45 mm stainless-steel electrodes on the top and bottom is sandwiched between a quartz plate and a Macor plate, forming a plane-to-plane double DBD. The inner dimensions of the rectangular quartz reactor section are 14 mm in height, 45 mm in width, and 152 mm in length. There is a wedge shaped 25.4 mm OD calcium fluoride window in the sidewall of the vacuum chamber to allow the mid-infrared laser beam to pass through the chamber

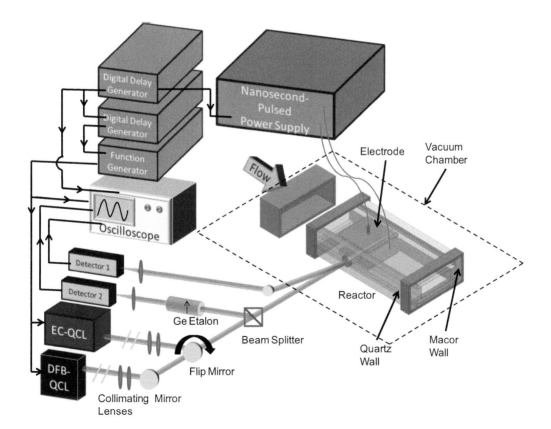

FIGURE 9.11 Schematic of the experimental setup for DBD flow reactor with TDLAS measurement system. In Chapter 5, the DFB-QCL laser is not included [24].

wall and into the Herriott cell for absorption measurements. The fuel (e.g., methane), oxygen, and diluent gases are metered using mass flow controllers (Brooks instruments) and are premixed prior to entering the plasma reactor. The lasers are co-aligned and coupled into a 24-pass Herriott cell such that either laser can be used individually by flipping a mirror, allowing convenient measurements without realignment of the cell. The Herriott cell (Figure 9.12a) is comprised of two opposed 12.7 mm OD, 50 mm focal length protected gold concave mirrors, which are located 30 mm downstream of the front edge of the electrodes. The laser light is coupled into the cell through a 2 mm hole in a stationary mirror in the quartz wall, and cell alignment is accomplished with the help of a two-dimensional tilt stage (Thorlabs KMS) fitted with a screw for axial translation, which is necessary for proper alignment of the Herriot cell, fixed to the Macor wall. Images of the beam profile of the Herriott cell are provided in Figure 9.12a for different pass numbers. The effective laser path length through the plasma is 1.08 m (24×0.045 m). The transmitted output beams are incident on an MCT detector (Vigo PVM-2TE10.6). In order to eliminate atmospheric water absorption outside the test section, the laser path is purged with nitrogen. The DBD plasma image is shown in Figure 9.12b.

The EC-QCL laser is scanned through the absorption lines by a 100 Hz sinusoidal signal sent to the controller (Thorlabs MDT694A) for the piezo-electric actuator in the laser cavity using a function generator (SRS DS345). In the time-dependent measurements, the laser scan was synchronized with the discharge trigger such that the absorption peak occurred at a controllable delay after the first plasma discharge pulse. The absorption signal used for quantification was averaged over 30–50

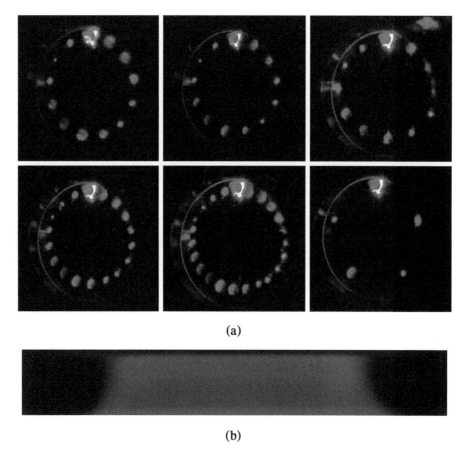

FIGURE 9.12 (a) Images of beam spots on Herriot cell mirror. The top left image is the beam profile used for all measurements [24]. (b) Two second exposure of 0.0625/0.1875/0.75 C2H4/O2/Ar plasma at pulse repletion frequency of 1,000 Hz.

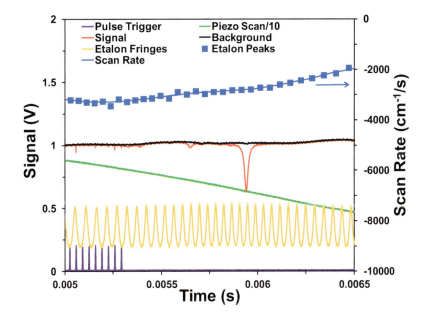

FIGURE 9.13 Collected signals in EC-QCL experiment with calculated laser scan rate. "Pulse trigger" refers to the signal sent to trigger the plasma discharge, "signal" is the absorption signal (of a water line), "etalon fringes" are the peaks and troughs measured after the laser is passed through the Ge etalon, "scan rate" is the rate of laser tuning calculated using the etalon fringes, "piezo scan" is the signal sent to the piezo-electric controller to tune the laser, and "background" is the signal from the EC-QCL in the absence of absorber [24].

plasma burst experiments. Figure 9.13 displays all of the signals collected for a single EC-QCL experiment, along with the calculated laser scan rate. To scan the DFB-QCL, the laser temperature is held constant while the current is swept by a sawtooth waveform at a rate of 15,000 Hz. This allowed all time-dependent CH_2O data points to be collected in a single scan, with an absorption measurement taken between every plasma pulse, and averaged over 100 experiments. The wavelength scan rate of both lasers is monitored using a 50.8 mm Germanium etalon (0.74 GHz at 1345 cm^{-1}). Due to inconsistency in scan rates for the EC-QCL, the etalon signal is collected for every experiment.

The direct absorption profile at different conditions of temperature and pressure can be modeled using data from the HITRAN database [25]. The pressure broadening coefficient was treated as a variable parameter, along with the absorber concentration, in a least squares nonlinear fitting algorithm for calculating the Voigt function [31,32]. The temperature was calculated by scanning over two different absorption lines of water or methane at the same delay after the pulse burst. Gases of known temperatures and concentrations of the absorbing species need to be used to validate the measurement technique, and a standard deviation of 3 ppm for methane in the range of 5–100 ppm was observed. Temperature measurements have a standard deviation of 5 K in the range of 300–500 K. Plots of these measurements are provided in Figure 9.14.

Figure 9.15 presents the mole fraction of measured and predicted CH_2O as a function of time at the same conditions. The comparison shows that the model under-predicts the peak concentration of CH_2O by a factor of 5, indicating major limitation in the model's predictive ability of this primary intermediate. Therefore, tunable diode laser absorption spectroscopy (TDLAS) is a very powerful time-resolved quantitative measurement method to probe plasma chemistry.

To collect a more complete set of product species, the plasma is run in continuous mode and temperature measurements were conducted by two-line absorption of methane. Because of the fast thermal conductivity rate of helium, the gas temperature will quickly reach a steady state in which the heat loss to the walls balances the heat addition by the plasma. Figure 9.16 presents the

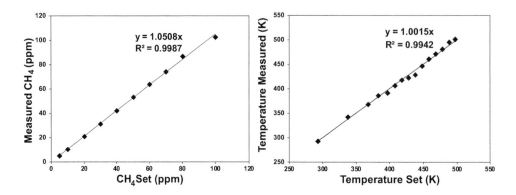

FIGURE 9.14 Left: Comparison of methane mole fraction measured by LAS vs. prepared mixture. Right: Comparison of temperature measured by two-line LAS of methane vs. temperature measured by thermocouples.

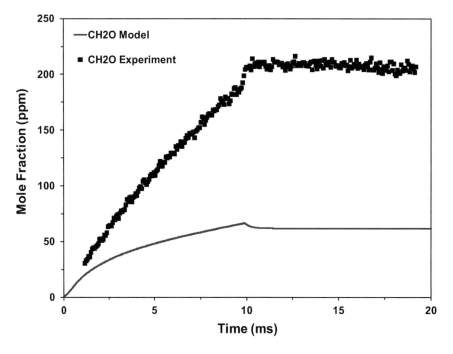

FIGURE 9.15 Formaldehyde measurements and model predictions during and after a 300 pulse burst at 30 kHz repetition rate and 8.76 kV peak voltage in a stoichiometric $CH_4/O_2/He$ mixture with 75% dilution.

measured and predicted gas temperature during steady-state NRP discharges at frequencies from 100 to 30,000 Hz. At lower frequencies (<10,000 Hz) the model and the measurement are in reasonable agreement. However, at higher frequencies, the model and the measurement diverge such that at 30,000 Hz there is a 50 K differential between the measurement and the model. This is due to wall heating in continuous plasma. More experimental data of time-resolved temperature and species data can be found in Figure 5.20.

Recently, Telfah et al. [33] applied time-resolved cavity ring down spectroscopy (CRDS) to measure absolute number densities of HO_2 in O_2–He mixtures excited by a repetitive ns pulse

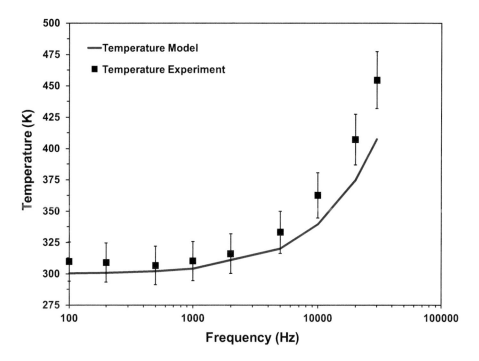

FIGURE 9.16 Temperature measurements and model predictions in a continuous plasma at 30 kHz repetition rate and 8.76 kV peak voltage in a stoichiometric $CH_4/O_2/He$ mixture with 75% dilution.

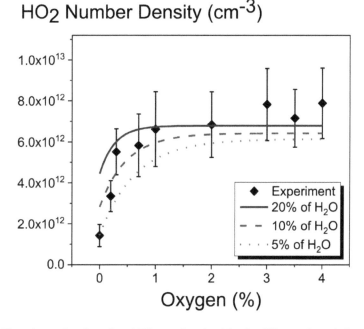

FIGURE 9.17 Experimental and predicted HO_2 number densities for different O_2 mole fractions in the mixture. $P = 85$ Torr, excitation by a 20-pulse, 10 kHz discharge burst. HO_2 is measured 100 μs after the last pulse in the burst. Modeling predictions are plotted for different water vapor mole fractions in the mixture [33].

discharge. Figure 9.17 shows the measured HO_2 dependence on oxygen concentration in plasma discharge at $P = 85$ Torr with excitation by a 20-pulse, 10 kHz discharge burst. Since hydroperoxyl radical is a key species in low-temperature plasma-assisted fuel oxidation [34–37] and plasma-liquid interactions, sensitive detection using TDLAS with CRDS provides critical information to understand reactive species generation and plasma chemistry.

More recently, Zhong et al. [38] applied TDLAS for *in situ* diagnostics of N_2O/NO_x in nonequilibrium plasma-assisted ammonia oxidation. The experimental setup is shown in Figure 5.25. Two DFB QCLs were used for NO and OH measurements at 1,906.73 and 3,568.41 cm^{-1}, respectively. An external-cavity QCL was also used for the detection of H_2O at 1,338.55 cm^{-1} and N_2O at 1,306.93 cm^{-1}. In addition, a sensitive off-axis integrated cavity output spectroscopy (ICOS) [39] was employed for NO_2 measurement at 6,640.4 cm^{-1}. The results are shown in Figure 5.26. Significant derivation was seen between experiments and model prediction. The experimental data contributed to develop a better predictive model in plasma-assisted ammonia oxidation [40].

Therefore, *in situ* time-resolved TDLAS provides accurate temperature measurements of plasma to validate and develop plasma kinetic models. Many other applications of TDLAS can be found in the literature [40,41].

9.2.4 Femtosecond LAS

Although mid-infrared TDLAS discussed above provides great advantage in sensitive detection molecules, the absorption cross-section is often smaller than that in UV. In addition, it requires scanning the laser in frequency domain to achieve time-resolved measurements. Recently, femtosecond (fs) lasers have provided new opportunities in laser diagnostics for combustion and plasma since they can generate pulses that feature femtosecond timescale in the time domain, and broadband spectrum in the frequency domain at kHz repetition rates [42,43]. The feature of the broadband spectrum can be utilized to develop various ultrafast laser diagnostics such as fs mid-infrared and UV LAS [44], fs cavity-enhanced absorption spectroscopy (fs-CEAS), and fs laser electronic excitation tagging [45]. Femtosecond lasers have a bandwidth of tens to hundreds cm^{-1} [46,47], which is much broader than that of narrow-line TDLAS [48–50] in the frequency domain. Therefore, the broadband nature of fs lasers can enable multi-parameter/species measurements [51,52] by simultaneously sensing many transitions of different species. Moreover, the multi-transition absorption spectroscopy can enable the absorption fitting of multi-line spectra and improve the experimental accuracy. Furthermore, the fs laser pulse duration provides a fs or ps time resolution to explore ultrafast plasma chemistry in breakdown and formation.

Recently, Ning et al. [51,52] reported a direct fs ultraviolet laser absorption spectroscopy (fs-UV-LAS) for simultaneous temperature and species measurements in plasma. The fs-UV-LAS was demonstrated based on the $X^2\Pi$-$A^2\Sigma^+$ transitions of hydroxyl (OH) radicals near 308 nm in low-temperature plasmas (LTPs) and flames. The results showed that with abundant absorption features within ~3.2 nm near 308 nm for spectral fitting, temperature, and OH concentration can be simultaneously measured with accuracies enhanced significantly by 29%–88% and 58%–91%, respectively, compared to those from past methods. Secondly, an ultrafast time resolution of ~120 ps was enabled in the measurements. Thirdly, with the large OH $X^2\Pi$-$A^2\Sigma^+$ transitions and absorption cross-section near 308 nm, a single-pass absorption configuration was possible for measurements in LTPs featuring low OH number density down to ~2×10^{13} cm^{-3}.

The experimental setup of fs-UV-LAS is shown in Figure 9.18 [52]. It consists of a fs laser system, a plasma reactor, a spectrometer, and a camera. The fs laser system (Coherent Astrella) combining a Ti:Sapphire oscillator, an ultrafast amplifier, and an optical parametric amplifier (OPA) was used to generate the fs UV laser pulses. The Ti:Sapphire oscillator and the ultrafast amplifier generated the laser pulses centered at 800 nm with a pulse duration of 80 fs at a repetition rate of 1 kHz. Then, these laser pulses were converted by the OPA into UV pulses with a bandwidth of ~3.2 nm near 308 nm and a pulse energy of 38 µJ.

Figure 9.19 demonstrates the advantage of fs-UV-LAS in enhancing accuracy over other TDLAS methods. Figure 9.19a–c shows the comparisons of temperature accuracies between fs-UV-LAS and past methods using simulations based on three sets of published data. Figure 9.19d–f shows the comparisons of OH concentration accuracies. Figure 9.19a shows that fs-UV-LAS can significantly

Plasma Diagnostics

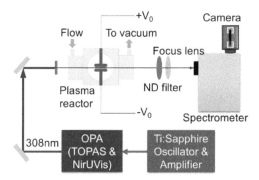

FIGURE 9.18 Experimental setup for fs-UV-LAS for OH detection in low-temperature plasma [52].

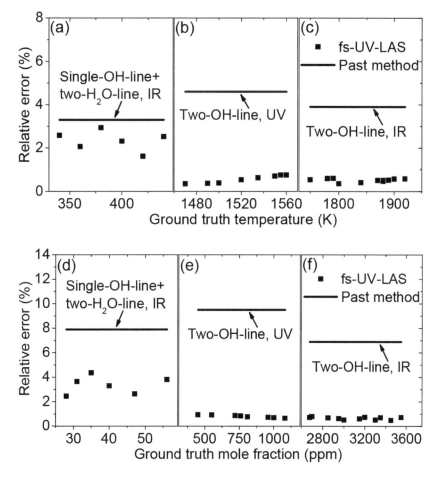

FIGURE 9.19 Comparison of fs-UV-LAS and past methods using simulations on the accuracies of (a–c) temperature, and (d–f) OH concentration [52].

reduce the measurement error of temperature by 29%–2.3% on average, compared to 3.3% obtained using the past method that combined single OH line with two H_2O lines in IR in a ns discharge plasma featuring a small absorbance of ~0.01 (~10^{13} cm^{-3} in OH number density). This significant error reduction was offered by the broadband sensing of fs-UV-LAS. In flames, Figure 9.19b and c show that fs-UV-LAS reduced the temperature error by 88% to 0.56% on average compared to

4.6% obtained by a past method using two OH lines in UV, and also by 86% to 0.51% on average compared to 3.9% by a two-OH-IR-line method. Figure 9.19d, fs-UV-LAS reduced the OH concentration error by 58% to 3.3% on average compared to 7.9% from the hybrid OH/H_2O-IR-line method with temperature obtained from two H_2O lines in plasmas with low OH number density. Also, as shown in Figure 9.19e and f, fs-UV-LAS substantially reduced the OH error by 91% to 0.8% on average in comparison with 9.5% from the two-OH-UV-line method, and also by 90% to 0.65% in comparison with 6.9% from the two-OH-IR-line method. In summary, this work not only demonstrated that fs-UV-LAS enabled simultaneous *in situ* measurements of temperature and species in plasma but also revealed three major benefits over previous TDLAS methods: (1) simultaneous temperature and OH measurements with significantly enhanced accuracies, (2) the ultrafast time resolution of ~120 ps, and (3) the simple single-pass absorption configuration for the measurements with low OH concentration down to ~2×10^{13} cm^{-3}, and the single-shot capability that would further extend fs-UV-LAS to the study of combustion/flow/plasma dynamics.

In plasma discharge, active species number density can be at ppb levels. To enhance the detection sensitivity, more recently, Ning et al. [51] further extended the fs-LAS method to a combined cavity-enhanced absorption spectroscopy (CEAS) method with femtosecond (fs) pulses to enable more sensitive single-shot measurements of OH concentration and temperature with a time resolution of ~180 ns in plasma. With the appropriately designed cavity, the results showed that an absorption gain of ~66 was achieved, enhancing the actual OH detection limit by ~55 times to the 10^{11} cm^{-3} level (ppb level).

Figure 9.20 shows the schematic of the fs-CEAS setup. The fs-CEAS setup was similar to that in Figure 9.18. It consisted of a fs laser, a plasma reactor bounded with cavity mirrors, a spectrometer, and a camera. The fs UV laser pulses were coupled into a plasma cell bounded with a pair of cavity mirrors. The plasma cell was largely similar to that introduced in [52,53] with only one difference on the electrodes. Two cylindrical electrodes with a 5.5-mm gap distance were fixed in parallel in the cell center. The electrodes were made of stainless steel, and one of them was covered by dielectric quartz with a 1.6-mm thickness. Each electrode had a diameter of 30 mm and a thickness of 5.5 mm, and their edges were smoothed to avoid the electric field concentration. The electrodes were

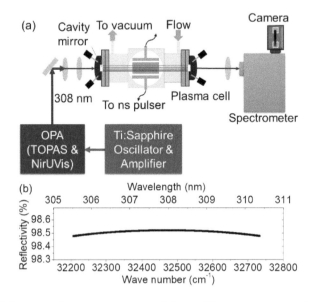

FIGURE 9.20 (a) Schematic of temperature and ppb-level OH measurements using fs cavity-enhanced absorption spectroscopy (fs-CEAS) for low-temperature plasma chemistry. (b) Reflectivity curve of cavity mirrors [51].

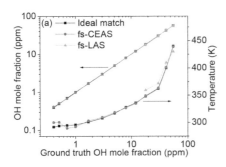 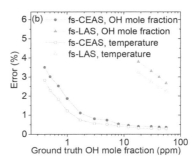

FIGURE 9.21 (a) Comparison of fs-CEAS and fs-LAS on the detection sensitivity of OH concentration and temperature using numerical simulations. (b) The relative errors associated to panel (a) [51].

connected to a ns discharge pulser (FID GmBH FPG) that operated in a burst mode with a burst rate of 1 Hz, a rate that could flush the plasma volume with new gas between two sequential bursts. Each burst had 600 pulses at a repetition rate of 30 kHz, and each pulse had a pulse width of 12 ns, and a peak voltage of 5 kV. The left and right parts of the plasma cell were a pair of high-finesse cavity mirrors (Rocky Mountain Inc) separated by 0.5 m distance, and each mirror had a radius of curvature of 6 m and a diameter of 20.3 mm. The reflectivity curve of cavity mirrors as a function of wavelength. The cavity mirrors featured a reflectivity of ~98.5% near 307.8 nm.

Figure 9.21 [51] shows the detection sensitivity comparison between fs-CEAS and fs-LAS using numerical simulations. Figure 9.21a shows the comparisons of OH mole fractions and temperatures at different OH mole fractions from 0.4 to 56 ppm. Figure 9.21b shows the OH mole fraction and temperature errors of these two methods relative to the ground truth data [54] of plasma. Due to the cavity enhancement, fs-CEAS extended the OH detection limit to sub-ppm level (~5×10^{11} cm^{-3}) with OH mole fraction within an acceptable accuracy of 3.5% and with temperature within 2.8%. By contrast, fs-LAS only detected OH at ~20 ppm (~1×10^{13} cm^{-3}) with accuracies of the OH mole fraction (within 3.8%) and temperature (within 3.2%). Therefore, the OH detection limit was improved by ~55× by fs-CEAS, compared to fs-LAS. In addition, fs-CEAS was able to maintain accuracies of OH mole fraction and temperature below 1% with OH mole fraction larger than 3 ppm. Therefore, fs-CEAS significantly increases the detection limit to ppb levels and improves the accuracy for plasma diagnostics.

9.3 FARADAY ROTATIONAL SPECTROSCOPY AND ELEMENTARY REACTION RATE MEASUREMENTS

9.3.1 Faraday Rotational Spectroscopy and Experimental Method

Intermediate radical species such as HO_2 and OH are critical for plasma chemistry. However, the detection of these species, especially HO_2, is challenging. For example, due to serious overlaps between spectra of HO_2, RO_2, H_2O_2, and HCO radicals with H_2O, fuel, and other species in infrared (IR) and ultraviolet (UV) regions, quantitative measurements of HO_2 radicals and their related reaction rates are extremely difficult with conventional TDLAS or laser-induced fluorescence (LIF) spectroscopy [37]. Although the near-IR and UV laser absorption has been used to observe HO_2, quantitative measurements are difficult due to complex IR spectra of hydrocarbon molecules and its structure-less UV absorption (200–250 nm) [33,55–58]. Similarly, the UV photo-fragmentation LIF method by breaking down the O-O bond of HO_2 and measuring OH suffers from the spectrum overlaps of HO_2, H_2O_2, aromatics, and OH [59].

Faraday rotational spectroscopy (FRS) is a highly selective and sensitive laser-based magneto-optical technique for paramagnetic gas species like HO_2, OH, NO, and HCO. In FRS, an external magnetic field is applied to the target species (e.g. HO_2 and OH), and the degeneracy of

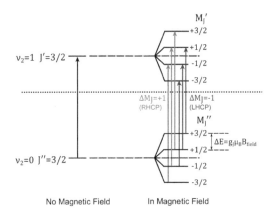

FIGURE 9.22 Zeeman splitting of HO$_2$ energy levels and $\Delta M = \pm 1$ transitions with a collinear laser propagation.

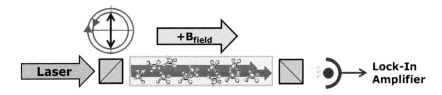

FIGURE 9.23 Schematic of laser beam, LHCP and RHCP polarization, Zeeman effect, and polarized angle change detection in FRS process.

the magnetic sublevels (electron spin) in the upper and (or) lower state of the optical transitions is broken due to Zeeman effect [36,37,60]. This results in splitting of the transitions into sets of transitions where the magnetic quantum number changes by $\Delta M = \pm 1$ or 0 (Figure 9.22). For a collinear laser propagation along the direction of magnetic field only transitions where $\Delta M = +1$ and -1 will interact with right and left-handed circularly polarized light (RHCP and LHCP), respectively (Figure 9.23). As a result, the applied magnetic field induces circular birefringence in the vicinity of the optical transition, and the plane of polarization is rotated when light passes through the mixture with HO$_2$ (Faraday rotation). This rotation of the light polarization is converted into light intensity changes by using a linear polarizer. The signal intensity changes from the lock-in amplifier output, $V_{RMS}(\tilde{v})$, is proportional to the product of,

$$V_{RMS}(\tilde{v}) = GP_0 \sin 2\theta \, \Theta_{RMS} \tag{9.19}$$

where G is the gain of the experimental setup including amplifier gain and transmission loss, P_0 is the optical power incident on the polarizer, θ is the uncrossing angle of second polarizer or offset angle, and Θ_{RMS} is the root mean square value for the polarization rotation angle due to the Faraday effect as a function of optical frequency, which is the function of the concentration of the detected species. Therefore, FRS is a "zero-background" detection method in which the signal is detected only when the laser frequency is scanned over the Zeeman split spectra. The absorption signals from non-paramagnetic species such as H$_2$O and other hydrocarbon molecules are strongly suppressed and do not contribute to the FRS signal.

To achieve higher sensitive detection of these radicals and O(^{1}D)/HO$_2$ reaction kinetics, a new FRS/LAS integrated photolysis flow reactor was developed in a collaborative effort [36,37,61,62] (Figure 9.24). The photolysis reactor consists of a pair of gold-coated spherical mirrors located at the ends of the reactor to form a Herriot multi-pass cell with 21 passes. As a spectroscopic light

Plasma Diagnostics 587

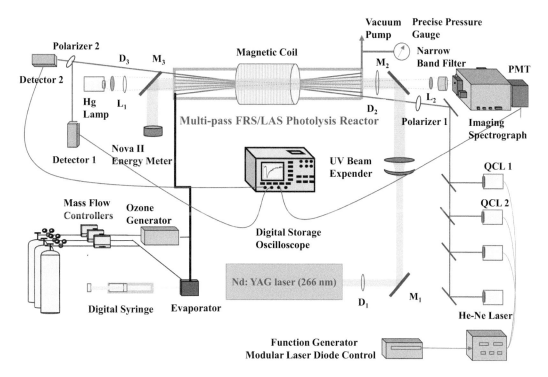

FIGURE 9.24 Top: Experimental layout of the mid-IR FRS/LAS integrated photolysis reactor and diagnostic system. Measurements: OH and HO_2 by FRS; $O(^1D)$, CH_2O, CO, H_2O and C_2H_2 by LAS. Bottom: Direct image of newly designed and tested FRS photolysis reactor.

source, a quantum cascade laser (QCL) operating at 7.1 μm, which was coupled into the multi-pass cell, was used. The central part of the spherical mirrors remains uncoated to allow coupling of the UV photolysis beam (266 nm), which generated $O(^1D)$ atoms from O_3 photolysis and enables the formation of radicals (HO_2 and OH) by reacting with fuels. A solenoid was placed around the photolysis reactor to generate an axial magnetic field (~200 G at the center of the reactor) that was utilized for FRS measurements. Accounting for 21 passes through the reactor and considering the overlap between the QCL and UV beam, the FRS signal was resulted from an effective path length of ~6.3 m (limited by solenoid length of 30 cm) while the TDLAS signal is attributed to a path length of ~7.5 m. The FRS/LAS diagnostic system with the QCL and DFB lasers provided the capability for time-resolved measurements of HO_2 (1,396.91 cm^{-1}) and OH (3,568.52 cm^{-1}) by FRS, $O(^1D)$, CH_2O, CO, H_2O, CH_3OH, C_2H_2, C_2H_4, O_3, and other species by TDLAS.

As shown in Figure 9.24, Polarizer #1 ensures linear polarization upon entering the magnetically induced circular birefringence (MCB) from paramagnetic molecules. Polarizer #2 was set to 45° with respect to the incident polarization angle, allowing roughly equal splitting of the p and s polarizations. Since commercially available auto-balancing detectors are limited to near-IR wavelengths, the current setup uses two separate mid-IR photodetectors and digital post-processing to obtain auto-balanced detection. During post-processing, the two orthogonal polarizations are digitally balanced to compensate for detector gain mismatch and any detector offsets. The differential signal extracted from digitally balanced channels (labeled as detectors 1 and 2 in Figure 9.24) is linearly proportional to the polarization rotation angle, which is used to determine the HO_2 and OH concentrations using FRS modeling.

The output FRS signal voltage (V_{RMS}) of the balanced detection setup in Figure 9.24 is given as [37]

$$V_{\text{RMS}}(\tilde{v}) = GP_0 \sin(2\theta)\Theta_{\text{RMS}}\left(\tilde{v}_0 + \Delta\tilde{v}_{\text{mod}}\cos(2\pi ft), n_{\text{HO}_2}, B_{\text{field}}\right) \quad (9.20)$$

Unlike in the AC-field FRS (Figure 9.23), the Zeeman splitting in the system is constant and set by the DC magnetic field. The DC-field balanced FRS signal measurement is performed via laser wavelength modulation (Δv_{mod}). The number density of HO_2 can be retrieved from Θ_{RMS} spectrum acquired over the laser-accessible spectral window by performing a spectral fitting that used the experimental parameters (pressure, temperature, magnetic field strength) and HITRAN line intensities.

Figure 9.25a shows the measured time-dependent HO_2 formation as UV photolysis of ozone in C_2H_2/O_2 mixture was activated at 2 ms using the experimental setup in Figure 9.24 [62]. The inset in Figure 9.25a shows a digitally balanced FRS spectrum with least-mean-squares spectral line fitting using HITRAN line parameters for HO_2. Note that due to the differential nature of the FRS measurement, prominent features from C_2H_2 absorption in the same spectral region (see Figure 9.25b inset) have been suppressed as common mode spectral interference and do not contribute to the FRS spectral retrieval. This result shows the selective detection of FRS for HO_2 even with the interference from H_2O and hydrocarbon molecules.

The FRS method has been applied to measure HO_2 in a low-temperature flow reactor [37], key elementary rate measurements of $HO_2+C_2H_5$ [36], $O(^1D)+C_2H_2$ [62], $O(^1D)$+alcohol [61], and $O(^1D)$+DME [63]. For example, measured OH and HO_2 time histories using TDLAS and balanced FRS for $O(^1D)+CH_3C_2OH$ reaction system [61] are shown in Figure 9.26. It is seen that the FRS detection can enable ppm level measurement of HO_2 in 200 kHz time resolution. Therefore, this sensitive and selective FRS diagnostic method provides an unprecedented capability for paramagnetic molecule diagnostics and has enabled the elementary rate measurements of $O(^1D)$+fuels and radicals, which are extremely challenging if not possible for other methods. Some of the measured reaction rates were described in Tables 5.10 and 5.11.

9.4 RAMAN SCATTERING AND THOMSON SCATTERING

In nonequilibrium plasma, the temperature of electrons, vibrational modes, and rotational modes can be very different. Therefore, it is necessary to probe the nonequilibrium temperatures. Although laser absorption can measure rovibrational temperature, it is a line-of-sight measurement and it requires the near equilibrium Boltzmann distributions of rotation and vibration modes, rendering it difficult to probe vibrational temperature and rotational temperature separately. Spontaneous Raman scattering has been developed to measure the molecular rotation and vibration distribution

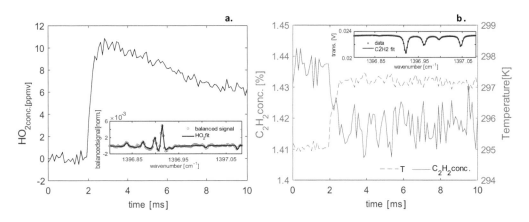

FIGURE 9.25 (a) Balanced FRS detection of HO_2. Inset demonstrates the FRS spectrum of HO_2 due to Faraday rotation in the presence of 200 G magnetic field. (b) TDLAS measurement of C_2H_2 and temperature during the photolysis reaction. Inset shows a measured C_2H_2 spectrum.

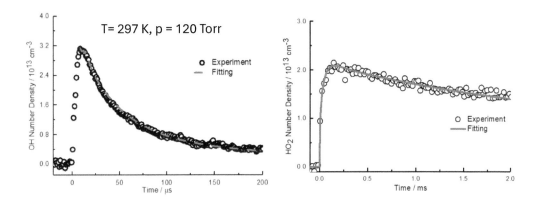

FIGURE 9.26 Left: OH time history measured by TDLAS at 2.8 μm with numerical fitting to experimental data for elementary rate determination. Right: HO2 time history measured by balanced FRS at 7.2 μm with numerical fitting to experimental data for elementary rate determination.

functions and their corresponding characteristic temperatures in combustion [64–66] and plasmas [67,68]. This technique has proven to be quite powerful in characterizing the vibrational energy transfer dynamics in a variety of plasmas.

To probe the electron temperature and density in plasma, laser Thomson scattering was developed to probe the elastic scattering of photons off of free electrons in a plasma [69,70]. Thomson scattering has become a staple diagnostic method for plasma properties [65,71–74]. The key advantage of Thomson scattering is that it can simultaneously provide both the EEDF and its characteristic temperature along with the electron number density. In the section below, the principles and applications of Raman scattering and Thomson scattering will be briefly discussed.

9.4.1 Raman Scattering

Raman scattering is an inelastic light scattering process where the photon exchanges energy with the internal energy of the molecule. As shown in Figure 9.1, the internal energy of a molecule can be divided into rotational, vibrational, and electronic energy modes, and these modes are quantized into discrete energy levels. If a photon interacts with a molecule and scatters, its energy will be scattered and sometimes its energy will change. As shown in Figure 9.27, if the scattered photon is unchanged, it is Rayleigh scattering. If it loses energy (ΔE), it is called Stokes Raman scattering. If it gains energy, then it is called anti-Stokes Raman scattering. The incident photon does not have to be resonant with a molecular energy level to the virtual state and Raman scattering can be performed with a wide variety of laser photon energies greater than ΔE.

Detailed quantum mechanical description of Raman scattering can be found in [64]. When an incident electric field, $\vec{E}$, the molecular polarizability, α, relates the electric field to the induced dipole moment:

$$\vec{p} = \alpha \varepsilon_0 \vec{E} \tag{9.21}$$

where ε_0 is the vacuum permittivity. The molecular polarizability can be linearly expanded in a normal coordinate of x such as a vibrational mode near the equilibrium position:

$$\alpha = \alpha_0 + \left(\frac{\partial \alpha}{\partial x}\right)_0 x \tag{9.22}$$

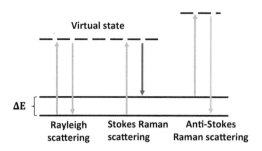

FIGURE 9.27 Raman and Rayleigh energy level diagram. ΔE is a molecular rovibrational energy level difference [65]. Courtesy of Dr. Timothy Chen.

The normal coordinate x can be expressed as an oscillator with frequency ω_v:

$$x = x_0 \cos(\omega_v t) \tag{9.23}$$

The electric field can also be expressed as a wave with frequency ω_0:

$$\vec{E} = \vec{E}_0 \cos(\omega_0 t) \tag{9.24}$$

After substitution and simplification using a trigonometric identity, the expression of the induced dipole moment, $\vec{p}$, becomes,

$$\vec{p} = \alpha_0 \, \varepsilon_0 \, \vec{E}_0 \cos(\omega_0 t) + \varepsilon_0 \left(\frac{\partial \alpha}{\partial x}\right)_0 \frac{x_0 \vec{E}_0}{2}\left[\cos(\omega_0 - \omega_v)t + \cos(\omega_0 + \omega_v)t\right] \tag{9.25}$$

In Eq. 9.25, the first term is responsible for scattered light at the incident frequency ω_0, i.e. Rayleigh scattering. The second and third terms are responsible for frequency-shifted scattered photons due to resonances in the scattering medium. The $\omega_0 - \omega_v$ shift corresponds to Stokes Raman scattering while the $\omega_0 + \omega_v$ shift corresponds to anti-Stokes Raman scattering. Eq. 9.25 indicates that a change in the polarizability of the molecule in the normal coordinate, x, of a rotational or vibrational mode is required for Raman scattering. However, Raman scattering does not need a permanent dipole moment. Therefore, homonuclear diatomic molecules such as N_2 can be measured. This distinguishes Raman scattering from photon absorption discussed in the previous section, where a permanent dipole moment is needed [65].

Raman scattering is linear to the number density of the scattering molecules. Absolute measurements can be calibrated using gas at a known pressure and temperature. Determining a temperature requires modeling the spectrum, particularly the distribution of rotational and vibrational states for a given temperature. For a rigid rotor, the quantized energy is given by the following expression:

$$E_r = hcBJ(J+1), \quad B = h/(8\pi^2 cI) \tag{9.26}$$

where J is the rotational quantum number, B is the rotational constant, h is Planck's constant, c is the speed of light, and I is the moment of inertia of the rigid rotor. Since molecules are not exactly rigid rotors, additional correction factors need to be made:

$$E_r = hc\left[BJ(J+1) - DJ^2(J+1)^2\right] \tag{9.27}$$

where D is the correction term for centrifugal distortion of the rotational energy determined by the rigid rotor approximation. This constant is typically much smaller than B, the rotational constant. The rotational and centrifugal constants in Eq. 9.27 can undergo additional modifications if the molecule is not in the vibrational ground state:

$$B_v = B_e - \alpha_e(v+1/2) \tag{9.28a}$$

$$D_v = D_e - \beta_e(v+1/2) \tag{9.28b}$$

where subscript e refers to the equilibrium position of the molecular nuclei and α_e and β_e are the corrections to the rotational and centrifugal distortion constants, respectively. This additional correction is necessary due to changes in the moment of inertia of the molecule as it simultaneously rotates and vibrates. This effect can be used to simultaneously measure both rotational and vibrational temperatures of diatomic molecules.

The vibrational energy of a harmonic oscillator can be expressed as follows:

$$E_v = h\nu(v+1/2) \tag{9.29}$$

where h is Planck's constant, ν is the frequency of the vibration of the mode and v is the vibrational quantum number. The harmonic oscillator approximation can be modified by considering anharmonicity:

$$E_v = hc\left[\omega_e(v+1/2) - \omega_e x_e(v+1/2)^2 + \omega_e y_e(v+1/2)^3\right] \tag{9.30}$$

where x_e and y_e are higher order anharmonicity constants and $\omega_e = \nu/c$. These constants can be found as part of a Dunham expansion in the literature such as the National Institute of Standards and Technology (NIST) Chemistry WebBook [75].

The Raman scattering signal involves a change in molecular energy levels following the selection rules of quantum mechanics. For pure rotational Raman scattering, they are $\Delta J = \pm 2$. For vibrational Raman scattering the selection rules are $\Delta v = \pm 1$, $\Delta J = 0, 2, -2$. Therefore, to calculate the Raman transition frequency, the difference between the initial and final state energies needs to be calculated.

Once the transition frequencies are known, the intensity of each transition needs to be determined to model the Raman spectrum. This requires the probability that a molecule is in a given energy state. From statistical mechanics [76], the probability is given as:

$$P_j = \frac{g_j e^{-E_j/(k_B T)}}{\sum_j g_j e^{-E_j/(k_B T)}} = \frac{g_j e^{-E_j/(k_B T)}}{Q} \tag{9.31}$$

where Q is the partition function, which can be factorized in terms of individual molecular energy modes of translational, rotational, vibrational, and electronic degree of freedom:

$$Q_{\text{total}} = Q_{\text{tr}} Q_{\text{rot}} Q_{\text{vib}} Q_{\text{elec}} \tag{9.32}$$

The probability of a molecule being in a state with vibrational and rotational quantum numbers (v, J) and its corresponding number density can be expressed as follows:

$$P_{v,j} = \frac{g e^{-E_{\text{rot}}/(k_B T_{\text{rot}})} e^{-E_{\text{vib}}/(k_B T_{\text{vib}})}}{Q_{\text{rot}} Q_{\text{vib}}} \tag{9.33}$$

Here, g is the total degeneracy including both the rotational and nuclear degeneracy. The separation of the rotational and vibrational temperatures in Eq. 9.33 suggests that they can have different values to consider nonequilibrium in plasmas. However, an important assumption in the formulation of Eq. 9.33 is that the energy level populations follow a Boltzmann distribution. In some cases, the vibrational state distribution is non-Boltzmann which may require a two-temperature distribution. Despite this, it will still be applicable if the interest is in the difference of rational and vibrational temperatures. In this case, the vibrational temperature is the temperature of vibrational ground state. In addition to Eq. 9.33, the information of the differential Raman cross-sections of photon-molecule interaction for scattered light polarized parallel to and perpendicular to the incident laser polarization is needed to calculate Raman spectra.

For a diatomic molecule, the pure rotational differential scattering cross-sections in parallel and vertical polarization can be given as [64,67],

$$\left(\frac{\partial \sigma}{\partial \Omega}\right)_{\parallel} = \frac{64\pi^4}{45\varepsilon_0^2} \tilde{v}^4 b_{J_i \to J_f} \gamma_0^2 \tag{9.34a}$$

$$\left(\frac{\partial \sigma}{\partial \Omega}\right)_{\perp} = \frac{4}{3}\left(\frac{\partial \sigma}{\partial \Omega}\right)_{\parallel} \tag{9.34b}$$

where ε_0 is the vacuum permittivity, $\tilde{v}$ is the wavenumber of the scattered photon, $b_{J_i \to J_f}$ is the so-called Placzek-Teller coefficient, and γ_0 is the anisoptoric polarizability. Note that Raman cross-section is proportional to the 4th power of the laser frequency. Higher laser frequency gives stronger Raman scattering signal.

For the diatomic molecular vibrational Stokes Q-branch ($\Delta J=0$, $\Delta v=+1$), the differential scattering cross-section can be written as [65],

$$\left(\frac{\partial \sigma}{\partial \Omega}\right)_{\parallel} \propto (v+1)\tilde{v}^4 \left(a'^2 + \frac{4}{45} b_{J_i \to J_f} \gamma'^2\right) \tag{9.35}$$

where v is the initial vibrational quantum number, a' and γ' are the first derivatives of the mean polarizability and polarizability anisotropy, respectively. The vibrational Raman scattering cross-section is proportional to the product of the 4th power of the wave number and the vibrational energy level. Therefore, vibrational Raman scattering is sensitive to the laser frequency and the high vibrational energy levels.

Plonjes et al. [77] studied nonequilibrium molecular plasma sustained by a combination of a continuous wave CO laser and a sub-breakdown radio frequency electric field. The plasma was sustained in a CO/N_2 mixture containing trace amounts of NO or O_2 at pressures of 0.4–1.2 atm. Figure 9.28 shows the measured N_2 Raman spectra at these conditions, taken with the RF voltage on and off. It is seen that with RF on, the N_2 vibrational temperature at 5,600 torr increased from 1,850 to 2,460 K. However, the translational temperature of the gas mixture at 5,600 torr, inferred from the Raman spectra, only increased from 360 to 540 K.

Bekerom et al. [78] further studied the vibration–vibration (V–V) relaxation (v_1 symmetric stretch, the doubly degenerate v_2 bending, and the v_3 asymmetric stretch) of in pulsed microwave CO_2 plasma using Raman scattering. Figure 9.29 shows the simulated CO_2 vibrational Raman spectrum [78] and the effect of anharmonicity on the vibrational excitation. It is seen that the Raman shifts in dependence of the vibrational excitation are dependent on the anharmonicity of the modes. In the high-energy branch, the v_1 and v_2 modes have a positive anharmonicity, thus the Raman shift increases for higher levels. Opposingly, v_3 vibrational level spacing decreases for higher vibrational quanta. Consequently, the lowest energy peaks that correspond to higher asymmetric stretch levels

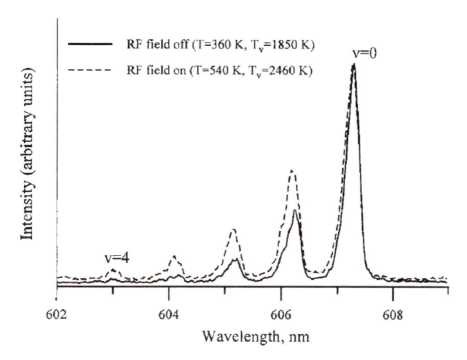

FIGURE 9.28 Raman spectra of nitrogen in the CO laser/rf field pumped 1% CO–99% N_2–150 ppm NO gas mixture at 5,600 torr. The spectra are normalized on the $v=0$ peak intensity [77].

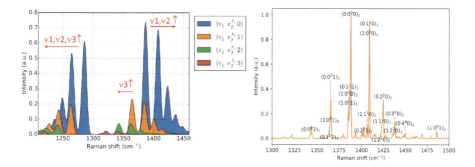

FIGURE 9.29 (Left) Synthetic spectrum of the vibrational Raman spectrum for $T_{v1}=T_{v2}=1,000$ K and $T_{v3}=3,000$ K. Anharmonicity in the v_1:$2v_2$ mode is positive for the 1,388 cm^{-1} band, causing levels with higher v_1 or v_2 quanta to appear at higher wavenumbers, while higher v_3 levels appear at lower wavenumbers so that v_3 excitation can be quantified. The arrows indicate the shift due to increasing quantum number. In the 1,285 cm^{-1} band, peak congestion occurs as vibrational excitation increases. (Right) Detail of the 1,388 cm^{-1} band at the same conditions but higher resolution of 1 cm^{-1}, with the vibrational level of the most intense lines labeled [78].

are well separated from higher v_1 and v_2 levels. Conversely, in the lower energy branch, all higher vibrational levels shift to lower energy, and peak congestion occurs. The higher energy branch (1,388 cm^{-1}) therefore offers an opportunity to quantify v_3 population. By using Raman scattering method, this study showed that in the pulsed microwave discharge, the asymmetric stretch mode of CO_2 was found to be in nonequilibrium with the symmetric stretch and bending modes. After 60 μs, these vibrational modes became equilibrated, while maintaining an overall nonequilibrium between vibrational and rotational/translational modes. This was the first experimental confirmation that nonequilibrium between asymmetric vibration of CO_2 and other vibrational modes could be maintained in a microwave plasma. The quantitative measurements of vibrational and rotational temperatures using Raman scattering clearly demonstrated the nonequilibrium energy coupling in plasma.

9.4.2 Thomson Scattering for Electron Property Measurements

To probe electron energy distribution and number density, elastic Thomson scattering is often preferred than OES. Unlike Rayleigh or Raman scattering which use molecular scatterers, Thomson scattering uses electrons as scatters. As a result, the Thomson scattering cross-section is larger than that of Rayleigh scattering if electron number density is relatively high (greater than $10^{13} cm^{-1}$). The differential Thomson scattering and the total Thomson scattering cross-section can be given as:

$$\frac{d\sigma_e}{d\Omega} = r_e^2 \left(1 - \sin^2\theta \cos 2\phi\right) \quad (9.35a)$$

$$\sigma_e = \int_0^{4\pi} \frac{d\sigma_e}{d\Omega} d\Omega = 6.65 \times 10^{-25} \, cm^2 \quad (9.35b)$$

where r_e is the classical radius of the electron, θ is the scattering angle, and ϕ is the angle between the scattering plane and the laser polarization (Figure 9.30). For typical scattering geometries such as $\theta = 90°$ and $\phi = 90°$ as shown in Figure 9.30, the differential Thomson scattering cross-section reduces to the square of the radius of the electron. The Rayleigh scattering cross-section of air at 532 nm is $5.17 \times 10^{-27} cm^2$, or two orders of magnitude smaller than the total Thomson scattering cross-section. However, the ionization fraction of nonequilibrium plasmas is typically on the order of 10^{-6} to 10^{-4}. Therefore, there is a 4–6 orders of magnitude difference in the electron number density compared to the molecular gas number density. This necessitates the use of special filtering of scattering, large laser fluences, and sensitive detection for Thomson scattering. Furthermore, rotational Raman scattering will also be present in molecular gas mixtures and further obscure the Thomson scattering signal.

If a laser is incident on a plasma with electron density n_e, the amount of scattered power per wavelength is linearly dependent on n_e [65] as,

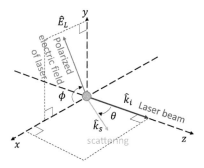

FIGURE 9.30 Laser scattering geometry for Thomson scattering. The laser polarization direction is $\vec{E}_L$, the laser propagation wavevector is $\vec{k}_i$, and the scattering wavevector is $\vec{k}_s$. The scattering angle is labeled as θ and the laser polarization angle is ϕ. Courtesy of Dr. Timothy Chen.

$$P_s(\lambda) = P_0 n_e \eta \frac{d\sigma_e}{d\Omega} \Delta\Omega S(\lambda) L \tag{9.36a}$$

where P_s is the scattered power per wavelength, P_0 is the incident laser power, η is the transmission efficiency of the detection optics, $\Delta\Omega$ is the collection solid angle, $S(\lambda)$ is the scattering spectral distribution, and L is the length of the detection volume. Note that $S(\lambda)$ is normalized such that $S(\lambda)d\lambda$ is 1. All parameters besides n_e in Eq. 9.36b are known or can be calibrated using rotational Raman scattering [65]. Thus, Thomson scattering can give absolute electron number densities. For example, the absolute electron number density in a nitrogen-diluted nonequilibrium plasma can be determined by using the ratio of the integrated Thomson scattering to the integrated N_2 rotational Raman scattering for calibration with the population f at the J rotational state.

$$n_e = \frac{\frac{\hat{P}_e}{\hat{P}_{N_2}} \left(\frac{d\sigma_{N_2}}{d\Omega} \right)_{\text{Raman, 532nm}}}{\frac{d\sigma_e}{d\Omega}} n_{N_2} f_J \tag{9.36b}$$

To detect the EEDF of plasma, the velocity distribution of a group of electrons can be probed based on the Doppler-shifted Thomson scattering spectrum. For a Maxwellian velocity distribution, this spectrum can be fitted with a Gaussian profile, and the width of the Gaussian is proportional to the square root of the electron temperature. In this case, the electron temperature can be determined using the following expression [65],

$$T_e = \frac{c^2 m_e}{8 k_B \sin^2(\theta/2)} \left(\frac{\Delta\lambda_{1/e}}{\lambda_i} \right)^2, \quad \Delta\lambda_{1/e} = \frac{2\lambda_i}{c} \sin\left(\frac{\theta}{2}\right) \left(\frac{2 k_B T_e}{m_e} \right)^{\frac{1}{2}} \tag{9.37}$$

where m_e is the mass of an electron, c is the speed of light, k_B is Boltzmann's constant, λ_i is the laser wavelength, θ is the scattering angle, and $\Delta\lambda_{1/e}$ is the 1/e half-width of the fitted Gaussian spectral shape. For a scattering angle of 90° and $\lambda_i = 532$ nm, this expression can be simplified to

$$T_e = 0.452 \, \Delta\lambda_{1/e}^2 \tag{9.38}$$

where $\Delta\lambda_{1/e}$ is in units of nm and T_e is in units of electronvolts (eV). Note that this expression can only be used in the case of incoherent scattering. For Debye lengths that are comparable to the wavelength of the incident laser light, the scattering medium is no longer individual electrons. Rather, the collection of charges within a Debye sphere (sphere of radius λ_D) will be the scatterer. This is called collective Thomson scattering and can be used to infer information about the ions [79].

9.4.3 Application of Thomson Scattering for Electron Property Measurements in Plasma

Chen et al. [73,80] developed a Thomson/Raman scattering experimental setup utilizing a single grating spectrometer paired with a physical mask to measure electron number density and temperature in a CH_4/He ns-DBD for CH_4 percentages (Figure 9.31). This design eliminates the second and third grating in a triple grating spectrometer. This work extends this detection strategy to moderate pressures and electron densities of 10^{12} cm^{-3}. Figure 9.31 shows the experimental setup. The experimental apparatus consists of three parts: the laser entrance optics, the plane-to-plane ns-DBD cell, and the detection optics.

A Quantel Q-Smart 850 Nd:YAG laser was used to generate a 6 ns FWHM laser pulse with a pulse energy of 230 mJ. The shot-to-shot variation in pulse energy during each measurement was

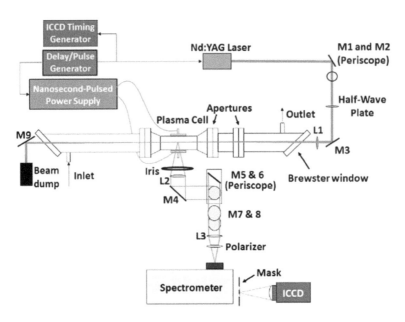

FIGURE 9.31 Experimental setup for Thomson and Raman scattering in a plane-to-plane ns-DBD. L and M are used to denote lenses and mirrors, respectively. Courtesy of Dr. Timothy Chen.

measured by a power meter. The laser beam was directed into the chamber using a periscope and a half-wave plate to ensure vertical polarization entering the chamber. An $f = 1,300$ mm AR-coated plano-convex lens was used to focus the beam to the plasma. Brewster windows at the entrance and exit of the chamber were used to reduce stray light. To further limit the stray light, before the discharge region, "subcritical" and "critical" apertures were placed.

The plasma was generated using 44.5 mm by 44.5 mm electrodes with a gap distance of 14 mm. The nanosecond pulses were generated by a FID power supply capable of delivering 12 ns FWHM and 32 kV peak applied voltage pulses at 30 kHz. The scattered light was collected and focused by two $f = 150$ mm 1-inch AR-coated achromatic lenses. The spectrometer (Acton 2500i, $f = 0.500$ m, 2,400 g/mm) was used to disperse the light into a spectrum. A slit width of 150 μm was used for Thomson scattering measurements, and a slit width of 250 μm was used for the vibrational Raman scattering measurements. At the focal plane of the spectrometer output, a blackened aluminum mask was aligned to block the stray light. The mask was mounted on two 1-axis micrometer translation stages for precise alignment in the wavelength and focal axes. The mask enabled many on-chip charge-coupled device (CCD) accumulations which improves the signal-to-noise ratio of the measurement. Any residual light that leaked through the mask was subtracted by a background measurement. With this arrangement, only a single camera lens was needed to relay the image from the mask to the ICCD (Princeton Instruments PIMAX 1300, UNIGEN II intensifier). 20 ns gate widths were used to account for any time jitter [73,80].

The calibration N_2 rotational Raman scattering and the Thomson scattering spectrum in ns-DBD plasma discharge of 1% CH_4/He mixture taken 100 ns after the pulse is shown in Figure 9.32. By using the N_2 rotational Raman scattering results and the Thomson scattering spectrum in Figure 9.32, the electron number density and temperature can be estimated using Eqs. 9.36 and 9.37.

Figure 9.33 shows the effect of CH_4 addition in helium on the electron number density. It is seen that the electron number density decreased with the increase of methane. At the same time, it was also observed that the electron temperature decreased. This result showed that as electron energy was transferred to dissociation and vibrational excitation of CH_4 molecules, the electron temperature and the overall available energy for ionization decreased. Thus, the overall electron density

Plasma Diagnostics 597

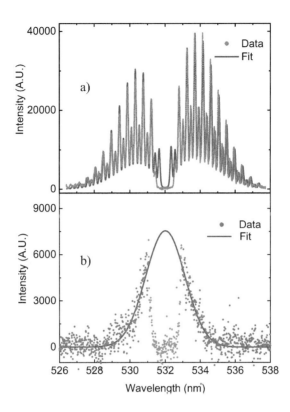

FIGURE 9.32 (a) Calibration N_2 rotational Raman scattering. (b) Thomson scattering spectrum in 1% CH_4/He mixture, taken 100 ns after the pulse[73, 80].

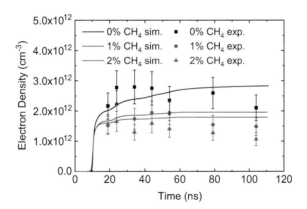

FIGURE 9.33 Time-resolved (a) electron temperature and (b) density with varying CH_4 addition. The discharge current is shown as a reference. 0 ns on the time axis refers to the start of the voltage pulse of plasma [65].

also decreased. This result implies that methane addition into He has a nonlinear effect on plasma properties, which is contrary to previous assumptions in the literature.

Figure 9.34 shows the measured dependence of vibrational temperature (by Raman scattering), electron temperature, and electron number density as a function of time as well as the effect of nitrogen dilution [73]. It is seen that the vibrational temperature increased to ~1,350 K after the electron temperature dropped below 1 eV, and decreased to 1,100 K after 50 μs. This long V-T relaxation

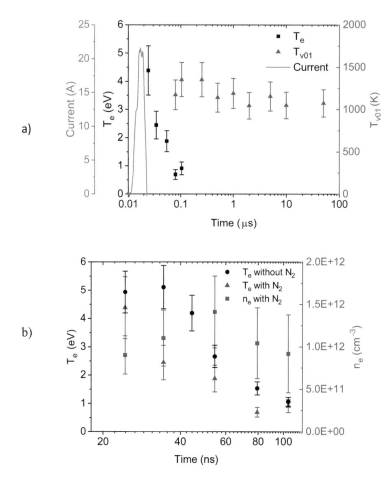

FIGURE 9.34 (a) Vibrational temperature and electron temperature of a $CH_4/N_2/He$ discharge after the 100th voltage pulse. (b) Electron density and temperature for the first 100 ns of the discharge at 60 Torr. 0 ns on the time axis refers to the start of the voltage pulse [73,80].

time allowed for vibrationally excited N_2 populations to build up. Figure 9.34b shows that the addition of N_2 dropped the electron temperature by more than 1 eV during the first 50 ns and the electron density on average dropped also. Furthermore, the electron temperature dropped to 1 eV faster than in the CH_4/He mixture.

9.5 ELECTRIC FIELD INDUCED SECOND HARMONIC GENERATION (E-FISH)

In nonequilibrium plasma, the electron energy distribution is a strong function of the reduced electric field, E/N. Thus, it is necessary to quantify the electric field. The E-FISH effect has been known since the early 1970s [81] and has been used since then to measure the nonlinear hyperpolarizabilities of gases and liquids [82]. Normally, like the second harmonic generation, the second-order optical nonlinear processes are forbidden in a centrosymmetric medium like argon [65]. However, the presence of an external electric field breaks the symmetry and converts the process into a third-order optical nonlinearity. The intensity of the second harmonic signal with an external electric field can be expressed as [83,84]:

$$I^{(2\omega)} = \left|P_i^{(2\omega)}\right|^2 = \left(\frac{3}{2}N\chi^{(3)}(-2\omega,0,\omega,\omega)\right)^2 E_{ext}^2 I_{pump}^2 = CN^2 E_{ext}^2 I_{pump}^2 \quad (9.38)$$

where P_i is the induced polarization characterized by the third-order nonlinear susceptibility term $\chi^{(3)}$. The E-FISH signal scales quadratically with the gas number density (N), pump laser intensity, I_{pump}, and external electric field, E_{ext}. If a known external electric field is used for calibration, the constant, C, can be determined. This calibration constant includes experimental optical transmission losses as well as the third-order nonlinear susceptibility of the probed gas, χ^3. The susceptibility is a constant that relates to the induced polarization of a medium by an electric field. The dependence of the polarization on the field and high-order nonlinear susceptibility can be written as [65,85]:

$$P(t) = \varepsilon_0\left(\chi^{(1)}E(t) + \chi^{(2)}E^2(t) + \chi^{(3)}E^3(t) + \cdots\right) \quad (9.39)$$

These nonlinear susceptibilities $\chi^{(i)}$ are specific to the medium being probed and vary for different gases. The second harmonic generation is a second-order optical nonlinearity since two photons mix to produce a third photon. As mentioned earlier, E-FISH is a third-order optical nonlinearity since it involves an applied electric field and two input pump photons. If this applied electric field varies slowly with respect to the laser pulse width, then it can be considered constant for the duration of this nonlinear optical process. Therefore, the applied electric field can be viewed as a "zero-frequency wave" (Eq. 9.38) and by energy conservation, the output photon is the second harmonic of the two input pump photons, 2ω.

Since the original work of [83,84], there has been rapid application of E-FISH technique for plasma diagnostics [83,86–89]. However, it should be noted that E-FISH is a line-of-sight averaged diagnostic method, although recent effort has been made to use two-beam E-FISH to improve the spatial resolution [86,89]. Moreover, if the calibration was performed at low gas temperatures, then the quadratic number density scaling in Eq. 9.39 can introduce uncertainties in the measurement. Furthermore, the change in gas chemical composition modifies the nonlinear susceptibility, which is a property of the gas mixture [65].

To simultaneously measure the electric field and the electron number density and temperature. Chen et al. [80] integrated the E-FISH setup in Figure 9.35 with the Thomson scattering in Figure 9.31 for plasma-assisted methane reforming. A schematic of the E-FISH setup is shown in Figure 9.35.

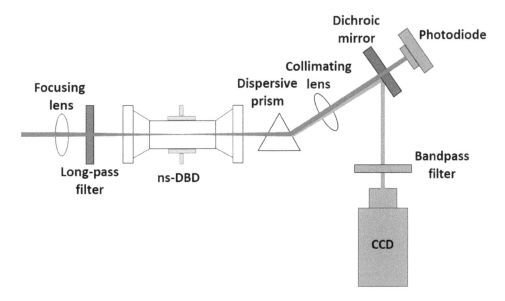

FIGURE 9.35 E-FISH experimental setup [65]. Courtesy of Dr. Timothy Chen.

A Spectra-Physics Solstice Ace laser was used as the pump source and produced a 50 fs laser pulse centered at 800 nm with a 15 nm bandwidth. The laser was focused into a sheet at the center of the dielectric barrier discharge using an f=400 mm cylindrical lens. The measurement length, which follows the laser confocal beam parameter, was 2 cm. After passing through the plasma region, the signal beam was collimated using a matched cylindrical lens and separated from the pump beam by a dispersive prism and a dichroic mirror. A PIXIS 512 CCD was used as the detector with a bandpass filter placed in front to limit interference from plasma emission. A 500 ms camera gate was used, effectively averaging 500 laser shots together. Measurements were performed with 1 ns temporal resolution, as the laser jitter with respect to the discharge was measured to be less than 500 ps.

Figure 9.36 shows the time evolution of the electric field, electron temperature, and electron density as a function of time in an Ar ns-DBD. It is seen that the electron temperature rapidly dropped from roughly 3 to 0.5 eV within the first 50 ns. Furthermore, the electron temperature did not continue decaying after reaching 0.5 eV. The electron number density grew after the breakdown. The detection limit is approximately 1×10^{12} cm^{-3} at 30 ns. The electric field decreased sharply after the applied electric field until breakdown occurred. The sharp drop in electric field corresponds to the establishment of a diffuse discharge. Following the initial drop, a smaller peak in electric field was observed near the end of the applied field pulse. The smaller peak indicates a response of the plasma to the decreasing applied field. A secondary reflected pulse appeared at 160 ns, with a small peak in the electric field at approximately 160 ns. Therefore, the combination of E-FISH with Thomson scattering provides a new platform to quantify time-resolved plasma property variation. More recently, two-beam E-FISH (Figure 9.37) has also been developed to improve the spatial resolution of electric field and reduce the effect of surface scattering in plasma-assisted detonation [86,89]. More details can be found in [86,89].

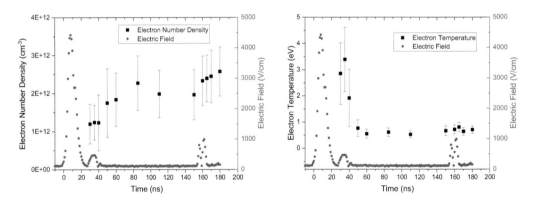

FIGURE 9.36 Electric field, electron temperature, and electron density as a function of time in an Ar ns-DBD [80].

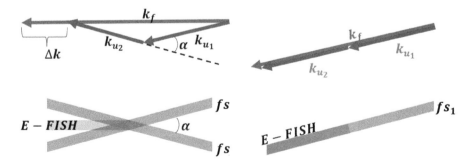

FIGURE 9.37 Phase matching vector diagram for the 2-beam (left) and 1-beam (right) cases [89].

9.6 HYBRID FEMTOSECOND/PICOSECOND COHERENT ANTI-STOKES RAMAN SCATTERING (FS/PS CARS)

While Raman scattering can detect molecular energy transfer between vibration and rotation states in plasmas, its main demerit is that the spontaneous Raman cross-section is three orders magnitude smaller than Rayleigh scattering [65]. However, high laser power can result in laser-induced breakdown and destroy the ability of plasma measurement. To solve this issue, CARS can enhance the Raman signal by several orders of magnitude [90] when using nanosecond pulse width lasers (ns CARS). This enabled the acquisition of spatially resolved one-dimensional Raman images in flames [91]. However, a major challenge for ns CARS is the modeling of the nonresonant background [92]. Moreover, appropriate consideration of collision quenching in ns CARS is also challenging at high pressures.

The development of ultrafast lasers, with pulse widths on the order of picoseconds down to tens of femtoseconds, enabled a new approach called "hybrid femtosecond/picosecond CARS" (fs/ps CARS) to generate a CARS spectrum free from the nonresonant background [93–95]. The key idea was to use a time-delayed spectrally narrow picosecond pulse to scatter off of the Raman coherence prepared by a spectrally broad femtosecond pulse. When the two pulses overlap, four-wave mixing occurs and generates the nonresonant background alongside the resonant CARS signal. The hybrid fs/ps CARS has the advantage that the probe delay can be fixed and the resultant CARS spectrum can be captured within a single laser shot. Additionally, the high intensity from the femtosecond and picosecond pulses significantly enhanced the ability to generate fs/ps CARS signals and the broadband femtosecond pulse eliminated the need for wavelength scanning from a dye laser [65]. Thus, fs/ps CARS has become a promising diagnostic for probing the thermo-chemical states such as temperature and compositions of reacting flows and plasma with high spatial and temporal resolution.

9.6.1 fs/ps CARS

The theory of fs/ps CARS has been described previously for rotation-vibration equilibrium [93,95–97] and nonequilibrium plasma environments [53,98–101]. As shown in Figure 9.38, CARS is a third-order optical nonlinear process that utilizes the difference frequency of the pump and Stokes (Raman) photons (ΔE) that is in resonant with a molecular rotational or vibrational transition. The time-delayed probe pulse scatters off the coherently excited molecules which generates the CARS signal at the anti-Stokes Raman frequency. From the energy conservation, the CARS signal wavenumber will be the summation of the wavenumbers of the pump and probe laser photons subtracted

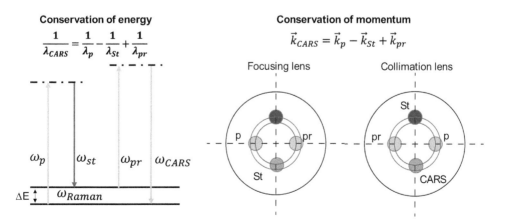

FIGURE 9.38 CARS energy level diagram, photon energy and momentum conservations, and laser beam locations on focusing and collimation lens. ΔE is a molecular rovibrational energy level difference. Courtesy from Dr. Ziqiao Chang, Princeton.

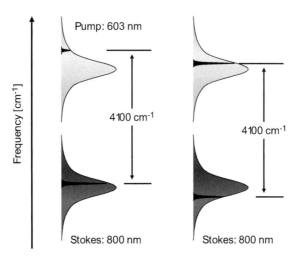

FIGURE 9.39 Schematic of broadband excitation at different energy states with pump and Stokes beams respectively at 603 and 800 nm. Courtesy from Dr. Ziqiao Chang, Princeton.

by that of the Stokes beam. From the conservation of momentum (Figure 9.38), one can align or arrange the pump, Stokes, probe, and CARS beams on the focusing lens and collimation lens in diagnostics.

Although the information from a fs/ps CARS spectrum is identical to that of spontaneous Raman scattering. Therefore, there are several differences between them. Firstly, spontaneous Raman scattering is quite weak when compared to Rayleigh scattering. Inducing a coherent excitation in the molecules prior to the probe pulse in fs/ps CARS allows for much stronger signals and therefore higher signal-to-noise ratio. Secondly, since femtosecond laser pulses contain broadband frequency spectra (Figure 9.39), these femtosecond pulses can supply both the pump and Stokes photons (for rotational CARS) [99] or excite molecules at many rovibrational states simultaneously. For a "transform-limited" pulse, the pulse width and frequency bandwidth are related via the following expression, for Gaussian pulses [102]:

$$\Delta\tau\Delta f \geq 0.44 \tag{9.40}$$

where $\Delta\tau$ is the pulse FWHM and Δf is the pulse frequency bandwidth. Given a certain frequency width in its spectrum, the laser pulse can be no shorter than that defined by Eq. 9.40. Therefore, a shorter laser pulse contains a broader frequency spectrum (Figure 9.39). For example, for a 50 fs Gaussian transform-limited pulse centered at ~800 nm, the corresponding FWHM bandwidth is 300 cm^{-1}, while a 5 fs pulse can span 3,000 cm^{-1}. For reference, 3,000 cm^{-1} is sufficient to access rotations and vibrations of most gas molecules. For example, vibrational stretch modes of N_2 are located at 2,330 cm^{-1} as and CH_4 at 2,960 cm^{-1}. As seen in Figure 9.39, with the pump and Stokes beams respectively at 603 and 800 nm, many rotational states can be excited due to the broadband fs laser beam. Therefore, the utility of using broadband fs pulses for spectroscopy has attracted significant attention not only for CARS but also for laser absorption [51], two-photon laser-induced fluorescence [103,104], and surface spectroscopy. The introduction of a fs pump/Stokes beam brings the minimum number of beams down to two rather than three [53] and still allows to detect both rotational and vibrational temperatures, which significantly simplifies the experimental setup. Furthermore, the two-beam arrangement enables simple 1-D and 2-D measurements [65].

From Eq. 9.39, the scaling for the three beam CARS signal intensity as a function of the intensity of pump, Stokes, and probe beams becomes:

$$I_{CARS} \sim \chi^{(3)} I_{pump} I_{Stokes} I_{probe} N^2 \tag{9.41}$$

Plasma Diagnostics

where N is the gas number density.

The use of a time-delayed probe beam introduces a new degree of freedom in the time domain not presented in traditional nanosecond CARS. This allows one to avoid the "nonresonant background" observed in nanosecond CARS from frequency mixing between the pump, Stokes, and probe beams. This background obscures the resonant CARS signal and could even be stronger than the resonant CARS signal. Since this nonlinear optical process can only occur if all three pulses overlap in time, delaying the probe pulse away from the fs pump/Stokes beam eliminates the nonresonant background as discussed in [93]. In the time domain, the intensity of the CARS signal is the function of the third-order molecular polarization $P^{(3)}(t)$ [105]:

$$I_{CARS} \sim \left| P^{(3)}(t) \right|^2 \tag{9.42}$$

which is dependent on the electric fields of the pump and probe beams as well as the conjugate electric field of the Stokes beam and the time delays between the pump and Stokes beams and between the Stokes and probe beams [65]. The frequencies of the rotational and vibrational CARS transitions of diatomic molecules are calculated in the same manner as in spontaneous Raman scattering. Therefore, the fs/ps CARS signal is now a function of both probe delay as well as the coherence lifetime of each Raman transition, although the initial and final rotational and vibrational states are also governed by the same selection rules as in spontaneous Raman scattering. Since the Raman linewidth is so critical for understanding the time-domain behavior of the fs/ps CARS signal, many studies measure these linewidths through probe beam delay scans [65].

9.6.2 Applications of fs/ps CARS

Dedic et al. [98] used four-beam (rotational and vibrational) hybrid fs/ps CARS for simultaneous, single-shot measurement of pure rotational and rovibrational energy distributions in the highly nonequilibrium environment of a DBD plasma. This approach allowed concise measurements of vibrational/rotational energy distributions in nonequilibrium environments at kHz rates that were free of nonresonant background and minimize interference from molecular collisions. Figure 9.40 shows the four-beam hybrid fs/ps CARS frequency and timing for the pump, Stokes, probe, and CARS signal pulses. To measure both rotational and vibrational temperature simultaneously, two pump beams (ω_{p1} and ω_{p2}) were used.

The DBD plasma (Figure 9.41a) [98] was generated using a 13.56 MHz radio frequency power source. The electrodes consisted of a stainless-steel plate (17.5 mm by 14 mm unobstructed portion) and a tungsten rod with a flat end and diameter of 2.38 mm. The dielectric, acting to limit current spikes and preventing the plasma from transitioning to a thermal arc, consisted of a 130-μm-thick quartz slide covering the stainless-steel plate. The tungsten rod was enclosed within a cylindrical ceramic shroud, and mixtures of N_2 and He flowed within the annulus. The gap between the top

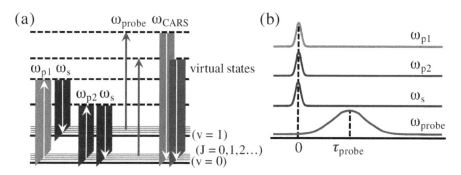

FIGURE 9.40 (a) Frequency and (b) timing diagrams for dual-pump fs/ps vibrational/rotational CARS [98].

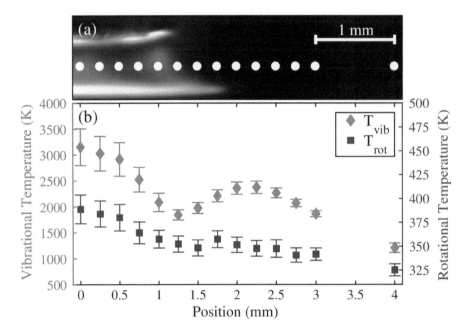

FIGURE 9.41 (a) Averaged image of the DBD and CARS measurement locations and (b) corresponding vibrational and rotational temperatures; symbols and bars represent the average and standard deviation of 400 single-shot measurements at each location [98].

electrode and the quartz slide varied between 670 and 760 μm, and the discharge was operated at atmospheric pressure. A heat sink was attached to the bottom electrode to prevent damage to the quartz plate due to overheating. The rotational and vibrational temperature were measured at different spatial locations across the DBD plasma and afterglow is shown in Figure 9.41b. The 0 mm location corresponded to the center of DBD (symmetric centerline). It is seen that the largest degree of rotational-vibrational nonequilibrium occurred at the center position, with a difference in the average rotational and vibrational temperature of 2,766 K. Moving away from the center toward the edge of the plasma, both the vibrational and rotational temperatures decreased. At 1.25 mm from the reactor center, however, the vibrational temperature begins to increase again, accompanied by a small initial increase in rotational temperature. The secondary peak in vibrational temperature occurred approximately 2.25 mm from the center. It was explained that this second peak was caused by the electronically excited $N_2(A)$ state, which subsequently underwent collisional relaxation to excited vibrational levels of the $N_2(X)$ and led to an increase of vibrational temperature. The results clearly demonstrated that the four-beam fs/ps CARS can simultaneously probe plasma nonequilibrium and help to understand the energy relaxation of the excited states.

Chen et al. [53] developed a two-beam pure rotational hybrid fs/ps CARS method for one-dimensional imaging of rotation-vibration nonequilibrium. This method is simple and enables simultaneous measurements of the spatial distribution of molecular rotation-vibration nonequilibrium in plasma only using 1 fs laser. This method takes advantage of the red shift of N_2 rotational levels by vibrational excitation (Eqs. 9.27 and 9.28) to determine the rotational and vibrational temperatures from the pure rotational spectrum and opens the possibility of simultaneous 2D imaging of rotation-vibration nonequilibrium.

The experimental setup is shown in Figure 9.42. The 65 ps probe beam was provided by a seeded ps regenerative amplifier that operated at 20 Hz with a pulse energy of 6 mJ. The oscillator of the seed laser for the probe beam was phase-locked to the oscillator of the femtosecond laser with <1 ps of jitter. The probe delay was controlled electronically by an electromechanical

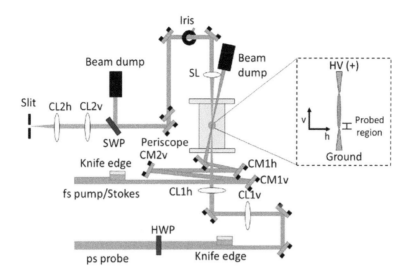

FIGURE 9.42 Experimental setup for fs/ps CARS imaging. The probed region included a 2 mm region above the cathode. CM1h, CM1v, CM2v: concave cylindrical mirrors with f=400 mm, 400 mm, and 200 mm. CL1v, CL1h, CL2v, CL2h: plano-convex cylindrical lenses with f=300 mm, 400 mm, 400 mm, 75 mm. SL: spherical plano-convex lens with f=400 mm. SWP: short-wave pass [53].

trombone. For these measurements, the probe delay was set to 60 ps to avoid the nonresonant background. The probe beam was focused into a sheet parallel to the electrode gap by a cylindrical lens (f=400 mm). A 1.5 mJ portion of a 1 kHz Ti:Sapphire regenerative amplifier with a pulse width of 50 fs (Legend Elite, Coherent) was split off and focused into a 1 m long hollow-core fiber with 310 μm core diameter filled with 400 Torr of Ar. The output energy of the hollow-core fiber was 0.6 mJ and the spectral profile of the pulse was significantly broadened by self-phase modulation within the fiber. The femtosecond beam was focused into a sheet by an additional silver concave cylindrical mirror (f=400 mm) to serve as the pump/Stokes beam. At the 5° intersection of the two sheets, the CARS signal was produced and propagated with the probe beam. The CARS signal was separated from the probe beam by an angle-tuned short-wave pass filter (Semrock) and imaged 1:1 onto the 100 μm wide slit of the spectrometer (iHR550, Horiba). The CARS signal was dispersed by a 2,400 g/mm grating and was detected by a water-cooled CCD with 13.5 μm pixels (Newton 920, Andor). The electrodes were machined to 500 μm cylindrical diameter tips and were separated by 1 cm. The 4 kV, 500 ns voltage pulse was supplied by a high-voltage switch at 20 Hz triggered by a delay generator. The gas mixture used was 40% CH_4 and 60% N_2 at 60 Torr.

The 1D measurement of rotation-vibration nonequilibrium distribution is presented in Figure 9.43 [53]. The CARS image was taken 20 μs after the voltage pulse. It is seen that there were significant rotation-vibration nonequilibrium and spatial gradients in both temperatures, where the peak rotational and vibrational temperatures along the laser sheet were 380 and 5,500 K, respectively. The vibrational temperature exhibited a strong spatial dependence within the first mm of the cathode with a peak at 860 μm from the cathode. This result indicates a possible abnormal glow discharge structure. Therefore, simultaneous measurements of 1D rotation-vibration nonequilibrium can be probed by using pure rotational fs/ps CARS in a CH_4/N_2 nanosecond-pulsed pin-to-pin discharge. The key advantage of this technique is that it enabled two-beam phase matching for straightforward implementation of 1D imaging with a spatial resolution of 40 μm.

A new simple and sensitive method for single-shot rotation-vibration nonequilibrium measurements using pure rotational hybrid femtosecond/picosecond coherent anti-Stokes Raman scattering (fs/ps CARS) coherence beating between vibrationally excited and ground state N_2. To our knowledge, these are the first measurements to use pure rotational fs/ps CARS coherence beating to simultaneously measure rotational and vibrational temperatures of diatomic molecules.

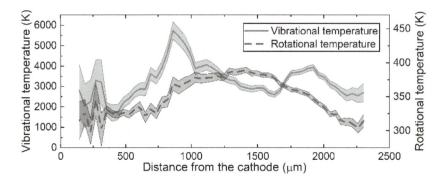

FIGURE 9.43 Spatial distribution of vibrational and rotational temperatures 20 μs after the 4 kV voltage pulse. Error bars are shaded [53].

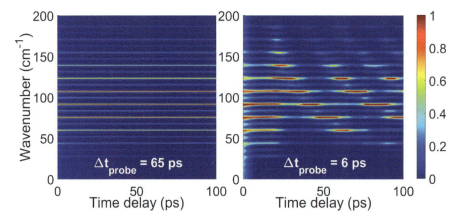

FIGURE 9.44 Modeled time delay scan of N_2 rotational CARS spectrum at $T_{rot}=400$ K and $T_{vib} = 3{,}500$ K with a 65 ps probe (left) and a 6 ps probe (right). Coherence beating was only observed for the 6 ps probe [65]. (Courtesy of Dr. Timothy Chen.

Rotation-vibration nonequilibrium is an important topic of study for research fields such as hypersonics, atmospheric chemistry, plasma-assisted materials synthesis, catalysis, and combustion, where nonequilibrium states evolve dynamically. To understand molecular energy transfer in these systems, rotation-vibration nonequilibrium must be determined simultaneously. Previous approaches required either a separate narrow linewidth picosecond laser to resolve the one-sided broadening of rotational lines or an OPA and two cameras to capture the rotational and vibrational CARS spectra simultaneously. We believe our method is a major improvement since it can quantify rotation-vibration nonequilibrium in a single spectrum using only two beams originating from the femtosecond laser at kHz rates. This makes single-shot rotation-vibration nonequilibrium measurements accessible to all fs/ps CARS practitioners since the only major pre-requisite is the femtosecond laser itself.

It is known that in fs/ps CARS, the time-domain interferences between ground state and vibrationally excited molecules induce complicated coherence beating (Figure 9.44) [98,99] and require to use longer probe beam pulse time to reduce this beating effect. For example, Figure 9.44 (left) [99] shows a demonstration of the advantage of using a long ps probe for the detection of rotation-vibration nonequilibrium. A time delay scan of the probe pulse with respect to the pump/Stokes beams was simulated for 60 Torr N_2 at a rotational temperature, T_{rot} of 400 K and a vibrational temperature, T_{vib}, of 3,500 K. In Figure 9.44, only the duration of the probe pulse was changed from 65 (left) to 6 ps (right). With the 65 ps probe beam, the 0.23 cm^{-1} spectral width (Eq. 9.40) was narrower than the 1 cm^{-1} spacing between adjacent vibrational levels. Each transition was probed separately without any spectral overlaps. With the 6 ps probe, however, beating was observed due to simultaneous sampling of neighboring vibrational levels, which can be detrimental for accurate rotational CARS thermometry.

To address this challenging beating issue of using short ps probe beam in fs/ps CARS, more recently, Chen et al. [99] developed a simple and sensitive two-beam hybrid pure rotational fs/ps CARS method to purposely induce this coherent beating non-Boltzmann distributions in the pure rotational spectra for simultaneously measurement of the rotational and vibrational temperatures of diatomic molecules (N_2). It is shown that this method can be more sensitive than a pure rotational fs/ps CARS approach by using a spectrally narrow probe pulse. The experimental setup is the same as Figure 9.42. The only change is the use of shorter ps probe beam time. A 75 Torr N_2 DC glow discharge was used as a testbed for experimentally demonstrating this technique and vibrational temperatures ranging from 1,000 to 3,600 K were measured on a single shot basis.

The main idea to use the beating information for rotational and vibrational temperature measurements is that the beating is a function of the vibrational temperature, and the interference becomes stronger with larger population of molecules in vibrationally excited states. The effect of the nonequilibrium vibrational temperature on the rotational CARS spectra can be modeled using Eqs. 9.27 and 9.28. Figure 9.45 [99] shows the experimental spectra of the 75 Torr N_2 DC glow discharge at different probe pulse delays as well as the nonequilibrium fits. The effects of coherence beating on the spectra are obvious from the comparison of the data and the rotation-vibration equilibrium simulation. Varying the probe delays induced different beat patterns in the rotational spectrum and these coherence beating effects were successfully fit by the CARS model. For the 92 ps delay, a clear deviation from a Boltzmann distribution of rotational states can be seen, while at 44 ps, the distortion is less obvious. This suggests that there should be an optimal probe delay to maximize the sensitivity of the measurement.

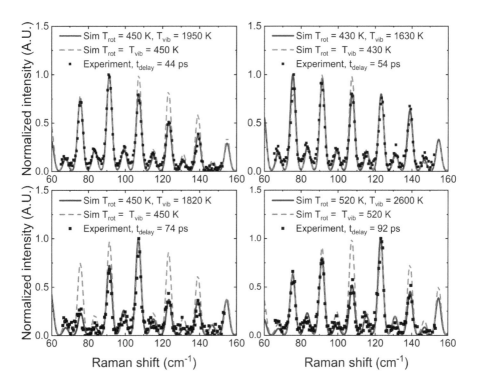

FIGURE 9.45 Single-shot fs/ps CARS spectra of a N_2 glow discharge and corresponding simulated fits for rotation-vibration nonequilibrium at different probe delays. Equilibrium spectra at the same rotational temperatures were plotted for reference [99].

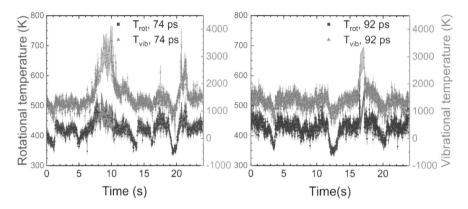

FIGURE 9.46 Time-dependent rotation-vibration nonequilibrium measured in a N_2 DC glow discharge with probe delays of 74 ps (left) and 92 ps (right). Error bars are shaded [99].

The time-dependent rotation-vibration nonequilibrium using probe delays of 74 and 92 ps were measured in the N_2 glow discharge and are shown in Figure 9.46. Both conditions show similar degrees of rotation-vibration nonequilibrium and temporal dynamics. The rotational temperature ranged from 350 to 500 K and the vibrational temperature was between 1,000 and 1,500 K. However, in both time traces, brief peaks in vibrational temperature up to 3,600 K were also observed. This likely was due to the optimal overlap between the glow discharge and the CARS probe volume. In addition, by scanning the probe delay across 300 ps with 10 ps steps, time-resolved CARS was performed and the Fourier transform was used to retrieve the beat frequencies. The measured and theoretical beat frequencies as well as the absolute difference between them [99] were less than 0.2 cm^{-1} between the theory and experiment. This study demonstrated that coherence beating in pure rotational hybrid fs/ps CARS can be used for simultaneous measurement of rotational and vibrational temperatures of molecules.

9.7 FEMOTSECOND TWO-PHOTON LASERS-INDUCED FLUORESCENCE

To probe nonequilibrium plasma chemistry, time, and spatial resolved detection of radical species such as H, O, N, OH, CH_3, and NH are important. In LIF, atoms or molecules are electronically excited at resonant laser wavelength defined by the target species. However, some atoms such as H, O, and N require deep UV lasers (<200 nm) for a single photon LIF, which is difficult to produce and leads ionization of air. TALIF provides a solution to excite these atoms to an electronically excited state using two low energy photons. Since its invention, TALIF has been used extensively in reactive flow and plasma [106,107]. Femtosecond TALIF not only provides broadband excitation of atomic species (Figure 9.47) but also provides higher laser power and temporal resolution for atomic species diagnostics for quantitative 2D measurements. Recently, femtosecond two-photon lasers-induced fluorescence (fs-TALIF) has been applied to nonequilibrium plasmas [104,108–110]. Therefore, with calibration, fs-TALIF enables sensitive, spatially, and temporally resolved absolute atomic number density measurements in plasmas.

9.7.1 TALIF AND CALIBRATION

A schematic of TALIF method with Xenon calibration to measure the absolute atomic O concentration produced by plasma is shown in Figure 9.47 [107]. Ground state atomic O is excited by absorbing two UV photons at 225.7 nm. The transition from the excited $3p^3P$ state to the $3s^3S$ state will release a single photon at 844.6 nm. Xenon can be excited from $5p^6{}^1S_0$ to $6p'[3/2]_2$ with two UV photons at 224.31 nm. The de-excitation from $6p'[3/2]_2$ to $6s'$ $[1/2]_1$ produces fluorescence at 834.91 nm.

Plasma Diagnostics

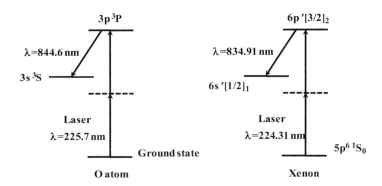

FIGURE 9.47 Schematic of atomic O and Xenon atom energy levels with two-photon absorption excitation and fluorescence schematic [51].

The TALIF fluorescence can be used to quantify the atomic number density of the O atom in the ground state by comparing it with the TALIF signal of Xe at a known concentration. In TALIF study, Xe or Kr is often used to calibrate the concentrations of O, H, and N atoms. The TALIF spectro-temporal integrated signal for an atomic species S_i is given as [111],

$$S_i = CT_i\eta_i \left(\frac{A}{A+Q}\right)_i \left(\frac{I_0}{h\nu}\right)_i^2 \sigma_i n_i \qquad (9.42)$$

where C is a constant to be determined by experimental setup such as focal length and optical collection efficiency. T_i represents the optical transmission coefficients of the experimental system and η_i is the detector efficiency. A is the natural radiative rate of the excited state and can be related to the natural lifetime ($\tau = 1/A$) [108]. Q is the quenching rate which determines the decay rate of the TALIF signal. σ_i is the two-photon absorption cross-section. I_0 is the incident laser intensity, ν is the incident photon frequency, and n_i is the number density of the fluorescence emitting atom.

From Eq. 9.42 and assuming the same laser intensity, the number density of atomic O can be estimated from the TALIF signal of Xe with a known concentration of n_{Xe} using the following equation,

$$n_O = \frac{S_O}{S_{Xe}} \frac{T_{Xe}}{T_O} \frac{\eta_{Xe}}{\eta_O} \frac{\sigma_{Xe}}{\sigma_O} \frac{\left(\frac{A}{A+Q}\right)_{Xe}}{\left(\frac{A}{A+Q}\right)_O} n_{Xe} \qquad (9.43)$$

As mentioned in the main text, the TALIF calibration for inferring the absolute number density of atomic species is very critical. First, the calibration gas, (e.g. Kr or Xe) needs to have similar excitation and fluorescence wavelengths (e.g. same optical configuration, same optical transmission, and detector efficiency) so that T_i, η_i, and I_i can be canceled in calibration [112]. Second, since TALIF requires high photon flux to create many excitations from the ground state, it needs to ensure that laser power does not cause breakdown of the mixture and the saturation of the TALIF signal. As shown in Figure 9.48, it is necessary to confirm the quadratic laser energy dependence of the TALIF signal as well as the linear relationship with the molecule number density (Eq. 9.42). In addition, to maximize the signal and check the lifetime of TALIF signal (quenching mechanism), the exponential decay of the TALIF signal needs to be confirmed (Figure 9.49). Both lifetimes and quenching rates are known for noble gases. Therefore, careful calibration of fs-TALIF is important for quantitative measurements. Recently, a method of independent calibration of H atom fs-TALIF was proposed for high-voltage plasmas in Xe-H_2 mixtures.

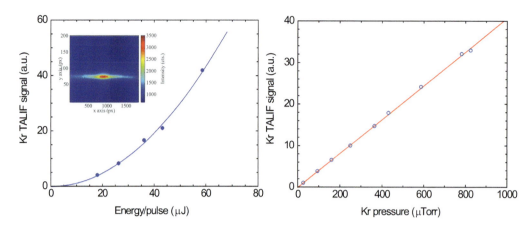

FIGURE 9.48 Conformation of quadratic dependence of Kr TALIF signal on laser energy and a linear relationship with the pressure. Courtesy from Dr. Arthur Dagoriu.

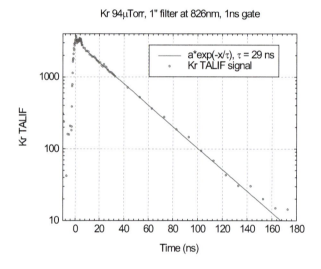

FIGURE 9.49 Conformation of exponential decay of Kr TALIF signal as a function of time. Courtesy from Dr. Arthur Dagoriu.

9.7.2 Femtosecond TALIF Applications in Plasma

There have been many studies using fs-TALIF for atomic species detection in plasma [104,108–110] for H, O, and N measurements. Recently, Liu et al. [109] conducted H and N measurements in N_2/H_2 DBD plasma for ammonia synthesis. The results showed that vibrational excitation of nitrogen and hydrogen molecules plays a critical role in ammonia production. Experimental setup is shown in Figure 9.50. The fs laser system (Coherent Astrella and Light Conversion TOPAS) generated UV fs laser pulses with a pulse energy of 4 µJ at central wavelengths of 205.08 nm for H and 206.65 nm for N, respectively, at a repetition rate of 1 kHz, by combining a Ti:Sapphire oscillator, an ultrafast amplifier, and an OPA. The fs laser beam was focused into the plate-to-plate DBD AC-DBD plasma reactor to excite the $1s^2S_{1/2} \rightarrow 3d^2D_{3/2,5/2}$ transition of H radical and the $2p^{34}S_{3/2} \rightarrow 3p^4S_{3/2}$ transition of N radical. The rectangular stainless-steel electrodes have the dimension of 55 mm×21 mm with a gap distance of 7 mm. A gas mixture of H_2 and N_2 was flowed into the plasma reactor at 100 Torr. The DBD image is shown in Figure 9.50.

Plasma Diagnostics 611

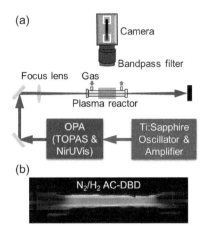

FIGURE 9.50 (a) Experimental setup for fs-TALIF and (b) Photo of N_2/H_2 AC-DBD plasma [109].

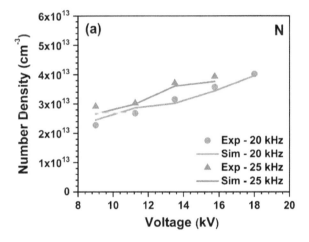

FIGURE 9.51 Experimental results and model predictions of N number density [109].

The H fluorescence signals emitted at 656.3 nm and the N fluorescence signals at 742–746 nm from the plasma were imaged by an ICCD camera (Princeton Instruments PIMAX-4). The camera was synchronized with the laser and the gate time was 80 ns. Each measurement used 2,000 shots on average. The H and N fs-TALIF measurements were calibrated using Krypton (Kr) fs-TALIF measurements as discussed above [113]. The calibration parameters including cross-section, natural lifetime, and quenching rate were taken from [104,108,111,114,115]. It is seen in Figure 9.51 [109] that N atom increases with the increase of discharge peak voltage. Moreover, the increase of AC frequency also led to an increase of N atom. Similar observation was seen for the H atoms. Readers can find more information from [109]. More literature reviews related to TALF can be found in [112]. Therefore, quantitative measurement of atomic species using fs-TALIF provides critical experimental data to understand plasma chemistry in electrified plasma manufacturing and to develop experimentally validated models to optimize the reactor design and efficiency.

In summary, various time- and spectral-resolved advanced diagnostics using tunable mid-IR lasers and fs/ps lasers have been developed over the past decade. Such diagnostics have provided valuable data to understand plasma physics and develop physical and chemical kinetic models. In future studies, time, space, state, and species-resolved quantitative diagnostic method for plasma chemistry in gas phase and on the surface of catalysts are needed.

REFERENCES

1. Devia, D., L. Rodriguez-Restrepo, and E.R. Parra, Methods employed in optical emission spectroscopy analysis: A review. *Ingeniería y ciencia*, 2015. **11**(21): pp. 239–267.
2. Bruggeman, P.J., et al., Gas temperature determination from rotational lines in non-equilibrium plasmas: a review. *Plasma Sources Science and Technology*, 2014. **23**(2): p. 023001.
3. Laux, C., et al., Optical diagnostics of atmospheric pressure air plasmas. *Plasma Sources Science and Technology*, 2003. **12**(2): p. 125.
4. Descoeudres, A., et al., Optical emission spectroscopy of electrical discharge machining plasma. *Journal of Physics D: Applied Physics*, 2004. **37**(6): p. 875.
5. Gottscho, R.A. and V.M. Donnelly, Optical emission actinometry and spectral line shapes in rf glow discharges. *Journal of applied physics*, 1984. **56**(2): pp. 245–250.
6. Wu, P., et al., Inductively coupled plasma optical emission spectrometry in the vacuum ultraviolet region. *Applied Spectroscopy Reviews*, 2009. **44**(6): pp. 507–533.
7. Bogaerts, A., R. Gijbels, and J. Vlcek, Collisional-radiative model for an argon glow discharge. *Journal of Applied Physics*, 1998. **84**(1): pp. 121–136.
8. Shukla, N., R.K. Gangwar, and R. Srivastava, Diagnostic of Ar-CO_2 mixture plasma using a fine-structure resolved collisional radiative model. Spectrochimica Acta Part B: *Atomic Spectroscopy*, 2021. **175**: p. 106019.
9. Laux, C.O., L. Pierrot, and R.J. Gessman, State-to-state modeling of a recombining nitrogen plasma experiment. *Chemical Physics*, 2012. **398**: pp. 46–55.
10. Luque, J., et al., Gas temperature measurement in CH_4/CO_2 dielectric-barrier discharges by optical emission spectroscopy. *Journal of Applied Physics*, 2003. **93**(8): pp. 4432–4438.
11. Ralchenko, Y., *Modern Methods in Collisional-Radiative Modeling of Plasmas*. 2016: Springer.
12. Zaplotnik, R., G. Primc, and A. Vesel, Optical emission spectroscopy as a diagnostic tool for characterization of atmospheric plasma jets. *Applied Sciences*, 2021. **11**(5): p. 2275.
13. Bruggeman, P., et al., Optical emission spectroscopy as a diagnostic for plasmas in liquids: Opportunities and pitfalls. *Journal of Physics D: Applied Physics*, 2010. **43**(12): p. 124005.
14. Rusterholtz, D., et al., Ultrafast heating and oxygen dissociation in atmospheric pressure air by nanosecond repetitively pulsed discharges. *Journal of Physics D: Applied Physics*, 2013. **46**(46): p. 464010.
15. Donnelly, V.M., Plasma electron temperatures and electron energy distributions measured by trace rare gases optical emission spectroscopy. *Journal of Physics D: Applied Physics*, 2004. **37**(19): p. R217.
16. Malyshev, M. and V. Donnelly, Trace rare gases optical emission spectroscopy: Nonintrusive method for measuring electron temperatures in low-pressure, low-temperature plasmas. *Physical Review E*, 1999. **60**(5): p. 6016.
17. Zhu, X.-M., et al., Measurement of the temporal evolution of electron density in a nanosecond pulsed argon microplasma: Using both Stark broadening and an OES line-ratio method. *Journal of Physics D: Applied Physics*, 2012. **45**(29): p. 295201.
18. Minesi, N., et al., Plasma-assisted combustion with nanosecond discharges. I: Discharge effects characterization in the burnt gases of a lean flame. *Plasma Sources Science and Technology*, 2022. **31**(4): p. 045029.
19. Herzberg, G., *Molecular Spectra and Molecular Structure*. 2 ed. 1950: Van Nostrand Reinhold Company.
20. Demtröder, W., *Laser Spectroscopy: Basic Concepts and Instrumentation*. 2013: Springer Science & Business Media.
21. Hanson, R.K., Applications of Quantitative Laser Sensors to Kinetics, Propulsion and Practical Energy Systems. *Proceedings of the Combustion Institute*, 2011. **33**(1): pp. 1–40.
22. Hanson, R.K., *Quantitative Laser Diagnostics for Combustion Chemistry and Propulsion*. 15 Lecture Short Course at Princeton University. 2013: Princeton University.
23. Miles, R.B., *MAE 521: Optics and Lasers*. 2010: Princeton University.
24. Lefkowitz, J.K., Plasma Assisted Combustion: Fundamental Studies and Engine Applications, in Mechanical and Aerospace Engineering. 2015: Princeton University.
25. Rothman, L.S., et al., The HITRAN 2008 molecular spectroscopic database. *Journal of Quantitative Spectroscopy and Radiative Transfer*, 2009. **110**(9–10): pp. 533–572.
26. Farooq, A., J.B. Jeffries, and R.K. Hanson, In situ combustion measurements of H_2O and temperature near 2.5 µm using tunable diode laser absorption. *Measurement Science and Technology*, 2008. **19**(7): p. 075604.
27. Lefkowitz, J.K., et al., In situ species diagnostics and kinetic study of plasma activated ethylene dissociation and oxidation in a low temperature flow reactor. *Proceedings of the Combustion Institute*, 2015. **35**(3): pp. 3505–3512.

28. Lefkowitz, J.K., et al., Species and temperature measurements of methane oxidation in a nanosecond repetitively pulsed discharge. *Philosophical Transactions of the Royal Society A*, 2015. **373**(2048): p. 20140333.
29. Lefkowitz, J.K., et al., Time Dependent Measurements of Species Formation in Nanosecond-Pulsed Plasma Discharges in $C_2H_4/O_2/Ar$ Mixtures, in *52nd Aerospace Sciences Meeting*. 2014: National Harbor, MD.
30. Wang, S., D.F. Davidson, and R.K. Hanson, High-temperature laser absorption diagnostics for CH_2O and CH_3CHO and their application to shock tube kinetic studies. *Combustion and Flame*, 2013. **160**(10): pp. 1930–1938.
31. Armstrong, B.H., Spectrum lines profiles: The voigt function. *Journal of Quantitative Spectroscopy and Radiative Transfer*, 1967. **7**: pp. 61–88.
32. Weideman, J.A.C., Computation of the complex error function. *SIAM Journal on Numerical Analysis*, 1994. **31**(5): pp. 1497–1518.
33. Telfah, H., et al., Formation and consumption of HO_2 radicals in ns pulse O_2–He plasmas over a liquid water surface. *Plasma Sources Science and Technology*, 2022. **31**(11): p. 115019.
34. Ju, Y. and W. Sun, Plasma assisted combustion: Dynamics and chemistry. *Progress in Energy and Combustion Science*, 2015. **48**: pp. 21–83.
35. Ju, Y., et al., Plasma assisted low temperature combustion. *Plasma Chemistry and Plasma Processing*, 2016. **36**(1): pp. 85–105.
36. Zhong, H., et al., Kinetic study of reaction $C_2H_5 + HO_2$ in a photolysis reactor with time-resolved Faraday rotation spectroscopy. *Proceedings of the Combustion Institute*, 2021. **38**: pp. 871–880.
37. Brumfield, B., et al., Direct in situ quantification of HO2 from a flow reactor. *The Journal of Physical Chemistry Letters*, 2013. **4**(6): pp. 872–876.
38. Zhong, H., et al., Understanding non-equilibrium N_2O/NO_x chemistry in plasma-assisted low-temperature NH_3 oxidation. *Combustion and Flame*, 2023. **256**: p. 112948.
39. Ombrello, T., et al., Flame propagation enhancement by plasma excitation of oxygen. Part II: Effects of O_2 (a 1 Δ g). *Combustion and Flame*, 2010. **157**(10): pp. 1916–1928.
40. Zhong, H., et al., Kinetic Studies of Low-Temperature Ammonia Oxidation in a Nanosecond Repetitively-Pulsed Discharge, in *AIAA SCITECH 2023 Forum*. Jan. 23-27, 2023, National Harbor, Maryland, 2023.
41. Goldenstein, C.S., et al., Infrared laser-absorption sensing for combustion gases. *Progress in Energy and Combustion Science*, 2017. **60**: pp. 132–176.
42. Xu, H., et al., Femtosecond laser ionization and fragmentation of molecules for environmental sensing. *Laser & Photonics Reviews*, 2015. **9**(3): pp. 275–293.
43. Apatin, V., et al., Decomposition of CF_2HCl molecules by femtosecond laser radiation. *Chemical Physics Letters*, 2005. **414**(1–3): pp. 76–81.
44. Tancin, R.J., et al., Ultrafast laser-absorption spectroscopy for single-shot, mid-infrared measurements of temperature, CO, and CH_4 in flames. *Optics Letters*, 2020. **45**(2): pp. 583–586.
45. Zhang, Y. and R.B. Miles, Femtosecond laser tagging for velocimetry in argon and nitrogen gas mixtures. *Optics Letters*, 2018. **43**(3): pp. 551–554.
46. Yalcin, S., Y. Wang, and M. Achermann, Spectral bandwidth and phase effects of resonantly excited ultrafast surface plasmon pulses. *Applied Physics Letters*, 2008. **93**(10): p. 101103.
47. Stauffer, H.U., et al., Broadband, background-free, single-laser-shot absorption. *Optica*, 2020. **7**(7): pp. 847–853.
48. Johnson, S., et al., Shock tube/laser absorption measurements of the isomerization rates of allene and propyne. *Combustion and Flame*, 2022. **238**: p. 111962.
49. Dong, L., et al., Compact CH_4 sensor system based on a continuous-wave, low power consumption, room temperature interband cascade laser. *Applied Physics Letters*, 2016. **108**(1): p. 011106.
50. Muraviev, A., et al., Quantum cascade laser intracavity absorption spectrometer for trace gas sensing. *Applied Physics Letters*, 2013. **103**(9): p. 091111.
51. Liu, N., et al., Sensitive and single-shot OH and temperature measurements by femtosecond cavity-enhanced absorption spectroscopy. *Optics Letters*, 2022. **47**(13): pp. 3171–3174.
52. Liu, N., et al., Femtosecond ultraviolet laser absorption spectroscopy for simultaneous measurements of temperature and OH concentration. *Applied Physics Letters*, 2022. **120,** 201103.
53. Chen, T.Y., et al., 1-D imaging of rotation-vibration non-equilibrium from pure rotational ultrafast coherent anti-Stokes Raman scattering. *Optics Letters*, 2020. **45**(15): pp. 4252–4255.
54. Rousso, A., et al., Kinetic studies and mechanism development of plasma assisted pentane combustion. *Proceedings of the Combustion Institute*, 2019. **37**(4): pp. 5595–5603.

55. Hippler, H., J. Troe, and J. Willner, Shock wave study of the reaction HO2+ HO2→H2O2+ O2: Confirmation of a rate constant minimum near 700 K. *The Journal of Chemical Physics*, 1990. **93**(3): pp. 1755–1760.
56. McAdam, K., B. Veyret, and R. Lesclaux, UV absorption spectra of HO 2 and CH 3 O 2 radicals and the kinetics of their mutual reactions at 298 K. *Chemical Physics Letters*, 1987. **133**(1): pp. 39–44.
57. Crowley, J., et al., The HO_2 radical UV absorption spectrum measured by molecular modulation, UV/diode-array spectroscopy. *Journal of Photochemistry and Photobiology A: Chemistry*, 1991. **60**(1): pp. 1–10.
58. Stone, D. and D.M. Rowley, Kinetics of the gas phase HO_2 self-reaction: Effects of temperature, pressure, water and methanol vapours. *Physical Chemistry Chemical Physics*, 2005. **7**(10): pp. 2156–2163.
59. Li, B., et al., Quantitative detection of hydrogen peroxide in an HCCI engine using photofragmentation laser-induced fluorescence. *Proceedings of the Combustion Institute*, 2013. **34**(2): pp. 3573–3581.
60. Brumfield, B. and G. Wysocki, Faraday rotation spectroscopy based on permanent magnets for sensitive detection of oxygen at atmospheric conditions. *Optics Express*, 2012. **20**(28): pp. 29727–29742.
61. Zhong, H., et al., Kinetic studies of excited singlet oxygen atom O (1D) reactions with ethanol. *International Journal of Chemical Kinetics*, 2021. **53**(6): pp. 688–701.
62. Yan, C., et al., The kinetic study of excited singlet oxygen atom O(1D) reactions with acetylene. *Combustion and Flame*, 2020. **212**: pp. 135–141.
63. Zhong, H., et al., Direct kinetic measurements and theoretical predictions of singlet oxygen atom reaction with dimethyl ether. *Applied Physics* Letters, 2022. **120**: p. 1201103.
64. Eckbreth, A.C., *Laser Diagnostics for Combustion Temperature and Species*. 2022: CRC press.
65. Chen, T.Y., In situ time-resolved laser diagnostics for plasma methane reforming, in Mechanical and Aerospace Engineering. 2021: Princeton University.
66. Drake, M.C. and G.M. Rosenblatt, Flame temperatures from Raman scattering. *Chemical Physics Letters*, 1976. **44**(2): pp. 313–316.
67. Lempert, W.R. and I.V. Adamovich, Coherent anti-Stokes Raman scattering and spontaneous Raman scattering diagnostics of nonequilibrium plasmas and flows. *Journal of Physics D: Applied Physics*, 2014. **47**(43): p. 433001.
68. Butterworth, T., et al., Quantifying methane vibrational and rotational temperature with Raman scattering. *Journal of Quantitative Spectroscopy and Radiative Transfer*, 2019. **236**: p. 106562.
69. Fiocco, G. and E. Thompson, Thomson scattering of optical radiation from an electron beam. *Physical Review Letters*, 1963. **10**(3): p. 89.
70. Gerry, E.T. and D. Rose, Plasma diagnostics by Thomson scattering of a laser beam. *Journal of Applied Physics*, 1966. **37**(7): pp. 2715–2724.
71. Roettgen, A., et al., Time-resolved electron density and electron temperature measurements in nanosecond pulse discharges in helium. *Plasma Sources Science and Technology*, 2016. **25**(5): p. 055009.
72. Muraoka, K. and A. Kono, Laser Thomson scattering for low-temperature plasmas. *Journal of Physics D: Applied Physics*, 2011. **44**(4): p. 043001.
73. Chen, T.Y., et al., Time-resolved characterization of plasma properties in a CH4/He nanosecond-pulsed dielectric barrier discharge. *Journal of Physics D: Applied Physics*, 2019. **52**(18): p. 18LT02.
74. Vincent, B., et al., A compact new incoherent Thomson scattering diagnostic for low-temperature plasma studies. *Plasma Sources Science and Technology*, 2018. **27**(5): p. 055002.
75. Linstrom, P.J. and W.G. Mallard, The NIST Chemistry WebBook: A chemical data resource on the internet. *Journal of Chemical & Engineering Data*, 2001. **46**(5): pp. 1059–1063.
76. McQuarrie, D., Statistical Mechanics; University Science: Sausalito, CA, 2000.
77. Plönjes, E., et al., Radio frequency energy coupling to high-pressure optically pumped nonequilibrium plasmas. *Journal of Applied Physics*, 2001. **89**(11): pp. 5911–5918.
78. Van Den Bekerom, D., et al., Mode resolved heating dynamics in pulsed microwave CO_2 plasma from laser Raman scattering. *Journal of Physics D: Applied Physics*, 2019. **53**(5): p. 054002.
79. Miles, J., et al., Time resolved electron density and temperature measurements via Thomson scattering in an atmospheric nanosecond pulsed discharge. *Plasma Sources Science and Technology*, 2020. **29**(7): p. 07LT02.
80. Chen, T.Y., et al., Impact of CH_4 addition on the electron properties and electric field dynamics in a Ar nanosecond-pulsed dielectric barrier discharge. *Plasma Sources Science and Technology*, 2023. **31**(12): p. 125013.
81. Levine, B. and C. Bethea, Molecular hyperpolarizabilities determined from conjugated and nonconjugated organic liquids. *Applied Physics Letters*, 1974. **24**(9): pp. 445–447.
82. Kaatz, P., E.A. Donley, and D.P. Shelton, A comparison of molecular hyperpolarizabilities from gas and liquid phase measurements. *The Journal of Chemical Physics*, 1998. **108**(3): pp. 849–856.

83. Goldberg, B.M., et al., 1D time evolving electric field profile measurements with sub-ns resolution using the E-FISH method. *Optics Letters*, 2019. **44**(15): pp. 3853–3856.
84. Dogariu, A., et al., Species-independent femtosecond localized electric field measurement. *Physical Review Applied*, 2017. **7**(2): p. 024024.
85. Boyd, R.W., A.L. Gaeta, and E. Giese, Nonlinear optics, in Springer *Handbook of Atomic, Molecular, and Optical Physics*. 2008: Springer. pp. 1097–1110.
86. Raskar, S., et al., Spatially enhanced electric field induced second harmonic (SEEFISH) generation for measurements of electric field distributions in high-pressure plasmas. *Plasma Sources Science and Technology*, 2022. **31**(8): p. 085002.
87. Chen, T.Y., et al., Single Shot Burst Imaging of Electric Fields Measured by Split Excitation Electric Field Induced Second Harmonic Generation (SEE-FISH). 2020: Sandia National Lab (SNL-NM), Albuquerque, NM (United States).
88. Rousso, A.C., et al., Time and space resolved diagnostics for plasma thermal-chemical instability of fuel oxidation in nanosecond plasma discharges. *Plasma Sources Science and Technology*, 2020. **29**(10): p. 105012.
89. Vorenkamp, M., et al., Suppression of coherent interference to electric-field-induced second-harmonic (E-FISH) signals for the measurement of electric field in mesoscale confined geometries. *Optics Letters*, 2023. **48**(7): pp. 1930–1933.
90. Tolles, W.M., et al., A review of the theory and application of coherent anti-Stokes Raman spectroscopy (CARS). *Applied Spectroscopy*, 1977. **31**(4): pp. 253–271.
91. Stufflebeam, J.H. and A.C. Eckbreth, CARS temperature and species measurements in propellant flames. *International Journal of Energetic Materials and Chemical Propulsion*, 1994. 3(1–6): pp. 115–131.
92. Eckbreth, A.C. and R.J. Hall, *CARS concentration sensitivity with and without nonresonant background suppression*. 1981.
93. Pestov, D., et al., Optimizing the laser-pulse configuration for coherent Raman spectroscopy. *Science*, 2007. **316**(5822): pp. 265–268.
94. Prince, B.D., et al., Development of simultaneous frequency-and time-resolved coherent anti-Stokes scattering Raman scattering for ultrafast detection of molecular Raman spectra. *The Journal of Chemical Physics*, 2006. **125**(4): p. 044502.
95. Kearney, S.P., D.J. Scoglietti, and C.J. Kliewer, Hybrid femtosecond/picosecond rotational coherent anti-Stokes Raman scattering temperature and concentration measurements using two different picosecond-duration probes. *Optics Express*, 2013. **21**(10): pp. 12327–12339.
96. Retter, J.E., G.S. Elliott, and S.P. Kearney, Dielectric-barrier-discharge plasma-assisted hydrogen diffusion flame. Part 1: Temperature, oxygen, and fuel measurements by one-dimensional fs/ps rotational CARS imaging. *Combustion and Flame*, 2018. **191**: pp. 527–540.
97. Dogariu, A. and A. Pidwerbetsky. Coherent Anti-Stokes Raman Spectroscopy for Detecting Explosives in Real Time, in *Proc. SPIE, Defense, Security, and Sensing*, 2012, Baltimore, Maryland, United States, 2012.
98. Dedic, C.E., T.R. Meyer, and J.B. Michael, Single-shot ultrafast coherent anti-Stokes Raman scattering of vibrational/rotational nonequilibrium. *Optica*, 2017. **4**(5): pp. 563–570.
99. Chen, T.Y., et al., Simultaneous single-shot rotation–vibration non-equilibrium thermometry using pure rotational fs/ps CARS coherence beating. *Optics Letters*, 2022. **47**(6): pp. 1351–1354.
100. Chen, T.Y., et al., Time-domain modelling and thermometry of the CH_4 v1 Q-branch using hybrid femtosecond/picosecond coherent anti-Stokes Raman scattering. *Combustion and Flame*, 2021. **224**: pp. 183–195.
101. Chen, T.Y., et al., One-dimensional imaging of rotation-vibration non-equilibrium from pure rotational fs/ps coherent anti-Stokes Raman scattering. *Optics Letters*, 2020. **45**(15): pp. 4252–4255.
102. Lazaridis, P., G. Debarge, and P. Gallion, Time–bandwidth product of chirped sech 2 pulses: Application to phase–amplitude-coupling factor measurement. *Optics Letters*, 1995. **20**(10): pp. 1160–1162.
103. Ding, P., et al., Temporal dynamics of femtosecond-TALIF of atomic hydrogen and oxygen in a NRP discharge-assisted methane-air flame. *Journal of Physics D: Applied Physics*, 2021. **54**: p. 275201.
104. Dogariu, A., et al., Neutral Atomic-Hydrogen Measurements in a Mirror/FRC Plasma Device using fs-TALIF, in *APS Annual Gaseous Electronics Meeting Abstracts*. Virtual conference, October 5–9, 2020.
105. Stauffer, H.U., et al., Time-and frequency-dependent model of time-resolved coherent anti-Stokes Raman scattering (CARS) with a picosecond-duration probe pulse. *The Journal of Chemical Physics*, 2014. **140**: p. 024316.

106. Sun, W., et al., Effects of non-equilibrium plasma discharge on counterflow diffusion flame extinction. *Proceedings of the Combustion Institute*, 2011. **33**(2): pp. 3211–3218.
107. Uddi, M., et al., Atomic oxygen measurements in air and air/fuel nanosecond pulse discharges by two photon laser induced fluorescence. *Proceedings of the Combustion Institute*, 2009. **32**(1): pp. 929–936.
108. Urdaneta, G., et al., Fs-TALIF for Low Pressure Interfacial Plasmas, in *AIAA SCITECH 2024 Forum*. Jan. 8-12, 2024, Orlando, 2024.
109. Liu, N., et al., Unraveling Non-equilibrium generation of atomic nitrogen and hydrogen in plasma aided ammonia synthesis. *ACS Energy Letters*, 2024. **9**(5): pp. 2031–2036.
110. Kulatilaka, W.D., et al., Photolytic-interference-free, femtosecond two-photon fluorescence imaging of atomic hydrogen. *Optics Letters*, 2012. **37**(15): pp. 3051–3053.
111. Niemi, K., V. Schulz-Von Der Gathen, and H. Döbele, Absolute calibration of atomic density measurements by laser-induced fluorescence spectroscopy with two-photon excitation. *Journal of Physics D: Applied Physics*, 2001. **34**(15): p. 2330.
112. Gazeli, K., et al., Progresses on the use of two-photon absorption laser induced fluorescence (TALIF) diagnostics for measuring absolute atomic densities in plasmas and flames. *Plasma*, 2021. **4**(1): pp. 145–171.
113. Liu, N., et al., Quantitative Femtosecond Two-Photon Absorption Laser Induced Fluorescence Measurements of Hydrogen and Nitrogen Atoms in an AC Dielectric Barrier Discharge, in *AIAA SCITECH 2023 Forum*. Jan. 23-27, 2023, National Harbor, MD, 2023.
114. Dumitrache, C., et al., Quantitative fs-TALIF in high-pressure NRP discharges: Calibration using VUV absorption spectroscopy. *Plasma Sources Science and Technology*, 2022. **31**(1): p. 015004.
115. Miles, R., A. Dogariu, and L. Dogariu, Localized time accurate sampling of nonequilibrium and unsteady hypersonic flows: Methods and horizons. *Experiments in Fluids*, 2021. **62**(12): p. 248.

Index

acoustic instability 495
activation energy 24, 191, 346
active radicals 15, 40, 288
adaptive chemistry 22
adiabatic flame speed 414, 423
adiabatic flame temperature 413
adsorption and dissociation 526
adsorption energy 522
advanced compression ignition 8
after top dead center 499
air pollution and smog 4, 6
air/fuel ratio 499
alternating current 16
alternative fuels 7
ammonia combustion 20, 354
ammonia synthesis 28, 515
anharmonicity 591
anisoptoric polarizability 592
arc 18, 44
arc jet 480
Arrhenius law 191, 347
Asia-Pacific Economic Cooperation 4
attachment reaction 40, 363
autoignition 43, 309, 403
autoignition temperature 317, 420

Balmer transition 569
barrier discharge 97
battery electrode materials 557
Beer-Lambert law 432, 574
binding energy 522
blowoff limit 492
Boltzmann constant 415, 574
Boltzmann distribution 21, 37, 574
Boltzmann equation 38, 106
Bronsted–Evans–Polanyi relation 522

carbon capture 12
carbon capture and storage (CCS) 553
carbon emissions 1
carbon index 7
carbon nanotubes 23
carbon nanotubes 561
Carnot efficiency 18
catalysis 521
catalytic reaction 522
cathode materials 30, 557
cavity flame 486
cavity ring down spectroscopy (CRDS) 580
cavity-enhanced absorption spectroscopy 20, 584
chain-branching reaction 42, 346
chain-initiation reaction 42, 346
chain-propagation reaction 42, 346
chain-termination reaction 346
Chapman–Jouguet (CJ) detonation 464
charge-coupled device (CCD) 85, 596
chemical heat release rate 403

chemical heat release time or excitation time 465
chemical looping 553
chemical potential 512
CHEMKIN 379
CJ detonation velocity 467
climate change 1, 12
CO_2 conversion 26, 546
CO_2 dissociation 24, 540
CO_2 emissions per capita 3
CO_2 hydrogenation 27, 547–551
CO_2 methanation 28
CO_2 reduction 24, 539
CO_2/CH_4 dry reforming 25
Coherent anti-Stokes Raman spectroscopy 20, 54, 566
collision time 16, 37
collisional excitation 16, 38
collisional–radiative models 568
combustion engines 8
Committee on Aviation Environmental Protection 19
compression ignition 8
compression ratio 8
contact surface 403
cool diffusion flame 458
cool flames 8
co-optimization 9
corona 16
correction term for centrifugal distortion 590
Coulomb force 16
critical flame radius 438
critical temperature gradient 465

Damkohler number 367, 420
deflagration to detonation transition 18, 464, 488
degeneracy of state 574
$DeNO_x$ 375
Department of Energy 9
detonation 18, 462
detonation engine 479, 488
detonation velocity 464
dielectric barrier discharge 16
diesel engines 8
diffusion flame 449
diffusion time 367
diffusivity 413
dipole moment 572
Dirac delta function 423
direct current 16
dissociative adsorption energy 523
dissociative ionization 37, 363
distributed feedback quantum cascade laser 373, 577
Doppler broadening 576
dry reforming 542

E-FISH 598
e-fuel 8
Einstein coefficients 575
electric field 16, 38

617

electric field induced second harmonic generation 598
electric vehicles 6
electrical charge 16
electrified materials manufacturing and recycling 30, 512
electrode metal oxide materials 558
electrolysis 12
electromagnetic radiation 572
electron beam 487
electron collision frequency 38, 415
electron energy 16, 38
electron energy distribution function 38, 379
electron energy exchange 38, 415
electron mass 415
electron mean free bath 16
electron number density 16, 569
electron temperature 16, 38, 569, 595
electron velocity 16, 81
electronic excitation 16, 38, 363
electron-impact molecular beam mass spectrometer 387
electron-impact reaction 86, 363
elementary reaction 191, 346
Eley–Rideal mechanism 22, 528
emission control 503
emissions of particulates 2
energy density 7, 40
Energy Information Association 2
energy loss fraction 362
Energy Policy Act 7
engine knock 8
engine thermal efficiency 8
enthalpy and entropy 513
Environmental Protection Agency 5
equilibrium constant 513
equilibrium plasma 17, 43
equivalence ratio 426–427
excited states 15, 40
expansive wave 403
extinction 409
extreme low-temperature combustion 391

Faraday rotation spectrometer 387, 566
fast ionization wave 81
Federal Aviation Administration 9
femotsecond two-photon lasers-induced fluorescence (fs-TALIF) 608
femto-second or fs 20
first-generation bioethanol 7
Fischer-Tropsch fuels 402
flame ball 438
flame location 425
flame propagation 412
flame speed 412
flame stabilization 19
flame structure 412
flame thickness 412
flammability limit 18, 422
flammable region 428
flow residence time 367
fluidized bed 520, 544
fossil fuel gasification 12
fractional loss of energy per collision 415
free energy diagram 530
fs cavity-enhanced absorption spectroscopy 582

fs/ps CARS 601
fs-UV-LAS 582
fuel consumption 6, 7
fuel reforming 12
full width at half maximum (FWHM) 72

gas chromatograph 368
gas turbine 9, 495
gasoline engines 8
Gaussian broadening 569
Gaussian line shape 576
G-curve 428
Gibbs energy 24, 512
gliding arc 16, 482
global warming 1
global warming potential 12
glow discharge 16, 96
greenhouse gas 12

Haber–Bosch 11
harmonic oscillator 591
heat addition 425
heat release 413
heating values 10
Heisenberg uncertainty principle 576
Herriott cell 367, 578
high-temperature ignition 308, 358
HITRAN database 572
HO_2 chemistry 350
homogenous charge compression ignition 8
hot flames 415
hot spot gradient 465
HP-Mech 348
Hugoniot curve 463
hybrid NSD/DC discharge 411
hydrogen 11
hydrogen explosion limit 350–351
hydrogen production 22, 537

ignition and ignition delay time 308, 402
ignition assisted flame propagation 422
ignition Damköhler number 422
ignition enhancement 410
ignition kernel 438
ignition power 440
ignition probability 444
ignition temperature 408
in situ diagnostic 20
induced absorption 575
induced dipole moment 590
induced emission 575
induction zone 464
inductively coupled plasma 569
inelastic neutron scattering 533
infrared (IR) 585
Intergovernmental Panel on Climate Change 1
intermediate-temperature chemistry 360
intermittency 13
internal combustion engine 498
International Civil Aviation Organization 19
ionic wind 19, 415
ionization reaction 363

jet-stirred reactor 387

K-curve 429
kinetic energy 16

Langmuir–Hinshelwood mechanism 22, 527
large activation energy assumption 404
laser absorption spectroscopy 20, 566, 572
laser ignition 487, 502
laser intensity 574
laser-induced fluorescence (LIF) 20, 585
lean-burn 8
Lewis number 413
line shape factor 575
line strength 575
local thermodynamic equilibrium 568
lock-in amplifier 586
Lorentzian broadening 569
low carbon fuels 8, 9
low speed pre-ignition 18
low-temperature chemistry 358
low-temperature combustion 8, 15
low-temperature ignition 358
low-temperature plasmas 20
LXCat database 106, 374

Mach number 463
machine learning 22
magnetic stabilized gliding arc 455
magneto-hydrodynamic discharge 17
Mars-van Krevelen mechanism 22, 528
mass burning rate 414
mass diffusivity 423
mass fraction 413
materials recycling 557
Mauna Loa observatory 1
methane emissions 12
methane oxidation kinetics 284, 352
methane pyrolysis 23
methane reforming 537
microchannel detonation 492
microwave 16, 486
mid-IR 367
mild combustion 19
million tons of oil equivalent 1
minimum ignition energy 438
minimum ignition power 444
molecular polarizability 589
molecular weight 413
molecular-beam mass spectrometer 533
molecule number density 16
multi-channel cascading spark 444
multi-channel nanosecond discharge 445
multi-fluid models 22
multiscale adaptive reduced chemistry solver for plasma-assisted combustion 20
MW discharge 70

nanosecond discharge 17, 59, 485
National Academies of Sciences 9
National Institute of Standards and Technology (NIST) 591
National Jet Fuels Combustion Program 9

National Oceanic and Atmospheric Administration 1
negative temperature coefficient 359
neutral molecules 15
nitric oxide 250, 535
nitrogen fixation 28, 535
non-Boltzmann distribution 21, 37
nonequilibrium manufacturing 22, 512–514
nonequilibrium plasma 8, 16, 38
nonequilibrium reaction kinetics 15, 40, 191
normalized chemical reaction time 465
normally stretched flame 428
NO_x chemistry 250, 355–358
NO_x emissions 9, 503

one-step chemistry 403
optical discharges 66
optical emission spectroscopy 566
organic chemicals 11
oxygenated fuel 354
ozone and ozone chemistry 389
ozone assisted combustion 429
ozonolysis effect 433
ozonolysis reaction 392

partially premixed compression ignition 8
particle-in-cell–Monte Carlo collision 22, 104
particulate matter 4
partition function 574, 591
path flux analysis 376
path length 574
P-branch 574
photoionization 72, 363
picosecond (ps) 20
Placzek-Teller coefficient 592
Planck's constant 572
plasma 15
plasma assisted Mars–van Krevelen mechanism 22, 29, 528–529
plasma assisted materials synthesis 557
plasma catalysis 15
plasma conductivity 415
plasma diagnostics 566
plasma thermal-chemical instability 20
plasma torch 480
plasma-assisted chemical manufacturing 13
plasma-assisted ignition and combustion 11, 37, 406
plasma-assisted recycling 557
plastic pyrolysis and recycling 517
polarizability 592
polarizer 587
pollutant emissions 19
polyatomic species 573
polycyclic aromatic hydrocarbons 506
polyethylene terephthalate 519
polypropylene 519
premixed flame 412
pressure gradient 465
pressure gradient and hydraulic resistance mechanism 466
primary energy consumption 2
primary ozonide 393
programmable heating and quenching 515
propagating spherical flame 438
pulse detonation engine 489

pulse repetition frequency 444, 448
pump/Stokes beam 605

quasi-steady state 350

radiation extinction 426
radiative heat loss 423
radio frequency 16, 486
Raman scattering 20, 566, 588
rapid compression machine 355
Rayleigh line 463
Rayleigh scattering 20
Rayleigh scattering cross-section 594
R-branch 574
reaction rate 37, 347
reactivity controlled compression ignition 8
reduced electric field 16
renewable electricity 11, 13
Renewable Fuel Standard 7
renewable fuels 13
renewable resources 2
residence time 520
residence time distribution 520
RF discharge 70
right and left-handed circularly polarized light (RHCP and LHCP) 586
rotational CARS 607
rotational energy transfer 569
rotational quantum number 574
rotational temperature 17, 568

scattering cross-sections 592
Schrodinger equation 572
scramjet engine 479
second explosion limit 350
second-generation biofuel 7
self-extinguishing flame 438
shock tube 308, 402
shock wave 19, 312
singlet oxygen assisted flame 435
singlet oxygen reaction 401–405
SKYACTIV-X 8
solid angle 594
soot emissions 19, 505
soot print, soot foil 492
spark 18, 62
spark plug 501
spark-assisted compression ignition 8
spark-assisted ignition 8
specific energy input 536
specific heat at constant pressure 413
spectral absorption coefficient 574
spectral absorptivity 574
spectral transmissivity 574
spherical flame 427
spontaneous emission 575
spontaneous ignition 465
stagnation plane 423–427
Stark broadening 570
steam-methane reforming 12
stoichiometric coefficient 512
strain rate 408, 423
streamer discharge 45
stretch extinction 426
stretched flame speed 425
stretched flames 422
supersonic combustion 480
supersonic ramjet engine 19
surface adsorption 524–527
surface charge 22
surface dielectric barrier discharge 407
surface site 524
surface vacancies 22
susceptibility 599
swirl stabilized burner 493

thermal conductivity 423
thermal plasma 17
thermodynamic equilibrium 512
third-body coefficient 348
Thomson Scattering 20, 566, 588
Thomson scattering cross-section 594
tunable diode laser absorption spectroscopy 368, 579
turnover frequency 522
two photon laser-induced fluorescence 20
two-line method 576

ultrahigh-temperature plasma 559
ultraviolet (UV) 585
unburned hydrocarbons 10
unimolecular surface adsorption 524
USC-MECH 373

vacuum permittivity 589
Van't Hoff equation 514
vibrational excitation 16, 38
vibrational quantum number 183, 574
vibrational temperature 17, 568
vibrationally excited species 16, 181
vibrational-rotational energy transfer 17, 186
vibrational-rotational energy transfer 210, 379
vibrational-vibrational energy transfer 210, 379
Voigt profile 576
Volatile Organic Compounds (VOC) 373

warm flame 8, 459
waste materials recycling 30
water gas shift 12
wavenumber 573
weak flame 428
weakly stretched flame 427
Wobbe Index 10

X-Ray diffraction analysis 558

ZDPlasKin 379
Zeldovich mechanism 536
Zeldovich numbers 442
Zeldovich, von Neumann, and Doring structure 464